# SPECTROSCOPY AND PHOTOCHEMISTRY OF PLANETARY ATMOSPHERES AND IONOSPHERES

The chemical composition of any planetary atmosphere is of fundamental importance in determining its photochemistry and dynamics in addition to its thermal balance, climate, origin, and evolution. Divided into two parts, this book begins with a set of introductory chapters, starting with a concise review of the Solar System and fundamental atmospheric physics. Chapters then describe the basic principles and methods of spectroscopy, the main tool for studying the chemical composition of planetary atmospheres and of photochemical modeling and its use in the theoretical interpretation of observational data on chemical composition. The second part of the book provides a detailed review of the carbon dioxide atmospheres and ionospheres of Mars and Venus and the nitrogen-methane atmospheres of Titan, Triton, and Pluto. Written by an expert author, this comprehensive text will make a valuable reference for graduate students, researchers, and professional scientists specializing in planetary atmospheres.

Now retired, VLADIMIR A. KRASNOPOLSKY was previously a research professor in the Department of Physics at the Catholic University of America, Washington, DC. An expert on spectroscopy and photochemical modeling, he is the author of 3 books, 6 book chapters, and 182 refereed publications. He is one of the most highly cited scientists working on planetary atmosphere research and was awarded the USSR State Prize in 1985 for his studies of Venus. He has worked on many space missions throughout his career and was the principal investigator of the airglow spectrometer on the Mars 5 spacecraft and the Venera 9 and 10 missions to Venus, the three-channel spectrometer on the Vega mission to Venus, and the infrared spectrometer on the Phobos 2 orbiter.

CAMBRIDGE PLANETARY SCIENCE

***Series Editors***

Fran Bagenal, David Jewitt, Carl Murray, Jim Bell, Ralph Lorenz, Francis Nimmo, Sara Russell

***Books in the Series***

1. *Jupiter: The Planet, Satellites and Magnetosphere*†
   Edited by Bagenal, Dowling, and McKinnon
   978-0-521-03545-3
2. *Meteorites: A Petrologic, Chemical and Isotopic Synthesis*†
   Hutchison
   978-0-521-03539-2
3. *The Origin of Chondrules and Chondrites*†
   Sears
   978-1-107-40285-0
4. *Planetary Rings*†
   Esposito
   978-1-107-40247-8
5. *The Geology of Mars: Evidence from Earth-Based Analogs*†
   Edited by Chapman
   978-0-521-20659-4
6. *The Surface of Mars*†
   Carr
   978-0-521-87201-0
7. *Volcanism on Io: A Comparison with Earth*†
   Davies
   978-0-521-85003-2
8. *Mars: An Introduction to its Interior, Surface and Atmosphere*†
   Barlow
   978-0-521-85226-5
9. *The Martian Surface: Composition, Mineralogy and Physical Properties*
   Edited by Bell
   978-0-521-86698-9
10. *Planetary Crusts: Their Composition, Origin and Evolution*†
    Taylor and McLennan
    978-0-521-14201-4
11. *Planetary Tectonics*†
    Edited by Watters and Schultz
    978-0-521-74992-3

12. *Protoplanetary Dust: Astrophysical and Cosmochemical Perspectives*[†]
    Edited by Apai and Lauretta
    978-0-521-51772-0
13. *Planetary Surface Processes*
    Melosh
    978-0-521-51418-7
14. *Titan: Interior, Surface, Atmosphere and Space Environment*
    Edited by Müller-Wodarg, Griffith, Lellouch, and Cravens
    978-0-521-19992-6
15. *Planetary Rings: A Post-Equinox View (Second edition)*
    Esposito
    978-1-107-02882-1
16. *Planetesimals: Early Differentiation and Consequences for Planets*
    Edited by Elkins-Tanton and Weiss
    978-1-107-11848-5
17. *Asteroids: Astronomical and Geological*
    Bodies Burbine
    978-1-107-09684-4
18. *The Atmosphere and Climate of Mars*
    Edited by Haberle, Clancy, Forget, Smith, and Zurek
    978-1-107-01618-7
19. *Planetary Ring Systems*
    Edited by Tiscareno and Murray
    978-1-107-11382-4
20. *Saturn in the 21st Century*
    Edited by Baines, Flasar, Krupp, and Stallard
    978-1-107-10677-2
21. *Mercury: The View after Messenger*
    Edited by Solomon, Nittler, and Anderson
    978-1-107-15445-2
22. *Chondrules: Records of Protoplanetary Disk Processes*
    Edited by Russell, Connolly, and Krot
    978-1-108-41801-0
23. *Spectroscopy and Photochemistry of Planetary Atmospheres and Ionospheres*
    Krasnopolsky
    978-1-107-14526-9

[†]Reissued as a paperback

# SPECTROSCOPY AND PHOTOCHEMISTRY OF PLANETARY ATMOSPHERES AND IONOSPHERES

## Mars, Venus, Titan, Triton, and Pluto

VLADIMIR A. KRASNOPOLSKY
Catholic University of America, Washington, DC

**CAMBRIDGE**
**UNIVERSITY PRESS**

University Printing House, Cambridge CB2 8BS, United Kingdom

One Liberty Plaza, 20th Floor, New York, NY 10006, USA

477 Williamstown Road, Port Melbourne, VIC 3207, Australia

314–321, 3rd Floor, Plot 3, Splendor Forum, Jasola District Centre, New Delhi 110025, India

79 Anson Road, #06–04/06, Singapore 079906

Cambridge University Press is part of the University of Cambridge.

It furthers the University's mission by disseminating knowledge in the pursuit of education, learning, and research at the highest international levels of excellence.

www.cambridge.org
Information on this title: www.cambridge.org/9781107145269
DOI: 10.1017/9781316535561



First published 2019

Printed in the United Kingdom by TJ International Ltd. Padstow Cornwall

*A catalogue record for this publication is available from the British Library.*

*Library of Congress Cataloging-in-Publication Data*
Names: Krasnopol'skiĭ, V. A. (Vladimir Anatol'evich), 1938– author.
Title: Spectroscopy and photochemistry of planetary atmospheres and ionospheres : Mars, Venus, Titan, Triton and Pluto / Vladimir Krasnopolsky (Catholic University of America).
Description: Cambridge, United Kingdom ; New York, NY : Cambridge University Press, 2019. | Series: Cambridge planetary science ; 23 | Includes bibliographical references and index.
Identifiers: LCCN 2018040321 | ISBN 9781107145269 (hardback : alk. paper)
Subjects: LCSH: Planets–Atmospheres. | Spectrum analysis.
Classification: LCC QB603.A85 K75 2019 | DDC 551.5099/2–dc23
LC record available at https://lccn.loc.gov/2018040321

ISBN 978-1-107-14526-9 Hardback

# Contents


*Preface* — *page* xiii
1 About the Book — xiii
2 About Me — xiii

1 The Solar System — 1
1.1 Objects and Sciences — 1
1.2 Planetary Atmospheres — 4
1.3 Outer Planets — 5
1.4 Asteroids, Transneptunian Objects, and Comets — 6
1.5 Formation of the Solar System — 8

2 Atmospheric Structure — 11
2.1 Barometric Formula and Its Versions — 11
2.2 Vertical Transport — 14
2.3 Thermal Balance — 18
2.4 Upper Atmosphere — 22
2.5 Escape Processes — 26

3 Spectroscopy — 30
3.1 Quantum Mechanics and Schroedinger Equation — 30
3.2 Hydrogen-like Atoms: Energy Levels and Quantum Numbers — 31
3.3 Radiation Types and Transition Probabilities — 33
3.4 Spectra of Hydrogen-like Atoms — 34
3.5 Multielectron Atoms — 36
3.6 Energy Levels and Selection Rules — 37
3.7 Spectra of Multielectron Atoms — 38
3.8 Rotational and Vibrational Levels of Diatomic Molecules — 41
3.9 Rotational and Rovibrational Spectra — 43
3.10 Electronic States of Diatomic Molecules — 46
3.11 Electronic Spectra of Diatomic Molecules — 47
3.12 Polyatomic Molecules — 51

4 Aerosol Extinction and Scattering 52
4.1 Spherical Particles: Mie Formulas 52
4.2 Some Approximations and Nonspherical Particles 54
4.3 Particle Size Distributions: Photometry, Polarimetry, and Nephelometry of Aerosol Media 55
4.4 On the Radiative Transfer 57
4.5 Aerosol Altitude Distribution 61

5 Quantitative Spectroscopy 65
5.1 Line Broadening 65
5.2 Line Equivalent Widths and Curves of Growth 67
5.3 Ground-Based Spatially Resolved High-Resolution Spectroscopic Observations 69
5.4 Equivalent Widths in the Observation of HF on Venus 70
5.5 Fitting of Observed Spectra by Synthetic Spectra 73

6 Spectrographs 78
6.1 CVF and AOTF Spectrometers 78
6.2 Grating Spectrographs 78
6.3 Echelle Spectrographs 80
6.4 Fourier Transform Spectrometers 82
6.5 Tunable Laser and Cavity Ring-Down Spectroscopy 83
6.6 Infrared Heterodyne Spectrometers 84

7 Spectroscopic Methods to Study Planetary Atmospheres 86
7.1 Spacecraft, Earth-Orbiting, and Ground-Based Observations 86
7.2 Nadir Observations to Measure Species Abundances 87
7.3 Vertical Profiles of Temperature from Nadir Observations of the $CO_2$ Bands at 15 and 4.3 μm 87
7.4 Vertical Profiles of Temperature and CO Mixing Ratio from CO Line Shapes in the Submillimeter Range 90
7.5 Vertical Profiles of Terrestrial Ozone from Nadir UV Spectra 90
7.6 Measurements of Rotational Temperatures and Isotope Ratios 94
7.7 Inversion of Limb Observations 96
7.8 Solar and Stellar Occultations 98
7.9 Some Other Applications of Spectroscopy 100
7.10 Mass Spectrometry and Gas Chromatography 100

8 Solar Radiation, Its Absorption in the Atmospheres, and Airglow 103
8.1 Structure of the Solar Atmosphere 103
8.2 Solar Spectrum 105
8.3 Airglow 108
8.4 Photodissociation and Photoionization of $CO_2$ and Related Dayglow 109

8.5 Resonance Scattering and Fluorescence 112
8.6 Photoelectrons and Energetic Electrons 115
8.7 Chemiluminescent Reactions 118

9 Chemical Kinetics 120
9.1 Double and Triple Collisions of Molecules 120
9.2 Thermochemical Equilibrium 121
9.3 Bimolecular Reactions 122
9.4 Unimolecular Reactions 132
9.5 Termolecular Association 136
9.6 Heterogeneous Reactions 137
9.7 Literature on Reaction Rate Coefficients, Absorption Cross Sections, and Yields 138

10 Photochemical Modeling 140
10.1 Continuity Equation and Its Finite Difference Analog 141
10.2 Solution of the Problem and Boundary Conditions 142
10.3 Example: Modeling of Global-Mean Photochemistry in the Martian Lower and Middle Atmospheres 144
10.4 Time-Dependent Models 148
10.5 Isotope Fractionation 149

11 Mars 155
11.1 History of Studies, General Properties, Topography, and Polar Caps 155
11.2 $CO_2$, Aerosol, and Temperature 158
11.3 Water Vapor, HDO, and Ice 167
11.4 Carbon Monoxide and Molecular Oxygen 173
11.5 Mass Spectrometric Measurements in the Lower Atmosphere and Martian Meteorites: Variability of Argon 178
11.6 Photochemical Tracers: Ozone, $O_2$ Dayglow at 1.27 μm, and Hydrogen Peroxide $H_2O_2$ 180
11.7 Methane 187
11.8 Some Upper Limits 194
11.9 Photochemistry of the Lower and Middle Atmosphere 197
11.10 Variations of Mars Photochemistry 201
11.11 Dayglow 204
11.12 Nightglow, Polar Nightglow, and Aurora 213
11.13 Upper Atmosphere and Ionosphere 219
11.14 Some Aspects of Evolution 232

12 Venus 238
12.1 General Properties and History of Studies 238
12.2 $CO_2$, $N_2$, Model Atmosphere below 100 km, Atmospheric Dynamics, and Superrotation 242

12.3 Noble Gases and Isotopes 248
12.4 Carbon Monoxide, Oxygen, and Ozone 251
12.5 Sulfur Species 257
12.6 Hydrogen-Bearing Species: $H_2O$, HCl, HF, HBr, $NH_3$, and Their D/H Ratios 266
12.7 Nitric Oxide and Lightning 275
12.8 Thermosphere 279
12.9 Ionosphere 290
12.10 Night Airglow 298
12.11 Day Airglow 310
12.12 Escape of H, O, and He and Evolution of Water 325
12.13 Clouds and Haze 330
12.14 Chemical Kinetic Model for Lower Atmosphere (0–47 km) 342
12.15 Photochemistry of the Middle Atmosphere (47–112 km) 350
12.16 Nightglow and Nighttime Chemistry at 80–130 km 362
12.17 Some Unsolved Problems 365

13 Titan 367
13.1 General Properties and Pre-Voyager Studies 367
13.2 Voyager 1 Observations 370
13.3 Ground-Based and Earth-Orbiting Observations 376
13.4 Observations from the Huygens Landing Probe 382
13.5 Cassini Orbiter Observations below 500 km 387
13.6 Cassini/UVIS Occultations and Airglow Observations 396
13.7 Ion/Neutral Mass Spectrometer Measurements 403
13.8 Ionosphere 409
13.9 Isotope Ratios 420
13.10 Photochemical Modeling of Titan's Atmosphere and Ionosphere 424
13.11 Unsolved Problems 441

14 Triton 443
14.1 General Properties and Pre-Voyager Studies 443
14.2 Interior and Surface 444
14.3 Atmosphere 449
14.4 Photochemistry 456
14.5 Triton's Atmosphere after the Voyager Encounter 460

15 Pluto and Charon 467
15.1 Discovery and General Properties 467
15.2 Interior and Surface 468
15.3 Atmosphere before New Horizons and ALMA Observations 473

15.4 Atmosphere: New Horizons and ALMA Observations 478
15.5 Haze 485
15.6 Photochemistry 489

*References* 497
*Index* 536

*Color plate section to be found between pages 304 and 305*

# Preface

## 1 About the Book

This book is based on the author's experience in the field and is intended for senior and graduate students and research scientists specializing in planetary atmospheres. The book comprises two parts, the first of which includes 10 chapters: three chapters (Chapters 1, 2, and 8) present some basic science on the Solar System, atmospheric physics, and effects of solar radiation in atmospheres; five chapters (Chapters 3, 4, 5, 6, and 7) are related to spectroscopy and spectroscopic tools to study atmospheres; and two chapters (Chapters 9 and 10) describe photochemical modeling.

Some chapters are sufficient for practical work in our science, while the others are brief introductions or reviews. For example, the review of different types of spectrographs in Chapter 6 may help in making a choice for a given task but is not supportive for detailed instrument design.

Five chapters of the second part of the book include present data on the structure and composition of the atmospheres and ionospheres of Mars, Venus, Titan, Triton, and Pluto, respectively. There are hundreds of research articles on the subject; my goal was to describe concisely the basic facts with indications of how they were obtained. Therefore only some papers are cited, and many are missing. However, I hope that these chapters on specific atmospheres are complete and will help the reader gain time in studying an atmosphere of interest. This book is not a substitute for the current research literature on specific problems that a researcher may study.

## 2 About Me

I was born in 1938 in Moscow, Russia (which was part of the Soviet Union at that time). In June 1941, Germany, which controlled most of continental Europe, attacked Russia, approaching the Moscow suburbs in November of that year. My father was an engineer in the aviation industry. The loss of Moscow seemed imminent, and so his institution, with all its employees and their families, moved east to Saratov, a city on the Volga River. However, during the following autumn in 1942, the Germans reached Volgograd

(Stalingrad), where the most critical battle of World War II was to take place. It was to the south of Saratov, and the German bombers appeared from time to time over our city. We returned to Moscow in 1943.

A hundred boys graduated from my school in 1955, and four of us were awarded gold medals and four silver medals. (The schools were separate for boys and girls. My gold medal was number one.) The school program covered the basic sciences and was rather broad, and our classrooms for physics and chemistry were well equipped.

I was a student of the Department of Physics, Lomonosov Moscow State University, in 1955–61. Besides attending lectures and seminars, we spent one day weekly in a physics practicum. For example, Millikan's experiment to measure electron charge was assigned to a student, and he or she was given a booklet with a description of the experimental idea and the equipment to study at home. Then the student had six hours to study the installation directly, make measurements, analyze them, and report the results to a tutor. That was during the first 3 years; later students chose their fields in physics and participated in research at proper university labs or at research centers in Moscow. I took some courses in theoretical physics from the famous theorist Lev Landau. His lectures made the exciting suggestion that physics could be created using only logic and math.

In 1961, I joined a lab headed by Professor Aleksandr Lebedinsky at the Nuclear Research Institute of Moscow State University. Lebedinsky was excited by perspectives of space studies of the Solar System, and our work was not related to nuclear physics. I was a radio amateur during my student years and had some experience in designing and adjusting electronic devices, which proved to be advantageous in developing scientific instruments for space missions. In June 1961 Lebedinsky told me that a mission to Venus with a landing on the planet had been planned for a launch in August 1962 – science fiction at that time. Lebedinsky visited M. V. Keldysh, president of the Academy of Sciences, and proposed a simple instrument to distinguish landing on the solid surface from landing on water. The instrument was based on a mercury level with platinum contacts. The circuit was on if the probe was horizontal within $\pm 3^\circ$ (water) and off on a solid surface with a mean deviation of $\approx 15^\circ$.

Keldysh approved the proposal, and Lebedinsky asked me to be technically responsible for the instrument. I was happy and proud to do that work. Finally, the instrument could also measure the period and amplitude of waves on water and gamma radiation of the surface rocks if the surface was solid. Its mass of 550 g looks reasonable even now. However, the fourth stage of the rocket failed, and the mission was lost. The Mariner 2 flyby of Venus in December 1962 indicated that the intense radio emission of Venus originates in the lower atmosphere, not in the ionosphere, and the hot lower atmosphere rules out liquid water on Venus.

Attempts to reach Venus and Mars continued in Russia, and Lebedinsky was principal investigator (PI) of a UV spectrometer (170–340 nm) onboard the Venera 2 (flyby) and of a photometer on Venera 3 (descent probe) missions. I was technically responsible for both instruments. (Here I exclude other of Lebedinsky's experiments that did not include my participation.) Both missions were launched in November 1965 and were lost before they approached Venus. A mission to Mars was planned to be launched in 1965 as well; however, the launch window was lost because of some delays, and the spacecraft was directed to the Moon as the Zond 3 mission. We had the UV spectrometer at Zond 3 and gathered UV spectra of the Moon's rocks. The UV spectrometer was also installed onboard two Cosmos orbiters that were launched in 1965 and 1966 and gave the first satellite data on the global ozone distribution. In June 1967, I defended my PhD thesis; two months later, Lebedinsky had a heart attack while swimming in the Black Sea and passed away.

Our team prepared a UV spectrometer for the unsuccessful Mars 1969 mission and a dayglow multiband photometer for one of the Cosmos orbiters. In 1971, I transferred to the Space Research Institute, to its planetary department headed by Professor Vasiliy I. Moroz. I proposed and became PI of the visible nightglow spectrometers for the Mars 5 and Venera 9 and 10 orbiters that reached Mars in February 1974 and Venus in October 1975, respectively. Mars' nightglow was not detected with some sensitive upper limits, while the observations of Venus revealed the nightglow spectra, their morphology, and some data on lightning and haze. Analysis and interpretation of the observations and photochemical modeling of the atmospheres of Mars and Venus (with V. A. Parshev), including two books (*Photochemistry of the Atmospheres of Mars and Venus*, Moscow: Nauka, 1982; Berlin: Springer, 1986; *Physics of the Planetary and Cometary Airglow*, Moscow: Nauka, 1987), took up most of my time through the mid-1980s. In June 1977, I defended my thesis for the degree doctor of physics and math, and in 1985, I was awarded the USSR State Prize (for studies of Venus).

In 1981, Russia initiated a complicated Vega mission that involved delivery of a balloon and a descent probe to Venus, while the remaining spacecraft moved to comet Halley using Venus' gravity. A few European countries and the United States participated in the scientific payload of the mission. I was Russian PI of a three-channel spectrometer to study the spatial distribution of various species in the coma of comet Halley. French and Bulgarian teams with their PIs, G. Moreels and M. Gogoshev, respectively, contributed very significantly to that instrument. Two twin spacecraft were launched in December 1984, reached Venus in June 1985, and flew through the coma at 8000 km from the nucleus of comet Halley in March 1986.

Detection of the $H_2O$ emission band at 1.38 μm and total production of water by the comet were among our results. (There had been indirect evidence of water in comets, but direct detection had been lacking.) At a conference on comet Halley in 1987, I was deeply impressed by reports by M. J. Mumma, H. P. Larson, and H. A. Weaver, who observed the $H_2O$ band at 2.7 μm using a high-resolution spectrometer at the Kuiper Airborne Observatory. Except for total production of water, they measured its temperature using the rotational line distribution, gas expansion velocity using the Doppler shift, and temperature of formation using the para-to-ortho hydrogen ratio, offering proof that high spectral resolution can be advantageous even relative to a close distance to the object during missions to planets.

We had an infrared spectrometer for solar occultations at Phobos 2 that orbited Mars for two months in 1989. Vertical profiles of water vapor and dust had been observed. The mission was complicated, and its goal to study Phobos at a close distance down to 50 m with laser evaporation and analysis of the surface material was not achieved.

Three Russian scientists, including me, were invited to be co-investigators for the Voyager 2 flyby of the Neptune system in August 1989. I joined the ultraviolet spectrometer (UVS) team, headed by A. L. Broadfoot, and analyzed the UVS solar occultations of Triton's atmosphere and the data on Triton's haze. Later a photochemical model of Triton's atmosphere and ionosphere was made (with D. P. Cruikshank).

Conditions in Russian science degraded significantly in the 1990s, and I transferred to the United States in 1991. There I was impressed by difficult but possible access to high-level Earth-orbiting observatories, from which some observations could give results that otherwise would require a special instrument on a mission to a planet.

We detected helium on Mars using the Extreme Ultraviolet Explorer (EUVE), and the related modeling showed that He on Mars originates from the captured solar wind alpha particles, not from the radioactive decay of uranium and thorium, as was previously supposed. Then we detected atomic deuterium and molecular hydrogen on Mars using the Hubble Space Telescope (HST) and the Far Ultraviolet Spectroscopic Explorer, respectively. A superior-quality spectrum of Mars dayglow was observed at 90–120 nm as well. Those results changed some aspects of the hydrogen photochemistry and hydrogen isotope fractionation that are related to the evolution of water on Mars. The first UV spectrum of Pluto at 180–255 nm was observed using the HST as well.

Our EUVE and CXO (Chandra X-ray Observatory) observations of comets and analyses of other CXO observations resulted in significant progress in understanding the nature of the unexpectedly bright and initially puzzling X-ray and EUV emissions from comets and in abundances of the solar wind heavy ions that originate these emissions.

Ground-based spatially resolved high-resolution spectroscopy is another powerful tool to study the atmospheres of planets. My long-term observations of Mars and Venus were conducted at the NASA Infrared Telescope Facility using CSHELL and TEXES spectrographs and at the Canada–France–Hawaii Telescope using FTS. The observations of Mars involved variations of the $O_2$ dayglow at 1.27 μm as a tracer of photochemistry, variations of CO as a tracer of the subpolar dynamics, dayglow of CO at 4.7 μm, detection of $CH_4$ as a tracer of possible microbial life, and variations of $HDO/H_2O$ related to the evolution of water. The observations of Venus referred to the cloud tops and included the first detections of NO and OCS at these altitudes and variations of CO, HCl, HF, $H_2O$, and $SO_2$; night airglow of $O_2$ at 1.27 μm and OH; dayglow of CO at 4.7 μm; and isotope D/H ratios in $H_2O$, HCl, and HF.

Significant efforts were made to photochemically model the atmospheres and ionospheres of Mars, Venus, Titan, Triton, and Pluto. Modeling of some other phenomena, e.g., hydrodynamic escape from Pluto, excitation of X-rays in comets by electron capture by the solar wind heavy ions, and excitation of oxygen emissions on the terrestrial planets, has been done as well. Using the British system, with all achievements equally divided among coauthors, my citation index is the best in the field of planetary atmospheres.

# 1
# The Solar System

## 1.1 Objects and Sciences

The Solar System presents a great variety of objects: the Sun, planets, satellites, rings, asteroids, comets, transneptunian or Kuiper belt objects, and interplanetary medium. Basic properties of the planets are given in Table 1.1. This table includes Pluto, which was recently assigned to the transneptunian objects by the International Astronomical Union. However, Pluto has an atmosphere and a few satellites, and this decision is not supported by some members of the planetary community.

There is a simple approximate formula for the heliocentric distances of the planets, a so-called Titius–Bode rule:

$$R_{TB}(\mathrm{AU}) = 0.4 + 0.3{*}2^n. \tag{1.1}$$

Here $n = -\infty$ for Mercury, 0, 1, 2 for Venus, Earth, and Mars, 3 for the asteroid belt, and 4, 5, 6, 7 for Jupiter, Saturn, Uranus, and Neptune, respectively. The calculated values are shown in Table 1.1 for comparison with the true heliocentric distances. The mean accuracy of the Titius–Bode approximation is 5.5%, which is very high for such a simple relationship. However, there is no physical base for this rule.

### *1.1.1 Two-Body Problem*

Two-body problem in the polar ($r$, $\theta$) coordinates for mass of the Sun $M_S$ exceeding mass of a planet by orders of magnitude results in the following equations for conservation of energy and angular momentum:

$$\frac{1}{2}\left[\left(\frac{dr}{dt}\right)^2 + \left(r\frac{d\theta}{dt}\right)^2\right] - \frac{\gamma M_S}{r} = C_0; \quad r^2\frac{d\theta}{dt} = C_1. \tag{1.2}$$

Here $\gamma = 6.67\times10^{-8}\ \mathrm{g^{-1}\ cm^3\ s^{-2}}$ is the gravitational constant and $C_0$ and $C_1$ are the initial energy and angular momentum that are conserved. The problem solution is an elliptic trajectory,

$$r = \frac{p}{1 + \varepsilon\cos\theta}, \tag{1.3}$$

Table 1.1 *Properties of planets*

| Parameter | Mercury | Venus | Earth | Mars | Jupiter | Saturn | Uranus | Neptune | Pluto |
|---|---|---|---|---|---|---|---|---|---|
| Mean distance, AU[a] | 0.387 | 0.723 | 1.0 | 1.524 | 5.204 | 9.582 | 19.19 | 30.07 | 39.53 |
| $R_{TB}$,[b] AU | 0.4 | 0.7 | 1.0 | 1.6 | 5.2 | 10.0 | 19.6 | 38.8 | – |
| Eccentricity | 0.206 | 0.0067 | 0.0167 | 0.0935 | 0.0488 | 0.0557 | 0.0472 | 0.0087 | 0.2488 |
| Orbital period[c] | 88.0 d | 224.7 d | 365.3 d | 687 d | 11.86 y | 29.42 y | 84.0 y | 164.8 y | 248 y |
| Rotational period[c] | 58.6 d | −243 d | 24 h | 24.6 h | 9.93 h | 10.57 h | −17.24 h | 16.11 h | 6.39 d |
| Obliquity | 0° | 177.4° | 23.5° | 24.0° | 3.1° | 26.7° | 97.77° | 28.3° | 122.5° |
| Mass/$M_E$[d] | 0.0554 | 0.815 | 1 | 0.107 | 318 | 95.2 | 14.54 | 17.15 | 0.00218 |
| Equatorial radius, km | 2440 | 6052 | 6378 | 3396 | 71,492 | 60,268 | 25,559 | 24,764 | 1190 |
| Density, g cm$^{-3}$ | 5.43 | 5.243 | 5.518 | 3.93 | 1.326 | 0.687 | 1.27 | 1.638 | 1.86 |
| Surface gravity, m s$^{-2}$ | 3.70 | 8.87 | 9.78 | 3.71 | 24.8 | 10.44 | 8.69 | 11.15 | 0.62 |
| Escape velocity, km s$^{-1}$ | 4.25 | 10.2[e] | 10.85[e] | 4.87[e] | 59.5 | 35.5 | 21.3 | 23.5 | 0.78 |

[a]One astronomic unit (AU) = the Sun–Earth distance = $1.496\times10^8$ km. [b]$R_{TB}$ is heliocentric distance predicted by the Titius–Bode rule. [c]Sidereal. [d]$M_E = 5.98\times10^{27}$ g is the Earth's mass. [e]At the exobase.

with the Sun in one of the foci. Here $\varepsilon$ is the eccentricity and $p/(1-\varepsilon)$, $p/(1+\varepsilon)$ are the orbit aphelion and perihelion, respectively. If a body enters the Solar System with a low velocity at the infinity (e.g., a comet from the Oort cloud), then $\varepsilon \approx 1$, the orbit is parabolic, and the perihelion is $p/2$. If the velocity at the infinity is significant, then $\varepsilon > 1$, and the orbit is hyperbolic.

### 1.1.2 Kepler Law

The solar gravity is balanced by the centrifugal force for circular orbits ($\varepsilon = 0$):

$$\frac{\gamma M_S}{r^2} = \omega^2 r; \quad \omega \equiv \frac{d\theta}{dt} = \frac{2\pi}{T}; \quad \frac{r^3}{T^2} = \frac{\gamma M_S}{4\pi^2}. \tag{1.4}$$

Here $\omega$ is the angular velocity and $T$ is the orbital period, that is, the length of year. The final relationship is also applicable to elliptic orbits as the Kepler law if $r$ is a semi-major axis, that is, a half sum of aphelion and perihelion. There are two obvious sequences of this law: (1) if a semi-major heliocentric distance is known, then the length of year may be calculated by scaling, e.g., the Earth's values, and (2) if both values are measured, then mass of the Sun can be determined.

Planets in the inner Solar System, the so-called terrestrial planets Mercury, Venus, Earth, and Mars, are much smaller but significantly denser than the outer or giant planets Jupiter, Saturn, Uranus, and Neptune.

### 1.1.3 Satellites; Roche Limit

All planets except Mercury and Venus have satellites that are orbiting them similar to the planets orbiting the Sun. Measurements of a satellite orbital radius and period may be used to derive mass of the planet. Furthermore, accurate measurements of spacecraft orbits near or around a planet make it possible to get some data on a mass distribution within the planet.

Tidal effects of a planet on its satellite may be strong and even result in its graduate destruction. Let $M$, $R$, $\rho_M$ be mass, radius, and mean density of a planet, $m$, $r$, $\rho_m$ of a satellite, $M >> m$, and $d$ is the distance between their centers. Then $\gamma M/d^2 = \omega^2 d$; however, a particle at the point of the satellite opposite to the planet is attracted weaker to the planet than the satellite center, while its centrifugal force is stronger. The particle can leave the satellite if this tidal force exceeds the satellite gravity. If they are equal, then

$$\omega^2(d+r) - \frac{\gamma M}{(d+r)^2} = \frac{\gamma m}{r^2}. \tag{1.5}$$

Substituting $\omega^2 = \gamma M/d^3$ and assuming $r << d$, one gets

$$d = r\left(\frac{3M}{m}\right)^{1/3} = R\left(\frac{3\rho_M}{\rho_m}\right)^{1/3}. \tag{1.6}$$

Table 1.2 *Properties of satellites with atmospheres and the Moon*

| Satellite | Planet | Mass[a] | Radius (km) | $R/R_p$[b] | Period[c] | Density (g cm$^{-3}$) | Gravity (m s$^{-2}$) | $V_E$ (kms$^{-1}$)[d] |
|---|---|---|---|---|---|---|---|---|
| Moon | Earth | 1 | 1737 | 60.3 | 27.32 | 3.34 | 1.62 | 2.38 |
| Io | Jupiter | 1.21 | 1822 | 5.9 | 1.77 | 3.53 | 1.80 | 2.56 |
| Titan | Saturn | 1.83 | 2576 | 20.3 | 15.95 | 1.88 | 1.35 | 2.64 |
| Triton | Neptune | 0.29 | 1353 | 14.3 | −5.88 | 2.06 | 0.78 | 1.46 |

[a]Times mass of the Moon, $7.35\times10^{25}$ g. [b]Ratio of orbit radius to radius of planet. [c]Sidereal, in days. [d]Escape velocity.

This is the Roche limit: if a satellite is closer to a planet than this limit, which is $\approx$1.5 times the planet radius, then weathering and other processes of gradual destruction result in loss of the satellite mass. The same effect occurs at the part of the satellite that is closest to the planet.

Three satellites have appreciable atmospheres; they are compared with the Moon in Table 1.2. The Moon and Io look rather similar, and their mean densities indicate that they are composed of rocks. Densities of Titan and Triton are smaller, suggesting comparable quantities of rocks and water ice.

## 1.2 Planetary Atmospheres

Three basic sciences are related to studies of the planets and satellites: geology, atmospheric science, and magnetospheric physics, including solar wind interactions. A subject of our interest is the atmospheric science. All planets (except Mercury) and three satellites have atmospheres. Their main properties are summarized in Table 1.3.

The atmospheres in the Solar System cover a great variety of conditions, with surface temperatures from 38 to 737 K and surface pressures varying within a factor of $10^{10}$. Another important topic is the chemical composition that determines photochemistry and dynamics of a planetary atmosphere, its thermal balance, climate, and reflects its origin and evolution. Abundances of three main gases in each atmosphere are given in Table 1.4.

Based on the chemical composition, the atmospheres in the Solar System can be divided into five types: the hydrogen–helium atmospheres of the outer planets, the nitrogen–methane atmospheres of Titan, Triton, and Pluto, the $CO_2$ atmospheres of Mars and Venus, the $N_2$–$O_2$ atmosphere of the Earth, and the $SO_2$ atmosphere of Io. This separation correlates with other properties of the bodies and their classification to the outer and inner planets and big satellites. The atmosphere of the Earth originated as a $CO_2$ atmosphere; however, carbon dioxide was dissolved in the ocean and formed carbonates. The atmosphere is significantly affected by life, photosynthesis, and currently by human activity. The atmosphere of Io is supported by intense volcanic eruptions stimulated by tidal forces from Jupiter. Volcanism on Io exceeds that on the Earth by orders of magnitude.

Table 1.3 *Properties of planetary atmospheres at the surface or 1 bar*

| Planet | $p$ (bar) | $T$ (K) | $H$ (km) | Dry lapse rate (K km$^{-1}$) |
|---|---|---|---|---|
| Jupiter | 1 | 165 | 27 | 1.8 |
| Saturn | 1 | 134 | 60 | 0.7 |
| Uranus | 1 | 76 | 19 | 0.8 |
| Neptune | 1 | 72 | 28 | 1.0 |
| Titan | 1.5 | 94 | 21.5 | 1.3 |
| Triton | $1.4\times10^{-5}$ | 38 | 13.4 | 0.74 |
| Pluto[a] | $1.2\times10^{-5}$ | 45 | – | – |
| Io | $\sim10^{-8}$ | 110 | 8 | 3.1 |
| Mars | 0.006 | 210 | 11 | 5.5 |
| Venus | 93 | 737 | 16 | 10.5 |
| Earth | 1 | 288 | 8.4 | 9.8 |

*Note.* $H$ is scale height. Updated from Yung and DeMore (1999).
[a] At the flyby of New Horizons in July 2015. Temperature gradient is very large near the surface, and $H$ and lapse rate are uncertain.

Table 1.4 *Three most abundant gases in each planetary atmosphere*

| | | | |
|---|---|---|---|
| Jupiter | $H_2$ (0.93) | He (0.07) | $CH_4$ ($3\times10^{-3}$) |
| Saturn | $H_2$ (0.96) | He (0.03) | $CH_4$ ($4.5\times10^{-3}$) |
| Uranus | $H_2$ (0.82) | He (0.15) | $CH_4$ (0.023) |
| Neptune | $H_2$ (0.80) | He (0.19) | $CH_4$ (0.01–0.02) |
| Titan | $N_2$ (0.94) | $CH_4$ (0.057) | $H_2$ (0.001) |
| Triton | $N_2$ (0.999) | CO ($7\times10^{-4}$) | $CH_4$ ($2\times10^{-4}$) |
| Pluto[a] | $N_2$ (0.995) | $CH_4$ ($4\times10^{-3}$) | CO ($5\times10^{-4}$) |
| Io | $SO_2$ (0.94) | SO (0.05) | O (0.01) |
| Mars | $CO_2$ (0.96) | $N_2$ (0.018) | Ar (0.019) |
| Venus | $CO_2$ (0.96) | $N_2$ (0.035) | $SO_2$ ($1.3\times10^{-4}$) |
| Earth | N2 (0.78) | O2 (0.21) | Ar (0.0093) |

*Note.* Mixing ratios are per volume and refer to the surface or 1 bar level for the outer planets. Updated from Yung and DeMore (1999).
[a] At the flyby of New Horizons in July 2015.

The $CO_2$ atmospheres of Mars and Venus and the $N_2$–$CH_4$ atmospheres of Titan, Triton, and Pluto are the main subject of this book and the author's research.

## 1.3 Outer Planets

Jupiter is the largest planet, with mass of 318 $M_E$, exceeding the total mass of all other planets by a factor of 2.5. The most detailed study of the Jupiter system was made by the

Galileo orbiter and probe. The probe studied the Jupiter environment down to a pressure of 23 bar, where temperature was equal to 426 K. (The first Venus probe Venera 4 operated in 1967 down to 22 bar as well.) A dense rocky and metal core retrieved from the orbit evolution constitutes 4%–14% of the Jupiter mass or 13–45 $M_E$ (Earth masses). Other data on the internal structure are obtained by modeling. Temperature in the core may reach 35,000 K. Metallic hydrogen is above the rocky core and extends to 78% of the Jupiter radius. Hydrogen above this level is in a supercritical fluid state, where liquid and gas are rather similar. It is assumed that hydrogen is gas down to 1000 km from the visible surface and it is liquid below this level. Jupiter has internal heat sources that are equal to 0.7 times the absorbed solar radiation. These sources are residual heat from the original accretion, the current compression, and penetration of the heavier helium into the deep interior. The Jupiter system involves 67 satellites and faint rings.

Mass of Saturn is 95 $M_E$ and smaller than that of Jupiter by a factor 3.3. The internal structure reminds that of Jupiter. The rocky and metal core is estimated at 9–22 $M_E$ with temperature reaching 12,000 K. The metallic hydrogen layer is thinner than that in Jupiter, while the layer of the supercritical fluid is thicker. Similar to that on Jupiter, the outer layer of 1000 km thick is gaseous. Saturn radiates heat more than it receives from the Sun by a factor of 1.8. Saturn has a well-developed ring system that extends to three Saturn radii and consists of mostly water ice, the big satellite Titan (Table 1.2), and numerous small satellites.

While Jupiter and Saturn are considered as the gas giants, Uranus and Neptune may be classified as the ice giants. Uranus was studied by Voyager 2 during its flyby in 1986. Its axis is near the orbit plane resulting very unusual seasons on Uranus. This axis position may be caused by a collision with an Earth-size body. Mass of Uranus is 14.5 $M_E$, ices are 9–13 $M_E$, rocks and metals are 0.5–3.5 $M_E$, and hydrogen and helium are 0.5–1.5 $M_E$. The outer 20% of the Uranus radius is gas, the remaining being a mantle of mostly water and ammonia with a rocky-metal core. Temperature in the core center is evaluated at 5000 K. Internal heat is low on Uranus, so that the radiated heat is near the balance with the absorbed solar flux. Uranus has 27 satellites, including Titania ($R$ = 790 km), and rings.

Neptune and Uranus are twins with similar sizes, masses, and internal structures. Neptune's mass is 17.2 $M_E$; the axis position is rather usual with obliquity of 29°; the internal heat flow is significant and equal to 1.6 times the absorbed solar flux. Neptune was studied by the Voyager 2 flyby in 1989. Its internal structure is rather similar to that of Uranus. Neptune has a large satellite, Triton (Table 1.2); 13 small satellites; and faint rings.

## 1.4 Asteroids, Transneptunian Objects, and Comets

### 1.4.1 Asteroids

Asteroids are the small bodies in the Solar System that consist of rocks and metals. Their orbits are mostly concentrated between the orbits of Mars and Jupiter. There are ~$10^6$ asteroids that size exceeds 1 km and ~200 asteroids exceeding 100 km. The largest asteroid is Ceres with diameter of 950 km. The asteroid size distribution may be approximated by

$$N \approx 10^6 D^{-2}. \tag{1.7}$$

Here $N$ is the number of asteroids with size exceeding $D$ km. Then the total mass of the asteroid belt is

$$M_A = \frac{4\pi\rho}{3} \int_{D_{min}}^{D_{max}} \left(\frac{D}{2}\right)^3 \frac{dN}{dD} dD = 3 \times 10^{24} \mathrm{g} = 0.04 M_{\mathrm{M}}, \tag{1.8}$$

where $\rho \approx 3$ g cm$^{-3}$ is the asteroid mean density. Therefore the total mass of the asteroid belt is rather low and equal to 4% of the Moon mass (Table 1.2), and Ceres constitutes one-third of the total mass.

There are three types of the asteroids: carbonaceous, silicate, and metallic. Their relative populations are ~75%, 17%, and 8%, respectively. Asteroids significantly evolved since their formation by melting from impacts, micrometeorite bombardment, and space weathering. Therefore they are not samples of the primordial Solar System. Impacts of asteroids exceeding ~0.1 km had and may have catastrophic sequences for the biosphere and humans. For example, an asteroid impact ~50 million years ago resulted in the end of the dinosaur era. Therefore there are a few astronomical facilities that watch the asteroid environment to detect asteroids with the potentially hazardous orbits.

### *1.4.2 Kuiper Belt*

It was suggested that the mean density in the Solar System cannot drop to zero beyond the Neptune orbit, and there should be a belt at 30–50 AU populated by comparatively small objects that are poorly seen at such large distances. This is a so-called Kuiper belt. Pluto is currently assigned as a transneptunian or Kuiper belt object (TNO or KBO), and Triton is believed to originate from the Kuiper belt as well. Another large KBO is Eris with radius of 1160 km at 96.3 AU. It is slightly smaller than Pluto but exceeds Pluto in mass. Currently the number of the observed KBOs exceeds a thousand, and the number of KBOs with a size over 100 km is estimated at ~$10^5$. This value may be coupled with the power index of $-3 \pm 0.5$ in relationship (1.7) derived for KBOs. The total mass of the Kuiper belt is evaluated at 0.04–0.1 Earth mass based on the observational data, while models for the Solar System formation predict ~10 Earth masses.

Densities of Pluto, Triton, and some KBOs that have satellites are ~2 g cm$^{-3}$, that is, KBOs consist of comparable quantities of rocks and ices. Ices of water, carbon monoxide and dioxide, methane, nitrogen, and ammonia dominate in the KBOs.

### *1.4.3 Comets*

Comets are small bodies ranging in size from 0.1 to 30 km. They consist of ice and dust, whose proportion varies significantly from comet to comet, while the mean quantities are comparable. Therefore comets are sometimes defined as dirty snowballs. Their density is ~0.5 g $cm^{-3}$, indicating a porous structure. Solid parts of comets are called nuclei, and they are black with reflectivity of ~0.05.

Comets reside in the Öpik-Oort cloud that extends from ~2000 to ~50,000 AU and ends at a quarter of the distance to the closest star. There is a spherical outer cloud at (2 to 5)$\times 10^4$ AU and a doughnut-shaped inner cloud at 2,000–20,000 AU. The number of comets in the outer cloud is ~$10^{12}$, and their total mass is near 5 Earth masses. The inner cloud is more populated and supplies comets to the outer cloud. Perturbations by passing stars, galactic tides, and the outer planets make some comets either leave the Solar System or move to its inner part. Typically 5–10 comets appear in the inner Solar System annually.

Intense solar radiation in the inner Solar System results in evaporation of the cometary ices. This process in vacuum accelerates the gas to ~1 km $s^{-1}$ and drags the dust. Gas molecules dissociate by the solar UV photons, and radial outflow of dust, parent and daughter molecules, and radicals forms a beautiful phenomenon called a coma. A coma extends typically to ~$10^5$ km from the nucleus. The solar light pressure and the solar wind affect trajectories of radicals and ions and form cometary tails.

Composition of ices in comets is studied by spectroscopy of comas. Water dominates and constitutes ~90% of the ice; the remaining ~10% is CO, $CO_2$, $CH_4$, $NH_3$, and primitive organics (HCN, hydrocarbons, formaldehyde, methanol, etc.).

The composition of comets reflects composition of the primordial nebula and conditions in a region where comets formed. An early model of the Solar System formation by Safronov (1969) predicted formation of comets between Jupiter and Neptune and then removal of them by gravity of the giant planets to a periphery of the Solar System where they formed the Öpik-Oort cloud.

Tests of this prediction were made by sensitive searches for Ne and Ar in comets (Krasnopolsky and Mumma, 2001; Weaver et al. 2002) using the Extreme Ultraviolet Explorer and the Far Ultraviolet Spectroscopic Explorer, respectively. If comets were formed very far from the Sun at extremely low temperatures of ~10 K, then their Ne and Ar abundances would be similar to the solar values Ne/O = 0.15 and Ar/O = 0.005. However, the observed upper limits show that Ne and Ar are depleted in comets relative to the solar abundances by more than factors of 1000 and 10, respectively. This means that comets were formed at temperatures exceeding 30 and 60 K, respectively, in favor of the conditions between Jupiter and Neptune.

However, mass spectrometry of comet 67P/Churyumov-Gerasimenko by the Rosetta spacecraft resulted in detection of Ar with the Ar/$H_2O$ ratio in the range of (0.1 to 2.3)$\times 10^{-5}$ (Balsiger et al. 2015), smaller than the protosolar ratio by a factor of more than 200. This argon could be adsorbed during formation of the cometary grains.

## 1.5 Formation of the Solar System

If laws that govern a system and its current state are known, then the past and future of the system can be recognized and predicted. This rule is exact for mechanical systems, where coordinates and velocities of the system elements at a given time determine its behavior at any time. For example, there is a website for the astronomic observations with a code that calculates position of any body in the Solar System as seen from any point on the Earth at any chosen time in the past or future.

While basic ideas of the Solar System formation from a gas and dust nebula appeared three centuries ago, there are significant changes in the understanding and modeling of this phenomenon in the last decades. First of all, some observational data on various phases of presolar nebula became available from observations of other stars and nebulae. Furthermore, the planet positions vary in the current models while they were previously adopted rather stable.

A nearby supernova triggered a gravitational collapse of a fragment of a giant molecular cloud 4.6 Byr ago. Mass $m$ in the collapsing nebula has angular momentum $L = m\omega r^2$, where $\omega$ and $r$ are the angular velocity and distance relative to the mass center. If the angular momentum is conserved in the collapse, then the centrifugal force is

$$F_C = m\omega^2 r = \frac{L^2}{mr^3}; \tag{1.9}$$

that is, it increases steeper than gravity that is proportional to $r^{-2}$. This results in a so-called rotational instability that flattens the nebula into a spinning protoplanetary disk. The density increases in the disk, and collisions convert kinetic energy into heat. The disk radius was ~100 AU with a hot and dense protostar in the center. That protosun radiated due to the gravitational contraction. It looks like viscosity of the gas near the protosun was sufficient to transfer the protosun angular momentum to the disk. The formation of the disk and the protosun took a short time, ~0.1 Myr. The Sun became a T Tauri star with a very strong solar wind. The gravitational contraction continued for ~50 Myr, when temperature in the Sun core became high enough to start hydrogen fusion. The fusion energy prevented further contraction and resulted in hydrostatic equilibrium. The Sun became a main sequence star, and its lifetime in this phase is evaluated at ~10 Byr, so that the Sun is currently in the middle of this period.

Basic elements of the protoplanetary disk are metals (mostly Fe, Al, Ni), rocks (silicates), ices ($H_2O$, $CO_2$, $NH_3$, etc.), and gases ($H_2$ and He). The gases are most abundant and remain in the gas phase at any temperature in the disk, the ices remain solid below ~100 K, and the metals and rocks remain solid at any temperature in the disk. Direct contacts between dust grains and condensation of ices made clumps of $\leq 200$ m, and their collisions formed planetesimals with a size of ~10 km. Planetesimals could form embryos of ~0.05 Earth masses, and those could form the planets.

The inner part of the protoplanetary disk inside ~4 AU was warm and did not contain ices. Therefore only metals and rocks participated in formation of the inner planets, and that is why their masses are comparatively low and densities high. The inner planets formed in a rather dense gas and dust environment and gradually dragged inward to their current orbits.

The giant planets formed beyond 4 AU in the protoplanetary disk where temperature was low enough for the ices to exist. They had much more material to accrete and grew to

~10 Earth masses, sufficient to accrete the surrounding gas. Saturn appeared a few Myr after Jupiter, and it had less gas to consume. Even later, Uranus and Neptune formed. By that time the very strong solar wind of the T Tauri Sun blew away the most of the disk gas, and Uranus and Neptune could accrete about one Earth mass of $H_2$ and He. They formed near Saturn and then migrated to their present orbits. This migration was pushed by a 2:1 resonance between the orbits of Saturn and Jupiter and took ~0.5 Byr to end. The growth of the planets ceased a few Myr after the beginning of the Solar System formation.

# 2

# Atmospheric Structure

Considering problems of the atmospheric structure, we will mostly discuss variations of atmospheric properties with altitude. Variations with local time, season, and from place to place refer to the atmospheric dynamics and will be mentioned but not discussed in detail. A more detailed description of the subject may be found, e.g., in Chamberlain and Hunten (1987).

## 2.1 Barometric Formula and Its Versions

Atmospheric pressure $p$ is the weight of the overhead gas per unit area:

$$p = mgN. \tag{2.1}$$

Here $m$ is the mean molecular mass, which is equal to mass $\mu$ in atomic units divided by the Avogadro number $A = 6.022\times10^{23}\ \mathrm{g}^{-1}$; $g$ is the gravity acceleration; and $N$ is the column abundance of molecules. Actually $m$ and $g$ may slightly vary, while the derivative of (2.1) by altitude $z$ is exact and may be coupled with the gas law:

$$\frac{dp}{dz} = mg\frac{dN}{dz}; \quad \frac{dN}{dz} = -n; \quad p = nkT; \quad \frac{dp}{p} = -\frac{mg}{kT}dz = -\frac{dz}{H}; \quad H = \frac{kT}{mg}. \tag{2.2}$$

Here $n$ is the number density, i.e., the number of molecules per $\mathrm{cm}^3$, $k = 1.3806\times10^{-16}\ \mathrm{erg\ K}^{-1}$ is the Boltzmann constant, and $H$ is the scale height. Solution of the differential equation in (2.2) gives

$$p(z) = p(z_0)\exp\left(-\int_{z_0}^{z}\frac{dz}{H(z)}\right) \approx p(z_0)e^{-(z-z_0)/H}. \tag{2.3}$$

Therefore pressure is exponentially decreasing in the atmospheres with a scale height $H$, which is an increment in altitude that reduces atmospheric pressure by a factor of $e = 2.718\ldots$ Scale height is therefore a vertical characteristic size of an atmosphere; for example, substitution of the values for the Earth's troposphere in $H = kT/mg$ gives $H \approx 8$ km. The highest mountain is of 8.9 km, and pressure on the summit is ~$1/e \approx 0.35$ bar.

$H$ is the scale height for pressure, and a scale height $H^*$ for density may be of some interest:

$$\frac{dp}{p} = \frac{d(nT)}{nT} = \frac{dn}{n} + \frac{dT}{T} = -\frac{dz}{H}; \quad \frac{dn}{n} = -dz\left(\frac{1}{H} + \frac{1}{T}\frac{dT}{dz}\right) \equiv -\frac{dz}{H^*}; \quad \frac{1}{H^*} = \frac{1}{H} + \frac{1}{T}\frac{dT}{dz}. \tag{2.4}$$

Then column abundance is

$$N(z) = -\int_z^{\infty} n(x)dx = -n(z)\int_z^{\infty} e^{-\frac{x}{H^*}}\, dx \approx n(z)H^*. \tag{2.5}$$

### 2.1.1 Barometric Formula and Boltzmann's Statistics

According to Boltzmann's statistics, the number of molecules, whose energy exceeds $E$, is

$$N(E) = N_0 e^{-\frac{E}{kT}}.$$

$E = mgz$, and this results in the barometric formula using $n = -dN/dz$.

### 2.1.2 Temperature Gradient

In some cases a temperature profile may be approximated by a linear function with a constant temperature gradient $\alpha = dT/dz$. Then

$$T(z) = T_0 + \alpha z; \quad \frac{dp}{p} = -\frac{mg}{k(T_0 + \alpha z)}dz; \quad \ln p = -\frac{mg}{k\alpha}\ln(T_0 + \alpha z) + C, \tag{2.6}$$

and finally,

$$p(z) = p_0\left(1 + \frac{\alpha z}{T_o}\right)^{-\frac{mg}{k\alpha}}. \tag{2.7}$$

### 2.1.3 Temperature Gradient in the Troposphere

The troposphere is the lowest part of the atmosphere near the surface of a planet. Solar radiation in the visible range dominates in the troposphere; the air is transparent in the visible that heats the surface. The air gets heat from the surface and moves up adiabatically:

$$pV^{\gamma} = C_0; \quad pV = RT; \quad p = C_1 T^{\frac{\gamma}{\gamma-1}}.$$

The second relationship is the gas law with the gas constant $R = 8.3143\times10^7$ erg $K^{-1}$ $mole^{-1}$, and its substitution to the adiabat gives the third relationship. The derivative

$$\frac{dp}{dz} = \frac{\gamma}{\gamma-1}\frac{p}{T}\frac{dT}{dz}$$

may be compared to $dp/dz = -p/H$ in (2.2) to give

$$\frac{dT}{dz} = -\frac{\gamma - 1}{\gamma}\frac{T}{H} = -\frac{\gamma - 1}{\gamma}\frac{mg}{k}. \tag{2.8}$$

The adiabat index is $\gamma = 1 + 2/f$, where $f$ is the number of degrees of freedom for a gas molecule: $f = 3$ for monatomic, 5 for diatomic and linear molecules, and 6 for polyatomic molecules, if their vibrational excitation (see Section 3.8.2) is negligible. Heat capacity of gas under constant pressure is

$$C_p = \left(1 + \frac{f}{2}\right)\frac{k}{m} = \frac{\gamma}{\gamma - 1}\frac{k}{m}, \text{ therefore } \frac{dT}{dz} = -\frac{g}{C_p}. \tag{2.9}$$

For example, the mean $m = 28.9/A$ in the Earth's troposphere, $f = 5$ for diatomic gases, and $dT/dz = -9.8$ K km$^{-1}$. This is the dry adiabatic lapse rate. However, condensation of water vapor in the upward motions of the air heats the atmosphere, and the wet adiabatic lapse rate is $-5$ K km$^{-1}$. The mean temperature gradient is between these values at $\approx -6.5$ K km$^{-1}$. Vibrational excitation of $CO_2$ is very low at the Martian temperatures, $f \approx 5$, and the dry adiabatic lapse rate is 5.5 K km$^{-1}$. The $CO_2$ heat capacity increases at the high temperatures of 400–700 K typical of the Venus troposphere and reduces the lapse rate (Table 1.3).

### *2.1.4 Troposphere of Saturated Vapor*

Air in troposphere may be in equilibrium with ice on the surface. Saturated vapor pressure is

$$p_{si} = a_i e^{-\frac{L_i}{RT}}. \tag{2.10}$$

Here $L_i$ is the latent heat of condensation per mole, and $a_i$ is a constant of species $i$. Generally $L_i$ varies with temperature, and more complicated relationships are used as well. Adiabatic gradient results in condensation of the air, while the condensation heats the atmosphere. Finally, the temperature profile corresponds to the saturation conditions. Then derivative of (2.10) may be compared with (2.2):

$$\frac{dp}{dz} = \frac{Lp}{RT^2}\frac{dT}{dz} = -\frac{mg}{kT}p.$$

Taking into account that molecular mass in atomic units is $\mu = Am$ and $R = Ak$,

$$\frac{dT}{dz} = -\mu g T/L. \tag{2.11}$$

Atmospheres of saturated vapor over the surface ice are on Triton, Pluto, and the Martian winter polar caps. Latent heats per mole are 47, 33, and 6.3 kJ mole$^{-1}$ for $H_2O$, $CO_2$, and $N_2$, respectively.

## 2.2 Vertical Transport

Vertical fluxes of species significantly affect their vertical density profiles.

### 2.2.1 Molecular Diffusion

Molecular diffusion is a dominant transport process in the upper atmospheres. Stable long-living species reach diffusive equilibrium in the upper atmospheres, and partial pressure $p_i$ of species $i$ is determined by the barometric formula with a scale height $H_i = kT/m_i g$, where mean molecular mass is replaced by molecular mass of the species.

Inhomogeneities in gas distribution form diffusion fluxes in horizontal and vertical directions

$$\Phi_{ix} = -D_i \frac{\partial n_i}{\partial x} \text{ and } \Phi_{iz} = -D_i \left( \frac{\partial n_i}{\partial z} + Z \right).$$

The horizontal flux just reflects the definition of diffusion, while the vertical flux includes the unknown $Z$ to account for gravity and a vertical decline in $n_i$. The vertical flux is a subject of our interest and will be used without the subscript $z$. To calculate $Z$, we consider a case of the diffusive equilibrium, where $\Phi_i = 0$ and $n_i(z) = n_{i\,0} \exp(-z/H_i^*)$. $H_i^*$ is similar to $H^*$ in (2.4), and the substitution gives

$$\Phi_i = -D_i \left[ \frac{\partial n_i}{\partial z} + n_i \left( \frac{1}{H_i} + \frac{1 - \alpha_i}{T} \frac{\partial T}{\partial z} \right) \right]. \tag{2.12}$$

Here we add thermal diffusion factor $\alpha_i$, which is equal to 0.25 for hydrogen and helium and zero for other gases.

The diffusion coefficient $D_i$ is a product of thermal velocity and mean free path of the molecules $i$ with coefficient of 0.42 that reflects geometry of the process. It is proportional to $T^{0.5}/n$ using the hard-sphere approximation. A better approximation is

$$D = \frac{b}{n} = a\, 10^{17} \frac{T^{\gamma}}{n} \text{cm}^2\text{s}^{-1}. \tag{2.13}$$

Data on $a$ and $\gamma$ for various gas pairs may be found in Marrero and Mason (1972). Some values are given in Table 2.1. Diffusion coefficient is proportional to $\mu^{-1/2}$, where $\mu$ is the reduced mass:

$$\frac{1}{\mu} = \frac{1}{m_1} + \frac{1}{m_2}. \tag{2.14}$$

This dependence may be used if a required diffusion coefficient is unknown.

Abundances of species are usually determined by their mole fractions or mixing ratios $f_i = n_i / n$. Then (2.12) and (2.13) may be combined to give

$$\Phi_i = -b_i \left[ \frac{\partial f_i}{\partial z} + f_i \left( \frac{1}{H_i} - \frac{1}{H} - \frac{\alpha_i}{T} \frac{\partial T}{\partial z} \right) \right]. \tag{2.12a}$$

Its identity to (2.12) may be checked by substitution of $f_i = n_i / n$ into (2.12a).

Table 2.1 *Coefficients of molecular diffusion of gas 1 through gas 2*

| Gas 1 | H | | | $H_2$ | | | He | | O | | Ne | | Ar | $CH_4$ |
|---|---|---|---|---|---|---|---|---|---|---|---|---|---|---|
| Gas 2 | $H_2$ | air | $CO_2$ | Air | $CO_2$ | $N_2$ (100 K) | $CO_2$ | $N_2$ | $CO_2$ | $N_2$ | $CO_2$ | air | $CO_2$ | $N_2$ |
| $a$ | 14.5 | 6.5 | 8.4 | 2.7 | 2.2 | 1.9 | 2.7 | 0.97 | 0.92 | 1.15 | 0.77 | 0.67 | 1.26 | 0.734 |
| $\gamma$ | 0.61 | 0.7 | 0.60 | 0.75 | 0.75 | 0.82 | 0.72 | 0.774 | 0.75 | 0.743 | 0.776 | 0.749 | 0.646 | 0.749 |

### 2.2.2 Eddy Diffusion

Molecular diffusion becomes ineffective in lower atmospheres, where number density $n$ is large. Winds and turbulent motions tend to mix uniformly an atmosphere, and this process is called eddy diffusion. The dimension of eddy diffusion coefficient $K$, cm$^2$ s$^{-1}$, is similar to that of the molecular diffusion coefficient $D$; however, $K$ is the same for all species. Unfortunately, $K$ is poorly predictable from general properties of an atmosphere, but can be retrieved from observations of atmospheric inhomogeneities: natural and artificial (sodium) clouds, aircraft traces, and vertical distributions of gases with different masses.

Each atmosphere has a *homosphere*, where $K >> D$ and all stable and long-living species are uniformly mixed with constant *mixing ratios* or *fractions* $f_i = n_i / n$, and a *heterosphere*, where $D >> K$ and all stable and long-living species and distributed with scale heights that are determined by masses of the species. Units for mixing ratios are percentage, parts per million (ppm), and parts per billion (ppb).

Diffusive separation and enrichment of light species occurs in heterosphere. *Homopause* is the boundary between homosphere and heterosphere, where $K \approx D$. Homopause is at 100 km in the Earth's atmosphere and at ~125 km on Mars and Venus. Eddy diffusion in the Martian troposphere is $K \approx 10^6$ cm$^2$ s$^{-1}$.

Flux of species $i$ for eddy diffusion is similar to that for molecular diffusion but with mean scale height $H$ instead $H_i$:

$$\Phi_i = -K\left[\frac{\partial n_i}{\partial z} + n_i\left(\frac{1}{H} + \frac{1}{T}\frac{\partial T}{\partial z}\right)\right]. \tag{2.15}$$

This relationship is identical to

$$\Phi_i = -Kn\frac{\partial f_i}{\partial z}, \tag{2.16}$$

and this may be checked by substitution. If species is well mixed, its fraction is constant and the flux is zero. If both molecular and eddy diffusion are taken into account, then

$$\Phi_i = -(K + D_i)\frac{\partial n_i}{\partial z} - n_i\left[K\left(\frac{mg}{kT} + \frac{1}{T}\frac{\partial T}{\partial z}\right) + D_i\left(\frac{m_i g}{kT} + \frac{1-\alpha_i}{T}\frac{\partial T}{\partial z}\right)\right]. \tag{2.17}$$

The vertical fluxes considered here can contain a vertical wind component $n_i V_z$. However, this component is zero in global-mean models.

Actually, flux is a vector with three projections, and eddy diffusion has three components as well. General circulation models (GCMs) simulate large-scale atmospheric dynamics, a significant component in the vertical eddy diffusion. However, winds at scales smaller than grid sizes in GCMs cannot be simulated, and eddy diffusion produced by them is assumed in the GCMs.

The above relationships are valid in regions where species $i$ does not condense. Formation of aerosol layers by condensation will be considered in Section 4.5.

### 2.2.3 Ambipolar Diffusion

Ions have typically short lifetimes near peaks of their productions, and their diffusion becomes significant only in the topside ionosphere. This diffusion conforms to quasi-neutrality of the atmosphere; therefore this is diffusion of ion–electron pairs, that is, ambipolar diffusion with diffusion coefficient

$$D_{ai} = \frac{k(T_e + T_i)}{m_i \sum_k \nu_{ik}}. \tag{2.18}$$

Electrons and ions get excess energies in ionization processes and ion reactions. The numbers of collisions during their short lifetimes are insufficient for the full thermalization; therefore electron and ion temperatures $T_e$ and $T_i$ may significantly exceed the atmospheric temperature; $\nu_{ik}$ is the collision frequency between ion $i$ and neutral $k$. This frequency is determined by the Langevin formula

$$\nu_{ik} = 2\pi e n_k \left(\frac{\beta_k}{\mu_{ik}}\right)^{1/2}. \tag{2.19}$$

Here $\beta_k$ is the polarizability, $e$ is the electron charge, and $\mu_{ik}$ is the reduced mass (2.14). The vertical flux for ambipolar diffusion looks like that for molecular diffusion (2.12) with $T$ in $H_i$ replaced by $(T_e + T_i)/2$. Polarizabilities of some species are given in Table 2.2.

### 2.2.4 Time of Mixing and Diffusion

Vertical velocity of species in the lower and middle atmosphere may be evaluated as

$$|V| \approx \frac{|\Phi|}{n} \approx K\left(\frac{1}{n}\frac{\partial n}{\partial z} + \frac{1}{H}\right) \approx \frac{K}{H}.$$

Evidently this is a very approximate evaluation, and we even cannot state if it is upward (positive) or downward (negative). Mixing time is the time to pass the scale height with this velocity, that is,

$$\tau_m \approx \frac{H^2}{K}. \tag{2.20}$$

For example, if $H = 10$ km and $K = 10^6$ cm$^2$ s$^{-1}$, then mixing time is $\tau_m \approx 10^6$ s $\approx 12$ days. Similarly, diffusion time is $\tau_d \approx H_i^2/D_i$.

Table 2.2 *Polarizabilities (in $10^{-24}$ cm$^3$) of some species*

| Species | O | $N_2$ | $CO_2$ | $H_2$ | H | He |
|---|---|---|---|---|---|---|
| $\beta$ | 0.79 | 1.76 | 2.6 | 0.82 | 0.67 | 0.21 |

### 2.2.5 *Elemental Conservation*

Elements conserve in chemical reactions; therefore vertical flux of a sum of all species containing a given element and scaled by numbers of the element atoms in molecules is constant. This flux is equal to the element production or loss in the surface rocks and the space. These processes are typically very much slower than chemical reactions in the atmosphere, and vertical flux of an element may be assumed equal to zero in the homosphere. Then, according to equation (2.16), *an element mixing ratio is constant throughout the homosphere.*

This rule does not hold above homopause (because of the diffusion separation and enrichment of light species) and above condensation level of a species containing the element (because (2.16) becomes invalid here). Atmospheric dynamics can make perturbations in the atmosphere and break this rule in some points. However, an excess of element in some points would mean its shortage in other regions, so that the global mean mixing ratio is constant with altitude.

## 2.3 Thermal Balance

Temperature is the key parameter that determines the atmospheric structure. If a temperature profile is known, then profiles of pressure and density are known as well. Therefore problems of thermal balance are of great importance in physics of the planetary atmospheres.

### 2.3.1 *Blackbody Radiation*

Blackbody radiation is the basic issue in thermal balance. Full flux of the blackbody thermal radiation is $\sigma T^4$, and full brightness is $\sigma T^4/\pi$; here $\sigma = 5.67\times10^{-5}$ erg cm$^{-2}$ s$^{-1}$ K$^{-4}$ is the Stefan–Boltzmann constant. The monochromatic brightness in wavelengths is

$$B_\lambda = \frac{C_1}{\lambda^5\left(\exp\left(\frac{C_2}{\lambda T}\right)-1\right)};\, C_1 = 2hc^2 = 1.19\times10^{-5}\text{erg cm}^2\text{s}^{-1}\text{sr}^{-1},\ C_2 = hc/k$$
$$= 1.4388 \text{ cm K}. \tag{2.21}$$

Here $h = 6.626\times10^{-27}$ erg s is the Planck constant, $c = 3\times10^{10}$ cm s$^{-1}$ is the light speed, and $\lambda$ is in cm. The similar relationship for wavenumbers $\nu = 1/\lambda$ (cm$^{-1}$) is

$$B_\nu = \frac{2hc^2\nu^3}{\exp\left(\frac{C_2\nu}{T}\right)-1}\text{erg}\left(\text{cm}^2\text{ s cm}^{-1}\text{ sr}\right)^{-1}. \tag{2.22}$$

Radiations of the real bodies are those in (2.21) and (2.22) scaled by emissivity $\varepsilon = 1 - r$; here $r$ is the reflectivity. Formulas (2.21) and (2.22) are the Planck law; if $x = C_2/\lambda T << 1$, then $e^x - 1 \approx x$; if $x >> 1$, then $e^x - 1 \approx e^x$ (the Rayleigh–Jeans and Wien approximations, respectively).

### 2.3.2 Monochromatic Radiative Equilibrium

Suppose an atmosphere is optically thick in thermal infrared, say, in the $CO_2$ band at $\lambda = 15$ μm, and apply optical depth $\tau$ as a vertical coordinate: $\tau = 0$ at the upper boundary and $\tau = \tau_s$ at the surface. The Sun heats the surface in the visible range, and the surface radiates at 15 μm and heats the atmosphere; the solar radiation at 15 μm is negligible in this problem. There are upward intensity $I^+$, downward intensity $I^-$, mean intensity $I = (I^+ + I^-)/2$, and net intensity $F = I^+ - I^-$ at each $\tau$ in the atmosphere. Each layer heats the above layer by $I^+$ and is heated from the above layer by $I^-$. The net intensity is created by $dB/d\tau$, so that

$$\frac{dB}{d\tau} = \frac{3}{4}F. \tag{2.23}$$

The concept of radiative equilibrium means that the net intensity $F$ is constant. Then $I^+ = B + 0.5\,F$ and $I^- = B - 0.5\,F$, in accord with $I^+ - I^- = F$. At the top of the atmosphere,

$$I_0^- \equiv 0 = B(T_0) - 0.5\,F, \tag{2.24}$$

while at the surface,

$$B(T_S) = B(T_1) + 0.5\,F. \tag{2.25}$$

Here $T_0$ and $T_1$ are the temperatures at the top ($\tau = 0$) and bottom ($\tau = \tau_s$) of the atmosphere, respectively. Combining (2.23–25),

$$B(\tau) = B(T_0)(1 + 1.5\,\tau). \tag{2.26}$$

### 2.3.3 Mean Effective Temperature of a Planet

The mean effective temperature of a planet is

$$F = \frac{S}{4R^2}\,(1 - A) = \varepsilon\sigma T_e^4. \tag{2.27}$$

Here $S = 1361$ W m$^{-2}$ is the solar constant, $R$ is the heliocentric distance in AU, $A$ is the effective reflectivity (bolometric albedo), $\varepsilon \approx 0.5$ to 1, and the factor of 4 in the denominator is the sphere-to-disk surface ratio. $A = 0.29$ for the Earth, and $T_e = 255$ K for $\varepsilon = 1$. Extrapolating intensities at 15 μm to the whole thermal emissions, one finds from (2.24) that a so-called skin temperature is $T_0 = T_e/2^{1/4} = 214$ K, close to that of the Earth's tropopause. The global-mean air temperature on the Earth is $T_1 = 287$ K; then using (2.25) $T_S = (287^4 + 214^4)^{1/4} = 307$ K, and the mean difference between the surface and air temperatures is 20 K.

The column abundance of $CO_2$ on Mars exceeds that on Earth by a factor of 70, and the optical depth at 15 μm is large. The temperature gradient for radiative equilibrium, which may be calculated using (2.26), is large and its absolute value exceeds the adiabatic gradient $g/C_p$ = 5.5 K km$^{-1}$ (2.11). The atmosphere becomes convectively unstable and initiates convection with the adiabatic temperature gradient. Absorption and radiation of dust further reduces the temperature gradient.

### *2.3.4 Thermal Balance Equation*

The thermal balance equation is

$$\rho c_p \frac{dT}{dt} + \frac{d\emptyset_c}{dz} + \frac{d\emptyset}{dz} = q; \quad \emptyset_c = -\rho c_p K\left(\frac{dT}{dz} + \frac{g}{c_p}\right); \quad \emptyset = -\kappa\frac{dT}{dz}; \quad \kappa = aT^s. \qquad (2.28)$$

The equation means that changes in gas internal energy (the first term) are caused by convection, heat conduction, and absorption and radiation of energy (the last term $q$). This is a one-dimensional version of the equation; the three-dimensional equations look similar and are basic for the general circulation modeling.

The convective heat flux is proportional to eddy diffusion and equal to zero if temperature gradient is at the adiabatic value of $-g/c_p$. Thermal conductivity $\kappa$ is proportional to both density and mean free path and therefore does not depend on density. Parameters of thermal conductivity for some species are given in Table 2.3.

### *2.3.5 Structure of the Earth's Atmosphere*

The Sun heats the ground in the visible region of the spectrum, and thermal radiation at 15 μm from the ground heats the *troposphere* and stimulates convection with the wet adiabatic lapse rate of −6.5 K km$^{-1}$. The mean difference between temperatures of the ground and the nearby air is ≈20 K, and this discontinuity forms a *boundary layer* with a complicated thermal structure and of a few hundred meters thick. The troposphere extends up to ~11 km with the mean temperature decreasing from 287 K near the ground to 217 K at the *tropopause* (Figure 2.1).

Temperature increases in the *stratosphere* above the tropopause because of heating by the solar UV radiation at 200–320 nm that is absorbed by ozone and in the Herzberg continuum of $O_2$ at 200–240 nm. The ozone column abundance is ~0.3 cm-atm; 1 cm-atm = $2.69\times10^{19}$ cm$^{-2}$, that is, the Loschmidt number. Altitude profile of the $O_3$ number density

Table 2.3 *Thermal conductivity* $\kappa = aT^s$ *(erg cm$^{-1}$ s$^{-1}$ K$^{-1}$) of some gases*

| Gas | H | $H_2$ | $H_2$ | He | O | $N_2$ | $N_2$ | $N_2$ | $CO_2$ | $CH_4$ | $H_2O$ |
|---|---|---|---|---|---|---|---|---|---|---|---|
| T (K) | | <200 | >200 | | | <100 | 100–200 | >200 | | | |
| a | 235 | 90 | 295 | 354 | 54 | 9.4 | 13.4 | 30 | 0.64 | 9.83 | 0.95 |
| s | 0.75 | 0.94 | 0.73 | 0.665 | 0.75 | 1.0 | 0.93 | 0.78 | 1.38 | 1.03 | 1.33 |

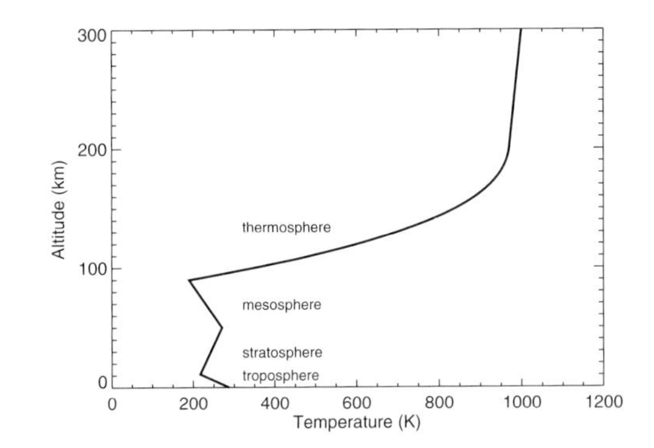


Figure 2.1 Structure of the Earth's atmosphere.

peaks near 25 km with a full width at half maximum of ≈20 km. The upper part of the layer gets the unattenuated solar radiation, refers to the atmosphere with the lower densities, and its heating effect is stronger than that from the lower part. The stratosphere ends near 50 km with temperature of 270 K.

Cooling by the $CO_2$ emission at 15 μm determines thermal balance of *mesosphere*, the atmosphere at 50 to 90 km. Collisions excite the upper state of this band, so that

$$\frac{[CO_2(1)]}{[CO_2(0)]} = \frac{g_1}{g_0} \exp\left(-\frac{h\nu}{kT}\right) \tag{2.29}$$

at the *local thermodynamic equilibrium* (*LTE*), *where collisions dominate and establish exact balance between all direct and inverse processes.* Here brackets mean number density of species inside, 0 and 1 refer to the ground and excited states for the emission of 15 μm, $g_0$ and $g_1$ are their statistical weights (see Chapter 3), and $h\nu$ is the photon energy.

The excited $CO_2(1)$ may either be quenched in collision and return its energy to heat the gas or radiate the photon. The former conforms to LTE, the latter has probability $A/(A + kn)$. Here $A = 1.35\ s^{-1}$ is the transition probability for the emission at 15 μm and $k = 2.5 \times 10^{-15}\ cm^3\ s^{-1}$ is the quenching rate coefficient for $CO_2(1)$ in the Earth's atmosphere. Both options become equally probable at $n = A/k = 5 \times 10^{14}\ cm^{-3}$, that is, near 80 km. This is a so-called level of vibrational relaxation. The $CO_2$ emission at 15 μm cools the mesosphere up to *mesopause* at 90 km, where heat conduction from *thermosphere* compensates for this cooling. Temperature of the mesopause is 190 K.

### 2.3.6 Water Vapor

Saturated pressure of $H_2O$ vapor is steeply decreasing with temperature with a mean rate of a factor of 3.4 per 10 K between 200 and 250 K. Its value over ice is

$$\ln p_{H2O}(\text{bar}) = -1.96250 - 5723.265/T + 3.53068\ \ln T - 0.00728332T \tag{2.30}$$

(Murphy and Koop 2005). The decreasing temperature in the troposphere results in a strong decline in the $H_2O$ mixing ratio from a few percent near the ground to a few parts per million (ppm) at the tropopause. The tropopause acts as a cold trap, and the $H_2O$ fraction is at a few ppm level up to the mesopause, where even these low abundances of $H_2O$ may condense and form *mesospheric or noctilucent clouds.*

## 2.4 Upper Atmosphere

### *2.4.1 Absorption of the Solar UV and EUV Radiation*

We divide the solar ultraviolet radiation into the near UV (400–300 nm), middle UV (300–200 nm), far UV (200–100 nm), and extreme UV (100–3 nm). This radiation will be considered in more detail in Chapter 8. Absorption of the UV radiation is mostly continuous with minor effects of absorption lines. Therefore the simple law of the exponential absorption may be applied, and formation of product $i$ from species with number density $n(z)$ is

$$P_i(z) = n(z)\sum\nolimits_{\lambda} I_\lambda \gamma_{i\lambda}\sigma_\lambda \exp\left(-\sec\theta \sum\nolimits_{K} \sigma_{k\lambda} n_k(z) H_k(z)\right). \quad (2.31)$$

Here $I_\lambda$ is the solar photon intensity at wavelength $\lambda$, $\gamma_{i\lambda}$ is the quantum yield of product $i$ in photodissociation of the given species at $\lambda$, $\sigma_\lambda$ is the cross section of the species at $\lambda$, $\theta$ is the solar zenith angle, $n_k$ and $H_k$ are the number densities and scale heights of species $k$.

Heating efficiency $\varepsilon$ is typically used to calculate heating of an upper atmosphere by the solar FUV or EUV radiation. This value is usually applied to the whole spectral range (FUV or EUV) and all species in the atmosphere, though all these details are taken into account when calculating $\varepsilon$. Then heating rate is

$$q(z) = \varepsilon\sum\nolimits_{\lambda} I_\lambda \sum\nolimits_{i} n_i(z)\sigma_{i\lambda} \exp\left(-\sec\theta \sum\nolimits_{k} \sigma_{k\lambda} n_k(z) H_k(z)\right). \quad (2.32)$$

Here $I_\lambda$ is in the energy units.

Both (2.31) and (2.32) are sums of the layers that are approximated by $A\tau e^{-\tau/\cos\theta}$ with $\tau = \tau_0 e^{-z/H}$. These so-called Chapman layers have maxima at $\tau/\cos\theta = 1$, their vertical integrals are $A \cos\theta$, and maximum values are $\frac{A \cos\theta}{2.72\, H}$.

### *2.4.2 Ionosphere*

If an effective EUV photon flux $I$ ionizes an atmosphere with density $n$, scale height $H$, and ionization cross section $\sigma$, then the ionization is balanced by recombination with coefficient $\alpha$:

$$I\sigma n\, e^{-\sigma n H \sec\theta} = \alpha n_e^2. \quad (2.32a)$$

Here $n_e$ is the electron density that is equal to the total density of positive ions. The ionization is maximum at $\sigma n H \sec\theta = 1$, and a peak electron density is

$$n_{e\ \max} = \left(\frac{I\cos\theta}{2.7\alpha H}\right)^{1/2}. \quad (2.32b)$$

Dissociative recombination of molecular ions is fast with $\alpha \approx 10^{-7}$ to $10^{-6}$ cm$^3$ s$^{-1}$, and plasma diffusion is neglected in (2.32a, b). This is an E-type ionosphere. Recombination of atomic ions is radiative and slow with $\alpha \approx 3\times10^{-12}$ cm$^3$ s$^{-1}$. If the neutral atmosphere is mostly atomic, then recombination proceeds in two steps: an ion-neutral reaction with a molecular species that forms a molecular ion that then recombines. For example, on the Earth above 150 km:

$$\mathrm{O} + h\nu(\lambda < 911\,\mathring{A}) \rightarrow \mathrm{O}^+ + \mathrm{e},$$

$$\mathrm{O}^+ + \mathrm{N}_2 \rightarrow \mathrm{NO}^+ + \mathrm{N},$$

$$\mathrm{NO}^+ + \mathrm{e} \rightarrow \mathrm{N} + \mathrm{O}.$$

The plasma density is proportional to the O/$N_2$ ratio and increases with altitude because of diffusive separation up to a level where ambipolar diffusion dominates. This is an F-type ionosphere.

### 2.4.3 Thermosphere

Thermosphere is a region above mesopause, where heating by absorption of the solar EUV radiation and heat conduction to the mesopause dominate in the thermal balance. Considering a steady state problem and neglecting temporal variations, the first term in the thermal balance equation (2.28) may be removed as well as the convection term, and the equation for thermosphere is

$$-\frac{d}{dz}\left(aT^s\frac{dT}{dz}\right) = q(z). \tag{2.33}$$

This equation is in the rectangular coordinates for a plane-parallel atmosphere, and the left side is the $z$-component of $-\mathrm{div}(\kappa\partial T/\partial \boldsymbol{r})$. Here $\boldsymbol{r}$ is a vector with three components. The spherical atmosphere is a better approximation to the problem; using the radial component of divergence, the equation is

$$-\frac{1}{r^2}\frac{d}{dr}\left(r^2 aT^s\frac{dT}{dr}\right) = q(r) \tag{2.34}$$

and altitude $z$ is replaced by distance from the planet center $r$. The boundary conditions are $T(r_0) = T_0$ and $dT/dr = 0$ at infinity, that is, no heat conduction to space (the closed boundary). The first integration and the condition at infinity give

$$aT^s\frac{dT}{dr} = -\frac{1}{r^2}\int_r^\infty q(x)x^2dx,$$

and the second integration is

$$\frac{a}{s+1}\left(T^{s+1} - T_0^{s+1}\right) = \int_{r_0}^r \frac{dr}{r^2}\int_r^\infty q(x)x^2dx. \tag{2.35}$$

The simplest test of this solution is assuming that $s = 1$ and the EUV heating $q(r)$ is concentrated in a narrow layer at $r_1$ with a column heating rate $Q$. Then the second integral is equal to zero for $r > r_1$ and $Qr_1^2$ for $r < r_1$, and the solution is

$$T(r \le r_1) = T_0\left[1 + \frac{2Qr_1^2}{aT_0^2}\left(\frac{1}{r_0} - \frac{1}{r}\right)\right]^{1/2} \text{ and } T(r > r_1) = T(r_1). \qquad (2.36)$$

Although the approximation (2.36) of a narrow heating layer is very simplified, it reflects main features of the thermospheric temperature profile: a strong temperature gradient in the lower thermosphere $r_0 < r < r_1$ and the isothermal upper thermosphere at $r > r_1$. Temperature of the upper thermosphere is called exospheric temperature $T_\infty$. The Earth's thermosphere extends up to ~400 km, and its upper isothermal part is not shown in Figure 2.1.

Numerical solution of (2.35) may be calculated by iterations. For example, a constant temperature profile $T(r) = T_0$ may be adopted as an initial approximation, and vertical profiles of number densities of main species in the atmosphere have to be calculated assuming constant mixing ratios below the homopause and diffusive separation with individual scale heights above the homopause. Then heating rate $q(r)$ is calculated using (2.32) with the solar EUV intensities $I_\lambda$ and cross sections $\sigma_{i\lambda}$ (see Chapter 8). The calculated $q(r)$ is put into (2.35) to get the first approximation to $T(r)$. This cycle is repeated until differences between the successive approximations become small.

The solar EUV radiation is nearly proportional to the solar activity index $F_{10.7}$ (Section 8.2), and the right side of equation (2.35) may be adopted proportional to $F_{10.7}$ as well. Then

$$T_\infty = \left(T_0^{s+1} + \beta F_{10.7}\right)^{\frac{1}{s+1}}. \qquad (2.36a)$$

Usually, $T_0$ is known, and even the only measurement of $T_\infty$ is sufficient to predict variations of exospheric temperature with solar activity. Relationship (2.36a) looks preferable to the widespread linear approximation $T_\infty = \alpha + \beta F_{10.7}$.

### 2.4.4 Thermal Effects of Eddy Diffusion

Atmospheric mixing by eddies in the lower thermosphere enhances heat conduction to the mesopause and cools the thermosphere with a rate $q_m$. However, dissipation of turbulent energy becomes effective near homopause and heats the lower thermosphere with a rate $q_d$. Their values are

$$q_m = -\frac{\partial}{\partial z}\left[\rho c_p K\left(\frac{\partial T}{\partial z} + \frac{g}{c_p}\right)\right], \quad q_d = \frac{\rho g K}{TR_{f0}}\left(\frac{\partial T}{\partial z} + \frac{g}{c_p}\right). \qquad (2.37)$$

The effects are opposite and partially compensating; therefore they are usually neglected. $R_{f0}$ is the Richardson number; it is ≈0.2 in the Martian middle atmosphere.

### 2.4.5 Exosphere

A vacuum is defined as a medium with a size that is equal or smaller than the mean free path. Density exponentially decreases in an atmosphere and reaches a level, where mean free path $l = 1/\sigma n = H$. $H$ is the characteristic size of the atmosphere, and this level is the upper boundary of the atmosphere and called exobase. Collision cross section $\sigma$ is typically $\approx 3 \times 10^{-15}$ cm$^2$, the vertical column abundance is $\approx 3 \times 10^{14}$ cm$^{-2}$, and the number densities are $\approx 10^7$–$10^8$ cm$^{-3}$ at exobases. Exosphere is the atmosphere above the exobase.

The thermosphere is isothermal to some extent below the exobase. It is usually assumed isothermal in the exosphere as well, though the term "temperature" is questionable in the collisionless exosphere. However, the Boltzmann statistics remains applicable, and probability for a molecule $i$ at the exobase $z_0$ to reach altitude $z$ in the exosphere is proportional to $\exp(-m_i g(z - z_0)/kT)$. Here $T$ is the temperature at the exobase; therefore the barometric formula holds and exosphere may be considered isothermal with temperature of the exobase.

Gravity $g(r) = \gamma M/r^2$ is the only variable value in the scale height formula for the exosphere. Here $\gamma = 6.670 \times 10^{-8}$ g$^{-1}$ cm$^3$ s$^2$ is the gravitational constant and $M$ is mass of a planet. Then the barometric formula transforms into

$$n_i(z) = n_i(z_0) \exp\left(-\int_{z_0}^{z} \frac{dz}{H_i}\right) = n_i(z_0) \exp\left(-\frac{\gamma M m_i}{kT} \int_{r_0}^{r} \frac{dr}{r^2}\right)$$

$$= n_i(z_0) \exp\left(\frac{\gamma M m_i}{kT}\left(\frac{1}{r} - \frac{1}{r_0}\right)\right),$$

and

$$n_i(\lambda) = n_i(\lambda_0) e^{\lambda - \lambda_0}; \quad \lambda = \frac{\gamma M m_i}{rkT} = \frac{r}{H_i}. \tag{2.38}$$

Here $\lambda$ is the structure parameter.

This relationship is valid for diffusive equilibrium without escape. It is modified using partition functions $\zeta(\lambda)$ for ballistic, satellite, and escaping particles:

$$\frac{n}{n_0} = \frac{\zeta(\lambda)}{\zeta(\lambda_0)} e^{\lambda - \lambda_0}. \tag{2.38a}$$

These functions were calculated by J. W. Chamberlain and may be found in Chamberlain and Hunten (1987).

### 2.4.6 Vertical and Limb Column Abundances

Vertical column abundance of species $i$ is $N_i = n_i H_i^*$ (2.5), and column abundance $N_{iL}$ on the limb at impact parameter $r$ from the planet center may be calculated by proper integration:

$$N_{iL} = n_i (2\pi r H_i)^{1/2}. \tag{2.39}$$

Hence the limb airmass $\beta = (2\pi r/H_i)^{1/2}$ is ~30 on Mars and ~80 on Venus. Therefore limb observations provide significant advantages to search for minor species. The decrease in gravity with $r$ results in the following corrections to these formulas for the exosphere:

$$N_i = n_i H_i \left(1 + \frac{2}{\lambda}\right) \text{ and } N_{iL} = n_i (2\pi r H_i)^{1/2} \left(1 + \frac{9}{8\lambda}\right). \tag{2.40}$$

## 2.5 Escape Processes

Escape processes determine evolution of the planetary atmospheres. Usually gases immediately fill vacuum, and all pumps utilize this phenomenon. It looks like the atmospheres do not conform to this expectation, do not escape to space and exist since the formation of the Solar System. This is because of the planet gravity that reduces the atmospheric density with scale height $H$ and prevents the gas to leave the planet. However, fast molecules may escape from exobase $r_e$ if their total energy is positive:

$$E = \frac{mv^2}{2} - \frac{\gamma M m}{r_e} > 0, \text{ that is, } \quad v > \left(\frac{2\gamma M}{r_e}\right)^{1/2} = v_e. \tag{2.41}$$

Here $v_e$ is the parabolic, or second space, or escape velocity (Tables 1.1 and 1.2).

### *2.5.1 Thermal Escape*

The Maxwell–Boltzmann velocity distribution in gas is

$$dn = 4\pi n \left(\frac{m}{2\pi kT}\right)^{3/2} \exp\left(-\frac{mv^2}{2kT}\right) v^2 dv, \tag{2.42}$$

and the mean velocity of gas molecules is $\left(\frac{8kT}{\pi m}\right)^{\frac{1}{2}}$. Evidently velocities of light molecules, e.g., H and $H_2$, may be high enough to exceed $v_e$ at the tail of the velocity distribution. For example, escape velocity on Mars is 5.0 km $s^{-1}$, while the mean velocity of hydrogen atom is 2.3 km $s^{-1}$ at the exospheric temperature of 250 K. Vertical flux is $d\Phi_e = vdn/4$, and the escaping flux is

$$\Phi_e = \frac{1}{4} \int_{Ve}^{\infty} v dn = \frac{n}{2} \left(\frac{2kT}{\pi m}\right)^{1/2} (1 + \lambda_e) e^{-\lambda_e} = nV_f. \tag{2.43}$$

This is the Jeans formula for thermal escape that is obtained by the simple integration of (2.42). The escaping flux decreases exponentially with $\lambda_e = \frac{\gamma M m}{r_e kT}$. For example, $\lambda_e = 5.7$ for H on Mars at $T_\infty = 250$ K, and the bulk flow velocity $V_f = 1300$ cm $s^{-1}$. For $H_2$ and He, $\lambda_e = 11.4$ and 22.8, with $V_f = 5.7$ and $9 \times 10^{-5}$ cm $s^{-1}$. Therefore thermal escape is significant only for very light species.

Donahue (1969) derived a relationship for the vertical profile of escaping species:

$$n(z) = n_e(z) \left(1 - \int_{z_0}^{z} \frac{\Phi(x) dx}{(D(x) + K(x)) n_e(x)}\right), \tag{2.43a}$$

where $n_e(z)$ is the diffusion equilibrium profile for $\Phi = 0$ and $z_0$ is a reference level that may be the homopause.

### 2.5.2 Nonthermal Escape

Photodissociation and photoionization processes, ion reactions, and collisions with the solar wind particles may release some excess energy into kinetic energy of their products that may be sufficient to escape. Dissociative recombination

$$O_2^+ + e \rightarrow O + O + 6.96\ \text{eV}$$

releases oxygen atoms in the ground and metastable states with a mean velocity of 5.5 km $s^{-1}$ that exceeds the escape velocity of 5.0 km $s^{-1}$ on Mars. This is the most important process of nonthermal escape on Mars. Thermal escape of O is negligible on Mars.

The hot O atoms collide with He atoms and $H_2$ molecules and push them to escape. Photo- and electron impact ionization above the ionopause (the upper boundary of ionosphere with the solar wind phenomena dominating above it, ~300 km on Mars) result in sweeping out of the ions by the solar wind. $O^+$ ions accelerated in the induced magnetosphere of Mars and Venus result in sputtering of species as well. These are processes of nonthermal escape.

### 2.5.3 Hydrodynamic Escape

We concluded in Section 2.4 that the vertical column is $N_e = 1/\sigma \approx 3\times10^{14}$ $cm^{-2}$ at the exobase. Using (2.40) and (2.38),

$$N_e = n_e H_e \left(1 + \frac{2}{\lambda_e}\right) = n_0 r_0 \lambda_0\, e^{-\lambda_0}\, C(\lambda_e);\quad C(\lambda_e) = \frac{e^{\lambda_e}}{\lambda_e^2}\left(1 + \frac{2}{\lambda_e}\right). \tag{2.44}$$

Here $r_0$ is any point in the isothermal part of thermosphere. The function $C(\lambda_e)$ is minimal and equal to 3.5 at $\lambda_e = 6^{1/2} = 2.45$; therefore if $n_0 r_0 \lambda_0 e^{-\lambda_0} > 10^{14}$ $cm^{-2}$, then the isothermal thermosphere does not have exobase. This means that rather high gas column densities exist at significant distances from a planet, where the gravitational energy is comparable to thermal energy of molecules that may escape to space as a hydrodynamic gas flow. The expanding gas cools and accelerates to the speed of sound at $\lambda = 2$. Gas flow from comets is the extreme case of hydrodynamic escape with a negligible gravity. Venus and Mars could lose most of their water by hydrodynamic escape in the first ~0.5 Byr.

Except for the existence of exobase (2.44) in the isothermal thermosphere, hydrodynamic escape may be tested by numerical solution of (2.35) for thermal balance in thermosphere:

$$\frac{a}{s+1}\left(T^{s+1} - T_0^{s+1}\right) = \int_{r_0}^{r} \frac{dr}{r^2} \int_{r}^{\infty} q(x) x^2 dx.$$

If iterations for (2.35) do not converge, this is an indication that the stable isothermal thermosphere cannot exist and hydrodynamic escape occurs. This test is more sensitive than that of the exobase.

To describe hydrodynamic escape, Watson et al. (1981) used dimensionless variables

$$\lambda = \frac{\gamma M m}{r k T_0}; \quad \tau = \frac{T}{T_0}; \quad \psi = \frac{m v^2}{k T_0}; \quad f^2 = \frac{F k}{4\pi \kappa_0 r_0 \lambda_0}; \quad \beta = S \frac{r_0 \lambda_0}{\kappa_0 T_0}. \tag{2.45}$$

Here $\lambda$ refers to variable $r$ but for the fixed $T_0$, $v$ is the bulk velocity, $F$ is the total outflow referred to $4\pi$ sr, $\kappa = aT^s$ is thermal conductivity, $S$ is the EUV and UV heating in erg cm$^{-2}$ s$^{-1}$, and subscript 0 refers to the lower boundary. Equations for the mass, momentum, and energy conservation are

$$F = 4\pi r^2 n v, \tag{2.46}$$

$$\frac{d\psi}{d\lambda} = 2\,\frac{\dfrac{d\tau}{d\lambda} + \dfrac{2\tau}{\lambda} - 1}{\dfrac{\tau}{\psi} - 1}, \tag{2.47}$$

$$\frac{\tau^s}{f^2}\frac{d\tau}{d\lambda} = \lambda - e - (1 + j/2)\tau - \psi/2. \tag{2.48}$$

Here $j$ is the number of degrees of freedom for a gas molecule ($j$ = 3, 5, 6 for atoms, linear and other molecules, respectively) and $e$ is the total energy per escaping molecule in $kT_0$,

$$e = \frac{4\pi}{F k T_0} \int_r^\infty q(x) x^2 dx, \tag{2.49}$$

and $q(x)$ is calculated using (2.32). If the solar energy is released in a narrow layer at $\lambda_1$, then $e = 0$ for $\lambda < \lambda_1$ and $e = \frac{\beta}{f^2 \lambda_1^2}$ for $\lambda > \lambda_1$.

The flow velocity is equal to speed of sound at $\tau = \psi$. To avoid the divergence of (2.47) at this sonic level (subscript $c$), its numerator should be zero. Then

$$f^2 = b\frac{1 - 2b}{1 - 4b}; \quad b = \frac{\tau_c}{\lambda_c}. \tag{2.50}$$

However, the ambiguity 0/0 at the sonic level in (2.47) affects stability of the numerical solution. The problem for Pluto was solved by McNutt (1989) and Krasnopolsky (1999) using some approximations and by Strobel (2008) and Zhu et al. (2014). The assumption that thermal conductivity does not depend on density becomes questionable at extremely low densities far from a planet in this problem. Direct Monte Carlo simulations of a fluid/kinetic model (Johnson 2010) were used to solve this problem for Pluto's atmosphere. However, the New Horizons observations rule out hydrodynamic escape from the current atmosphere of Pluto (Chapter 15).

### 2.5.4 Diffusion Limit

Here we will compare a vertical profile $n_{ie}(z)$ of species $i$ in equilibrium to that $n_i(z)$ for escaping flux $\Phi_i$. Using (2.4),

$$\frac{1}{H_{ie}^*} = \frac{1}{H_{ie}} + \frac{1-\alpha_i}{T}\frac{dT}{dz} = -\frac{1}{n_{ie}}\frac{dn_{ie}}{dz}; \qquad \frac{1}{H_i^*} = -\frac{1}{n_i}\frac{dn_i}{dz}.$$

As in (2.4), $H^*$ is the density scale height. Then relationships (2.16) and (2.17) may be written as

$$\Phi_i = -D_i\left[\frac{dn_i}{dz} + n_i\left(\frac{1}{H_i} + \frac{1-\alpha_i}{T}\frac{dT}{dz}\right)\right] - Kn\frac{df_i}{dz} = -D_i n_i\left[\frac{1}{n_i}\frac{dn_i}{dz} - \frac{1}{n_{ie}}\frac{dn_{ie}}{dz}\right] - Kn\frac{df_i}{dz}$$

$$= -b_i f_i\left(\frac{1}{H_{ie}^*} - \frac{1}{H_i^*}\right) - Kn\frac{df_i}{dz}.$$

Here $b_i$ is from $D_i = b_i/n$, and $n$ is the total number density. A light species should be enriched in the upper atmosphere above the homopause, and a limiting flow corresponds to no enrichment with $df_i/dz = 0$ and $H_i^* = H^* \approx H$. Then

$$\Phi_i^{\max} = b_i f_i\left(\frac{1}{H} - \frac{1}{H_{ie}^*}\right) = \frac{b_i f_i}{H}\left(1 - \frac{m_i}{m}\right) \approx \frac{b_i f_i}{H}. \tag{2.51}$$

This is a so-called diffusion limit. Escape flux at any conditions cannot exceed the diffusion limit that depends on a species mixing ratio in the homosphere and does not depend on eddy diffusion and position of homopause.

Consider the case of $H_2$ in $CO_2$ on Mars:

$$\begin{aligned} D_{H_2/CO_2} &= 2.27\times10^{17}T^{0.75}\,e^{-11.7/T}/n\ \text{cm}^2\text{s}^{-1};\ f_{H_2} = 1.7\times10^{-5},\ T\approx120\ \text{K},\\ \Phi_{H_2}^{\max} &= 4.2\times10^{13}\frac{e^{-11.7/T}}{T^{0.25}}f_{H_2} = 1.14\times10^{13}f_{H_2} = 2\times10^{8}\ \text{cm}^{-2}\text{s}^{-1}. \end{aligned} \tag{2.52}$$

The observed escape flux is $\Phi_H + 2\,\Phi_{H2} = 2\times10^8$ cm$^{-2}$ s$^{-1}$, that is, half the diffusion limit.

# 3

# Spectroscopy

## 3.1 Quantum Mechanics and Schroedinger Equation

The physical base of spectroscopy is quantum mechanics. Its main principle is the unity of the wave and the corpuscular properties of the matter (wave particle dualism). According to this principle, the motion of any particle is associated with a wave of length $\lambda = h/p$. Here $h = 6.626\times10^{-27}$ erg s is the Planck constant and $p$ is the particle momentum. The wave motion is described by a so-called wave function of coordinates and time $\Psi(\boldsymbol{r}, t)$; bold symbols represent vectors. It gives the probability of $\Psi\Psi^*$ that a particle is located at a given point $\boldsymbol{r}$ in space at time $t$. Here the asterisk denotes the complex conjugate. We remind that complex numbers include $i = \sqrt{-1}$ and are $a + bi$ or $ae^{i\varphi} = a(\cos\varphi + i \sin\varphi)$; the complex conjugates are $a - bi$ or $ae^{-i\varphi}$, and products of the complex and complex conjugate numbers are $a^2 + b^2$ or $a^2$, respectively.

Some formulas from classic mechanics can be carried over to quantum mechanics, if quantities are substituted by operators. For example, operators of momentum and energy are

$$\hat{P} = -i\,\hbar\nabla,\ \ \nabla = \boldsymbol{i}\frac{\partial}{\partial x} + \boldsymbol{j}\frac{\partial}{\partial y} + \boldsymbol{k}\frac{\partial}{\partial z},\ \ \hat{E} = i\hbar\frac{\partial}{\partial t}\ . \tag{3.1}$$

Here $\hbar = h/2\pi$, $\nabla$ is the vector differential operator, and $\boldsymbol{i}, \boldsymbol{j}, \boldsymbol{k}$ are the basic vectors of the rectangular coordinate system. Hamilton's function in classic mechanics, which describes the total particle energy $H = p^2/2m + V$ (where $V$ is potential energy), results in the Schroedinger wave equation by substitution of the operators:

$$i\hbar\frac{\partial\Psi}{\partial t} = -\frac{\hbar^2}{2m}\nabla^2\Psi + V(\mathrm{r}, t)\Psi,\ \ \nabla^2 = \frac{\partial^2}{\partial x^2} + \frac{\partial^2}{\partial y^2} + \frac{\partial^2}{\partial z^2}\,. \tag{3.2}$$

Usually potential energy does not depend on time; then the variables may be separated: $\Psi(\boldsymbol{r},t) = \varphi(t)\ \psi(\boldsymbol{r})$, and we have after substitution

$$\frac{i\hbar}{\varphi}\frac{\partial\varphi}{\partial t} = -\frac{\hbar^2\nabla^2\psi}{2m\psi} + V(\mathrm{r}) = E. \tag{3.3}$$

$E$ does not depend on either $\boldsymbol{r}$ or $t$, that is, it is a constant, a scalar integral of motion, a conserved quantity. Of the three integrals of motion (momentum, angular momentum, and energy), only energy is a scalar, therefore $E$ is energy. Then the equation for $\varphi$ is $i\hbar\frac{d\varphi}{dt} = E\varphi$ (or $\hat{E}\varphi = E\varphi$); its solution is

$$\varphi(t) = \varphi_0 \exp\left(-\frac{iEt}{\hbar}\right). \quad (3.4)$$

The equation for $\psi$ is

$$\left(\frac{\hbar^2}{2m}\nabla^2 + E - V(r)\right)\psi = 0. \quad (3.5)$$

Its solution depends on the type of the potential energy $V(\boldsymbol{r})$.

An important sequence of quantum mechanics is the uncertainty principle that links the quantities in the operators:

$$\Delta p \times \Delta r \approx \hbar, \quad \Delta E \times \Delta t \approx \hbar\,. \quad (3.6)$$

These relations have analogs in the classic wave motion:

$$\Delta k \times \Delta x \approx 1(k = 2\pi/\lambda), \ \Delta\omega \times \Delta t \approx 1(\omega \text{ is the angular frequency}).$$

## 3.2 Hydrogen-like Atoms: Energy Levels and Quantum Numbers

The problem of electron motion in a central electric field is among the simplest tasks of quantum mechanics. This is a two-body problem that has an exact solution and makes it possible to test the laws that control the system. These are a motion of a body in the gravitational field in classic mechanics, the spectrum and structure of the hydrogen atom in atomic physics, and the deuterium nucleus in nuclear physics.

This problem for the hydrogen atom and hydrogen-like ions $He^+$, $Li^{++}$, etc. is solved substituting $V(r) = -\frac{ze^2}{r}$ and $\nabla^2$ in the spherical coordinates into equation (3.5); $e = 4.8\times10^{-10}\ g^{1/2}\ cm^{3/2}\ s^{-1}$ is the electron charge. Then the spatial variables are separated using $\psi(r,\theta,\varphi) = R(r)\Theta(\theta)\Phi(\varphi)$. This gives three equations that determine three components of the wave function, three motion integrals with three quantum numbers.

1 Energy is

$$E_n = -\frac{z^2e^4\mu}{2\hbar^2n^2} \approx -Ry\,\frac{z^2}{n^2}\,. \quad (3.7)$$

Here $n \geq 1$ is an integer and the main quantum number. The negative values for energy correspond to the bound electron and are discrete. The solution for $E > 0$ is continuous and refers to a free electron. Hence the ionization potential of hydrogen is

$$E_i \approx Ry = \frac{e^4m_e}{2\hbar^2} = 109{,}737\ \text{cm}^{-1} \approx 13.6\ \text{eV}. \quad (3.8)$$

$Ry$ is the Rydberg constant that is used as an energy unit in atomic physics. A small difference between $E_i$ and $Ry$ reflects the difference between the hydrogen reduced mass $\mu = m_e m_p/(m_e + m_p)$ and $m_e$; $m_e = 9.11\times10^{-28}$ g is the electron mass and $m_p = 1836\, m_e$ is the proton mass.

2 Modulus of the electron angular momentum (we will call it "angular momentum" or just "momentum" for short):

$$M = \hbar\sqrt{l(l+1)}\,, \tag{3.9}$$

where $l$ is an integer in the range $n > l \geq 0$, with $n$ possible values.

3 Projection of momentum in the $z$-direction, $M_z = \hbar m$, where $|m| \leq l$ is an integer. Each $l$ gives $2l + 1$ values of $m$, and each $n$ results in $n^2$ different combinations of $l$ and $m$.

Electron energy does not depend on $l$ and $m$. The latter is caused by spherical symmetry of the field and therefore trivial, while the former contradicts the classic mechanics that connects energy and momentum.

For each value of energy $E_n$ there are $n^2$ solutions that correspond to $n^2$ electronic states. Such a level is called $n^2$-degenerated, or having statistical weight $g_n = n^2$. Statistical weights are important in probability processes. For example, the ratio of populations of atoms at levels $E_i$ and $E_k$ at thermodynamic equilibrium is

$$\frac{N_i}{N_k} = \frac{g_i}{g_k}\exp\left(-\frac{E_i - E_k}{kT}\right). \tag{3.10}$$

Transition probabilities from the $j$-level to $i$ and $k$ are also proportional to $g_i$ and $g_k$.

The above solution does not account for the electron spin and the relativistic dependence of the electron mass on velocity. These corrections are small and can be made by perturbation tools:

$$E_{nj} = -Ry\,\frac{z^2}{n^2}\left[1 + \frac{\alpha^2 z^2}{n}\left(\frac{1}{j+1/2} - \frac{3}{4n}\right)\right]. \tag{3.11}$$

This formula describes the fine structure of hydrogen-like atoms, and $\alpha = \dfrac{e^2}{\hbar c} \approx \dfrac{1}{137}$ is the dimensionless fine-structure constant; $c = 3\times10^{10}$ cm s$^{-1}$ is the speed of light. The correction in (3.11) is ~$10^{-4}$ for $z = 1$. It depends on the interaction between the electron spin and momentum and is determined by total momentum $M_j = \hbar\sqrt{j(j+1)}$ with $j = l + s$ and $s = \pm1/2$ are two possible spin projections.

Taking the spin into consideration, the set of quantum numbers for an electron's state is $n$, $l$, $m_l$, $j$, $m_j$, and $s$. However, only four of them are independent, and we choose $n$, $l$, $j$, and $m_j$. The number of $m_j$ values is equal to the statistical weight of a level $g_j = 2j + 1$, the number of states equals $2(2l + 1)$ for a given combination of $n$ and $l$, and $2n^2$ for a given $n$. The largest value of $j$ is $n - 1/2$, the smallest is $1/2$, i.e., there is a total of $n$ values. Electrons with $l = 0, 1, 2, 3, 4, \ldots$ are designated by $s, p, d, f, g, \ldots$, and electron states are indicated by $nl_j$. For example, $2p_{3/2}$ means the state with $n = 2$, $l = 1$, and $j = 3/2$. The projection $m_j$ is essential only in an external field and therefore not fixed.

## 3.3 Radiation Types and Transition Probabilities

According to classic electrodynamics, radiation can be induced by some change in the electric or magnetic dipole, or quadrupole, etc., moment of a system. Let us recall that a quadrupole is a system of two identical but oppositely directed dipoles. In practice it is necessary to examine only electric and magnetic dipole radiation, and electric quadrupole radiation.

Transitions between levels $i$ and $k$ can be forced, with probabilities $\rho(v)B_{ki}$ and $\rho(v)B_{ik}$ for absorption and stimulated emission, respectively, and spontaneous with a transition probability $A_{ik}$ ($i$ is the upper level). Here $\rho(v)$ is the incident photon flux at the transition frequency $v$. There are following relations between the Einstein coefficients $A$ and $B$:

$$g_i B_{ik} = g_k B_{ki},\ A_{ik} = 8\pi h\,\lambda^{-3} B_{ik}. \tag{3.12}$$

The lifetime of an excited $i$-state is

$$\tau_i = \left(\sum\nolimits_k A_{ik}\right)^{-1}. \tag{3.13}$$

Along with transition probabilities $A_{ik}$, some other quantities are used. The oscillator strength $f_{ki}$ (in absorption) is a dimensionless value that gives an effective number of electrons participating in the transition:

$$f_{ki} = \frac{g_i}{g_k}\frac{m_e c}{8\pi^2 e^2}\lambda^2 A_{ik} = 1.5\,\frac{g_i}{g_k}\,\lambda^2 A_{ik}\ (\lambda \text{ is in cm}). \tag{3.14}$$

The absorption line strength

$$S_{ki} = \int_0^\infty \sigma(v)dv = \frac{\pi e^2}{m_e c^2} f_{ki} = 8.85 \times 10^{-13} f_{ki} = \frac{\lambda^2}{8\pi c}\frac{g_i}{g_k} A_{ik}, \tag{3.15}$$

where $\sigma(v)$ is the cross section at wavenumber $v$ and $S_{ki}$ is in cm.

The transition dipole moment is

$$P_{ik} = e\int \psi_i^* \mathbf{r}\psi_\mathbf{k} dv, \quad \text{and} \quad A_{ik} = \frac{32\pi^3}{3\hbar}\,\lambda^{-3}|P_{ik}|^2 \tag{3.16}$$

for the electric dipole transitions. $\boldsymbol{P_{ik}}$ is substituted by the magnetic dipole moment $\boldsymbol{\mu_{ik}}$ in (3.16) for the magnetic dipole transitions, and

$$A_{ik}^Q = \frac{16\pi^6}{5\hbar}\,\lambda^{-5}|Q_{ik}|^2 \tag{3.17}$$

for the quadrupole transitions; $Q_{ik}$ is the transition quadrupole moment. Substituting the numerical values, one gets

$$A_{ik} = 3 \times 10^{29}\lambda^{-3}|P_{ik}|^2,\ A_{ik}^M = 3 \times 10^{29}\lambda^{-3}|\mu_{ik}|^2,\ A_{ik}^Q = 9 \times 10^{29}\lambda^{-5}|Q_{ik}|^2. \tag{3.18}$$

These values may be estimated assuming $P = ea$, $Q = ea^2$, where $a = 0.1$ nm is a typical size of atom, $\mu = \dfrac{e\hbar}{2m_e c}$ is the Bohr magneton, and $\lambda = 500$ nm. Then $A_{ik} = 6\times10^7\ \mathrm{s}^{-1}$, $A_{ik}^M = 250\ \mathrm{s}^{-1}$, and $A_{ik}^Q = 8\ \mathrm{s}^{-1}$.

Thus electric dipole transitions are more intense than magnetic dipole and electric quadrupole transitions by many orders of magnitude. Such transitions are possible for $\boldsymbol{P_{ik}} \neq 0$ and are called *allowed.* If $\boldsymbol{P_{ik}} = 0$, then the transition can occur if $\boldsymbol{\mu_{ik}}$ or $Q_{ik} \neq 0$ and is called *forbidden*.

### 3.3.1 Selection Rules

Selection rules determine combinations of quantum numbers of the initial and final state that result in allowed transitions. The spatial variables are separated in the wave functions for hydrogen-like atoms: $\psi(\boldsymbol{r}) = R(r)\Theta(\theta)\Phi(\varphi)$ (Section 3.2). Therefore the integral $\boldsymbol{P_{ik}}$ in (3.16) is the product of three integrals along three coordinates, and each of those should not be equal to zero. Therefore a transition is forbidden if a selection rule is not obeyed for even one of the quantum numbers. The simplest case is for $m$, since $\Phi_m(\varphi) = e^{im\varphi}$ :

$$P_{ik}^x = e\int_0^{2\pi} x\ \exp\left(i\varphi(m_i - m_k)\right)d\varphi = er\sin\theta\int_0^{2\pi} \exp\left(i\varphi(m_i - m_k + 1)\right)d\varphi,$$

$$P_{ik}^y = e\int_0^{2\pi} y\ \exp\left(i\varphi(m_i - m_k)\right)d\varphi = er\sin\theta\int_0^{2\pi} \exp\left(i\varphi(m_i - m_k - 1)\right)d\varphi,$$

$$P_{ik}^z = e\int_0^{2\pi} z\ \exp\left(i\varphi(m_i - m_k)\right)d\varphi = er\cos\theta\int_0^{2\pi} \exp\left(i\varphi(m_i - m_k)\right)d\varphi.$$

These integrals are not equal to zero only if their exponential indices are zero. Hence the selection rule is $\Delta m = 0, \pm1$. For the other quantum numbers, $\Delta l = \pm1$, $\Delta j = 0, \pm1$, and $\Delta s = 0$. There is no restriction for $n$ that results in series in the spectra.

## 3.4 Spectra of Hydrogen-like Atoms

A diagram of the hydrogen atom energy levels is shown in Figure 3.1. The transition wavenumbers and wavelengths are accurately described by

$$\nu_{ik} = \frac{1}{\lambda_{ik}} = z^2 Ry\left(\frac{1}{n_k^2} - \frac{1}{n_i^2}\right), \tag{3.19}$$

with $i$ and $k$ being the upper and lower level, respectively. Transitions with the same $k$ form series (Lyman, Balmer, Paschen, Brackett series for $k = 1, 2, 3, 4$ etc., respectively), and lines in a series are designated $\alpha$, $\beta$, $\gamma$, $\delta$, … For example, $L_\alpha$ is at 121.6 nm for H and 30.4 nm for $He^+$.

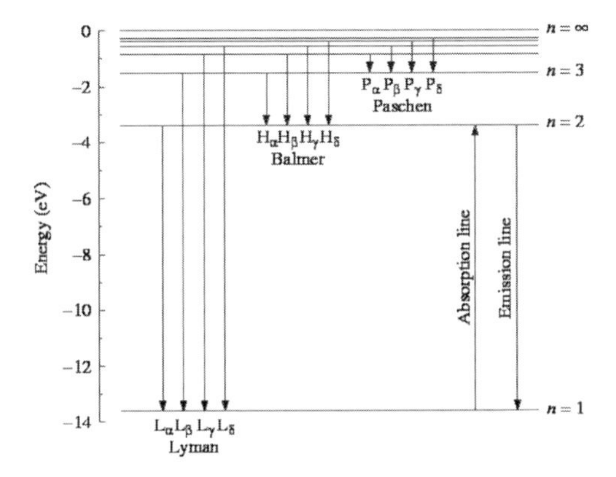


Figure 3.1 Energy levels and transitions in the hydrogen atom.

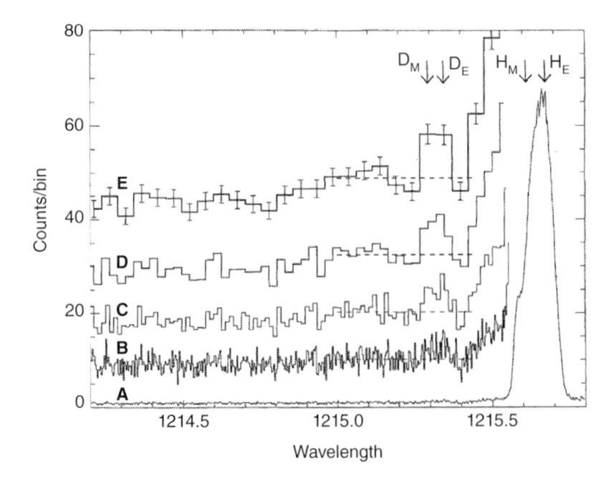


Figure 3.2 Search for atomic deuterium on Mars using the Goddard High-Resolution Spectrograph at the Hubble Space Telescope. The spectrum reveals four $L_\alpha$ lines: the Earth's $H_E$, the Martian Doppler-shifted $H_M$ seen as a shoulder, the isotopically shifted $D_E$, and both isotopically and Doppler-shifted $D_M$. Five curves reflect different averaging of the data with proper scaling. From Krasnopolsky et al. (1998).

A weak dependence of the reduced mass on the nucleus mass results in a small shift of isotope lines (isotopic shift). This shift is maximal for deuterium and equal to 0.033 nm for the deuterium $L_\alpha$ (Figure 3.2).

Figure 3.1 does not show the fine structure of the lines that is depicted for $H_\alpha$, that is, Balmer-alpha, in Figure 3.3. All transitions shown in the figure conform to the selection rules. All transitions $n_i\ l_i$–$n_k\ l_k$ form a multiplet. Intensities of separate transitions of a multiplet obey an addition rule: the sum of intensities of transitions from the same initial or final state is proportional to the statistical weight $2j + 1$ of that state. Let us apply this rule to the $2p$–$3d$ multiplet in Figure 3.3. It consists of three lines, because $2p_{1/2}$–$3d_{5/2}$ is forbidden ($\Delta j = 2$). Then $i_1/(i_2 + i_3) = 6/4$, $(i_1 + i_2)/i_3 = 4/2$, and $i_1 : i_2 : i_3 = 9 : 1 : 5$.

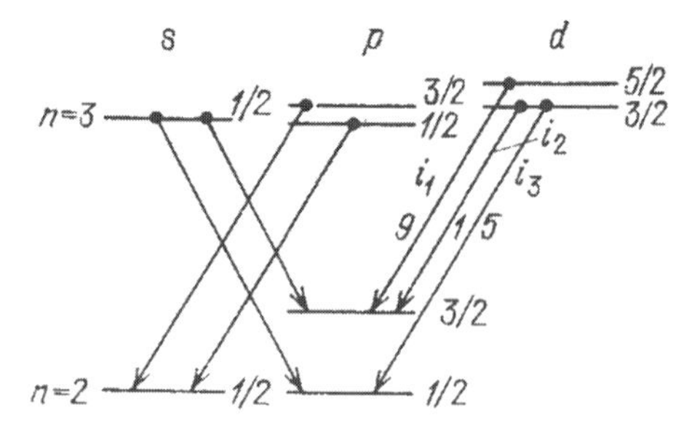


Figure 3.3 Transitions in the $H_\alpha$ line at 656.3 nm. Line intensities of the $2p$–$3d$ multiplet are given.

Besides fine structure, there also exists superfine structure of spectral lines that is induced by the interaction of electron with the spin of the nucleus. Furthermore, there is a so-called Lamb shift of levels with identical $n$ and $j$:

$$E\left(ns_{1/2}\right) - E\left(np_{1/2}\right) = 8Ry\,\frac{\alpha^3 z^4}{n^3}\,C(n,z)\,, \tag{3.20}$$

where $C$ depends weakly on $n$ and $z$. This shift is explained by quantum electrodynamics.

Magnetic and electric fields remove the $m_j$ degeneracy and result in line splitting (the Zeeman, Paschen–Back, and Stark effects).

Nine lines of the H Lyman series down to $L_\kappa$ 91.8 nm were observed in the Martian spectrum measured by the Far Ultraviolet Spectroscopic Explorer (FUSE) with resolution of 0.02 nm (Krasnopolsky and Feldman 2002).

The $He^+$ line at 30.4 nm was detected on Venus (Bertaux et al. 1981; Krasnopolsky and Gladstone 2005) and in spectra of comets (Krasnopolsky and Mumma 2001). Emissions of the H-like ions $C^{5+}$, $O^{7+}$, and $Ne^{9+}$ are revealed in the X-ray spectra of Mars (Dennerl et al. 2006) and comets (Krasnopolsky 2015 and references therein).

## 3.5 Multielectron Atoms

An exact description of a many-body quantum system is very complicated. However, there are some assumptions, approximations, and rules that make it possible to characterize the system qualitatively. Here a basic rule is the Pauli exclusion principle that all electrons in an atom must be in different quantum states $nljm_j$. This principle explains the periodic table of the chemical elements that was created by D. I. Mendeleev in 1869, long before the Pauli principle, from an analysis of chemical properties of the elements.

Electrons with the same $n$ form a shell with up to $2n^2$ electrons, while a subshell $nl$ can contain up to $2(2l + 1)$ electrons. However, the energy of a multielectron atom depends on both $n$ and $l$, and this results in a more complicated structure of the periods IV, V, and more. Electron configuration is the full set of $nl$ for all electrons in an atom, e.g., $1s^2 2s^2 2p^6 3s$ for Na. Here the superscripts give the numbers of electrons in subshells. The total momentums of a filled shell and a filled subshell are zero, and the chemical and spectroscopic properties of atoms are determined mainly by the electrons in the outer shells that are not full, i.e., valence electrons.

There is a strong interaction between the orbital momentums of electrons in the light atoms that is described by $L = \Sigma l$, and a strong interaction between the spins with $S = \Sigma s$. An interaction between $L$ and $S$ is weak; therefore an atom energy depends strongly on $L$ and $S$ and weakly on $J = L + S$. For example, a configuration of two electrons $s$ and $p$ is $L = 1$ and $S = 0, 1$, so that $J = 1$ for $S = 0$ and $J = 2, 1, 0$ for $S = 1$; two terms ${}^1P_1$ and ${}^3P_{2,1,0}$ correspond to this configuration.

The term is designated ${}^{\kappa}L_J$, where $\kappa$ is the number of possible projections of vector $S$ on $L$ that is called *multiplicity* $\kappa = 2S + 1$. Generally speaking, this number should be $2L + 1$ for $L < S$; nevertheless, $\kappa = 2S + 1$ even in this case, although the real number of components is smaller. The term designation may be supplemented by $nl$ of the subshell that originates the term, e.g., $3s\,{}^2S_{1/2}$ ($n = 3$, one $s$-electron, $L = 0$, $J = ½$, $\kappa = 2$, $S = ½$) or the full electron configuration, e.g., $1s^2 2s^2 2p^6 3s\,{}^2S_{1/2}$ for Na. If $J$ is determined unambiguously (e.g., ${}^1D_2$) or not of interest, it may be skipped (e.g., ${}^1D$).

For heavy atoms a strong interaction is between $l$ and $s$ for each electron and a weak connection between $j$ of different electrons. This case is of low importance for the planetary atmospheres where the heavy atoms are rare.

Since a filled subshell has zero momentum and $2(2l + 1)$ electrons, the quantity and type of terms for $k$ and $2(2l + 1) - k$ equivalent electrons are identical. For example, for $p$-electrons identical terms have configurations $p$ and $p^5$ (elements B and F), $p^2$ and $p^4$ (C and O).

## 3.6 Energy Levels and Selection Rules

Within the same configuration $nl$, the energy is smallest for large $S$, and then for large $L$. The distance between neighbor levels of a term with $J$ and $J - 1$, that is, the multiplet splitting, is equal to $\Delta E_{J,J-1} = aJ$ (the interval rule). If the number of electrons in a subshell $k < 2l + 1$, that is, the subshell is less than half filled, then $a > 0$, and the lowest level corresponds to the smallest $J$. For $k > 2l + 1$, $a < 0$ and the level with the maximum $J$ has the lowest energy; $a \approx 0$ for $k = 2l + 1$, and the multiplet splitting is very small.

Excitation and transitions of only one electron are important in the planetary atmospheres, although the multielectron excitation is possible in powerful laser beams and by electron impact. Two-electron transitions occur if a one-electron transition is strongly forbidden.

### *3.6.1 Selection Rules*

Selection rules are the following: $\Delta J = 0, \pm 1$ except for $0 \to 0$. The parity conservation requires $u \leftrightarrow g$. A state is even ($g$) if its wave function does not change sign under inversion, i.e., under a sign change of its spatial coordinates $x, y, z$. In the opposite case the state is odd ($u$). The parity of states is determined by the parity of the arithmetic sum of the momentums of electrons, and odd terms are sometimes indicated by a superscript °: $^2P^o_{1/2}$. The required change in parity is caused by the parity conservation, since a dipole photon has $l = 1$. A quadrupole photon has $l = 2$, and transitions between terms of the same configuration $nl$ can be only quadrupole, that is, forbidden.

Other selection rules are $\Delta S = 0$ and $\Delta L = 0, \pm 1$ except for $0 \rightarrow 0$. Radiation and absorption of a photon in allowed transitions change a state of only one electron, which has $\Delta l = \pm 1$. This restriction should be kept in mind as well.

## 3.7 Spectra of Multielectron Atoms

The alkali metals (group I in the periodic table) have one electron in an outer shell with a ground state $ns^2S_{1/2}$. Transitions related to this state form the principal series $n'p^2P_{1/2,\ 3/2} - ns^2S_{1/2}$ that is seen in absorption and emission. The emission series are $n's^2S_{1/2} \rightarrow np^2P_{1/2,\ 3/2}$ (the sharp series), $n'd^2D_{3/2,\ 5/2} \rightarrow np^2P_{1/2,\ 3/2}$ (the diffuse series), and $n'f\ ^2F_{5/2,\ 7/2} \rightarrow nd^2D_{3/2,\ 5/2}$ (the fundamental series). The electrons of closed shells screen the electric field of the nucleus, and the energy levels are $E_{nl} = -Ry/(n - \Delta_l)^2$; here $\Delta_l$ is the quantum defect (Table 3.1).

A level diagram for the helium atom is shown in Figure 3.4. Here energies of the ground states are assigned to zero. Two electrons of the helium atom in the ground state form a strong filled shell $1s^2\ ^1S_0$. The next configuration $1s2s$ gives two metastable terms $^1S_0$ and $^3S_1$. (All spontaneous transitions are forbidden from the metastable terms.) All configurations of two electrons form either singlet ($S = 0$) or triplet ($S = 1$) terms, transitions between them are forbidden by the selection rule $\Delta S = 0$, and systems of singlet and triplet lines (called sometimes parahelium and orthohelium systems) occur.

The lowest allowed transition of He ($1s2p\ ^1P \rightarrow 1s^2\ ^1S$) at 58.4 nm was observed on Venus and Mars (Kumar and Broadfoot 1975; Bertaux et al. 1981; Krasnopolsky and Gladstone 2005) and in comets (Krasnopolsky et al. 1997). Emissions of the He-like ions $C^{4+}$, $N^{5+}$, $O^{6+}$, and $Ne^{8+}$ were revealed in X-ray spectra of Mars (Dennerl et al. 2006) and comets (Krasnopolsky 2015a and references therein).

Terms of the alkaline Earth elements (group II in the periodic table) is similar to that of helium with the following differences: (1) the filled subshell $ns^2$ is not so strong as the shell $1s^2$, and the excitation energies are not very high; (2) terms with the same $n$ as in the

Table 3.1 *Quantum defect in the alkali metal spectra*

| Element | $s$ | $p$ | $d$ | $f$ |
|---|---|---|---|---|
| Li | 0.41 | 0.04 | 0.00 | 0.00 |
| Na | 1.37 | 0.88 | 0.01 | 0.00 |
| K | 2.23 | 1.77 | 0.15 | 0.01 |
| Rb | 3.20 | 2.72 | 1.23 | 0.01 |
| Cs | 4.13 | 3.67 | 2.45 | 0.02 |

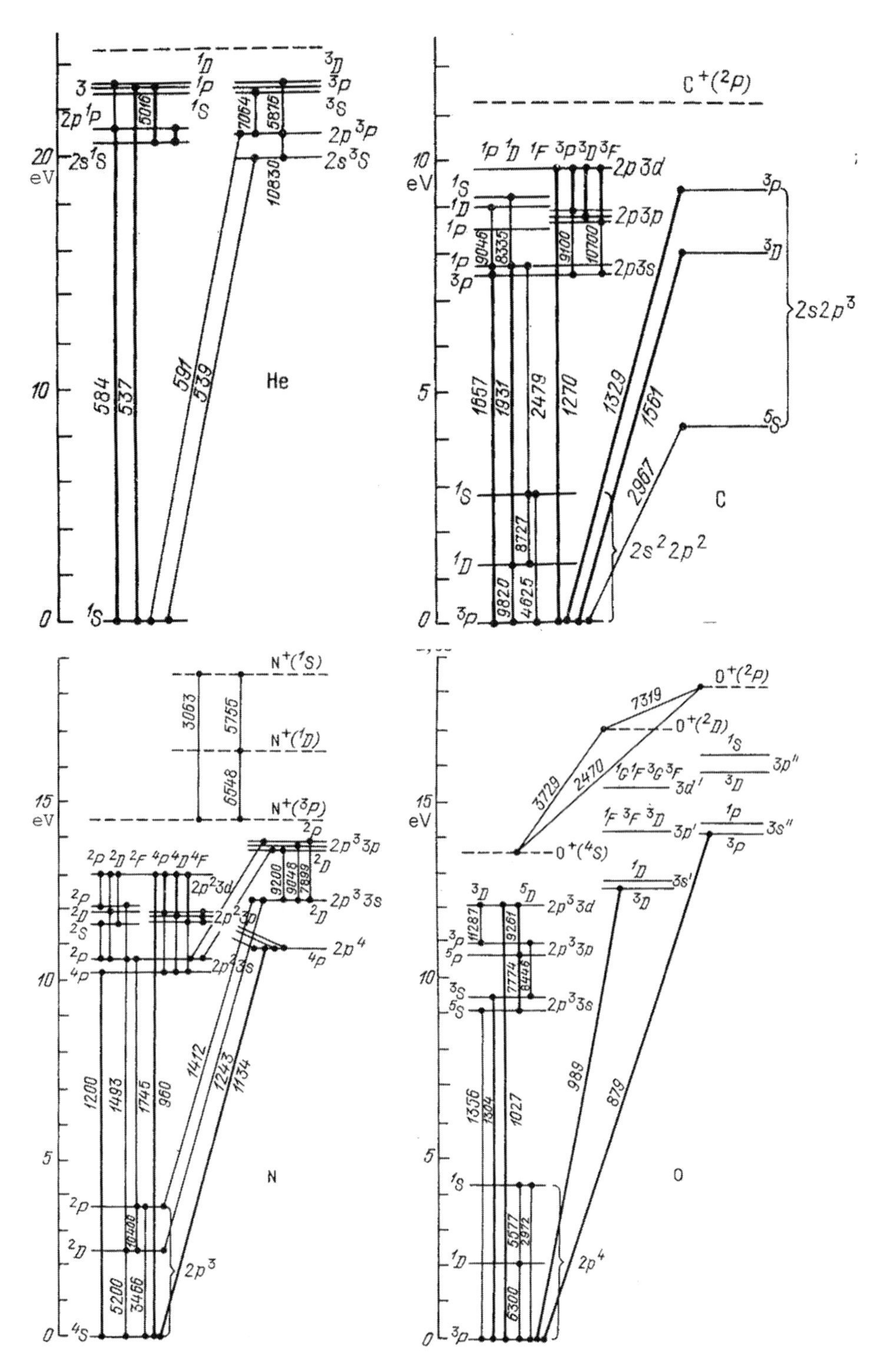

Figure 3.4 Energy levels and transitions of He, C, N, and O. Transitions are in angstroms, and ground state energies are zero. Allowed transitions are shown in bold.

ground state appear and have low excitation energies; (3) the intercombination prohibition $\Delta S = 0$ is not so strict, and singlet-triplet transitions become more intense as $z$ increases.

### *3.7.1 Atomic Oxygen*

Terms of the elements of groups IV and VI are similar, and we consider here the energy levels of the O atom (Figure 3.4). Using the Pauli exclusion principle, terms of $2p^4$ configuration are $^3P_{2,1,0}$, $^1D$, and $^1S$. The lowest term is $^3P_{2,1,0}$ as having the greatest spin, and the lowest component is $J = 2$, because more than half of the subshell are populated (4 electrons in the subshell of 6 electrons). The second term is $^1D$ with the greater $L$ than that of $^1S$. All transitions between these terms are parity-forbidden. The observed transitions are:

$^1D_2 \rightarrow {}^3P_{2,1}$ : 630/636 nm, the O red line, $^3P_0$ as the lower state is additionally forbidden because of $\Delta J = 2$; the line is additionally spin-forbidden and its radiative time is $\tau \approx 120$ s.

$^1S_0 \rightarrow {}^3P_1$ : 297 nm, the O UV line; the lower states $^3P_{0,\ 2}$ are additionally forbidden because of $J = 0 \rightarrow 0$ and $\Delta J = 2$, respectively; the line is spin-forbidden with $\tau \approx 20$ s.

$^1S_0 \rightarrow {}^1D_2$ : 558 nm, the O green line. Comparing with the UV line, the additional violation $\Delta L = 2$ is overcompensated by the spin conservation, and $\tau \approx 1$ s.

The next configuration is $1s^2\, 2s^2\, 2p^3\, 3s$ with terms $^5S_2$ and $^3S_1$. $^5S_2$ is the lowest because of the large spin, and all transitions from $^5S_2$ are spin-forbidden. The O ($^5S_2 \rightarrow {}^3P_{2,1}$) emission at 136 nm is prominent in the airglow on the terrestrial planets. The allowed triplet $^3S_1 \rightarrow {}^3P_{2,1,0}$ at 130 nm is used to measure abundances of atomic oxygen in the upper atmospheres. Excitations of the $2s$-electrons result in allowed transitions at 99 and 88 nm (Figure 3.4).

### *3.7.2 Atomic Nitrogen*

Atomic nitrogen is the most important of the group V elements for our science. Its configuration is $1s^2\, 2s^2\, 2p^3$ that makes terms $^4S$, $^2D_{3/2,\ 5/2}$ and $^2P_{1/2,\ 3/2}$. The lowest term has the largest spin, and the next term has the largest momentum. The multiplet splitting is very small, near zero for the $p^3$ configuration. The N ($^2D \rightarrow {}^4S$) transition at 520 nm is parity-, spin-, and momentum-forbidden, and its radiative time is ~$10^5$ s.

Excitation of the $2p$-electrons to the $2p^2\, 3s$ configuration forms terms $^4P$ and $^2P$ and allowed transitions $^4P \rightarrow {}^4S$ at 120 nm and $^2P \rightarrow {}^2D$ at 149 nm. Excitation of the $2s$-electrons to the $2p^4$ configuration forms a term $^4P$ and an allowed transition $^4P \rightarrow {}^4S$ at 113 nm. Here are two $^4P$ terms that refer to the different configurations.

### *3.7.3 Metastable States O($^1$D) and N($^2$D)*

Metastable states O($^1D$) and N($^2D$) are chemically active in the planetary atmospheres. The O($^1D$) energy is 1.97 eV relative to the ground state O($^3P$), and the radiative lifetime is

120 s. This state is formed by photolyses of $O_3$, $O_2$, and $CO_2$ with significant quantum yields. $O(^1D)$ is effectively quenched in collisions with almost all molecules. However, a small part of the $O(^1D)$ production in the terrestrial atmospheres reacts with $H_2O$, $H_2$, and $CH_4$. The reaction with $H_2O$ initiates the odd hydrogen chemistry in the Earth's stratosphere and significantly contributes to this chemistry on Mars. The reactions with $H_2$ and $CH_4$ are the main losses of these species on Mars and Earth.

Quantum yield of $N(^2D)$ is significant in almost all processes of the formation of atomic nitrogen. The $N(^2D)$ energy is 2.38 eV, and its radiative lifetime is $10^5$ s. $N(^2D)$ reacts with $O_2$ and $CO_2$ to form NO and initiates the odd nitrogen chemistry in the atmospheres of Earth, Mars, and Venus. $N(^2D)$ reacts with hydrocarbons and $H_2$ and forms species with the NH and CN bonds in the atmospheres of Titan, Triton, and Pluto.

## 3.8 Rotational and Vibrational Levels of Diatomic Molecules

The complexity of energy states and wave functions is determined to a great extent by symmetry of a system. In the case of atoms, the nucleus field is spherically symmetric. For a diatomic molecule, the field has an axis of symmetry along the line that connects the nuclei. The linear triatomic molecules are axial symmetric as well; however, this symmetry is broken under any bending of the molecule. All triatomic molecules have planes of symmetry, some molecules have symmetry axes of different order, etc. Here is a path along which the spectral structure becomes more and more complicated. We will consider below the spectra of diatomic molecules. Main features of these spectra are applicable to the more complicated molecules.

### *3.8.1 Rotational Levels*

Aside from the motion of electrons, there are motions of the nuclei relative to the center of mass. These motions are rotation of the molecule around two axes that are normal to the molecule axis and each other, and vibration relative to some equilibrium position. Both rotational axes are equivalent, all three motions are approximately independent, and the wave function of the system can be presented as product of the wave functions for each motion (electronic, vibrational, and rotational):

$$\psi = \psi_e \psi_v \psi_r; \text{ then } E = E_e + E_v + E_r. \tag{3.21}$$

The angular momentum of a rotating molecule looks similar to those of atoms:

$$M = I\omega = \hbar\sqrt{J(J+1)}. \tag{3.22}$$

Here $I = \mu r^2$ is the moment of inertia, $\mu = \dfrac{m_1 m_2}{m_1 + m_2}$ is the reduced mass, $r$ is the distance between the nuclei, $\omega$ is the angular velocity, and $J = 0, 1, 2, 3, \ldots$ is the rotational quantum number. Hence the rotational term in wavenumbers $1/\lambda$ is

$$F(J) = \frac{M^2}{2I}/hc = BJ(J+1), \tag{3.23}$$

and

$$B = \frac{h}{8\pi^2 c\mu r^2}$$

is the rotational constant.

Two additional factors affect the rotation: centrifugal force that increases the distance between nuclei $r$, and a weak dependence of the rotational constant on the vibrational number v (see below). The proper corrections result:

$$F_v(J) = B_v\, J(J+1) - D_v\, J^2(J+1)^2, \quad B_v = B_e - \alpha_e(\mathrm{v}+1/2). \tag{3.24}$$

Here $B_e$ is the rotational constant for the minimum of the potential energy curve $r = r_e$, and

$$D_v = D_e + \beta_e\left(\mathrm{v}+\frac{1}{2}\right),\ D_e = 4B_e^3\, c^2/v. \tag{3.25}$$

Here $v$ is the vibrational frequency, and $\alpha_e$, $\beta_e$ are small constants.

### 3.8.2 Vibrational Levels

Vibrational levels appear in a molecule to describe displacements of the nuclei relative to the equilibrium internuclear distance $r_e$. The equilibrium position requires a minimum of the potential energy $V(r)$ (Figure 3.5). The lower part of the curve may be approximated by the parabola $V(r) = k\,(r - r_e)^2/2$, and in this case (an approximation of the harmonic oscillator),

$$v = \frac{1}{2\pi}\left(\frac{k}{\mu}\right)^{1/2},\ E_\mathrm{v} = hv(\mathrm{v}+1/2),\ G(\mathrm{v}) = \omega_e(\mathrm{v}+½). \tag{3.26}$$

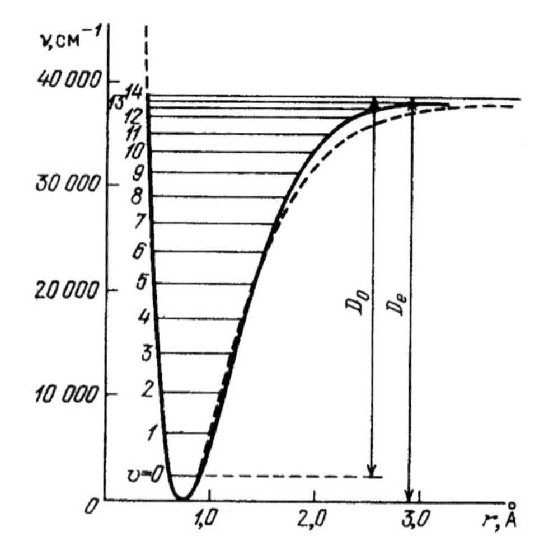


Figure 3.5 Potential energy and vibrational levels of the $H_2$ molecule. Fitting of the curve by the Morse function is shown by the dashed line.

Here $G(\mathrm{v})$ is the vibrational term in wavenumbers, $\omega_e = \nu/c$, and the vibrational quantum number v = 0, 1, 2, 3, . . . The parabolic approximation of the potential energy curve becomes invalid for large v (Figure 3.5), and the Morse-function approximation can be used:

$$V(r) = D_e\left[1 - e^{-\beta(r-r_e)}\right]^2, \quad \beta = 1.218 \times 10^{-7}\omega_e\left(\frac{\mu}{D_e}\right)^{1/2}, \tag{3.27}$$

where $D_e$ is the dissociation energy measured from the curve minimum (as opposed to $D_0$ measured from the v = 0 level; do not confuse with $D_e$ in (3.25)). Now

$$G(\mathrm{v}) = \omega_e(\mathrm{v} + 1/2) - \omega_e x_e(\mathrm{v} + 1/2)^2 + \omega_e y_e(\mathrm{v} + 1/2)^3, \tag{3.28}$$

and $x_e$, $y_e$ are small constants.

### 3.8.3 *Comparison of the Electronic, Vibrational, and Rotational Energies*

Thus the rotational constant is proportional to $\mu^{-1}$ and the vibrational constant to $\mu^{-1/2}$. These relationships can be applied to an order-of-magnitude estimate of the motion of an electron in a molecule or atom to get

$$E_e : E_\mathrm{v} : E_r \approx 1 : \sqrt{\frac{m_e}{M}} : \frac{m_e}{M}, \text{ and } E_e >> E_\mathrm{v} >> E_r. \tag{3.29}$$

The rotational levels are typically ~100 $\mathrm{cm}^{-1}$, the vibrational energy is ~2000 $\mathrm{cm}^{-1}$, and the electronic energy is ~$5\times10^4$ $\mathrm{cm}^{-1}$, 1 eV = 8065 $\mathrm{cm}^{-1}$, and the rotational energy refers to the far infrared and millimeter regions, the vibrational energy to the near- and mid-infrared, and the electronic energy to the visible and UV ranges.

## 3.9 Rotational and Rovibrational Spectra

The selection rule for rotational transitions is $\Delta J = \pm 1$, hence the transition frequency in $\mathrm{cm}^{-1}$ is

$$F(J) - F(J-1) = 2BJ - 4DJ^3. \tag{3.30}$$

Therefore a rotational spectrum is a set of the almost equally spaced lines, because $D << B$. However, the second term becomes significant at large $J$.

The selection rule for the harmonic oscillator is $\Delta v = \pm 1$, with no restriction on $\Delta v$ in the general case. Actually transitions $\Delta v = \pm 1$ are much more intense than others.

Homonuclear molecules (made up of identical atoms) do not have a dipole moment independent of the distance between atoms and the rotation of the molecule. Therefore rotational and vibrational transitions are forbidden for them.

Vibration numbers v′ and v″ determine the spectral band (v′,v″) or (v′−v″) that consists of lines corresponding to the set of allowed rotational transitions between $J'$ and $J''$. Single and double primes refer to the upper and lower states, respectively.

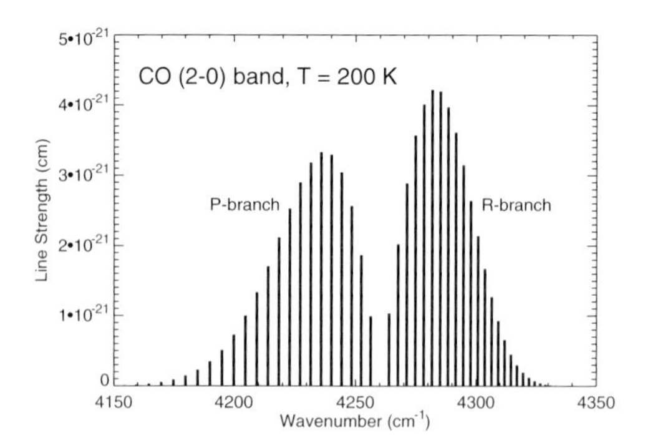


Figure 3.6 Spectrum of the $^{12}C^{16}O$ (2–0) band.

The selection rule for a molecule momentum is the same as for an atom: $\Delta J = 0, \pm 1$, except $0 \leftrightarrow 0$. Therefore, if the projection of the momentum of an electronic state equals zero ($\Lambda = 0$, $\Sigma$-state, see below), then only $\Delta J = \pm 1$ transitions are allowed. This gives two branches of a band:

$$\text{R-branch for } J' = J'' + 1 \text{ and } \nu = \nu_0 + (B' + B'')(J'' + 1) + (B' - B'')(J'' + 1)^2 \quad (3.31)$$

$$\text{P-branch for } J' = J'' - 1 \text{ and } \nu = \nu_0 - J''(B' + B'') + J''^2(B' - B''). \quad (3.32)$$

The lines are designated by their $J''$, that is, the CO (2–0) R6 line is a transition between $J' = 7$ and $J'' = 6$ of the CO band with $v' = 2$ and $v'' = 0$. The R-branch starts with R0, while the P-branch with P1. Usually $B' - B'' << B' + B''$, and a rovibrational band is a set of the lines without $\nu_0$ (Figure 3.6). Typically $B'$ is smaller than $B''$ (see 3.24), and the line spacing is reduced to the blue.

If the electronic state is $\Lambda > 0$ ($\Pi$, $\Delta$ states, see below), then rovibrational transitions $\Delta J = 0$ are allowed and form the Q-branch

$$\nu = \nu_0 + (B' - B'')\, J(J + 1). \quad (3.33)$$

$B' - B''$ is usually small; therefore this branch is very dense and favorable for detection and observation by instruments with a moderate spectral resolution (Figure 3.7).

### 3.9.1 Rotational Temperature

Population of rotational levels is $N_J \sim g \exp(-\alpha F(J)/T)$ at thermodynamic equilibrium, and $g = 2J + 1$ is the statistical weight of the level. This relationship can be applied to get line strength

$$S = A(2J + 1)\, \exp(-\alpha BJ(J + 1)/T). \quad (3.34)$$

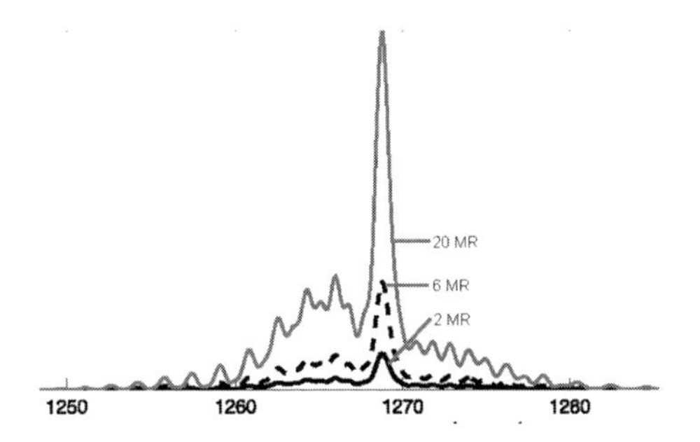


Figure 3.7 Synthetic spectra of the $O_2(a^1\Delta_g \rightarrow X^3\Sigma_g^-)$ (0–0) band at 175 K and resolving power $\nu/\delta\nu = 2200$. The wavelength scale is in nm. The spectra are calculated for the total emission of 2, 6, and 20 MR (mega-rayleighs). The prominent central peak is the Q-branch. From Fedorova et al. (2006).

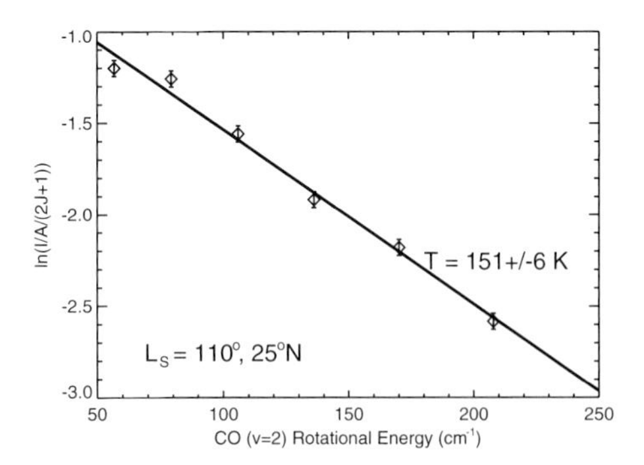


Figure 3.8 Rotational temperature of the CO (2–1) dayglow at 4.7 μm on Mars. Logarithm of the line intensities corrected for transition probabilities and statistical weights versus rotational energy gives a straight line with the slope of $-\alpha/T$. The retrieved temperature refers to the dayglow maximum near 50 km. From Krasnopolsky (2014a).

Here $A$ is the transition probability and $\alpha = hc/k = 1.439$ cm K. The line strength increases at small $J$ and then decreases exponentially at large $J$ (Figure 3.6). Temperature significantly affects line distributions in the absorption and emission bands, and accurate measurements of these distributions make it possible to determine temperature of the absorbing or emitting layers (Figure 3.8). Thermodynamic equilibrium is applicable if a lifetime of the absorbing or emitting state is much longer than the time between collisions. Temperatures retrieved by this method are called *rotational temperatures.*

### *3.9.2 Spectra of Isotopologues*

Both vibrational and rotational constants depend on the reduced mass of the molecule. For example, if masses of two atoms in a molecule are rather similar and the isotopes differ by the

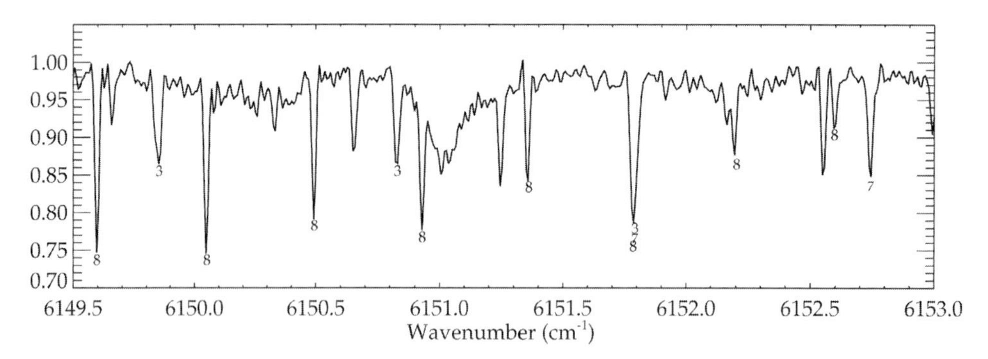


Figure 3.9 A small part (1.2%) of the observed spectrum of Mars that shows lines of $^{13}CO_2$, $C^{18}OO$, $C^{17}OO$ (marked 3, 8, 7, respectively), and $CO_2$. Abundances of these isotopes are depleted relative to the basic isotope by factors of 85, 238, and 1310, respectively, on Mars. The resolving power $\nu/\delta\nu = 3.5\times10^5$. From Krasnopolsky et al. (2007).

unit mass (e.g., $^{12}CO$ and $^{13}CO$), then the vibrational frequency ratio is $\nu_{M+1}/\nu_M \approx 1 - \frac{1}{2M}$. The isotopic shift in the atomic spectra is $\nu_{M+1}/\nu_M = 1 + \frac{m_e}{M(M+1)}$, that is, smaller by a factor of $M/2m_e \approx 10^4$. Therefore detection of molecular isotopologues is a more feasible spectroscopic task than that for atoms (Figure 3.9).

## 3.10 Electronic States of Diatomic Molecules

In an atom $n$ and $l$ are the main quantum numbers of an electron. There is an axis of symmetry and an electric field between the nuclei in a diatomic molecule. The electron energy depends on the projection $m_l$ on the axis, and the electron state, or orbital, is given as $\lambda = |m_l|$. Orbitals with $\lambda = 0, 1, 2, \ldots$ are called $\sigma, \pi, \delta, \ldots$ orbitals. Orbitals with $\lambda \neq 0$ correspond to two values of $m_l = \pm\lambda$ and are doubly degenerate. An electron with a given $l$ can have various orbitals, e.g., $d\sigma$, $d\pi$, and $d\delta$ orbitals for a $d$ electron ($l = 2$).

The sum of orbitals with signs taken into account gives the angular momentum of the molecule $\Lambda = 0, 1, 2, \ldots$ with designations $\Sigma, \Pi, \Delta, \ldots$ The total spin $S$ and multiplicity $\kappa = 2S + 1$ are similar to those in atomic spectroscopy.

Homonuclear molecules ($H_2$, $O_2$, $N_2$, ...) have centers of symmetry, and their wave functions (for electrons and the whole molecule) can be even or odd (Section 3.6) with indication by subscript $g$ or $u$. The parity of electron wave functions is determined by the parity of $l$; since an orbital can be formed by an electron with a different $l$, there exist $\sigma_g$, $\pi_g$, $\delta_g$, ... and $\sigma_u$, $\pi_u$, $\delta_u$, ... orbitals. The parity of a molecular state depends on whether the number of $u$ orbitals is even or odd, that is, as for an atom, it is equal to the parity of the arithmetic sum $\Sigma\, l_i$.

For example, a configuration of two electrons $\sigma_g\pi_u$ gives two states $^3\Pi_u$ and $^1\Pi_u$. The configuration $\pi\pi$ gives states $^1\Sigma$, $^3\Sigma$, $^1\Delta$, and $^3\Delta$. The values of $\Lambda$ may be described in the following way (arrows indicate projections of electron momentum onto the molecular axis): $\leftarrow\rightarrow$, $\rightarrow\leftarrow$, $\rightarrow\rightarrow$ and $\leftarrow\leftarrow$. Strictly speaking, the first and the second combinations are not equivalent, and this symmetry, typical only of $\Sigma$ states, is indicated by signs $+$ or $-$. Therefore, while projection of angular momentum and spin are applicable to all diatomic molecule states, parity and sign are indicated only for homonuclear molecules and $\Sigma$ states, respectively.

The Pauli exclusion principle restricts the number of possible states. For example, there are three of six states for the configuration $(2p\pi)^2$: $^3\Sigma^-$, $^1\Delta$, and $^1\Sigma^+$. (This is the case of the $O_2$ molecule.)

The interaction of rotational and orbital moments removes the degeneration for $\Lambda \neq 0$. Rotational levels double in this case ($\Lambda$ doubling) and begin with $J = \Lambda$ instead of zero. States with $\Lambda \neq 0$ are split into $2S + 1$ components that are defined by the total projection of momentum and spin onto the internuclear axis: $\Omega = \Lambda + \Sigma$ and $\Sigma = S, S - 1, \ldots, -S$.

Above we examined the formation of various molecular states with an unchanged quantum number $n$. States that arise in a transition of a loosely bound electron to orbitals with large $n$ are called Rydberg states. They converge to the ionization potential of the molecule.

Ground states are designated X, e.g., $O_2(X^3\Sigma_g^-)$, and excited states with the same spin and increasing energy are marked $a, b, c, \ldots$ and $A, B, C, \ldots$ for another spin, e.g., $O_2(a^1\Delta_g)$ and $O_2(A^3\Sigma_u^+)$.

## 3.11 Electronic Spectra of Diatomic Molecules

The wave function of a molecular state is $\psi = \psi_e \psi_v \psi_r$; all three multipliers can change in the electronic transitions, and some new selection rules intrinsic to $\psi_e$ should be observed: $\Delta\Lambda = 0, \pm 1$ and $\Delta S = 0$. Transitions of the same sign are allowed for $\Sigma$ states ($\Sigma^+ \leftrightarrow \Sigma^+$, $\Sigma^- \leftrightarrow \Sigma^-$), and $g \leftrightarrow u$ for homonuclear molecules.

All transitions between two electronic states form a band system, e.g., the CO($a^3\Pi \rightarrow X^1\Sigma^+$) Cameron band system and the CO($A^1\Pi \rightarrow X^1\Sigma^+$) fourth positive system. Names of the band systems may be related to researchers who discovered and studied them (e.g., the Fox–Duffedack–Barker system of $CO_2^+$). Many of the band systems were discovered in the gas discharge, and their names may reflect their positions near anode or cathode.

To understand how a band system looks like, we consider the harmonic oscillator approximation for terms in both the lower and upper electronic states:

$$F'' = \omega_e''(\mathrm{v}'' + ½),\ F' = \nu_0 + \omega_e'(\mathrm{v}' + ½),\ \omega_e' = \omega_e'' + \alpha$$
$$\nu(\mathrm{v}', \mathrm{v}'') = F' - F'' = \nu_0 + \omega_e''(\mathrm{v}' - \mathrm{v}'') + \alpha(\mathrm{v}' + ½). \tag{3.35}$$

Bands that originates from a fixed $\mathrm{v}'$ or $\mathrm{v}''$ form progressions with almost equal spaces between the bands, because typically $\alpha << \omega_e$. For example, usually only the $\mathrm{v}'' = 0$ progressions are present in absorption, since almost all molecules are in the ground state with $\mathrm{v}'' = 0$ at low and moderate temperatures. The $\mathrm{v}' = 0$ progressions from three

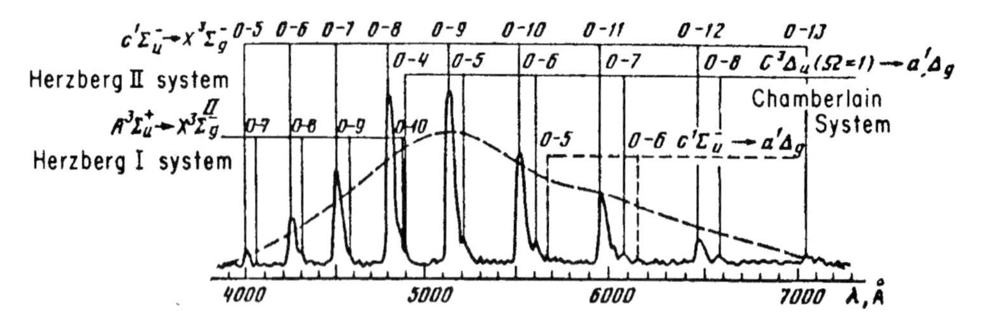


Figure 3.10 Spectrum of the Venus nightglow in the visible range observed from the Venera 9 and 10 orbiters. The spectrum consists of the v′ = 0 progressions from three excited electronic states. The dashed curve is the instrument sensitivity. From Krasnopolsky (1983a).

excited electronic states are seen in the $O_2$ nightglow on Venus (Figure 3.10). Bands with equal $\Delta v = v' - v''$ form closely located groups that are called sequences.

The spaces between sequences with negative and positive $\Delta v$ are $\omega_e''$ and $\omega_e'$, respectively, and a change in the spaces at $\Delta v = 0$ makes it possible to identify the zero sequence and the (0–0) band in the observed spectrum. The band rotational structure is similar to that for rovibrational transitions, although $B' - B''$ can be both positive and negative and is typically significantly larger than those for rovibrational transitions. Spaces between the lines decrease to the blue for $B' < B''$, and an intensity peak is formed from the large number of densely arranged lines in the R-branch – a high-frequency band head. It is said of such a band that it is shaded to the red. A low-frequency band head is formed for $B' > B''$, and the band is shaded to the blue. For large $J$ the space between lines passes through zero and changes sign.

### 3.11.1 Franck–Condon Principle

Motions of the electrons are much faster than those of the nuclei in a molecule, and displacements of the nuclei are negligible during an electron transition (the Franck-Condon principle). This means that probability of the electron transition is proportional to square of the overlapping vibrational wave functions:

$$q_{v'v''} = \left| \int \psi_{v'}^* \psi_{v''} dr \right|^2 .$$

This is the Franck–Condon factor of the v′–v″ band, and a band strength equals $R_{ik}^2 q_{v'v''}$ for

$$R_{ik} = \int \psi_{ei}^* M_e \psi_{ek} dv. \tag{3.36}$$

Here $\psi = \psi_e\ \psi_v$, and $\psi_e, \psi_v$ are the electronic and vibrational components of the wave function. $R_{ik}$ is the electronic moment of the transition, $M_e$ is the electronic component of

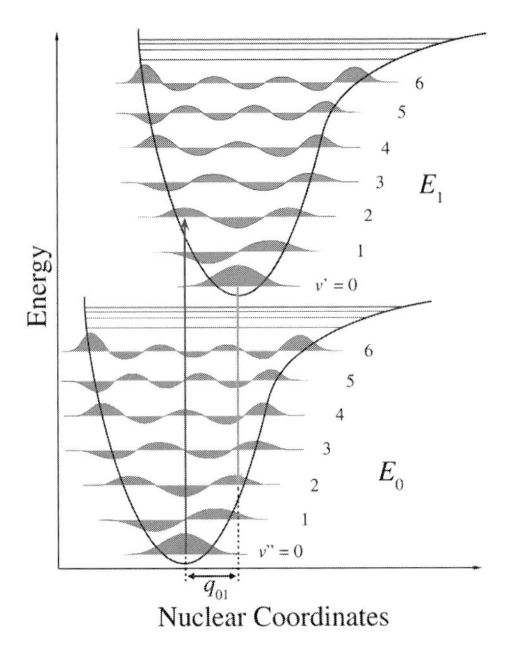


Figure 3.11 Potential energy curves and vibrational wave functions for two electronic states of a molecule. Electronic transitions are vertical, i.e., at the same internuclear distance, and their probabilities are proportional to products of the proper wave functions for the lower and upper states. The strongest absorption and emission lines (from v″ = 0 and v′ = 0, respectively) are shown.

the dipole moment operator, *dv* involves all spatial coordinates except the internuclear distance *r*.

Potential energy curves and vibrational wave functions for two electronic states of a molecule are shown in Figure 3.11. The electronic transitions occur with an unchanged internuclear distance, i.e., vertically. The most probable transitions from the lower and upper states at v = 0 are shown. The photon energy in emission (from the upper state) may be significantly smaller than that in absorption. For example, emission from v′ = 3 is maximal at v″ = 0 and 6, and this two-peak distribution is typical of the vibrational bands for electronic transitions.

### 3.11.2 Electronic States of $O_2$

Energy level diagram for $O_2$ is shown in Figure 3.12. Three lowest states of $O_2$ are even: $X^3\Sigma_g^-$, $a^1\Delta_g$, and $b^1\Sigma_g^+$. The $a \rightarrow X$ band system is forbidden by the momentum ($\Delta\Lambda = 2$), spin, and parity ($g \rightarrow g$), and the radiative lifetime is $\tau = 4460$ s. Fortunately, quenching of $O_2(a^1\Delta_g)$ in collisions with other molecules is very slow as well, and this so-called infrared atmospheric band system of $O_2$ is the brightest in the airglow on the terrestrial planets. Almost all system intensity (~98%) is concentrated in the (0–0) band at 1.27 μm (Figure 3.7).

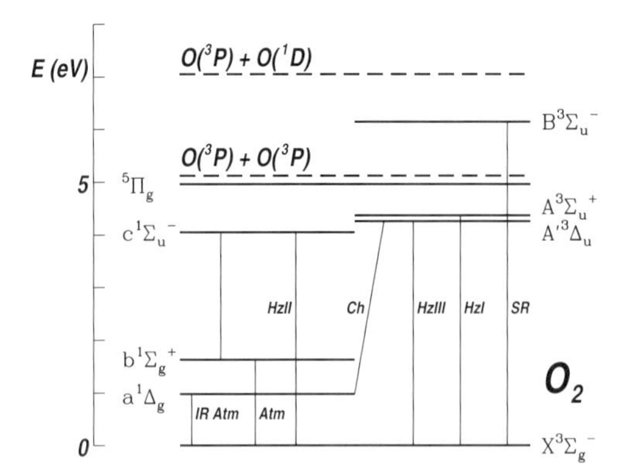


Figure 3.12 Energy level diagram for $O_2$. Hz, Ch, and SR are the Herzberg, Chamberlain, and Schumann–Runge band systems, respectively.

The $b \rightarrow$ X band system is forbidden by spin, sign, and parity, and its radiative lifetime is $\tau = 13$ s. Again, most of the system intensity is in the (0–0) band at 762 nm, the so-called atmospheric band of $O_2$. The band is significant in Earth's airglow but strongly quenched by $CO_2$ on Venus and Mars.

Three odd states of $O_2$, so-called Herzberg states $c^1\Sigma_u^-$, $A'^3\Delta_u$, and $A^3\Sigma_u^+$, have close energies of 4.05, 4.26, and 4.34 eV, respectively. Their transitions to the ground state give the Herzberg II, III, and I band systems that were observed by Herzberg at 240–260 nm in absorption. Continuous absorption occurs for $\lambda < 240$ nm (the Herzberg continuum).

The Herzberg II $c \rightarrow$ X band system is spin-forbidden with $\tau$ = 5–7 s for various v. It is prominent in the visible nightglow on Venus (Figure 3.10). The $c \rightarrow b$ system is sign-forbidden and was observed in Earth's nightglow. The $A'^3\Delta_u$ state emits the Chamberlain system $A' \rightarrow a$ that is spin-forbidden with $\tau$ = 2–4 s. The Herzberg III system is momentum-forbidden ($\Delta\Lambda = 2$). The Herzberg I system A $\rightarrow$ X is sign-forbidden with $\tau = 0.14$ s. The Herzberg I and Chamberlain band systems are prominent in Earth's nightglow and present in the Venus nightglow (Figure 3.10).

All these states converge to the dissociation limit of 5.11 eV for O($^3P$) + O($^3P$). There is a shallow state $^5\Pi_g$ with a depth of 0.15 eV that converges to this limit as well. The internuclear distance for this state is longer than for the other states by a factor of ~1.5.

Productions of the electronic states in the formation of $O_2$ in three-body collisions O + O + M (M is any molecule) are approximately proportional to their statistical weights times square of the internuclear distances. Statistical weights are equal to multiplicity for Σ-states and twice multiplicity for Π- and Δ-states because of Λ-doubling. Then two-thirds of the $O_2$ molecules are formed in the $^5\Pi_g$ state, and energy transfer from this state adds to the populations of the other states.

The $B^3\Sigma_u$-state originates a so-called Schumann–Runge band system B – X. The transitions are allowed and observed in absorption at 200 to 175 nm. The B-state approaches the dissociation limit of 7.08 eV for O($^3P$) + O($^1D$) with a continuous absorption shortward of 175 nm (the Schumann–Runge continuum). Energies of the vibrational levels of the B-state correspond to levels of the continuum in dissociation to $O(^3P)$ + $O(^3P)$ that makes it possible radiationless transitions between the levels, a so-called Auger effect. If the Auger effect results dissociation, it is called predissociation. Probabilities of predissociation of the B-state are smaller than the transition probabilities at low v and dominate at $v > 2$.

## 3.12 Polyatomic Molecules

If a molecule is made up of $N$ atoms, then it has $3N$ degrees of freedom. Of them, three degrees of freedom give the coordinates of the mass center, and three are related to rotation around three major axes of the ellipsoid of the moment of inertia. The remaining $3N - 6$ degrees of freedom are related to so-called normal vibrations.

An important example is the linear triatomic $CO_2$ molecule with three normal vibration modes: symmetric $v_1$ 1388 cm$^{-1}$, bending $v_2$ 667 cm$^{-1}$, and antisymmetric $v_3$ 2349 cm$^{-1}$. All three vibration numbers can change or remain unchanged in a transition. The symmetric mode does not initiate a dipole moment and is therefore forbidden. However, $v_1$ can change simultaneously with $v_2$ and $v_3$. Bending of $CO_2$ is possible in two orthogonal directions, and this is equivalent to an elliptic rotation of the molecule with some angular momentum $l$. Frequency of the symmetric mode is close to the doubled frequency of the bending mode, and this results perturbation of vibrational levels of both modes, a so-called Fermi resonance. The $CO_2$ vibration states are designated either $v_1\ v_2^l\ v_3$ or $v_1\ v_2\ l\ v_3$ r. The latter five-digit designation locates the level r in the Fermi resonance (1 is the upper level, 2, 3,... (if exist) are the lower levels).

$CO_2$ lines are used to measure mixing ratios of species on Venus and Mars, where $CO_2$ constitutes 96% of the atmospheric composition. Comparison of a species line with a nearby $CO_2$ line significantly reduces uncertainties in the derived mixing ratio. The $CO_2$ fundamental band $v_2$ at 15 μm is a standard tool to study vertical thermal structures of the atmospheres of Earth, Mars, and Venus.

The electronic states of linear molecules are similar to the electronic states of diatomic molecules. However, molecules that are linear in the ground states can be nonlinear in excited states. Numerous vibrational bands and the Auger transitions between vibrational levels can make spectra look like continua in some cases. Some spectra retain clear structure, e.g., the $CO_2^+$ ($A^2\Pi \rightarrow X^2\Pi$) emission has a developed vibrational structure only in the symmetric mode $v_1$ with two subsystems: ($v'00$–$v''00$) and ($v'00$–$v''02$).

# 4

# Aerosol Extinction and Scattering

Aerosol is a significant component of any atmosphere, and its composition, size distribution, vertical profile, variations, processes of formation, and removal are important for understanding the atmosphere. Aerosol affects optical observations in an atmosphere that should be corrected to the aerosol scattering and extinction. On the other hand, optical observations are a powerful tool to study aerosol properties. There are three basic problems related to aerosol extinction and scattering: (1) how separate particles scatter the light, (2) what populations of particles look like, and (3) radiative transfer in aerosol with known properties.

## 4.1 Spherical Particles: Mie Formulas

Scattering of electromagnetic waves with wavelength $\lambda$ by a spherical particle with radius $r$ and complex refractive index $m = m_r - i\, m_i$ was studied by Gustav Mie, who obtained analytical formulas for this process. The formulas look complicated but may be easily computed. The real part of $m$, $Re\ m$, is just refractive index, while the imaginary part is $m_i = \frac{k\lambda}{4\pi}$, and $k$ is the absorption coefficient of the particle material. The computation is made using complex numbers.

A parallel unpolarized beam with intensity $I_0$ illuminates the particle, and intensities of two components of the light scattered at angle $\theta$ and distance $D$ from the particle are calculated. The components refer to the light polarized normal to and in the scattering plane. The input parameters are $x = 2\pi r/\lambda$, $m$, and $\mu = \cos\theta$. Then

$$I_\perp = \frac{I_0\lambda^2}{8\pi^2 D^2} i_1(x, m, \mu) = \frac{I_0\lambda^2}{8\pi^2 D^2} \left| \sum_{n=1}^{\infty} \frac{2n+1}{n(n+1)} [a_n \pi_n(\mu) + b_n \tau_n(\mu)] \right|^2; \quad \frac{d\sigma_1}{d\theta} = i_1 \frac{\lambda^2}{8\pi^2} \tag{4.1}$$

$$I_\parallel = \frac{I_0\lambda^2}{8\pi^2 D^2} i_2(x, m, \mu) = \frac{I_0\lambda^2}{8\pi^2 D^2} \left| \sum_{n=1}^{\infty} \frac{2n+1}{n(n+1)} [b_n \pi_n(\mu) + a_n \tau_n(\mu)] \right|^2; \quad \frac{d\sigma_2}{d\theta} = i_2 \frac{\lambda^2}{8\pi^2}. \tag{4.2}$$

The auxiliary functions are

$$\pi_0 = 0,\ \pi_1 = 1,\ \pi_2 = 3\mu, \ldots,\quad \pi_n = \frac{2n-1}{n-1}\mu\pi_{n-1} - \frac{n}{n-1}\pi_{n-2}, \tag{4.3}$$

$$\tau_0 = 0,\ \tau_1 = \mu,\ \tau_2 = 6\mu^2 - 3.\ldots, \tau_n = \mu(\pi_n - \pi_{n-2}) - \left(1-\mu^2\right)(2n-1)\pi_{n-1} + \tau_{n-1}, \tag{4.4}$$

$$w_{-1} = \cos x - i\sin x,\ w_0 = \sin x + i\cos x, \ldots,\ w_n = \frac{2n-1}{x}w_{n-1} - w_{n-2}, \tag{4.5}$$

$$A_0(mx) = \cot mx, \ldots,\ A_n(mx) = -\frac{n}{mx} + \left[\frac{n}{mx} - A_{n-1}(mx)\right]^{-1}. \tag{4.6}$$

Here, for $z = x + i\,y$, $\cot z = \dfrac{2\sin 2x + i(e^{-2y} - e^{2y})}{e^{-2y} + e^{2y} - 2\cos 2x}$. The coefficients in $i_1$ and $i_2$ are

$$a_n(m,x) = \frac{\left[\dfrac{A_n(mx)}{m} + \dfrac{n}{x}\right] Re\ w_n - Re\ w_{n-1}}{\left[\dfrac{A_n(mx)}{m} + \dfrac{n}{x}\right] w_n - w_{n-1}} \tag{4.7}$$

$$b_n(m,x) = \frac{\left[mA_n(mx) + \dfrac{n}{x}\right] Re\ w_n - Re\ w_{n-1}}{\left[mA_n(mx) + \dfrac{n}{x}\right] w_n - w_{n-1}}. \tag{4.8}$$

Extinction and scattering coefficients (efficiencies) are

$$K_e = \frac{\sigma_e}{\pi r^2} = \frac{2}{x^2}\sum\nolimits_{n=1}^{\infty}(2n+1)Re\ (a_n + b_n) \tag{4.9}$$

$$K_s = \frac{\sigma_s}{\pi r^2} = \frac{2}{x^2}\sum\nolimits_{n=1}^{\infty}(2n+1)\left(|a_n|^2 + |b_n|^2\right), \tag{4.10}$$

and absorption coefficient $K_a = K_e - K_s$. The number $n$ may be restricted at $n_{\max} \approx x$; if $x \leq 5$, then $n_{\max} \approx 5$ may be used.

Thus, the Mie formulas make it possible to calculate the extinction and scattering cross section of a spherical particle, its scattering phase function, and polarization. Phase function of scattering is the scattering intensity versus scattering angle with the integral over the sphere normalized to either 1 or $4\pi$; the latter is used below. Scattering efficiency functions of the size parameter $x$ and imaginary refractive index $m_i$ are shown in Figure 4.1. The peak value at $x \approx 4$ exceeds the geometric cross section $\pi r^2$ by a factor of ~4, and an asymptotic value for very large nonabsorbing particles is 2. There are a few approximations for the Mie formulas, and some of them are considered below.

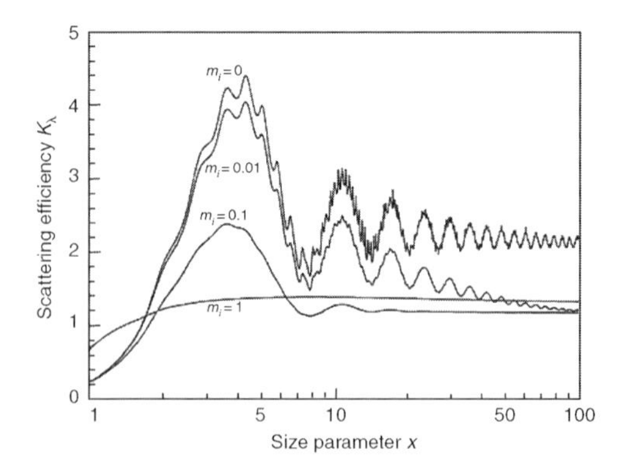


Figure 4.1 Scattering efficiency as a function of size parameter and imaginary refractive index for $m_r = 1.33$.

## 4.2 Some Approximations and Nonspherical Particles

### 4.2.1 Rayleigh Scattering by Very Small Particles and Gas Molecules

If $x << 1$, then the phase function, scattering and absorption efficiencies are

$$\gamma(\theta) = \frac{3}{4}\left(1 + \cos^2\theta\right); K_s = \frac{8}{3}x^4\left|\frac{m^2 - 1}{m^2 + 2}\right|; \quad K_a = -4x\, Im\, \frac{m^2 - 1}{m^2 + 2}. \tag{4.11}$$

The Rayleigh scattering is proportional to $x^4$, that is, to $\lambda^{-4}$, and steeply increases to the blue.

Scattering by gas molecules is the important case of the Rayleigh scattering, and the scattering cross section is

$$\sigma_s = \frac{32\pi^3\delta}{3\lambda^4}\left(\frac{m - 1}{n}\right)^2. \tag{4.12}$$

Here $\delta \approx 1$ is the depolarization factor, $n$ is the gas number density, and $n$ is equal to the Loschmidt number $2.69\times10^{19}$ cm$^{-3}$, if the refractive index $m$ is measured at the standard conditions. For example, the vertical optical depth for molecular scattering in the Earth's atmosphere is equal to 1.2 at $\lambda = 0.3$ μm.

### 4.2.2 Weakly Absorbing Particles

The ratio of $K_s/K_e = \omega$ is a so-called single-scattering albedo of a particle. If this value is close to 1, that is, $K_a << K_s$, then $1 - \omega = \frac{2}{3}kr$ ($k$ is the absorption coefficient of the particle material), and

$$K_s = 2 - \frac{4}{y}\sin y + \frac{4}{y^2}(1 - \cos y); \tag{4.13}$$

$$y = \frac{4\pi r}{\lambda}(m_r - 1). \tag{4.14}$$

The phase function is symmetric for very small particles with equal forward ($\theta < \pi/2$) and back ($\theta > \pi/2$) scattering, and forward scattering dominates for medium and especially large ($x >> 1$) particles.

### 4.2.3 Nonspherical Particles

There are some tools to calculate light scattering by the particles with a more complicated geometry. For example, haze particles on Titan are aggregates of a few thousand very small spherical monomers with $r = 0.02$–$0.1$ μm. Rannou et al. (1999) created a code to calculate the particle extinction cross section, single-scattering albedo, and scattering asymmetry factor (see below) using the monomer radius, number, and complex refractive index. A few types of nonspherical particles are considered by Mishchenko (2000, 2014).

According to van der Hulst (1981), light scattering by the nonspherical randomly oriented convex particles is identical to that by the spheres of the same surface area. This makes it possible to apply the Mie formulas with some effective radius to nonspherical particles.

### 4.2.4 Solid Surfaces

Hapke (1993) adjusted the Mie theory to calculate reflectance of solid surfaces that consist of particles and grains with given properties.

## 4.3 Particle Size Distributions: Photometry, Polarimetry, and Nephelometry of Aerosol Media

Aerosol particles of the same origin form a population with a size distribution $\frac{dn}{dr}$. Factors that affect a particle size distribution are the strong dependence of particle precipitation on their size, coagulation and preferential evaporation of small liquid particles etc. There are a few analytic presentations for the particle size distributions. Their basic parameters are particle number density, effective mean particle radius $a$, and effective width of the size distribution. The gamma distribution (Figure 4.2) is

$$\frac{dn}{dr} = Anr^{\frac{1}{b}-3}e^{-r/ab};\ A = \frac{(ab)^{2-\frac{1}{b}}}{\Gamma\left(\frac{1}{b}-2\right)};\ a = \frac{\int_0^\infty r^3 \frac{dn}{dr} dr}{\int_0^\infty r^2 \frac{dn}{dr} dr}; b = \frac{\int_0^\infty \left(1-\frac{r}{a}\right)^2 r^2 \frac{dn}{dr} dr}{\int_0^\infty r^2 \frac{dn}{dr} dr}. \quad (4.15)$$

Here $\Gamma$ is the gamma function, the distribution is maximum at $r_{max} = a\ (1 - 3b)$, and the standard deviation is $\sigma = a\ [b\ (1 - 2b)]^{1/2}$.

The Gaussian distribution is

$$\frac{dn}{dr} = \frac{n}{(2\pi)^{1/2}\sigma} e^{-\frac{1}{2}\left(\frac{r-a}{\sigma}\right)^2}. \quad (4.16)$$

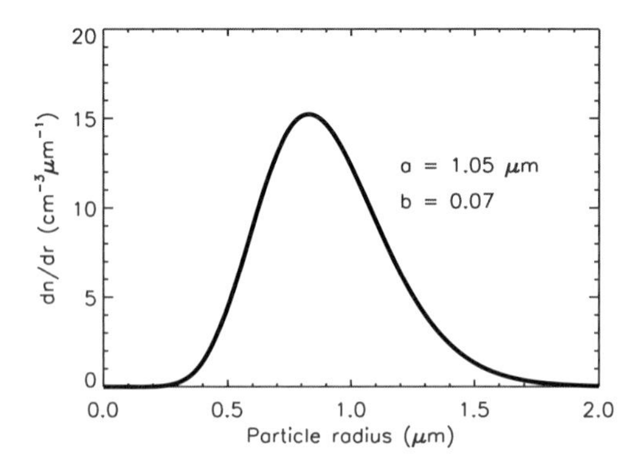


Figure 4.2 Gamma distribution of the sulfuric acid particles on Venus at 65 km, mode 2.

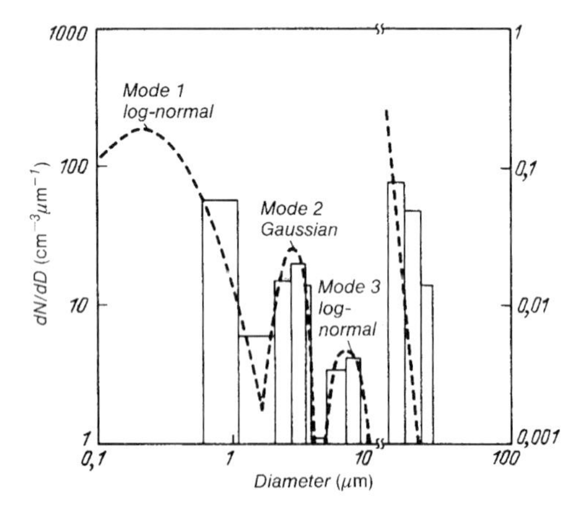


Figure 4.3 Fitting of the observed particle size distribution in the middle cloud layer at 54 km on Venus by three particle modes. The Pioneer Venus data are from Knollenberg and Hunten (1980).

The log-normal distribution is

$$\frac{dn}{dr} = \frac{n}{(2\pi)^{1/2} r \ln \sigma} e^{-\frac{1}{2}\left(\frac{\ln r/a}{\ln \sigma}\right)^2}. \tag{4.17}$$

The above distributions involve three parameters. The general gamma distribution is determined by four parameters,

$$\frac{dn}{dr} = Anr^{\alpha} \exp\left[-\frac{\alpha}{\gamma}\left(\frac{r}{r_m}\right)^{\gamma}\right], \tag{4.18}$$

which gives an additional degree of freedom for fitting to observational data.

An example of this fitting is shown in Figure 4.3 for the Pioneer Venus observations by the particle size spectrometer in the middle cloud layer at 54 km. Three different particle modes were detected and their quantitative parameters derived.

Photometry and polarimetry at various phase angles are sensitive to size distribution and refractive index of the observed aerosol. Therefore the remote observational data may be fitted by models to retrieve these parameters. In situ aerosol measurements may be made by nephelometers with a small aerosol volume illuminated by a laser beam and the scattered light measured in a few directions. For example, if four directions are used, then refractive index, particle number density, their mean size and width of the size distribution can be retrieved.

## 4.4 On the Radiative Transfer

Radiative transfer is a broad field of science that cannot be discussed here in detail. Two cases and some recommendations are considered below.

### *4.4.1 Optically Thin Atmosphere: Single Scattering Approximation*

The problem becomes simple if the multiple scattering is neglected (Figure 4.4). Then a layer $d\tau$ adds to the observed radiation

$$dI = I_0 e^{-\tau\left(\frac{1}{\cos\theta} + \frac{1}{\cos\varphi}\right)} \frac{\omega d\tau}{\cos\varphi} \frac{\gamma(\psi)}{4\pi}. \tag{4.19}$$

Here $I_0$ is the solar radiation, ω is the single-scattering albedo of the medium, and $\gamma(\psi)$ is the phase function. Brightness of a surface with reflectivity $A$ for the Lambert reflection without the atmosphere is

$$I_S = A\frac{I_0}{\pi}\cos\theta \tag{4.20}$$

and does not depend on the viewing angle $\varphi$. One gets after the integration

$$\rho = \frac{\pi I}{I_0\cos\theta} = \frac{\omega\gamma(\psi)}{4(\cos\theta + \cos\varphi)}\left[1 - e^{-\tau\left(\frac{1}{\cos\theta} + \frac{1}{\cos\varphi}\right)}\right] + Ae^{-\tau\left(\frac{1}{\cos\theta} + \frac{1}{\cos\varphi}\right)}, \tag{4.21}$$

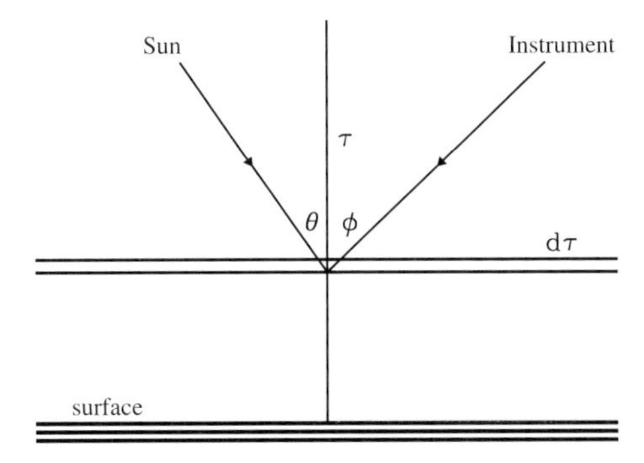


Figure 4.4 Single scattering in an atmosphere observed from space.

where $\rho = \frac{\pi I}{I_0 \cos\theta}$ is the definition of the brightness coefficient, and the brightness coefficient should be symmetric relative to the incidence $\theta$ and viewing $\varphi$ angles everywhere. The airmass factor $m = \frac{1}{\cos\theta} + \frac{1}{\cos\varphi}$ becomes invalid near the limb ($\cos\varphi \to 0$) and terminator ($\cos\theta \to 0$), and more complicated functions for the spherical atmosphere have to be applied. The approximation (4.21) may be used for the Martian atmosphere in the periods without dust storms.

### *4.4.2 Optically Thick Atmosphere*

A convenient tool for radiative transfer in the optically thick atmospheres is the Henyey–Greenstein phase function

$$\mu = \cos\varphi;\ \gamma(\mu) = \frac{1 - g^2}{(1 + g^2 - 2g\mu)^{3/2}};\ g = \frac{1}{2}\int_{-1}^{1} \gamma(\mu)\mu d\mu. \tag{4.22}$$

Here $g$ is the scattering asymmetry factor that is equal to zero for, e.g., isotropic and Rayleigh scattering and approaching 1 and −1 for very strong forward and back scattering, respectively. True phase functions can be approximated by the Henyey–Greenstein phase function using the relationship (4.22) for $g$. Dependences of $g$ on the effective size parameter $2\pi a/\lambda$ and refractive index are shown in Figure 4.5.

Optical observations in the range of 0.32–1.2 μm at the Venera 8 to 14 landing probes and at 0.63 μm from the Pioneer Venus large probe are the appropriate examples. The

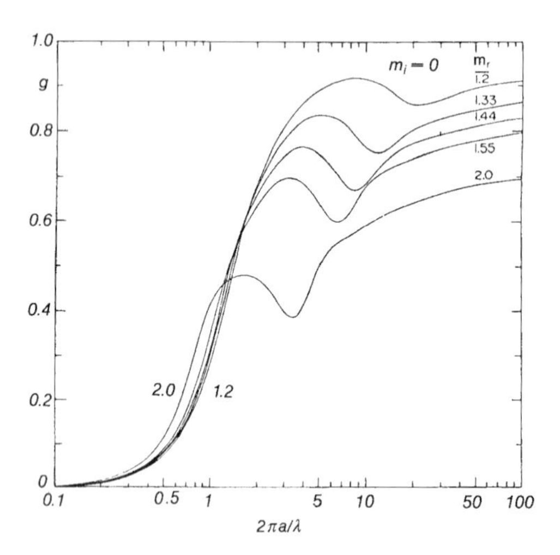


Figure 4.5 Asymmetry parameter $g$ as a function of effective size parameter $2\pi a/\lambda$ and refractive index for the gamma distribution with $b = 0.07$ (Hansen and Travis 1974).

results will be discussed in Chapter 12, and here we will consider two methods that were applied to the Venera data.

### 4.4.3 Inverse Problem

The difference between downward and upward light fluxes $F_\downarrow - F_\uparrow$ at the altitude $z$ is a flux absorbed by the atmosphere below $z$, the so-called net flux. A derivative of the net flux is the energy absorbed per unit length:

$$\frac{d(F_\downarrow - F_\uparrow)}{dz} = 2(F_\downarrow + F_\uparrow)\sigma(1-\omega)\,, \tag{4.23}$$

where $\sigma(1 - \omega)$ is the true absorption coefficient and $\sigma$ is the extinction coefficient. The factor of 2 is because the mean cosine of the incident angle is ½.

The net flux decreases volume luminosity in the low lying atmosphere, and the decrease is proportional to the scattering coefficient $\sigma\omega$ and scattering asymmetry:

$$\frac{d(F_\downarrow + F_\uparrow)}{dz} = A(g)\sigma\omega(F_\downarrow - F_\uparrow). \tag{4.24}$$

Ustinov (1977) found that $A(g) = \frac{3}{2}(1-g)$ for $\omega \approx 1$. Angular distribution of radiation in an optically thick atmosphere is approximated by

$$I(z, \cos\varphi) = \frac{1}{2} f_0(z) + \frac{3}{2} f_1(z) \cos\varphi\,. \tag{4.25}$$

Then integration over the upper and lower hemispheres yields

$$F_\downarrow + F_\uparrow = \pi f_0, \quad F_\downarrow - F_\uparrow = 2\pi f_1, \tag{4.26}$$

and the substitution gives

$$\sigma(1-\omega) = \frac{1}{f_0}\frac{df_1}{dz}, \quad \sigma(1-g) = \frac{1}{3f_1}\frac{df_0}{dz}\,. \tag{4.27}$$

This solution of the inverse problem of sounding in an optically thick atmosphere was obtained by Ustinov (1977) as a partial case of a more complicated matrix equation. If brightnesses in two directions are measured at varying altitude in an optically thick atmosphere, then extinction coefficient $\sigma(z)$ and single-scattering albedo $\omega(z)$ can be extracted from the data for an adopted asymmetry factor (e.g., $g \approx 0.75$ in the Venus clouds and zero below the clouds in the Rayleigh-scattering atmosphere).

### 4.4.4 Direct Problem

Here $\sigma$, $\omega$, and $g$ as functions of altitude are known, and the problem is to calculate downward and upward fluxes of the diffuse light at each altitude, that is, transmittance $T_i$ of

Table 4.1 K($\mu_0$, g)

| $\mu_0$ | $g = 0$ | 0.25 | 0.5 | 0.75 | 0.875 |
|---|---|---|---|---|---|
| 0 | 0.433 | 0.424 | 0.395 | 0.333 | 0.272 |
| 0.1 | 0.540 | 0.534 | 0.516 | 0.484 | 0.463 |
| 0.3 | 0.711 | 0.708 | 0.701 | 0.692 | 0.688 |
| 0.5 | 0.872 | 0.870 | 0.869 | 0.869 | 0.869 |
| 0.7 | 1.028 | 1.028 | 1.030 | 1.033 | 1.035 |
| 0.9 | 1.182 | 1.184 | 1.188 | 1.192 | 1.194 |
| 1 | 1.259 | 1.261 | 1.265 | 1.270 | 1.271 |

the atmosphere above level $i$ and reflectance $R_i$ of the atmosphere below this level. The atmosphere is divided into a few levels, and it is necessary to calculate transmittances $t_i$ and reflectances $r_i$ of all layers and then add them. If a layer $r_i$, $t_i$ is between levels $i$ and $i+1$ and $i$ increases with altitude, then (van der Hulst 1980)

$$R_{i+1} = r_i + \frac{R_i t_i^2}{1 - r_i R_i}; \quad T_i = \frac{T_{i+1} t_i}{1 - r_i R_i} \quad . \tag{4.28}$$

Reflectance $r$ and transmittance $t$ of an isolated layer can be calculated using the following formulas (Danielson et al. 1969):

$$1 - \omega = \frac{k^2}{3(1-g)}; \quad m = \frac{8k}{3(1-g)}; \quad l = e^{-\frac{1.42k}{1-g}}; \quad f = \exp\left[-k\left(\tau + \frac{1.42}{1-g}\right)\right]; \tag{4.29}$$

$$r = 1 - K(\mu_0)m\left[\frac{1}{2}\, l^{1/2} + \frac{f^2}{1-f^2}\right]; \quad t = \frac{K(\mu_0)mf}{1-f^2} \quad . \tag{4.30}$$

Here $K(\mu_0) = 1$ for all internal layers; $K$ for the upper layer at various cosines of the incidence angle $\mu_0 = \cos\theta$ and $g$ are given in Table 4.1.

For conservative scattering $\omega = 1$, formulas (4.30) require the evaluation of the indeterminate 0/0 and result in

$$r(\mu_0) = 1 - t(\mu_0) = 1 - \frac{4K(\mu_0)}{3(\tau(1-g) + 1.42)} \quad . \tag{4.31}$$

For example, the Venus reflectivity is 0.83 in the visible range, and one gets the cloud optical depth of 25 using (4.31) and neglecting the surface reflection. Here $g \approx 0.75$ for the sulfuric acid clouds with $m_r = 1.44$ (Figure 4.5). This easily obtained value is in reasonable agreement with the detailed studies by the Venera and Pioneer Venus landing probes.

Van der Hulst (1980) considered various approximations and calculated detailed tables for numerous cases of radiative transfer in the multiple scattering. Those tables can be interpolated and the results coadded using (4.28) for many practical problems. For example, single-scattering albedo ω and asymmetry factor $g$ may be combined into

$$s = \sqrt{\frac{1-\omega}{1-\omega g}}, \tag{4.32}$$

and spherical albedo (ratio of total flux scattered by a planet to the total incoming flux) of a semi-infinite atmosphere is

$$A = (1-s)\frac{1-0.139s}{1+1.17s}, \tag{4.33}$$

with accuracy of ~0.001. The spherical harmonic discrete ordinate method (Evans 2007) may be used for various applications as well.

## 4.5 Aerosol Altitude Distribution

Here we consider some cases for the aerosol altitude distribution. They are based on equilibrium between the particle precipitation and eddy diffusion that tends to mix the particles uniformly in the atmosphere. Evidently this equilibrium is not applicable to some dynamic phenomena, such as active phases of dust storms.

Precipitation velocity $V$ of a spherical particle is given by the Stokes law with corrections by Davies for particle radii $r$ comparable or smaller than the mean free path $l$:

$$V = \frac{2}{9}\rho g\frac{r^2}{\eta}\left[1+\frac{l}{r}\left(1.257+0.4e^{-1.1r/l}\right)\right]. \tag{4.34}$$

Here $\rho$ is particle material density, $g$ is gravity, and $\eta$ is gas viscosity.

### *4.5.1 Incondensable Aerosol*

Its flux is

$$\Phi = -K\left(\frac{dn}{dz}+\frac{n}{H^*}\right) - nV\,. \tag{4.35}$$

Here $H^*$ is the scale height for the gas number density, and $V$ is the absolute (positive) value of the precipitation velocity.

For example, solar occultation observations of the dust extinction at 1.9 and 3.7 μm from the Phobos orbiter of Mars (Korablev et al. 1993) resulted in particle radii and number densities for nine occultations. Their mean values and variances were $r = 1.26 \pm 0.2$ μm and $n = 0.32 \pm 0.15$ cm$^{-3}$ at 15 to 25 km. The observed profiles were fitted using (4.35), and the derived mean values were $K = (1.1 \pm 0.5)\times10^6$ cm$^2$ s$^{-1}$ below 25 km, $\Phi = 0.4 \pm 0.2$ cm$^{-2}$ s$^{-1}$, the mean altitude of the dust influx was 22 ± 4 km, and the mean total dust optical depth was 0.24 ± 0.06 at low latitudes near the spring equinox.

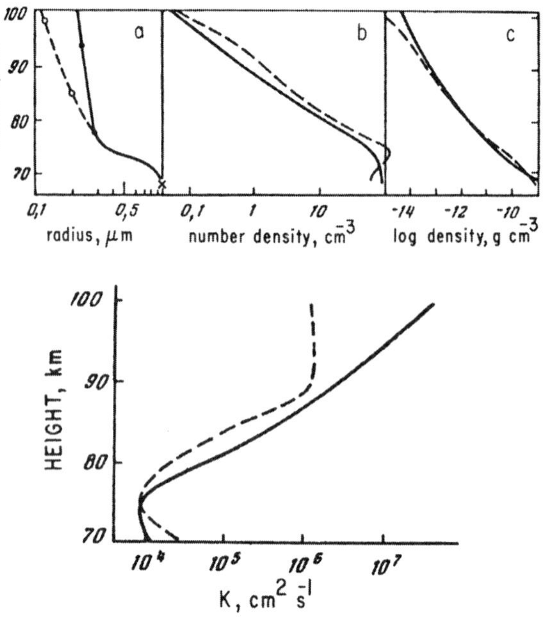


Figure 4.6 Vertical profiles of the particle radius, number density, mass loading, and eddy diffusion coefficient retrieved from the limb spectra of the sulfuric acid haze in the visible and NUV ranges observed by the Venera 9 and 10 orbiters. From Krasnopolsky (1980, 1983a).

Properties of the sulfuric acid haze above the clouds of Venus were retrieved from spectra on the limb in the visible and near-UV ranges observed from the Venera 9 and 10 orbiters (Krasnopolsky 1983a). Particle sizes and limb column abundances were obtained by fitting the observed spectra using the Mie formulas for the sulfuric acid refractive index of 1.46 at $T \approx 200$ K. Then the limb abundances were converted into particle number densities (Section 7.7) and mass loading (Figure 4.6). $\Phi = 0$ in (4.35) at equilibrium, and

$$-\frac{1}{n}\frac{dn}{dz} = \frac{1}{H_a} = \frac{1}{H^*} + \frac{V}{K}, \quad K = \frac{VH^*H_a}{H^* - H_a}, \tag{4.36}$$

where $H_a$ is the aerosol scale height. The retrieved profiles of eddy diffusion are shown in Figure 4.6 as well.

### 4.5.2 Condensation Layer

If a condensable species is mixed below a condensation level $z_0$ and its chemical production and loss are weak, then flux of the species in two phases (gas and aerosol) is zero for the equilibrium above $z_0$:

$$\Phi = -K\left[\frac{d(n_s + n_a)}{dz} + \frac{n_s + n_a}{H^*}\right] - n_a V = 0\,, \tag{4.37}$$

where $n_s$ is the saturated vapor number density and $n_a$ is the number density in the aerosol phase. The former may be approximated by

$$n_s = n_{s0} e^{(T-T_0)/\beta} \tag{4.38}$$

using, for example, the data from Fray and Schmitt (2009). Then the solution of the differential equation (4.37) for $T(z) = T_0 - \gamma\,(z - z_0)$ is

$$n_a = n_{s0} \frac{\dfrac{\gamma}{\beta} - \dfrac{1}{H^*}}{\dfrac{V}{K} - \dfrac{\gamma}{\beta} + \dfrac{1}{H^*}} \left[ e^{-(z-z_0)\gamma/\beta} - e^{-(z-z_0)(V/K+1/H^*)} \right]. \tag{4.39}$$

The layer maximum can be found using $\dfrac{dn_a}{dz} = 0$ :

$$z_{\max} = z_0 + \frac{\ln\left(\dfrac{V}{K} + \dfrac{1}{H^*}\right) - \ln\dfrac{\gamma}{\beta}}{\dfrac{V}{K} + \dfrac{1}{H^*} - \dfrac{\gamma}{\beta}}. \tag{4.40}$$

The condensate column abundance may be obtained by integration of (4.39):

$$N_a = n_{s0} \frac{1 - \dfrac{\beta}{\gamma H^*}}{\dfrac{V}{K} + \dfrac{1}{H^*}}. \tag{4.41}$$

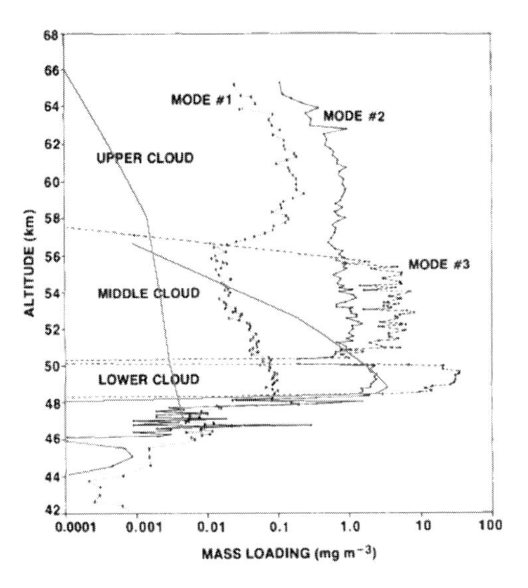


Figure 4.7 Vertical profile of the sulfur aerosol mass loading (the bottom solid curve) calculated using the method in Section 4.5.2 compared with the Pioneer Venus particle size spectrometer observations. The upper curve is the sulfur aerosol that is formed in the mesosphere of Venus. From Krasnopolsky (2016a).

Furthermore, equation (4.37) can be easily solved numerically, without these analytic expressions.

For instance, this approach was used to calculate a layer of sulfur aerosol in the clouds of Venus (Figure 4.7) using the $S_8$ vapor model abundance of 2.5 ppm at 47 km. The sulfur saturation vapor pressure, particle radius, temperature, pressure, and eddy diffusion data were taken from the literature.

Modeling of coagulation and other processes that determine particle size distribution may be found in Gao et al. (2014) and references therein.

# 5
# Quantitative Spectroscopy

The main task of quantitative spectroscopy is retrieval of quantities of absorbing or emitting species from the observed spectra. These problems are discussed, for example, in the *Journal of Quantitative Spectroscopy and Radiative Transfer*.

## 5.1 Line Broadening

Evidently all lines are of some width, and line widths and shapes are determined by the chaotic motion of molecules, their collisions, and radiative lifetimes of the upper states.

### *5.1.1 Thermal (Doppler) Broadening*

If pressure and therefore the collision rate are low, then thermal velocities of molecules determine the line shape. Mean thermal velocity is $\boldsymbol{v}_t = \sqrt{\frac{2kT}{m}}$, and the associated Doppler shift is

$$\delta\nu_D = \frac{\nu_0 v_t}{c} = 4.3 \times 10^{-7}\nu_0 \left(\frac{T}{\mu}\right)^{1/2} \text{cm}^{-1}. \tag{5.1}$$

Here $\mu$ is the molecular mass in the atomic units. The shape of the absorption cross section is a Gaussian,

$$\sigma_\nu = \frac{\sigma}{\sqrt{\pi}\,\delta\nu_D} e^{-[(\nu-\nu_0)/\delta\nu_D]^2} \text{cm}^2, \tag{5.2}$$

and the absorption line is $\frac{I_\nu}{I_0} = e^{-\sigma_\nu mN}$ (Figure 5.1). Here $\sigma = \int_0^\infty \sigma_\nu d\nu$ is the line strength, $m$ is the airmass, and $N$ is the absorber column abundance. For example, $\delta\nu_D \approx 0.005$ cm$^{-1}$ for the CO lines at 2.35 μm and $T$ = 200 K, so that $\delta\nu_D/\nu_0 \approx 10^{-6}$.

### *5.1.2 Collisional or Pressure (Lorentz) Broadening*

The line shape is collisionally induced at moderate and high pressure with a line width

$$\delta\nu_c = \frac{1}{2\pi c\tau_c} = \frac{pb}{\pi c(2mkT)^{1/2}} = \delta\nu_{c0}(p/p_0)(296/T)^\alpha. \tag{5.3}$$

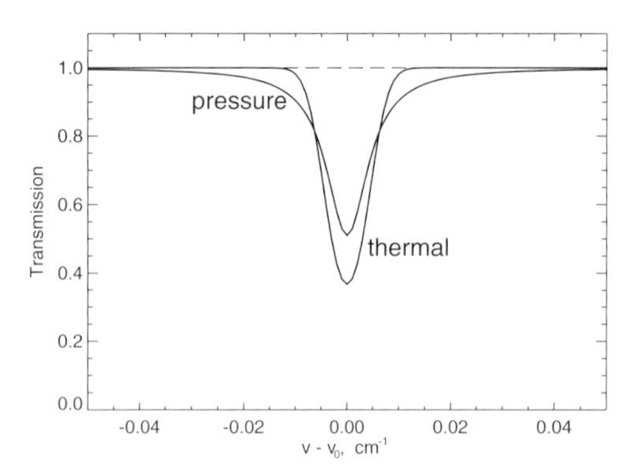


Figure 5.1 Line shapes for thermal and pressure broadening. The line widths and absorber abundances are the same.

Here $\tau_c$ is the mean time between the collisions, $b$ is the collisional cross section, and $m$ is the molecular mass. The collisionally induced half-widths are tabulated as $\delta\nu_{c0}$ at the standard conditions $p_0$ = 1 bar and $T_0$ = 296 K with index $\alpha$ for the temperature dependence. If a line is allowed and the radiative time is very small, then $\delta\nu_{rad} = A/4\pi c$ is to be added to $\delta\nu_c$; $A$ is the transition probability. The absorption cross section is

$$\sigma_\nu = \frac{\sigma\delta\nu_c}{\pi\left[(\nu - \nu_0)^2 + \delta\nu_c^2\right]}. \tag{5.4}$$

Typically $\delta\nu_{c0} \approx 0.1$ cm$^{-1}$ and $\delta\nu_c$ exceeds $\delta\nu_D$ at $p > 50$ mbar. The line shapes for thermal and pressure broadenings with the same line widths and absorber abundances are compared in Figure 5.1. The wings of the pressure-broadened line are much wider than those for thermal broadening.

### 5.1.3 Voigt Line Shape

Usually both thermal and pressure broadenings affect the spectral lines, and both effects are combined by the Voigt line shape:

$$\mathrm{a} = \delta\nu_\mathrm{c}/\delta\nu_\mathrm{D}; \quad \sigma_\nu = \frac{\sigma}{\pi^{1/2}\delta\nu_D} H(a,q); \quad H(a,q) = \frac{a}{\pi}\int_{-\infty}^{\infty} \frac{e^{-x^2}dx}{a^2 + (q-x)^2}; \quad q = \frac{\nu - \nu_0}{\delta\nu_D}. \tag{5.5}$$

Though the function $H(a,q)$ looks complicated, it is immediately returned by, e.g., IDL for any reasonable combination of the parameters. The Voigt shape may be approximated by

$$\sigma_\nu \approx \frac{\sigma}{\pi^{1/2}\delta\nu_D}\left[e^{-q^2} + \frac{a}{\pi^{1/2}q^2}\right]; \tag{5.6}$$

that is, it approaches thermal broadening near the line center and pressure broadening at the wings.

## 5.2 Line Equivalent Widths and Curves of Growth

### 5.2.1 Equivalent Width

Equivalent width is a quantitative assessment of the absorption effect:

$$W = \int_0^\infty \frac{I_0(v) - I(v)}{I_0(v)}\, dv = \int_0^\infty \left(1 - \frac{I(v)}{I_0(v)}\right) dv = 2\int_{v_0}^\infty \left(1 - e^{-\sigma_v mN}\right) dv, \quad (5.7)$$

$I_0(v)$ is the continuum intensity that is measured outside the line and interpolated inside the line, e.g., the dashed line for the thermal broadening in Figure 5.1. This interpolation is a source of uncertainty; furthermore, it becomes impossible if the line and the nearby continuum are contaminated by other lines or affected by some instrument features. On the other hand, equivalent width does not generally depend on the instrument sensitivity and spectral resolution and does not require the instrument absolute calibration. However, spectrographs with high sensitivity and spectral resolution provide better accuracy in measuring equivalent widths.

Here $m = \frac{1}{\cos\theta} + \frac{1}{\cos\varphi}$ is the airmass, and $N$ is the absorber column abundance. If the line is weak and $\sigma_v mN \le 0.3$, even in the line center, then $1 - e^{-\sigma_v mN} \approx \sigma_v mN$, and $W \approx \sigma mN$. This is the weak line approximation. For strong lines, substitution of $\sigma_v$ by the relationship for the Voigt broadening (5.5) and $dv$ by $dq = dv/\delta v_D$ to (5.7) results in

$$\frac{W}{\pi^{1/2}\delta v_D} = \frac{2}{\pi^{1/2}} \int_0^\infty \left[1 - \exp\left(-\frac{\sigma mN}{\pi^{1/2}\delta v_D} H(a,q)\right)\right] dq = F\left(\frac{\sigma mN}{\pi^{1/2}\delta v_D}, a\right). \quad (5.8)$$

### 5.2.2 Curves of Growth

These functions $F$ are so-called curves of growth, and relationship (5.8) can be used to calculate them (Figure 5.2). A curve of growth connects the input absorber amount $\sigma mN$

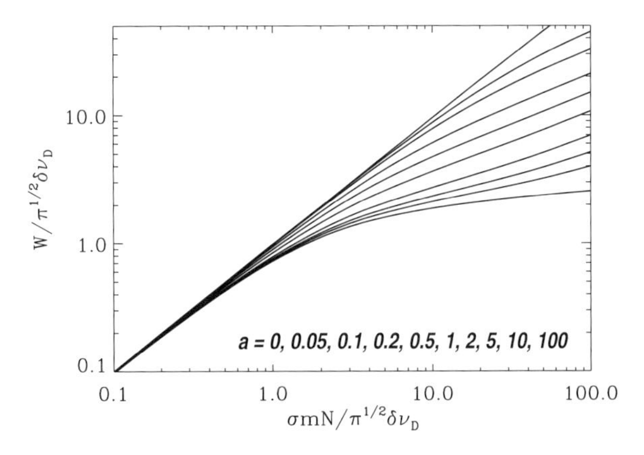


Figure 5.2 Curves of growth for various parameter $a = \delta v_c/\delta v_D$. The abscissa is optical depth in the line center.

with the equivalent width, both in units of $\pi^{1/2}\delta\nu_D$. The abscissa $x$ is optical depth in the line center for thermal broadening.

The pure thermal broadening ($a = 0$) results in saturation at $x \approx 3$ (Figure 5.2) and a weak sensitivity to further increase of $x$. For example, the equivalent widths for $x = 10$ and 100 differ just by a factor of 1.36. The next two curves with $a = 0.05$ and 0.1 are typical of the conditions in the **Martian atmosphere** for the near-infrared range, and the preferable option for observations is for $x \approx 1$ to 2, where the line equivalent width is rather large but still sensitive to variations of the absorber abundance. The value of $a \geq 0.5$ is typical of the light reflected by the clouds on **Venus**, without significant restrictions to the line strength.

### *5.2.3 Analysis of Measured Equivalent Widths*

Atmospheric pressure and temperature vary with altitude, and some effective values are to be used for the curves of growth. If an observed equivalent width refers to an atmosphere with low aerosol optical depth and pressure $p_0$ at the surface, then the equivalent width for a uniformly mixed absorber may be approximated by a uniform layer of the gas with the same vertical abundance $N$ at pressure $p_0/2$ and temperature $T(p_0/2)$ at this level. This is a version of the Curtis–Godson approximation, and $p_0/2$ is a level of line formation.

Being applied to Mars in the periods without global dust storms, the mean $p_0 \approx 6$ mbar and $T(p_0/2) \approx 200$ K, and true values for various conditions may be found in the TES database (thermal emission spectrometer, Smith 2004) and the Mars Climate Database (www-mars.lmd.jussieu.fr/mcd_python/). Line strength $\sigma$ is proportional to $\exp(-\alpha E_0/T)$ with $\alpha = 1.439$ cm K and the rotational energy $E_0$ of the ground state. Evidently variations of $\sigma$ with temperature are low for small $E_0$ and may be otherwise significant. Numerical tests show that the error of this approximation is ~1% for $E_0 \approx 200\ \text{cm}^{-1}$ and ~5% for $E_0 \approx 1000\ \text{cm}^{-1}$, that is, proportional to $E_0$.

If a line is collisionally broadened, then the simple integration results

$$\tau_\nu = \frac{\sigma m N}{4\pi\delta\nu} \ln\left[1 + \left(\frac{2\delta\nu}{\nu - \nu_0}\right)^2\right], \tag{5.9}$$

where $\delta\nu$ is the collisional half-width for $p_0/2$ and $T(p_0/2)$, and the equivalent width,

$$W = 2\int_{\nu_0}^{\infty} (1 - e^{-\tau_\nu})d\nu,$$

can be easily calculated. There is a divergence of (5.9) at $\nu = \nu_0$, and $\tau_{\nu 0}$ may be adopted equal to the neighbor value. These relationships can be applied to telluric lines of uniformly mixed species in the ground-based observations.

## 5.3 Ground-Based Spatially Resolved High-Resolution Spectroscopic Observations

Below we will consider two basic methods for interpretation of spectroscopic data: analysis of equivalent widths and fitting of observed spectra by synthetic spectra. This will be done using existing observations with a spectrograph CSHELL at NASA IRTF. Interpretation of the ground-based observations is typically more complicated than those by spacecraft and Earth-orbiting observatories because of contamination by the absorptions and emissions in the Earth's atmosphere.

The NASA Infrared Telescope Facility is based on a telescope with diameter of 3 m on the summit of Mauna Kea, Hawaii, with elevation of 4.2 km, pressure of 0.6 bar, and mean overhead water vapor of 2 precipitable mm. This is a location with the best astroclimate, and the low $H_2O$ abundance and atmospheric pressure are favorable for spectroscopy of the planetary atmospheres.

The cryogenic echelle spectrograph CSHELL (Greene et al. 1993) extracts a narrow spectral interval of $0.0023\nu_0$, and the central wavenumber $\nu_0$ can be chosen in a wide spectral range from 1800 to 9000 $cm^{-1}$ (5.6 to 1.1 μm, respectively). The instrument detector is an InSb array of $256\times150$ pixels that is cooled to 30 K. Some parts of the instrument are cooled by liquid nitrogen as well. The pixel size is $9\times10^{-6}\,\nu_0$ in the dispersion direction and 0.2 arcs in the aspect direction. The total slit length is $0.2\times150 = 30$ arcs, and each point of the slit results in a spectrum of 256 pixels with resolving power $\nu/\delta\nu = 4\times10^4$ for the slit width of 0.5 arcs. Spatial resolution of the spectrograph and telescope combination is ~1 arcs.

Latitudinal variations of a species can be measured by placing the slit along the central meridian, variations with local time are preferable with the slit along the equator, and mapping of a species can be achieved by a sequence of the slit positions with some increment on the disk of a planet.

The observation involves acquiring of spectral frames of a target, flat field, and dark current. If the target is significantly smaller than the slit size of 30 arcs, then spectra of the telluric foreground are just in the target spectral frames. For example, Mars diameter is typically 10 arcs in our observations, and the remaining 20 arcs is the atmospheric foreground. In the opposite case it is better to observe the atmospheric foreground separately near the target. For example, Venus diameter is ~25 arcs in our observations, and the foreground is measured 30 arcs off the Venus disk.

Effects of the foreground may be analyzed in the following way. Absorption of monochromatic light $I$ from an external source in the Earth's atmosphere is accompanied by thermal emission:

$$dI = -I(\tau)\,d\tau + B(\tau)d\tau.$$

Here $\tau$ is the slant optical depth in the Earth's atmosphere and $B(\tau)$ is the blackbody thermal emission at temperature $T(\tau)$. Integration of this differential equation results in

$$I(\tau) = I_0 e^{-\tau} + e^{-\tau}\int_0^{\tau} B(t)e^t dt. \tag{5.10}$$

The second term is the foreground emission, and subtraction of the foreground spectrum from the target spectrum corrects the latter for the thermal emission in the Earth's atmosphere, while the extinction remains. The night airglow emissions are deleted by this subtraction as well.

The flat field is a spectral frame observed with a continuous and uniformly distributed source of light. A ratio of the target minus the foreground to the flat field minus the dark current is corrected for the foreground, dark current, and variations of the pixel sensitivities. Then it is necessary to correct bad pixels replacing them by the means of their two neighbors.

If absolute calibration is required, then the observing sequence includes a nearby standard calibrated infrared star from a catalog at the NASA IRTF website. Star brightness is tabulated in stellar magnitude at various wavelengths in the catalog. To convert stellar magnitudes into the energy units, a detailed spectrum of a zero-magnitude star (Vega) from 115 nm to 35 μm can be found in Engelke et al. (2010). Rayner et al. (2009) published a spectral library of 210 stars at 0.8 to 5 μm.

## 5.4 Equivalent Widths in the Observation of HF on Venus

**Co-adding of atmospheric layers** is the most accurate tool for interpretation of the observed equivalent widths. Here we will illustrate this method using the observations of HF on Venus (Krasnopolsky 2010c).

Ground-based observations of Venus are the most convenient at the phase (Sun–target–observer) angle near 90° (Figure 5.3). This corresponds to a maximum elongation (Sun–observer–target angle) of 47° and maximal geocentric velocity of 13 km $s^{-1}$ that results in the maximum Doppler shift relative to telluric lines.

The observations were made using the CSHELL spectrograph at NASA IRTF. Optimal positions of the spectrograph slit to observe the Venus dayside and nightside are shown in

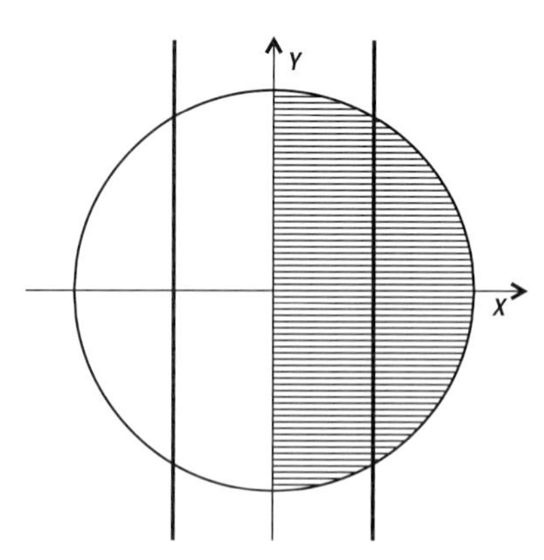


Figure 5.3 Ground-based observations of Venus at phase angle of 90° using a long-slit spectrograph. Positions of the slits for the dayside and nightside observations are shown by the vertical bars.

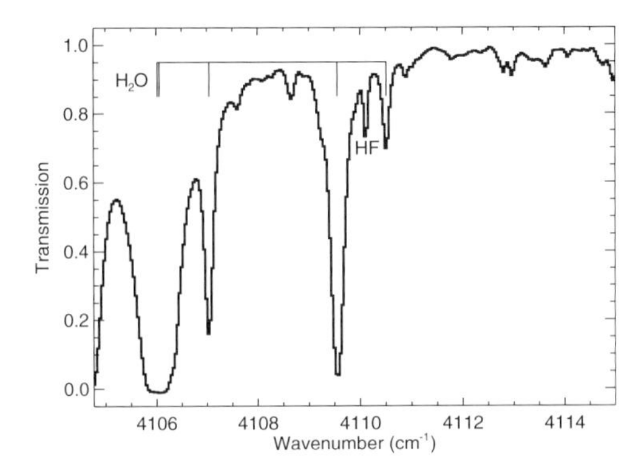


Figure 5.4 Ground-based observation of HF on Venus using CSHELL spectrograph at NASA IRTF with resolving power $\nu/\delta\nu = 4\times10^4$. The HF (1–0) R3 line at 4109.94 $cm^{-1}$ is Doppler-shifted by 0.174 $cm^{-1}$ and well seen in the spectrum. From Krasnopolsky (2010c).

Figure 5.3. Each point of the slit gives a spectrum, and the latitudinal coverage is from 60°S to 60°N in this geometry.

The observed spectrum near the equator (Figure 5.4) includes solar and telluric (mostly $H_2O$) lines and the HF (1–0) R3 line at 4109.94 $cm^{-1}$ from the Venus atmosphere. The line is Doppler-shifted to the blue by 0.174 $cm^{-1}$, in accord with the predicted geocentric velocity of Venus during the observation. All **parameters of targets for the Solar System observations** can be found at http://ssd.jpl.nasa.gov/horizons.cgi for any time in the past or future.

To test for a possible contamination of the HF line by the solar lines, **high-resolution solar infrared spectra** observed by the ATMOS (Farmer and Norton 1989) and ACE (Hase et al. 2010) orbiters with resolution of 0.01 and 0.02 $cm^{-1}$, respectively, can be used. The latest version of the ATMOS spectrum is prepared by Kurucz (http://kurucz.harvard.edu/sun/atmos) and covers 605–4800 $cm^{-1}$; the ACE range is 700–4430 $cm^{-1}$. These spectra show no contamination of the HF line by the solar lines.

Then a possible contamination by the telluric and Venusian lines should be verified using, for example, the **HITRAN 2016 spectroscopic database** (Gordon et al. 2017) and reasonable evaluations of species abundances in the Earth's and Venus atmosphere. This database includes parameters of spectral lines of 47 molecules. The test confirms that the HF line is clean.

The observed light is reflected by the clouds and absorbed in the atmosphere above the clouds on the way from the Sun and to the observer. A complicated process of the cloud reflection may be substituted by a simple surface reflection at some effective altitude. To determine this altitude, spectra of $CO_2$ lines in a nearby spectral region were observed as well (Figure 5.5). Properties of the clouds are rather similar at the wavenumbers in Figures 5.4 and 5.5, and the effective altitudes are the same.

Of six $CO_2$ lines in the spectrum, only R30 is clean. R36 can be used as well after a correction for some contamination by the telluric $CH_4$, and that is done by scaling of other $CH_4$ lines in the spectrum.

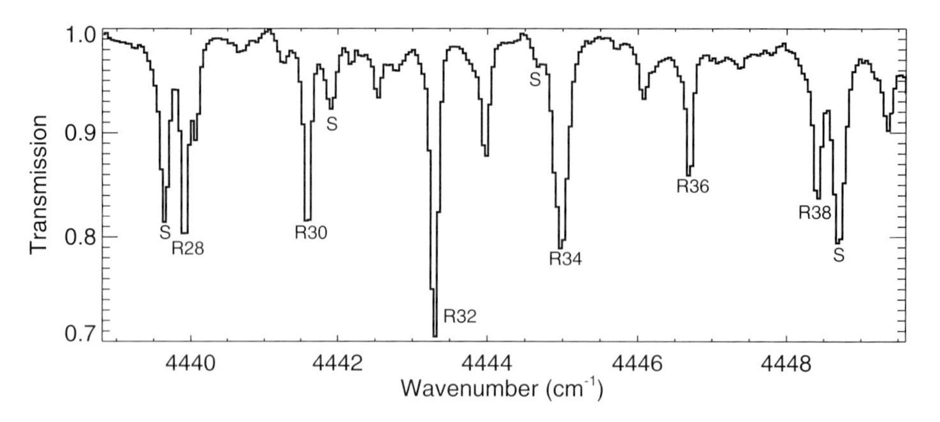


Figure 5.5 Spectrum of the $CO_2$ lines on Venus observed just after the observation of HF.

Next we need cosines of the incidence and viewing angles ($\theta$ and $\varphi$, respectively) for all points along the slit. A point on the Venus sphere (Figure 5.3) has coordinates $x$, $y$, $(1 - x^2 - y^2)^{1/2}$ times the Venus radius. Then cosine of angle between two vectors is

$$\cos \psi = x_0 x_1 + y_0 y_1 + \left(1 - x_0{}^2 - y_0{}^2\right)^{1/2}\left(1 - x_1{}^2 - y_1{}^2\right)^{1/2}. \tag{5.11}$$

For the case in Figure 5.3, the solar vector is [−1, 0, 0], the observer vector is [0, 0, 1], and a point on the slit is $[-0.5, y, (1 - 0.5^2 - y^2)^{1/2}]$; then $\cos \theta = 0.5$ for all points and $\cos \varphi = (0.75 - y^2)^{1/2}$.

Now we apply altitude profiles of temperature and pressure from the Venus International Reference Atmosphere (VIRA; Seiff et al. 1985) up to 100 km with a step of 2 km. Then absorption by the $CO_2$ line is calculated using the Voigt line shape (5.5) with a wavenumber step of 0.001 $cm^{-1}$ at each altitude. When the calculated line equivalent width exceeds the measured value, the two last points are interpolated to the measured value to give the effective altitude of the cloud reflection. The results for the two $CO_2$ lines are very similar, confirming reliability of the method.

Next we repeat this procedure for the HF line with some adopted HF mixing ratio and coadding the layer absorption down to the effective altitude of the cloud reflection from the $CO_2$ line analysis. This is done using the Voigt line shape with the collisional broadening of HF by $CO_2$ from Shaw and Lovell (1969). The calculated HF line equivalent width is compared to the measured value, and the input HF mixing ratio is adjusted to fit it.

This process is repeated for each of 113 spectra of Venus along the slit and gives a latitudinal distribution of the HF mixing ratio on Venus (Figure 5.6). A similar analysis was applied to the data obtained 2 years earlier to compare the HF abundances in the morning and in the afternoon. Both distributions are rather flat with the variations that are comparable to the uncertainties of the results. Thus one may conclude that the HF mixing ratio is equal to $3.3 \pm 0.3$ parts per billion (ppb) and does not vary with latitude and local time. The random uncertainty is 0.2 ppb, and we increase this value to account for possible systematic uncertainties.

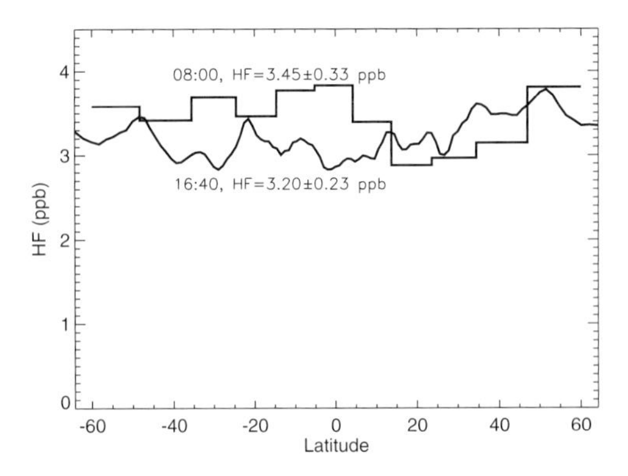


Figure 5.6 Latitudinal variations of the HF mixing ratio at 68 km on Venus observed in the morning and in the afternoon with a 2-year interval. The observed variations are comparable with the observational uncertainty and favor a uniform distribution of HF on Venus. From Krasnopolsky (2010c).

Our analysis of the $CO_2$ lines near 2.3 μm shows that the effective altitude of the cloud reflection is near 64 km with pressure of ~100 mbar. Then the level of line formation is at 50 mbar, that is, at 68 km. While we adopted a constant HF mixing ratio for all altitudes, our results should be referred to the level of line formation at 68 km.

## 5.5 Fitting of Observed Spectra by Synthetic Spectra

This is the basic tool for analysis of spectral data from the planetary atmospheres. Here we will consider the recent observations of variations of the CO mixing ratio on Mars using CSHELL/IRTF (Krasnopolsky 2015c).

The CO mean lifetime is a ratio of its column abundance and its column production rate by photolysis of $CO_2$, that is, a few years. It is much longer than the vertical mixing time $\tau_m = H^2/K \approx 1$ week and the global mixing time of half year (Krasnopolsky et al. 2004b; Lefèvre and Forget 2009). Therefore one may expect that CO is uniformly mixed in the atmosphere both vertically and geographically. However, the seasonal condensation and sublimation of $CO_2$ in the polar regions are of comparable duration with the global mixing time and therefore induce significant variations of long-living incondensable species on Mars ($N_2$, Ar and noble gases, $O_2$, CO, $H_2$, $CH_4$, etc.). These variations are essential to study the polar and subpolar processes on Mars.

### *5.5.1 Choice of the CO and $CO_2$ Lines*

The observation is aimed to measure latitudinal variations of the CO and $CO_2$ abundances at the same places and get their ratios. The CO lines are significant in the solar spectrum, and careful corrections for those lines is required using either the ATMOS or ACE solar spectra (Section 5.4). This restricts a choice of the CO band to (2–0), because the

(3–0) band is not covered by the solar spectra and the (1–0) band is very sensitive to thermal structure of the atmosphere.

The CSHELL spectral interval is 10 $cm^{-1}$ at the CO (2–0) band (Figure 3.6), and the lines to observe should be carefully chosen. We concluded in Section 5.2 and Figure 5.2 that the optimal optical depth in the line center is $x = \frac{\sigma m N}{\pi^{1/2}\delta\nu_D} \approx 1.5$ on Mars. Using the mean column $CO_2$ abundance of $2.3\times10^{23}$ $cm^{-2}$, the airmass of ~3, and the CO mixing ratio of ~$10^{-3}$, the best line strength is $\sigma \approx 3\times10^{-23}$ cm for CO and ~$3\times10^{-26}$ cm for $CO_2$.

Temperature varies from 150 to 270 K in the Martian lower atmosphere. Although these variations may be accounted for using the MGS/TES data (Smith 2004) and Mars Climate Database (www-mars.lmd.jussieu.fr/mcd_python/), it is preferable to use the lines that are weakly sensitive to temperature variations, that is, those with low ground state energy $E = BJ(J + 1)$; $B$ is the rotational constant. However, the CO (2–0) lines with low $J$ have strengths exceeding the desired value by two orders of magnitude. This contradiction is resolved by choosing the $^{13}CO$ isotopologue, and the $^{13}CO$ (2–0) P4, P5, and P6 lines have the proper strengths and low ground state energies. Furthermore, they are weakly contaminated by other lines.

The similar requirements are applied to the $CO_2$ lines. We choose seven lines of the $CO_2$ (31103–00001) band between P32e 4565.849 $cm^{-1}$ and P20e 4575.396 $cm^{-1}$.

### 5.5.2 Processing of Observed Spectra

The Martian abundances of $CO_2$ and CO exceed those on Earth by orders of magnitude, and a significant Doppler shift is not required for the observations. However, it was critical for some other aspects of the observational program, and Mars was observed at a reasonable combination of its angular diameter ~10 arcs and geocentric velocity ~15 km $s^{-1}$ with the CSHELL slit along the central meridian to cover latitudes from 60°S to the north pole.

As described in Section 5.3, spectral frames for the target, foreground, flat field, and dark current were acquired, and the ratio of target minus foreground to flat field minus dark current gives spectra corrected for foreground and variations of the pixel sensitivity. Then bad pixels are replaced by mean values of their two neighbors.

Even small errors in wavenumber result in significant differences between the observed and synthetic spectra. Therefore our technique for fitting the observed spectra by synthetic spectra require conversion of the observed spectra with the pixel size of ~0.04 $cm^{-1}$ to a wavenumber scale with a step of 0.001 $cm^{-1}$. We apply for this purpose a parabolic fitting of three adjacent pixels that transforms a pixel value to eight sampling points and keeps the sum of these points at the pixel value. (This is better for our analysis than the standard parabolic fitting that fixes the middle sampling point at the pixel value.) Then wavenumber scales are determined using identified lines in the spectra, and the spectra are linearly interpolated to the step of 0.001 $cm^{-1}$.

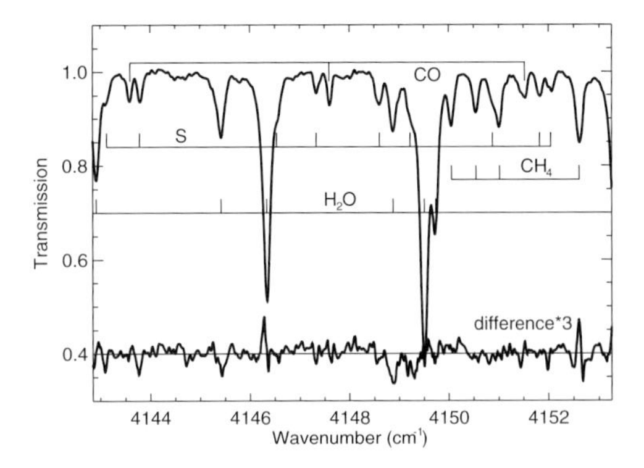


Figure 5.7 Observed and processed spectrum of CO on Mars at the subsolar latitude of 24°N and $L_S$ = 110°. All spectral features are identified by the telluric and solar (S) lines and three Martian CO lines. Difference between the observed and synthetic spectra is shown scaled by a factor of 3. From Krasnopolsky (2015c).

Both the instrument sensitivity and the Martian reflectivity may slightly vary with wavenumber within a spectrum, and we use a parabola $1 + ai + bi^2$ ($i$ is the point number) to compensate for these variations. Optical interactions between some parts of the instrument make a weak sinusoidal component in the spectrum that is subtracted using three parameters (amplitude, period, and phase). This sinusoidal has a period exceeding the resolution element by a factor of ~15 and amplitude of ~1%. Its subtraction helps getting a more accurate fitting of the spectral continuum. Small errors in wavenumbers result in significant differences between the observed and synthetic spectra. Therefore we apply wavenumber corrections at the edges and in the middle of the observed spectrum. Overall, eight parameters are used to adjust each observed spectrum.

One of the observed and processed spectra of CO is shown in Figure 5.7. All spectral features are identified, as well as the three chosen CO lines. The similar processing is applied to the $CO_2$ spectra (Figure 5.8).

### 5.5.3 Synthetic Spectra

Synthetic spectra are based on the solar high-resolution ATMOS spectrum that is Doppler-shifted by a sum of geocentric and heliocentric velocities of Mars. Telluric water is approximated by its abundance, mean temperature and pressure. Two parameters, abundance and temperature, are used for telluric methane, and its mean pressure is equal to half the pressure of 0.6 bar at Mauna Kea. We find that the Voigt line shape gives results for the telluric $H_2O$ and $CH_4$ lines that are rather similar to those for the collisional broadening, and use the latter.

The Martian CO lines are calculated using the Voigt line shape and CO line broadening by $CO_2$ from Sung and Varanasi (2005). Half surface pressures and temperatures at these

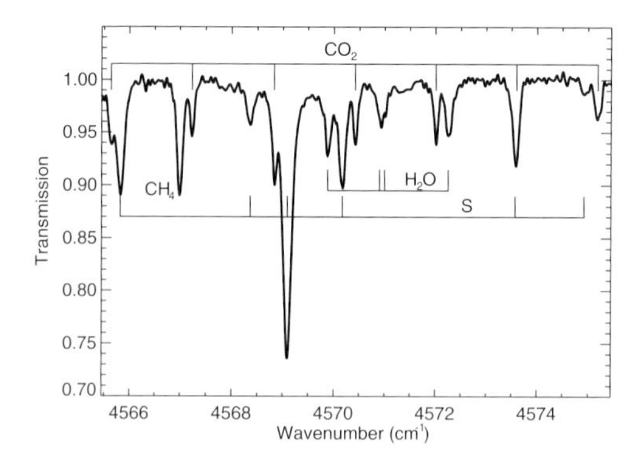


Figure 5.8 Observed and processed spectrum of $CO_2$ at the same location on Mars as the CO spectrum in Figure 5.7. The solar lines, telluric $H_2O$ and $CH_4$ lines, and seven Martian $CO_2$ lines are identified. From Krasnopolsky (2015c).

levels are required for these calculations, and the data were taken from the MGS/TES (Smith 2004) for proper seasons, latitudes, and longitudes at three Martian years and averaged. One more parameter is the instrument spectral resolution that is approximated by a Gaussian with a variable width. Overall we have eight parameters to adjust the observed spectrum and seven parameters to calculate the synthetic spectrum. One more parameter results in the best scaling of the processed observed spectrum $OS_i^0$ to the synthetic spectrum $SS_i$:

$$OS_i = OS_i^0 \frac{\sum_i SS_i^2}{\sum_i (OS_i^0 \times SS_i)}. \tag{5.12}$$

Each parameter is adjusted to provide the least square difference between the observed and synthetic spectra $\sum_i (OS_i - SS_i)^2$. Standard deviation of the fit is

$$\sigma = \left(\frac{n-1}{n-p}\right)^{1/2} \left(\frac{1}{m}\sum_i (OS_i - SS_i)^2\right)^{1/2}. \tag{5.13}$$

Here $n$ is the number of pixels in the spectrum, $p$ is the number of fitting parameters, and $m$ is the number of sampling points.

The least square fit is a partial case of the $\chi^2$ fit with equal observational uncertainties for all points. If the photon noise dominates in uncertainties, then the uncertainty is proportional to square root of the signal and the $\chi^2$ fit should be applied. However, we have not found significant advantages of this method for our observational data.

The number of free parameters is much smaller than the number of degrees of freedom (256 pixels in the spectrum). The synthetic spectrum that fits the observed spectrum in Figure 5.7 is not shown, because both spectra are very similar with a mean deviation of 0.6%. Their difference is shown scaled by a factor of 3. The $CO_2$ spectra at 4565–4576 $cm^{-1}$ (Figure 5.8) are processed similarly.

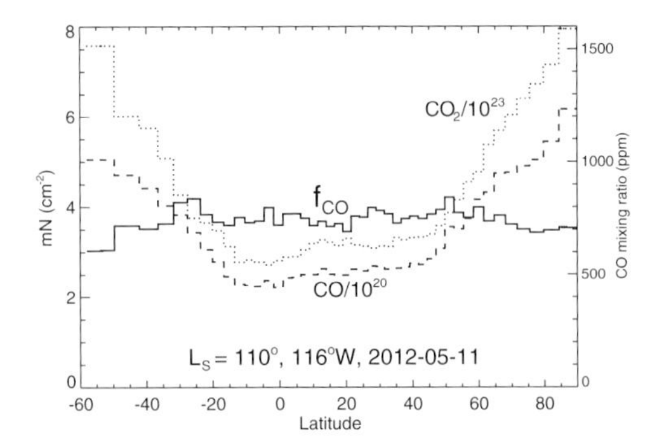


Figure 5.9 Retrieved vertical abundances of CO and $CO_2$ times airmass factors in the observed latitude range from 60°S to the north pole. While the abundances vary within a factor of 2 to 2.5, their ratio $f_{CO}$ is constant at 747 $\pm$ 42 ppm. From Krasnopolsky (2015c).

### 5.5.4 Results

We do not account for aerosol scattering in our analysis. Aerosol slightly reduces apparent abundances of species in spectroscopic observations of the Martian atmosphere. This reduction is rather similar for CO and $CO_2$, which are measured at close wavenumbers, and compensated in the $CO/CO_2$ ratios.

Retrieved column CO and $CO_2$ abundances times airmass $m$ are shown in Figure 5.9. Both curves demonstrate strong limb brightening that disappears in their ratios. The ratios are corrected for the presence of $N_2$ and Ar, so that $CO_2$ constitutes 0.96 of the atmospheric composition, and then for $^{13}CO/CO$ = 1.023 times the terrestrial $^{13}C/^{12}C$ isotope ratio of 0.0112. The observed CO mixing ratios are rather constant with a mean value of 747 ppm and standard deviation of 42 ppm. Correction for seasonal variations of the $CO_2$ atmospheric abundances results in an annually mean CO mixing ratio of 684 ppm.

These results were obtained at four seasonal points in late spring and the first half of the northern summer on Mars, and the CO mixing ratio does not vary with latitude from the north pole to 60°S in these seasons, in accord with the Mars Climate Database. Significant variations of the CO mixing ratio are expected at these seasons to the south of 60°S.

# 6
# Spectrographs

## 6.1 CVF and AOTF Spectrometers

Interferential light filters may be designed with a variable transmission wavelength, and the most convenient shape of these filters is circular. These circular variable filters (CVF) can serve as a dispersion element for spectrometers with low resolution. For example, the infrared spectrometer IKS (Combes et al. 1988) to study comet Halley at the VEGA mission had two CVFs to cover the ranges of 2.5 to 5.0 μm and 6 to 12 μm with resolving power of $\lambda/\delta\lambda \approx 50$. (VEGA 1 and 2 was the Russian–ESA mission that deployed two landing probes and two balloons on Venus in 1985 and then reached and studied comet Halley in 1986 during flybys at 8000 km.)

The acousto-optical tunable filter (AOTF) consists usually of a tellurium dioxide ($TeO_2$) crystal with refraction that is different for the light polarized in two orthogonal directions relative to the crystal axes. A piezoelectric transducer generates ultrasound waves in the crystal that make waves of the refractive index. They induce interferential phenomena, which restrict the crystal transmission to a wavelength that is determined by the wave frequency and can be changed by changing this frequency. The near-infrared channel of the SPICAM instrument (Korablev et al. 2006) at the Mars Express orbiter covers the range of 1.0 to 1.7 μm with resolving power of $\lambda/\delta\lambda \approx 1700$ (Figure 6.1). The instrument mass is just 0.75 kg. Spectral range of the AOTF is restricted to the visible and near-infrared because of the transmission of $TeO_2$.

## 6.2 Grating Spectrographs

The diffraction grating (Figure 6.2) is the mostly used dispersion element in spectrographs. Its formula is

$$d(\sin\varphi + \sin\psi) = k\lambda; \quad \lambda/\delta\lambda \le kN. \tag{6.1}$$

Here $d$ is the grating constant, that is, width of the groove; $\varphi$ and $\psi$ are the incidence and viewing angles relative to the normal to the grating and in the plane normal to the grooves; $k$ is the diffraction order; and $N$ is the total number of grooves. The groove density may be very large, up to 5000 grooves per mm. A slope of the grooves can be optimized to

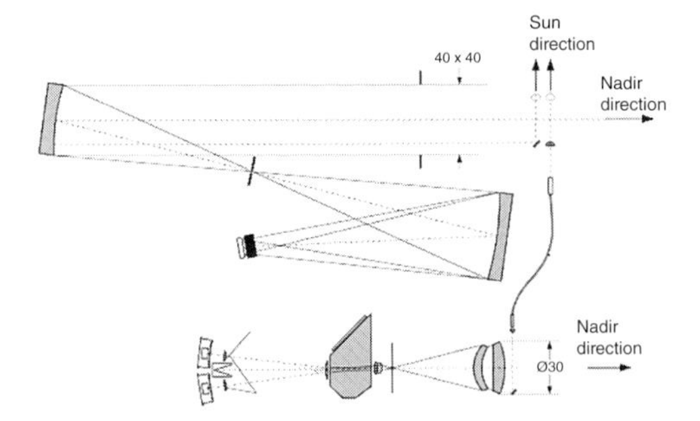


Figure 6.1 Optical scheme of the UV (top) and IR (bottom) channels of the SPICAM instrument at the Mars Express orbiter. The UV channel is a grating spectrometer, and the IR channel is based on the AOTF crystal. Both channels are designed for the nadir and solar occultation observations. From Bertaux et al. (2006).

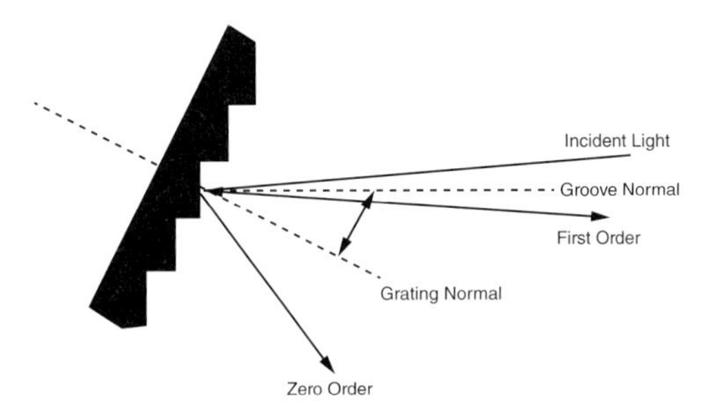


Figure 6.2 Geometry of reflecting diffraction grating.

reflection at a chosen spectral interval. Gratings are mostly reflective and may have different shape: flat, spherical, and toroidal. The spherical and toroidal gratings combine functions of the dispersive and optical elements; the toroidal gratings compensate for some aberrations in spectrographs.

The flat gratings can be copied with quality of the copies (replicas) approaching the original grating. Holographic gratings are made by photolithography of holographic interference pattern. These gratings have sinusoidal grooves and cannot be blazed at a chosen diffraction angle but make much less stray light.

There are numerous designs of the grating spectrographs. Spectrographs for the visible and UV ranges operate usually in the first diffraction order $k = 1$. For example, the UV channel of the SPICAM instrument (Bertaux et al. 2006) is shown in Figure 6.1. It consists of the off-axis parabolic mirror as the instrument telescope, the slit of variable width, the toroidal holographic grating with 290 grooves per mm, and the intensified CCD detector. The instrument covers a spectral range from 118 to 320 nm with resolving power of $\lambda/\delta\lambda \approx 200$.

Structure of the intensified CCD detector is shown in Figure 6.3. The $MgF_2$ window is transparent down to Lyman-alpha, and an additional sapphire window is transparent down to 150 nm and prevents overlapping of the diffraction orders and removes stray light from

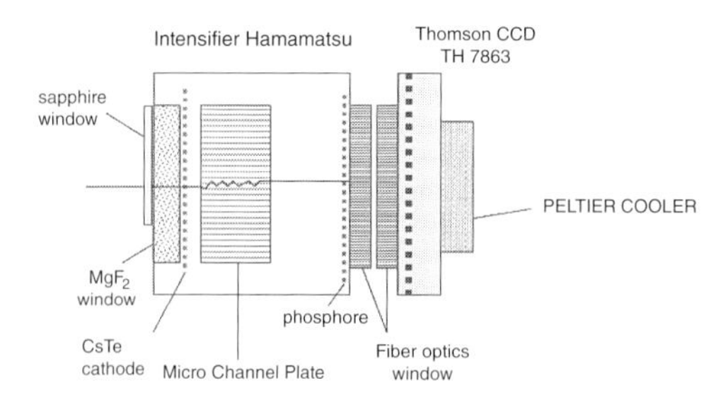


Figure 6.3 SPICAM UV detector. From Bertaux et al. (2006).

Lyman-alpha. The solar-blind CsTe photocathode is insensitive to light beyond 320 nm. The microchannel plate intensifies the photoelectron flux by a few orders of magnitude, and the amplification factor can be changed by varying the voltage within 500 to 900 volts. The phosphor screen forms a visible image of the observed UV spectrum that is transmitted by fiber optics to the Peltier-cooled charge-coupled device (CCD) with 288×384 pixels in the aspect and dispersion directions, respectively.

## 6.3 Echelle Spectrographs

An echelle grating has a low groove density and a groove shape that is effective at high angles and in high diffraction orders. A short side of the groove becomes a reflecting surface, and the diffraction orders are large, up to thousands. Assuming $\sin\varphi \approx \sin\psi \approx 1$, it follows from (6.1) that

$$L \geq R\lambda/2, \quad \text{or} \quad L \approx R\lambda. \tag{6.2}$$

Here $L = Nd$ is the echelle size, $R = \lambda/\delta\lambda$ is the resolving power, and the first relationship is for the diffraction limit while the second one is more practical. This means that echelles for high-resolution infrared spectrographs are large: for example, if $\lambda$ = 10 μm and $R = 10^5$, then the echelle is 1 m long.

Extraction of a selected diffraction order and removal of all other orders is achieved in the echelle spectrographs using prisms, interferential filters, and diffraction gratings. High-resolution infrared instruments for ground-based astronomy are mostly the echelle spectrographs. Their optical schemes are rather complicated and not considered here. For example, the CSHELL spectrograph (Section 5.3) includes a CVF for the order separation.

Using large detector arrays with great numbers of pixels, it is possible to observe simultaneously a few diffraction orders by so-called cross-dispersed echelle spectrographs. The order separation in those spectrographs is achieved by the second grating, whose grooves are normal to the echelle grooves. Then a continuum spectrum of a point-like source (star) looks like a few lines on the array detector, which correspond to different diffraction orders. There are spaces on the array between the diffraction orders that are determined by dispersion of the second grating. These spaces may be large enough to observe extended objects like planets. Here we briefly mention some outstanding infrared cross-dispersed echelle spectrographs for the planetary astronomy.

The TEXES spectrograph (Lacy et al. 2002) is designed for a spectral range from 5 to 25 μm with spectral coverage of $0.005\nu_0$ ($\nu_0$ is the central wavenumber) and resolving power of $\nu/\delta\nu \approx 10^5$ with a slit of 1.5×8 arcs. The instrument is used at the NASA IRTF (Section 5.3) and the Gemini North observatory with the telescope diameter of 8.1 m, both on the summit of Mauna Kea, Hawaii (elevation 4.2 km, mean overhead water vapor of 2 pr. mm).

The echelle of TEXES is 915 mm long with a groove spacing of 7.62 mm and blazed at $\tan \varphi = 10$ ($\varphi = 84.3°$). The second cross-dispersed echelle has 31.6 grooves per mm, and the detector is a SiAs array of 256×256 pixels. The detector and all radiating parts of the instrument are cooled by liquid helium.

The CRIRES spectrograph (Käufl et al. 2004) was designed for the Very Large Telescope with diameter of 8.2 m at Cerro Paranal (Chile) with elevation of 2.6 km and mean overhead water vapor of 2.8 pr. mm. VLT is a part of the European Southern Observatory. A spectral range of CRIRES is from 0.95 to 5.2 μm with resolving power of $10^5$; the slit length is 40 arcs. CRIRES has the echelle grating of 40×20 cm with 31.6 grooves/mm blazed at $\varphi = 63.5°$ ($\tan \varphi = 2$) and the ZnSe prism as a predisperser. The detector is a combination of four InSb arrays with 4096×512 pixels total. The optical parts are cooled to 65 K and the detector to 25 K.

NIRSPEC (McLean et al. 1998) is another outstanding near-infrared echelle spectrograph for the range of 0.95 to 5.5 μm with resolving power of $2.5\times10^4$. Its detector is an InSb array of 1024×1024 pixels, and a few target spectra are measured simultaneously by the array. The instrument is used at the Keck telescope with diameter of 10 m on the summit of Mauna Kea.

CSHELL at NASA IRTF is described in Section 5.3. Recently the immersion grating echelle spectrograph iSHELL was commissioned. Comparison of the normal reflecting grating with the immersion grating is shown in Figure 6.4. Dispersion of the latter is greater by a factor of the refractive index that is equal to 3.4 for the silicon grating in iSHELL. This made it possible to significantly reduce the size of the instrument. Its resolving power is 80,000 at 1.25–2.5 μm and 67,000 at 2.8–5.5 μm (Tokunaga et al. 2008). The detector array is 2048×2048 InSb pixels with 0.083 arcs per pixel. Three changeable cross-disperser gratings cover the instrument spectral range. Up to nine diffraction orders can be observed

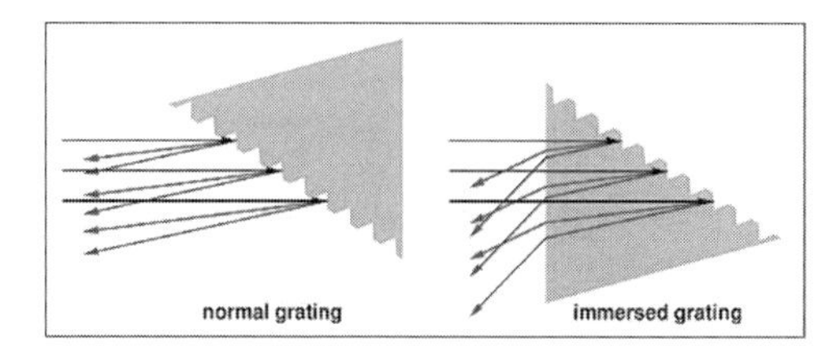


Figure 6.4 Comparison of the normal and immersed gratings. Dispersion of the latter is higher by a factor of the refractive index.

simultaneously. For example, the echelle orders 167–175 cover a range of 2826–2954 nm, that is, $0.044\nu_0$ compared to $0.0023\nu_0$ for CSHELL. Here $\nu_0$ is the central wavenumber.

## 6.4 Fourier Transform Spectrometers

The optical scheme of the Fourier transform spectrometer (FTS), or the Michelson interferometer, is shown in Figure 6.5. Light from a source is directed by the beamsplitter to two mirrors, and a sum of the reflected light from both mirrors is measured by the detector. The path difference $p = l_1 - l_2 = 2l$ results in interference between the beams, and the signal depends on the translating mirror displacement $l$:

$$I(l,\nu) = I(\nu)(1 + \cos 4\pi\nu l). \tag{6.3}$$

The measured signal as a function of $l$ changing from 0 to $L$ is an interferogram

$$I(l) = \int_0^\infty I(\nu)(1 + \cos 4\pi\nu l)d\nu. \tag{6.4}$$

Using the Fourier transform,

$$I(\nu) = 8\int_0^L F(l)\left[I(l) - \frac{I(0)}{2}\right]\cos 4\pi\nu l\, dl \tag{6.5}$$

is the measured spectrum, and $F(l) = 1$. The instrument line shape function, that is, the instrument response to a monochromatic line is (Figure 6.6)

$$ILS = \frac{\sin 4\pi(\nu - \nu_0)L}{4\pi(\nu - \nu_0)L}. \tag{6.6}$$

This function is designated $\operatorname{sinc} x = \frac{\sin x}{x}$, where $x = 4\pi(\nu - \nu_0)L$. The instrument spectral resolution is

$$\delta\nu = \frac{1}{2L}. \tag{6.7}$$

The Fourier transform spectroscopy was a breakthrough in the planetary astronomy in 1960s, and the instruments at ground-based observatories had $L$ = 50 cm and $\delta\nu \approx 0.01$ cm$^{-1}$, that is, resolving power of $\nu/\delta\nu \approx 3\times10^5$ near 3 μm.

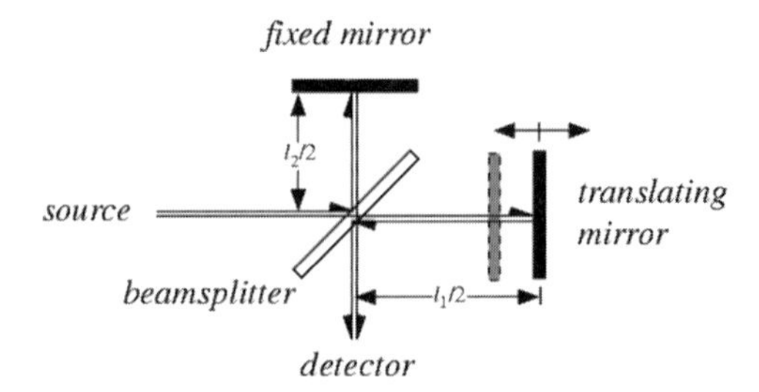


Figure 6.5 Fourier transform spectrometer: optical scheme.

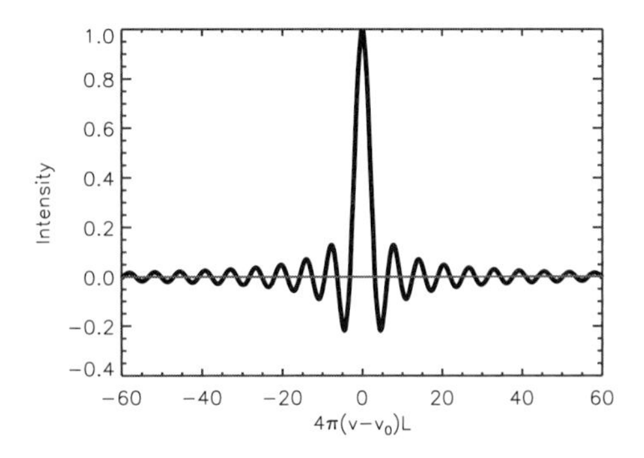


Figure 6.6 FTS line shape function sinc $x = \frac{\sin x}{x}$, $x = 4\pi(v - v_0)L$.

The photon noise, i.e., random fluctuations of the photon flux, contributes significantly to the observed signal-to-noise ratio. It may be reduced using an interferential filter that restricts a spectral range to an interval of the interest.

The FTS design looks simple; however, there are severe requirements to the parallel displacements of the translating mirror and accurate measurements of these displacements. Therefore FTSs are rather complicated instruments.

The secondary peaks and minima in the sinc function distort the spectrum and reduce its quality. They can be significantly reduced or even eliminated by apodization. The apodization function $F(l) = 1$ at $l = 0$ and gradually decreases to $l = L$. The steeper is the decrease, the smaller are the secondary features and the greater is degradation of the resolving power. Norton and Beer (1976) derived optimal functions for weak, medium, and strong apodization that reduce the resolving power by factors 1.2, 1.4, and 1.6, respectively. Usually even the weak apodization may be sufficient.

Apart from the high spectral resolution, the FTSs were much more sensitive than the slit instruments (grating monochromators), because the whole spectrum is measured at each time and the whole planet (or its significant part) is in the field of view instead of a narrow slit. Later the progress in multielement detector arrays made the former insignificant, while the latter does not support the spatially resolved spectroscopy.

Fourier transform spectrometers were used to study Mars at the Mariner 9 (Hanel et al. 1972), Mars Global Surveyor (Christensen et al. 1998), and Mars Express (Formisano et al. 2005) orbiters, at the Venera 15 orbiter (Moroz et al. 1990), and at the Voyager Grand Tour mission.

## 6.5 Tunable Laser and Cavity Ring-Down Spectroscopy

Spectroscopy of the planetary atmospheres is mostly related to absorption of the sunlight, absorption and emission of the thermal radiation, and airglow observations. The first case of the active spectroscopy, when the instrument included a light source and a cell with an atmospheric gas, was the ISAV experiment (Bertaux et al. 1996) at the VEGA entry probes with a xenon light bulb, a gas cell, and a UV spectrometer.

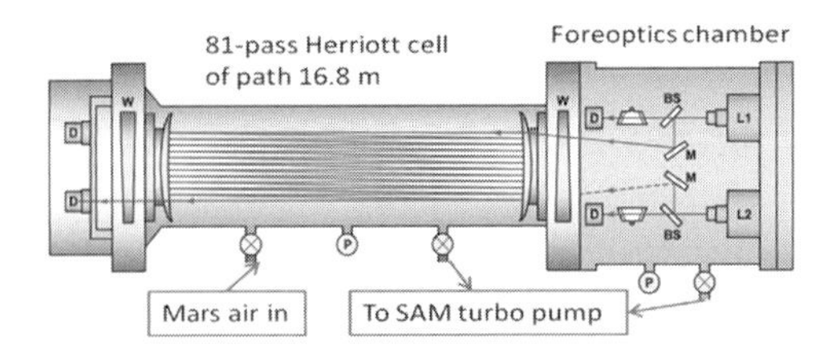


Figure 6.7 Tunable laser spectrometer (Webster et al. 2015a) at the Curiosity rover.

Currently an active spectroscopic study of the Martian atmosphere is carried out by the tunable laser spectrometer (Webster et al. 2013, 2015a) at the Curiosity rover. While the spectral instruments usually provide a wavelength selection, here the light source is selective, and two tunable lasers L1 and L2 (Figure 6.7) scan almost monochromatically two narrow spectral intervals at 3058 and 3594 $cm^{-1}$ to search for methane and measure the $CO_2$ and $H_2O$ isotopologues with resolving power $\nu/\delta\nu \approx 10^7$. The laser beam is controlled using a calibrated gas cell and beamsplitter BS in the foreoptics chamber. The main beam is directed into the 81-pass Herriot cell with a length of 20 cm and a path of 16.8 m. The cell is either empty using the SAM (Sample Analysis at Mars instrument suite) turbo pump or filled by Mars air. The absorption spectra are recorded by two detectors D. There were enrichment sessions for this instrument, when methane was enriched by a factor of 23 ± 1 by chemical removal of $CO_2$, and the achieved uncertainty in the observed methane mixing ratio was 0.1 ppb (parts per billion). It is better than those in the ground-based observations (Krasnopolsky 2012a; Villanueva et al. 2013) by a factor of ~25, a value similar to the enrichment factor.

Mirrors with very high reflectivity exceeding 99.9% in a narrow spectral range can be made using interferential coatings. A cavity ring-down spectrometer (CRDS) includes a gas cell with two parallel mirrors of this type. A monochromatic laser beam passes a few thousand times through this cell with a total path of a few km. The light intensity in the cell is controlled through a weak leak. Then the laser is switched off, and the light is gradually decreasing with a time constant $\tau$ because of the loss on the mirrors. If a gas in the cell adds to this loss, the decrease becomes steeper with a smaller time constant. Extremely low absorber abundances can be measured by this method. However, the measurements require long time, because the individual measurement covers an extremely small spectral interval, and the number of these measurements is large. This technique has not been used to study atmospheres of planets other than Earth. Recently, Chen et al. (2015) developed a CRDS that is aimed to detect both $^{12}CH_4$ and $^{13}CH_4$ on Mars and measure their ratio.

## 6.6 Infrared Heterodyne Spectrometers

The highest spectral resolution can be achieved in remote spectroscopy using infrared heterodyne spectrometers. The heterodyne principle was initially developed for the short-wave broadcasting receivers, where a signal is mixed with a signal from a local oscillator and transformed to an intermediate frequency that is equal to difference between

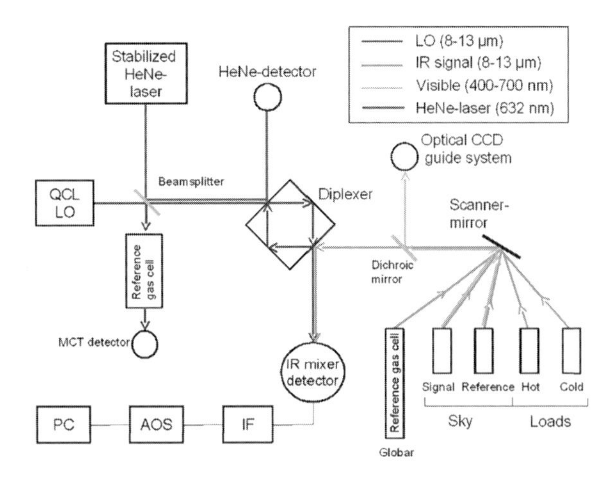


Figure 6.8 Structure of the Tunable Heterodyne Infrared Spectrometer (Sonnabend et al. 2008). See explanation in the text.

frequencies of the signal and the local oscillator. The intermediate frequency is much lower than the signal frequency and more convenient for analysis.

Currently there are three infrared heterodyne spectrometers: Heterodyne Instrument for Planetary Wind And Composition (HIPWAC; Kostiuk et al. 2001), Tunable Heterodyne Infrared Spectrometer (THIS; Sonnabend et al. 2008), and the heterodyne spectrometer for near-infrared (Rodin et al. 2014). A scheme of THIS is shown in Figure 6.8. The instrument is designed for observations from ground-based telescopes. Signals from a sky target and a reference source, hot and cold blackbody imitators, and a reference gas cell for absolute frequency calibration can be chosen using a scanning mirror. Position of the field of view (typically with diameter of 1 arcs) on the sky is controlled by a CCD guide system using a dichroic mirror.

The local oscillator is a quantum-cascade laser (QCL LO) with a complicated control system that includes a reference gas cell with a mercury–cadmium–telluride detector and a diplexer. The diplexer consists of two elliptic mirrors and two beamsplitters, and one of those has a piezo actuator. A very stable helium–neon laser is used to tune the system, and the achieved stability of the local oscillator is ~1 MHz.

The mercury–cadmium–telluride detector provides the intermediate frequency signal up to 3 GHz. Then the signal is amplified and shifted to the optimum range of 3.5–6.5 GHz for the acousto-optical spectrometer (AOS). The acousto-optical spectrometer operates similar to the acousto-optical tunable filter (Section 6.1) and disperses the signal at a linear CCD with 6000 pixels with a resolution bandwidth of ~1 MHz.

A set of the quantum-cascade lasers covers a range of 8 to 13 μm, and the achieved resolving power is $10^7$. This makes it possible to resolve line shapes in the planetary atmospheres and measure temperatures at levels of line formation and even temperature profiles near these levels. Doppler shifts by winds can be measured as well to give wind speeds.

# 7

# Spectroscopic Methods to Study Planetary Atmospheres

## 7.1 Spacecraft, Earth-Orbiting, and Ground-Based Observations

Spacecraft missions to planets are currently the main source of observational data on the planetary atmospheres. Actually the Solar System planets were rediscovered since the beginning of the space era. The possibility to observe a planet in situ and at close distances without absorption and emission from the terrestrial atmosphere gives obvious advantages to the planetary missions.

On the other hand, the planetary missions are rather expensive and comparatively sparse. Implementation of a new idea related to a planetary mission requires very significant efforts that can take a decade and even more. Furthermore, spacecraft instruments are always restricted in mass, size, and power consumption, and their properties are inevitably moderate.

Earth-orbiting observatories are primarily designed for astrophysical observations of stars, galaxies, nebulae, etc. Their spectroscopic equipment is mostly aimed at the spectral regions that are not accessible from the Earth. For example, the Space Telescope Imaging Spectrograph (STIS) at the Hubble Space Telescope (HST) is a multipurpose instrument with sets of apertures, gratings, echelles, and detectors that make it possible to cover a spectral range from 115 to 1030 nm with resolving power up to $10^5$.

The Far Ultraviolet Spectroscopic Explorer (FUSE) operated from 1999 to 2007. It covered a range from 90 to 120 nm with resolving power of 24,000 for point sources and 5000 for extended sources like Mars.

The Extreme Ultraviolet Explorer (EUVE) was in operation from 1992 to 2001 and had four cameras and three spectrometers that covered a range from 7 to 76 nm. Spectral resolving power was equal to 200 for point sources and degraded to 100 for targets with diameter of 4 arcmin.

The Chandra X-ray Observatory (CXO) and the X-ray Multi-Mirror Mission (XMM) are advanced observatories that have been operating since 1999. They have sets of instruments for spectroscopy and imaging in X-rays starting from 6 nm, that is, from the photon energy $E = 200$ eV and higher. The intrinsic resolving power of the CCD detectors is $E/\delta E \approx 7.5$ at 500 eV; the Reflection Grating Spectrometers (RGS) at XMM have resolution of 3.3 eV in the range of 330 to 2000 eV.

These Earth-orbiting observatories (except XMM) were used by the author to observe planets and comets. Evidently the list of the observatories is not full; besides, there are observatories designed for specific goals, e.g., to search for the extrasolar planets.

We briefly discussed in Section 6.3 some spatially resolved high-resolution spectrographs for the near- and mid-infrared ranges at the ground-based observatories. Overall, both Earth-orbiting and ground-based observatories have significant capabilities to solve numerous tasks in the studies of the planetary atmospheres. Obviously this way is much cheaper than that related to a planetary mission and much shorter, of 1 or 2 years between an idea and its implementation.

## 7.2 Nadir Observations to Measure Species Abundances

This is a typical geometry and method of the spectroscopic observations. It is considered in Section 5.4 for the observations of HF above the cloud tops on Venus, and coadding the atmospheric layers and the equivalent width technique were used for analysis of the spectroscopic data. Another case with fitting of observed spectra by synthetic spectra is described in Section 5.5 for variations of the CO mixing ratio on Mars. Simultaneous observations of $CO_2$ lines in nearby spectral intervals are key features in both cases on Venus and Mars to compensate for errors related to the simplified schemes of line formation that avoid detailed calculations of aerosol scattering.

If the nearby lines $CO_2$ are not available, then the radiative transfer technique should be applied. This is the case of spacecraft observations, when a whole unresolved band is measured and simultaneous observations of an appropriate $CO_2$ band may be lacking. Then aerosol scattering is taken into account using, e.g., the spherical harmonic discrete ordinate method (Evans 2007). Absorption by a band is simulated using line-by-line calculations.

## 7.3 Vertical Profiles of Temperature from Nadir Observations of the $CO_2$ Bands at 15 and 4.3 μm

Nadir observations of a shape of the $CO_2$ band at 15 μm are a standard tool to study thermal balance and dynamics of the lower and middle atmospheres of the Earth, Mars, and Venus. Fourier transform spectrometers (Figures 7.1 and 7.2) are used for these purposes. The band is the $CO_2$ bending mode, which is very strong. Its full width including all isotopologues is ~200 cm$^{-1}$ on Mars and Venus, and the instruments with a moderate spectral resolution of ~10 cm$^{-1}$ observe ~20 points along the band shape.

Monochromatic thermal radiation $I_\nu$, directed to the space at angle $\theta$ to the local zenith, is absorbed by a layer $d\tau_\nu$ and enhanced by the thermal emission from this layer:

$$dI_\nu = I_\nu \sec\theta d\tau_\nu - B_\nu(T(\tau_\nu)) \sec\theta\, d\tau_\nu; \quad \mu\frac{dI_\nu}{d\tau_\nu} = I_\nu - B_\nu(T(\tau_\nu)). \tag{7.1}$$

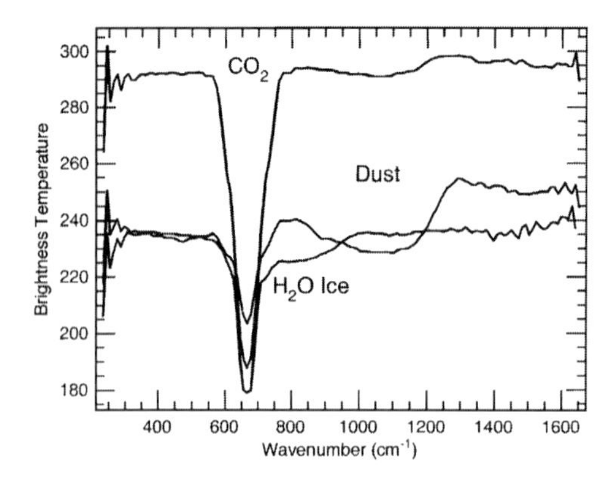


Figure 7.1 MGS/TES spectra of Mars at 250–1650 cm$^{-1}$ (6–40 μm) with resolution of ~10 cm$^{-1}$. The main feature is the $CO_2$ band at 670 cm$^{-1}$ (15 μm). The upper spectrum refers to a warm region near noon. Two other spectra are from regions with dust storm and a thick $H_2O$ cloud. From Smith et al. (2000).

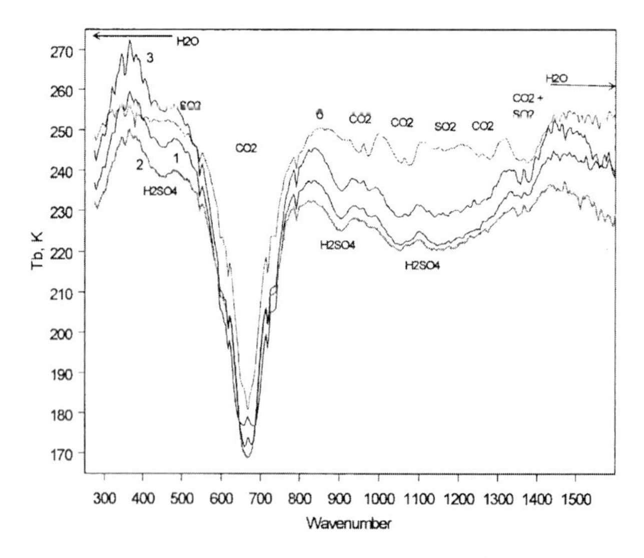


Figure 7.2 Venera 15 spectra of Venus with resolution of ~5 cm$^{-1}$ at low latitudes in the morning, after sunset, and near noon (1, 2, 3, respectively) and at the north polar region (6). From Zasova et al. (2004).

Here $\tau$ is increasing from zero at the infinity down to the surface; therefore $d\tau$ is opposite to the direction of the thermal emission and negative, $\mu = \cos\theta$, and $B_\nu$ is the blackbody radiation (2.22). Relationship (7.1) is a linear differential equation, and its solution is

$$I_\nu = \int_0^\infty B_\nu(T(\tau_\nu))e^{-\tau_\nu/\mu}\frac{d\tau_\nu}{\mu}. \tag{7.2}$$

Evidently the measured thermal emission depends on the temperature profile $T(\tau_\nu)$. To understand this dependence, we assume that the atmosphere is isothermal and the absorption cross section $\sigma_\nu$ does not depend on pressure. Then the column abundance $N(z) = N_0\, e^{-z/H}$, $\tau_\nu = \sigma_\nu N$, and

$$d\tau_\nu = -\frac{\sigma_\nu N(z)}{H}dz; \quad I_\nu = \int_0^\infty B_\nu(T)\frac{\sigma_\nu N(z)}{\mu H}\exp\left(-\frac{\sigma_\nu N(z)}{\mu}\right)dz. \tag{7.3}$$

$B_\nu(T)$ is constant here, and the integrand

$$\frac{x}{H}e^{-x}, \; x = \frac{\sigma_\nu N(z)}{\mu} = \frac{\tau_\nu}{\mu} \tag{7.4}$$

is a so-called weighting function. It is actually the Chapman function (Section 2.4) that is maximal at $x = 1$ with a full width at half maximum of 2.5 $H$. Therefore the observed intensity $I_\nu$ is mostly sensitive to temperature at a level of $\tau_\nu/\mu = 1$ and in the nearby region of $\pm H$. This level corresponds to $N(z) = \mu/\sigma_\nu$, and the effective absorption cross section $\sigma_\nu$ varies within the band by orders of magnitude, providing vertical sounding of the atmospheric temperature.

Transmission $t_\nu = \exp\left(-\frac{\tau_\nu}{\mu}\right)$ may be used in (7.2) instead of $\tau_\nu/\mu$ to give

$$I_\nu = \int_\infty^0 B_\nu(T(t_\nu))dt_\nu\,. \tag{7.5}$$

The TES data (Figure 7.1) are presented with a constant increment in ln $p$ (Smith 2004); then $dt_\nu$ in (7.5) may be replaced by

$$dt_\nu = \frac{dt_\nu}{d\ln p}d\ln p.$$

The TES temperature profiles retrieved from the nadir observations extend from the surface to 0.1 mbar (~37 km), and those from the limb observations (see below) end at 0.01 mbar (~55 km). The increment is $\Delta \ln p = 0.25$. The tabulated functions $T(p)$ may be converted to the altitude scale using

$$\frac{z - z_0}{1 + (z + z_0)/R} \approx z - z_0 = -\frac{k}{mg_0}\int_{p_0}^{p} T(p)\frac{dp}{p} = -\frac{k}{mg_0}\int_{\ln p_0}^{\ln p} T(p)d\ln p. \tag{7.5a}$$

Temperature profiles observed by the Venera 15 spectroscopy of the $CO_2$ band at 15 μm cover altitudes from 55 to 100 km (Zasova et al. 2004, 2006).

The $CO_2$ band at 4.3 μm (the antisymmetric mode) can be also used for temperature sounding of the atmospheres of the terrestrial planets. The observations were conducted using the VIRTIS-H spectrograph with resolving power of 1700 at the Venus Express orbiter. Miglirioni et al. (2012) analyzed the nightside observations, because the dayside observations are significantly affected by the solar light. The retrieved temperature profiles cover the altitudes from 65 to 80 km.

## 7.4 Vertical Profiles of Temperature and CO Mixing Ratio from CO Line Shapes in the Submillimeter Range

The submillimeter spectral range contains rotational lines of various molecules. The heterodyne technique, which is feasible in this range, makes it possible to achieve very high spectral resolution. A large one-dish submillimeter telescope is the James Clerk Maxwell Telescope (JCMT) on the summit of Mauna Kea, Hawaii. Its diameter is 15 m, the beam size is 14 arcs, and resolving power is up to $10^7$. The beam size is sufficient to make spatially resolved observations of the Venus nightside near the inferior conjunction, when the Venus diameter is ~60 arcs. Even larger is the IRAM telescope for the millimeter range in Spain with diameter of 30 m. The greatest instrument for the millimeter and submillimeter range is ALMA, which is a combination of 66 dishes with diameter of 12 and 7 m.

The $CO_2$ molecule does not have dipole moment in the ground state, and all $CO_2$ pure rotational lines are forbidden. Thermal sounding can be made using the very strong $^{12}CO$ ($J = 2 \rightarrow 3$) line at 346 GHz. Retrieval of a temperature profile from an observed line shape is generally similar to that for the $CO_2$ band at 15 μm, and relationship (7.2) and its versions are valid as well. However, the optical depths depend on a vertical profile of the CO mixing ratio, and this problem is lacking for the $CO_2$ band.

To resolve the situation, Clancy et al. (2012a) observed both $^{12}CO$ and $^{13}CO$ lines, the latter at 330 GHz and optically thin. The $^{13}CO/^{12}CO$ ratio may be adopted equal to the terrestrial carbon isotope ratio of 0.0112 and constant with altitude. Vertical profiles of both temperature and CO are retrieved from the shapes of two lines. The blackbody radiation in (7.2) is just a function of altitude in these observations, because frequency varies insignificantly within the line shape. The retrieval is made using the Voigt line broadening in each altitude interval.

The observed $^{12}CO$ line shape is shown in Figure 7.3a in full; its central part is in Figure 7.3b. The circles depict the model fitting. The strong absorption feature in the line center reflects a steep depletion in the nighttime temperature above 95 km. The $^{13}CO$ line observed at the same conditions is shown in Figure 7.4. Weighting functions of the $^{12}CO$ line for temperature and of the $^{13}CO$ line for the CO mixing ratio are shown in Figure 7.5. The CO mixing ratio is significantly depleted below 85 km, and additional weighting for this region can be taken from the far wings of the $^{12}CO$ line. The retrieved CO varies from 60 ppm at 76 km to 2000 ppm at 111 km. The derived temperature profile is Z-shaped with 200 K at 78 km, 175 K at 88 km, 188 K at 97 km, and 127 K at 113 km.

The submillimeter range is also used for detection and measurements of minor species in the planetary atmospheres. Then a background atmosphere is adopted from models, and the species mixing ratios or their vertical profiles (depending on quality of the data) are fitting parameters.

## 7.5 Vertical Profiles of Terrestrial Ozone from Nadir UV Spectra

Ozone $O_3$ is among basic products of the oxygen photochemistry on the terrestrial planets. Its dissociation proceeds mostly in the near-UV range (Figure 7.6) with a peak cross

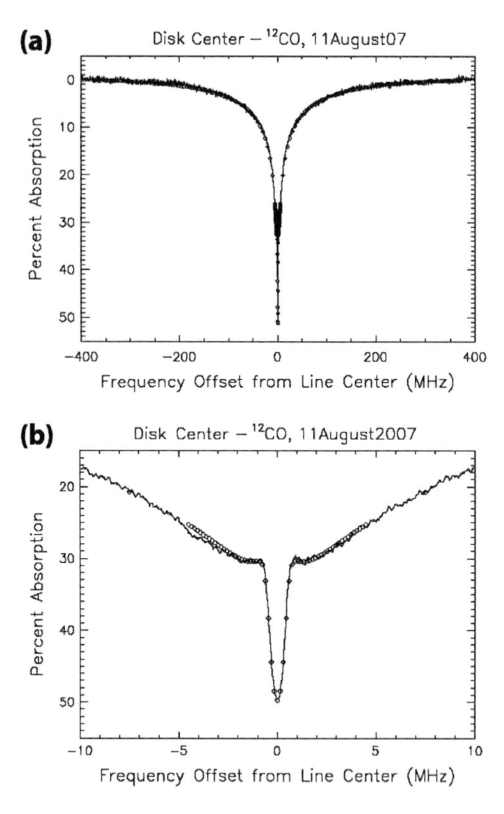


Figure 7.3 The observed $^{12}$CO 346 GHz line shape in two scales at the Venus midnight equator. The model fitting is shown by small circles. The strong absorption feature in the line center reflects a steep decrease in temperature above 95 km. From Clancy et al. (2012a).

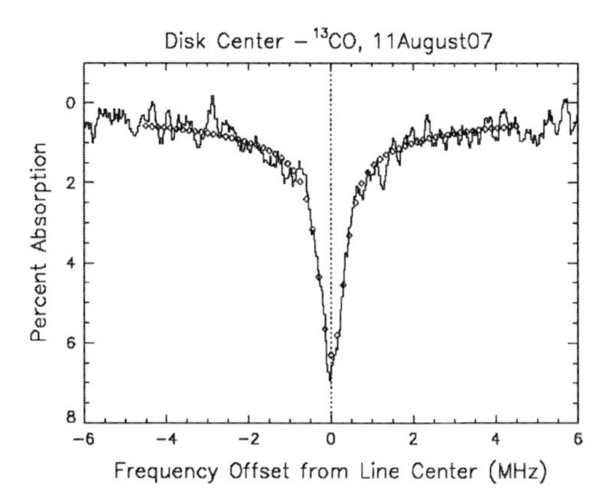


Figure 7.4 The observed $^{13}$CO 330 GHz line shape at the Venus midnight equator. Circles are the model fit. The line shape reflects the CO vertical profile. From Clancy et al. (2012a).

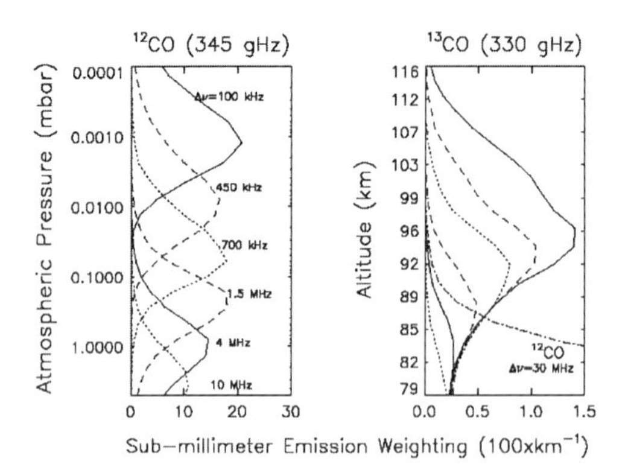


Figure 7.5 Weighting functions of the $^{12}CO$ and $^{13}CO$ lines for different offsets from the line center. The $^{12}CO$ functions are for temperature and those of $^{13}CO$ are for the CO mixing ratio. The latter is steeply decreasing below 85 km, and additional weighting for CO here can be obtained at the far wings of the $^{12}CO$ line. From Clancy et al. (2012a).

section of $1.1\times10^{-17}$ cm$^2$ at 255 nm. Dissociation energy of ozone is low (1.05 eV); therefore ozone is chemically active, direct methods of its measurement are complicated, and low-resolution UV spectroscopy is the main tool for observations of ozone.

Ozone in the Earth's atmosphere forms a layer with a maximum density near 25 km and column abundance of ~0.3 cm-atm. This layer protects living organisms from hard solar UV radiation. Ozone abundances and vertical distributions vary because of variations of the natural conditions in the atmosphere and environmental effects of the human activity. Therefore control of the ozone layer and keeping of its stability are among the priority tasks related to atmospheric chemistry.

The nadir UV spectroscopy at 260–310 nm from space is used to sound vertical profiles of the stratospheric ozone. The ozone layer is opaque in this range, and the tropospheric ozone is not seen. Absorption by other gases is negligible in this range. We concluded in Section 2.3 that there is a cold trap for water vapor at the tropopause and its abundance is rather constant at a few ppm in the stratosphere that has a positive temperature gradient and extends up to 50 km. Therefore the conditions in the stratosphere are not favorable for condensation of water vapor. The positive temperature gradient prevents significant vertical flow, and dust from the surface is almost lacking in the stratosphere, while the abundances of the cosmic and meteorite dust are low.

Therefore reflectivity of the atmosphere at 260–310 nm is determined by the Rayleigh scattering of the air and the ozone absorption. Optical depth of the atmosphere for Rayleigh scattering is

$$\tau_\lambda^R = 1.2p(300/\lambda)^4\,. \tag{7.6}$$

Here $p$ is pressure in bar and $\lambda$ is the wavelength in nm. Using the ray geometry from Figure 4.4 with $p$ instead of $\tau$ and the Rayleigh scattering phase function $\gamma(\psi) = \frac{3}{4}(1 + \cos^2\psi)$ (Section 4.2),

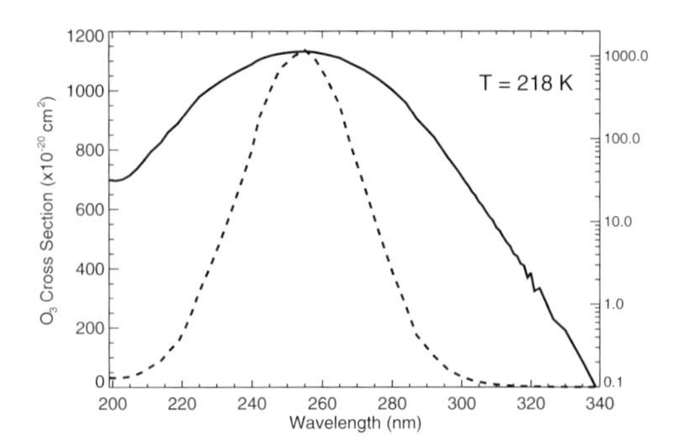


Figure 7.6 Absorption cross section of ozone at 218 K in the linear and logarithmic scales. Data from Burkholder et al. (2015).

$$dI_\lambda = I_{0\lambda} \frac{3 \times 1.2}{16\pi} \left(\frac{300}{\lambda}\right)^4 \left(1 + \cos^2\psi\right) \frac{dp}{\cos\varphi} e^{-\sigma_\lambda N(p)\left(\frac{1}{\cos\theta} + \frac{1}{\cos\varphi}\right)}. \tag{7.7}$$

Here $\sigma_\lambda$ is the ozone absorption cross section (Figure 7.6), and $N(p)$ is the column ozone abundance above a level $p$. Then integration gives

$$\rho_\lambda = \frac{\pi I_\lambda}{I_{0\lambda} \cos\theta} = \frac{3 \times 1.2}{16} (300/\lambda)^4 \frac{1 + \cos^2\psi}{\cos\theta \cos\varphi} \int_0^\infty e^{-\sigma_\lambda N(p)\left(\frac{1}{\cos\theta} + \frac{1}{\cos\varphi}\right)} dp, \tag{7.8}$$

and $\rho_\lambda$ is the observed brightness coefficient (Section 4.4). Assign

$$p_\lambda = \frac{16}{3 \times 1.2} \rho_\lambda (\lambda/300)^4 \frac{\cos\theta \cos\varphi}{1 + \cos^2\psi} \text{ and } \mu = \frac{1}{\cos\theta} + \frac{1}{\cos\varphi}. \tag{7.9}$$

Here $\mu$ is the two-way airmass, and

$$p_\lambda = \int_0^\infty e^{-\sigma_\lambda \mu N(p)} dp. \tag{7.10}$$

This is the Laplace-type integral equation that makes it possible to retrieve the ozone vertical profile $N(p)$ using the measured function $p_\lambda$. To get weighting function, we assume a constant ozone mixing ratio

$$f = \frac{N}{ap}, \tag{7.11}$$

where $a = 2.1 \times 10^{21}$ cm$^{-2}$ bar$^{-1}$ is the coefficient for conversion of pressure in bar into the column abundance of the air molecules on the Earth. Then, using $dp = -\frac{p}{H} dz$ from the barometric formula,

$$p_\lambda = \int_\infty^0 e^{-\sigma_\lambda \mu a f p(z)} \frac{p(z)}{H} dz, \tag{7.12}$$

and the weighting function is the Chapman layer $xe^{-x}$, where $x = \sigma_\lambda\ \mu a f p(z) = \sigma_\lambda \mu N$. The function is maximum at $x = 1$, that is, at ozone column $N = \frac{1}{\sigma_\lambda \mu}$, and has a full width at half maximum of 2.5 *H*. The ozone layer sounding starts at $N \approx 3$ μm-atm, that is, ~0.001 of the total abundance.

Using $x = \frac{1}{\sigma_\lambda \mu}$ and $dp = \frac{dp}{dN} dN$, equation (7.10) becomes

$$p_x = \int_0^\infty e^{-\frac{N}{x}} \frac{dp}{dN} dN. \tag{7.13}$$

Assuming

$$\frac{dp}{dN} = a_0 + a_1 N + a_2 N^2 + \ldots + a_i N^i + \ldots \tag{7.14}$$

and taking into account that

$$\int_0^\infty t^k e^{-\frac{t}{y}} dt = k!\, y^{k+1},$$

the integration gives

$$p_x = a_0 x + a_1 x^2 + 2a_2 x^3 + \ldots + a_i i! x^{i+1} + \ldots \tag{7.15}$$

Thus, the observed UV spectrum at 260–310 nm is transformed into $p_x$ using (7.8), (7.9), and $x = \frac{1}{\sigma_\lambda \mu}$. Then the polynomial fit to $p_x$ using, e.g., the *poly_fit* IDL procedure results in the coefficients $a_i$ and finally

$$p(N) = a_0 N + \frac{a_1}{2} N^2 + \frac{a_2}{3} N^3 + \ldots + \frac{a_i}{i+1} N^{i+1} + \ldots \tag{7.16}$$

The function $p(N)$ relates pressure with column ozone abundance and can be transformed into a vertical profile of ozone number density $[O_3] = -\frac{dN}{dz}$ using $p(z)$ from a model atmosphere.

The ozone layer becomes transparent at 310–340 nm, and this spectral interval can be used to derive column ozone abundances. The problem of sounding from space of the vertical ozone profile and its total abundance was solved by Krasnopolsky (1966), and the first orbiter study of the global ozone distribution was made at the Cosmos 65 orbiter (Krasnopolsky et al. 1966; Iozenas and Krasnopolsky 1970).

## 7.6 Measurements of Rotational Temperatures and Isotope Ratios

Distribution of the rotational line intensities (relationship 3.34) and retrieval of rotational temperature from the observed distribution was discussed in Section 3.9 and Figure 3.8. Rotational excitation and quenching are very effective in the molecular collisions. Rotational temperature is equal to the ambient mean temperature of the absorbing or emitting layer, if radiative lifetime is much longer than time between collisions, that is, if the conditions of local thermodynamic equilibrium hold.

Rotational temperatures of the $O_2$ ($a^1\Delta_g \rightarrow X^3\Sigma_g^-$) nightglow at 1.27 μm on Venus are shown in Figures 7.7–7.9. This nightglow on Venus was discovered by Connes et al.

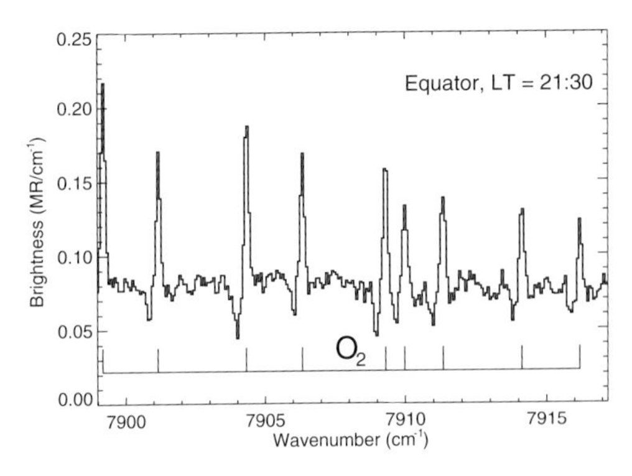


Figure 7.7 One of the spectra of Venus $O_2$ nightglow at 1.27 μm. Nine emission lines exactly fit their expected Doppler-shifted positions. From Krasnopolsky (2010a).

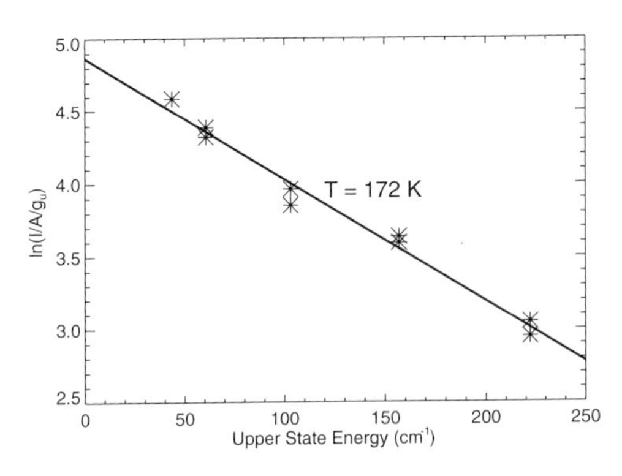


Figure 7.8 Logarithm of the upper state rotational population versus upper state rotational energy for nine lines in Figure 7.7 results in rotational temperature of 172 K. From Krasnopolsky (2010a).

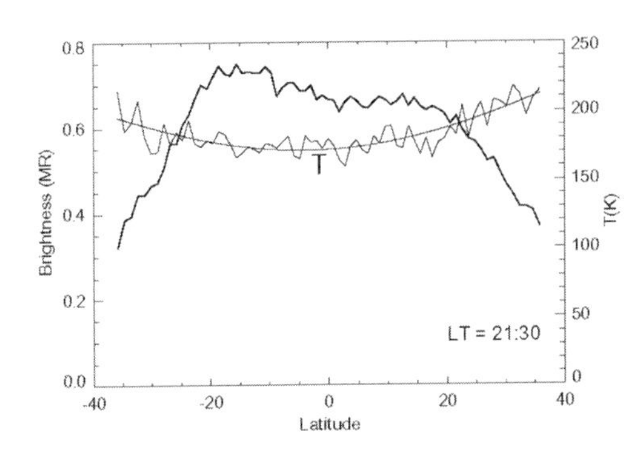


Figure 7.9 Variations of the vertical $O_2$ nightglow brightness and rotational temperature on Venus. The temperatures refer to $\approx$97 km. From Krasnopolsky (2010a).

(1979), who obtained its mean rotational temperature of 185 ± 15 K. Figure 7.7 reproduces one of the nightglow spectra observed using IRTF/CSHELL. Retrieval of rotational temperature from this spectrum is in Figure 7.8, and latitudinal variations of the nightglow vertical intensity and rotational temperature are in Figure 7.9. Scatter of temperatures reflect local variations and random errors. If they contribute equally, then uncertainties of the retrieved temperatures are 5 K.

Detailed study of the nightglow morphology was made at Venus Express (Piccioni et al. 2009). The $O_2$ nightglow at 1.27 μm forms a layer of ~15 km thick with a maximum at 97 ± 2.5 km. Resolution of the VITRIS spectrograph at Venus Express was insufficient to observe the nightglow rotational structure. The radiative time is very long (4460 s; Section 3.11), and the rotational temperatures reflect ambient temperatures near 97 km on Venus.

High-resolution spectroscopy may be applied to measure isotope ratios in molecular species in the planetary atmospheres (Section 3.9; Figure 3.9) that are important clues to their evolution. While D/H fractionation is typically large, those for the other elements are of a few percent and require very high accuracy. Therefore strengths of the chosen lines should be known with uncertainties of less than 1%. Next, thermal structure of an atmosphere in the observed region should be taken into account and the results have to be properly averaged over the observed region. Finally, the large number of the lines should be measured. For example, if an isotopologue abundance can be extracted from a line with uncertainty of 10%, then the uncertainty can be reduced to 1% by observing of 100 lines.

For example, the study of carbon and oxygen isotopes in $CO_2$ on Mars (Krasnopolsky et al. 2007) covered a spectral range of 6022–6308 $cm^{-1}$ with resolving power of $3.5\times10^5$ (Figure 3.9) that included 475 lines of the main isotope, 184 lines of $^{13}CO_2$, 181 lines of $CO^{18}O$, and 119 lines of $CO^{17}O$. (Lines with strengths exceeding $10^{-27}$ cm at 218 K are considered here.) Of those, the most accurate 229 lines of the main isotope, 76 lines of $^{13}CO_2$, and 80 lines of $CO^{18}O$ were chosen for analysis.

## 7.7 Inversion of Limb Observations

Vertically resolved limb observations reflect atmospheric vertical structure that can be retrieved from those observations. The simplest and practically important case is airglow in the transparent atmosphere (Figure 7.10). At first we find a relationship between volume emission rate $i(h)$ (in photons $cm^{-3}$ $s^{-1}$) and limb emission $I(h)$ (in photons $cm^{-2}$ $s^{-1}$), and both values refer to $4\pi$ steradians. Then

$$dI = i(h^*)dx;\ \ h^* = \frac{R+h}{\cos\theta} - R;\ \ x = (R+h)\tan\theta;\ \ dx = (R+h)\frac{d\theta}{\cos^2\theta}, \tag{7.17}$$

and

$$I(h) = 2(R+h)\int_0^{\frac{\pi}{2}} i\left(\frac{R+h}{\cos\theta} - R\right)\frac{d\theta}{\cos^2\theta}. \tag{7.18}$$

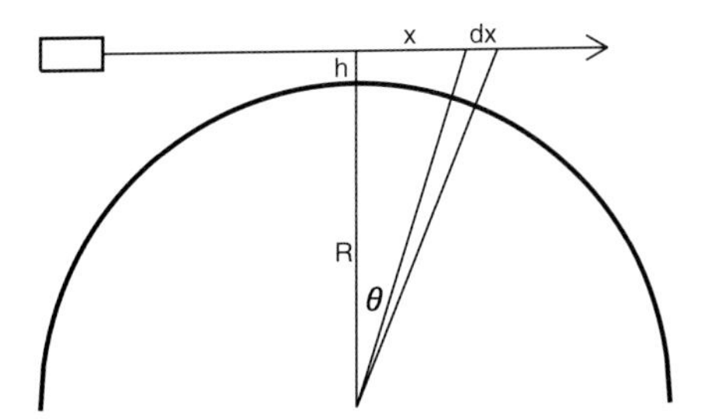


Figure 7.10 Parameters to deduce the airglow limb intensity.

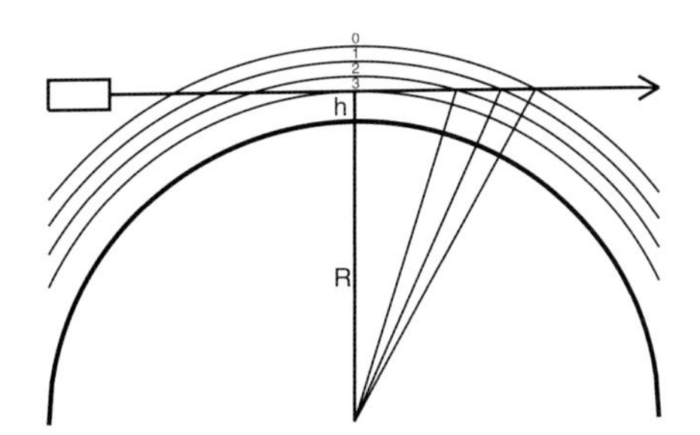


Figure 7.11 The onion peel technique for limb observations and solar and stellar occultations.

Using (7.18), the limb intensity $I(h)$ can be calculated if volume emission rate $i(h)$ is known. Changing properly the variables $h$ and $\theta$, (7.18) becomes the Abel integral equation, and its solution is

$$i(h) = \frac{1}{\pi(R+h)^3} \frac{d}{dy} \int_0^y \frac{I_1(t)dt}{\sqrt{t(y-t)}} . \tag{7.19}$$

Here $y = (R+h)^{-2}$, and this variable transforms $I(h)$ into $I_1(y)$. The direct derivative of the right side is a $\infty - \infty$ ambiguity. To avoid it, the integral should be calculated and then differentiated.

A more convenient solution of the problem is a so-called onion peel technique (Figure 7.11). Consider at first some auxiliary relationships.

The limb intensity of a uniform layer with thickness $H$ is

$$I = 2i\sqrt{(R+h+H)^2 - (R+h)^2} \approx i\sqrt{8RH}; \qquad \mu = \sqrt{8R/H}. \tag{7.20}$$

Here $\mu$ is the airmass. The limb intensity of an exponentially decreasing layer can be integrated to give

$$I = i\sqrt{2\pi RH}; \qquad \mu = \sqrt{2\pi R/H} . \tag{7.21}$$

The limb airmass is typically ~30 for Mars and ~50 for Venus. The limb intensity of a uniform layer with thickness $H$ as the instrument optical axis is displaced downward by $\delta$ is

$$I = I_0\left(\sqrt{1+\delta/H} - \sqrt{\delta/H}\right). \tag{7.22}$$

Now there is a set of limb intensities $I_0, I_1, \ldots, I_n$ with altitude decreasing with the number. We are looking for a solution using a steplike function $i_i$. Then the simple geometry in Figure 7.11 gives

$$i_0 = I_0[2\pi H(R+h_0)]^{-1/2}, \tag{7.23}$$

$$i_1 = (I_1 - A_1)[8(R+h_0)(h_0-h_1)]^{-1/2},$$

$$i_2 = \left[I_2 - A_2 - \sqrt{8(R+h_0)}\left(i_1\left(\sqrt{h_0-h_2} - \sqrt{h_1-h_2}\right)\right)\right][8(R+h_0)(h_1-h_2)]^{-1/2},$$

$$\cdots\cdots\cdots\cdots\cdots\cdots\cdots\cdots\cdots\cdots\cdots\cdots\cdots\cdots$$

$$i_n = \left[I_n - A_n - \sqrt{8(R+h_0)}\sum_{k=1}^{n-1} i_k\left(\sqrt{h_{k-1}-h_n} - \sqrt{h_k-h_n}\right)\right][8(R+h_0)(h_{n-1}-h_n)]^{-1/2}.$$

Here

$$A_k = I_0\left(\sqrt{1+(h_0-h_k)/H} - \sqrt{(h_0-h_k)/H}\right). \tag{7.24}$$

If $I_0$ is rather low and $A_n$ may be neglected, then the solution for a constant step $\Delta = h_n - h_{n+1}$ is

$$i_n = \frac{I_n}{\sqrt{8\Delta(R+h_0)}} - \sum_{k=1}^{n-1} i_k\left(\sqrt{n-k+1} - \sqrt{n-k}\right). \tag{7.25}$$

This problem was solved by Krasnopolsky (1970, 1974) and applied to the dayglow limb observations from the Cosmos 224 orbiter.

Limb observations of scattering and absorption phenomena can be also used to retrieve their vertical structure. For example, this may be applied to haze. However, the sounding becomes uncertain if the slant optical depth exceeds ~3, that is, the vertical depth exceeds ~0.1.

Limb observations of the $CO_2$ band at 15 μm were made on a regular basis using the MGS/TES (Smith 2004). The retrieved temperature profiles in the Martian atmosphere from those measurements covered altitudes from 15–20 to 55–60 km, while those from the nadir observations extended up to 35–40 km. Limb observations are the standard technique for the Mars Climate Sounder (McCleese et al. 2010) at the Mars Reconnaissance Orbiter.

### 7.8 Solar and Stellar Occultations

Solar and stellar occultations are a powerful technique to study vertical profiles of absorbing species in atmosphere. They employ the limb observational geometry

(Figure 7.11) with the instrument directed to the Sun or a star. The atmosphere is scanned and the impact altitude *h* changes while the orbiter is moving to or from the shadow. The method has important advantages:

1 Vertical profiles of species are retrieved from the observations. Volume extinction coefficient is extracted similar to the volume emission rate in relationships (7.23) and (7.25).
2 This is a pure absorption technique that does not require radiative transfer.
3 The observations are relative, scaled to the spectra observed at high altitudes, and do not require absolute calibration.
4 The large airmass is favorable for detection of minor species.
5 The signal is so high in the case of the Sun that the problems of sensitivity and signal-to-noise ratio become insignificant.

However, the method requires a tracking system to keep the Sun center or the star at the instrument optical axis. The vertical sounding is restricted down to slant optical depth of ~3 and does not cover, e.g., the lowest 10–15 km on Mars. The solar occultations refer to dusk and dawn and do not reproduce the diurnal cycle in full. The occultations cover a limited latitude range that depends on the orbit. The stellar occultations are mostly possible on the nightside. Anyway, the advantages overweight some restrictions, and the solar and stellar occultations are a standard tool for atmospheric studies from orbiters and flyby missions. Vertical profiles of $H_2O$ and HDO mixing ratios observed by the solar occultations using the SOIR spectrograph at Venus Express are shown in Figure 7.12. Stellar occultations were applied at Mars Express to observe nighttime variations of ozone (Figure 7.13).

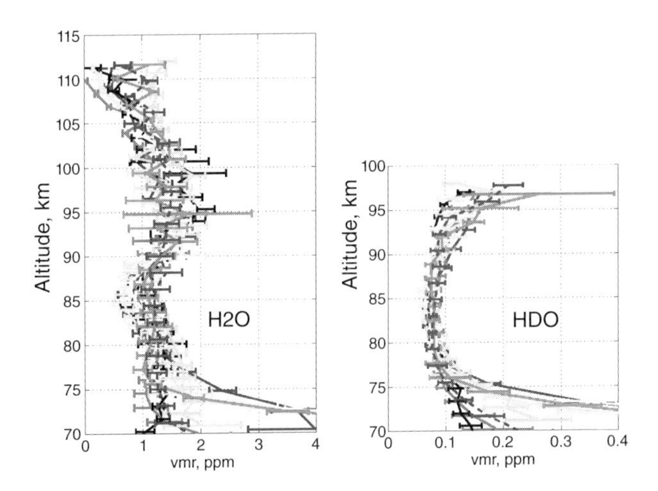


Figure 7.12 Vertical profiles of the $H_2O$ and HDO volume mixing ratios (vmr) from eight solar occultations observed by the SOIR spectrograph at Venus Express. From Fedorova et al. (2008).

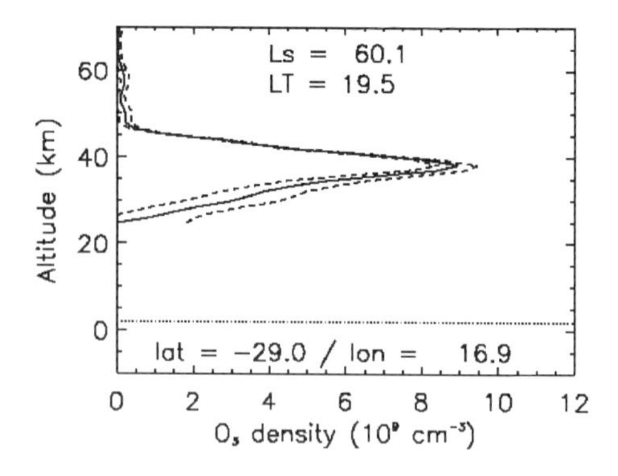


Figure 7.13 One of the $O_3$ nighttime profiles observed by SPICAM using Mars Express stellar occultations (Lebonnois et al. 2006). The retrieved uncertainties are shown by the dashed curves.

There is another type of the stellar occultations that is based on broadband photometry in the visible or near-infrared ranges, where absorption processes are weak. It is aimed to measure densities and temperatures using refraction of light in the atmospheres. Ground-based and Earth-orbiting observatories are used for these purposes. The observations are essential for the atmospheres of Triton and Pluto.

## 7.9 Some Other Applications of Spectroscopy

Observations of the solar light scattered in various directions at descent probes are a method to study aerosol media in the atmospheres. These observations were made in the Venus atmosphere from the Venera 8 to 14 (Figure 7.14) and Pioneer Venus descent probes, and some aspects of those measurements were discussed in Section 4.4. Detailed optical observations of aerosol and methane in the lower atmosphere on Titan (Tomasko et al. 2008a) were conducted at the Huygens descent probe.

Optical observations of the sky near zenith or the Sun may be done from Mars landing probes. (Only short-time observations of this type can be made from the other planets because of the extremely low or high surface temperatures.) These observations can monitor some absorbing species and give coarse data on their vertical distributions.

## 7.10 Mass Spectrometry and Gas Chromatography

Mass spectrometry and gas chromatography are the alternative methods to analyze atmospheric composition.

### *7.10.1 Mass Spectrometry*

A sample of an atmospheric gas can be ionized by an electron beam. The behavior of ions in electric and magnetic fields is determined by their mass to the electric charge ratio.

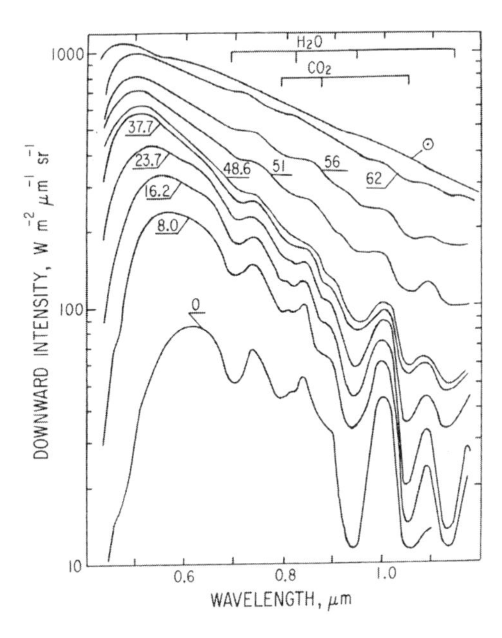


Figure 7.14 Nadir spectra of the solar light at various altitude (labeled in km) in the Venus atmosphere observed at the Venera 11 descent probe (Ekonomov et al. 1983a).

Therefore various types of electrostatic and magnetic devices can be made to analyze the mass spectrum of an atmospheric gas. The measured mass spectrum reflects chemical composition of the atmosphere. Mass spectrometry is a powerful and universal tool to study the chemical and isotopic composition of planetary atmospheres. For example, the first mass spectrometer study of the Martian atmosphere at the Viking landers (Owen et al. 1977) revealed mixing ratios of $N_2$, Ar, Ne, Kr, and Xe, the terrestrial isotope ratios for C and O, and isotope anomalies for N, Ar, and Xe. The measured $^{132}$Xe mixing ratio was 30 ppb, indicating the high sensitivity of the method. Mass spectrometric measurements of the isotope compositions are typically more accurate than those by spectroscopy. Nitrogen and the noble gases are poorly detectable by spectroscopy, and only far UV emissions of $N_2$ and Ar have been observed in the Martian upper atmosphere (see, e.g., Krasnopolsky and Feldman 2002). Therefore mass spectrometry is a key tool in the studies of chemical and isotope composition of the atmospheres.

Species dissociate in mass spectrometers, and analyses of the observed mass spectra require detailed laboratory modeling using duplicates of the flight instruments. This helps to resolve species of the same mass. For example, $N_2$ and CO can be separated using a peak of N at mass 14.

Mass spectrometers are also used to measure ion composition in the ionospheres. Then the internal electron beam is not required, and a sample of an ionosphere is directly analyzed. Mass spectrometers can achieve very high sensitivity in this mode. For instance, ion densities down to ~$10^{-4}$ cm$^{-3}$ were measured in Titan's ionosphere (Vuitton et al. 2007).

### 7.10.2 Gas Chromatography

A sample of atmospheric gas can be inserted into a tube filled by a sorbent, that is, a powder with great gaseous adsorption on the surface of particles. After the inlet is closed, a flow of a carrier gas, e.g., helium, moves the atmospheric mixture to the end of the tube. The time of this transport, the so-called retention time, depends on properties of the sorbent and the gas. Therefore, a separation of components of the atmospheric mixture occurs at the output of the device. The instrument detector should be insensitive to the carrier gas but sensitive to the atmospheric gases. Ionization potential of He is 24.6 eV and exceeds those of other gases by a factor of ~2. Therefore ionization by soft electrons is effective for the atmospheric gases and may be insignificant for helium. Then the output ion current is a measure of a quantity of a gas, while the retention time indicates its chemical nature. A function of ion current versus retention time is called a gas chromatogram, and the described instrument is a gas chromatograph. This idea was used for the first time by the Russian scientist M. S. Tzwet in 1903.

Though the idea looks simple, gas chromatographs are rather complicated devices with many valves, separators, chromatographic columns (tubes with sorbents), gas sample systems, and detectors. The detection limits are of ~100 ppb for the total pressure of ~1 bar and become worse at low pressure. The analysis duration is of a few minutes, and gas species do not decompose in the instrument. Some gases cannot be separated by a given sorbent, and a few columns with different sorbents are used. Gas chromatographs do not separate isotopes. Gas chromatographs may be combined with mass spectrometers, for instance, at the Viking landers (Owen et al. 1977).

# 8

# Solar Radiation, Its Absorption in the Atmospheres, and Airglow

## 8.1 Structure of the Solar Atmosphere

The Sun, its structure, and related physics are a fascinating science that is beyond our scope. Here we will briefly discuss the solar atmosphere and corona (Figure 8.1), which radiate photons and charged particles that are essential for the planetary atmospheres.

The photosphere, the visible surface of the Sun, is a layer of a few hundred kilometers thick with a mean temperature of 5800 K. It consists of mostly H atoms and ~3% of $H^+$ (protons) with the density of $\sim 10^{17}$ cm$^{-3}$, that is, smaller than that of the air by a factor of 300. Temperature is decreasing with altitude in the photosphere, and this results in limb darkening on the Sun. Electron attachment and detachment $H + e \rightleftarrows H^- + h\nu$ are the basic processes of emission and absorption. (Electron affinity of H is 0.73 eV.)

Temperature is decreasing above the photosphere to a minimum of 4100 K at ~500 km. This temperature is rather low for some molecules to exist, and CO and $H_2O$ absorption lines with high rovibrational excitation are formed. The chromosphere is a layer with a positive temperature gradient that extends to 2000 km above the minimum. Temperature reaches 20,000 K at the top of the chromosphere, and this temperature is sufficient for partial ionization of helium.

Temperature jumps by a factor of ~50 in a thin transition region that is ~200 km thick above the chromosphere. The corona is the next layer with typical temperatures of $(1\text{–}2)\times 10^6$ K, which extent is comparable with the Sun radius of $7\times 10^5$ km. The hottest regions of the corona can reach up to $2\times 10^7$ K. A significant part of even heavy elements like C, N, O, and Ne is ionized to the bare nuclei in the corona.

The energetic plasma forms the solar wind in the heliosphere that extends from ~0.1 AU to 50–70 AU. The solar wind at 1 AU is a flow with a mean speed of 400 km s$^{-1}$ and density of ~5 cm$^{-3}$, so that the flux is $2\times 10^8$ cm$^{-2}$ s$^{-1}$. It consists of protons $H^+$, alpha particles $He^{++}$ (~5%) and heavy ions $C^{+6}$, $C^{+5}$, $O^{+8}$, $O^{+7}$, etc. (~0.1% total). Local speed, strength and direction of magnetic field are the other properties of the solar wind.

Composition of the solar wind is measured by mass spectrometers, while abundances of $C^{+6}$, $C^{+5}$, $N^{+6}$, $O^{+8}$, $O^{+7}$, $Ne^{+10}$, and $Ne^{+9}$ can be retrieved from X-ray spectra of

Figure 8.1 The solar atmosphere and corona during a total eclipse.

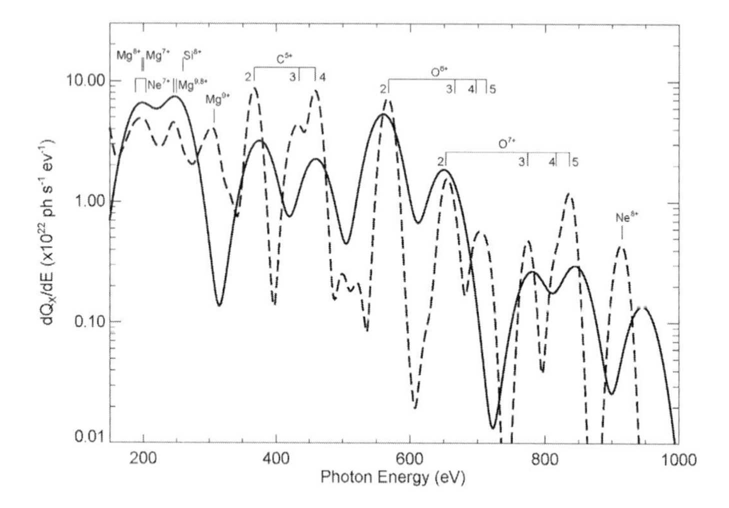


Figure 8.2 Emission lines of the solar wind heavy ions in the X-ray spectrum of comet McNaught–Hartley observed using the Chandra X-ray Observatory. The lines are excited by electron capture from the cometary molecules. The observed and processed spectrum is compared with a synthetic spectrum (dashed line). Emission lines and their positions are identified. From Krasnopolsky et al. (2002)

comets (Figure 8.2). X-ray emissions are excited in comets by the solar wind heavy ions that capture electrons from the cometary molecules, e.g.

$$O^{+8} + H_2O \rightarrow O^{+7} + H^+ + OH + 853 \text{ eV}.$$

Most of the released energy results in excitation of the $O^{+7}$ ion at $n$ = 2, 3, 4, 5 that emit Lyman-$\alpha$, $\beta$, $\gamma$, $\delta$ (Figure 8.2). Heavy ion abundances in the solar wind were derived from the spectra of nine comets measured by the Chandra X-ray Observatory (Krasnopolsky 2015a).

The outer boundary of the heliosphere is the heliopause, where the solar wind pressure becomes comparable with that of the stellar wind. The Voyager 1 and 2 Grand Tour probes intersected and detected the heliopause. The heliopause is not a boundary of the Solar System, and the Öpik–Oort cloud of comets extends to $5\times10^4$ AU (Section 1.4).

## 8.2 Solar Spectrum

A solar spectrum from the soft X-rays to the microwave range is shown in Figure 8.3. The spectrum is well fitted by the blackbody spectrum at 5785 K in a wide spectral range from the blue (400 nm) to ~1 cm. Radiation from the photosphere dominates in this range.

The solar constant $S$ is the total radiation from the Sun at 1 AU and equal to 1361 W m$^{-2}$. The solar energy absorbed by a planet is radiated (Section 2.3):

$$\frac{S}{4R^2}(1 - A) = \sigma T_e^4.$$

Here $R$ is the heliocentric distance in AU, $A$ is the effective (bolometric) albedo (reflectivity), and $T_e$ is the effective (bolometric) temperature. Using approximate albedo values for the terrestrial planets, their effective temperatures are calculated in Table 8.1.

Though the solar radiation at the Martian orbit is weaker than that at the Venus orbit by a factor of 4.5, their effective temperatures are rather similar because of the much higher reflectivity of the Venus clouds. While the energy is conserved, the entropy $E/T$ increases in the process of reradiation of the solar energy by a factor of ~25. Furthermore, the solar flux is almost parallel, while the planet flux is isotropic with a higher degree of disorder, and this adds to the entropy.

The solar spectrum from 1 to 10 μm is shown in Figure 8.4. It consists of the continuum and the Fraunhofer absorption lines. The continuum intensity drops within the spectrum by a factor of 3000. The solar radiation beyond 10 μm is weak, much weaker than thermal emission from the planets and typically of low importance for them. Detailed solar high-resolution spectra are required for infrared spectroscopy of planetary atmospheres, and those spectra were measured by the ATMOS (Farmer and Norton 1989; http://kurucz.harvard.edu/sun.html) and ACE (Hase et al. 2010) orbiters.

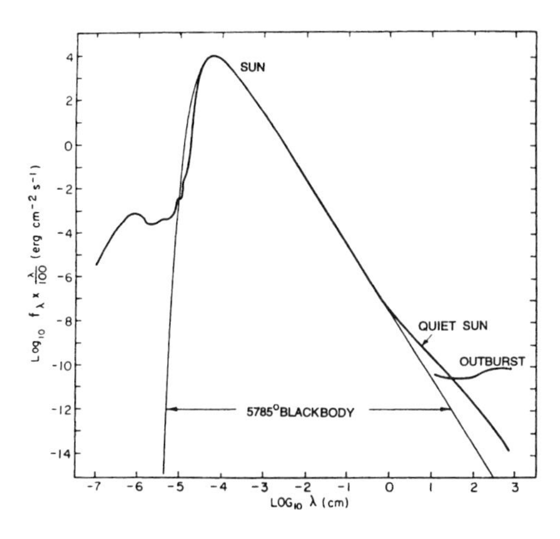


Figure 8.3 Solar radiation at 1 AU from X-rays to 10 m range and its fit by the blackbody emission at 5785 K.

Table 8.1 *Effective temperatures of the terrestrial planets*

| Planet | Venus | Earth | Mars |
|---|---|---|---|
| $R$, AU | 0.72 | 1.0 | 1.52 |
| $A$ | 0.8 | 0.3 | 0.15 |
| $T_e$, K | 220 | 255 | 217 |

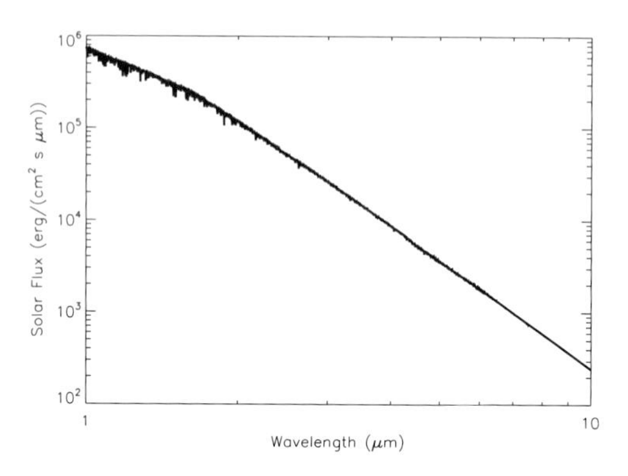


Figure 8.4 Solar flux from 1 to 10 μm based on http://kurucz.harvard.edu/sun.html.

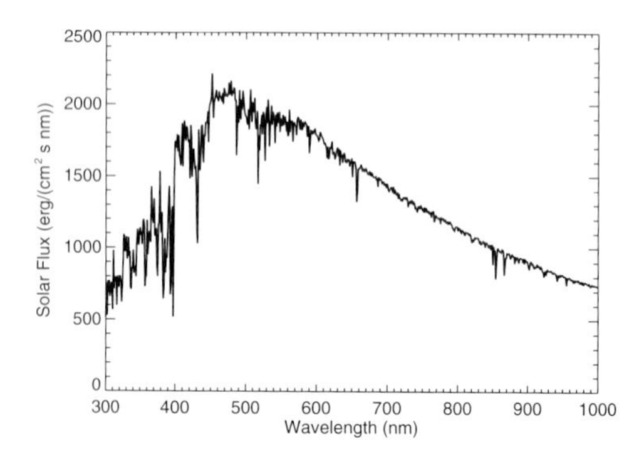


Figure 8.5 Solar flux from 300 to 1000 nm averaged within intervals of 1 nm and calculated using http://kurucz.harvard.edu/sun.html.

The spectra cover ranges of 605–4800 and 700–4430 $cm^{-1}$ with resolution of 0.01 and 0.02 $cm^{-1}$, respectively.

Absorption lines become deeper in the visible and UV ranges (Figure 8.5 and 8.6). Structure of some deep lines is complicated with, e.g., three minima separated by two maxima for the $Ca^+$ 393 nm. Intensity at each point of the line shape reflects temperature at a level of $\tau \approx 1$ at this wavelength, and the line structure reproduces a temperature profile in some altitude range in the solar atmosphere.

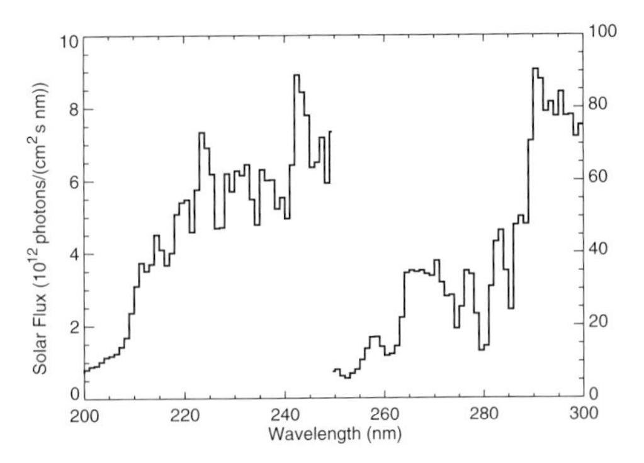


Figure 8.6 Solar flux from 200 to 300 nm averaged within intervals of 1 nm (based on Woods et al. 1996).

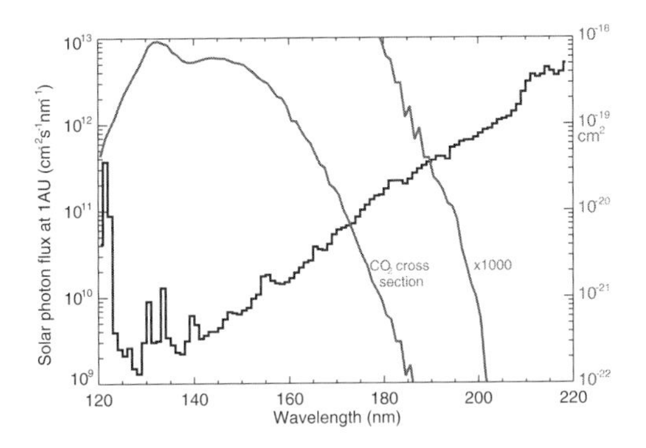


Figure 8.7 Solar spectrum at 120–220 nm averaged within intervals of 1 nm (based on Woods et al. 1996; mean solar activity) and $CO_2$ absorption cross section at 195 K (based on Parkinson et al. 2003). The right curve is the $CO_2$ cross section at 184–202 nm scaled by a factor of 1000.

The solar light below 300 nm dissociates and ionizes atmospheric species, and the solar fluxes are given in the photon numbers in Figures 8.6–8.8. The decrease in the spectral intensity from 300 to 200 nm is of a factor of 100 (Figure 8.6). This decline continues down to 140 nm (Figure 8.7) and looks exponential (linear in the log scale).

Emission lines dominate in the spectrum shortward of 140 nm. Here are the C 133 nm, O 130 nm, and especially the bright and important H Lyman-alpha at 122 nm (Figure 8.7). The extreme ultraviolet (EUV) spectrum (Figure 8.8) is a sum of quasi-continua and strong emission lines. The quasi-continua present mostly weak unresolved lines.

Solar activity changes with time, and the variations have random and regular components, i.e., cycles. The most prominent are those with periods of 11 years and 28 days. There are a few parameters that are used to characterize the solar activity. The most convenient is the solar radiation at 10.7 cm measured in $10^{-22}$ W m$^{-2}$ Hz$^{-1}$. Then $F_{10.7} < 100$ at 1 AU refers to low solar activity, 100–170 is medium, and $F_{10.7} > 170$ is high solar activity.

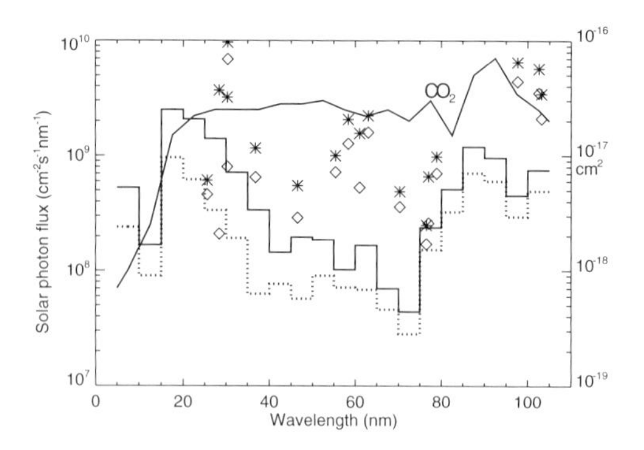


Figure 8.8 Solar EUV flux at low and high solar activity ($F_{10.7}$ = 80 and 200, respectively). Weak and unresolved lines are averaged in the 5 nm intervals (dotted and solid lines), and strong lines are shown by diamonds and asterisks, respectively. Based on Richards et al. (1994). The $CO_2$ cross sections are shown as well.

The solar UV and especially EUV radiation is typically tabulated for low and high solar activity, e.g., at $F_{10.7}$ = 80 and 200 in Figure 8.8, and the linear interpolation is usually applied to evaluate the solar emissions at a given level of solar activity.

The solar EUV lines originate in either the chromosphere or corona, and their variations with solar activity are different. For example, the $He^+$ 30.4 nm line is chromospheric and varies by a factor of 1.4 from $F_{10.7}$ = 80 to 200, while the line at 30.3 nm is coronal and varies by a factor of 4.

## 8.3 Airglow

Dayglow is a phenomenon related to excitation of atomic and molecular species in the daytime atmospheres of planets. Rayleigh and aerosol scattering, extinction, and refraction do not refer to dayglow, though they can affect the dayglow intensities. Dayglow emissions are measured in rayleighs, and 1 R is equal to the emission of $10^6$ photons per cm$^2$, s, and $4\pi$ sr along a given line of sight. The dayglow emissions are almost isotropic, and the observed brightness $I$ per steradian is extrapolated for the whole sphere as $4\pi I$. The dayglow emissions are often optically thin; then observed intensities can be converted to vertical (nadir or zenith) intensities being divided by cosine of observing angle.

Dayglow intensities present integrated volume emission rates that are in photons per cm$^3$ and s or in R km$^{-1}$. Vertical profiles of volume emission rates reveal altitude distributions of the excitation and quenching processes. Retrieval of volume emission rates from the observed dayglow intensities can be made using limb observations (Section 7.7). This is also done by sounding rockets in the Earth's atmosphere, when the observed intensity reflects a position of the instrument in the dayglow layer.

Excitation processes occur in the nighttime as well, though their nature may be very different with dominance of chemiluminescent reactions. This is nightglow. Airglow near sunset and sunrise is called twilight airglow. Airglow is an interesting phenomenon and a tool for diagnostics of the atmospheres.

## 8.4 Photodissociation and Photoionization of $CO_2$ and Related Dayglow

$CO_2$ is the most abundant gas in the atmospheres of Mars and Venus, and its photodissociation and photoionization are of practical importance for those atmospheres. This is also of interest as an example of the unimolecular reaction. Photodissociation and photoionization of $CO_2$ excite the bright dayglow that was studied for the first time from Mariners 6, 7, and 9 (Figure 8.9).

We discussed in Section 2.4 that formation of product $i$ from species with number density $n(z)$ is

$$P_i(z) = n(z)\sum\nolimits_{\lambda} I_\lambda \gamma_{i\lambda}\sigma_\lambda \exp\left(-\sec\theta\sum\nolimits_{k}\sigma_{k\lambda}n_k(z)H_k(z)\right),$$

where $I_\lambda$ is the photon intensity of the solar radiation at wavelength $\lambda$, $\gamma_{i\lambda}$ is the quantum yield of product $i$ at $\lambda$, $\sigma_\lambda$ is the absorption cross section of the species at $\lambda$, $\theta$ is the solar

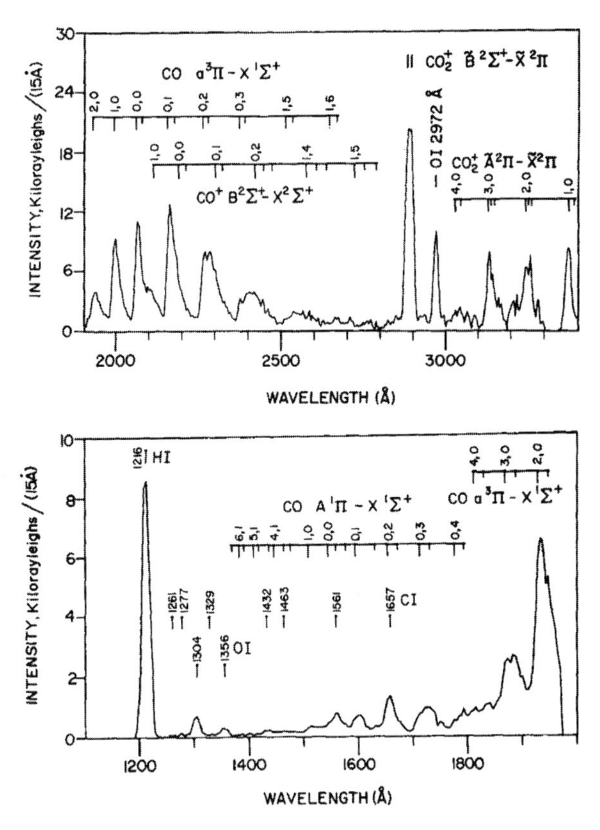


Figure 8.9 Mars dayglow observed by the UV spectrometers onboard Mariner 6, 7, and 9. These are averaged spectra of the Martian limb at 170 km with resolution of 1.5 nm. From Barth et al. (1972).

zenith angle, $n_k$ and $H_k$ are the number densities and scale heights of species $k$. If one species dominates, then

$$P_i(z) = n(z)\sum\nolimits_{\lambda} I_\lambda \gamma_{i\lambda}\sigma_\lambda e^{-\sigma_\lambda n(z)H(z)\sec\theta}.$$

The CO=O bond energy is 5.45 eV and corresponds to the dissociation limit at 227 nm. The dissociation into the ground states

$$CO_2(X^1\Sigma_g{}^+) + h\nu \rightarrow CO(X^1\Sigma^+) + O(^3P)$$

is spin-forbidden, because the spins of $CO_2$ and CO are zero and that of O is one. Therefore some excess over the Rayleigh scattering cross section

$$\sigma^R_{CO_2} = 8\times 10^{-25}\left(\frac{200\text{ nm}}{\lambda}\right)^4 \text{cm}^2 \qquad (8.1)$$

appears only near 205 nm and increases by a factor of $10^4$ at 180 nm (Figure 8.7). The $CO_2$ cross sections at the Martian temperatures ~200 K are smaller than those at the room temperatures by a factor of 1.5–2 that varies with wavelength.

The photodissociation can form $O(^1D)$ and $O(^1S)$ below 167 and 129 nm, respectively, and these processes are spin-allowed. $O(^1D)$ is immediately quenched by $CO_2$; however, its quenching by $H_2O$ and $H_2$ results in dissociation of these molecules. The quantum yield of $O(^1S)$ is shown in Figure 8.10. The solar Lyman-alpha emission dominates in the production of $O(^1S)$ with a yield of ~0.15. The $O(^1S\rightarrow{}^3P)$ emission at 297 nm is prominent in the Martian dayglow (Figure 8.9). Another branch of the $CO_2$ photolysis that forms $CO(a^3\Pi)$ is possible below 108 nm (Figure 8.10). It originates the $CO(a^3\Pi\rightarrow X^1\Sigma^+)$ Cameron band system that is the most outstanding feature in the Martian UV dayglow (Figure 8.9).

Photoionization of $CO_2$ occurs below 90 nm, and the limits for excitation of the $CO_2{}^+$ electronic states *A*, *B*, and *C* are 72, 69, and 64 nm, respectively. Yields of the states in the range of $h\nu$ = 20–40 eV (62–31 nm) are shown in Figure 8.11. $CO_2{}^+(C^2\Sigma_g{}^+)$ completely predissociates to CO + $O^+$ for $v = 0$ and partially to $CO^+$ + O for $v > 0$. The total yields of the channels are 80% and 20% (Fox and Dalgarno 1979b).

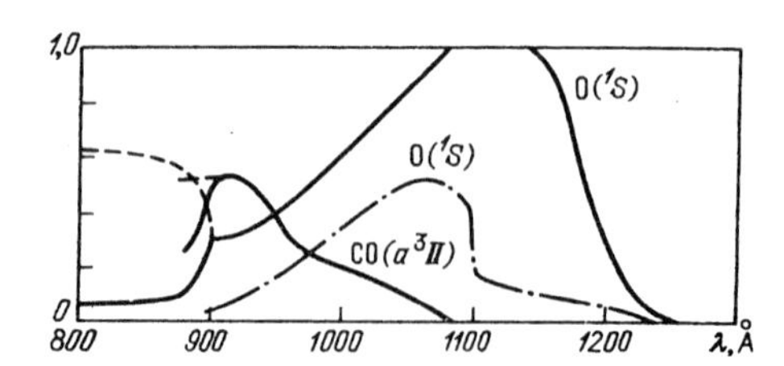


Figure 8.10 Yields of $O(^1S)$ and $CO(a^3\Pi)$ in photodissociation of $CO_2$. The dashed lines show the yields below the photoionization limit of 902 Å. The dash-dotted line is the yield in photodissociation of $O_2$ scaled by a factor of 5, i.e., 0.1 at 1060 Å. From Lawrence (1973).

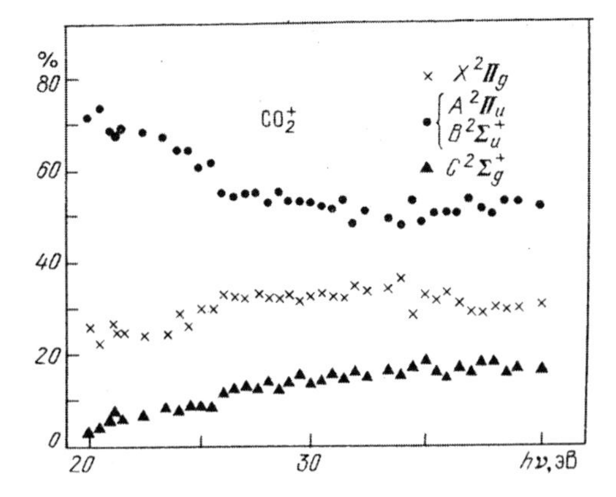


Figure 8.11 Photoionization yields of the $CO_2^+$ electronic states (Gustaffson et al. 1978).

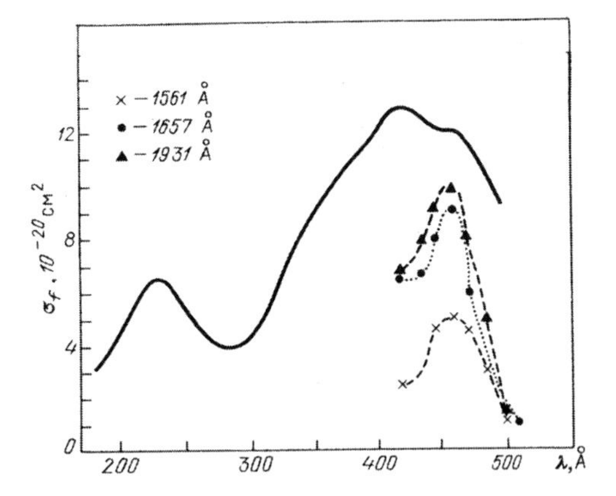


Figure 8.12 Cross sections for excitation of atomic carbon in photoabsorption by $CO_2$. The solid line is for total luminescence at 160–183 nm. From Wu et al. (1978).

The $CO_2^+$ ($A \rightarrow X$) and ($B \rightarrow X$) band systems at 300–400 nm and 289 nm, respectively, are prominent in the dayglow spectrum of Mars (Figure 8.9). The $B \rightarrow X$ so-called doublet system consists actually of the only (0–0) band at 289 nm. Almost one-half of the produced B-states populate the A-state via the Auger effect (Section 3.11). Therefore relative yields of the A and B states measured by photoelectron spectroscopy are inapplicable to the interpretation of the Martian dayglow in these band systems, and laboratory studies of the $CO_2^+$ fluorescence are preferable. For example, Lee and Judge (1972) measured yields of the A and B vibrational states from 72 to 46 nm.

Dissociation of $CO_2$ to $C + O_2$ does not proceed, while the atomization to $C + O + O$ is possible below 75 nm. Excitation of the C 166 nm line has a threshold at 52 nm. Cross sections for the carbon emissions (Figure 8.9) in this process are shown in Figure 8.12. Using the $CO_2$ total cross section of $3 \times 10^{-17}$ cm$^2$ at 40–50 nm (Figure 8.8), the yield of the C 166 emission peaks at 0.3% near 46 nm.

## 8.5 Resonance Scattering and Fluorescence

### 8.5.1 Absorption, Spontaneous, and Stimulated Emissions between Two Levels

Consider a quantum system with two levels $l$ and $u$ (lower and upper). An external radiation $I_\nu$ at the transition frequency $h\nu = E_u - E_l$ populates $u$ from $l$ and stimulates transitions from $u$ to $l$ that can also proceed spontaneously with probability $A$:

$$i_{lu} = n_l B_{lu} \int I_\nu d\omega; \quad i_{ul} = n_u \left( A + B_{ul} \int I_\nu d\omega \right). \tag{8.2}$$

Here $n_l$, $n_u$ are the populations of the lower and upper levels, $B$ is the probability of the induced transition that does not depend on the external radiation, and the integration is over the sphere. The relationships may be applied to thermodynamic equilibrium where

$$\frac{n_u}{n_l} = \frac{a_u}{a_l} \exp\left(-\frac{h\nu}{kT}\right), \quad i_{lu} = i_{ul}, \tag{8.3}$$

and $a_l$ and $a_u$ are the statistical weights of the lower and upper states, respectively. Then $I_\nu$ is the Planck function for the photon number, and

$$\int I_\nu d\omega = 4\pi \times \frac{2\nu^2}{c^2 (e^{h\nu/kT} - 1)} = \frac{f_\nu}{e^{h\nu/kT} - 1}; \quad f_\nu = \frac{8\pi\nu^2}{c^2} = \frac{8\pi}{\lambda^2}. \tag{8.4}$$

Therefore

$$n_l B_{lu} \frac{f_\nu}{e^{h\nu/kT} - 1} = n_u \left( A + B_{ul} \frac{f_\nu}{e^{h\nu/kT} - 1} \right) \tag{8.5}$$

and

$$a_l B_{lu} \frac{f_\nu}{e^{h\nu/kT} - 1} = a_u \left( A + B_{ul} \frac{f_\nu}{e^{h\nu/kT} - 1} \right) e^{-h\nu/kT} \tag{8.6}$$

are analogs of (8.3). Equation (8.6) should be valid for all temperatures, that is, it must be an identity. One may check that substitution of

$$B_{lu} = \frac{a_u A}{a_l f_\nu} = \frac{\lambda^2}{8\pi} \frac{a_u}{a_l} A, \quad a_l B_{lu} = a_u B_{ul} \tag{8.7}$$

transforms (8.6) into identity. (This problem was solved by A. Einstein in 1916.)

### 8.5.2 Implications for Planetary Atmospheres

$\int I_\nu d\omega$ for the solar radiation is just the solar photon flux. It is usually given at the wavelength scale, and $I_\nu = \frac{\lambda^2}{c} I_\lambda$. Typically, $h\nu >> kT$, and $n = n_l + n_u \approx n_l$. Then

$$i_{lu} = n \frac{\lambda^4}{8\pi c} \frac{a_u}{a_l} A I_\lambda = n g_{lu}. \tag{8.8}$$

Here $g_{lu}$ is the emission rate factor, and

$$g_{lu}\left(\frac{\text{photons}}{s}\right)=\frac{\lambda^4}{8\pi c}\frac{a_u}{a_l}AI_\lambda(\lambda\text{ in cm})=1.33\times10^{-21}\lambda^4_{\mu m}\frac{a_u}{a_l}AI_\lambda\left(\frac{\text{photons}}{\text{cm}^2\text{s nm}}\right)$$
$$=6.67\times10^{-10}\lambda^5_{\mu m}\frac{a_u}{a_l}AI_\lambda\left(\frac{\text{erg}}{\text{cm}^2\text{s nm}}\right). \tag{8.9}$$

The solution becomes trivial if line strength $S$ and absorption oscillator strength $f$ are used (Section 3.3):

$$S=\int_0^\infty\sigma_\nu d\nu=\frac{\pi e^2}{mc}f=0.02654f\text{ cm}^2\text{s}^{-1} \tag{8.10}$$

$$g=0.02654fI_\nu=8.85\times10^{-14}f\lambda^2_{\mu m}I_\lambda\left(\frac{\text{photons}}{\text{cm}^2\text{s nm}}\right). \tag{8.11}$$

The emission from the upper level is

$$i_{ul}=n_u\left(A+g_{lu}\frac{a_l}{a_u}\right).$$

The second term for the solar radiation is typically much smaller than $A$ and may be neglected. Therefore, if a system has terms $i$ between $l$ and $u$, then

$$g_i=g_{lu}\frac{A_i}{k_qn_q+\sum A_i}, \tag{8.12}$$

and $A_i$ is probability of the $ui$ transition. Here the denominator includes quenching of the upper level by species $q$ with number density $n_q$ and quenching rate coefficient $k_q$. Transitions to intermediate levels are called fluorescence.

Phase function of resonance scattering and fluorescence is similar to that for Rayleigh scattering (Section 4.2):

$$\gamma(\theta)=\frac{3}{4}\left(1+\cos^2\theta\right). \tag{8.13}$$

Asymmetry parameter is zero for this function, similar to that for isotropic scattering, and the approximation of isotropic scattering is sometimes applied to resonance scattering and fluorescence.

### *8.5.3 Retrieval of Atmospheric Abundances*

Some emission rate factors are given in Table 8.2. These emissions may coincide with absorption lines of the same species in the solar spectrum, and the solar intensity is smaller in these cases than those of the nearby continua. This may result in a significant scatter in the published emission rate factors.

Dayglow emissions from resonance scattering and fluorescence are directly converted into abundances of the radiating species, if the emission is optically thin. Volume emission

Table 8.2 *Emission rate factors for some species at 1 AU*

| Species | λ (nm) | g ($s^{-1}$) |
|---|---|---|
| $He^+$ | 30.4 | 9.2–6[a,b] |
| He | 58.4 | 6.5–6[g] |
| H | 121.6 | 1.2–3[b] |
| O | 130.4 | 5.7–6[b] |
| C | 165.7 | 4.4–5[f] |
| | 156.1 | 1.1–5[f] |
| Na | 589.0/9.6 | 0.8[c] |
| Li | 670.8 | 16[c] |
| CO(A→X) | | |
| Total | 130–180 | 7.5–7[d] |
| (14–5) | 139.2 | 2.9–9[e] |
| (14–3) | 131.7 | 3.5–9[e] |
| NO | 214.9 | 7.7–6[b] |
| $N_2^+$ | 391.4 | 1.43[b] |

[a]9.2–6 = $9.2\times10^{-6}$. [b]From Meier (1991) at solar minimum. [c]From Chamberlain and Hunten (1987). [d]Fox and Dalgarno (1979b). [e]Durrance (1981) [f]Paxton (1985). [g]$F_{10.7}$ = 100.

rate $i$ (photon $cm^{-3}$ $s^{-1}$) is just equal to $gn$, and integration of $i$ along a line of sight gives the dayglow intensity.

The interplanetary dust and meteors are decomposed near the mesopauses in the atmospheres of Earth, Venus, and Mars and form populations of metals and other atomic species that radiate by resonance scattering. The most prominent emission is Na 589.0/9.6 nm with a mean intensity of ~ 2 kR on the Earth, where sodium forms a layer of ~10 km thick with a maximum near 90 km. The Na densities of ~100 $cm^{-3}$ can be measured by the observations of resonance scattering.

Lithium has the highest emission rate factor. Lithium is scarce in the interplanetary dust and meteors, and the Li 670.8 nm emission is ~10 R on the Earth. However, significant quantities of lithium are formed in the fusion reaction, and observations of the lithium line were used for diagnostics of the thermonuclear tests in the atmosphere.

The carbon emissions are significant in the UV dayglow spectra of Mars (Figure 8.9) and Venus. However, resonance scattering is among a few excitation mechanisms for these emissions, and extraction of carbon densities from the observed emissions is not straightforward.

A similar situation is for the fourth positive band system of CO ($A^1\Pi \rightarrow X^1\Sigma$) at 140–180 nm (Figure 8.9), and self-absorption of the hot (0–0) and (1–0) bands by thermal atmospheric CO was used to retrieve the CO abundances in the Martian atmosphere (Mumma et al. 1975; Krasnopolsky 1981). However, the CO (14–0) band coincides with the strong solar Lyman-alpha emission with the excitation factor of $3.9\times10^{-8}$ $s^{-1}$ at 1 AU (Durrance 1981). The least contaminated band in the v′ = 14 progression is (14–5) at 139.2 nm, and contribution of other processes to excitation of this band is negligible. Therefore the CO band at 139.2 nm may be used to measure CO abundances.

Analyses of the H, He, and O emissions are more complicated because the atmospheres are optically thick in the line centers. An excited atom emits the photon almost isotropically, and thermal speed of the atom changes the photon frequency depending on the photon direction. Complete or partial frequency redistributions are the standard assumptions for the scattering process. A more detailed discussion of the problem may be found in Meier (1991).

## 8.6 Photoelectrons and Energetic Electrons

Ionization of atmospheric species by the solar hard UV photons $\lambda \leq 30$ nm forms photoelectron with energies of dozens and hundreds eV that can excite, dissociate, ionize, and heat the atmosphere. Typically the photoelectron mechanisms add to other basic processes; however, they may dominate and be even only in some cases. For example, excitation of the triplet band systems of $N_2$ can only be made by photoelectrons. Here we will consider a simplified technique for calculation of photoelectron production and loss that, however, is applicable for many tasks and is accurate within $\approx$20%.

It is suggested that photoelectrons appear and disappear at the same point in the atmosphere. This assumption is reasonable for many tasks but becomes questionable at high altitudes with low atmospheric density and long path of the photoelectrons.

Electron energy loss processes $j$ in the species $i$ with cross sections $\sigma_{ij}(E)$ are

$$\frac{dE}{dx} = -n_i \sum\nolimits_j \sigma_{ij}(E)\Delta E_{ij} = -n_i L_i(E). \tag{8.14}$$

Here $E$ is the electron energy, and $\Delta E_{ij}$ is the loss of electron energy in the process $ij$. This relationship is actually a definition of the electron energy loss function $L_i(E)$ in species $i$. The number of events $j$ in species $i$ from appearance $a$ to stop $b$ of the photoelectron is

$$\int_a^b \sigma_{ij}(E) n_i dx = n_i \int_0^E \frac{\sigma_{ij}(\varepsilon) d\varepsilon}{\sum_k n_k L_k(\varepsilon)}, \tag{8.15}$$

because

$$dx = -\frac{dE}{\sum_k n_k L_k(E)}$$

from (8.14). Each photoelectron with energy exceeding $E$ is decelerated to $E$, and the number of photoelectrons with energy $E$ formed per cm$^3$ s is

$$P(E) = \int_E^\infty p(\varepsilon) d\varepsilon. \tag{8.16}$$

Here $p$ and $P$ are the differential and integral spectra of photoelectron generation, respectively. Photoelectron flux spectrum is

$$\Phi(E) = \frac{P(E)}{\sum_k n_k L_k(E)} \text{ cm}^{-2} \text{ s}^{-1} \text{ eV}^{-1}. \tag{8.17}$$

This spectrum is measured by some instruments. Total excitation rate is

$$q_{ij} = n_i \int_0^\infty \frac{P(\varepsilon)\sigma_{ij}(\varepsilon)d\varepsilon}{\sum_k n_k L_k(\varepsilon)}. \tag{8.18}$$

Mean loss of the photon energy for ionization and prompt excitation of species like $CO_2$, CO, $N_2$, $CH_4$, and O is $\approx$16 eV. This assumption helps avoiding detailed analyses of production of ions in various excited states. Then

$$p(E) = I(E + 16\text{eV})\sum_i n_i \,\sigma_i(E + 16\text{eV}) \,\exp\left(-\sum_j N_j \,\sigma_j(E + 16\text{eV})/\cos\theta\right). \tag{8.19}$$

Here $I(E)$ is the solar EUV spectrum (Figure 8.4), $\sigma_i(E)$ is the photoionization cross section, $N_j$ is the column abundance of species $j$, and $\theta$ is the solar zenith angle.

Cross sections of the elastic collisions of electrons with gas species that consist of O, C, N atoms may be approximated by $5\times10^{-16}$ cm$^2$ times the number of atoms. The electron mass is very much smaller than the molecule mass; therefore the electron energy loss is very low in the elastic collisions with the gas species. More important is a change of the electron direction. The simplest assumption is the isotropic electron scattering in the elastic collisions.

Inelastic electron collisions result in rotational and vibrational excitation at $E \leq 1$ and 3 eV, respectively. Dissociative electron attachment (Munro et al. 2012) is significant for some species, e.g., $O_2$, $H_2O$, $Cl_2$, at energies of 3–10 eV:

$$H_2O + e \rightarrow H^- + OH.$$

Then the increasing electron energy is lost for excitation, dissociation, and ionization of gases.

Cross sections of the major electron energy loss processes in $CO_2$ are shown in Figure 8.13. The calculated electron energy loss function for $CO_2$ is in Figure 8.14, along with the dissociation and ionization components of this function. Sometimes it is helpful to get the number of chosen events from a photoelectron of a given energy. For example, the number of excitation of the CO ($a^3\Pi$) state as a function of electron energy is shown in Figure 8.15 for two degrees of ionization. Similar figures were calculated by Fox and Dalgarno (1979a) for other photoelectron impact products in $CO_2$.

Photon energy loss per ionization is ~16 eV, and that for the low energy electrons is ~40 eV. These values can be combined to give an effective quantum yield of photoionization at $\lambda < 40$ nm:

$$\gamma_\lambda = 1 + \frac{h\nu - 16 \text{ eV}}{40 \text{ eV}} = 0.6 + \frac{31 \text{ nm}}{\lambda} \tag{8.20}$$

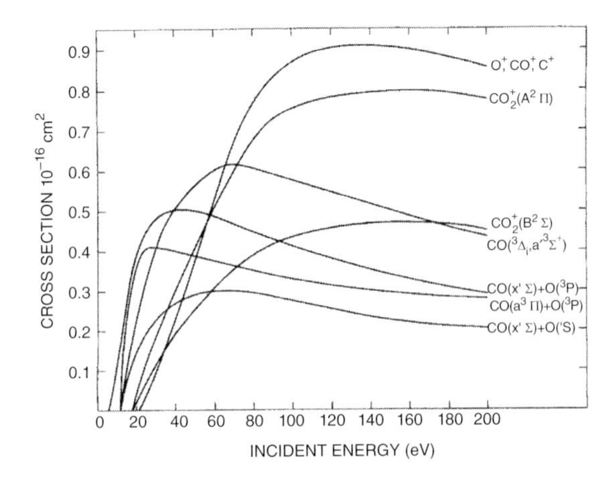


Figure 8.13 Cross sections for the electron impact excitation, dissociation, and ionization of $CO_2$ (Fox and Dalgarno 1979a).

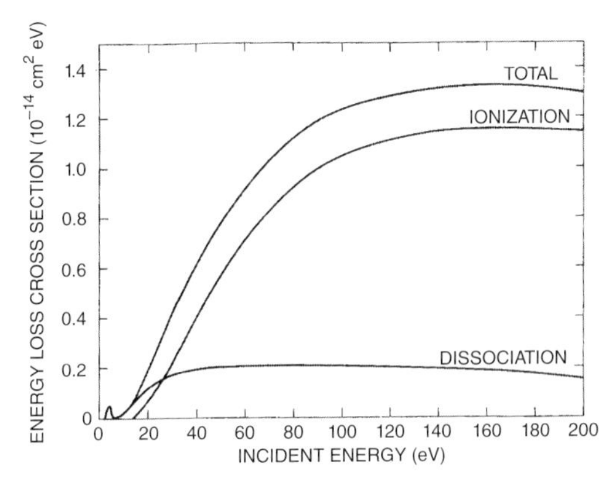


Figure 8.14 Electron energy loss function in $CO_2$ and its components for dissociation and ionization (Fox and Dalgarno 1979a).

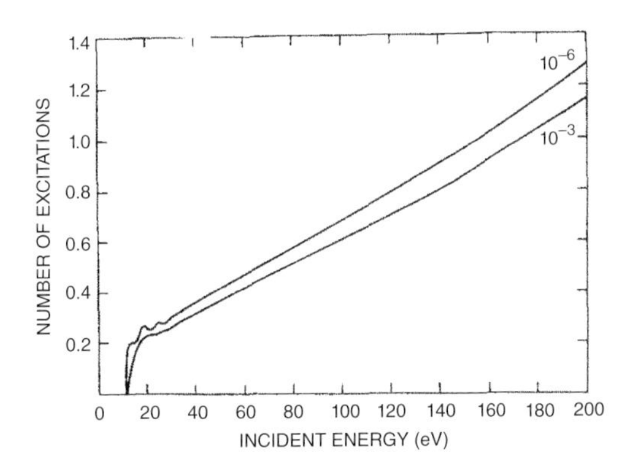


Figure 8.15 The number of excitations of CO ($a^3\Pi_u$) from electron impact on $CO_2$ for two values of fractional ionization (Fox and Dalgarno 1979a).

For example, the effective ionization yield is 1.6 for the solar emission of $He^+$ 30.4 nm and 5 at $\lambda = 7$ nm. Ionization by photoelectrons forms secondary electrons. Their mean energy is ~5 eV.

Magnetospheric processes may produce electron precipitation in an atmosphere. These electrons affect the atmosphere similarly to photoelectrons. If electron energy is significantly higher than those of photoelectrons, then the calculated cross sections and energy loss function may be extrapolated using the Born approximation.

## 8.7 Chemiluminescent Reactions

While these reactions are numerous, we consider here only five of them that are essential for our science.

Dissociative recombination of $O_2^+$ is responsible for escape of atomic oxygen from Mars. The required escape velocity of 5.03 km s$^{-1}$ corresponds to the energy of O atoms of 2.1 eV, and the first and second channels of the reaction provide this energy (Table 8.3). The total recombination rate coefficient is $2 \times 10^{-7}(300/T_e)^{0.7}$ and $7.4 \times 10^{-8}(1200/T_e)^{0.56}$ cm$^3$ s$^{-1}$ for electron temperature $T_e$ lower and higher than 1200 K, respectively (Mehr and Biondi 1969). This reaction is also essential in excitation of the green and red oxygen lines.

Dissociative recombination of $CO_2^+$ is significant in the ionospheres of Mars and Venus as well and contributes to excitation of the CO Cameron ($a^3\Pi \rightarrow X^1\Sigma^+$) and fourth positive ($A^1\Pi \rightarrow X^1\Sigma^+$) band systems. The measured recombination rate coefficient is $6.4 \times 10^{-7}(300/T_e)^{0.8}$ cm$^3$ s$^{-1}$ (Seiersen et al. 2003) with the following channel yields

$$
\begin{aligned}
CO_2^+ + e \rightarrow\ & CO(X^1\Sigma^+) + O(^3P) + 8.3\ \text{eV} \\
& CO(a^3\Pi) + O(^3P) + 2.29\ \text{eV};\ \gamma = 0.3 \pm 0.1\ \text{(Skrzypkowski et al. 1998)} \\
& CO(A^1\Pi) + O(^3P) + 0.27\ \text{eV};\quad 0.05\ \text{(Gutcheck and Zipf 1973)} \\
& C(^3P) + O_2(X^3\Sigma_g^-) + 2.3\ \text{eV};\quad 0.09 \pm 0.03\ \text{(Seiersen et al. 2003)}
\end{aligned}
$$

Table 8.3 *Yields for various channels in recombination of* $O_2^+(v)$

| Channel | $v = 0$ | $v = 1$ | $v = 2$ |
|---|---|---|---|
| $O(^3P) + O(^3P) + 6.98$ eV | 0.265 | 0.073 | 0.020 |
| $O(^3P) + O(^1D) + 5.03$ eV | 0.473 | 0.278 | 0.764 |
| $O(^1D) + O(^1D) + 3.05$ eV | 0.204 | 0.510 | 0.025 |
| $O(^1D) + O(^1S) + 0.83$ eV | 0.058 | 0.139 | 0.211 |

*Note.* From Petrignani et al. (2005).

Table 8.4 *Yields of the $O_2$ electronic states in termolecular association at 180 K*

| State | $X^3\Sigma_g^-$ | $a^1\Delta_g$ | $b^1\Sigma_g^+$ | $c^3\Sigma_u^-$ | $A'^3\Delta_u$ | $A^3\Sigma_u^+$ | $^5\Pi_g$ |
|---|---|---|---|---|---|---|---|
| Yield | 0.08 | 0.05 | 0.02 | 0.03 | 0.12 | 0.04 | 0.66 |

Channels with some other excited states of CO and O are energetically possible and not ruled out by the existing laboratory studies.

Three-body association of oxygen is among the main processes of photochemical relaxation in the terrestrial atmospheres:

$$O + O + N_2 \rightarrow O_2{}^* + N_2; k = 3 \times 10^{-33}(300/T)^{3.25}\mathrm{cm^6\ s^{-1}}$$

(Smith and Robertson 2008).

The expected efficiency of $CO_2$ as a third body is higher than that of $N_2$ by a factor of 2.5. This reaction is the main source of the oxygen airglow, and yields of seven electronic states of $O_2$ in this reaction were calculated by Wraight (1982) and Smith (1984). They are rather similar, do not depend on the third body, and are shown in Table 8.4.

The NO nightglow is excited on the terrestrial planets by the following set of reactions:

$$N + O \rightarrow NO(C^2\Pi, v = 0)\ ;\ k = 1.9 \times 10^{-17}(300/T)^{1/2}\mathrm{cm^3\ s^{-1}}(\text{Dalgarno et al. 1992})$$

$$NO(C^2\Pi, v = 0) \rightarrow NO(X^2\Pi) + \delta - \text{bands};\ \ \gamma = 2/3$$

$$NO(C^2\Pi, v = 0) \rightarrow NO(A^2\Sigma^+, v = 0) + 1.22\ \mu\text{m};\ \ \gamma = 1/3$$

$$NO(A^2\Sigma^+, v = 0) \rightarrow NO(X^2\Pi) + \gamma - \text{bands}.$$

Excitation of the OH vibrational bands is caused by the hydrogen-ozone reaction

$$H + O_3 \rightarrow OH(v) + O_2; k = 1.4 \times 10^{-10}e^{-470/T}\mathrm{cm^3\ s^{-1}}$$

(Sander et al. 2011).

Yields of the vibrational states v = 9, 8, 7, 6, 5, 4 are equal to 0.35, 0.29, 0.19, 0.07, 0.05, 0.05, respectively, and to zero for v $\leq$ 3 (García Muñoz et al. 2005).

# 9
# Chemical Kinetics

## 9.1 Double and Triple Collisions of Molecules

Evidently collision rates for these processes, that is, the numbers of collisions per $cm^3$ and s, are equal to

$$q_2 = k_2 n_1 n_2, \; q_3 = k_3 n_1 n_2 n_3. \tag{9.1}$$

Here $n_i$ is the number density of a colliding species and $k_2$ ($cm^3$ $s^{-1}$), $k_3$ ($cm^6$ $s^{-1}$) are the collision rate coefficients. For double (binary, or two-body) collisions

$$k_2 = \sigma V, \; V = \sqrt{\frac{8kT}{\pi\mu}}, \; \frac{1}{\mu} = \frac{1}{m_1} + \frac{1}{m_2}, \tag{9.2}$$

where $\sigma$ is the collision cross section, $V$ is the mean relative velocity, and $\mu$ is the reduced mass. Using the hard sphere approximation, $\sigma = \pi(r_1 + r_2)^2$, $r \approx 2\times10^{-8}$ cm, and

$$k_2 \approx 10^{-11} T^{1/2}\ \mathrm{cm^3\ s^{-1}}. \tag{9.3}$$

To calculate triple collisions, one may assume that complexes (1–2), (1–3), and (2–3) exist in the mixture, and their number density

$$n_{12} = q_{12}\tau_{12} = V_{12}\pi(r_1 + r_2)^2 n_1 n_2 \frac{r_1 + r_2}{V_{12}} = n_1 n_2 \pi (r_1 + r_2)^3. \tag{9.4}$$

Here $\tau_{12} = \frac{r_1 + r_2}{V_{12}}$ is the collision time. The collision rate of ((1–2) – 3) is

$$q_{(12)3} = V_{(12)3}\pi(r_{12} + r_3)^2 n_{12} n_3.$$

Taking into account all complexes,

$$k_3 \approx 1500\, r^5 V_{123} \approx 10^{-32} T^{1/2}\ \mathrm{cm^6\ s^{-1}}. \tag{9.5}$$

There are two sequences from these relationships: (1) rate coefficients of the double and triple collisions are proportional to $r^2$ and $r^5$, respectively, and (2) rate of triple collisions becomes comparable to the rate of double collisions at rather high densities $n_3 \approx k_2/k_3 \approx 10^{21}$ $cm^{-3}$. This value is proportional to $r^{-3}$ and may significantly vary.

The collisional cross section $\sigma = \pi(r_1 + r_2)^2$ is called gas kinetic. Sometimes this term refers to $\sigma = \pi a_0^2 = 0.88 \times 10^{-16}$ cm$^2$, where $a_0 = 0.53\times10^{-8}$ cm is the radius of the hydrogen atom. However, the collisional cross section exceeds typically $\pi a_0^2$ by an order of magnitude.

## 9.2 Thermochemical Equilibrium

We will use this term for thermodynamic equilibrium in chemical reactions. All direct and inverse processes are balanced under thermochemical equilibrium, e.g., $H + H \rightleftarrows H_2$. Enthalpy of formation $H$ is the energy content of a molecule. Energy is measured relative to an adopted level, and the enthalpy of $H_2$ is adopted zero; then that of H is 218 kJ mol$^{-1}$ or 52.1 kcal mol$^{-1}$ (1 cal = 4.184 J, 1 eV = 96.48 kJ mol$^{-1}$). The system tends to minimal energy at the equilibrium, and this favors $H_2$ to dominate in the mixture, because destruction of $H_2$ requires 436 kJ mol$^{-1}$, much higher than the collisional energy at moderate temperatures.

However, there is another principle of maximum entropy *S*, that is, maximum disorder, and it tends the equilibrium to H + H, because two separate H atoms have more disorder than when they are bound in the molecule: $S_{H2}$ = 130.7 J mol$^{-1}$ K$^{-1}$, $S_H$ = 114.7 J mol$^{-1}$ K$^{-1}$. Both principles are combined into the principle of minimal Gibbs free energy

$$G = H - TS = \min. \tag{9.6}$$

For example, the change of free energy in the reaction H + H = $H_2$ at 300 K is

$$0 - 0.1307{*}300 - 2(218 - 0.1147{*}300) = -406.4 \text{ kJ mol}^{-1}$$

and at 6000 K is

$$0 - 0.1307{*}6000 - 2(218 - 0.1147{*}6000) = 156.2 \text{ kJ mol}^{-1}.$$

Therefore hydrogen is molecular at the room conditions and atomic on the Sun.

Consider the general case

$$\alpha_1 A_1 + \alpha_2 A_2 + \ldots + \beta_1 B_1 + \beta_2 B_2 + \ldots = \gamma_1 C_1 + \gamma_2 C_2 + \ldots + \delta_1 D_1 + \delta_2 D_2 + \ldots \tag{9.7}$$

Here A, C are gases and B, D are solids and liquids with saturated vapor pressures in the system. Then the equilibrium constant is

$$K_p(T) = \frac{k_f}{k_r} = \frac{P_{C1}^{\gamma 1}\, P_{C2}^{\gamma 2} \cdots}{P_{A1}^{\alpha 1}\, P_{A2}^{\alpha 2} \cdots} \tag{9.8}$$

$$\begin{aligned} RTlnK_p(T) &= -\Delta G(T) \\ &= \alpha_1 G_{A1} + \alpha_2 G_{A2} + \ldots + \beta_1 G_{B1} + \beta_2 G_{B2} + \ldots - \gamma_1 G_{C1} - \gamma_2 G_{C2} \cdots \\ &\quad - \delta_1 G_{D1} - \delta_2 G_{D2} - \ldots \end{aligned} \tag{9.9}$$

Here $k_f$ and $k_r$ are the rate coefficients of the forward and reverse reactions, respectively.

Thermochemical equilibria can be easily calculated if thermodynamic properties of participating species are known. Enthalpies and entropies of formation of simple molecules are given in Tables 9.1 and 9.2. The values are also recorded in http://webbook.nist.gov/chemistry/form-ser.html or in Chase (1998). Thermochemical equilibria hold if chemical lifetimes of participating species are shorter than mixing time (Section 2.2.4). However, thermodynamics does not provide a tool to get chemical lifetimes that may exceed the age of the Universe in some cases. Usually high temperatures and pressures stimulate fast chemistry, and thermochemistry holds in the planet interiors and probably near the surface of Venus. It may be applicable to some other processes, e.g., thermal decomposition of sulfuric acid below the clouds of Venus:

$$H_2SO_4 \leftrightarrow H_2O + SO_3.$$

Thermochemistry makes it possible to calculate reaction rate coefficients using (9.8), if rate coefficients of the reverse reactions are known.

Consider for example the equilibrium between HDO and DCl near the surface of Venus at $T = 735$ K: HDO + HCl = $H_2O$ + DCl ; the calculation gives $K = 0.37$. Then

$$\mathrm{DCl/HCl} = 0.37\ \mathrm{HDO}/\mathrm{H_2O} \text{ and } (\mathrm{D/H})_{\mathrm{HCl}} = 0.74\ (\mathrm{D/H})_{\mathrm{H_2O}}.$$

## 9.3 Bimolecular Reactions

Bimolecular, or binary, or two-body reactions are the basic type of reactions, and unimolecular and termolecular reactions may involve bimolecular reactions as intermediate steps. Evidently not all collisions of reactants result in the products. Each reaction forms or breaks a chemical bond $D$. This process occurs in a very short collision time $V/(r_1 + r_2)$ via a very small contact area, and the flow of energy is

$$DV/(r_1 + r_2)^3 \approx 100 \text{ MW cm}^{-2} \text{ for } D \approx 5 \text{ eV}, \tag{9.10}$$

while 100 megawatts are close to the production of a power plant. This energy transfer is usually less probable than a decay of a complex AB to the initial reactants A and B.

At moderate temperatures typical of the planetary atmospheres the second term in the Gibbs free energy $G = H - TS$ is small and

$$\frac{k_f}{k_r} \approx \frac{a_2}{a_1} e^{\Delta E/kT},$$

where $\Delta E$ is the energy yield of the reaction and $a_1$, $a_2$ are the statistical weights of reactants and products. If $\Delta E \approx 0.5$ eV, then the reverse reaction is weaker at the room temperature than the forward reaction by a factor of ~$10^8$. Therefore endothermic reactions are usually negligible in the planetary atmospheres.

Another restriction is the spin conservation. Reactions that obey this rule are typically much faster than those violating this rule. For example,

$$\mathrm{N}(^4S) + \mathrm{CO_2}(\mathrm{X}^1\Sigma_g{}^+) \rightarrow \mathrm{NO}(\mathrm{X}^2\Pi) + \mathrm{CO}(\mathrm{X}^1\Sigma^+), \quad k < 10^{-16}\mathrm{cm}^3\ \mathrm{s}^{-1}$$
$$\mathrm{N}(^2D) + \mathrm{CO_2}(\mathrm{X}^1\Sigma_g{}^+) \rightarrow \mathrm{NO}(\mathrm{X}^2\Pi) + \mathrm{CO}(\mathrm{X}^1\Sigma^+), \quad k = 3.6 \times 10^{-13}\ \mathrm{cm}^3\ \mathrm{s}^{-1}.$$

Table 9.1 *Enthalpies of some molecules*

| Species | $\Delta H_f$ (298 K) (kcal mol$^{-1}$) | Species | $\Delta H_f$ (298 K) (kcal mol$^{-1}$) | Species | $\Delta H_f$ (298 K) (kcal mol$^{-1}$) | Species | $\Delta H_f$ (298 K) (kcal mol$^{-1}$) |
|---|---|---|---|---|---|---|---|
| H | 52.1 | $C_2H_4$ | 12.45 | $FNO_2$ | $-26 \pm 2$ | $CFCl_3$ | $-68.1$ |
| $H_2$ | 0.0 | $C_2H_5$ | 28.4 | $FONO_2$ | $2.5 \pm 7$ | $CF_2Cl$ | $-64 \pm 3$ |
| O | 59.57 | $C_2H_6$ | $-20.0$ | CF | | $CF_2Cl_2$ | $-117.9$ |
| $O_2$ | 0.0 | $CH_2CN$ | $57 \pm 2$ | $CF_2$ | $-44 \pm 2$ | $CF_3Cl$ | $-169.2$ |
| $O_3$ | 34.1 | $CH_3CN$ | 15.6 | $CF_3$ | $-112 \pm 1$ | $CHFCl_2$ | $-68.1$ |
| HO | 9.3 | $CH_2CO$ | $-11 \pm 3$ | $CF_4$ | $-223.0$ | $CHF_2Cl$ | $-115.6$ |
| $HO_2$ | $3 \pm 1$ | $CH_3CO$ | $-5.8$ | $CHF_3$ | $-166.8$ | COFCl | $-102 \pm 2$ |
| $H_2O$ | $-57.81$ | $CH_3CHO$ | $-39.7$ | $CHF_2$ | $-58 \pm 2$ | $CH_3CH_2F$ | $-63 \pm 2$ |
| $H_2O_2$ | $-32.60$ | $C_2H_5O$ | $-4.1$ | $CH_2F_2$ | $-107.2$ | $CH_3CHF$ | $-17 \pm 2$ |
| N | 113.00 | $CH_2CH_2OH$ | $10 \pm 3$ | $CH_2F$ | $-8 \pm 2$ | $CH_2CF_3$ | $-124 \pm 2$ |
| $N_2$ | 0.0 | $C_2H_5OH$ | $-56.2$ | $CH_3F$ | $-55.9 \pm 1$ | $CH_3CHF_2$ | $-120 \pm 1$ |
| NH | 85.3 | $CH_3CO_2$ | $-49.6$ | FCO | $-41 \pm 14$ | $CH_3CF_2$ | $-71 \pm 2$ |
| $NH_2$ | 45.3 | $C_2H_5O_2$ | $-6 \pm 2$ | $COF_2$ | $-153 \pm 2$ | $CH_3CF_3$ | $-179 \pm 2$ |
| $NH_3$ | $-10.98$ | $CH_3COO_2$ | $-41 \pm 5$ | $CF_3O$ | | $CF_2CF_3$ | $-213 \pm 2$ |
| NO | 21.57 | $CH_3OOCH_3$ | $-30.0$ | $CF_3O_2$ | | $CHF_2CF_3$ | $-264 \pm 2$ |
| $NO_2$ | 7.9 | $C_3H_5$ | 39.4 | $CF_3OH$ | | $CH_3CF_2Cl$ | $-127 \pm 2$ |
| $NO_3$ | $17 \pm 2$ | $C_3H_6$ | 4.8 | $CF_3OOCF_3$ | | $CH_2CF_2Cl$ | $-75 \pm 2$ |
| $N_2O$ | 19.61 | n-$C_3H_7$ | $22.6 \pm 2$ | $CF_3OOH$ | | $C_2Cl_4$ | $-3.0$ |
| $N_2O_3$ | 19.8 | i-$C_3H_7$ | $19 \pm 2$ | $CF_3OF$ | | $C_2HCl_3$ | $-1.9$ |
| $N_2O_4$ | 2.2 | $C_3H_8$ | $-24.8$ | Cl | 28.9 | $CH_2CCl_3$ | $17 \pm 2$ |
| $N_2O_5$ | $2.7 \pm 2$ | $C_2H_5CHO$ | $-44.8$ | $Cl_2$ | 0.0 | $CH_3CCl_3$ | $-34.0$ |
| HNO | 23.8 | $CH_3COCH_3$ | $-51.9$ | HCl | $-22.06$ | $CH_3CH_2Cl$ | $-26.8$ |
| HONO | $-19.0$ | $CH_3COO_2NO_2$ | $-62 \pm 5$ | ClO | 24.4 | $CH_2CH_2Cl$ | $22 \pm 2$ |
| $HNO_3$ | $-32.3$ | S | 66.22 | ClOO | $23 \pm 1$ | $CH_3CHCl$ | $17.6 \pm 1$ |
| $HO_2NO_2$ | $-11 \pm 2$ | $S_2$ | 30.72 | OClO | $23 \pm 2$ | Br | 26.7 |
| C | 170.9 | HS | $34 \pm 1$ | $ClO_2$ | 21.3 | $Br_2$ | 7.39 |
| CH | 142.0 | $H_2S$ | $-4.9$ | $ClOO_2$ | 16.7 | HBr | $-8.67$ |

Table 9.1 (*cont.*)

| Species | $\Delta H_f$ (298 K) (kcal mol$^{-1}$) | Species | $\Delta H_f$ (298 K) (kcal mol$^{-1}$) | Species | $\Delta H_f$ (298 K) (kcal mol$^{-1}$) | Species | $\Delta H_f$ (298 K) (kcal mol$^{-1}$) |
|---|---|---|---|---|---|---|---|
| $CH_2$ | 93 ± 1 | SO | 1.3 | $ClO_3$ | 52 ± 4 | HOBr | −19 ± 2 |
| $CH_3$ | 35 ± 0.2 | $SO_2$ | −70.96 | $Cl_2O$ | 19.5 | BrO | 30 |
| $CH_4$ | −17.88 | $SO_3$ | −94.6 | $Cl_2O_2$ | 31 ± 3 | BrNO | 19.7 |
| CN | 104 ± 3 | HSO | −1 ± 3 | $Cl_2O_3$ | 34 ± 3 | BrONO | 25 ± 7 |
| HCN | 32.3 | $HSO_3$ | −92 ± 2 | FOCl | 18 ± 3 | $BrNO_2$ | 17 ± 2 |
| $CH_3NH_2$ | −5.5 | $H_2SO_4$ | −176 | NOCl | 12.36 | $BrONO_2$ | 12 ± 5 |
| NCO | 38.0 | CS | 67 ± 2 | ClNO | 12.4 | BrCl | 3.5 |
| CO | −26.42 | $CS_2$ | 28.0 | $ClNO_2$ | 3.0 | $CH_2Br$ | 40 ± 2 |
| $CO_2$ | −94.07 | $CS_2OH$ | 26.4 | ClONO | 13 | $CHBr_3$ | 6 ± 2 |
| HCO | 10 ± 1 | $CH_3S$ | 33 ± 2 | $ClONO_2$ | 5.5 | $CHBr_2$ | 45 ± 2 |
| $CH_2O$ | −26.0 | $CH_3SOO$ | 0.0 | $ClNO_3$ | 6.28 | $CBr_3$ | 48 ± 2 |
| COOH | −53 ± 2 | $CH_3SO_2$ | −57 | FCl | −12.1 | $CH_2Br_2$ | −2.6 ± 2 |
| HCOOH | −90.5 | $CH_3SH$ | −5.5 | $CCl_2$ | 57 ± 5 | $CH_3Br$ | −8.5 |
| $CH_3O$ | 4 ± 1 | $CH_2SCH_3$ | 36 ± 3 | $CCl_3$ | 18 ± 1 | $CH_3CH_2Br$ | −14.8 |
| $CH_3O_2$ | 4 ± 2 | $CH_3SCH_3$ | −8.9 | $CCl_4$ | −22.9 | $CH_2CH_2Br$ | 32 ± 2 |
| $CH_2OH$ | −6.2 | $CH_3SSCH_3$ | −5.8 | $C_2Cl_4$ | −2.97 | $CH_3CHBr$ | 30 ± 2 |
| $CH_3OH$ | −48.2 | OCS | −34 | $CHCl_3$ | −24.6 | I | 25.52 |
| $CH_3OOH$ | −31.3 | F | 18.98 | $CHCl_2$ | 23 ± 2 | $I_2$ | 14.92 |
| $CH_3ONO$ | −15.6 | $F_2$ | 0.0 | $CH_2Cl$ | 29 ± 2 | HI | 6.3 |
| $CH_3NO_2$ | −17.86 | HF | −65.34 | $CH_2Cl_2$ | −22.8 | $CH_3I$ | 3.5 |
| $CH_3ONO_2$ | −28.6 | HOF | −23.4 ± 1 | $CH_3Cl$ | −19.6 | $CH_2I$ | 52 ± 2 |
| $CH_3NO_3$ | −29.8 | FO | 26 ± 5 | ClCO | −5 ± 1 | IO | 41.1 |
| $CH_3O_2NO_2$ | −10.6 ± 2 | $F_2O$ | 5.9 ± 0.4 | $COCl_2$ | −52.6 | INO | 29.0 |
| $C_2H$ | −133 ± 2 | $FO_2$ | 6 ± 1 | CHFCl | −15 ± 2 | $INO_2$ | 14.4 |
| $C_2H_2$ | 54.35 | $F_2O_2$ | 5 ± 2 | $CH_2FCl$ | −63 ± 2 | | |
| $C_2H_2OH$ | 30 ± 3 | FONO | −15 ± 7 | CFCl | 7 ± 6 | | |
| $C_2H_3$ | 72 ± 3 | FNO | −16 ± 2 | $CFCl_2$ | −22 ± 2 | | |

*Note.* From Yung and DeMore (1999).

Table 9.2 *Entropies of some molecules*

| Species | $S^\circ$ (298 K) ($cal^{-1}$ $mol^{-1}$ $deg^{-1}$) | Species | $S^\circ$ (298 K) ($cal^{-1}$ $mol^{-1}$ $deg^{-1}$) | Species | $S^\circ$ (298 K) ($cal^{-1}$ $mol^{-1}$ $deg^{-1}$) | Species | $S^\circ$ (298 K) ($cal^{-1}$ $mol^{-1}$ $deg^{-1}$) |
|---|---|---|---|---|---|---|---|
| H | 27.4 | $C_2H_4$ | 52.5 | $FNO_2$ | 62.3 | $CFCl_3$ | 74.0 |
| $H_2$ | 31.2 | $C_2H_5$ | 58.0 | $FONO_2$ | 70.0 | $CF_2Cl$ | 68.7 |
| O | 38.5 | $C_2H_6$ | 54.9 | CF | 50.9 | $CF_2Cl_2$ | 71.8 |
| $O_2$ | 49.0 | $CH_2CN$ | 58.0 | $CF_2$ | 57.5 | $CF_3Cl$ | 68.3 |
| $O_3$ | 57.0 | $CH_3CN$ | 58.2 | $CF_3$ | 63.3 | $CHFCl_2$ | 70.1 |
| HO | 43.9 | $CH_2CO$ | 57.8 | $CF_4$ | 62.4 | $CHF_2Cl$ | 67.2 |
| $HO_2$ | 54.4 | $CH_3CO$ | 64.5 | $CHF_3$ | 62.0 | COFCl | 66.2 |
| $H_2O$ | 45.1 | $CH_3CHO$ | 63.2 | $CHF_2$ | 61.7 | $CH_3CH_2F$ | 63.3 |
| $H_2O_2$ | 55.6 | $C_2H_5O$ | 65.3 | $CH_2F_2$ | 58.9 | $CH_3CHF$ | |
| N | 36.6 | $CH_2CH_2OH$ | | $CH_2F$ | 55.9 | $CH_2CF_3$ | 71.8 |
| $N_2$ | 45.8 | $C_2H_5OH$ | 67.5 | $CH_3F$ | 53.3 | $CH_3CHF_2$ | 67.6 |
| NH | 43.3 | $CH_3CO_2$ | | FCO | 59.4 | $CH_3CF_2$ | 69.9 |
| $NH_2$ | 46.5 | $C_2H_5O_2$ | 75.0 | $COF_2$ | 61.9 | $CH_3CF_3$ | 68.6 |
| $NH_3$ | 46.0 | $CH_3COO_2$ | | $CF_3O$ | | $CF_2CF_3$ | 81.6 |
| NO | 50.3 | $CH_3OOCH_3$ | 74.1 | $CF_3O_2$ | | $CHF_2CF_3$ | |
| $NO_2$ | 57.3 | $C_3H_5$ | 62.1 | $CF_3OH$ | | $CH_3CF_2Cl$ | 68.7 |
| $NO_3$ | 60.3 | $C_3H_6$ | 63.8 | $CF_3OOCF_3$ | | $CH_2CF_2Cl$ | |
| $N_2O$ | 52.6 | n-$C_3H_7$ | 68.5 | $CF_3OOH$ | | $C_2Cl_4$ | 81.4 |
| $N_2O_3$ | 73.9 | i-$C_3H_7$ | 66.7 | $CF_3OF$ | 77.1 | $C_2HCl_3$ | 77.5 |
| $N_2O_4$ | 72.7 | $C_3H_8$ | 64.5 | Cl | 39.5 | $CH_2CCl_3$ | 80.6 |
| $N_2O_5$ | 82.8 | $C_2H_5CHO$ | 72.8 | $Cl_2$ | 53.3 | $CH_3CCl_3$ | 76.4 |
| HNO | 52.7 | $CH_3COCH_3$ | 70.5 | HCl | 44.6 | $CH_3CH_2Cl$ | 65.9 |
| HONO | 59.6 | $CH_3COO_2NO_2$ | | ClO | 54.1 | $CH_2CH_2Cl$ | |
| $HNO_3$ | 63.7 | S | 40.1 | ClOO | 64.0 | $CH_3CHCl$ | |
| $HO_2NO_2$ | | $S_2$ | 54.5 | OClO | 61.5 | Br | 41.8 |
| C | 37.8 | HS | 46.7 | $ClO_2$ | 63.0 | $Br_2$ | 58.6 |
| CH | 43.7 | $H_2S$ | 49.2 | $ClOO_2$ | 73.0 | HBr | 47.4 |
| $CH_2$ | 46.3 | SO | 53.0 | $ClO_3$ | 67.0 | HOBr | 59.2 |

Table 9.2 (*cont.*)

| Species | $S°$ (298 K) ($cal^{-1}$ $mol^{-1}$ $deg^{-1}$) | Species | $S°$ (298 K) ($cal^{-1}$ $mol^{-1}$ $deg^{-1}$) | Species | $S°$ (298 K) ($cal^{-1}$ $mol^{-1}$ $deg^{-1}$) | Species | $S°$ (298 K) ($cal^{-1}$ $mol^{-1}$ $deg^{-1}$) |
|---|---|---|---|---|---|---|---|
| $CH_3$ | 46.4 | $SO_2$ | 59.3 | $Cl_2O$ | 64.0 | BrO | 56.8 |
| $CH_4$ | 44.5 | $SO_3$ | 61.3 | $Cl_2O_2$ | 72.2 | BrNO | 65.3 |
| CN | 48.4 | HSO | | $Cl_2O_3$ | | BrONO | |
| HCN | 48.2 | $HSO_3$ | | HOCl | 56.5 | $BrNO_2$ | |
| $CH_3NH_2$ | 58.0 | $H_2SO_4$ | 69.1 | NCCl | 62.5 | $BrONO_2$ | |
| NCO | 55.5 | CS | 50.3 | ClNO | 62.5 | BrCl | 57.3 |
| CO | 47.3 | $CS_2$ | 56.9 | $ClNO_2$ | 65.1 | $CH_2Br$ | |
| $CO_2$ | 51.1 | $CS_2OH$ | | ClONO | | $CHBr_3$ | 79.1 |
| HCO | 53.7 | $CH_3S$ | 57.6 | $ClCNO_2$ | | $CHBr_2$ | |
| $CH_2O$ | 52.3 | $CH_3SOO$ | | $ClNO_3$ | | $CBr_3$ | 80.0 |
| COOH | 61.0 | $CH_3SO_2$ | | FCl | 52.1 | $CH_2Br_2$ | 70.1 |
| HCOOH | 59.4 | $CH_3SH$ | 61.0 | $CCl_2$ | 63.4 | $CH_3Br$ | 58.7 |
| $CH_3O$ | 55.0 | $CH_2SCH_3$ | | $CCl_3$ | 71.0 | $CH_3CH_2Br$ | 68.6 |
| $CH_3O_2$ | 65.3 | $CH_3SCH_3$ | 68.4 | $CCl_4$ | | $CH_2CH_2Br$ | |
| $CH_2OH$ | 58.8 | $CH_3SSCH_3$ | 80.5 | $C_2Cl_4$ | 81.4 | $CH_3CHBr$ | |
| $CH_3OH$ | 57.3 | OCS | 55.3 | $CHCl_3$ | 70.7 | I | 43.2 |
| $CH_3OOH$ | 67.5 | F | 37.9 | $CHCl_2$ | 66.5 | $I_2$ | 62.3 |
| $CH_3ONO$ | 68.0 | $F_2$ | 48.5 | $CH_2Cl$ | 58.2 | HI | 49.3 |
| $CH_3NO_2$ | 65.7 | HF | 41.5 | $CH_2Cl_2$ | 64.6 | $CH_3I$ | 60.6 |
| $CH_3ONO_2$ | 72.1 | HOF | 54.2 | $CH_3Cl$ | 56.1 | $CH_2I$ | |
| $CH_3NO_3$ | 76.1 | FO | 51.8 | ClCO | 63.3 | IO | 58.8 |
| $CH_3O_2NO_2$ | | $F_20$ | 59.1 | $COCl_2$ | 67.8 | INO | 67.6 |
| $C_2H$ | 49.6 | $FO_2$ | 61.9 | CHFCl | | $INO_2$ | 70.3 |
| $C_2H_2$ | 48.0 | $F_2O_2$ | 66.3 | $CH_2FCl$ | 63.3 | | |
| $C_2H_2OH$ | | FONO | 62.2 | CFCl | 62.0 | | |
| $C_2H_3$ | 56.3 | FNO | 59.3 | $CFCl_2$ | 71.5 | | |

*Note.* From Yung and DeMore (1999).

$N(^4S)$ and $N(^2D)$ are the ground and metastable states of atomic nitrogen. Spins are zero for $CO_2$ and CO, 1/2 for $N(^2D)$ and NO, and 3/2 for $N(^4S)$. The first reaction is spin-forbidden and much slower than the second spin-allowed reaction.

### 9.3.1 Two-Body Association

$$A + B \rightarrow AB + h\nu.$$

Considering the problem of two atoms at their mass center, the product AB is formed in the continuum above the dissociation limit. It may either dissociate or emit a photon and get a bound state. Therefore the reaction rate coefficient is

$$k = \pi(r_A + r_B)^2 V^* \frac{r_A + r_B}{V} A \frac{a_{AB}}{a_{A+B}} = \pi(r_A + r_B)^3 \frac{a_{AB}}{a_{A+B}} A. \tag{9.11}$$

Here $A$ is the transition probability, $a$ is the statistical weight, and $(r_A + r_B)/V$ is the collision time. The rate coefficient is ~$10^{-16}$ $cm^3$ $s^{-1}$ for $A \approx 10^6$ $s^{-1}$.

The collision time is close to a period of the A–B vibration in the diatomic molecule. If AB is a polyatomic molecule, then the excess energy can be distributed between a few vibrational modes of the molecule, and numerous vibration periods are required to get the excess energy in a mode that breaks. The lifetime of the AB* complex is much longer in this case, the number of optical transitions to stabilize it is greater; however, their transition probabilities are typically smaller. Overall, rate coefficients of two-body association are greater for polyatomic molecules.

### 9.3.2 Exchange Reactions and Transition States

Most of the bimolecular reactions result in exchange of chemical bonds:

$$A + BC \rightarrow AB + C. \tag{9.12}$$

This reaction proceeds via a transient triatomic complex ABC* (transition state) that may either return the reactants A + BC or form the products AB + C. Figure 9.1 shows potential curves of A + BC and AB + C that form a potential barrier for the reaction. Though the energy of AB + C is lower than that of A + BC and the reaction is exothermic, some excess energy called activation energy $E_a$ is required for the reaction:

$$k = Ae^{-E_a/kT}. \tag{9.13}$$

This is the Arrhenius form of reaction rate coefficient. The preexponential factor $A$ is a product of the collision rate coefficient and a geometric factor that is a share of the collisions with the proper geometry. For example, the reaction does not occur if A impacts C in BC.

Usually data on the activation energies and preexponential factors are measured in the laboratory. There are some complicated methods to calculate them, e.g., the

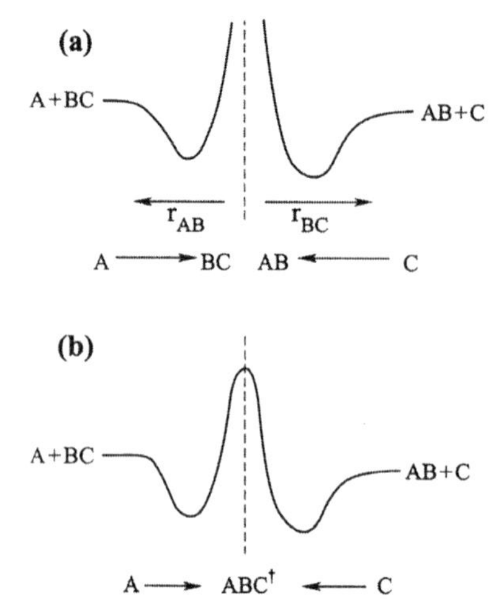


Figure 9.1 Potential curves of A + BC and AB + C in (a) form a potential barrier for formation of AB + C (b). From Yung and DeMore (1999).

Rice–Ramsperger–Kassel–Marcus (RRKM) theory that accounts for contributions to reaction of various vibrational and rotational modes.

### 9.3.3 Attraction Fields

Here we will consider formation of a complex by two molecules with attraction potential $V(r) = -A/r^n$. Then, for

$n = 1$: charge–charge interaction, $V = -e^2/r$;
$n = 2$: charge–dipole, $V = -pe/r^2$, $p$ is the dipole moment;
$n = 3$: dipole–dipole, $V = -p_1p_2/r^3$;
$n = 4$: charge–neutral with polarizability $\beta$: $V = -\beta\, e^2/2r^4$;
$n = 6$: dipole–neutral, $V = -\beta\, p^2/r^6$.

A reasonable approximation for the collisional cross section $\pi r_0^2$ in this case is $|V(r_0)| = kT$. Then the substitution of $r_0$ gives

$$k = \left(\frac{8\pi}{\mu}\right)^{1/2} A^{2/n}(kT)^{\frac{1}{2}-\frac{2}{n}}\,. \tag{9.14}$$

### 9.3.4 Centrifugal Barrier Model

Interaction of two particles with the attraction field between them may form a rotating complex. The system at the mass center has angular momentum $M = \mu\, v_0\, a$; here

$\mu$ is the reduced mass, $v_0$ is the relative velocity before the collision, and $a$ is the impact parameter. Then

$$av_0 = rv, \; E = \frac{\mu v_0^2}{2} = \frac{\mu v^2}{2} - \frac{A}{r^n}, \; \frac{\mu v^2}{r} = \frac{dV(r)}{dr} = n\frac{A}{r^{n+1}} \,. \tag{9.15}$$

The first and second relationships are the angular momentum and energy conservations, and the third one is the balance between centrifugal and attraction forces. Their solution is

$$r = a\left(1 - \frac{2}{n}\right)^{1/2}. \tag{9.16}$$

Energy of the system prevents formation of the complex, similar to a potential barrier. Its maximum is at

$$\frac{dE}{dr} = 0; \text{ then } a_{\max} = \left(\frac{nA}{\mu v_0^2}\right)^{1/n}\left(1 - \frac{2}{n}\right)^{\frac{1}{n}-\frac{1}{2}}. \tag{9.17}$$

The rate coefficient is $k = \pi a_{\max}^2 v_0$; $v_0 = \left(\frac{8kT}{\pi\mu}\right)^{1/2}$, and

$$k = \frac{\pi}{1 - 2/n}\left(\frac{n-2}{\mu}A\right)^{2/n}\left(\frac{8kT}{\pi\mu}\right)^{\frac{1}{2}-\frac{2}{n}} = \pi n(n-2)^{\frac{2}{n}-1}\frac{A^{2/n}}{\mu^{1/2}}\left(\frac{8kT}{\pi}\right)^{\frac{1}{2}-\frac{2}{n}}. \tag{9.18}$$

This relationship is rather similar to (9.14) for the approximation $|V(r_0)| = kT$, and both are

$$k \approx 5\mu^{-1/2}A^{2/n}(kT)^{\frac{1}{2}-\frac{2}{n}}. \tag{9.19}$$

The calculated rate coefficient is negative and does not make sense for the charge–charge interaction ($n = 1$) that results in a hyperbolic orbit without any complex. The rate coefficient for the charge-dipole ($n = 2$) is

$$k = \frac{\pi^{3/2}}{(2\mu kT)^{1/2}}\, ep \approx 2 \times 10^{-9}\ \mathrm{cm^3\ s^{-1}} \tag{9.20}$$

for $\mu \approx 10$, $T \approx 300$ K, and $p \approx 10^{-18}$ CGSE = 1 debye. Here $(n-2)^{\frac{2}{n}-1} \to 1$ for $n \to 2$. The dipole-dipole interaction ($n = 3$) gives

$$k \approx 3 \times 10^{-10}(p_1 p_2)^{2/3}\ \mathrm{cm^3\ s^{-1}}, \tag{9.21}$$

where $p_1$ and $p_2$ are in debyes.

The case of charge-neutral interaction results in a beautiful relationship,

$$k = 2\pi e(\beta/\mu)^{1/2}, \tag{9.22}$$

which is known as the Langevin formula. This formula is applicable to a broad class of the ion-molecular reactions. Polarizability $\beta$ is

$$\beta = \frac{m-1}{2\pi n}. \tag{9.23}$$

Here $m$ is the refractive index, and $\beta \approx 10^{-24}$ cm$^3$ (Section 2.2.3; Table 2.2). The typical rate coefficients are ~$10^{-9}$ cm$^3$ s$^{-1}$.

The dipole-neutral interactions ($n$ = 6) have rate coefficients

$$k \approx 10^{-10} \sqrt[3]{\beta p^2} \text{ cm}^3 \text{ s}^{-1} \tag{9.24}$$

for $\beta$ in $10^{-24}$ cm$^3$ and $p$ in debyes.

All above relationships for the electric fields between interacting molecules are in reasonable agreement with the laboratory studies of these processes.

### 9.3.5 Dissociative and Radiative Recombination

Though the ion–electron interaction cannot form a complex, this interaction is strong because of the electric field and the high mobility of electron. For the molecular ion, a level of the incoming electron is in the dissociation continuum of the neutral molecule, and the radiationless transition results in dissociative recombination with rate coefficients of $10^{-7}$ – $10^{-6}$ cm$^3$ s$^{-1}$ (Section 8.7). This is the main loss process in the planetary ionospheres.

Recombination of atomic ions is radiative, e.g.,

$$O^+ + e \rightarrow O + h\nu, \; k = 3 \times 10^{-12} \text{ cm}^3 \text{ s}^{-1}. \tag{9.25}$$

This process is similar to the radiative association (Section 9.3.1); however, the cross section is much larger because of the electric field, and the emitted photon is in the EUV range with high transition probability.

While the rate coefficient of radiative recombination is larger than that of radiative association by five orders of magnitude, it is slower than that of dissociative recombination by the similar factor. Therefore ionospheres with dominating atomic composition ($F$-type ionospheres, e.g., that on Triton) are much denser than ionospheres with molecular ions ($E$-type, e.g., those on Titan and Pluto).

### 9.3.6 Chemical Lifetime

For number densities of the reacting species $n_1$ and $n_2$

$$\frac{dn_1}{dt} = -kn_1n_2; \;\; n_1(t) = n_1(0)e^{-kn_2t} = n_1(0)e^{-t/\tau_1} \tag{9.26}$$

and the time constant for the decay of $n_1$ is $\tau_1 = (kn_2)^{-1}$. (Actually this formulation is valid if $n_2 >> n_1$ and $n_2$ is therefore constant. However, we do not apply this restriction.) If a species with number density $n$ reacts with a few species $n_i$ and the reaction rate coefficients are $k_i$, then the decay time constant is

$$\tau = \left(\sum_i k_i n_i\right)^{-1}. \tag{9.27}$$

This time constant is called chemical lifetime of species $n$. This lifetime is local and refers to a given altitude. It is actually equal to a ratio of the species density to its loss rate. This formulation may be applied to a mean chemical lifetime of species in the atmosphere that is equal to a ratio of the species column abundance to its column loss or production rate, because column loss and production rates are typically balanced.

For example, the CO mixing ratio on Mars is $7\times10^{-4}$, and the atmosphere of $p = 6$ mbar consists mostly of $CO_2$ and contains $N = p/mg = 2.3\times10^{23}$ molecules per cm$^2$. (The gravity acceleration is $g = 370$ cm s$^{-2}$.) The CO global-mean column production rate is equal to the $CO_2$ column photolysis rate that is a quarter of the solar photon flux for $\lambda < 200$ nm, i.e., $7.5 \times 10^{11}$ cm$^{-2}$ s$^{-1}$. (The quarter is the disk-to-sphere surface ratio.) Then the CO mean chemical lifetime is

$$7 \times 10^{-4} * 2.3 \times 10^{23}/7.5 \times 10^{11} = 2.15 \times 10^8 \text{ s} = 6.8 \text{ yr}.$$

### *9.3.7 Some General Considerations*

Energy barriers in chemical reactions result in the Arrhenius form of their rate coefficients with the exponential term (9.13), while interaction fields between reactants require a power index to temperature (9.14, 9.18). Therefore a general form for the reaction rate coefficient is

$$k = a(T/T_0)^{\alpha} e^{-A/T}, \tag{9.28}$$

which is given by three parameters $a$, $\alpha$, and $A$.

Reaction rate coefficient strongly depends on changes of chemical bonds in the reaction. For example,

$$\begin{aligned}
&N^+ + O_2 \rightarrow NO^+ + O, && k = 2.6 \times 10^{-10},\\
&O^+ + N_2 \rightarrow NO^+ + N, && k = 2.4 \times 10^{-12},\\
&O_2{}^+ + N_2 \rightarrow NO^+ + NO, && k < 10^{-16},
\end{aligned}$$

all in cm$^3$ s$^{-1}$. The first reaction is breaking the double bond in $O_2$, the second refers to the triple bond in $N_2$, and the rate coefficient is smaller by a factor of ~100. Both $O_2{}^+$ and $N_2$ bonds are broken in the third reaction, which is very slow.

Consider the following reactions:

$$\begin{aligned}
&OH + O \rightarrow O_2 + H && k = 2.2 \times 10^{-11} e^{120/T} && k(300\text{ K}) = 3.3 \times 10^{-11},\\
&OH + O_3 \rightarrow HO_2 + O_2 && 1.5 \times 10^{-12} e^{-880/T} && 8 \times 10^{-14},\\
&OH + H_2 \rightarrow H_2O + H && 3.3 \times 10^{-13} (T/300)^{2.7} e^{-1150/T} && 7 \times 10^{-15},\\
&H_2 + O_2 \rightarrow HO_2 + H && 6.3 \times 10^{-11} (T/300)^{0.17} e^{-28000/T} && 10^{-51},\\
&CO + O_2 \rightarrow CO_2 + O && 4.2 \times 10^{-12} e^{-24000/T} && 10^{-46}.
\end{aligned}$$

All rate coefficients are in $cm^3\ s^{-1}$. The first reaction is between two radicals, and its rate coefficient is ~$10^{-11}\ cm^3\ s^{-1}$. The second reaction is between a radical and a weakly bound ozone molecule with the bond energy of 1.05 eV; it is slower by two orders of magnitude. Even slower is the reaction with $H_2$ with the bond of 4.48 eV. The explosive mixture of $H_2$ and $O_2$ is very exothermic; however, the initial reaction in this mixture given here is endothermic and so slow at the room temperature that the mixture lifetime exceeds the age of the universe by orders of magnitude. Yet this conclusion refers to the exothermic oxidation of CO as well. Overall, the range of the reaction rate coefficients is extremely wide.

## 9.4 Unimolecular Reactions

These are reactions of breaking of chemical bonds: AB → A + B. Actually, the process is more complicated and involves three steps:

$$\mathrm{AB + M \rightarrow AB^* + M} \qquad k_1,$$

$$\mathrm{AB^* + M \rightarrow AB + M} \qquad k_2,$$

$$\mathrm{AB^* \rightarrow A + B} \qquad k_3.$$

The effective reaction rate coefficient is

$$k = \frac{k_1 k_3 [M]}{k_2 [M] + k_3}. \tag{9.29}$$

The reaction has low- and high-pressure limits:

$$k_0 = k_1[M] \text{ for } k_2[M] \ll k_3 \text{ and } k_\infty = \frac{k_1 k_3}{k_2} \text{ for } k_2[M] \gg k_3. \tag{9.30}$$

Thermal energies are typically much smaller than chemical bonds, and the collisionally induced thermal dissociation is scarce in the planetary atmospheres.

### *9.4.1 Photodissociation of Diatomic Molecules*

Photodissociation is the basic process that initiates photochemistry in the planetary atmospheres. Dissociation energies of some molecules are given in Table 9.3. The process is relatively simple in diatomic molecules. For example, the $\mathbf{O_2}$ molecule has three low-energy even ($g$) electronic states, three odd ($u$) states, and $^5\Pi_g$ state. All these states approach the dissociation limit to $O(^3P) + O(^3P)$ at 5.12 eV (see Section 3.11.2 and Figure 3.12). Photons absorbed above this limit $\lambda < 242$ nm result in dissociation to two ground-state atoms in the so-called Herzberg continuum (Figure 9.2).

The Schumann–Runge bands at 176–200 nm involve the $B^3\Sigma_u^-$ state, which has the dissociation limit $O(^3P) + O(^1D)$. The absorption above this limit forms the Schumann–Runge continuum at $\lambda < 176$ nm. Furthermore, energy levels of the Schumann–Runge

Table 9.3 *Dissociation energies (eV) of some molecules*

| $H_2$ | HF | HCl | CO | $N_2$ | NO | $O_2$ | $S_2$ | $Cl_2$ | CH | NH | OH | CN | $C_2$ | ClO | SO | $H_2O$ |
|---|---|---|---|---|---|---|---|---|---|---|---|---|---|---|---|---|
| 4.48 | 5.86 | 4.43 | 11.12 | 9.76 | 6.50 | 5.12 | 4.37 | 2.48 | 3.47 | 3.54 | 4.39 | 7.85 | 6.11 | 2.75 | 5.34 | 5.12 |
| $H_2S$ | HCN | $CO_2$ | OCS | $NO_2$ | $O_3$ | $SO_2$ | $HO_2$ | $CH_2$ | $NH_2$ | $C_3$ | $NH_3$ | $PH_3$ | $C_2H_2$ | $H_2CO$ | $H_2O_2$ | |
| 3.91 | 5.20 | 5.45 | 3.12 | 3.12 | 1.05 | 5.65 | 2.00 | 4.36 | 4.10 | 7.31 | 4.40 | 3.40 | 5.38 | 3.70 | 2.15 | |
| $COCl_2$ | $C_2N_2$ | $CH_3$ | $NO_3$ | $SO_3$ | $CH_4$ | $CH_2CO$ | HCOOH | $HC_3N$ | $HNO_3$ | $ClONO_2$ | $C_2H_4$ | $C_2H_6$ | | | | |
| 3.30 | 5.58 | 4.69 | 2.10 | 3.55 | 4.48 | 3.32 | 4.00 | 6.21 | 2.07 | 1.15 | 4.80 | 4.36 | | | | |

*Note.* Dissociation energies are relative to v = 0 ($D_0$; Section 3.8.2).

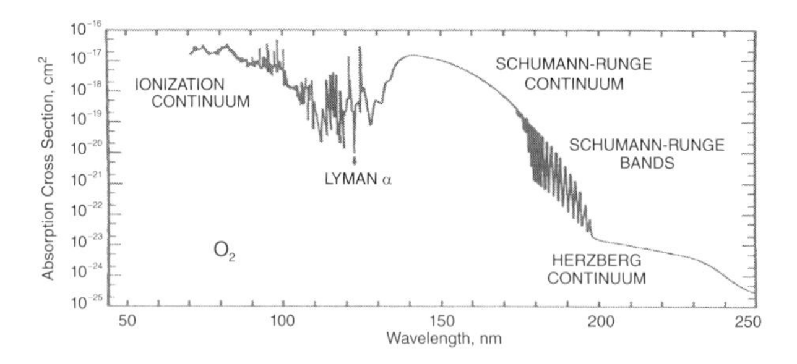


Figure 9.2 Absorption spectrum of $O_2$.

bands coincide with the Herzberg continuum for two $O(^3P)$ atoms, and the Auger transitions between them result in predissociation of $O_2$ that dominates at vibration levels $v' > 2$.

The $\mathbf{N_2}$ dissociation limits to $N(^4S) + N(^4S)$ and $N(^4S) + N(^2D)$ are 9.76 eV and 12.14 eV, that is, 127 nm and 102 nm, respectively. However, the dissociative continua of $N_2$ are very weak, while a few absorption band systems appear below 150 nm and become prominent at 80–100 nm. The bands at 80–100 nm predissociate to $N(^4S) + N(^2D)$. Shapes of the lines in these bands are poorly known, and accurate calculations of the $N_2$ absorption are problematic. Krasnopolsky and Cruikshank (1995) approximated the $N_2$ mean optical depth at 80–100 nm by

$$\tau = 1.1 \times 10^{-8}\{N_2\}^{0.434}. \tag{9.31}$$

Here $\{N_2\}$ is the $N_2$ column abundance. Then the $N_2$ global-mean predissociation rate is

$$p(z) = -\frac{I}{2}e^{-2\tau}\frac{d\tau}{dz}, \tag{9.32}$$

where $I$ is the solar photon flux at 80–100 nm and the factors 1/2 and 2 allow for the nightside and mean airmass, respectively. Photoionization of $N_2$ occurs below 80 nm (see Section 9.4.3).

Similar to $N_2$, the **CO** molecule has prominent absorption bands at 88.5–112 nm and the photoionization continuum below 88.5 nm. The absorption bands result in predissociation of CO. The process can be calculated using the data from Fox and Black (1989).

The $\mathbf{H_2}$ molecule is almost the simplest (the simplest is $H_2^+$), and the selection rules are very strict for $H_2$. Its bond energy is 4.48 eV ($\lambda$ = 277 nm); however, photodissociation to the ground-state atoms

$$H_2\left(X^1\Sigma_g^{\ +}\right) + h\nu \rightarrow H(1s) + H(1s)$$

is parity-forbidden: $H_2$ and H(1$s$) are even, while the dipole photon is odd. Therefore the parity-allowed photodissociation involves H(1$s$) + H(2$p$) with the dissociation limit at 84.5 nm. This limit is near the ionization limit at 80.4 nm, and photodissociation of $H_2$ is insignificant.

The Lyman band system of $H_2(B^1\Sigma_u^+ \rightarrow X^1\Sigma_g^+)$ lies between 111 and 85 nm. Fluorescence is proportional to $\lambda^4$ and the solar photon flux (Section 8.5.2), and both values are low for the $H_2$ Lyman bands. The best conditions are for the (6–0) P1 absorption line that coincides with the strong H Lyman-beta emission at 102.6 nm in the solar spectrum. This absorption originates three emission lines P1 (6–1, 2, 3) that were detected on Mars with a total intensity of 0.88 R (Krasnopolsky and Feldman 2001) using the Far Ultraviolet Spectroscopic Explorer (FUSE). The retrieved $H_2$ abundance extrapolated into the Martian lower atmosphere was 17 ppm.

### *9.4.2 Photodissociation of Polyatomic Molecules*

Photodissociation of polyatomic molecules may be similar for some species to that of diatomic molecules. The important case of $\mathbf{CO_2}$ was considered in Section 8.4. The photodissociation begins near 200 nm, various electronic states of CO + O are the only products, C + $O_2$ cannot form, and atomization C + O + O is possible below 75 nm but with a low yield. Population of the bending mode of $CO_2$ $\nu_2 = 667$ cm$^{-1}$ (Section 3.12) is proportional to $\exp(-\alpha\nu_2/T)$; $\alpha$ = 1.439 cm K. This population is not negligible and depends on temperature. Absorption from the $\nu_2 = 1$ level affects the $CO_2$ absorption cross sections and explains their significant temperature dependence.

Photodissociation of $\mathbf{CH_4}$ starts at 140 nm, while the $CH_3$–H bond corresponds to the dissociation limit at 277 nm. This is due to the very high symmetry of the molecule that is almost spherical. For example, photolysis of $\mathbf{C_2H_2}$ begins at 200 nm, though the $C_2H$–H bond is stronger than that of $CH_3$–H. This is because $C_2H_2$ is a linear molecule, which is very symmetric but not as symmetric as $CH_4$. The solar Lyman-alpha emission at 122 nm makes the bulk contribution in the dissociation of $CH_4$.

However, more complicated photodissociation processes with different products are possible as well. Figure 9.3 shows a case where the dissociating upper electronic state has a repulsive potential energy surface with an intermediate maximum. If the maximum is lacking, then the molecule decays during one vibrational period or less. Otherwise, the maximum serves as a potential barrier, resulting in indirect dissociation that can range up to thousands of the vibrational periods. There are many aspects of the problem, including dependences of the dissociation cross section on the photon energy and yields of various products and their electronic, vibrational, and rotational states. The system may be controlled by three tunable lasers: the first laser selects a vibrational level of the molecule ground state, the second laser dissociates the molecule, and the third laser measures the products using the laser-induced fluorescence of species.

### *9.4.3 Photoionization and Dissociative Photoionization*

These processes are generally similar to photodissociation. They occur in the EUV range of the solar spectrum, and the solar photon fluxes in this range and their variations with solar activity were discussed in Section 8.2 and Figure 8.8. Absorption and ionization cross

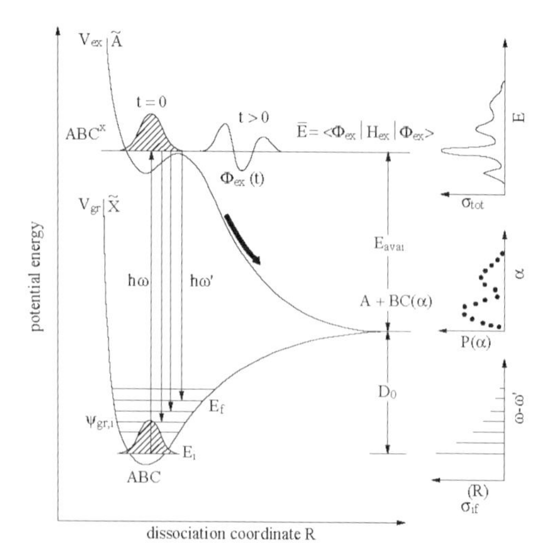


Figure 9.3 Schematic representation of the photodissociation of a molecule ABC into products A + BC($\alpha$). $\Psi gr;i$ is the initial wave function in the ground electronic state $X$ and $\Phi_{ex}(t)$ is the wave packet in the excited electronic state with average energy $E$. $\sigma_{tot}$ is the energy-dependent total dissociation cross section, $P(\alpha)$ is the probability for filling quantum state $\alpha$ of the products, and $\sigma_{if}^{(R)}$ is the Raman cross section for transition $i \rightarrow f$ in the lower state via resonant excitation of the upper electronic state. From Schinke (1995).

sections of various species and yields of dissociative ionization products are compiled in http://amop.space.swri.edu.

## 9.5 Termolecular Association

Emission of a photon is the only means to stabilize a product in the two-body association (Section 9.3.1). Collision of the excited complex AB* with a third body M is another way of the stabilization. This results in termolecular or three-body association:

$$\mathrm{A} + \mathrm{B} + \mathrm{M} \rightarrow \mathrm{AB} + \mathrm{M}; \; k. \tag{9.33}$$

The reaction sequence is

$$\begin{aligned} &\mathrm{A} + \mathrm{B} \rightarrow \mathrm{AB}^* && k_1, \\ &\mathrm{AB}^* \rightarrow \mathrm{A} + \mathrm{B} && k_2, \\ &\mathrm{AB}^* + \mathrm{M} \rightarrow \mathrm{AB} + \mathrm{M} && k_3. \end{aligned}$$

The reaction rate is

$$\frac{k_1[A][B]k_3[\mathrm{M}]}{k_2 + k_3[\mathrm{M}]},$$

where [X] is the X number density. A low-pressure limit is for $k_2 \gg k_3$ [M], and the reaction rate is

$$k_0[\mathrm{A}][\mathrm{B}][\mathrm{M}];\ k_0 = k_1 k_3/k_2. \tag{9.34}$$

A high-pressure limit is for $k_2 \ll k_3$[M], and the reaction rate is

$$k_\infty[\mathrm{A}][\mathrm{B}];\ k_\infty = k_1;\ [\mathrm{M}_\infty] = k_\infty/k_0 = k_2/k_3. \tag{9.35}$$

The reaction rate for arbitrary [M] is $k$ [A][B][M], and

$$k = \frac{k_0}{1 + [\mathrm{M}]/[\mathrm{M}_\infty]}. \tag{9.36}$$

However, experimental data for intermediate pressures are better fitted by the Troe approximation:

$$k = \frac{k_0}{1 + [\mathrm{M}]/[\mathrm{M}_\infty]} \times 0.6^{\left(1+X^2\right)^{-1}} \text{ and } X = lg\left(\frac{[\mathrm{M}]}{[\mathrm{M}_\infty]}\right). \tag{9.37}$$

The Troe approximation fits the requirements $k = k_0$ for low [M] and $k = k_\infty$/[M] for high [M].

It was discussed in Section 9.3.1 that the lifetime of AB*, $\tau = 1/k_2$, is comparatively long, up to thousands of vibrational periods, if AB* is polyatomic and has many vibrational modes. Therefore $k_0$ may be rather large for polyatomic molecules. Yung and DeMore (1999) give the following examples:

| | | |
|---|---|---|
| $O + O + M \to O_2 + M$ | 1 vibrational mode, $k_0 =$ | $4.8 \times 10^{-33}$ cm$^6$ s$^{-1}$at 300 K, |
| $H + CH_3 + M \to CH_4 + M$ | 8 | $2.4 \times 10^{-29}$, |
| $CH_3 + CH_3 + M \to C_2H_6 + M$ | 18 | $2.2 \times 10^{-26}$ |

(two last $k_0$ are updated). High-pressure limits for the last two reactions are near the gas-kinetic values, while $[\mathrm{M}_\infty] = 4 \times 10^{17}$ and $10^{14}$ cm$^{-3}$, respectively. Each rate coefficient exceeds the preceding value by a few orders of magnitude, and rate coefficients of termolecular association can be very large.

## 9.6 Heterogeneous Reactions

Gas molecules hit a surface of a planet or aerosol particle with a rate (in cm$^{-2}$ s$^{-1}$):

$$q = \frac{1}{4} nV; \qquad V = \sqrt{\frac{8kT}{\pi m}}. \tag{9.38}$$

The factor of 1/4 arises because one-half of the molecules are moving to the surface and integration of cos $\varphi$ around the hemisphere gives 1/2; $V$ is the mean thermal velocity. The molecules may be either reflected by the surface or adsorbed, the adsorption may be reversible or irreversible (reaction), and the process may have saturation. Number density of gas molecules dissolved in a liquid is equal at equilibrium to the gas partial pressure times a so-called Henry constant for the gas–liquid couple. All these processes along with detailed interactions at the atomic levels are the subject of heterogeneous chemistry.

Uptake, accommodation, or sticking coefficient $\alpha$ is the probability of the impacting gas molecule to be captured. Irreversible loss, or reaction probability, is denoted by $\gamma$. The proper rate is $\alpha$ or $\gamma$ times $q$ calculated from equation (9.38). Aerosol may be described by the surface area per unit volume, e.g., $A = 4\pi a^2 n_a$. Here $a$ is the mean particle radius and $n_a$ is the aerosol particle number density. If the particles are large, $2\pi a/\lambda \gg 1$, and weakly absorbing, then $A$ is twice the extinction coefficient $d\tau/dz$. Therefore a heterogeneous volume loss rate is

$$L = \frac{\alpha}{4} AnV = \alpha\pi a^2 n_a nV. \tag{9.39}$$

## 9.7 Literature on Reaction Rate Coefficients, Absorption Cross Sections, and Yields

Photochemical modeling involves a few dozen species and hundreds of reactions, and searching for their absorption cross sections, yields, and rate coefficients in original publications is very time consuming. Fortunately, there are some compilations where these data are collected.

NIST Chemical Kinetics Database (http://kinetics.nist.gov/) is the most complete source of data on reaction rate coefficients for neutral gases, including some metastable states.

Burkholder, J.B., et al., 2015. *Chemical Kinetics and Photochemical Data for Use in Atmospheric Studies*. Evaluation no. 18. JPL Publication 15–10 (http://jpldataeval.jpl.nasa.gov/). This compilation includes numerous reaction rate coefficients for neutral gases (with some metastable states), photodissociation cross sections and yields, and data on heterogeneous chemistry.

Atkinson, R., et al., 2004. Evaluated kinetic and photochemical data for atmospheric chemistry: Volume I – gas phase reactions of $O_X$, $HO_X$, $NO_X$ and $SO_X$ species. *Atmos. Chem. Phys.* 4, 1461–1738 (www.atmos.chem-phys.org/acp/4/1461/).

Atkinson, R., et al., 2006. Evaluated kinetic and photochemical data for atmospheric chemistry: Volume II – gas phase reactions of organic species. *Atmos. Chem. Phys.* 6, 3625–4055 (www.atmos.chem-phys.net/6/3625/2006/).

Atkinson, R., et al., 2007. Evaluated kinetic and photochemical data for atmospheric chemistry: Volume III – gas phase reactions of inorganic halogens. *Atmos. Chem. Phys.* 7, 981–1191 (www.atmos.chem-phys.net/7/981/2007/).

Atkinson, R., et al., 2008. Evaluated kinetic and photochemical data for atmospheric chemistry: Volume IV – gas phase reactions of organic halogen species. *Atmos. Chem. Phys.* 8, 4141–4496 (www.atmos.chem-phys.net/8/4141/2008/).

Crowley, J.N., et al., 2010. Evaluated kinetic and photochemical data for atmospheric chemistry: Volume V – heterogeneous reactions on solid substrates. *Atmos. Chem. Phys.* 10, 9059–9225 (www.atmos.chem-phys.net/10/9059/2010/)

www.atmosphere.mpg.de/enid/2295: this website contains photolysis cross sections and yields for numerous species.

www.lisa.univ-paris12.fr/GPCOS/SCOOPweb/: photolysis cross sections and yields for hydrocarbons and nitriles.

http://amop.space.swri.edu/ : photolysis and photoionization cross sections and yields.

UMIST Database for Astrochemistry 2012 (http://udfa.ajmarkwick.net): neutral and ion reaction rate coefficients.

McEwan, M.J., Anicich, V.G., 2007. Titan's ion chemistry: A laboratory perspective. *Mass Spectrom. Rev.* 26, 281–319. Detailed collection of ion-neutral reaction rate coefficients.

All these compilations greatly simplify search for chemical kinetic and photoabsorption data for photochemical modeling of planetary atmospheres.

# 10
# Photochemical Modeling

If parent species, expected photochemical products, their chemical reactions, rate coefficients, and transport processes in an atmosphere are known, then the chemical structure of the atmosphere can be calculated by means of photochemical modeling. Photochemical modeling is a powerful tool to study chemical composition of an atmosphere. In the case of Titan (Krasnopolsky 2014b), this makes it possible to calculate vertical profiles of 83 neutral and 33 ion species from the surface to the exosphere at 1600 km using very limited input data on the atmosphere: densities of the parent species $N_2$ and $CH_4$ near the surface and temperature and eddy diffusion profiles in the atmosphere.

The huge difference between the amounts of the input and output data on the atmosphere requires numerous auxiliary data: the solar UV and EUV spectra; fluxes of magnetospheric electrons, protons, and oxygen ions; meteorite influx of $H_2O$; and cosmic rays. Absorption of the solar UV and EUV photons is calculated interactively for the atmospheric gases using their absorption cross sections and product yields. Radiative transfer by aerosol is calculated using the aerosol observations from the Huygens probe. Vertical transport in the atmosphere is simulated by eddy, molecular, and ambipolar diffusion. The atmospheric chemistry involves 420 chemical reactions and their rate coefficients, despite the overall intent to reduce the numbers of species and their reactions to some reasonable minimum.

Below we will consider methods for calculating one-dimensional steady state and time-dependent photochemical models. Then a Martian global-mean model for the atmosphere below 80 km will be described as a relatively simple example.

There are self-consistent photochemical models that predict chemical structure of the whole atmosphere or its significant part, and partial models that are aimed at calculating of some species while the background atmosphere is assumed known. For example, models of ionospheric composition are typically based on the adopted neutral atmospheric composition being therefore the partial models.

Currently elements of photochemistry are included in some general circulation models. In the case of Mars, its photochemistry is included almost completely in the LMD GCM (Lefèvre and Forget 2009). That was a breakthrough in the modeling of variations of photochemistry in the Martian atmosphere. The model data are available online as the Mars Climate Database (www-mars.lmd.jussieu.fr/mcd_python/).

Methods for development of GCMs are beyond our scope. There are some restrictions to the GCMs. For example, currently the complicated photochemistries of Titan and Venus have not been included in full in their GCMs. For Mars, long-living photochemical products $H_2$, $O_2$, and CO require simulations for ~1000 years to get their equilibrium abundances, and this is beyond current computational capabilities. Therefore the Mars Climate Database simulates variations of these species, while their total abundances are taken from the observations. Furthermore, atmospheric mixing by eddy diffusion at the scales smaller than a grid size should be assigned in GCMs. Anyway, the one-dimensional photochemical modeling remains the basic tool to study the problems of chemical composition in the planetary atmospheres.

## 10.1 Continuity Equation and Its Finite Difference Analog

The mathematical base of photochemical modeling is the continuity equation

$$\frac{\partial n_i}{\partial t} + \text{div}\ \Phi_\text{i} = P_i - n_i L_i. \tag{10.1}$$

This means that a change in a species number density $n_i$ is caused by its production $P_i$, loss $n_iL_i$, and a net influx div$\Phi_{i;;}$ $\Phi_i$ is the species flux. The steady state one-dimensional version of this equation in the spherical coordinates is

$$\frac{1}{r^2}\frac{d(r^2\Phi_i)}{dr} = P_i - n_i L_i\ , \tag{10.2}$$

and according to (2.17), the flux $\Phi$ is

$$\Phi_i = -(K + D_i)\ \frac{\partial n_i}{\partial r} - n_i\left[K\left(\frac{mg}{kT} + \frac{1}{T}\frac{\partial T}{\partial r}\right) + D_i\left(\frac{m_i g}{kT} + \frac{1-\alpha_i}{T}\frac{\partial T}{\partial r}\right)\right]. \tag{10.3}$$

The combination of (10.2) and (10.3) is an ordinary second-order nonlinear differential equation. The problem of photochemical modeling results in a set of these equations, and their number is equal to the number of species in the problem.

The set of the continuity equations is solved numerically using their finite difference analogs for a variable vertical step

$$s_i = ab^i,\ \text{and}\ r_i = r_0 + a\frac{b^i - 1}{b - 1} \tag{10.4}$$

Here the subscript $i$ refers to altitude layer, while it designated species in (10.1–10.3). The step $s_i$ is usually constant ($b$ = 1) in models for the spatially thin atmospheres of Mars and Venus with exobase altitudes that are much smaller than the planet radii. It may be variable for the extended atmospheres of Titan, Triton, and Pluto. Then

$$\frac{dn_i}{dr} = \frac{n_{i+1} - n_i}{s_i}\ \text{and}\ \frac{d^2 n_i}{dr^2} = \frac{n_{i+1} - 2n_i + n_{i-1}}{s_i^2}\ . \tag{10.5}$$

Using the designations

$$p_i = P_i\left(\frac{r_i}{r_0}\right)^2,\ l_i = L_i\left(\frac{r_i}{r_0}\right)^2,\ k_i = K_i\left(\frac{r_i}{s_i r_0}\right)^2,\ d_i = D_i\left(\frac{r_i}{s_i r_0}\right)^2, \tag{10.6}$$

$$k1_i = k_i\left[2 - \frac{m_a g_0 s_i}{kT_i(1 + r_i/r_0)^2} - \frac{T_{i+1}}{T_i}\right],$$

$$d1_i = d_i\left[1 - \frac{m g_0 s_i}{kT_i(1 + r_i/r_0)^2} - (1 - \alpha)\left(\frac{T_{i+1}}{T_i} - 1\right)\right], \tag{10.7}$$

the continuity equation (10.2, 10.3) can be transformed to

$$p_i = -n_{i+1}(k_i + d_i)b + n_i[(k1_i + d1_i)b + k_{i-1} + d_{i-1} + l_i] - n_{i-1}(k1_{i-1} + d1_{i-1}). \tag{10.8}$$

Here the subscript 0 refers to the lowest level in the model, and $m_a$ is the mean molecular mass. Gravity $g_0/(1 + r_i/r_0)^2$ is corrected for the radial dependence.

The one-dimensional steady state photochemical problem results in equations (10.8) that are solved by iterations. First of all, it is necessary to establish species in the problem and their reactions and to find their rate coefficients, absorption cross sections, and quantum yields of products. Solar UV and EUV photon fluxes, other sources of photochemistry, diffusion coefficients, and vertical profiles of temperature and eddy diffusion are the input data of the model as well. Then vertical profiles of all species in the model are adopted arbitrarily but reasonably using the known data on the atmosphere. These profiles are the zeroth approximation. The first approximation for a chosen species can be calculated using (10.8). All values in (10.8) are known, except $n_{i+1}$, $n_i$, $n_{i-1}$. Two of them originate from boundary conditions, and the third is found from the equation.

## 10.2 Solution of the Problem and Boundary Conditions

To solve equation (10.8), auxiliary parameters $\alpha_i$ and $\beta_i$ are calculated from $i = u - 1$ to 0:

$$\alpha_i = \frac{(k1_i + d1_i)(b\alpha_{i+1} + l_{i+1})}{b\alpha_{i+1} + l_{i+1} + k_i + d_i},\ \beta_i = \frac{(k_i + d_i)\left(b\beta_{i+1} + p_{i+1}\right)}{b\alpha_{i+1} + l_{i+1} + k_i + d_i}. \tag{10.9}$$

Here $i = u$ is the upper boundary in the problem. All $\alpha_i$ and $\beta_i$ can be calculated if $\alpha_{u-1}$, $\beta_{u-1}$ are known. To find them the upper boundary conditions are used.

There are three basic types of the boundary conditions for photochemical modeling: densities $n$, fluxes $\Phi$, and bulk velocities $V = \Phi/n$. To use the modeling capabilities in full, data in the boundary conditions should be restricted to some minimum. For example, either an observed value of CO or $\Phi_{CO} = 0$ may be used as a lower boundary condition for the Martian atmosphere. Then in the first case the CO abundance is mostly assigned, not predicted by the model. The second case means that CO does not react with the surface rocks, the assumption is reasonable, the calculated CO in the model is of interest and may be compared with the observations.

To establish requirements to the boundary conditions, Krasnopolsky (1995) applied the element conservation in chemical reactions and deduced some rules for different cases. Generally, the number of boundary conditions given by densities or nonzero velocities should be equal at least to the number of elements in the system, and the boundary conditions should account for balance of the elements.

In the case of the $CO_2$–$H_2O$ chemistry on Mars, this means that at least two densities ($CO_2$ and $H_2O$) plus one nonzero velocity should be specified. For example, the model with the $CO_2$ and $H_2O$ densities near the surface and fluxes for all other conditions does not have a unique solution, while the same model with one flux replaced by a nonzero velocity has. (The zero velocity means the zero flux and does not change the situation.) The nonzero velocity may be the escape velocity of hydrogen on Mars. However, the loss of hydrogen should be accompanied by a loss of oxygen, otherwise oxygen is accumulated and the steady state is broken. Therefore the loss of oxygen should be specified as either a nonthermal loss at the upper boundary ($\Phi_O = \Phi_{H2} + \Phi_H/2$) or heterogeneous loss at the surface.

Conditions at the upper boundary specify $\alpha_{u-1}$ and $\beta_{u-1}$:

$$\text{If } n_u \text{ is given, then } \alpha_{u-1} = k1_{u-1} + d1_{u-1};\ \beta_{u-1} = n_u(k1_{u-1} + d1_{u-1})\,. \tag{10.10}$$

$$\text{If } \Phi_{u-1} \text{ is given, then } \alpha_{u-1} = 0;\ \beta_{u-1} = -\frac{\Phi_{u-1}}{s_{u-1}}\left(\frac{r_{u-1}}{r_0}\right)^2. \tag{10.11}$$

$$\text{If } V_{u-1} \text{ is given, then } \alpha_{u-1} = \frac{V_{u-1}}{s_{u-1}}\left(\frac{r_{u-1}}{r_0}\right)^2;\ \beta_{u-1} = 0. \tag{10.12}$$

Now all $\alpha_i$ and $\beta_i$ can be calculated down to $i = 0$ using (10.9), and the density at the lower boundary $n_0$ is either specified or equal to the following:

$$\text{If } \Phi_0 \text{ is given, then } n_0 = \frac{\beta_0 + \Phi_0/a}{\alpha_0}\,. \tag{10.13}$$

$$\text{If } V_0 \text{ is given, then } n_0 = \frac{\beta_0}{\alpha_0 - V_0/a}\,. \tag{10.14}$$

All other $n_i$ up to $i = u$ can be calculated using

$$n_{i+1} = \frac{(k1_i + d1_i - \alpha_i)n_i + \beta_i}{k_i + d_i}\,. \tag{10.15}$$

As usual, upward fluxes and velocities are positive. The best convergence of the system is for the next iteration equal to the mean of the calculated iteration and the previous iteration.

All relationships above are given without their derivations, which are actually not complicated. The initial density profiles (the zeroth iteration) are arbitrary and may result in negative values in the successive iterations. To avoid this situation, the condition “if $n_i$ less than $10^{-40}$, then $n_i = 10^{-40}$” may be applied. It may be adopted that the system is balanced if the most long-living photochemical product does not vary within, say, 1% per thousand iterations. It is difficult to predict the required number of iterations. For example, the very complicated photochemistry of Titan converges much faster than the comparatively simple photochemistry of Mars, which requires near a million iterations.

## 10.3 Example: Modeling of Global-Mean Photochemistry in the Martian Lower and Middle Atmospheres

Photochemistry of the Martian atmosphere is comparatively simple, and here we will consider it as an example of photochemical modeling. The model discussed below was calculated by Krasnopolsky (2010b).

The atmosphere consists of $CO_2$ and its products CO, O, $O_2$, and $O_3$; $H_2O$ and its products H, OH, $HO_2$, $H_2O_2$, and $H_2$; $N_2$ and its products, noble gases, and $CH_4$. The bulk photochemistry occurs in the lower and middle atmospheres, and our model will cover the altitude range from the surface to 80 km. The model is a global mean, that is, it refers to the mean heliocentric distance of 1.517 AU, the dayside mean cosine of solar zenith angle of 0.5, and solar flux scaled by a factor of 0.5 to allow for the nightside. The global mean solar conditions are similar to those at $L_S \approx 165°$ (end of northern summer) and middle latitudes near 45°N. A temperature profile for the model was obtained by averaging of the MGS/TES data (Smith 2004) at this season and latitude. Using this profile and the global mean atmospheric pressure of 6.1 mbar, the $CO_2$ density profile was calculated (Figure 10.1). The eddy diffusion profile was adopted at $K = 10^6$ cm$^2$ s$^{-1}$ for $z < 20$ km

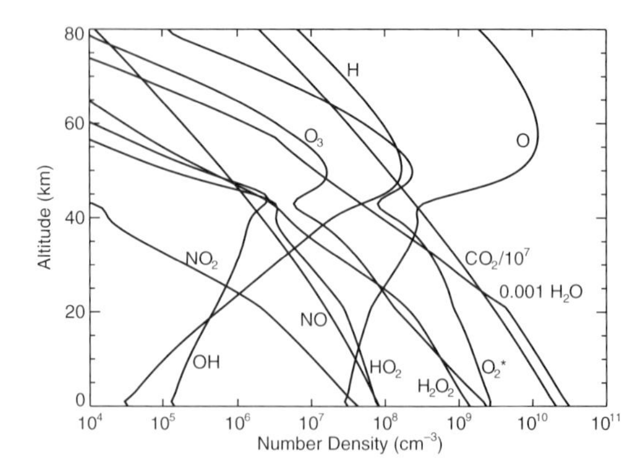


Figure 10.1 A global-mean photochemical model for Mars at 0–80 km. The calculated long-living species $O_2$ (900 ppm), CO (80 ppm at 20 km), and $H_2$ (15 ppm) are not shown. The better version of the model gives 1600, 120, and 20 ppm, respectively. From Krasnopolsky (2010b).

and $K = 10^6(n(20\ \text{km})/n(z))^{1/2}$. Then $K \approx 4\times10^7$ cm$^2$ s$^{-1}$ at 80 km, while the diffusion coefficient of, say, O in $CO_2$ (Table 2.1) is $1.7\times10^5$ cm$^2$ s$^{-1}$ at this altitude. Evidently, molecular diffusion can be neglected below 80 km.

Photoionization of $CO_2$ begins at 90 nm, the optical depth for the ionizing photons exceeds 100 at 80 km, and ionization processes with related ionospheric chemistry may be neglected below 80 km. Both neglects simplify the problem significantly.

The mean methane abundances are of a few ppb, and their effects on the bulk atmospheric chemistry are also negligible. The noble gases do not react with other species. $N_2$ does not dissociate below 80 km; however, N and NO are formed in almost equal quantities in the upper atmosphere. They are lost in the reaction N + NO $\rightarrow$ $N_2$ + O, and nitrogen chemistry below 80 km crucially depends on which of these species dominates in the downward flow from the upper atmosphere. If abundances of N exceed those of NO, then nitrogen chemistry is negligible in our model; otherwise, models (e.g., Krasnopolsky 1995) predict the NO mixing ratio of 0.6 ppb in the lower atmosphere. Then 3 of 33 reactions of nitrogen chemistry determine a balance between NO and $NO_2$ and affect oxygen species. We use these three reactions in our model (r33–34 in Table 10.1).

The total number of species is 15, and those include $CO_2$, $H_2O$, and their products. NO, $NO_2$, and two metastable species $O(^1D)$ and $O_2(^1\Delta_g)$. A choice of reactions in the model is generally a heuristic problem. The chosen reactions should cover all significant processes of production and loss for each species in the model. In our case reactions of the $CO_2$–$H_2O$ chemistry on Mars are well known and repeated with some modifications in many models since McElroy and Donahue (1972). They are given in Table 10.1.

Seven species dissociate in this model: $CO_2$, $O_2$, $H_2O$, $H_2O_2$, $HO_2$, $O_3$, and $NO_2$. The $CO_2$ absorption cross sections (Figure 8.3) may be taken from Parkinson et al. (2003) at 195 K and corrected for the Rayleigh scattering cross sections ($8\times10^{-25}(200/\lambda)^4$ cm$^2$, $\lambda$ is in nm) near 200 nm. The $O_2$ photolysis is complicated in the Schumann–Runge bands at 175–200 nm, and the photolysis frequencies may be derived from Nair et al. (1994) using their figure 8 and the calculated $O_2$ densities. The $H_2O$ cross sections were measured by Chung et al. (2001), and those for $H_2O_2$, $HO_2$, $O_3$, and $NO_2$ are compiled in Burkholder et al. (2015). (Actually the model was developed in 2009, when a previous version of that compilation was available.)

The solar photon fluxes are given in Woods et al. (1996; Figure 8.7) and can be used to calculate the $CO_2$ photolysis rate using relationship (2.31). Absorptions by the other species are much smaller than that of $CO_2$ in the $CO_2$ photolysis range $\lambda < 200$ nm and may be neglected. We do not allow for the absorption by dust and heterogeneous chemistry on the ice aerosol in this model. The $H_2O$ photolysis frequencies $\Sigma_\lambda I_\lambda(z)\sigma_{H2O}(\lambda)$ are calculated as well, and the $H_2O$ photolysis rate at altitude $z$ is a product of the photolysis frequency and $H_2O$ number density at this altitude.

Photolyses of $H_2O_2$, $HO_2$, $O_3$, and $NO_2$ occur mostly in the near-UV range $\lambda > 200$ nm where the atmosphere is transparent. Therefore their photolysis frequencies do not depend on altitude, and the calculated values are given in Table 10.1.

Table 10.1 *Photochemical reactions in the lower and middle atmosphere of Mars, their rate coefficients, and the calculated column rates*

| # | Reaction | Rate coefficient | Column rate |
|---|---|---|---|
| 1 | $CO_2 + h\nu \rightarrow CO + O$ | – | 7.44 + 11 |
| 2 | $CO_2 + h\nu \rightarrow CO + O(^1D)$ | – | 1.18 + 10 |
| 3 | $O_2 + h\nu \rightarrow O + O$ | – | 4.74 + 10 |
| 4 | $O_2 + h\nu \rightarrow O + O(^1D)$ | – | 3.14 + 9 |
| 5 | $H_2O + h\nu \rightarrow H + OH$ | – | 8.79 + 9 |
| 6 | $HO_2 + h\nu \rightarrow OH + O$ | $2.6 \times 10^{-4}$ | 1.41 + 10 |
| 7 | $H_2O_2 + h\nu \rightarrow OH + OH$ | $4.2 \times 10^{-5}$ | 2.45 + 10 |
| 8 | $O_3 + h\nu \rightarrow O_2(^1\Delta) + O(^1D)$ | $3.4 \times 10^{-3}$ | 2.36 + 12 |
| 9 | $O(^1D) + CO_2 \rightarrow O + CO_2$ | $7.4 \times 10^{-11}e^{120/T}$ | 2.38 + 12 |
| 10 | $O(^1D) + H_2O \rightarrow OH + OH$ | $2.2 \times 10^{-10}$ | 5.58 + 8 |
| 11 | $O(^1D) + H_2 \rightarrow OH + H$ | $1.1 \times 10^{-10}$ | 2.96 + 7 |
| 12 | $O_2(^1\Delta) + CO_2 \rightarrow O_2 + CO_2$ | $10^{-20}$ | 1.84 + 12 |
| 13 | $O_2(^1\Delta) \rightarrow O_2 + h\nu$ | $2.24 \times 10^{-4}$ | 5.22 + 11 |
| 14 | $O + CO + CO_2 \rightarrow CO_2 + CO_2$ | $2.2 \times 10^{-33}e^{-1780/T}$ | 2.21 + 7 |
| 15 | $O + O + CO_2 \rightarrow O_2 + CO_2$ | $1.2 \times 10^{-32}(300/T)^2$ | 4.38 + 9 |
| 16 | $O + O_2 + CO_2 \rightarrow O_3 + CO_2$ | $1.4 \times 10^{-33}(300/T)^{2.4}$ | 2.38 + 12 |
| 17 | $H + O_2 + CO_2 \rightarrow HO_2 + CO_2$ | $1.7 \times 10^{-31}(300/T)^{1.6}$ | 8.69 + 11 |
| 18 | $O + HO_2 \rightarrow OH + O_2$ | $3 \times 10^{-11}e^{200/T}$ | 7.26 + 11 |
| 19 | $O + OH \rightarrow O_2 + H$ | $2.2 \times 10^{-11}\ e^{120/T}$ | 1.41 + 11 |
| 20 | $CO + OH \rightarrow CO_2 + H$ | $1.5 \times 10^{-13}$ | 7.56 + 11 |
| 21 | $H + O_3 \rightarrow OH + O_2$ | $1.4 \times 10^{-10}e^{-470/T}$ | 2.33 + 10 |
| 22 | $H + HO_2 \rightarrow OH + OH$ | $7.3 \times 10^{-11}$ | 1.36 + 10 |
| 23 | $H + HO_2 \rightarrow H_2 + O_2$ | $1.3 \times 10^{-11}(T/300)^{0.5}e^{-230/T}$ | 4.54 + 8 |
| 24 | $H + HO_2 \rightarrow H_2O + O$ | $1.6 \times 10^{-12}$ | 2.97 + 8 |
| 25 | $OH + HO_2 \rightarrow H_2O + O_2$ | $4.8 \times 10^{-11}e^{250/T}$ | 8.16 + 9 |
| 26 | $HO_2 + HO_2 \rightarrow H_2O_2 + O_2$ | $2.3 \times 10^{-13}e^{600/T}$ | 2.50 + 10 |
| 27 | $OH + H_2O_2 \rightarrow HO_2 + H_2O$ | $2.9 \times 10^{-12}e^{-160/T}$ | 4.40 + 8 |
| 28 | $OH + H_2 \rightarrow H_2O + H$ | $3.3 \times 10^{-13}(T/300)^{2.7}e^{-1150/T}$ | 2.54 + 8 |
| 29 | $O + O_3 \rightarrow O_2 + O_2$ | $8 \times 10^{-12}e^{-2060/T}$ | 2.22 + 7 |
| 30 | $OH + O_3 \rightarrow HO_2 + O_2$ | $1.5 \times 10^{-12}e^{-880/T}$ | 6.85 + 6 |
| 31 | $HO_2 + O_3 \rightarrow OH + O_2 + O_2$ | $10^{-14}\ e^{-490/T}$ | 8.57 + 7 |
| 32 | $H_2O_2 + O \rightarrow OH + HO_2$ | $1.4 \times 10^{-12}\ e^{-2000/T}$ | 3.87 + 6 |
| 33 | $NO_2 + h\nu \rightarrow NO + O$ | 0.0037 | 4.25 + 10 |
| 34 | $NO_2 + O \rightarrow NO + O_2$ | $5.6 \times 10^{-12}\ e^{180/T}$ | 1.40 + 10 |
| 35 | $NO + HO_2 \rightarrow NO_2 + OH$ | $3.5 \times 10^{-12}\ e^{250/T}$ | 5.65 + 10 |
| 36 | diffusion-limited flux of $H_2$ | $V = 1.14\times10^{13}/n_{79}$ | 1.70 + 8 |
| 37 | loss of $H_2O_2$ at the surface | $V = 0.02$ | 2.81 + 7 |
| 38 | loss of $O_3$ at the surface | $V = 0.02$ | 4.71 + 7 |

*Note.* Photolysis rates and $k_{13}$ are in $s^{-1}$, velocities $V$ are in cm $s^{-1}$, and second- and third-order reaction rate coefficients are in $cm^3\ s^{-1}$ and $cm^6\ s^{-1}$, respectively. Photolysis rates for $HO_2$, $H_2O_2$, $O_3$, and $NO_2$ refer to the lower atmosphere and are calculated for $\lambda > 200$ nm at 1.517 AU. Column reaction rates are in $cm^{-2}\ s^{-1}$, summed up from 1 to 79 km and multiplied by $(1 + h/R)^2$. $7.44 + 11 = 7.44 \times 10^{11}$.

The reaction rate coefficients in Table 10.1 were mostly taken from Burkholder et al. (2015). The exceptions are reactions 12, 23, 28, and four termolecular associations 14–17. The latter are given in Burkholder et al. (2015) for the air molecules (mostly $N_2$) as a third body. Rate coefficients for these reactions were chosen from the NIST Chemical Kinetics Database. Sometimes an assumption is used that $CO_2$ is more effective as a third body than $N_2$ by a factor of 2.5.

It is believed that water vapor is uniformly mixed in the Martian atmosphere up to a saturation level (hygropause) and then follows the saturation densities that may be found in (2.30) and Fray and Schmitt (2009). Actually the nighttime condensation of water vapor at the surface and in the boundary layer breaks this altitude distribution and extends even in the daytime (Krasnopolsky and Parshev 1977). Furthermore, significant supersaturation of water vapor is observed by the Mars Express solar occultations, especially near the perihelion in southern summer (Maltagliati et al. 2013). However, these deviations are usually neglected in the photochemical models for Mars, and we apply the standard assumption and vary the $H_2O$ density near the surface to fit a column $H_2O$ abundance of 9.5 precipitable μm ($3.2 \times 10^{19}$ cm$^{-2}$) adopted in the model. The calculated profile of $H_2O$ is shown in Figure 10.1.

The model is calculated for a constant altitude step of 1 km. The code for this model starts with a procedure that calculates $\alpha_i$ and $\beta_i$ using given $\alpha_{79}$, $\beta_{79}$, $p_i$, $l_i$, and relationships (10.9). Then a density profile is calculated using relationships (10.13–10.15). The mean profile of the calculated and previous iteration is the procedure output.

At first the $CO_2$ and $H_2O$ profiles and then their photolysis rates are calculated as described above and profiles of $k_i$ and $kl_i$ using (10.6). Then profiles of all species except $CO_2$ and $H_2O$ are arbitrarily chosen as the zeroth iteration (Section 10.2). The next step is calculation of vertical profiles of all reaction rates using the given density profiles and rate coefficients.

Our boundary conditions are the $CO_2$, $H_2O$, and NO densities near the surface that we have used above and an upward bulk velocity of $H_2$ that corresponds to its diffusion limit (Zahnle et al. 2008). Using (2.52), this velocity is

$$V_{H_2}^{max} = 0.96 \times 1.14 \times 10^{13} / [CO_2]_{80} \text{ cm s}^{-1}. \quad (10.16)$$

Here the $CO_2$ fraction of 0.96 in the total number density is involved. To compensate for the loss of hydrogen, Zahnle et al. (2008) suggested losses of oxygen via heterogeneous oxidation of the rocks by $O_3$ and $H_2O_2$ with velocity of 0.02 cm s$^{-1}$ (lines 37–38 in Table 10.1).

All other boundaries are closed for all species, that is, their fluxes are zero. Generally it would be better to specify downward fluxes of CO and O from photolyses of $CO_2$ and $O_2$ above 80 km and an upward velocity of $O_2$ to account for its photolysis above 80 km. This was done in another version of the model, while here we follow the conditions in Zahnle et al. (2008).

The next phase in our code is to specify $\alpha_{79}$, $\beta_{79}$, $p_i$, and $l_i$ for each component using the calculated profiles of the reaction rates, the low boundary condition by (10.13–10.14), and apply the procedure to calculate a new iteration for the component. When the new iteration is calculated for all species, the cycle is repeated. It is necessary to control the solution by checking some key values, e.g., mixing ratios of $O_2$, CO, and $H_2$ at 20 km, at, say, each thousandth cycle, and to stop the code when the convergence is satisfactory, say, better than 1%.

The final phase is the calculations of column reaction rates

$$CR = \sum_1^{79} R_i \left(\frac{r_i}{r_0}\right)^2. \tag{10.17}$$

Here $R_i$ is the reaction rate at the level $i$, and the sum is from 1 to 79 km because the model does not include reactions at the boundaries. Column reaction rates of production and loss for each species should be equal within the model accuracy. For example, the production of $NO_2$ in r35 is balanced by its loss in r33 and r34, the diffusion-limited escape of $H_2$ (r36) is equal to the loss at the surface of $H_2O_2$ (r37) plus that of $O_3$ (r38) times a factor of 3. $H_2O_2$ can be considered as $H_2O + O$, $H_2O$ is fixed at the surface, and r37 does not require scaling. More complicated balances for other species may be checked using the column reaction rates from Table 10.1.

## 10.4 Time-Dependent Models

One-dimensional time-dependent models are aimed to simulate variations of atmospheric photochemistry with local time and sometimes with season. Species do not vary with local time if their chemical lifetimes (Section 9.3.6) are significantly longer than the day duration. Therefore, in the case of the Martian lower and middle atmosphere, vertical profiles of the long-living photochemical products $O_2$, CO, and $H_2$ are fixed in the time-dependent models, while diurnal variations of the observable photochemical tracers $O_3$, $H_2O_2$, and $O_2(^1\Delta_g)$ are the main objective of these models (Figure 10.2).

A code for these models is similar to that for the steady state models with the following changes in the production and loss terms in (10.6):

$$p_i = \left(P_i + \frac{n_i^k}{2t}\right)\left(\frac{r_i}{r_0}\right)^2, \quad l_i = \left(L_i + \frac{1}{2t}\right)\left(\frac{r_i}{r_0}\right)^2. \tag{10.18}$$

Here $t$ is the time step and $n_i^k$ is the number density at $r_i$ and time $kt$. These values are used to calculate the next iteration $n_i$ and then

$$n_i^{k+1} = \frac{n_i^k + n_i}{2}. \tag{10.19}$$

However, the photolysis rates and some other parameters vary with local time, and this makes the problem more complicated. Furthermore, the time step should be significantly shorter than the shortest photolysis lifetime, and it was equal to 30 s for the model in Figure 10.2, that is, ~3000 time intervals per the Martian day. The problem is solved when the calculated diurnal behaviors of all species in the model do not change from day to day.

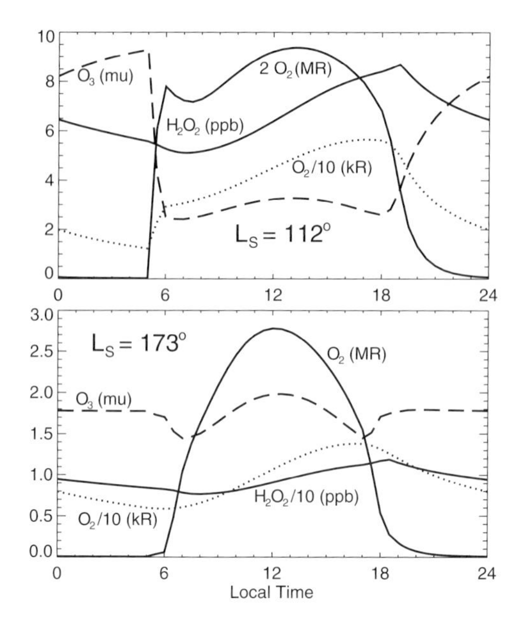


Figure 10.2 Diurnal variations of photochemical tracers in the Martian atmosphere: column ozone in μm-atm, column peroxide in ppb, the $O_2(^1\Delta_g)$ airglow at 1.27 μm in MR, and its component from the termolecular association of oxygen in kR. The models are near the fall equinox ($L_S$ = 173°, equator) and near the peak of northern summer ($L_S$ = 112°, 20°N). From Krasnopolsky (2006b).

## 10.5 Isotope Fractionation

### *10.5.1 Isotope Ratios*

While properties of isotopologues of a species are rather similar, some minor differences can be accumulated for a long time and result in measurable variations of isotope ratios. These variations may be keys to understanding evolution of an atmosphere or some of its aspects. Isotope ratios can be measured by spectroscopy (Sections 3.9 and 7.6) and mass spectrometry (Section 7.10.1). Measured isotope ratios are scaled to the terrestrial standards (Table 10.2).

The isotope ratio standards are the Standard Mean Ocean Water (SMOW) for H and O, Pee Dee Belemnite (PDB) for C, and the air for N and noble gases. Isotope ratios for heavy elements in the planetary atmospheres are typically very similar to the terrestrial values and given as deviations from the terrestrial standard. For example, $\delta^{13}C = -18‰$ means that $^{13}C/^{12}C$ is equal to (1 – 0.018) times the standard ratio.

### *10.5.2 Isotopes of Noble Gases*

Although the cosmic abundance of helium is very large (~8%), its fraction in the Earth's atmosphere is just 5 ppm, and it is produced by radioactive decay of $^{238}U$, $^{235}U$, and $^{232}Th$

Table 10.2 *Standard isotope ratios*

| Isotopes | D/H | $^{18}O/^{16}O$ | $^{17}O/^{16}O$ | $^{13}C/^{12}C$ | $^{15}N/^{14}N$ | $^{3}He/^{4}He$ | $^{22}Ne/^{20}Ne$ | $^{21}Ne/^{20}Ne$ | $^{38}Ar/^{36}Ar$ | $^{40}Ar/^{36}Ar$ |
|---|---|---|---|---|---|---|---|---|---|---|
| Ratio | 1.56–4 | 2.005–3 | 3.73–4 | 1.124–2 | 3.65–3 | 1 34–6 | 0.102 | 0.003 | 0.19 | 299 |
| Inverse ratio | 6420 | 499 | 2680 | 89 | 274 | 7.46+5 | 9.78 | 335 | 5.35 | 0.00335 |
| Standard | SMOW | SMOW | SMOW | PDB | air | air | air | air | air | air |

with 8, 7, and 6 alpha particles per the decay chains, respectively. $^3$He is mostly formed by the nuclear fusion in the stars and the Sun, and its abundance in the solar wind and interplanetary medium is higher than on the Earth.

$^{40}$Ar is another radiogenic isotope that is produced by the decay of $^{40}$K. Both $^4$He and $^{40}$Ar are formed in the crust and released into the atmosphere mostly by volcanism. Losses of $^{40}$Ar from the atmospheres of the inner planets are low, and its abundance may be used for calibration of the outgassing of helium.

The other isotopes of argon and the isotopes of neon and krypton reflect their abundances in the solar nebula and in the solar wind that may be partially captured by a planet. Krypton has five isotopes, and $^{84}$Kr is the most abundant. There are nine isotopes of xenon; $^{129}$Xe is radiogenic and originates from $^{129}$I, while $^{132}$Xe is the most abundant nonradiogenic isotope.

### *10.5.3 Fractionation in Chemical Reactions*

Isotope fractionation in a chemical reaction depends on small differences in bond energies of the reactant and product that may change the energy barrier. Other factors are the collision rate and the collision time, and the former is proportional to relative velocity $V = \sqrt{\frac{8kT}{\pi\mu}}$, while the latter is proportional to $1/V$. If a reaction is without energy barrier and a rate coefficient for an isotopologue is unknown, then it is typically adopted similar to that for the main isotopologue or that scaled by the changed reduced mass $\mu^{-1/2}$.

Photo-induced isotope fractionation may be significant. Miller and Yung (2000) argued that photolysis cross sections of isotopologues are similar with a small shift in the energy scale that is equal to difference in zero-point energies. The zero-point energy is equal to the vibrational term

$$G(v) = \omega_e\left(v + \frac{1}{2}\right) - \omega_e x_e\left(v + \frac{1}{2}\right)^2 + \omega_e y_e\left(v + \frac{1}{2}\right)^3 \tag{10.20}$$

for $v = 0$. It is a sum of the vibrational terms at $v = 0$ of all vibrational modes for polyatomic molecules. The most significant difference is for hydrides because the mass ratio D/H of 2 is the greatest. For example, the observed absorption cross sections of $H_2O$ and HDO (Chung et al. 2001) result in a difference in their photolysis frequencies by a factor of 2.5 on Mars. The similar difference in photolysis frequencies of HCl and DCl on Venus is a factor of 6 (Bahou et al. 2001).

Isotope fractionation may be induced by condensation processes. Deuterium fractionation in condensation of water vapor was studied by Merlivat and Nief (1967) at 230–273 K, and the data were approximated by

$$\left(\frac{D}{H}\right)_{\text{ice}} / \left(\frac{D}{H}\right)_{\text{vapor}} = \exp\left(\frac{16288}{T^2} - 0.0934\right). \tag{10.21}$$

Being extrapolated to the Martian temperatures, this ratio is $\approx$1.5 at $T \approx$ 180 K typical of the Martian water ice haze. Unfortunately, there is a great scatter in the laboratory data on the deuterium fractionation between water ice and vapor at low temperatures typical of the Martian atmosphere. Even temperature trends in decreasing or increasing of the fractionation are opposite in the data (Lecuyer et al. 2017). Evidently this badly affects the model predictions.

### *10.5.4 Fractionation by Atmospheric Escape*

Atmospheric escape results in irreversible isotope fractionation. Its effect during the age of the Solar System may be very significant. For example, the D/H ratio in water vapor on Venus exceeds that on the Earth by two orders of magnitude. There are two basic reasons for the large isotope fractionation in the escape. First of all, it is diffusive separation in the upper atmospheres that builds enrichment of lighter isotopes. Second, escape is very preferential for light isotopes. This can be easily evaluated for thermal escape using relationship (2.43).

Isotope fractionation of light species by thermal and nonthermal escape in the Martian atmosphere was calculated by Krasnopolsky (2010b), and the results are given in Table 10.3. The nonthermal escape involves charge exchange with the solar wind protons

Table 10.3 *Flow velocities of light species for various escape processes at 300 km on Mars*

| Process | H | | $H_2$ | | D | | HD | | He | |
|---|---|---|---|---|---|---|---|---|---|---|
| $p_{SW}+\alpha_{SW}$ | 23.9 | 36.4 | 0.12 | 0.17 | 10.6 | 15.6 | 0.08 | 0.11 | 0.02 | 0.03 |
| $e_{SW}$ | 1.91 | 2.91 | 0.85 | 1.25 | 0.79 | 1.16 | 0.54 | 0.78 | 0.09 | 0.12 |
| $h\nu$ | 1.39 | 4.62 | 0.66 | 2.38 | 0.62 | 1.98 | 0.42 | 1.50 | 0.25 | 1.27 |
| $P_{ex}$ | 19.3 | 77.0 | – | – | 3.90 | 4.10 | – | – | – | – |
| $O_{hot}$ | 3.82 | 12.0 | 4.14[a] | 12.2[a] | 3.16 | 9.30 | 3.80[a] | 10.3[a] | 2.74 | 6.79 |
| Total nonthermal | 50.3 | 133 | 5.77 | 16.0 | 19.1 | 32.1 | 4.84 | 12.7 | 3.10 | 8.21 |
| Thermal | 387 | 2513 | 0.48 | 22.5 | 0.48 | 22.5 | 5e-4 | 0.18 | 6e-7 | 1.5e-3 |
| Total | 437 | 2646 | 6.25 | 38.5 | 19.6 | 54.6 | 4.84 | 12.9 | 3.10 | 8.21 |

*Note.* $p_{SW}+\alpha_{SW}$ is charge exchange with protons and α particles of the solar wind, $e_{SW}$ is ionization by electrons of the solar wind, $h\nu$ is photoionization above the ionopause calculated for $F_{10.7}$ = 27 and 90, $P_{ex}$ is production above the exobase, and $O_{hot}$ is collisions with hot oxygen. Values for each species and process are given at 300 km for two models with exospheric temperatures of 200 and 300 K. Velocities are in cm $s^{-1}$, and 5e-4 = $5 \times 10^{-4}$. From Krasnopolsky (2010b).

[a] Calculations by Gacesa et al. (2012) for $T_\infty$ = 240 K account for anisotropic scattering and vibrational excitation of $H_2$ and HD and give smaller flow velocities by factors of $\approx$10 for $H_2$ and $\approx$2 for HD.

and alpha particles, photoionization and ionization by energetic electrons of the solar wind and the induced magnetosphere above the ionopause that is near 300 km on Mars. Then almost all ions above the ionopause are swept away by the solar wind. Two other processes are the formation of species by exothermic reactions above the exobase that is near 200 km, because a very low energy is required for the light species to escape, and collisions with hot oxygen formed by recombination of $O_2^+$ (Section 2.5.2). The results are presented for two exospheric temperatures. Thermal escape dominates for H and becomes comparable with nonthermal escape for $H_2$ and D, while the latter dominates for HD and He.

### *10.5.5 Variations of Isotope Ratios*

For the sake of simplicity, we will consider fractionation of heavy $H$ and light $L$ isotopes between condensate $C$ and vapor $V$, though the approach is applicable to other isotope fractionations. There are three limiting cases in this consideration.

A closed system means that all released vapor is always in equilibrium with condensate. Adopt $L >> H$ and $C_0 = 1$ with the isotope ratio $R_{C0} = H_{C0}/L_{C0}$ at the beginning of evaporation, when $V_0 = 0$, and fractionation factor is $F = R_V/R_C$. Then abundances of the isotopes in the system are

$$L = C + V = 1, \qquad H = CR_C + VR_V, \qquad \frac{H}{L} = R_{C0} = R_C[C + (1 - C)F]. \tag{10.22}$$

The isotope enrichment in the condensate and ratio in the vapor are

$$\varepsilon = R_C/R_{C0} = \frac{1}{C + (1 - C)F}, \qquad R_V = \frac{R_{C0}}{1 - C(1 - 1/F)}. \tag{10.23}$$

For example, if $F = 0.99$, then $\varepsilon = 1$ ($\delta_C = 0$) at the beginning ($C = 1$), while $\varepsilon = 1.01$ ($\delta_C = 10‰$) at the end of evaporation ($C = 0$); $\delta_V = -10‰$ and 0, respectively. The closed system is typical of the terrestrial clouds, where both vapor and condensate are in the same medium for a long time.

An open system is when vapor leaves the system and does not affect the remaining condensate. Then

$$\frac{dH_C}{dt} = F\frac{H_C}{L_C}\frac{dL_C}{dt}, \tag{10.24}$$

and its solution is

$$\varepsilon = R_C/R_{C0} = \left(\frac{L_{C0}}{L_C}\right)^{1-F}. \tag{10.25}$$

This is a so-called Rayleigh distillation. $L_{C0}/L_C$ is the ratio of the initial to remaining condensate abundances, and $\delta$ = 23, 47, and 72‰ for this ratio of 10, 100, and 1000, respectively, for $F = 0.99$. Isotope fractionation in the open system is more significant than that in the closed system.

We considered above the case when a condensate is well mixed and all its parts have the same isotope ratio. However, if a condensate is solid (ice) and evaporation is rather intense, then diffusion in the ice cannot compensate for the depletion of the light isotope on the surface, and the ice surface is enriched in the heavy isotope. According to the Raoult law, this enrichment increases evaporation of the heavy isotope until both isotopes sublime without fractionation. This is the bulk evaporation.

A more complicated case with a major reservoir having a fixed isotope ratio $R_{C0}$ and a smaller buffer of a constant size may be of practical interest (Krasnopolsky 2000). A species evaporates and escapes with fractionation factor

$$F = \frac{\Phi_H/\Phi_L}{(H/L)_C} = \frac{\Phi_H}{\Phi_L R_C} \tag{10.26}$$

relative to the isotope ratio in the buffer. The loss of the buffer is compensated from the major reservoir, so that $L_C = a$ is constant. Then

$$\frac{dH_C}{dt} = \Phi_L(t)R_{C0} - \Phi_H(t) = \frac{\Phi_L(t)H_{C0}}{a}(1 - F\varepsilon)\ . \tag{10.27}$$

Here enrichment $\varepsilon = R_C/R_{C0} = H_C/H_{C0}$, and the substitution gives

$$\frac{d\varepsilon}{dt} = \frac{\Phi_L(t)}{a}(1 - F\varepsilon). \tag{10.28}$$

Solution of this differential equation is

$$\varepsilon(t) = \frac{1}{F} - \left(\frac{1}{F} - 1\right)e^{-\alpha(t)F}\ . \tag{10.29}$$

Here $\alpha$ is the ratio of the total escape of the species to its buffer abundance. The solution approaches steady state,

$$\varepsilon = \frac{1}{F}\ , \tag{10.30}$$

if $\alpha F \gg 1$; otherwise, it may be given as

$$\alpha = \frac{1}{F} ln \frac{1 - F}{1 - F\varepsilon}\ . \tag{10.31}$$

For example, this model may be applicable to water on Mars, where the north polar cap may be considered as a buffer that supplies water to the atmosphere with the subsequent escape. The loss of water from the north polar cap is compensated by water from the regolith.

# 11
# Mars

## 11.1 History of Studies, General Properties, Topography, and Polar Caps

### *11.1.1 History of Studies*

Mars is the neighbor planet with a comparatively thin atmosphere that is transparent to observe the surface. Early observations established polar caps, seasonal variations of the Martian brightness, and even so-called straight channels. The seasonal variations were ascribed to vegetation, and attempts were made to detect the chlorophyll bands and other "bands of life." G. A. Tikhov, a Russian scientist, suggested the term astrobotany in 1945 for that new field and published a book with this title in 1949 and followed it in 1953, with a book entitled *Astrobiology*. It was later recognized that the observed seasonal variations are caused by windblown dust with global dust storms in some perihelion periods.

The channels were explained as an indication of intelligent life, and even the Martian satellite Phobos was hypothesized as artificial, hollow, and populated inside. Actually the channels are lacking on Mars, and some canyons and combinations of craters could stimulate imagination of the observers.

A modern period in the studies of Mars started by Kuiper (1949), who compared near-infrared spectra of Mars and the Moon and revealed $CO_2$ bands at 1.57 and 1.6 μm. Later attempts were made to retrieve both $CO_2$ column abundance $N$ and atmospheric pressure $p$ from a combination of a weak band at 870 nm and the strong bands near 1.6 μm. The weak band gave a reasonable vertical $CO_2$ abundance of $N \approx 2\times10^{23}$ cm$^{-2}$ (Kaplan et al. 1964; Owen 1964). Equivalent widths of strong bands are proportional to $\sqrt{pN}$ ; however, the derived pressure was between 10 and 20 mbar with $N_2$ as a possible major component, exceeding the true value of $\approx$6.1 mbar from the Mariner 9 radio occultations (Kliore et al. 1973). Another significant result of that period was the detection of water vapor (Spinrad et al. 1963).

Spacecraft studies of Mars started by flybys of Mariner 4 in 1965 and Mariners 6 and 7 in 1969. Three orbiters, Mariner 9 and Russian Mars 2 and 3, were deployed at the end of 1971. The Russian spacecraft had landers; however, their landings were unsuccessful. The Viking 1 and 2 spacecraft that reached Mars in 1976 included two orbiters, two landers, and in situ measurements in the upper atmosphere. That was a breakthrough in the study of Mars atmosphere.

Currently six orbiters (NASA's Mars Odyssey, Mars Reconnaissance Orbiter (MRO), and Mars Atmosphere and Volatile Evolution mission (MAVEN), ESA's Mars Express, Roscosmos/ESA ExoMars Trace Gas Orbiter (TGO), Indian Mars Orbiter) and two of NASA's rovers (Opportunity and Curiosity) are operating on Mars.

### *11.1.2 Basic Properties*

Main properties of Mars are given in Table 11.1. Mars mean insolation is 0.43 that on the Earth, and it varies at the elliptic orbit from aphelion to perihelion by a factor of 1.45. This complicates the seasonal cycle that otherwise could be similar to that on the Earth because of the similar obliquity. Rotation periods are similar on both planets as well, while the Martian year is longer by almost a factor of 2.

Martian solar longitude $L_S$ defines seasons on Mars, so that $L_S = 0°$ and 180° are the vernal and fall equinoxes, $L_S = 90°$ and 270° are the northern summer and winter solstices. The seasonal variations are significant, and $L_S$ is given for each observation in the atmosphere. Mars aphelion is at $L_S = 71°$ and perihelion at $L_S = 251°$. Therefore the southern summer is warmer than the northern summer, while the southern winter is colder than the northern winter.

Mars has a crust of 50 km thick that consists mostly of O, Si, Fe, Mg, Ca, and K. The red color of the planet is caused by iron oxides. A silicate mantle below the crust extends down to $R \approx 1800$ km. The mantle is currently dormant without volcanic phenomena on Mars. A core below the mantle consists primarily of Fe, FeS, and Ni and involves ≈25% of the total mass.

Table 11.1 *Basic properties of Mars*

| | |
|---|---|
| Aphelion (AU) | 1.6660 |
| Perihelion (AU) | 1.3814 |
| Semi–major axis (AU) | 1.5237 |
| Eccentricity | 0.0934 |
| Obliquity | 25.19° |
| Mean radius (km) | 3390 |
| Mass (g) (km) | $6.42 \times 10^{26}$ |
| Mass/Earth's mass | 0.107 |
| Mean density (g cm$^{-3}$) | 3.93 |
| Surface gravity (cm s$^{-2}$) | 371 |
| Sidereal year (Earth days) | 687 |
| Sidereal year (sols) | 668.6 |
| Sidereal period of rotation (hours) | 24.6 |
| Escape velocity at exobase (km s$^{-1}$) | 4.87 |

### 11.1.3 Mars Topography

Mars topography (Figure 11.1) was carefully studied using the Mars Orbiter Laser Altimeter (MOLA) on board Mars Global Surveyor (MGS). There is a striking contrast between the flat plains with low (negative) elevation in the northern hemisphere and the highlands with numerous craters in the southern hemisphere. However, there are four prominent ancient volcanos in the north – including Olympus Mons (altitude of 21.2 km), while the Hellas impact basin found in the south is 7–8 km deep (below the standard topographical datum) and ~2000 km in diameter.

### 11.1.4 Polar Caps

Water ice may migrate on Mars as vapor and should be accumulated in regions with minimal insolations, that is, on the polar caps. The two Martian polar caps are not similar, because, for example, the northern winter is warmer than the southern winter on Mars and the northern cap is on a deep plain while the southern cap is on highlands. (There are significant differences between Arctics and Antarctic on the Earth as well.) The northern cap has a radius of 500 km in summer and consists of water ice and some dust frozen into the ice. Sublimation rate in summer is partially controlled by the dust abundance. A mean thickness of the northern cap is 2 km, and the amount of water in the cap is estimated as a global ocean of 10 m deep (Zuber et al. 1998). Radius of the southern cap is $\approx$180 km in summer with a thickness of $\approx$3 km. Water ice is also buried in the adjacent layer deposits, and its total amount in the southern cap is estimated at 11 m of the global-equivalent layer (Plaut et al. 2007). The polar caps are sinks of water vapor for most of the Martian year and sources of water vapor in the periods near the summer solstices.

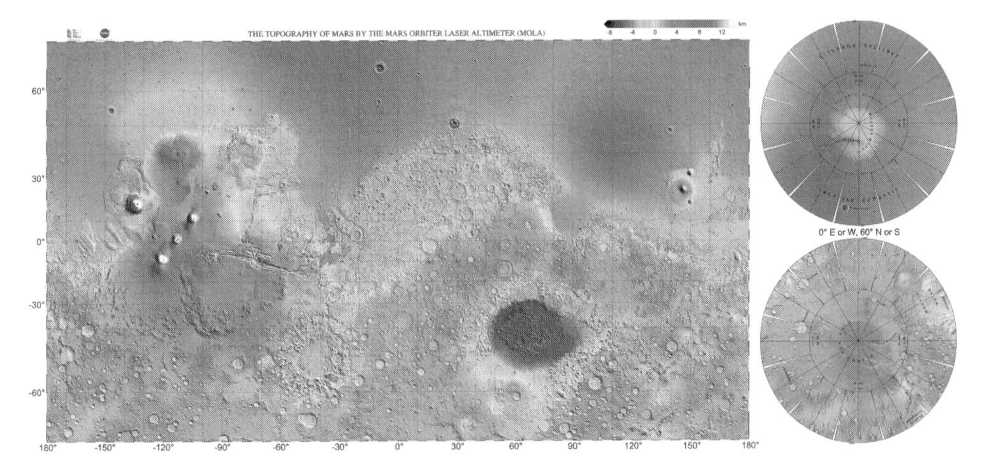


Figure 11.1 Mars elevation map observed by MOLA/MGS. (A black and white version of this figure will appear in some formats. For the color version, please refer to the plate section.)

Advection is insufficient to prevent condensation of $CO_2$ during the polar nights. Latent heat of the $CO_2$ condensation supports a temperature of 143–145 K that corresponds to the $CO_2$ saturated pressure of 4–8 mbar typical of the Martian atmosphere. The accumulated layer of $CO_2$ ice is $\approx 1$ m at the end of winter on the northern cap that is is completely lost in spring. The southern cap has a permanent $CO_2$ ice that is estimated to be 8 m thick in summer.

## 11.2 $CO_2$, Aerosol, and Temperature

### *11.2.1 $CO_2$ and Atmospheric Pressure*

Comparison of spectroscopic observations of the $CO_2$ abundance on Mars with atmospheric pressure measured by Mariner 4, 6, and 7 radio occultations was in favor of $CO_2$ as the only main component of the atmosphere. A lack of nitrogen emissions in the dayglow spectra observed by Mariner 6 and 7 (see below) confirmed that conclusion. Therefore atmospheric pressure uniquely determines the $CO_2$ abundance.

Numerous observations of pressure by the Mariner 9 radio occultations (Kliore et al. 1973) revealed a mean value near 6.1 mbar. This value corresponds to the triple point of water where three phases (solid, liquid, and gas) can coexist. If pressure is smaller than 6.1 mbar, then liquid water is unstable. For example, if temperature is higher than or equal to the melting point of 273 K, then water boils below 6.1 mbar. Actually variations of pressure are significant, an accurate globally and annually mean pressure is not known even now, and we will adopt the value of 6.1 mbar.

Annual variations of pressure on Mars were measured for the first time by the Viking landers (Figure 11.2). The curves contain data averaged over each sol (Martian

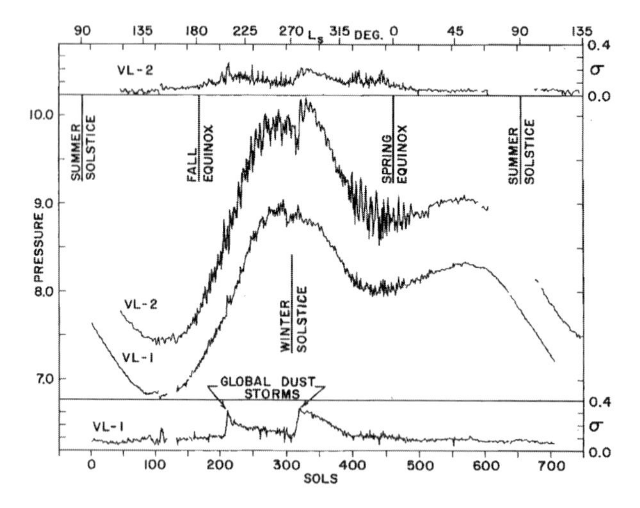


Figure 11.2 Seasonal variations of atmospheric pressure observed by Viking landers. VL1 was at 22.5°N and 48°W with elevation of −3.66 km. VL2 was at 48°N and 134°E with elevation of −4.44 km. From Hess et al. (1980).

day). Day-to-day variability is significantly greater at VL2 (48°N) than at VL1 (22°N). The observed seasonal variations are caused by sublimation and condensation of $CO_2$ at the polar caps. Each curve has two minima and two maxima that reflect differences between the opposite processes at the opposite polar caps. Annually mean pressure is 7.9 mbar at VL1, and peak-to-peak variations are $\pm 1.1$ mbar. The main minimum and maximum are at $L_S = 150°$ and $270°$, respectively. Similar observations at the later landers and rovers confirm the curves in Figure 11.2. These curves are important tests for general circulation models that simulate thermal balance and dynamics of the Martian atmosphere.

### *11.2.2 Aerosol*

Aerosol on Mars includes mineral dust and fog, clouds, and detached layers of water ice and sometimes of $CO_2$ ice. Dust on Mars significantly changes the atmospheric temperature and therefore abundance of water vapor and atmospheric photochemistry. The condensed aerosol reflects temperature conditions in the atmosphere. Aerosol affects optical observations of the atmosphere and surface that, on the other hand, may be used to study the aerosol properties (see Chapter 4).

#### *11.2.2.1 Viking Data*

Observations from landers are restricted to their locations but their geometry is preferable to study the Martian aerosol. Long-term Viking lander camera observations (Pollack et al. 1977, 1979) applied brightnesses of the Sun and Phobos to measure the daytime and nighttime aerosol optical depth, respectively. The observations at VL1 show a gradual increase from $\tau \approx 0.6$ at $L_S = 100°$ to $\tau \approx 1$ at $L_S = 200°$ with a jump to $\tau = 3$ at $L_S = 210°$ and a linear decrease to $\tau \approx 1$ at 270°, a jump to $\tau \approx 3$ at 270°–300°, a decrease to $\tau \approx 1.5$ at 320°, and a gradual decrease to $\tau \approx 0.5$ at 30°. The VL2 optical depths are $\tau \approx 0.3$ at 120°–170°, $\tau \approx 1$ at 175°–205°, $\tau \approx 2$ at 215° decreasing to 1.3 at 235°, $\tau \approx 2$ at 260°–305°, and 1.2 at 340° decreasing to 0.5 at 30°. The event at $L_S = 210°$ is the most prominent and seen at both landers being a global dust storm. The data refer to an effective wavelength of 0.67 μm.

Observed brightness distributions on the sky make it possible to derive properties of the aerosol particles. The cameras had six narrow band filters that covered the range of 0.4 to 1.1 μm. The retrieved properties averaged over this range are cross section weighted mean particle radius of 2.5 μm, single scattering albedo $\omega = 0.86$, asymmetry parameter $g = 0.79$, and extinction efficiency of 2.74. The real refraction index is adopted at 1.5, and the derived imaginary index has a minimum of 0.03 at 0.8 μm increasing to 0.09 at 0.49 and 1.05 μm.

Nighttime observations of Phobos at VL2 in northern summer revealed appearance of fog with $\tau \approx 0.1$ near 02:00 and its evaporation near noon. The particle radius was ~2 μm and the fog depth equaled ~0.4 km.

Each Viking orbiter had a photometer that was sensitive at 0.3–3.0 μm with an effective mean wavelength of 0.67 μm. Observations of a chosen location at different angles resulted in an emission phase function that made it possible to determine scattering properties of the particles. One hundred ten observations of these functions were analyzed by Clancy and Lee (1991). They found $\omega$ = 0.92 and $g$ = 0.55 for dust that differ from the values by Pollack et al. (1979). The observed north polar spring clouds had $g$ = 0.55 as well, while $g$ = 0.66 for midlatitude fall clouds, and $\omega$ = 1.0 was adopted for all clouds.

Jaquin et al. (1986) analyzed 116 limb images observed by the Viking orbiters. Dust was not seen above 50 km, and haze layers of water ice extended up to 90 km. The maximum observed aerosol elevation at low latitudes between $\pm 30°$ is approximated by a sine function of season with the minimum of 40 km at aphelion and maximum of 75 km at perihelion.

#### *11.2.2.2 MGS/TES Observations*

Detailed study of the dust and water ice aerosol distributions, temperature profiles, and water vapor abundances were made using the Thermal Emission Spectrometer (TES) on board the Mars Global Surveyor orbiter (MGS). The near-circular Sun-synchronous nearly polar orbit was fixed at local time of 14:00 on the dayside and 02:00 on the nightside. The observations began in July 1998 and ended in January 2006. The instrument spectral range is from 250 to 1650 $cm^{-1}$ (40 to 6 μm, respectively) with resolution of either 12.5 or 6.25 $cm^{-1}$, and both nadir and limb observations were made. Some observed spectra are shown in Figure 7.1. They include the $CO_2$ band at 670 $cm^{-1}$ that is used to retrieve temperature profiles (Section 7.3), the $H_2O$ bands at 250 and 1600 $cm^{-1}$, the dust band at 1075 $cm^{-1}$, and the $H_2O$ ice band at 825 $cm^{-1}$.

Optical depths of dust and water ice and column abundances of water vapor were retrieved from each TES spectrum and collected in the MGS/TES database. The data for three Martian years are presented as functions of $L_S$ with a step of 5°, latitude (3°), and longitude (7.5°). The TES optical depths are typically smaller than those in the visible range by a factor of ≈2.5. Seasonal variations of the globally mean dust are shown in Figure 11.3. The data for three Martian years are almost identical, except for a period from

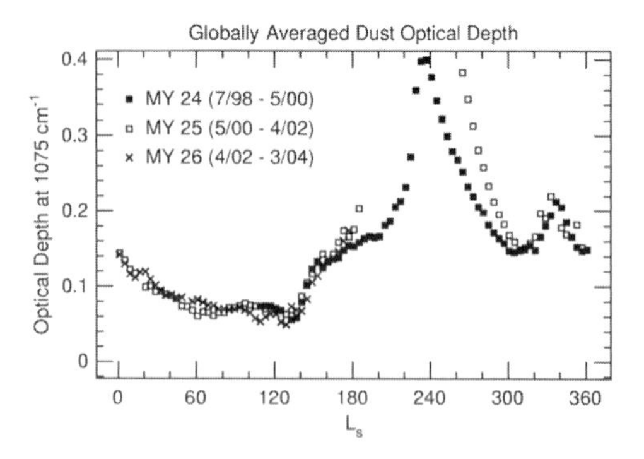


Figure 11.3 Globally averaged dust optical depth observed for three Martian years by MGS/TES (Smith 2004). The dust optical depths at 1075 $cm^{-1}$ are smaller than those in the visible range by a factor of ≈2.5.

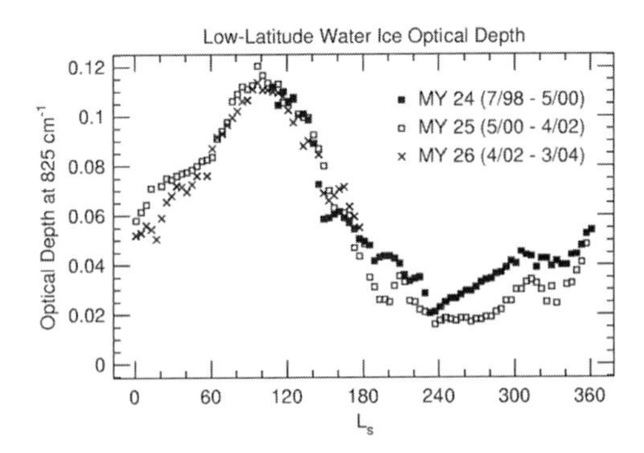


Figure 11.4 Variations of the water ice optical depth at low latitudes within $\pm 30°$ and local time of 14:00 observed by MGS/TES for three Martian years (Smith 2004). The $H_2O$ ice optical depths at 825 $cm^{-1}$ are smaller than those in the visible range by a factor of $\approx 2.5$.

$L_S$ = 180° to 300° near perihelion, when global dust storms appear and their strengths are different.

The dust decreases by a factor of $\approx 2$ in spring from 0° to 60° and then is constant up to midsummer. The steep increase at 140°–160° with subsequent global dust storms and the secondary maximum at 335° are the main features of the dust seasonal behavior.

Variations of the $H_2O$ ice aerosol at low latitudes are shown in Figure 11.4. The aerosol is maximal near the early northern summer ($L_S \approx 100°$) and smaller in the winter period of $L_S$ = 200°–340° by a factor of ~4. Other detailed data on the aerosol seasonal and latitudinal distribution may be found in the MGS/TES database.

Pankine et al. (2013) studied nighttime dust and water ice opacities from the MGS/TES data. They found some similarity between the nighttime and daytime opacities, with a nighttime increase by a factor of $\approx$ 1.3 for dust and up to a factor of $\approx 2$ for water ice.

### *11.2.2.3 THEMIS*

The Thermal Emission Imaging System (THEMIS) onboard the Mars Odyssey orbiter had a set of 10 filters that covered a range from 6.5 to 15 μm. The observations from February 2002 to December 2008 (Smith 2009) gave detailed maps of dust, water ice aerosol, and temperature at the 0.5 mbar level (~25 km).

### *11.2.2.4 Mars Express Observations*

Each spacecraft on or near Mars had/has tools to observe aerosol in the atmosphere, and we consider here only some of them. Two channels of SPICAM, UV and NIR, observed simultaneously 20 solar occultations, and properties of the aerosol in those occultations were analyzed by Fedorova et al. (2014).

The observations were conducted at $L_S$ = 56° to 97° and covered latitudes 43°–66°N and 32°–60°S. Ten NIR intervals at 1.0 to 1.7 μm and those UV at 200, 250, and 300 nm were chosen at wavelengths with minimal gaseous absorption. Data on real and imaginary

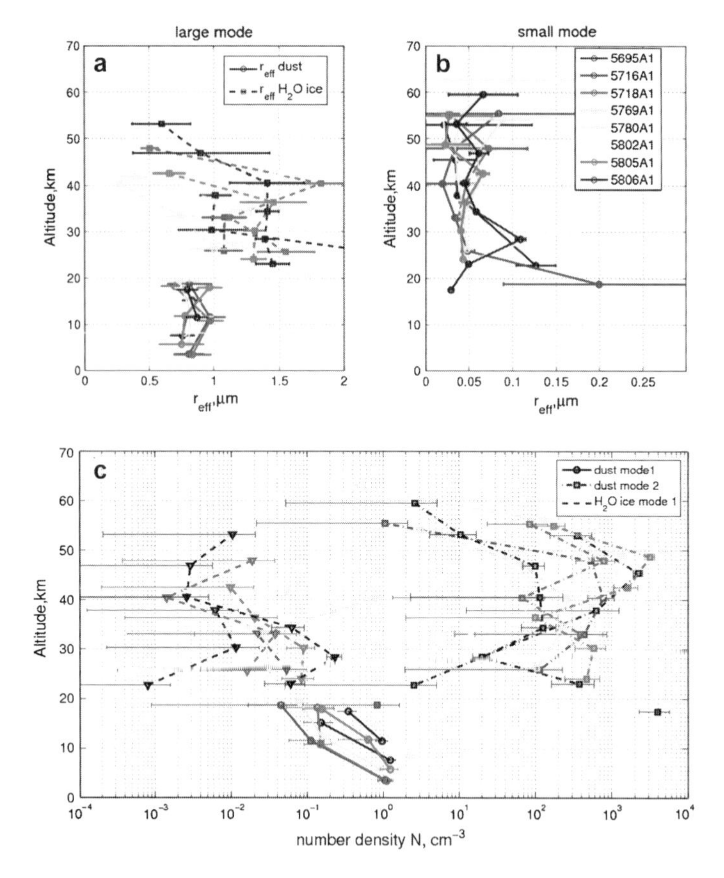


Figure 11.5 Aerosol properties from eight solar occultations at 200–300 nm and 1.0–1.7 μm. Vertical profiles of effective particle radius of the large (a) and small (b) modes and their number densities (c) are shown. The data refer to $L_S$ = 83°–97° and 32°–50°S. From Fedorova et al. (2014).

refractive indices of water ice and dust were taken from the literature, and the observed spectral transmittances of the atmosphere at various altitudes were fitted to get effective radii of particles and their number densities using the Mie formulas for spheres. Results for eight occultations at latitudes of 32°–50°S near the summer solstice ($L_S$ = 83°–97°) are shown in Figure 11.5.

The data require two populations of the aerosol. A large particle mode has effective radius of 0.8 μm below 20 km and ≈1.3 μm at 20–40 km decreasing to ≈0.5 μm near 50 km. Though the spectral data do not clearly distinguish water ice and dust particles, it is reasonable to anticipate that dust and water ice dominate in this mode below and above 20 km, respectively. Their number density diminishes from ~1 $cm^{-3}$ below 10 km to ~0.01 $cm^{-3}$ near 40 km. The second mode is very small particles with effective radius ≈0.05 μm that is almost constant from 30 to 60 km with number densities of a few hundred per cubic cm. The authors conclude that the meteor flux is insufficient to support the observed population of very small particles.

The $CO_2$ ice aerosol is observed as clouds at the winter polar caps or tiny hazes at the mesopause near 90 km. The mesospheric $CO_2$ haze was also observed spectroscopically using PFS (Aoki et al. 2018). The haze is seen as a daytime emission at 4.25 μm in the broad $CO_2$ absorption band at 4.3 μm.

### 11.2.2.5 MRO/MCS

Mars Climate Sounder (MCS) is a radiometer with a broad band in the visible-NIR (0.3 to 3.0 μm) and eight bands that cover 12 to 50 μm. Each band has 21 detectors for limb observations with resolution of 5 km in altitude. Detectors A5 with a band pass of 400–500 $cm^{-1}$ is used to observe the dust band at 463 $cm^{-1}$, and the results are shown in Figure 11.6 (Kleinboehl et al. 2015).

Dust extends up to 15 to 40 km depending on season and latitude, and no dust is observed in the middle atmosphere up to some levels where aerosol appears again. Though that aerosol is assigned to dust, it is impossible for dust to have a global-wide gap in the middle atmosphere. Probably water ice contributes to the band at 463 $cm^{-1}$, and its contribution is not completely compensated by the retrieval code. The band at 463 $cm^{-1}$ is much weaker than that at 1075 $cm^{-1}$ (Figure 7.1). Anyway, vertical distributions of dust at different seasons and latitudes in the lower atmosphere and their lack in the middle atmosphere are important results of this study.

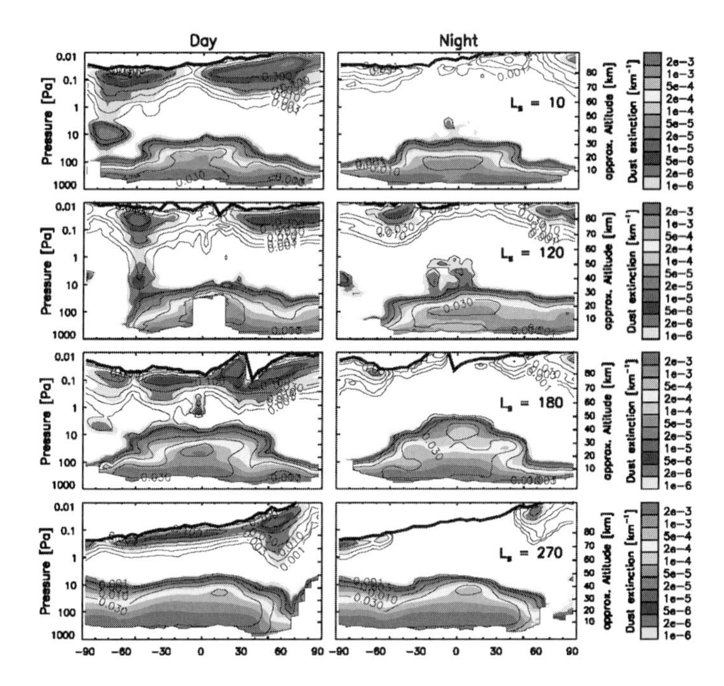


Figure 11.6 Vertical profiles of dust at various latitudes in four seasons observed by MRO/MCS (Kleinboehl et al. 2015). Solid lines are the upper boundaries of the observations. (A black and white version of this figure will appear in some formats. For the color version, please refer to the plate section.)

### 11.2.3 Temperature

Temperature determines vertical profiles of density and pressure, and measurements of those by the orbiter radio occultations and accelerometers at the descent probes give data on the atmospheric temperatures. However, the most convenient tool to study temperatures in the Martian low and middle atmosphere is spectroscopy of the $CO_2$ band at 15 μm (Section 7.3). The observations were made at Mars orbiters Mariner 9 (IRIS), Viking/IRTM, MGS/TES, Mars Express (PFS), and MRO/MCS.

#### 11.2.3.1 MGS/TES

The most detailed data were retrieved from MGS/TES (Smith 2004) and presented as a database for 3 Martian years with steps in $L_S$, latitude, and longitude of 5°, 3°, and 7.5°, respectively. The nadir data include daytime and nighttime surface temperatures and pressures and temperature profiles from the surface to 0.1 mbar (≈37 km) as functions of pressure with a step of $\delta \ln p = 0.25$ (2–3 km on Mars). The limb data give daytime and nighttime temperature profiles from 0.8 to 0.01 mbar (≈55 km). Feofilov et al. (2012) extended the temperature retrievals to 90 km using the total emission at 15 μm. The data are near the noise level at ≈90 km at the limb; however, averaging dozens of limb profiles, they got reasonable results up to this altitude.

A temperature profile in the atmosphere that is based on the TES data and approximates the daytime annually and globally mean conditions is shown in Figure 11.7. The mean heliocentric distance of 1.52 AU is reached near $L_S \approx 160°$, and regions with zero elevation at ≈40°N are close to the mean conditions at this season. The profile reflects main processes that determine thermal balance of the lower and middle atmosphere: heating of the surface that initiates convection, radiative transfer in the $CO_2$ band at 15 μm, absorption and radiation by dust, UV and NIR heating by the $CO_2$ absorption of the sunlight.

Seasonal variations of the globally averaged daytime surface temperature are shown in the right upper panel of Figure 11.8. The temperature minimum of 237 K is at aphelion at 1.67 AU and maximum of 260 K is at $L_S$ = 310° and 1.44 AU. Scaling the minimum

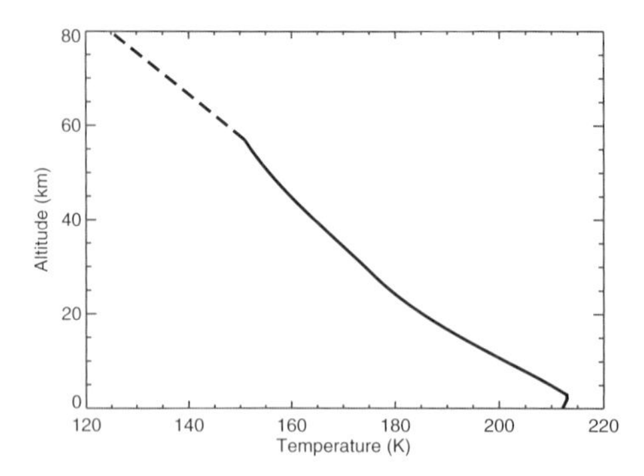


Figure 11.7 A temperature profile that approximates the daytime globally and annually mean conditions. The profile is based on the MGS/TES data. The profile is extrapolated above 57 km to fit $p$ = 1.2 nbar at 125 km according to the Mariner 9 radio occultation and dayglow observations.

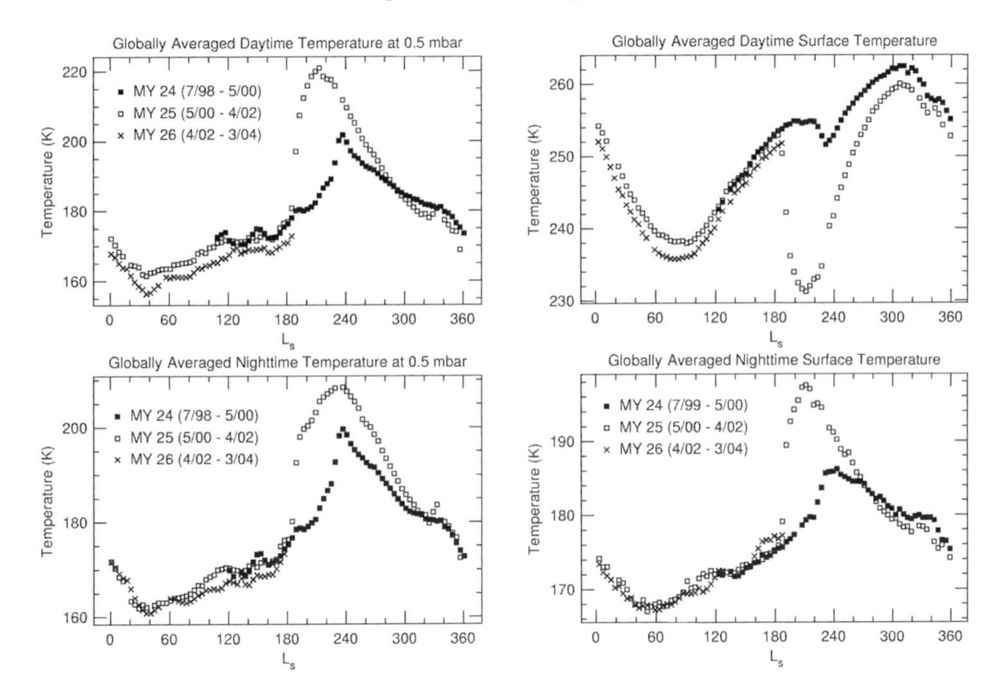


Figure 11.8 Seasonal variations of the atmospheric temperature at 0.5 mbar (~25 km) and surface temperature for 3 Martian years (Smith 2004).

temperature by the heliocentric distances gives 255 K at 310°. The difference can be explained if the albedo is higher at $L_S = 70°$ than that at 310° by a factor of 1.08 because of the water ice clouds. There are significant temperature minima near the perihelion that are caused by dust storms and depend on their strength.

The nighttime mean surface temperature at 02:00 has a minimum of 168 K at $L_S = 60°$. The temperature maxima correlate with the dust storms. The optically denser and warmer atmosphere during the dust storms either heats the surface or reduces its thermal loss to space. The difference between the daytime and nighttime temperatures is $\approx$70 K, much greater than that in the denser atmosphere of the Earth.

Contrary to the surface, the day–night variations are low in the atmosphere (left panels of Figure 11.8). Dust heats the atmosphere and increases its temperature during dust storms. The level of 0.5 mbars in the figure is near 25 km. The minimum is at $L_S = 40°$, that is, before aphelion.

### *11.2.3.2 MRO/MCS*

Arrays of 21 detectors A1 at 16.5 μm, A2 at 15.9 $\mu$m, A3 at 15.4 μm, and B1 at 31.7 μm are used to observe a limb structure of the $CO_2$ band at 15 μm and retrieve temperature profiles from the surface to 80–90 km (McCleese et al. 2010). Results of the observations are presented as zonally mean temperatures in a color code versus pressure and latitude at eight seasons with a step of $\delta L_S = 45°$ for day and night conditions. The observed temperature and pressure fields make it possible to calculate zonal gradient winds that are given by McCleese et al. (2010) in the same format.

### *11.2.3.3 MER Mini-TES*

The Spirit and Opportunity Mars Exploration Rovers (MER) had miniature versions of Thermal Emission Spectrometer (Mini-TES) that operated in up- and down-looking modes. The instrument covered a range from 340 to 2000 $cm^{-1}$ (5–29 μm) with a spectral sampling of 10 $cm^{-1}$. The observations made it possible to study variations of temperature in the boundary layer from 1 m to 2000 m above the surface with local time and during a Martian year (Smith et al. 2006). Variations of dust and water vapor were observed as well. Temperature profiles at various local time (Figure 11.9) show variations of the air near the surface by 54 K that are smaller than the variations of the surface temperature by 80 K at the same season in Figure 11.8. The atmosphere becomes more stable with altitude, and the observed temperature variations at 2 km are equal to 23 K.

Seasonal variations of the afternoon temperature at 1, 100, and 1000 m are shown in Figure 11.10. The data from both figures are essential for modeling of the boundary layer on Mars.

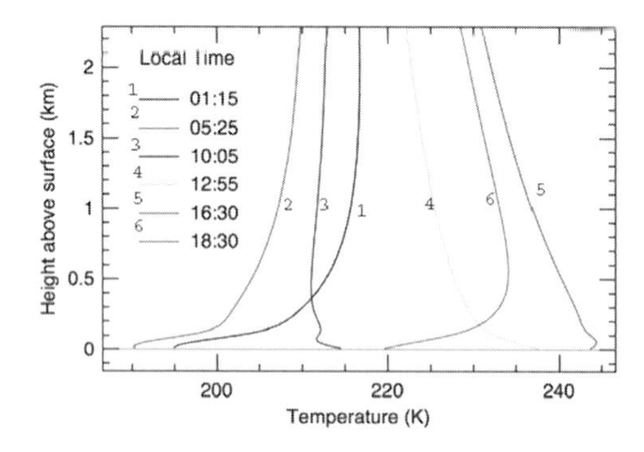


Figure 11.9 Variations of temperature in the boundary layer with local time observed by the mini-TES at the Spirit rover, $L_S$ = 5°, 15°S, and 185°W (Smith et al. 2006).

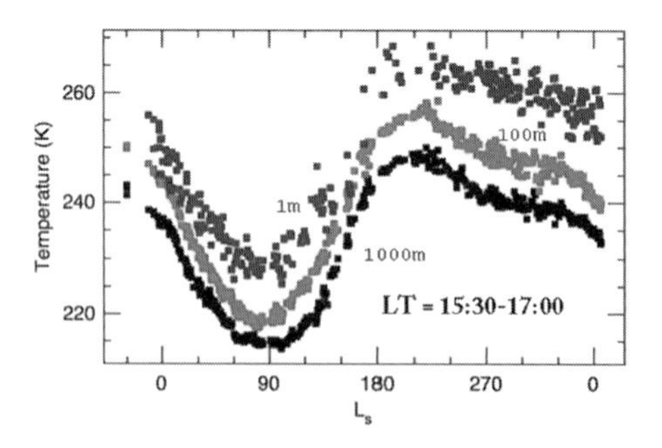


Figure 11.10 Seasonal variations of temperature at 1, 100, and 1000 m observed by the mini-TES at the Spirit rover (Smith et al. 2006).

## 11.3 Water Vapor, HDO, and Ice

### *11.3.1 Variations of Water Vapor Column Abundance*

Water vapor is the second gas detected in the Martian atmosphere. Spinrad et al. (1963) observed weak Doppler-shifted lines of the $H_2O$ band at 820 nm; the estimated $H_2O$ vapor abundance was 5–10 precipitable μm, a reasonable value that agrees with the later studies. (Column abundances of water vapor on Mars are usually measured in precipitable μm, and 1 pr. μm = $\frac{A\times10^{-4}}{\mu} = 3.3\times10^{18}$ cm$^{-2}$. Here $A$ = 6.02×10$^{23}$ is the Avogadro number, $\mu$ = 18 is the molecular mass in atomic units, and 1 μm = 10$^{-4}$ cm.) Water vapor is a sensitive tracer of thermal balance and dynamics of the atmosphere. Furthermore, water vapor and its products are catalysts in the atmospheric photochemistry, and global and vertical distribution of water vapor and its variations are among the key problems in the Martian science.

The first detailed study of water vapor on Mars was made by Jakosky and Farmer (1982) using the Mars Atmospheric Water Detector (MAWD) at the Viking 1 and 2 orbiter. The instrument was a grating spectrometer with five detectors. Three of those were centered at two strong and one weak $H_2O$ lines of the band at 1.38 μm, while two detectors measured background between the lines. Fedorova et al. (2010) revised the MAWD data using HITRAN 2004 and improved atmospheric models.

MGS/TES collected data on the $H_2O$ column abundances from July 1998 to January 2006 that are presented in the MGS/TES database (Smith 2004) as functions of season ($L_S$), latitude, and longitude for each Martian year. The $H_2O$ bands at 6.3 and 40 μm were used for the retrievals.

Spatial distribution of water vapor on Mars was measured for 5 Martian years using MEX/SPICAM IR spectrometer (Trokhimovsky et al. 2015). The $H_2O$ band at 1.38 μm was observed with resolving power $\lambda/\delta\lambda$ = 1700. The retrieved $H_2O$ column abundances were corrected for multiple scattering by dust and water ice aerosol measured by THEMIS at the Mars Odyssey orbiter (Smith 2009). Averaged seasonal and latitudinal variations of water vapor are shown in Figure 11.11. Maximum retrieved abundances are 60–70 pr. μm in the northern summer and 20 pr. μm in the southern summer. The seasonal cycle of water vapor

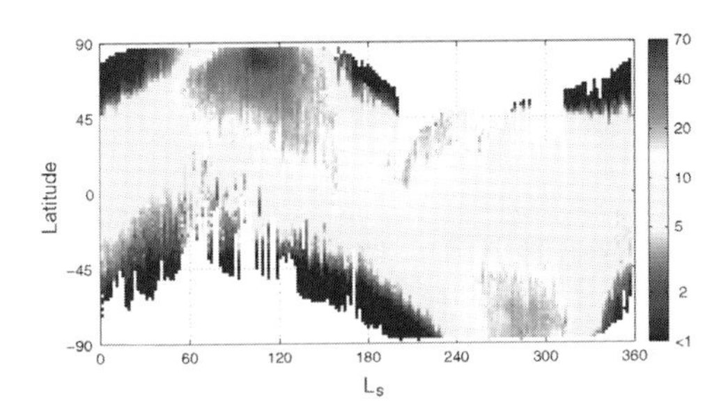


Figure 11.11 Seasonal and latitudinal variations of water vapor (in pr. μm) observed by MEX/SPICAM IR and averaged for 5 Martian years (Trokhimovsky et al. 2015). (A black and white version of this figure will appear in some formats. For the color version, please refer to the plate section.)

looks very stable, and interannual variations are comparatively low. The zonally averaged data are presented as files with a grid of 2° in $L_S$ and latitude for each of 5 Martian years observed and the mean of those data. Calculated altitudes of condensation are given as well.

### *11.3.2 $H_2O$ Vertical Distribution*

This is a key problem for photochemistry, because, for example, due to the $CO_2$ absorption, photolysis of $H_2O$ is more efficient at 40 km than near the surface by a factor of 400.

It was believed in the beginning of 1970s that the Martian troposphere has the adiabatic gradient $-5$ K km$^{-1}$ (Section 2.1.3) from $T = 225$ K near the surface to 150 K at 15 km and is isothermal above 15 km. Saturated vapor pressure of water is steeply decreasing with the decreasing temperature (Section 2.3.6, relationship (2.30)), and almost all water vapor should be in the lowest 5 km in this atmosphere. Therefore photolysis of $H_2O$ and production of odd hydrogen were weak in the early photochemical models, and high eddy diffusion was applied to enhance the odd hydrogen chemistry.

Later more reliable data on the atmospheric temperature profile became available. The Viking MAWD observations of the same location at different angles favored a uniformly mixed water vapor in the lowest 10 km (Davies 1979) The steady state photochemical models also predict uniformly mixed water vapor up to a saturation level.

However, the very low nighttime temperatures on the surface and in the boundary layer (Figures 11.8 and 11.9) result in condensation of water, a downward flow of water vapor, and accumulation of the condensate during night and some time after the sunrise when temperature in the boundary layer is still rather low. Later the condensate evaporates, and the total water mixing ratio (vapor plus condensate) in the lowest few km (Figure 11.12) exceeds the $H_2O$ mixing ratio above 10 km (by a factor of ≈2 in the calculations by Krasnopolsky and Parshev 1977).

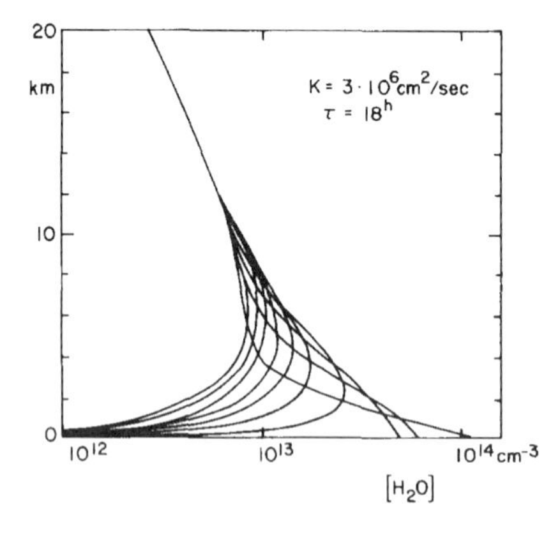


Figure 11.12 Variations of the water vapor number density with local time induced by the nighttime condensation on the surface and in the boundary layer. Time interval between the curves is 2.4 hours. The surface $H_2O$ is maximum at LT = $12^h$, the next curves are for $14.4^h$ and $16.8^h$, and the curve for $19.2^h$ shows the beginning of condensation. The last curve refers to $9.6^h$. From Krasnopolsky and Parshev (1977).

The calculated variations are significant and may be observed as variations of the $H_2O$ column abundance with local time. This problem was discussed by Trokhimovsky et al. (2015, section 5.2). They mentioned that Mariner 9/IRIS, Viking/MAWD, and Phobos-2/ISM show variations of the $H_2O$ column with local time by a factor of 2–3. However, mini-TES did not detect these variations, and neither did SPICAM IR on board Mars Express. Unfortunately, the conclusion by Trokhimovsky et al. (2015) on the lack of the $H_2O$ variations with local time is based on the SPICAM IR observations near the summer peak ($L_S$ = 108°–120°) at latitudes of 60°–80°N, that is, mostly during the polar day without significant temperature minimum at the surface. Therefore their conclusion on the lack of the local time variability of $H_2O$ refers to those specific conditions and is inapplicable to the middle and low latitudes.

Vertical profiles of the $H_2O$ mixing ratio were observed using MEX/SPICAM IR solar occultations (Maltagliati et al. 2013; Fedorova et al. 2018). The $H_2O$ band at 1.38 μm, the $CO_2$ band at 1.43 μm, and 11 continuum wavelengths were applied for the retrievals. Two observational campaigns covered $L_S$ = 50°–110° and 205°–270° and a broad latitude range of ±70°. The major results are shown in Figures 11.13 and 11.14. Both plots in both figures demonstrate large scatter of the observed profiles. However, the mean profile in the

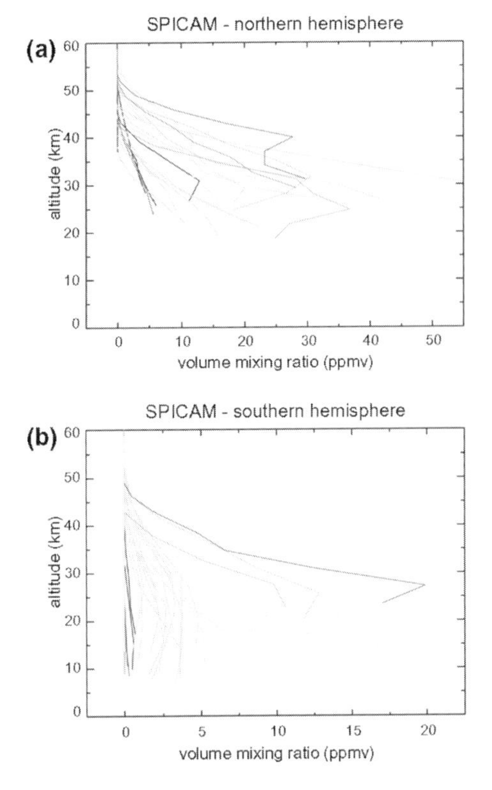


Figure 11.13 Vertical profiles of water vapor observed using SPICAM IR solar occultations at $L_S$ = 50°–110° (Maltagliati et al. 2013).

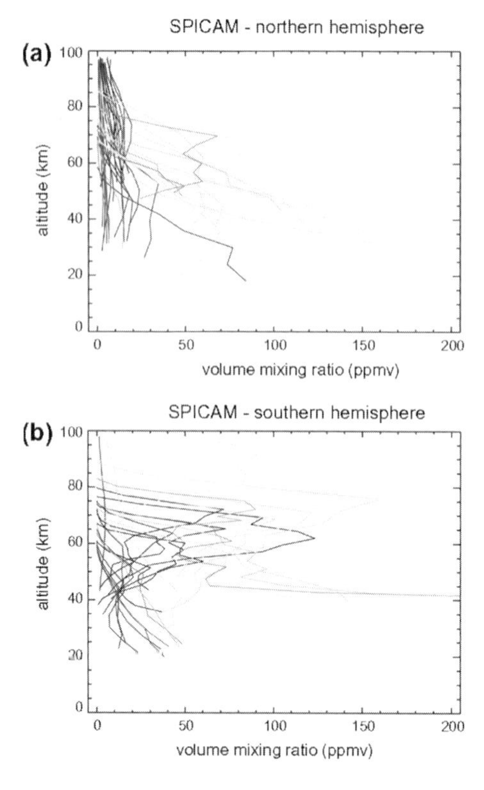


Figure 11.14 Vertical profiles of water vapor observed using SPICAM IR solar occultations at $L_S$ = 205°–270° (Maltagliati et al. 2013).

northern hemisphere at $L_S$ = 50°–110° show a decrease from ≈20 ppm at 25 km to ≈3 ppm at 45 km. The values for the southern hemisphere are ≈5 ppm at 15 km and ≈1 ppm at 40 km. Much higher $H_2O$ abundances were observed near perihelion at $L_S$ = 205°–270°: ≈60 ppm at 40 km and ≈3 ppm at 90 km in the northern hemisphere. The profiles in the southern hemisphere are mostly layers with ≈30 ppm at 40 km, peaks of ≈70 ppm near 70 km and reduction to ≈30 ppm near 80 km. The simultaneous retrievals of the aerosol structure favor exchange of water between vapor and the ice haze with some events of supersaturation up to a factor of 2.

The $H_2O$ abundances at high altitudes near perihelion are much greater during dust storms (Figure 11.15), because dust is heated by the sunlight and heats the atmosphere, preventing condensation of water vapor. Significant abundances of water in the lower thermosphere during global dust storms increase production and escape of atomic hydrogen.

### *11.3.3 HDO/$H_2O$ Ratio and Its Variations*

Preferential escape of hydrogen relative to deuterium through the history of Mars resulted in enrichment of the D/H ratio in water. This enrichment is among key values to study evolution of water on Mars. It was discovered by Owen et al. (1988) who observed

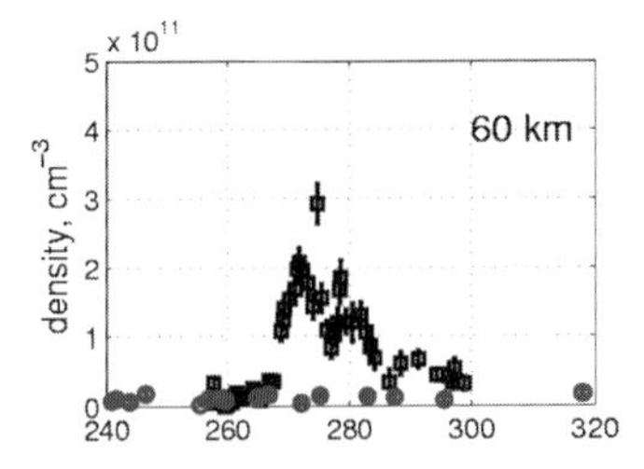


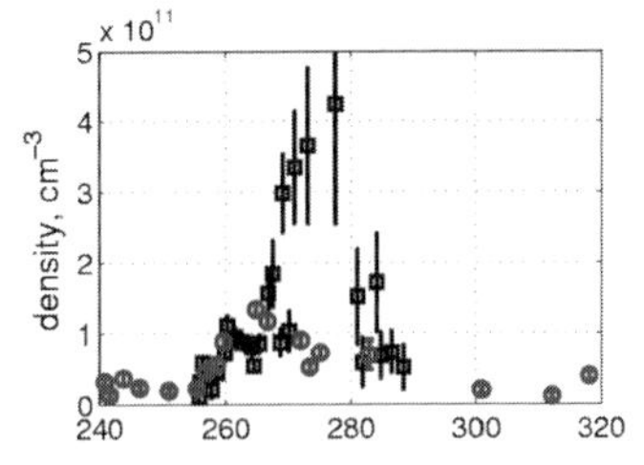


Figure 11.15 $H_2O$ densities at 60 km in the northern (left) and southern (right) hemispheres during the global dust storm (MY28, square symbols) and without global dust storm (MY 32, circular symbols). From Fedorova et al. (2018).

the HDO lines near 2722 $cm^{-1}$ (3.67 μm) using the FTS at the Canada–France–Hawaii Telescope. The observed D/H in the Martian water vapor was equal to 5.8 ± 2.6 times that of $1.56\times10^{-4}$ in the Standard Mean Ocean Water (SMOW). Similar values of 5.2 ± 0.2 and 5.5 ± 2 were measured by Bjoraker et al. (1989) and Krasnopolsky et al. (1997) using the FTSs at the Kuiper Airborne Observatory and the Kitt Peak National Observatory, respectively.

In situ observations of the D/H ratio in the Martian water vapor was made using the Tunable Laser Spectrometer at the Curiosity rover (Webster et al. 2013). Measurements for the first ≈100 days of the rover operation give D/H = 6 ± 1. However, water vapor abundances in the instrument cell were much higher, up to 1%, than those expected in the atmosphere (≈100 ppm), and a reason of this enrichment is not well identified. Of all other Mars-orbiting and landing spacecraft, only the Trace Gas Orbiter has capabilities to observe HDO on Mars.

M. J. Mumma and R. Novak initiated a ground-based study of variations of $HDO/H_2O$ using spatially resolved spectroscopy with CSHELL/IRTF. Their best results are maps of $HDO/H_2O$ at four seasons (Villanueva et al. 2015) observed using CRIRES/VLT. A map of $HDO/H_2O$ at $L_S = 113°$ was extracted from the EXES observations aboard the airborne observatory SOFIA (Encrenaz et al. 2016). Variations of $HDO/H_2O$ along the central meridian were observed in six seasons by Krasnopolsky (2015d) using CSHELL/IRTF and in two seasons by Aoki et al. (2015) using IRCS/Subaru.

One of the observed HDO spectra at 2722 $cm^{-1}$ is in Figure 11.16, and spectra of $H_2O$ were observed at 2994 $cm^{-1}$. Spectral properties of the dust and ice aerosol are rather similar at both wavenumbers, and their effects cancel out in the $HDO/H_2O$ ratios. Latitudinal variations of the measured $HDO/H_2O$ at $L_S = 70°$ are shown in Figure 11.17. The retrievals by synthetic spectra and by spectra of the Moon observed simultaneously using the same instrument gave similar results (Figure 11.17). Averaging of the data for six seasons give a mean $HDO/H_2O$ = 4.6 ± 0.7 times the terrestrial ratio. The ratio in vapor released by the north polar cap is 6.2 ± 1.4, and the ratio in the north polar cap ice is 7.1 ± 1.6.

An early version of the Mars Climate Database was used by Montmessin et al. (2005) to calculate two models assuming either rapid homogenization or Rayleigh distillation of HDO in the ice aerosol particles, and the results differ just by a few percent. They pointed out that the model does not account for adsorption and desorption of water vapor by the regolith and possible isotope fractionation in these processes. The residual north polar cap

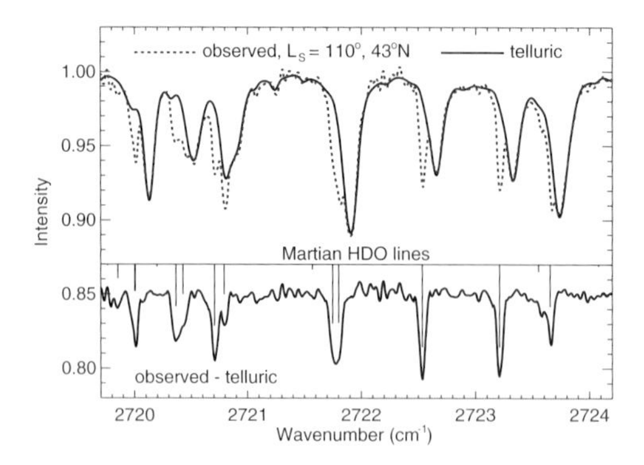


Figure 11.16 Observed spectrum of Mars near 3.67 μm is compared with the best-fit telluric spectrum. Their difference reveals Martian HDO lines; their strengths and Doppler-shifted positions are indicated. From Krasnopolsky (2015d).

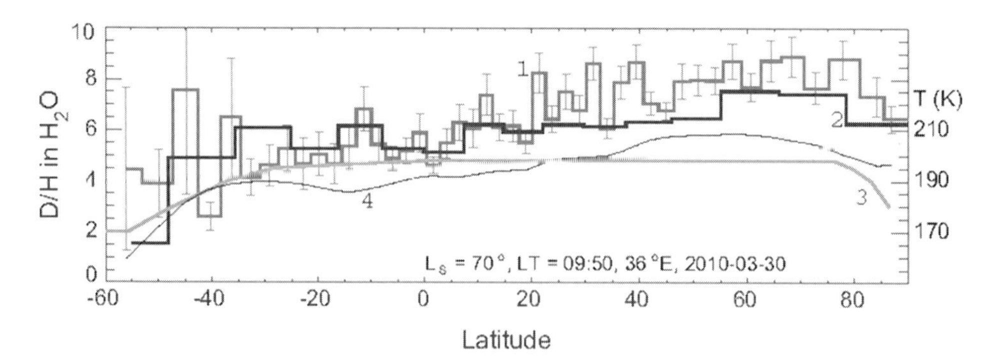


Figure 11.17 Variations of $HDO/H_2O$ (times the terrestrial ratio) retrieved using synthetic spectra (1) and spectra of the Moon (2) are compared with the model (3) by Montmessin et al. (2005). Temperature at 7 km is also shown (4) using the TES observations. From Krasnopolsky (2015d). (A black and white version of this figure will appear in some formats. For the color version, please refer to the plate section.)

is a source of water vapor in the model with recycling of water between the cap and the atmosphere, while the residual south polar cap is its permanent sink in the current epoch.

Unfortunately, there is a great scatter in the laboratory data on the deuterium fractionation between water ice and vapor at low temperatures typical of the Martian atmosphere. Even temperature trends in decreasing or increasing of the fractionation are opposite in the data (Lecuyer et al. 2017). Evidently this badly affects the model predictions.

### *11.3.4 Water Ice in the Regolith*

Large amount of water ice are concentrated in the north polar cap and the south polar deposits (Section 11.1.4). These amounts are evaluated at 10 m of

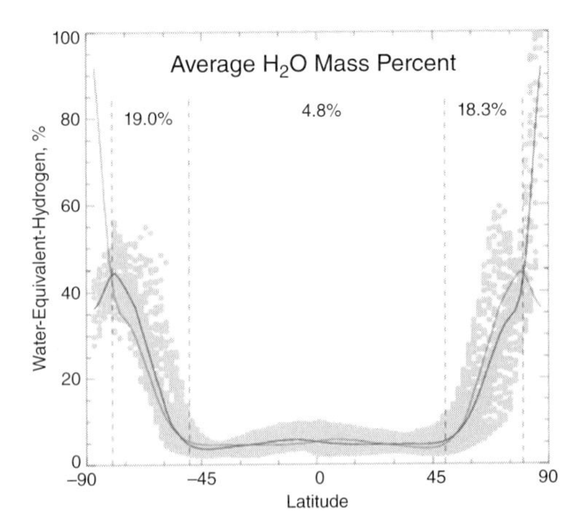


Figure 11.18 Zonally averaged water-equivalent hydrogen in the upper regolith layer of 1 m thick. Thin line is the same curve with inverted signs of latitude to compare the water contents in both hemispheres. From Feldman et al. (2004).

global equivalent layer in the north polar cap using the Mars Orbiter Laser Altimeter (Zuber et al. 1998) and 11 m in the south polar deposits using radar observations from Mars Express (Plaut et al. 2007).

A projectile loses its energy most efficiently in collisions with targets of the equal mass. Therefore hydrogen atoms in the regolith reduce its neutron reflectivity, and observations of the solar neutrons reflected by Mars may be used to measure hydrogen in the upper surface layer of $\approx$1 m thick. Hydrogen may be in water ice, $H_2O$ and OH bound in the rocks.

Two instruments onboard the Mars Odyssey orbiter (Feldman et al. 2004; Mitrofanov et al. 2007) measured Mars neutron reflectivity, and maps of water content in the regolith were retrieved from the observations. Zonally averaged data from Feldman et al. (2004) are shown in Figure 11.18. The mean $H_2O$ mass ratio in the regolith is typically 5% at the low and middle latitudes within $\pm 50^\circ$ and steeply increases to 45% at $\pm 80^\circ$. The different behavior near the poles reflects the different structures of the polar caps. A global mean water content is 14 cm in the upper regolith layer of 1 m thick.

## 11.4 Carbon Monoxide and Molecular Oxygen

### *11.4.1 CO Detection and First Observations*

CO was the third species detected in the Martian lower atmosphere. Kaplan et al. (1969) analyzed the first high-resolution spectra of Mars by Connes et al. (1969) and compared equivalent widths of the CO P6 lines of the (3–0) and (2–0) bands. The (2–0) line is stronger than the (3–0) line by a factor of 160 and therefore is sensitive to pressure. Both CO column abundance and pressure were retrieved from the data and resulted in the CO mixing ratio of 800 ± 300 ppm. Krasnopolsky (2007) pointed out that the CO (3–0) P6 line at 6325.80 $cm^{-1}$ is contaminated by a weak $CO_2$ line at 6325.77 $cm^{-1}$, and the retrieved CO fraction should be reduced by ≈10%. Anyway, the observed CO abundance was unexpectedly low, much smaller than predictions of the dry $CO_2$ photochemistry. This raised a problem of the $CO_2$ stability on Mars.

Ground-based high-resolution spectroscopy of the CO (1–0) band at 4.7 μm gave the CO mixing ratio of 550 ± 150 ppm (Billebaud et al. 1992). Encrenaz et al. (1991) derived 660 ± 220 ppm from the observed (2–0) band at 2.35 μm and ≈800 ppm from the submillimeter observations of the $^{12}CO$ and $^{13}CO$ J = 2 → 3 lines. Observations of the J = 1 → 2 lines of both isotopologues gave 800 ± 200 ppm (Lellouch et al. 1991). Observations of these lines by Clancy et al. (1996) confirmed this value with a better uncertainty: 800 ± 100 ppm.

### *11.4.2 Spatially Resolved Observations of CO and Variability of Long-Living Species on Mars*

Lellouch et al. (1991) observed Mars with a beam width equal to the Mars angular radius, and the measured variations of CO were within the claimed uncertainty. Krasnopolsky (2003) used a long-slit high-resolution spectrograph IRTF/CSHELL for mapping of CO on Mars. A chosen spectral interval of 6379–6394 $cm^{-1}$ includes lines of the CO (3–0) band and $CO_2$ lines. Their equivalent widths were converted into CO and $CO_2$ abundances that gave CO mixing ratios. The observation was made near the peak of northern summer ($L_S$ = 112°). The extracted CO mixing ratios showed an increase from 800 ppm in the northern hemisphere to ≈1200 ppm at 50°S. The increase was explained by condensation of $CO_2$ in the south polar regions that results in enrichments in mixing ratios of all incondensable long-living species. Studies of this phenomenon are related to dynamics of the polar processes on Mars.

Inclusion of photochemistry in the LMD general circulation model of the Martian atmosphere and further developments of the model (Lefèvre et al. 2004; Lefèvre and Forget 2009) transformed this model into the Mars Climate Database with significant prediction capabilities (MCD; www-mars.lmd.jussieu.fr/mcd_python/) that are used for comparison with observations and other models.

Two Mars Express instruments were used to observe variations of CO. The Omega low-resolution imaging spectrograph observed the CO band at 2.35 μm at the deepest Hellas basin (~6 km deep) that extends at 30°–50°S and 50°–80°E. The measured CO mixing ratio was 1150 ppm during the southern summer and fall ($L_S$ = 250°–360°–50°) and twice this value near 135° (Encrenaz et al. 2006). MCD at that time predicted doubling of mixing ratios of the incondensable species in Hellas relative to the nearby regions in northern summer. To check this prediction, Encrenaz et al. (2006) compared an observation at Hellas (40°S) with that at the north of Hellas at 23°S and confirmed the twofold difference. The observation by Krasnopolsky (2003) covered Hellas as well, and the similar difference was a factor of 1.75.

Billebaud et al. (2009) studied variations of CO using observations of the CO (1–0) band at 4.7 μm by the Planetary Fourier Spectrometer (PFS) at Mars Express. The observations covered seasons from $L_S$ = 330° to 50°. CO at $L_S$ = 330°–360° was constant at ≈1200 ppm from 40°N to 10°S decreasing to 750 ppm at 65°N. The data at $L_S$ = 0°–30° were constant at ≈1100 ppm from 20°S to 30°N decreasing to 450 ppm at 60°S and 800 ppm at 50°N.

The MEX/PFS observations of the CO (2–0) band in all seasons for 2.5 Martian years were analyzed by Sindoni et al. (2011). CO at $L_S$ = 90°–120° varies in that study from 600 ppm at 80°N to 750 ppm at 40°N, 1000 ppm at the equator, and 2000 ppm at 20°S. The variations at $L_S$ = 240°–270° are from 900 ppm at 40°N to 600 ppm at the equator and 520 ppm at 60°S.

CO on Mars is also observed using the low-resolution imaging spectrometer CRISM (Smith et al. 2009) at the Mars Reconnaissance Orbiter (MRO). The observations of the CO (2–0) band at 2.35 μm for 2.5 Martian years were analyzed by Toigo et al. (2013). The observed latitude range varies with season from 35°S–90°N at $L_S$ = 90° to 90°S–40°N at $L_S$ = 270°. The observed so-called enhancement factor for CO at 30°–45°S shows a linear decrease from 1.5 at $L_S$ = 0° to 1.0 at 340° with a jump to 1.5 at 360°. A peak of 2.0 centered at $L_S$ = 150° with a full width at half maximum of 50° overlaps this trend. Strong minima of CO in the subpolar regions of 60°–90° were observed on the south at $L_S$ = 200°–300° and on the north at $L_S$ = 60°–150° with typical mixing ratios of ≈200 and ≈350 ppm, respectively. More detailed data on the MRO/CRISM observations of CO in Smith et al. (2018) confirm the curve in Figure 11.20 and give a globally and annually mean CO mixing ratio ≈800 ppm.

Thus one may conclude that the Mars Express and MRO observations also indicate the increase in CO mixing ratio at southern latitudes in northern summer as observed by Krasnopolsky (2003). However, there are significant quantitative differences between the data from different teams and instruments.

Surprisingly, the latitudinal coverage of the spacecraft observations of CO is even smaller than that of the ground-based observations (for example, down to 35°S and 50°S, respectively, at $L_S$ ≈ 90°). While the observations from the Mars orbiters have obvious advantages in spatial resolution and regularity, the much better spectral resolution in the CSHELL/IRTF observations gives some rewards as well.

The latest CSHELL/IRTF observations of variations of CO on Mars (Krasnopolsky 2015c) were significantly improved using the synthetic spectra fitting instead of the equivalent widths and by a proper choice of the CO and $CO_2$ lines (Section 5.5). The results for four seasonal points in northern spring and summer, that is, near the expected maximum of CO in the southern subpolar regions, are shown in Figure 11.19. No significant variations have been observed from the north pole down to 55°S. The calculated standard deviations of 7% reflect both true CO variations and random errors, and both are low. Comparison of a few observations and the MCD data at $L_S$ =110° is shown in Figure 11.20. Comparing all combinations of the curves, the best agreement is between the observation in Krasnopolsky (2015c) and the MCD

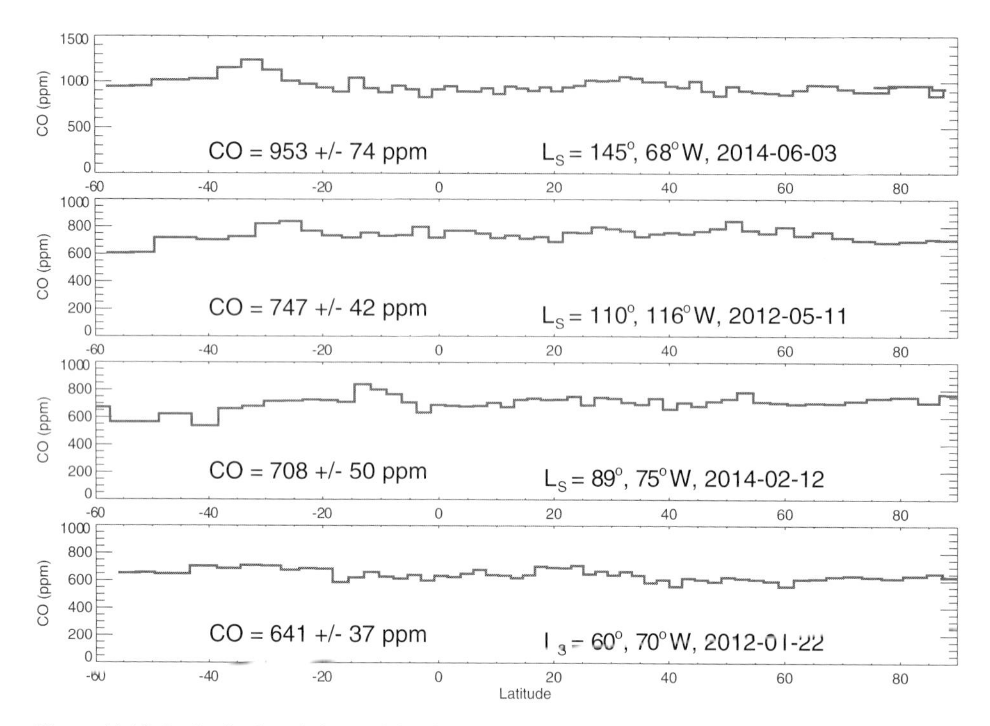


Figure 11.19 Latitudinal variations of the CO mixing ratio observed at four seasons. Mean CO and their standard deviations that reflect the CO variations are also given. From Krasnopolsky (2015c).

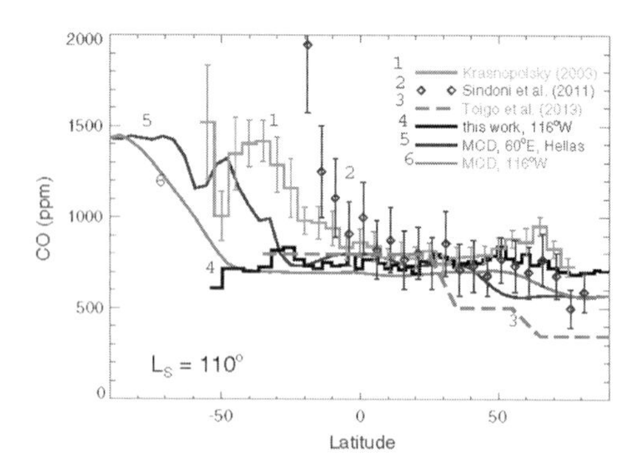


Figure 11.20 Comparison of the observational and model data on latitudinal variations of the CO mixing ratio at $L_S$ = 110°. The MCD predictions are shown for two longitudes. From Krasnopolsky (2015c). (A black and white version of this figure will appear in some formats. For the color version, please refer to the plate section.)

beyond Hellas. The twofold excess in the CO mixing ratio over Hellas affected the conclusion of Krasnopolsky (2003), because Hellas was heavily weighted in his derived latitude dependence of CO. Therefore the increase in the CO mixing ratio in the southern fall and winter does not extend to the middle and low latitudes to the north of 55°S.

Total amounts of CO and other incondensable long-living gases are not changed by the $CO_2$ condensation/sublimation. Therefore the observed CO mixing ratios may be corrected for the seasonal phase of the $CO_2$ cycle (Figure 11.2) to get an annually mean CO mixing ratios that are equal to 667, 693, and 684 ppm for the observing sessions at $L_S = 60°$, $89°$, and $110°$, respectively (Figure 11.19). The mass spectrometer at the Curiosity rover observed a CO mixing ratio of 747 ± 2.6 ppm (Franz et al. 2017) at $L_S = 184°$, and the annually mean CO abundance is 672 ± 2.6 ppm (Krasnopolsky 2017b).

### *11.4.3 Molecular Oxygen*

Homonuclear diatomic molecules of $O_2$ do not have dipole moment, and vibrational and rotational transitions are strongly forbidden for them. Transitions between seven lowest electronic states are forbidden as well (Section 3.11.2), and the allowed Schumann–Runge band system is affected by predissociation and screened by the $CO_2$ absorption. The $O_2$ ($b^1\Sigma_g^+ \rightarrow X^3\Sigma_g^-$, 0–0) band at 762 nm with a radiative lifetime of 12 s is the most convenient for detection of $O_2$. High spectral resolution and signal-to-noise ratio are required to detect weak Martian Doppler-shifted satellites to the strong telluric $O_2$ lines in the ground-based observations (Figure 11.21).

The first detections of $O_2$ on Mars were made by Barker (1972) and Carleton and Traub (1972), and later by Trauger and Lunine (1983), who employed a triple pressure-scanned Fabri–Perot interferometer with a resolving power $\lambda/\delta\lambda = 2.4\times10^5$. Figure 11.21 shows the very strong telluric absorption line of $O_2$, the much weaker lines of $^{18}OO$ and $^{17}OO$, and the faint traces of the Martian line. Barker (1972), Carleton and Traub (1972), and Trauger and Lunine (1983) retrieved $O_2$ abundances of 9.5 ± 0.6, 10.4 ± 1.0, and 8.5 ± 0.1 cm-atm, respectively, and an $O_2$ mixing ratio of $1.2\times10^{-3}$.

Hartogh et al. (2010) observed the $O_2$ line at 773.84 GHz (0.388 mm) using the Herschel submillimeter orbiting observatory. The measured $O_2$ mixing ratio was $(1.40 \pm 0.12)\times10^{-3}$ with some indications of possible variations with altitude.

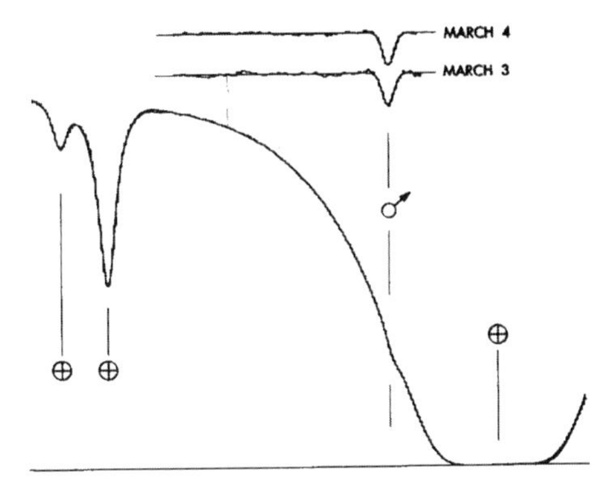


Figure 11.21 Fabry–Perot spectrum of the $O_2$ Q9 line at 7635.192 Å on Mars. Lines to the left of the Q9 line are those of $^{17}OO$, $^{18}OO$, and the Doppler-shifted Martian line. Its shapes corrected for the telluric absorption are shown above for two observing days. From Trauger and Lunine (1983).

Measurements using the quadrupole mass spectrometer at the Curiosity rover result in the $O_2$ mixing ratio of $(1.74 \pm 0.06)\times10^{-3}$ (Franz et al. 2017) at $L_S = 184°$ and the annual mean abundance of $(1.56 \pm 0.054)\times10^{-3}$ (Krasnopolsky 2017b). Actually the aerosol extinction was neglected in the retrievals of $O_2$ from the ground-based observations. According to Davies (1979), the similar neglect in the Viking MAWD retrievals from the $H_2O$ band at 1.38 μm results in an underevaluation by a factor of 1.3. Then the ground-based data corrected by this factor give $1.56\times10^{-3}$ and agree with the Curiosity value.

The measured $O_2$/CO ratio is ≈2 on Mars, while the expected value is 1/2 for the $CO_2$ photolysis products. This is an indication of a long-term accumulation of $O_2$ from a rather weak $H_2O$ photolysis.

## 11.5 Mass Spectrometric Measurements in the Lower Atmosphere and Martian Meteorites: Variability of Argon

### *11.5.1 Mass Spectrometric Measurements*

There have been three mass spectrometric studies of the Martian lower atmosphere: at the Viking 1 and 2 landers (Owen et al. 1977), Phoenix (Niles et al. 2010), and the Curiosity rover (Mahaffy et al. 2013; Wong et al. 2013; Franz et al. 2017; Conrad et al. 2016). Furthermore, the atmospheric gases are trapped in the Martian meteorites, which are Martian rocks ejected from Mars and captured on the Earth. The Martian meteorites are studied in the laboratory to reveal evolution of the atmosphere, because the trapped gases in different meteorites and different fractions refer to different epochs on Mars, and those in the glass fraction reflect the current atmosphere. Summary of the results on the Martian meteorites may be found in Bogard et al. (2001). The measured data on the atmospheric composition are in Table 11.2.

Absolute abundances of krypton and xenon are not given in the Curiosity data, while their isotope ratios are discussed in detail (Conrad et al. 2016). $Kr/^{84}Kr = 1.85$ and $Xe/^{132}Xe = 5.27$ in those data, and the abundances of Kr and Xe extracted from the meteorites are calculated here using these values, Ar from the Curiosity results, $^{40}Ar/^{36}Ar = 1800$, $^{36}Ar/^{132}Xe = 900$, and $^{84}Kr/^{132}Xe = 20.5$ from the meteorite data (Bogard et al. 2001).

Table 11.2 *Composition of Mars lower atmosphere from mass spectrometric measurements*

| Species | $N_2$ | Ar | Ne | Kr | Xe | $O_2$ | CO |
|---|---|---|---|---|---|---|---|
| Viking | 2.7% | 1.6% | 2.5 ppm | 300 ppb | 80 ppb | $(1\text{–}4)\times10^{-3}$ | – |
| Curiosity[a] | 2.5 | 1.9 | – | – | – | $1.56\times10^{-3}$ | 673 ppm |
| Meteorites | – | – | – | 460 | 70 | – | – |

[a] The measured mixing ratios from Franz et al. (2017) are scaled by 0.899 to reproduce annual mean values (Krasnopolsky 2017b).

Table 11.3 *Isotope ratios of atmospheric species on the Earth and Mars*

| Isotope ratio | Earth | Viking | Curiosity | Meteorites |
|---|---|---|---|---|
| $^{16}O/^{18}O$ | 499 | ≈500 | 476 | – |
| $^{12}C/^{13}C$ | 89 | ≈90 | 85 | – |
| $^{14}N/^{15}N$ | 274 | 168 | 173 | >180 |
| $^{40}Ar/^{36}Ar$ | 292 | 3000 | 1900 | 1800 |
| $^{36}Ar/^{38}Ar$ | 5.35 | 5.5 | – | ≤3.9 |
| $^{129}Xe/^{132}Xe$ | 0.97 | 2.5 | 2.57 | 2.5 |

Isotope ratios of oxygen and carbon in $CO_2$ were measured by the Viking landers equal to the terrestrial values (Table 11.3) with uncertainty of 5%. These ratios correspond to $\delta^{18}O = 48 \pm 5$ ‰ and $\delta^{13}C = 46 \pm 4$ ‰ in the Curiosity measurements, still within the uncertainties of the Viking values. The heavy isotopes in the atmosphere are enriched by nonthermal escape and sputtering, and depleted by fractionation with solid-phase reservoirs. The enrichment and depletion are almost balanced for oxygen and carbon on Mars.

The significant enrichment in heavy nitrogen ($^{14}N/^{15}N = 173 \pm 11$, Table 11.3) clearly indicates a large loss of nitrogen by nonthermal escape with isotope fractionation.

Radiogenic $^{40}Ar$ originates from a decay of $^{40}K$ by electron capture and emission of a gamma-photon at 1.46 MeV. (Another decay branch of $^{40}K$ is $\beta$-decay to $^{40}Ca$ with a total half-life of 1.25 Byr.) $^{40}Ar$ on Mars is enriched relative to the terrestrial ratio by a factor of 6.5, while two other isotopes of argon that were captured from the protosolar nebula have the same ratio as on the Earth. Radiogenic isotope $^{129}Xe$ is a product of $^{129}I$ and also enriched on Mars relative to the Earth. Though a few processes affected the Martian Ar and Xe isotope ratios, the most significant was a loss of the primordial Martian atmosphere by intense meteorite erosion in the first 0.8 Byr (Melosh and Vickery 1989).

Five isotopes of Kr and eight nonradiogenic isotopes of Xe are generally similar to those in the solar wind (Conrad et al. 2016). Corrections for fractionation in the solar wind and some minor deviations from the solar isotope ratios are discussed by the authors.

### *11.5.2 Seasonal-Latitudinal Variations of Ar*

$^{40}Ar$ can capture the solar neutron, and this capture is followed by the $\beta$-decay of $^{41}Ar$ to a metastable state of $^{41}K$ that emits a $\gamma$-photon at 1.294 MeV. This emission was measured by the Gamma Ray Spectrometer (GRS) at the Mars Odyssey orbiter (Sprague et al. 2012). The observed emission was averaged over latitude bands of 15° and converted into argon mixing ratios. Seasonal variations of argon are the most prominent near the South Pole (Figure 11.22) and reach a factor of 5.5 in southern winter. The effect is much smaller to the north of 60°S including the North Pole. Along with the CO observations (Section 11.4.2), these are the basic data on dynamics of the polar processes. The measurements of Ar near the South Pole are compared in Figure 11.22 with predictions of four general circulation models, and the differences are significant.

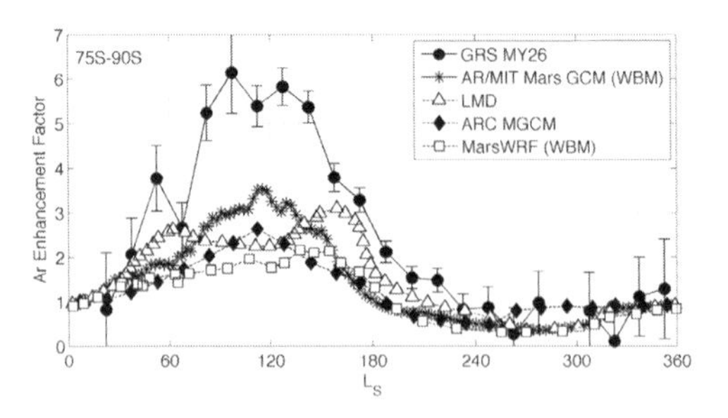


Figure 11.22 Seasonal variations of Ar in the southern subpolar regions observed by GRS and predicted by GCMs. Enhancement factor is the observed Ar abundance divided by the Viking value of 1.6%. From Sprague et al. (2012).

## 11.6 Photochemical Tracers: Ozone, $O_2$ Dayglow at 1.27 μm, and Hydrogen Peroxide $H_2O_2$

### *11.6.1 Ozone*

Similar to CO and $O_2$, ozone $O_3$ is a product of the $CO_2$ chemistry. While lifetimes of CO and $O_2$ are years and decades, lifetime of ozone as odd oxygen ($O_3$ + O) component is of the order of a day, and ozone is a sensitive tracer of photochemistry. Dissociation energy of ozone is low (1.05 eV), therefore ozone is chemically active.

Ozone is usually measured using its UV absorption in the so-called Hartley band (Figure 7.6):

$$O_3 + h\nu(\lambda < 310\,\text{nm}) \rightarrow O_2\left(a^1\Delta_g\right) + O\left(^1D\right). \tag{11.1}$$

$O_2(a^1\Delta_g)$ is either quenched by $CO_2$ with a rate coefficient of $\approx 10^{-20}$ cm$^3$ s$^{-1}$ or radiates a photon at 1.27 μm with radiative time of 4500 s. The latter dominates above ≈20 km on Mars. $O(^1D)$ is mostly quenched by $CO_2$; however, it also reacts with $H_2O$ and $H_2$ and dissociates these species.

The Hartley band (Figure 7.6) is maximal of $1.1\times10^{-17}$ cm$^2$ at 255 nm with a half width of 40 nm. The solar light varies within the band by a factor of ≈15, and the surface reflection and dust properties may vary as well. This restricts the $O_3$ detection, and the Mariner team that detected ozone for the first time from the Mariner 7 flyby (Barth and Hord 1971) and then studied its distribution in more detail from the Mariner 9 orbiter (Barth et al. 1973; Figure 11.23), evaluated the detection limit at 3 μm-atm. (Ozone on Mars is measured in μm-atm, and 1 μm-atm = $2.7\times10^{15}$ cm$^{-2}$.)

The Mariner 9 observations revealed very strong latitudinal and seasonal variations of ozone. The greatest ozone abundances ≈50 μm-atm were detected at the subpolar latitudes 50°–75°N at the end of northern winter. Anticorrelation between ozone and water vapor was another important conclusion from the Mariner 9 observations.

Later ground-based measurements of ozone became possible using the infrared heterodyne technique (Section 6.6). A selected Doppler-shifted line of the $O_3$ band at 9.6 μm was observed typically at eight locations along the dayside limb of Mars to gain in airmass. The observations (Espenak et al. 1991; Fast et al. 2006, 2009) were made at various seasons

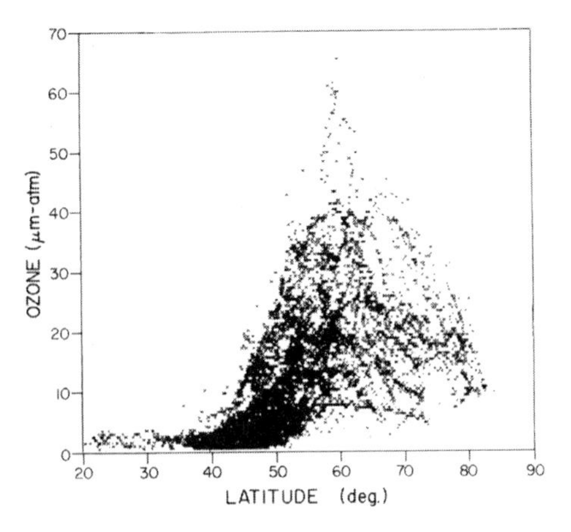


Figure 11.23 Mariner 9 observations of ozone at the end of northern winter (Traub et al. 1979).

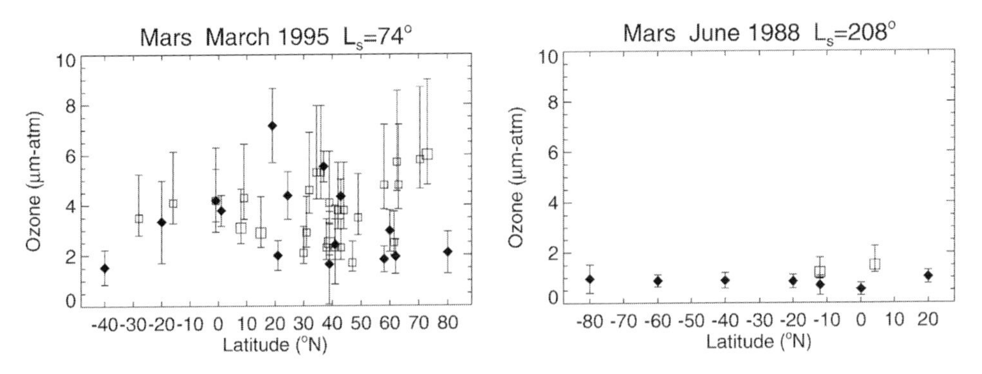


Figure 11.24 Ground-based (black symbols; Fast et al. 2006) and HST (open symbols; Clancy et al. 1999) observations of ozone near aphelion and perihelion.

using NASA IRTF. The instrument resolving power was up to $5\times10^6$, and the field of view was ~1 arcs with a typical angular diameter of Mars of 10 arcs.

Despite the very high spectral resolution, the ozone lines were weak and measured with significant uncertainties. On the other hand, the ground-based detection of the Martian ozone at the low and middle latitudes not covered by the Mariner 9 observations resulted in a significant progress in the problem. Furthermore, the ozone retrieval did not depend on spectral properties of the rocks and dust that affect the UV observations. Some results near aphelion and perihelion are shown in Figure 11.24 and demonstrate significant seasonal variability of ozone.

Clancy et al. (1999) observed ozone on Mars using a UV spectrometer at the Hubble Space Telescope. The observations were made near the dayside limb to improve the detection limit to ≈1 μm-atm. There are some similarities and differences (Figure 11.24) between the ozone abundances measured by Fast et al. (2006) and Clancy et al. (1999). For example, the data for aphelion at 60–80°N are ≈2.5 and 5 μm-atm, respectively.

The most detailed observations of seasonal and latitudinal variations of ozone were made using MEX/SPICAM (Section 6.2) in the UV range. The data for 1 Martian year

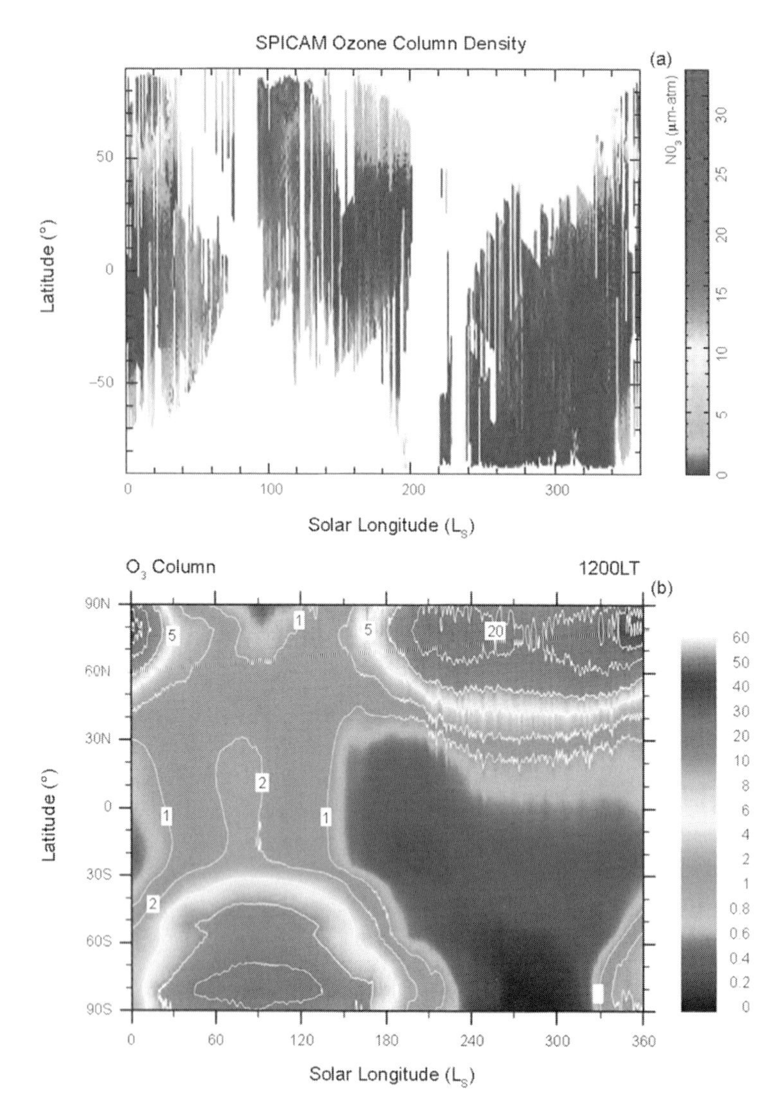


Figure 11.25 Seasonal variations of the daytime $O_3$ column (μm-atm) observed by MEX/SPICAM (upper panel; Perrier et al. 2006) and from the Mars Climate Database (lower panel; updated from Lefèvre et al. 2004). From Krasnopolsky and Lefèvre (2013). (A black and white version of this figure will appear in some formats. For the color version, please refer to the plate section.)

were analyzed by Perrier et al. (2006) and are shown in Figure 11.25. The observed spectra were divided by those measured at Olympus Mons with elevation of 22 km, where effects of ozone, atmospheric Rayleigh and dust scattering are minimal. This operation compensates significantly for uncertainties in the instrument calibration and the solar spectrum convolved by the instrument. The ratio was fitted assuming a linear albedo between 210 and 300 nm, the Rayleigh scattering atmosphere, a relative vertical profile of ozone from the Mars Climate Database, and dust with $\omega = 0.6$, $g = 0.88$, and $\tau_d$, all independent of

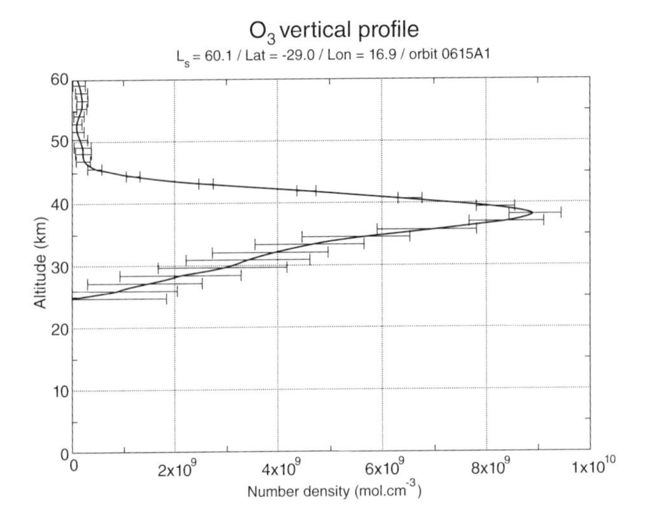


Figure 11.26 Ozone nighttime vertical profile observed at $L_S = 60°$ and 29°S (Lebonnois et al. 2006).

$\lambda$. Therefore there were four fitting parameters: surface albedo at 210 and 300 nm, $O_3$ column, and $\tau_d$. The approach is very reasonable, though the retrieved ozone abundances $\approx$1 μm-atm and smaller should be inevitably very uncertain. These small abundances were measured in the most of the area covered by the observations (Figure 11.25). The MCD map of ozone reproduces the main features of the observed distribution, is better seen in detail, and covers all latitudes including those at the polar night.

Photolysis is switched off at night, and a significant part of atomic oxygen at 30–50 km is converted into ozone and forms detached layers that are prominent near aphelion and much weaker near perihelion. This was predicted by a time-dependent model by Krasnopolsky (2006b, figures 7, 10, 11), by MCD, and observed by SPICAM stellar occultations (Figure 11.26; Lebonnois et al. 2006).

### *11.6.2 $O_2$ Dayglow at 1.27 μm*

It was mentioned above that the UV photolysis of ozone excites the $O_2(a^1\Delta_g)$ metastable state that emits photons at 1.27 μm. Evidently vertical intensity of the dayglow is

$$4\pi I\ (\mathrm{MR}) = \frac{10^{-12}J}{r_h^2}\int_0^\infty \frac{[\mathrm{O_3}]dh}{1+\tau k[\mathrm{CO_2}]}. \qquad (11.2)$$

Here $10^{-12}$ is the conversion of photons to mega-rayleighs, $J$ = 0.0082 s$^{-1}$ is the $O_3$ photolysis frequency at 1 AU, $r_h$ is the heliocentric distance in AU, $\tau$ = 4460 s is the radiative lifetime, and $k \leq 2\times10^{-20}$ cm$^3$ s$^{-1}$ is the quenching rate coefficient.

The dayglow was discovered by Noxon et al. (1976) and observed by Traub et al. (1979) in three latitude bands (north, equator, and south). The ground-based observations are possible at significant Doppler shifts relative the telluric $O_2$ lines and by means of high-resolution spectroscopy.

Clancy and Nair (1996) argued that the atmosphere at low and middle latitudes is clean and therefore cold near aphelion and dusty and warm near perihelion, with increase of the solar flux by a factor of 1.45 and heating by dust. Therefore water vapor extends much higher near perihelion and depletes high-altitude ozone above ≈15 km that may be significant near aphelion.

This stimulated Krasnopolsky (2013c) to start regular ground-based observations of the $O_2$ dayglow at 1.27 μm that is a sensitive tracer of high-altitude ozone and photochemistry in the middle atmosphere. Thirteen observing sessions were made in the period of 1997–2012 using IRTF/CSHELL.

An example of the observed spectrum is shown in Figure 11.27. The observed shapes of the telluric $O_2$ lines were fitted by sums of Gaussian and parabola, and differences between the observations and fits at the expected Doppler-shifted positions of the Martian lines were attributed to the dayglow. A summary of those long-term observations is the map of the dayglow seasonal and latitudinal variations (Figure 11.28) that is compared with a model of those variations (Krasnopolsky 2009b). The observed dayglow vertical intensities varied from 0.4 to 17 MR, and the mean intensities at low latitudes are 5 MR near aphelion and 1 MR near perihelion.

The most detailed observations of the $O_2$ dayglow were made by Guslyakova et al. (2016) using MEX/SPICAM IR. The observations covered 6 Martian years, and the data for the first year of the observations are shown in Figure 11.29a. They refer to different local time, and this is discussed by the authors. Limb observations of the dayglow were made as well (Guslyakova et al. 2014). The instrument field of view was ≈30 km on the limb, therefore the observational data were compared with the MCD results convolved with the instrument field of view. Gagne et al. (2012) applied MCD to calculate seasonal and latitudinal variations of the $O_2$ dayglow (Figure 11.29b).

Limb observations of the $O_2$ dayglow using MRO/CRISM in 2009–2016 were analyzed by Clancy et al. (2017). The retrieved vertical profiles were fitted by the latest MCD data with variable water vapor abundance and the quenching rate coefficient by $CO_2$. Summary

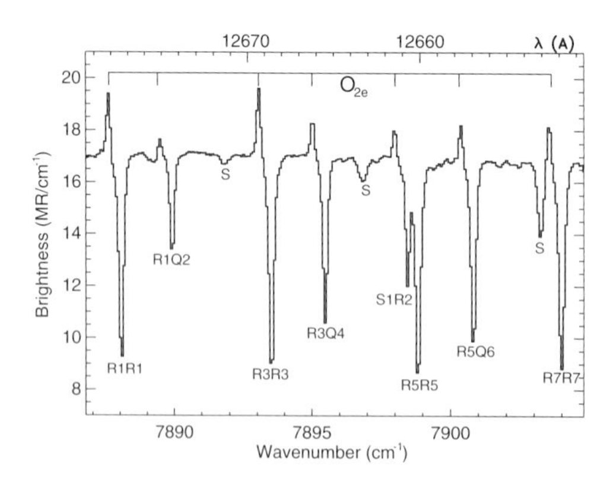


Figure 11.27 IRTF/CSHELL spectrum of the $O_2$ dayglow at 1.27 μm. The dayglow emission lines $O_{2e}$ are Doppler-shifted to the red from the telluric $O_2$ absorption lines. Solar lines are marked S. From Krasnopolsky (2007).

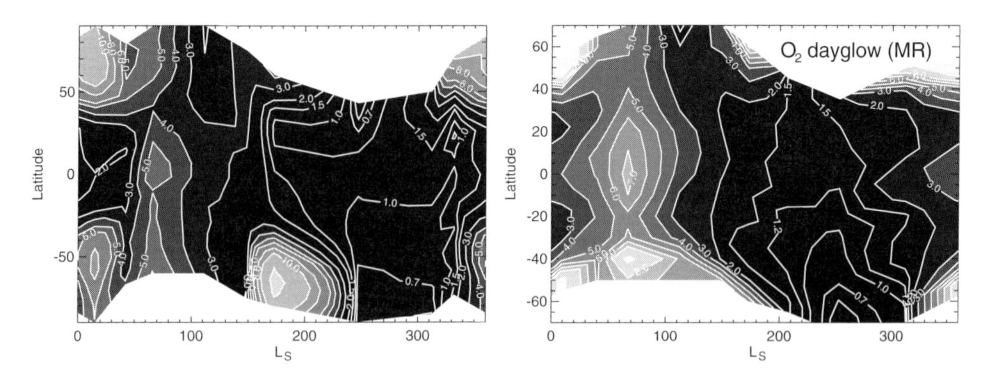


Figure 11.28 Seasonal-latitudinal variations of the $O_2$ dayglow at 1.27 μm: observations and model (left and right panels, respectively). From Krasnopolsky (2013c). (A black and white version of this figure will appear in some formats. For the color version, please refer to the plate section.)

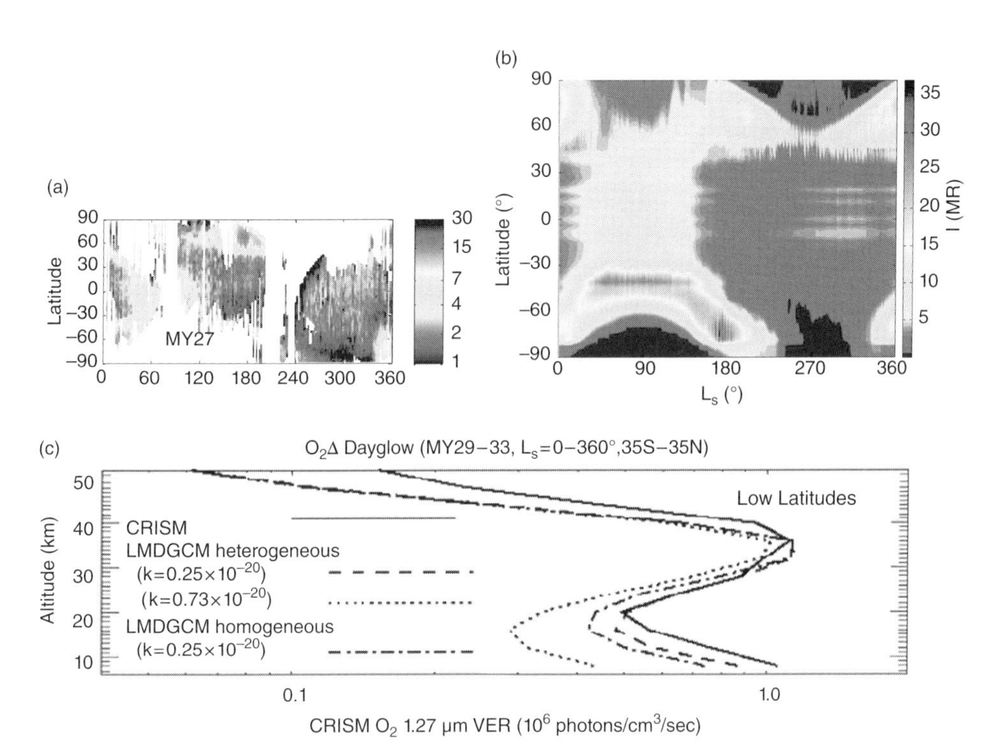


Figure 11.29 (a) SPICAM IR observations of the $O_2$ dayglow at 1.27 μm for 1 Martian year (Guslyakova et al. 2016). (b) Seasonal evolution of the dayglow at LT = 12 based on the MCD data (Gagne et al. 2012). (c) Mean vertical profile of the dayglow volume emission rate (VER) observed using CRISM/MRO at low latitudes compared with three model profiles (Clancy et al. 2017). (A black and white version of this figure will appear in some formats. For the color version, please refer to the plate section.)

of all data at the low latitudes ±35° is shown in Figure 11.29c and compared with MCD fittings. The preferable $k$ is $2.5\times10^{-21}$, smaller than $7.3\times10^{-21}$, and $5\times10^{-21}$ $cm^3$ $s^{-1}$ recommended by Guslyakova et al. (2016) and Krasnopolsky (2009b), respectively.

### 11.6.3 Peroxide $H_2O_2$

Peroxide $H_2O_2$ is a rather abundant product of the odd hydrogen chemistry on Mars. After a few attempts to detect that gave only upper limits, peroxide was detected almost simultaneously by Clancy et al. (2004) using the submillimeter JCMT telescope at 362 GHz and Encrenaz et al. (2004) using IRTF/TEXES (Section 6.3) at 8 μm (Figure 11.30). The TEXES observations provide a spectral resolving power of $8\times10^4$ and are spatially resolved with a mapping capability using various slit positions. However, the task is difficult even for such an advanced instrument as TEXES (Figure 11.30). The line depth is 0.5% for $H_2O_2$ = 15 ppb, and a very high signal-to-noise ratio of 400 is required to get a two-sigma detectivity of 10 ppb.

The last summary of the long-term observations of $H_2O_2$ at low latitudes from Encrenaz et al. (2015a) is shown in Figure 11.31. It includes two upper limits, six TEXES and one JCMT detections, and data of four models. The MCD model (Lefèvre et al. 2008) involves heterogeneous loss of OH and $HO_2$ on water ice particles, while the model by Krasnopolsky (2006b, 2009b) is based on the TES data and includes heterogeneous loss of $H_2O_2$ on water ice. Encrenaz et al. (2015a) argue that their last points, 15 ppb at $L_S$ = 96° and 30 ppb and $L_S$ = 156°, favor MCD. However, their figures 6 and 12 do not support the claimed values, and the smaller $H_2O_2$ abundances look more appropriate. The upper limits of 3 ppb at 77° (Herschel/HIFI; Hartogh et al. 2010) and 10 ppb at 112° (TEXES; Encrenaz et al. 2002) agree better with Krasnopolsky (2009b; Figure 11.32).

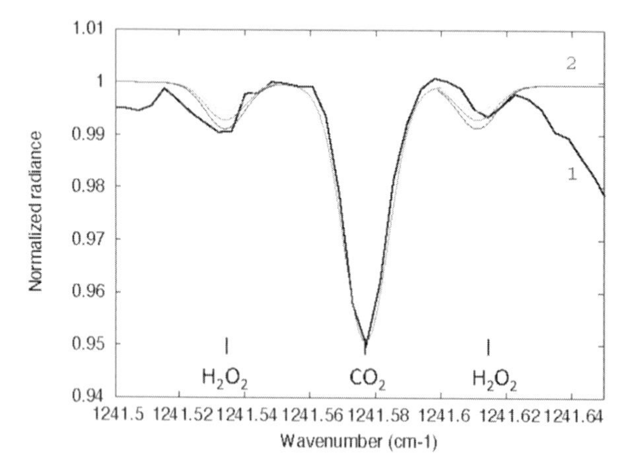


Figure 11.30 A narrow interval in TEXES spectrum of Mars that includes the $CO_2$ line at 1241.58 $cm^{-1}$ between two $H_2O_2$ lines. These lines are used to retrieve mixing ratio of $H_2O_2$. The spectrum is averaged over nine spatial pixels. The true spectrum is (1), a corrected spectrum is (2), and the other lines are spectral fits with $H_2O_2$ = 15, 20, and 25 ppb. The observed value is 20 ± 5 ppb. From Encrenaz et al. (2015a).

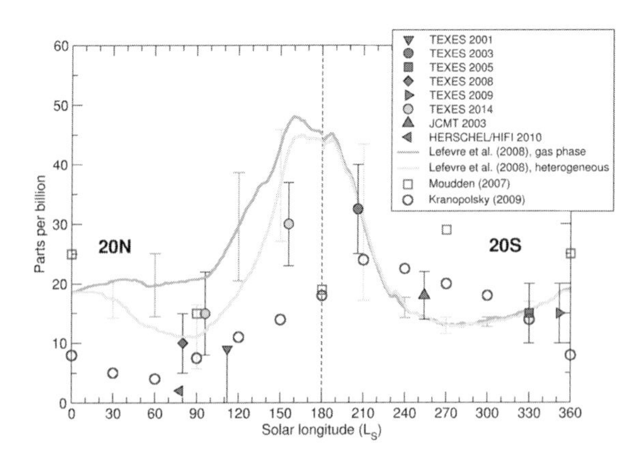


Figure 11.31 Seasonal variations of $H_2O_2$ at low latitudes (20°N at $L_S$ = 0°–180° and 20°S at 180°–360°). Two curves are based on MCD (Lefèvre et al. 2008) with and without heterogeneous loss of odd hydrogen (the lower and upper curves, respectively). Open circles are from the 1D model by Krasnopolsky (2009b), and squares at 0°, 90°, 180°, 270°, and 360° are from the 3D model by Moudden (2007). From Encrenaz et al. (2015a).

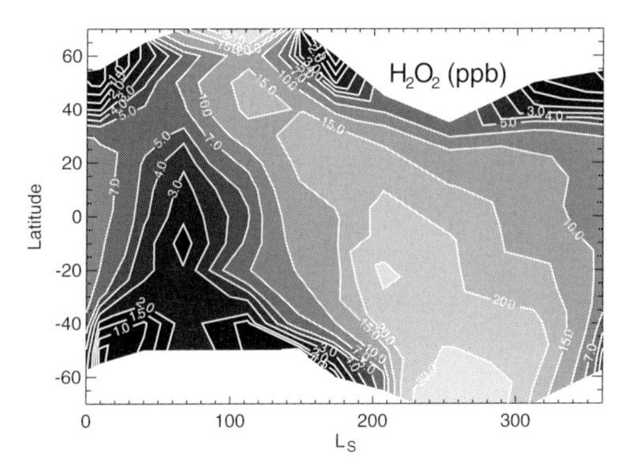


Figure 11.32 Seasonal and latitudinal variations of the $H_2O_2$ mixing ratio in the model by Krasnopolsky (2009b). (A black and white version of this figure will appear in some formats. For the color version, please refer to the plate section.)

Overall, the detection and long-term TEXES observations of $H_2O_2$ is a significant contribution to the Martian science.

## 11.7 Methane

It is currently thought that if microbial life exists on Mars, then methanogenesis is a likely pathway of its metabolism in net reactions

$$4\,H_2 + CO_2 \rightarrow CH_4 + 2\,H_2O + 1.53\text{ eV}$$

$$4\,CO + 2\,H_2O \rightarrow CH_4 + 3\,CO_2 + 3.17\text{ eV}.$$

The probable existence of subsurface liquid water would provide protected habitats for such organisms, similar to the methanogenic bacteria beneath the Earth's surface. A mass spectrometer at the British Beagle 2 lander was strictly aimed to detect methane and therefore life on Mars. That lander was a part of the Mars Express mission; however, its landing failed in December 2003.

### *11.7.1 Observations*

Krasnopolsky et al. (1997) found some indications of methane in their high-resolution spectrum. However, methane could be a by-product in that spectrum, and the results stimulated Krasnopolsky et al. (2004a, 2004b) to make special efforts to search for methane. The observations were made in January 1999 using the FTS at the Canada–France–Hawaii Telescope (CFHT). The P-branch of the $CH_4$ band at 3.3 μm was observed with the apodized resolving power of $1.8\times10^5$. The observation revealed methane slightly above the detection limit with a mixing ratio of $10 \pm 3$ ppb. The abstract to the European Geosciences Union (Krasnopolsky et al. 2004a) with the results of the observation was submitted on January 9, 2004, just at the beginning of the Mars Express science operations. That was the first publication on the detection of methane on Mars. Mumma et al. (2003) published an abstract on their observations of methane in 2003 without any results.

The PFS team at Mars Express (Formisano et al. 2004) observed methane using the Q-branch at 3018 $cm^{-1}$ of the $CH_4$ band at 3.3 μm (Figure 11.33). The instrument resolving power was $\nu/\delta\nu = 1500$ near 3 μm, and they averaged thousands of spectra to improve signal-to-noise ratio. However, averaging does not remove systematic errors, and the retrieved $CH_4$ absorption was comparable with the difference between the observed and the best-fit synthetic spectra (Figure 11.33). The retrieved methane abundances varied from 0 to 35 ppb. The latest PFS methane paper (Geminale et al. 2011) is a summary of

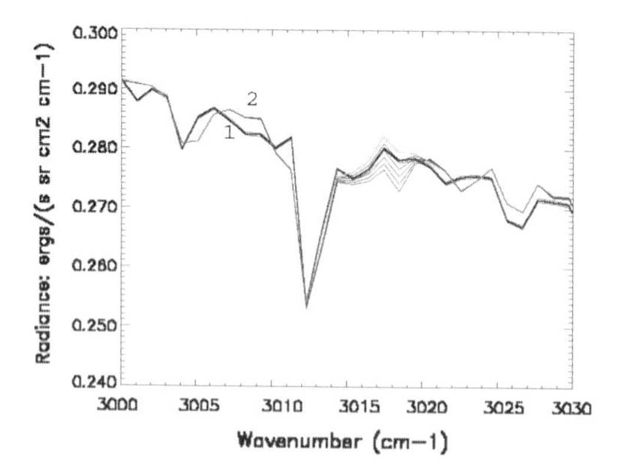


Figure 11.33 Average of 2931 PFS/MEX spectra (1) near the $CH_4$ Q-branch at 3018 $cm^{-1}$. Synthetic spectra (2) are calculated for 0, 10, 20, 30, 40, and 50 ppb of methane. The best fit is ≈15 ppb. From Formisano et al. (2004).

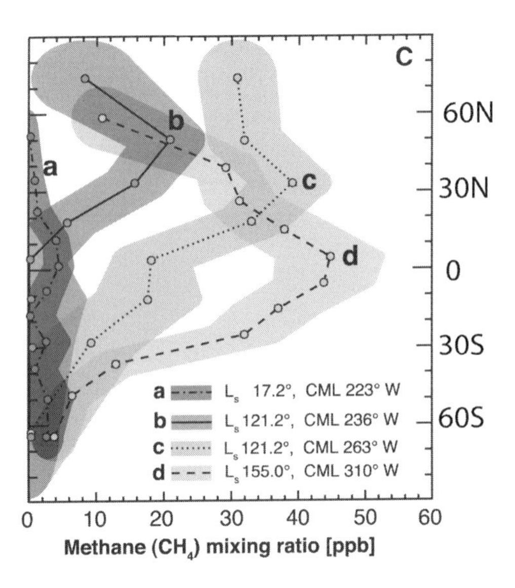


Figure 11.34 Latitudinal variations of methane in four observing sessions by Mumma et al. (2009). CML is central meridian longitude.

6 years of the observations. The authors presented a zonally averaged seasonal-latitudinal map of methane and maps for four seasons in the latitude range of $\pm 60°$ with a grid of $20° \times 20°$. The $CH_4$ mixing ratio in the maps varies from 0 to 70 ppb with a global-mean value of 15 ppb. Some regions with very abundant methane are in vicinity of those with very low methane, and this should cause an intense exchange of methane between them.

The best lines for ground-based search for methane on Mars are R0 3028.75 $cm^{-1}$ and R1 3038.50 $cm^{-1}$. They are strong, singlet, and without significant contamination by other lines. However, telluric methane is more abundant than that on Mars by a factor of $\approx 2 \times 10^4$, and its detection is a difficult task. Mumma et al. (2009) observed methane using IRTF/CSHELL in January and March 2003 and February 2006 (Figure 11.34). The abundances observed in 2003 varied from 0 to 45 ppb.

Fonti and Marzo (2010) analyzed the MGS/TES spectra that cover the $CH_4$ band at 1306 $cm^{-1}$. The TES spectral resolving power is $\approx 100$ and $\approx 200$, and averaging of 11,000 and 3000 spectra, respectively, is required to achieve a detection limit of 10 ppb, respectively (Fonti and Marzo 2010). They chose $3 \times 10^6$ dayside spectra in the latitude range of $\pm 60°$ and then selected to a methane cluster $9.3 \times 10^5$ spectra that favored the detection of methane. Eleven seasonal maps of methane that cover 3 Martian years are the final products of this study. Although the maps show numerous details, one should bear in mind that those 11 maps represent 103 detection points using the above values ($9.3 \times 10^5/11{,}000 \approx 85$; correction for regions observed with the resolving power of 200 results 103 points). Therefore the mean number of the effective points is $\approx 9$ per map, and four maps are based on $\approx 4$ points per map. Their approach is not standard, and it is not clear that the clustering does not break the true statistics of the signal near 1306 $cm^{-1}$.

Krasnopolsky (2011b, 2012a) observed the $CH_4$ R0 and R1 lines in February 2006, December 2009, and March 2010 using IRTF/CSHELL. The measurable signal at ≈9 ppb (Figure 11.35) was from a region near the deepest canyon Valles Marineris. Spectra of the Moon measured almost simultaneously by the same instrument were applied to improve extraction of the Martian methane in the observations in 2009 and 2010. However, no methane had been observed with upper limits of 8 ppb in both sessions (Figure 11.36).

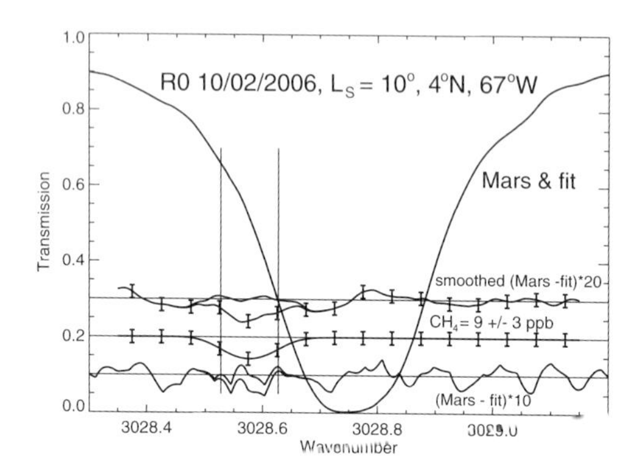


Figure 11.35 The best spectral fit to the observed CH4 R0 line. The observed and synthetic spectra are not distinguishable, and their difference is shown at the bottom scaled by a factor of 10 with and without the Martian methane. Interval of 0.1 cm$^{-1}$ centered at the expected position of the Martian methane is marked by the vertical lines. These curves smoothed within 0.1 cm$^{-1}$ are shown with their corrected standard deviation. The contribution of methane absorption is shown separately. From Krasnopolsky (2012a).

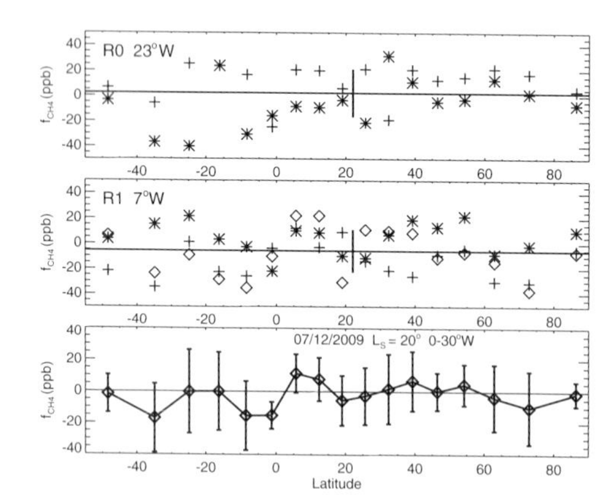


Figure 11.36 Two observations of the $CH_4$ R0 line and three observations of the R1 line in the session on December 7, 2009, reveal no methane with an upper limit of 8 ppb (Krasnopolsky 2012a). Horizontal lines and vertical bars in the upper and middle panels are the mean values and standard deviations. Lower panel shows the sum of the observations.

Villanueva et al. (2013) searched for methane using a few advanced instruments: VLT/CRIRES, IRTF/CSHELL, and Keck/NIRSPEC (Section 6.3). However, the observations resulted in only upper limits (8 ppb in January 2006 and 7 ppb in November 2009 and April 2010).

All above observations of methane were made by the spectral instruments of general purpose that were not specially designed to search for methane on Mars. The Mars Science Laboratory at the Curiosity rover includes a tunable laser spectrometer (TLS; Section 6.5) to measure the $CH_4$ R3 triplet at 3057.7 $cm^{-1}$ with resolving power $\nu/\delta\nu \approx 1.5\times10^7$ (Webster et al. 2015a). A multipass gas cell provides a 16 meter pass through the Martian air, and sessions with removal of $CO_2$ enhance abundances of the remaining gases by a factor of 23 $\pm$ 1. The measurements in the period of 2013–2017 are shown in Figure 11.37.

The observations reveal a background component of methane with regular seasonal variations from 0.24 ppb near aphelion to 0.65 ppb in late northern summer and temporary enhancements of ~7 ppb. The authors conclude that "the large seasonal variation in the background and occurence of higher temporary spikes (~7 ppb) are consistent with small localized sources of methane released from martian surface or subsurface reservoirs."

Search for methane on Mars using EXES at the SOFIA airborne observatory in March 2016 resulted in upper limits of 1 to 4 ppb at different locations (Aoki et al. 2017). EXES is a version of TEXES adjusted to the conditions of the SOFIA observatory. The observations were made at the altitude of $\approx$13 km with much lower pressure and overhead methane than, say, at NASA IRTF at elevation of 4.2 km.

While the data based on the spacecraft (PFS and TES) observations favor a permanent presence of highly variable methane, both ground-based and TLS observations indicate episodic injections of methane (Figure 11.38). It looks like there was a strong emission of

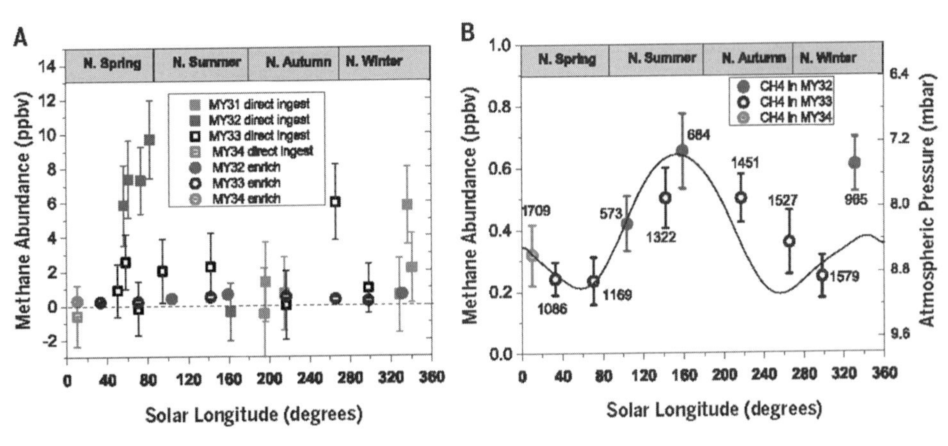


Figure 11.37 A: all TLS measurements of methane in 2013–2017; those direct are squares, with enrichment are circles. B: background variations observed with enrichment. The solid line is the mean pressure variation in the inverted scale. From Webster et al. (2018).

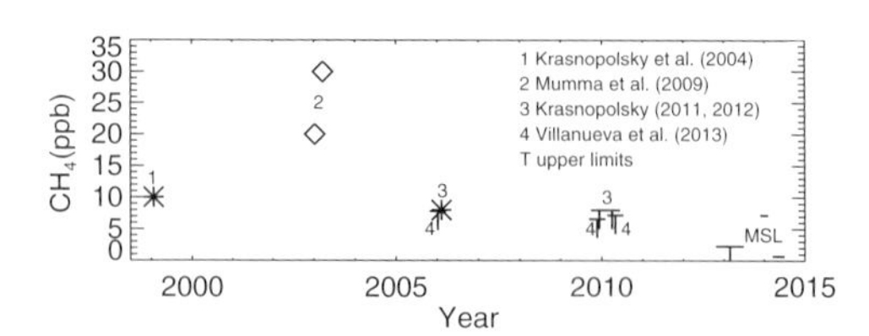


Figure 11.38 Comparison of the ground-based and TLS/MSL observations of methane. Four and five ground-based observations resulted in detections and upper limits for methane, respectively.

methane in 2003, and both points from Mumma et al. (2009) with the two-month interval refer probably to the same episode. Four detections and five upper limits result in the measurable abundances of methane in ≈40% of the ground-based observations.

Uncertainties of the TLS measurements without enrichment are similar to those of the ground-based observations, ≈2.5 ppb. The great disadvantages of the abundant telluric methane, the much lower spectral resolution, and the smaller brightness in the ground-based observations are compensated by the longer path in the Martian air, which is equal to 2–3 scale heights and longer than that in TLS by three orders of magnitude.

### *11.7.2 Variability*

There are three reactions that remove methane from the Martian atmosphere (Krasnopolsky et al. 2004b):

$$CH_4 + h\nu(1216\ \text{Å}) \rightarrow \text{products},$$

$$CH_4 + OH \rightarrow CH_3 + H_2O,$$

$$CH_4 + O(^1D) \rightarrow CH_3 + OH.$$

$CH_3$ reacts with atomic oxygen and forms formaldehyde $H_2CO$ that dissociates in a few hours. Therefore the above reactions are irreversible sinks of $CH_4$ with a total lifetime of ≈300 years. Other useful characteristic times are those of vertical mixing, ≈10 days, global mixing, ≈0.5 years, diffusion in the regolith from depths of 10 m and 1 km, 0.3 and 500 years, respectively (Krasnopolsky et al. 2004b). Comparing these times, methane should be uniformly mixed both vertically and horizontally and might vary within centuries. However, the observations show significant local (place-to-place) and temporal variabilities that require effective heterogeneous chemistry to be explained.

Krasnopolsky (2006c) concluded that the heterogeneous effect of dust is smaller than that of the surface rocks if they are of the same composition. He considered kinetic data on reactions of $CH_4$ with metal oxides, superoxide ions, and in the classic Fischer–Tropsch process used for production of synthetic gasoline. Trainer et al. (2011) established upper limits to trapping of methane by the polar ice analogs. All these reactions are extremely slow at the Martian temperatures and cannot explain the observed variations of methane. We do not consider effects of electrochemistry in dust devils that were overestimated by orders of magnitude by applying a steady state model to those transient events. Search for organics by the Viking instruments resulted in restrictive upper limits (Biemann et al. 1976).

Jensen et al. (2014) studied a loss of methane in collision between quartz ($SiO_2$) particles. Tumbling quartz grains of 125–1000 μm in size in pure methane at 600 mbar for 115 days, they found that 3% of the grain surface constitute $(Si–O)_3Si–CH_3$ complexes. Therefore the collisions activate the surface that attracts methyl and dissociates methane. However, quantitative assessment of this effect for the Martian conditions is difficult, and the authors confused the photochemical loss of methane of 270 tons $yr^{-1}$ from Krasnopolsky et al. (2004b) with its global amount of $10^5$ tons for 10 ppb. Furthermore, the effect in the extremely diluted (by a factor of a billion) methane on Mars would be very much weaker (maybe, by the same factor) than that in the pure methane at a pressure that exceeds the Martian pressure by a factor of 100. Anyway, this mechanism deserves further study.

Lefèvre and Forget (2009) and Mischna et al. (2011) simulated variations of methane using their GCMs. The only variations obtained with the conventional gas-phase chemistry are those induced by condensation and sublimation of $CO_2$ at the polar caps, similar to those for CO and Ar. The GCMs have tested the behavior of methane under various assumptions on its sources and heterogeneous sinks, their locations, and temporal variations, without reconciling current observations of methane and the known atmospheric chemistry and dynamics. Therefore, the problem remains unsolved and needs further study.

### *11.7.3 Origin of Methane*

The methanogenic bacteria are a distinct and exciting explanation for methane on Mars, and a possible methane cycle (Krasnopolsky 2012a) is

$$4\,CO + 2\,H_2O \rightarrow CH_4 + 3\,CO_2 + 3.17\text{ eV (metabolism)},$$
$$CH_4 + O \rightarrow CH_2 + H_2O + 0.34\text{ eV (heterogeneous process)},$$
$$CH_2 + O_2 \rightarrow CO + H_2O,$$
$$\text{Net } 3\,CO + O + O_2 \rightarrow 3\,CO_2.$$

Methane cannot be formed photochemically on Mars, and statements on this possibility are erroneous. Methane can be brought to Mars by comets that contain ≈1% of $CH_4$ in their ices. However, it does not matter how much methane is in comets, because thermochemistry in the cometary fireball in the Martian atmosphere determines the methane delivery. Calculations by Krasnopolsky (2006c) resulted in a methane yield of 0.25% of the comet mass. Using the published data, Krasnopolsky et al. (2004b) evaluated the mass influx from comets to Mars at $10^9$ g $yr^{-1}$. Then production of methane by impacts of comets is a negligible value of ≈2000 $cm^{-2}$ $s^{-1}$.

Accreted influx of carbon from the interplanetary dust particles on Mars is estimated at $2.4\times10^8$ g $yr^{-1}$ (Flynn 1996), and H/C ≈ 0.7 in those particles. The solar NUV radiation can release methane from organics in the particles, and the methane yield is restricted by the hydrogen abundance, that is, smaller than 0.7/4 = 0.17 relative to carbon. Most probably, $C_XH_Y$ and H radicals are formed and quickly react with O, $O_3$, and $O_2$, reducing the yield to a few percent. Anyway, an upper limit to the production of methane from organics in the interplanetary dust particles is a negligible value of $5\times10^4$ $cm^{-2}$ $s^{-1}$.

Volcanism is among geological sources of methane. However, traces of the latest volcanism are a few million years old. $SO_2$ is more abundant than methane in the released volcanic gases, and its upper limit of 0.3 ppb (see below) rules out this possibility.

Some papers favor serpentinization, that is, high-temperature reactions between hydrothermal water and some rocks that release $H_2$ and are followed by

$$4\,H_2 + CO_2 = CH_4 + 2\,H_2O.$$

According to Lyons et al. (2005), methane may be released for $\approx10^4$ yr with a rate of $\approx3\times10^8$ g $yr^{-1}$ after solidification of a dike of 100 $km^3$ at 870 K and gradually cooling to 520 K. However, the surface temperature above this dike is at least 0°C, and this source of endogenic heat should be detected by THEMIS, especially at night. Furthermore, this production of methane is smaller than the estimated global production by a factor of 400.

Oze and Sharma (2005) proposed serpentinization in a layer at 4–15 km. Among other problems, they overestimated the rate of the reaction between $H_2$ and $CO_2$ by orders of magnitude (Krasnopolsky 2006c). This reaction is effective at high temperatures, and only the bottom of the proposed layer may form methane.

Methane clathrate $CH_4$*5.75 $H_2O$, a hydrate that may exist at high methane pressure of a few bars and low temperatures, was also proposed as a source of methane (Chassefiere 2009). However, the origin of methane is substituted here by an origin of methane clathrate.

There are three major restrictions to the geological methane: (1) lack of the current volcanism, (2) lack of the places with endogenic heat release in the THEMIS observations (Christensen 2003), (3) a stringent upper limit of 0.3 ppb to $SO_2$. These facts are not in favor of geological methane.

## 11.8 Some Upper Limits

### *11.8.1 Ethane $C_2H_6$*

According to Allen et al. (2006, figure 2), typically $CH_4/C_2H_6 > 300$ and $^{12}CH_4/^{13}CH_4 > 1.05$ times the standard $^{12}C/^{13}C = 89$ in the biological sources with the opposite signs for the geological sources. Therefore measurements of these ratios are a way to establish the origin of methane.

Krasnopolsky (2012a) searched for the ethane subband at 2976.79 $cm^{-1}$ on Mars using IRTF/CSHELL. The telluric ethane was detected with a mixing ratio of 0.3 ppb, and an upper limit of 0.2 ppb was established for the Martian mixing ratio. Villanueva et al. (2013) observed upper limits of 0.2–0.6 ppb.

Comparing this limit to $CH_4 \approx 1$ ppb based on the TLS measurements, one gets a nonrestrictive ratio $CH_4/C_2H_6 > 5$. Furthermore, this is the abundance ratio, and it should be corrected for the species lifetimes to get the source ratio. Photochemical lifetime of ethane on Mars may be much shorter than that of methane because of the faster reaction with OH (Krasnopolsky 2006c), while its heterogeneous loss is generally unknown. Perspectives to improve the achieved upper limit by orders of magnitude are poor. Other hydrocarbons should be less abundant than ethane.

### 11.8.2 Sulfur Dioxide $SO_2$

Except for $H_2O$ and $CO_2$, $SO_2$ is the most abundant in the terrestrial volcanic outgassing, and its abundance may be a measure of outgassing on Mars. $SO_2$ was searched using IRTF/TEXES observations of its lines of the band at 7.35 μm with resolving power $\nu/\delta\nu = 8\times10^4$ (Krasnopolsky 2005, 2012a; Encrenaz et al. 2011). Small spectral intervals near 16 and 9 $SO_2$ lines were summed by two teams and revealed no $SO_2$ with upper limits of 0.3 ppb.

Sulfur chemistry on Mars was calculated by Krasnopolsky and Lefèvre (2013) for $f_{SO2}$ = 0.1 ppb and shown in Figure 11.39. Sulfur dioxide is lost via formation of $SO_3$ in reactions

$$SO_2 + OH + M \rightarrow HSO_3 + M \qquad \text{column rate} = 1.93\times10^6\,\text{cm}^{-2}\text{s}^{-1},$$

$$HSO_3 + O_2 \rightarrow SO_3 + HO_2 \qquad 1.93\times10^6,$$

$$SO_2 + O + M \rightarrow SO_3 + M \qquad 1.80\times10^5,$$

$$SO_2 + H_2O_2 + \text{ice} \rightarrow H_2SO_4 + \text{ice} \qquad 2.34\times10^5.$$

The last reaction is heterogeneous. Photolyses of $SO_3$ and $H_2SO_4$ (as well as of $SO_2$ and SO) were included in the model, though their effects appeared negligible. $SO_3$ is almost completely converted into $H_2SO_4$ in the reaction with two $H_2O$ molecules. $H_2SO_4$ vapor is lost by sticking to the ice particles. Liquid or solid sulfuric acid reacts with the surface rocks, and this is an irreversible loss of atmospheric sulfur in the model.

The calculated lifetime of sulfur species in the Martian atmosphere is 0.3 yr, and the observed upper limit of $f_{SO2} < 0.3$ ppb corresponds to production of $SO_2$ less than

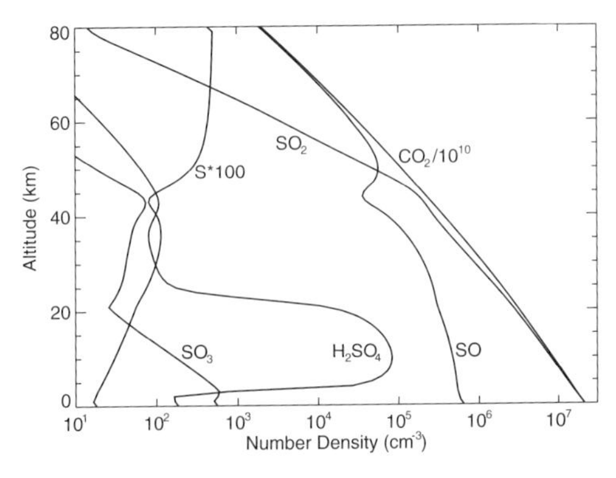


Figure 11.39 Calculated profiles of sulfur species assuming 0.1 ppb of $SO_2$ in the atmosphere (Krasnopolsky and Lefèvre 2013).

$7\times10^6$ cm$^{-2}$ s$^{-1}$ that is equal to 17 ktons yr$^{-1}$ of sulfur. It is smaller than the volcanic production of 5 megatons of sulfur per year on the Earth (Yung and DeMore 1999) by a factor of 300. Krasnopolsky (2005) estimated delivery of sulfur by comets at 35 tons yr$^{-1}$ and by meteorites and interplanetary dust at 170 tons yr$^{-1}$ of sulfur. Therefore the external sources are weak, while seepage of gases from the interior with no eruption of lava may dominate in production of sulfur in the Martian atmosphere.

The low upper limit to the $SO_2$ abundance does not support geological sources of methane.

### *11.8.3 Nitric Oxide NO*

It was discussed in Section 10.3 that odd nitrogen chemistry on Mars originates in the upper atmosphere with comparable productions of N and NO. These species recombine in $N + NO \rightarrow N_2 + O$, and an excess of either N or NO moves downward into the middle and lower atmosphere. In the first case, N recombines via the NO airglow:

$$\begin{aligned} &N + O \rightarrow NO^*, \\ &NO^* \rightarrow NO + h\nu, \\ &NO + N \rightarrow N_2 + O, \\ &\text{Net } N + N \rightarrow N_2 + h\nu. \end{aligned}$$

If NO dominates, then the NO airglow is excited only on the dayside by fluorescent scattering (Section 8.5) and odd nitrogen chemistry is initiated in the lower and middle atmosphere with a mixing ratio of NO of $\approx$0.6 ppb. Current photochemical models are uncertain in this aspect and adopt a moderate odd nitrogen chemistry with the above NO abundance (Section 10.3).

A sensitive search for the NO line at 1900.76 cm$^{-1}$ using IRTF/TEXES (Krasnopolsky 2006a) resulted in an upper limit of 1.7 ppb to the NO mixing ratio. This limit does not solve the problem of odd nitrogen chemistry in the lower and middle atmosphere of Mars.

### *11.8.4 Hydrogen Chloride HCl*

This species initiates chlorine chemistry that is very important on Venus. Effects of chlorine chemistry may be compared with those of odd hydrogen chemistry using the equilibrium

$$H + HCl = Cl + H_2.$$

The equilibrium constant is calculated using relationship (9.9) with enthalpies and entropies from Tables 9.1 and 9.2. This results in

$$\frac{[Cl]}{[H]} = \frac{[HCl]}{[H_2]} \times 0.52\, e^{574/T}.$$

Hartogh et al. (2010) searched for the HCl multiplet at 1876 GHz using Herschel/HIFI submillimeter orbiting observatory and established an upper limit of 0.2 ppb to the HCl mixing ratio on Mars. Comparing to the $H_2$ abundance of 17 ppm at $T \approx 200$ K, $[Cl]/[H] < 10^{-4}$, and chlorine chemistry is negligible on Mars.

Upper limits to $CH_3OH$, $H_2CO$, $NH_3$, and HCN (Table 11.4) are of some interest as well.

## 11.9 Photochemistry of the Lower and Middle Atmosphere

A global mean model of the Martian atmosphere from the surface to 80 km was considered in Section 10.3. The model includes $CO_2$ and its products CO, O, $O_2$, $O_3$, $O(^1D)$, $O_2(a^1\Delta_g)$, $H_2O$ and its products H, OH, $HO_2$, $H_2O_2$, $H_2$, NO and $NO_2$, that is, 15 species that participate in 38 reactions. Their rate coefficients and the calculated column rates are given in Table 10.1, and the calculated vertical profiles of all species are shown in Figure 10.1. Time-dependent models that simulate variations of minor species and photochemical tracers with local time were considered in Section 10.4. The calculated variations of the photochemical tracers are shown in Figure 10.2 at low latitudes for two Martian seasons. However, we concentrated in those sections on the general technique of the model calculations. Here we will briefly consider basic ideas of the Martian photochemistry.

Time of the global-scale mixing is ~0.5 year in Mars' atmosphere. Species with lifetimes exceeding this time are well mixed, and their variations are only caused by condensation and sublimation of $CO_2$ at the polar caps. $O_2$, CO, and $H_2$ are the long-living photochemical products, and the global-mean 1D models are currently the only means to calculate these species. Short-living photochemical products are variable on Mars and cannot be studied in detail by the global-mean models.

### *11.9.1 $CO_2$ Stability Problem and Basic $CO_2$–$H_2O$ Chemistry*

The observed low abundances of CO and $O_2$ ~$10^{-3}$ were initially puzzling, because the $CO_2$ dry (without $H_2O$) chemistry predicts their mole fractions of 0.08 and 0.04 (Nair et al. 1994) in reactions

$$CO_2 + h\nu(\lambda < 200\ \text{nm}) \rightarrow CO + O,$$
$$O + O + M \rightarrow O_2 + M,$$
$$O_2 + h\nu \rightarrow O + O,$$
$$O + O_2 + M \rightarrow O_3 + M,$$
$$O_3 + h\nu \rightarrow O_2 + O,$$
$$O + CO + M \rightarrow CO_2 + M.$$

Spins of O, CO, and $CO_2$ are 1, 0, and 0, respectively, and the spins do not conserve in the last reaction; therefore it is very slow giving rise to CO and $O_2$. That was a so-called $CO_2$ stability problem. The problem was solved by McElroy and Donahue (1972) and Parkinson and Hunten (1972) using a catalytic effect of odd hydrogen (H, OH, $HO_2$, and

Table 11.4 *Chemical composition of the Martian lower and middle atmosphere*

| Species | Mixing ratio | Comments and references |
|---|---|---|
| $CO_2$ | 0.96 | Kuiper (1949); global and annually mean pressure 6.1 mbar (Kliore et al. 1973) |
| $N_2$ | 0.025[a] | Owen et al. (1977), Franz et al. (2017) |
| Ar | 0.019[a] | Owen et al. (1977), Franz et al. (2017) |
| Ne | 2.5 ppm | Owen et al. (1977) |
| Kr | 300 ppb | Owen et al. (1977) |
| Xe | 80 ppb | Owen et al. (1977) |
| $O_2$ | $1.56\times10^{-3}$ [a] | Barker (1972), Carleton and Traub (1972), Trauger and Lunine (1983), Hartogh et al. (2010), Franz et al. (2017) |
| CO | 673 ppm[a] | Kaplan et al. (1969), Krasnopolsky (2015c), Franz et al. (2017) |
| $H_2O$ | 0–70 pr. μm | Spinrad et al. (1963), Jakosky and Farmer (1982), Smith (2004), Smith et al. (2009), Trokhimovsky et al. (2015) |
| $O_3$ | 0–60 μm-atm | Barth and Hord (1971), Barth et al. (1973), Clancy et al. (1999), Fast et al. (2006, 2009), Perrier et al. (2006), Lebonnois et al. (2006) |
| $O_2(^1\Delta_g)$ | 0.6–35 μm-atm[b] | Noxon et al. (1976), Krasnopolsky (2013c), Guslyakova et al. (2016), Clancy et al. (2017) |
| He | 10 ppm | Krasnopolsky et al. (1994), Krasnopolsky and Gladstone (2005) |
| $H_2$ | 17 ppm | Krasnopolsky and Feldman (2001) |
| $H_2O_2$ | 0–40 ppb | Clancy et al. (2004), Encrenaz et al. (2004, 2015a), Hartogh et al. (2010) |
| $CH_4$ | 0–40 ppb | Krasnopolsky et al. (2004b), Formisano et al. (2004), Mumma et al. (2009), Fonti and Marzo (2010), Geminale et al. (2011), Krasnopolsky (2011c, 2012a), Villanueva et al. (2013), Webster et al. (2018), Aoki et al. (2017) |
| $C_2H_6$ | <0.2 ppb | Krasnopolsky (2012a), Villanueva et al. (2013) |
| $SO_2$ | <0.3 ppb | Encrenaz et al. (2011), Krasnopolsky (2012a) |
| NO | <1.7 ppb | Krasnopolsky (2006c) |
| HCl | <0.2 ppb | Hartogh et al. (2010) |
| $H_2CO$ | <3 ppb | Krasnopolsky et al. (1997), Villanueva et al. (2013) |
| $NH_3$ | <5 ppb | Maguire (1977) |
| $CH_3OH$ | <7 ppb | Villanueva et al. (2013) |
| HCN | <2 ppb | Villanueva et al. (2013) |

*Note.* First detections along with the latest data are given.
[a]Corrected to annual mean conditions (Krasnopolsky 2017b). [b]$O_2(^1\Delta_g)$ dayglow at 1.27 μm is measured in mega-rayleighs; 1 MR = 1.67 μm-atm of $O_2(^1\Delta_g)$ for this airglow.

$H_2O_2$ due to its fast photolytic conversion to OH). Odd hydrogen is formed by photolysis of $H_2O$ and reactions of $O(^1D)$:

$$H_2O + h\nu(\lambda < 195\text{ nm}) \rightarrow OH + H,$$

$$O(^1D) + H_2O \rightarrow OH + OH,$$

$$O(^1D) + H_2 \rightarrow OH + H.$$

All these reactions and those discussed below, their rate coefficients, and column rates in the atmosphere are given in Table 10.1. O($^1D$) is the lowest metastable state of O that is formed on Mars by photolysis of ozone with minor contributions from photolyses of $CO_2$ and $O_2$ and is strongly quenched by $CO_2$.

Odd hydrogen drives the basic photochemical cycle:

$$H + O_2 + M \rightarrow HO_2 + M,$$

$$HO_2 + O \rightarrow OH + O_2,$$

$$CO + OH \rightarrow CO_2 + H,$$

$$\text{Net} \quad O + CO \rightarrow CO_2.$$

$HO_2$ is formed by the termolecular reaction; therefore H is the most abundant odd hydrogen species above ~35 km, while $HO_2$ and $H_2O_2$ dominate below this altitude. Odd hydrogen is not lost in the cycle and therefore acts as a catalyst. OH is also formed by photolyses of $H_2O_2$ and $HO_2$ and in reactions of H + $O_3$ and H + $HO_2$. Reaction of CO + OH is the major process of formation of the CO=O bonds that balances breaking of these bonds by photolysis of $CO_2$. Balances of key reactions of formation and breaking bonds of the basic species are convenient for analyses of photochemistry.

Another pathway of loss of odd oxygen (O and $O_3$) is formation of the O=O bonds:

$$O + O + M \rightarrow O_2 + M$$

$$O + OH \rightarrow O_2 + H.$$

Breaking of the O=O bonds occurs by photolyses of $O_2$, $H_2O_2$, $HO_2$, and $NO_2$.

Odd hydrogen is lost in reactions

$$OH + HO_2 \rightarrow H_2O + O_2,$$

$$H + HO_2 \rightarrow H_2 + O_2,$$

$$H_2O + O,$$

$$OH + H_2O_2 \rightarrow H_2O + HO_2.$$

Production of $H_2$ in H + $HO_2$ is balanced by its loss in reactions with OH, O($^1D$), and flow to the upper atmosphere with subsequent dissociation and escape of H and $H_2$. This escape is ~$2\times10^8$ cm$^{-2}$ s$^{-1}$ (Anderson 1974). Hydrogen escape should be balanced by loss of O from the atmosphere in proportion 2 : 1 in the steady state global mean models. This proportion is that in $H_2O$ and presumes a surface reservoir of water that replenishes its loss from the atmosphere. If this proportion is broken, accumulation of either H or O would break the steady state. According to McElroy (1972), the required loss of O occurs by nonthermal escape of fast oxygen atoms formed by dissociative recombination of $O_2^+$. Early calculations by Fox (1993) did not support that suggestion, and oxidation of the surface rock is an alternative solution. Model results are insensitive to this choice, and typically a flow of oxygen equal to ~$10^8$ cm$^{-2}$ s$^{-1}$ is adopted as the upper boundary condition. Current models of the oxygen escape induced by dissociative recombination of $O_2^+$ (Fox and Hac 2010) demonstrate a significant scatter of the results for various initial conditions and do not rule out the oxygen escape of ~$10^8$ cm$^{-2}$ s$^{-1}$. Their recent calculations (Fox and Hac 2018) result in the mean oxygen escape of $2\times10^7$ cm$^{-2}$ s$^{-1}$.

### 11.9.2 *Published Global-Mean Models*

Mars' photochemistry critically depends on the catalytic effect of $H_2O$ and its products. Photolysis of $H_2O$ is very sensitive to absorption by the overhead $CO_2$ that restricts the effective spectral interval to ~10 nm centered at 190 nm. Therefore the $H_2O$ photolysis is more efficient at 40 km than near the surface by a factor of 400.

The early models (e.g., Kong and McElroy 1977a) were aimed to simulate the observed CO and $O_2$ and adopted the $H_2O$ layer in the lowest 4 km with a steep cutoff above the layer. The odd hydrogen production was therefore low and occurred near the surface, and strong eddy diffusion $K \approx 3\times10^8$ cm$^2$ s$^{-1}$ was required to move odd hydrogen up and reduce CO and $O_2$.

Water vapor is uniformly mixed up to a saturation level near 20–25 km and moderate eddy diffusion $\approx10^6$–$10^7$ cm$^2$ s$^{-1}$ is adopted in the later models (Krasnopolsky 1993b, 1995, 2006b; Nair et al., 1994). However, the calculated CO abundances are much smaller than those observed (by a factor of 5.5 in the model in Figure 10.1). Some suggestions to improve the situation have not been confirmed by later laboratory studies, and the problem remains unsolved.

### 11.9.3 *Some Model Results*

Density profiles calculated by the model from Section 10.3 are shown in Figure 10.1. The $H_2O$ densities are restricted by the saturation values. The calculated abundances of the major photochemical products are 1600 ppm for $O_2$, 20 ppm for $H_2$, 0.9 μm-atm for $O_3$, 1.5 MR or 2.5 μm-atm for $O_2(^1\Delta_g)$, and 7.6 ppb for $H_2O_2$, in good agreement with the observations (Table 11.4). The CO mixing ratio varies from 120 ppm near the surface to 900 ppm at 80 km.

Mean lifetime of a species is equal to a ratio of its column abundance to its column production/loss rate (Table 10.3). The lifetimes of $O_2$ and $H_2$ are 60 and 370 years, respectively. The lifetime of the calculated CO is $\approx$1 year while a ratio of the observed column CO to its production is 7 years.

Photolysis of $CO_2$ and the subsequent termolecular association form CO and $O_2$ in proportion 1:0.5, and $O_2$ significantly exceeds this proportion for both observed and calculated CO. This means that $O_2$ is a photochemical product of mostly $H_2O$, though the $H_2O$ photolysis is weaker than that of $CO_2$ by two orders of magnitude.

Nitrogen chemistry may be essential in Mars' lower and middle atmosphere if a downward flow of NO from the upper atmosphere exceeds that of atomic nitrogen. This chemistry was considered in detail by Nair et al. (1994) and Krasnopolsky (1995). The calculated NO + $NO_2$ mole fraction was 0.6 ppb, below the observed upper limit to NO of 1.7 ppb (Table 11.4). Basic reactions of the nitrogen chemistry on Mars are

$$NO + HO_2 \rightarrow NO_2 + OH,$$
$$NO_2 + O \rightarrow NO + O_2,$$
$$NO_2 + h\nu \rightarrow NO + O.$$

The first reaction actually breaks the O=O bond and forms OH that reacts with CO, the second reaction restores the bond, and the third reaction ends the O=O breaking in the first reaction. Therefore it looks like the photolysis of $NO_2$ breaks the O=O bond, though the $NO_2$ structure is O=N=O. Nineteen percent of the O=O bonds are removed by the nitrogen chemistry (Table 10.1), and its role is neither critical nor negligible.

## 11.10 Variations of Mars Photochemistry

### *11.10.1 Steady State Models for Local Conditions*

Changes in temperature and dynamics induce variations of the $H_2O$ abundance and its vertical distribution, and $H_2O$ controls Mars' photochemistry. Variations of the Martian photochemistry may be studied by 1D models assuming that local productions and losses of species exceed their net delivery/removal by horizontal winds. These models are aimed to reproduce odd oxygen and odd hydrogen, while the long-living products $O_2$, CO, and $H_2$ may be adopted at their observed values. Using the data from Figure 10.1 and Table 10.1, lifetimes of odd oxygen and odd hydrogen are ~$10^5$ and ~$2\times10^5$ s, that is, of the order of the day, and these species may vary on a daily scale. This estimate agrees with the observations.

Liu and Donahue (1976) calculated a model with various abundance of water vapor. They proved anticorrelation between water and ozone that, however, was insufficient to explain the observed abundances of ozone up to 40–60 μm-atm in the subpolar regions at the end of winter. Photolysis of the abundant ozone enhances production of odd hydrogen in the reaction of $O(^1D) + H_2$ preventing a further increase of $O_3$ even in the lack of $H_2O$. Kong and McElroy (1977b) involved condensation of $H_2O$ and then $H_2O_2$, reduced $O_3$ and $H_2O_2$ photolysis rates at large airmass and high ozone, and influx of atomic oxygen from middle latitudes. They succeeded to explain variations of ozone (Figure 11.40) observed by Mariner 9.

Clancy and Nair (1996) argued that the enhanced solar heating near perihelion ($L_S$ = 251°) stimulates dust storms, and the atmosphere at low and middle latitudes is dust-free and cold near aphelion ($L_S$ = 71°) and dusty and therefore warm near perihelion. This changes a condensation level of $H_2O$ from ~10 to ~40 km, respectively, and induces significant seasonal variations of odd hydrogen (by one and two orders of magnitude at 20 and 40 km, respectively) and the overall photochemistry. The $O_3$ column varied from 1.0 μm-atm at perihelion to 3.2 μm-atm at aphelion at 30°N in their model. The growth is mostly due to ozone near 20 km that increases by an order of magnitude. Their model motivated Krasnopolsky (1997) to initiate regular ground-based observation of the $O_2$ dayglow at 1.27 μm that reflects ozone above ≈15 km.

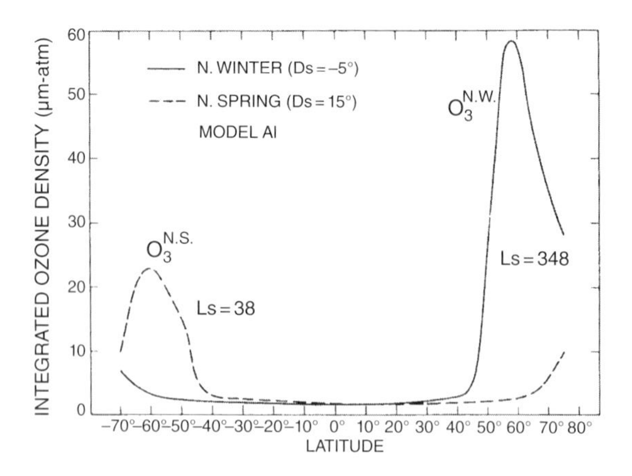


Figure 11.40 Latitudinal variations of ozone at two seasons in the model by Kong and McElroy (1977b).

Krasnopolsky (2006b) argued that contrary to the claimed anticorrelation of $O_3$ with $H_2O$, both the $O_2$ dayglow and ozone observations near aphelion ($L_S \approx 40°–130°$) show rather constant values at latitudes from 20°S to 60°N, while the water vapor abundance increases by an order of magnitude in this latitude range. Variations of the condensation level appear to be insufficient to explain this behavior, and heterogeneous loss of $H_2O_2$ at the water ice particles was suggested and added to the gas-phase chemistry from Table 10.1 to simulate the observed seasonal and latitudinal variations of Mars photochemistry. The adopted reaction probability was $5\times10^{-4}$ and $3\times10^{-4}$ in two versions of the model (Krasnopolsky 2006b, 2009b), much smaller than the uptake coefficient recommended from laboratory studies. However, the measured uptake of $H_2O_2$ on ice is reversible; therefore the adopted reaction probabilities (irreversible sinks) look plausible.

Krasnopolsky (2006b, 2009b) combined $T(z)$, $H_2O$, and aerosol observations from the MGS/TES database with variable heliocentric distance, daytime duration, and mean cosine of solar zenith angle to calculate photochemistry at ~100 seasonal-latitudinal points. The latitudinal coverage was ±50° centered at the subsolar latitude, because $H_2O$ becomes smaller than ~1 pr. μm and could not be measured by TES beyond this range. The calculated $O_3$, $O_2(^1\Delta_g)$, and $H_2O_2$ columns were interpolated to make seasonal-latitudinal maps (Figures 11.41, 11.28b, and 11.32, respectively). The model generally agrees with the observations.

### *11.10.2 Photochemical General Circulation Models and Mars Climate Database*

An attempt to account for the seasonal and latitudinal variations of Mars photochemistry was made in a two-dimensional zonal-mean model by Moreau et al. (1991). Recently the increase in computer power allowed the development of full general circulation models (GCM) of Mars with photochemistry. In addition to an adequate representation

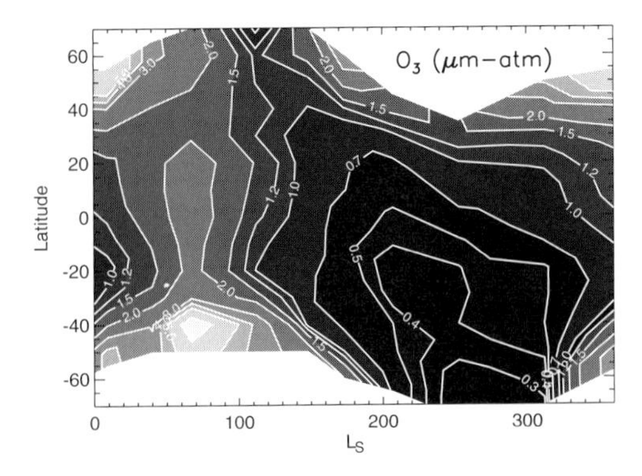


Figure 11.41 Seasonal and latitudinal variations of ozone from the model by Krasnopolsky (2009b). (A black and white version of this figure will appear in some formats. For the color version, please refer to the plate section.)

of atmospheric transport, Martian GCMs are able to provide a realistic description of the three-dimensional field of water vapor and its variations at all scales, which is a crucial advantage for constraining properly the fast chemistry of the lower atmosphere. Here we consider photochemical aspects of those models.

Seasonal variations on Mars are complicated by both the obliquity and the elliptic orbit and very prominent because of the comparatively thin atmosphere. Mars' photochemistry is relatively simple and may be involved in GCMs in its full extent. That is why the importance and success of the photochemical GCMs are so evident and impressive for Mars. These Martian models are the most advanced tool to simulate variations of the basic atmospheric properties.

Currently there are two Martian photochemical GCMs: Lefèvre et al. (2004, 2008) and Lefèvre and Forget (2009) and that of Moudden and McConnell (2007) and Moudden (2007). Because of the long computation time involved, photochemical GCMs have up to now been essentially applied to the analysis of short-lived species. Lefèvre et al. (2004) characterized the three-dimensional variations of ozone in the atmosphere of Mars, and later demonstrated the importance of heterogeneous chemistry to reproduce the observations of ozone or $H_2O_2$ (Lefèvre et al. 2008). The model included the heterogeneous loss of OH and $HO_2$ on the ice aerosol with uptake coefficients of 0.03 and 0.025, respectively.

The model is known as the Mars Climate Database, and its website (www-mars.lmd .jussieu.fr/mcd_python) may be used for various applications. This model is currently a standard tool for comparison with observations and other models. Any model is an approximation of the reality; evidently MCD reflects laws, principles, and assumptions in its initial data. As a GCM it generates dynamics of the atmosphere including its mixing that affects minor species. However, dynamics smaller that the grid size cannot be produced by MCD, and this requires eddy diffusion coefficient, similar to that in the one-dimensional models.

The $H_2$ lifetime is 370 years (Section 11.9.3), and a Martian model should run for more than 1000 years to get equilibrium in long living species. This is currently impossible for GCMs, and MCD adopts absolute abundances of $O_2 = 1.3\times10^{-3}$, CO = 800 ppm, and $H_2$ = 15 ppm (Lefèvre et al. 2004). This does not precule MCD to simulate vertical profiles of these species and their seasonal and latitudinal variations induced by condensation and sublimation of $CO_2$ on the polar caps.

## 11.11 Dayglow

### *11.11.1 Mariner, Mars Express, and MAVEN Observations*

Spectroscopy of the UV dayglow at 120–400 nm was among basic tasks of the Mariner 6 and 7 flybys and the Mariner 9 orbiter. Most of the observed emissions are excited by the solar photons and photoelectrons in dissociation and ionization of $CO_2$. These emissions and their excitation are discussed in Section 8.4, and the observed spectrum is in Figure 8.9. Vertical profiles of four major emission at 180–400 nm are depicted in Figure 11.42. Stewart (1972) calculated six models using various assumptions on the EUV heating efficiency and for either $CO_2^+$ or $O_2^+$ dominating ionospheres. The observed scale height of the CO ($a$ →X) Cameron bands was 19 km near 170 km. The bands are mostly excited in photodissociation of $CO_2$, and the scale height indicates an exospheric temperature $T_\infty \approx$ 315 K (high solar activity $F_{10.7} \approx 85$). Comparing the models with the observations, Stewart (1972) concluded that $CO_2^+$ constitutes ≈30% of the topside ionosphere. $CO_2^+$ is lost in reaction with O and dissociative recombination, and abundance of O at the ionospheric peak at 135 km required by this proportion is 2%. He concluded that the ionosphere below 200 km is not modified by the solar wind.

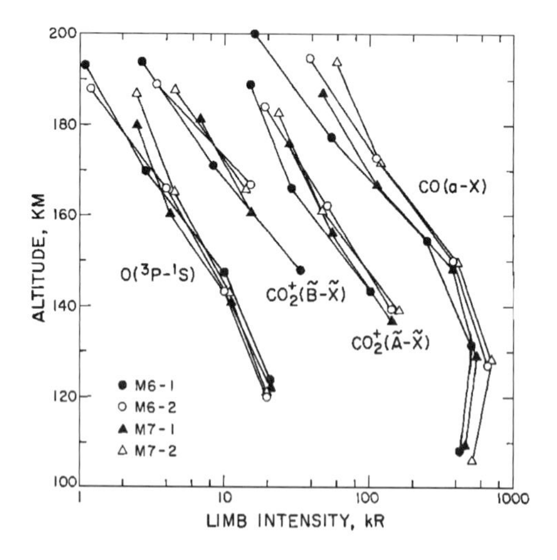


Figure 11.42 Limb intensity profiles of four basic UV dayglow emissions in the Mariner 6 and 7 observations (Stewart 1972).

The $CO_2^+$ (A→X) and (B→X) band systems are the brightest emissions at 280–400 nm. They are excited by photoionization of $CO_2$ and fluorescent scattering by $CO_2^+$. Vibrational distributions are different for these mechanisms, and Krasnopolsky (1975) analyzed the observed vibrational distribution and extracted vertical profiles $CO_2^+$ and $CO_2$ from these data. The retrieved $CO_2^+$ profile required significant deviations from the diffusion equilibrium profile of O. The $CO_2$ profile also showed deviations from the classic thermospheric temperature profile (Section 2.4.3) that were explained by effects of eddy diffusion and gravity waves.

Detailed analysis of the UV dayglow observations was made by Fox and Dalgarno (1979b) using the Viking mass spectrometer (Nier and McElroy 1977) and retarding potential analyzer (Hanson et al. 1977) data, some modeling, and laboratory data on the excitation processes.

SPICAM/MEX observations of the Martian dayglow at 120–300 nm (Leblanc et al. 2006) confirmed the main dayglow features and revealed the $N_2$ ($A^3\Sigma_u^+ \rightarrow X^1\Sigma_g^+$) Vegard-Kaplan bands (0–5) 261 nm and (0–6) 276 nm. The observed nadir intensity of the (0–6) band is 6 R (Leblanc et al. 2007), smaller than 20 R calculated by Fox and Dalgarno (1979b). The observations were conducted in summer ($L_S$ = 100°–170°) and winter ($L_S$ = 289°–321°), and solar activity varied from $F_{10.7} \approx 33$ to 60. Exospheric temperatures derived from scale heights of the $CO_2^+$ band at 289 nm were 195 ± 7 K and 207 ± 13 K, respectively.

The Imaging Ultraviolet Spectrograph (IUVS) aboard the Mars Atmosphere and Volatile Evolution (MAVEN) orbiter observes the Martian dayglow in the range of 115–340 nm. Exospheric temperatures obtained from scale heights of the $CO_2^+$ band at 289 nm were equal to 300 K at $F_{10.7} = 82$ and 250 K at $F_{10.7} = 49$ (Jain et al. 2015).

### *11.11.2 H Lyman-Alpha 122 nm*

H Lyman-alpha 122 nm emission is very bright and presents a convenient tool to study hydrogen coronae of planets. Resonance scattering of the solar Lyman-alpha photons in the line center is the only significant excitation mechanism. The flyby geometry is advantageous to measure the corona at large distances from Mars, and the Mariner 6 and 7 observations (Anderson and Hord 1971) remain important even now, a half-century after they were conducted.

The observed corona brightness distribution (Figure 11.43) is fitted by a model that is determined by density of H at the exobase $[H]_e$ and exospheric temperature. Vertical profile of hydrogen density is calculated using the partition functions (Section 2.4.5) for adopted $[H]_e$ and $T$, while their combination gives thermal escape of hydrogen. Then radiative transfer for resonance scattering in spherical atmosphere is calculated. The emission rate factor $g = 0.001$ s$^{-1}$ agrees with that in Table 8.2, taking into account the greater heliocentric distance compensated by the higher solar activity ($F_{10.7} \approx 90$ at Mars orbit during the Mariner flybys). The best-fit values are $[H]_e = 3\times10^4$ and $2.5\times10^4$ cm$^{-3}$ for Mariner 6 and 7, respectively, and $T \approx 350$ K. The large extent of the observations is

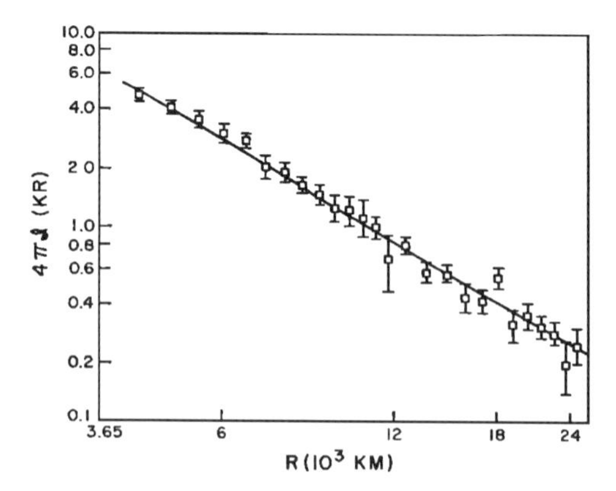


Figure 11.43 Brightness of the hydrogen Lyman-alpha corona: Mariner 6 observations and model (Anderson and Hord 1971).

very essential: for example, the data for $R < 7000$ km in Figure 11.43 show a slope that is much smaller than the final one.

The observed nadir intensities (Anderson 1974) depend on the column abundance of H above 80 km. The $CO_2$ optical depth is ≈1 at 80 km, and the atmosphere is adopted black below 80 km and pure scatteing without absorption above 80 km. Retrieved optical depths of hydrogen above 80 km were 2.2 for Mariner 6 and 7 and 5 for Mariner 9.

Photochemical interpretation of the data involved the $H_2$ mixing ratio of 20 ppm at 80 km, the $CO_2^+$ profile equal to 0.3 of the electron density profile from the radio occultations, and eddy diffusion $K = 5\times10^8$ cm$^2$ s$^{-1}$. A source of H is the $H_2$ dissociation in

$$H_2 + CO_2^+ \rightarrow CO_2H^+ + H$$

$$CO_2H^+ + e \rightarrow CO + O + H,$$

and atomic hydrogen moves either up and escapes or down into the middle atmosphere. Difference between the downward fluxes of H at the Mariner 6 and 7 versus Mariner 9 conditions was just 20% and induced the twofold difference in the H column abundances. Escape flux of H is $1.7\times10^8$ cm$^{-2}$ s$^{-1}$ and varies insignificantly from Mariner 6 and 7 to Mariner 9. This analysis that was made by Anderson (1974) at the beginning of the Mars exploration properly reflects main features of the problem.

The hygrogen corona was observed using SPICAM onboard Mars Express on seven orbits in 2005 (Chaufray et al. 2008). Mars was near the fall equinox ($L_S \approx 185°$) at medium solar activity with $F_{10.7}$ = 40–50 at Mars orbit, and the dayglow was observed up to 2100 km on two orbits with solar zenith angle ≈30° at the tangential points and up to 4000 km and SZA ≈ 90° on five orbits. The retrieved $[H]_e = (1–4)\times10^5$ cm$^{-3}$ and $T$ = 200–250 K and agree with the photochemical models. The authors considered also a two-component interpretation with hydrogen density $10^5$ cm$^{-3}$ at $T \approx 200$ K near 200 km and hot hydrogen density $2\times10^4$ cm$^{-3}$ at $T > 500$ K. However, escape of hot hydrogen dominates in this case, while it is 5%–10% according to Table 10.3. Therefore we do not support this version.

The observations were repeated in July, August, and December 2007 (Chaffin et al. 2014) at $L_S$ = 271°, 296°, and 2°, respectively. The analysis was made using best-fit temperatures that reach 1200 K and adopted temperatures of 300 and ≈190 K. The escape fluxes for 300 K were $10^9$, $4\times10^8$, and $10^8$ cm$^{-2}$ s$^{-1}$, respectively. The authors relate the high escape rate in July and August to the global dust storm that may heat the lower and middle atmosphere and prevent condensation of water vapor.

HST observations of the Martian hydrogen corona (Clarke et al. 2014; Bhattacharyya et al. 2015) were made at three dates in 2007 and five dates in 2014 that refer to $F_{10.7} \approx 70$ and 144 at the Earth, respectively. Similar to the Mars flybys, the HST observations cover the Martian hydrogen corona at large distances from Mars. However, the data need correction for the strong geocoronal emission. One-component fitting (Figure 11.44) resulted in temperatures of 360–440 K, while that with two components gave 170 and 200–240 K for the thermal component in 2007 and 2014, respectively, with 800 K adopted for hot hydrogen. Using the data in Figure 11.60 (Section 11.13.3), the expected temperatures are 190 and 280 K, and the derived temperatures for one- and two-component fittings look too high and low, respectively. The observed escape fluxes show some seasonal behavior (Figure 11.44) that, however, is not corrected for the solar activity. We will discuss variations of the hydrogen escape later in Section 11.13.4.

HST high-resolution observation of H Lyman-alpha is shown in Figure 11.45. Four components are seen in the spectrum: telluric $H_E$, Doppler-shifted Martian $H_M$ observed as a shoulder, and deuterium telluric and Martian lines $D_E$ and $D_M$. The Martian deuterium line intensity was 23 ± 6 R, and the instrument was pointed at the Martian limb. Modeling of the emission at the observational geometry resulted in [D] ≈ 450 cm$^{-3}$ at 250 km. The observations were conducted near aphelion ($L_S$ = 67°) at low solar activity $F_{10.7}$ = 25.

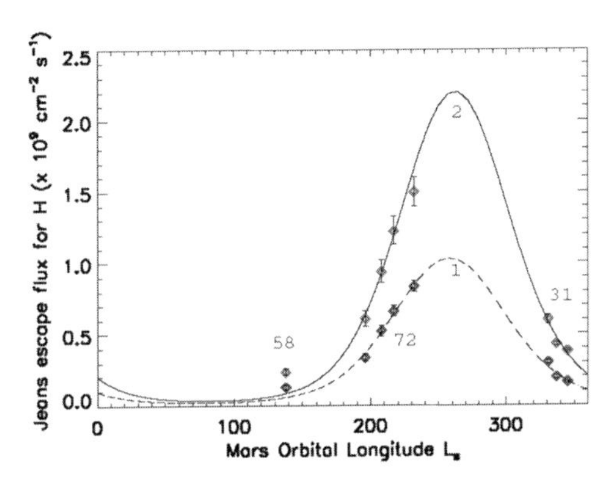


Figure 11.44 Hydrogen escape retrieved from HST observations of the Martian hydrogen corona using one- and two-component distributions (1 and 2, respectively). The other numbers are $F_{10.7}$ on Mars during the observations. Modified from Bhattacharyya et al. (2015).

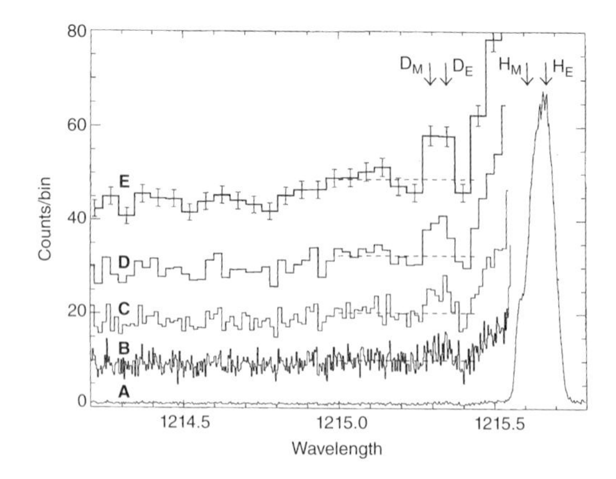


Figure 11.45 HST high-resolution observation of H Lyman-alpha. $H_E$ is the terrestrial emission; $H_M$ is the Doppler-shifted Martian emission seen as a shoulder. B, C, D, and E are the spectra scaled by a factor of 10 and averaged by one, two, four, and eight pixels. They reveal the Martian and telluric deuterium components of almost equal intensity. From Krasnopolsky et al. (1998).

IUVS/MAVEN has an echelle channel to observe H Lyman-alpha with resolving power of $1.5 \times 10^4$, similar to that in the HST observation. Significant variations of both H and D components and their ratio have been observed during the first Martian year of the MAVEN operation (Clarke et al. 2017).

### 11.11.3 Atomic Oxygen Triplet at 130 nm

Excitation of $2p$ electron to $3s$ in atomic oxygen results in configuration $1s^2 2s^2 2p^3 3s$. The lowest state $^5S_2$ has the largest spin, all spontaneous transitions are spin-forbidden, and the strongest emission is to the ground state $^5S_2 \rightarrow {}^3P_{2,1}$. This is the doublet at 136 nm that is seen in the Martian dayglow (Figure 8.9).

The next state is $^3S_1$ with allowed transitions to and from the ground state $^3P_{2,1,0}$ that form a resonance triplet 130.2, 130.5, and 130.6 nm with relative intensities of 5 : 3 : 1 in emission. This triplet (Figure 8.9) is used to measure abundances of atomic oxygen in the upper atmospheres.

The Mariner 9 nadir observations of the O 130 nm intensities (Stewart et al. 1992) were simulated by a model that included vertical profiles of O and $CO_2$ for exospheric temperature of 250 K. Resonance scattering in the optically thick atmosphere and photoelectron exctitation (Figure 11.46) of the triplet emission was calculated using the Monte Carlo method and assuming partial frequency redistribution with the $O/CO_2$ ratio at 125 km ($p$ = 1.2 nbar) as a fitting parameter. The results are shown in Figure 11.47. The mean $O/CO_2$ is $\approx$0.8%, $L_S = 290°$ to $350°$, and the mean solar activity was $F_{10.7} = 54$ on Mars orbit.

SPICAM/MEX observed limb intensities of the oxygen triplet (Chaufray et al. 2009) at tangential altitudes from 150 to 400 km at $L_S = 101°$ to $164°$ and $F_{10.7} = 32$ to 54. Atomic oxygen density at 200 km and temperature were fitting parameters. The best-fit temperature

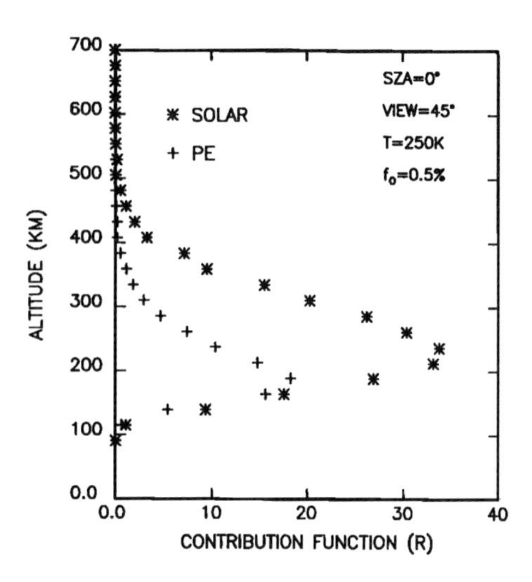


Figure 11.46 Contribution of resonance scattering and photoelectron excitation to the O 130 nm emission (Stewart et al. 1992).

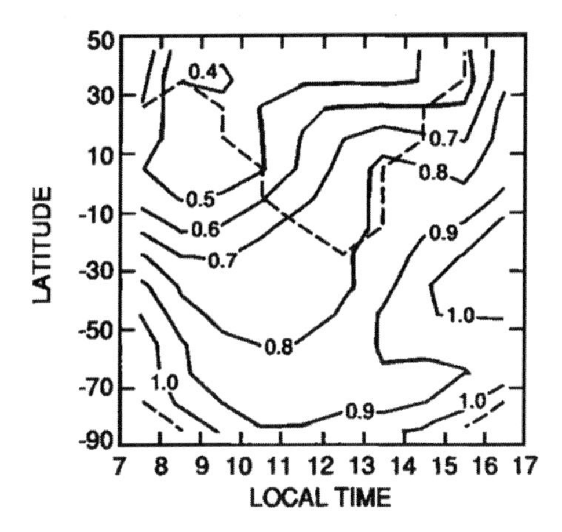


Figure 11.47 Distribution of $O/CO_2$ (%) at the ionospheric peak near 130 km retrieved from the Mariner 9 observations of O 130 nm. Regions poleward of the dashed lines are poorly covered by the observations. From Stewart et al. (1992).

is 300 K, and the measured densities correspond to $O/CO_2$ = 0.6%–1.2% at 135 km. The retrieved temperature is higher than that from other observed dayglow emissions. This is an indication of hot oxygen, and both cold and hot oxygen populations are expected to have equal densities $\approx$2000 cm$^{-3}$ near 500 km.

The hot oxygen corona was observed at 130 nm using IUVS/MAVEN (Deighan et al. 2015). The measured oxygen densities are 300 cm$^{-3}$ at 400 km, 40 cm$^{-3}$ at 700 km, and

5 $cm^{-3}$ at 3500 km for SZA = 40°–80°. The oxygen scale height corresponds to its mean energy of 1 eV.

### 11.11.4 CO ($A^1\Pi \rightarrow X^1\Sigma^+$) Fourth Positive System

This band system at 130–180 nm (Figure 8.9) is allowed and can be directly excited by the solar radiation. However, the solar light is weak in this range, and other excitation mechanisms, including photo- and electron impact dissociation of $CO_2$, electron impact excitation of CO, and dissociative recombination of $CO_2^+$, are much stronger than the CO fluorescence.

The observed relative intensities of the (1–0) and (0–0) bands are smaller in the Martian dayglow than those in the laboratory spectra because of absorption by CO. The self-absorption of the (1–0) band at 150 nm is ≈60% (transmission ≈40%) at 150 km. Rotational temperature of the band for all excitation mechanisms is much higher than the ambient temperature, reducing the self-absorption. Mumma et al. (1975) observed self-absorption by CO of the (1–0) and (0–0) bands excited by electron impact of $CO_2$ at 30 eV. The required self-absorption corresponds to CO = $5\times10^{15}$ $cm^{-2}$ that means [CO] = $1.5\times10^8$ $cm^{-3}$ at 150 km assuming pure absorption from the tangential point to the instrument. More careful integration of the emission and absorption and inclusion of the $CO_2$ absorption results in $CO/CO_2 \approx 1\%$ at 135 km (Krasnopolsky 1981).

### 11.11.5 Far UV (900–1200 Å) and EUV Dayglow

Dayglow in the range of 900–1200 Å (Krasnopolsky and Feldman 2001, 2002) was observed using the Far Ultraviolet Spectroscopic Explorer (FUSE). A large aperture of FUSE made it possible to collect light from the whole Martian disk, providing high sensitivity at the excellent spectral resolution of 0.2 Å. Two fragments of the observed spectrum are shown in Figure 11.48. The measured spectrum includes lines of all atomic species H, O, C, N, Ar, and He, three ions $C^+$, $N^+$, and even $Ar^+$, and bands of $N_2$ and CO. The measured intensities of 70 observed lines, multiplets, and bands are tabulated. For example, the H Lyman-gamma 973 Å, which was previously observed as a broad peak, clearly shows four components (Figure 11.48, upper panel) with the hydrogen line and the oxygen triplet well resolved.

Figure 11.48 (lower panel) includes the strongest detected $H_2$ (6–1) P1 line at 1071.6 Å with intensity of 0.47 R. Fluorescence is proportional to $\lambda^4$ and the solar spectral intensity (Section 8.5, relationship 8.9). Both factors are unfavorable for the $H_2$ Lyman band system at 845–1108 Å. The best case is for the (6–0) P1 line that coincides with the rather bright solar H Lyman-beta at 1025.7 Å, and three of four detected $H_2$ lines originate from P1 v′ = 6. The measured intensity resulted in an $H_2$ column abundance $\{H_2\}$ = $(1.17 \pm 0.13)\times10^{13}$ $cm^{-2}$ at 140 km. The observations were made at mean solar activity

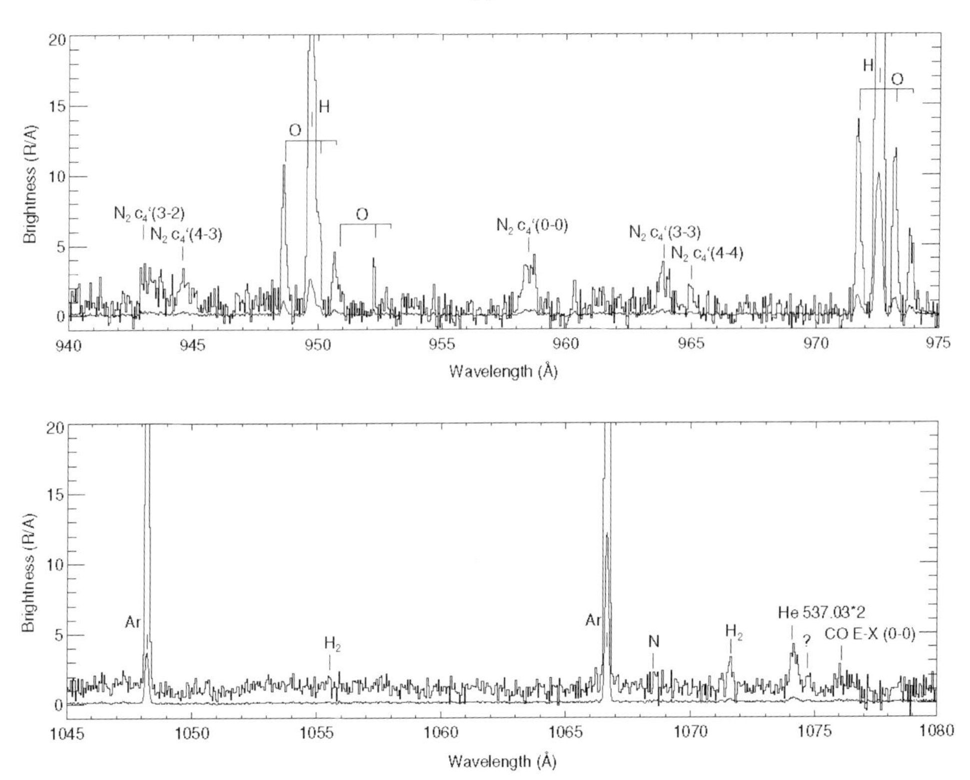


Figure 11.48 Two fragments that are a quarter of the observed FUSE spectrum. Four components of the emission near 973 Å are well resolved, while they looked like a single emission in the previous spectra. The $H_2$ lines, two bright Ar lines, and even a faint He 537 Å emission in the second diffraction order are observed. From Krasnopolsky and Feldman (2002).

$F_{10.7} = 61$ and near the fall equinox ($L_S = 160°$). Their interpretation will be discussed in Section 11.13.3.

Argon lines at 1048 and 1067 Å are bright with intensities of 6.4 and 21.8 R (Figure 11.48). The retrieved argon mixing ratio is 1.5% at 150 km. Two helium lines at 584 and 537 Å were observed in the second diffraction order (Figure 11.48). The He 584 Å brightness is 58 ± 20 R after corrections for N 1168.42/54 Å, the geocoronal foreground, and the second-order effective area.

Mars was observed at 80–780 Å (Krasnopolsky and Gladstone 2005) using the Extreme Ultraviolet Explorer (EUVE) at $F_{10.7} = 43$ and $L_S = 146°$. The only detected emission was He 584 Å with intensity of 41 ± 7 R scaled to the Martian disk. The EUVE observation resulted in the first detection of helium on Mars, and the EUVE intensity of the helium line is more reliable than that of FUSE because calibration of the latter was rather uncertain in the second diffraction order. Interpretation of the observation will be considered below (Section 11.13.3).

### 11.11.6 CO Dayglow at 4.7 μm

CO dayglow at 4.7 μm was discovered by Billebaud et al. (1991) using FTS at the Canada–France–Hawaii Telescope (CFHT). The dayglow is emitted by the CO (2–1) and (1–0) bands with intensities of $2.4^{+2}_{-1}$ MR and $1.3^{+1.1}_{-0.6}$ MR, respectively, that referred to the whole disk. They identified absorption of the solar light by the CO (1–0), (2–0), and (3–0) bands at 4.7, 2.35, and 1.58 μm, respectively, and the UV photolysis of $CO_2$ as the excitation mechanisms.

Much later spatially resolved high-resolution spectroscopy of six (2–1) and three (1–0) dayglow lines was made using CSHELL/IRFT (Krasnopolsky 2014a). A small fragment of the spectrum is shown in Figure 11.49. It includes the telluric CO (1–0) P2 line, the Doppler-shifted Martian absorption and emission lines, and the Martian (2–1) R4 emission line. Observed latitudinal variations of the dayglow intensity are depicted in Figure 11.50. Intensities of the six (2–1) lines at each latitude were used to retrieve rotational temperature that should be equal to the ambient temperature near the calculated dayglow maximum at 46 km (Figure 11.51). The retrieved temperatures are compared with those from the TES and MCD data for the similar conditions.

Vertical profile of the CO (2–1) dayglow was calculated for various CO abundances and solar zenith angles. Excitation by the UV photolysis of $CO_2$ appeared much weaker than that in Billebaud et al. (1991), because the solar Lyman-alpha is only effective in the excitation. The (2–1) dayglow is quenched by $CO_2$ ($k_0 = 1.4\times10^{-12} \exp\left(-\frac{1157}{T}+\frac{70900}{T^2}\right)$ $cm^3\ s^{-1}$) and CO ($k_1 = 5.6\times10^{-11}\ T^{-1/2}\ e^{-39/T}$ $cm^3\ s^{-1}$). Using these data, CO mixing ratios were retrieved from the observed dayglow intensities

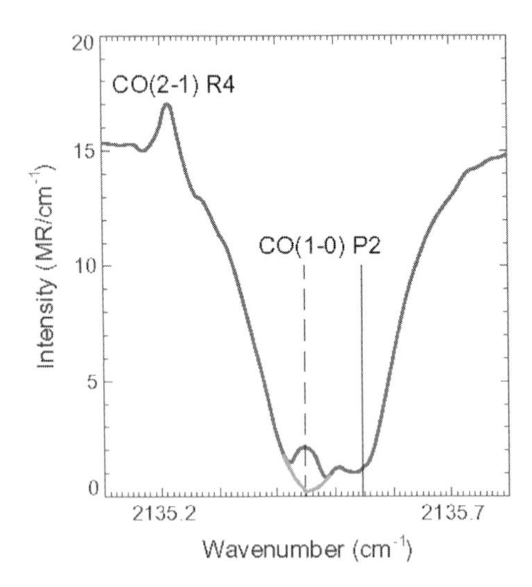


Figure 11.49 Observed structure of the CO (1–0) P2 line includes the telluric absorption line and the Doppler-shifted Martian absorption and emission lines. The CO (2–1) R4 line of the Martian dayglow is on the shoulder. From Krasnopolsky (2014a).

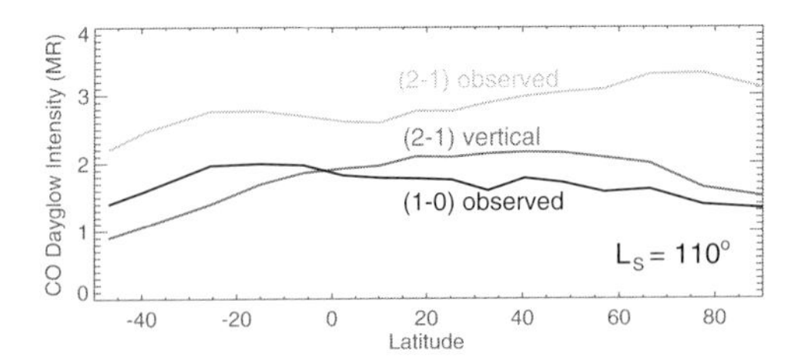


Figure 11.50 Latitudinal variations of the CO dayglow at 4.7 μm (Krasnopolsky 2014a).

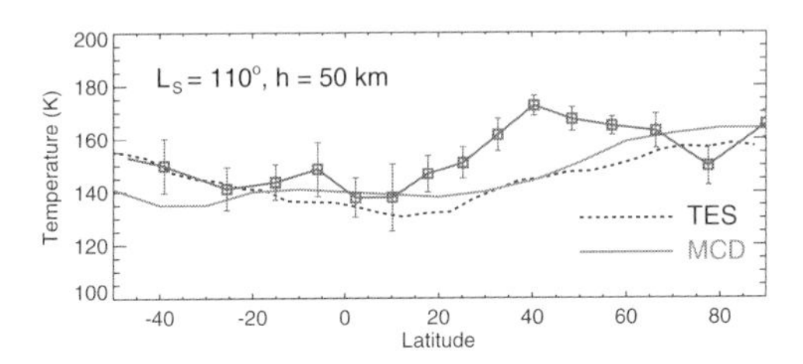


Figure 11.51 Retrieved temperatures at 50 km are compared with those from TES and MCD (Krasnopolsky 2014a).

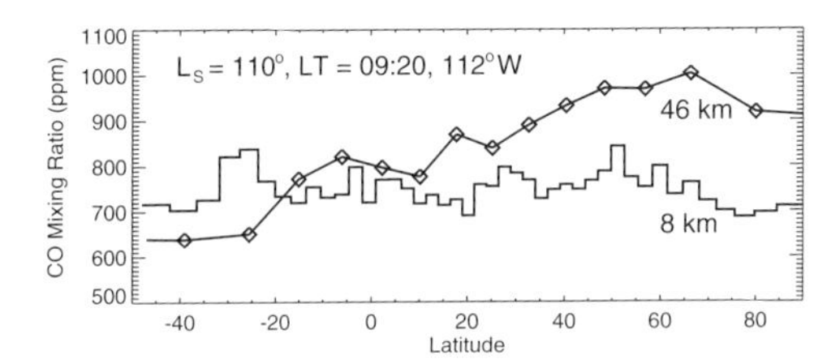


Figure 11.52 CO mixing ratios at 46 km are compared with those observed simultaneously at 8 km (Krasnopolsky 2014a, 2015c).

(Figure 11.52). These mixing ratios refer to the expected dayglow peak at 46 km, where $k_0[CO_2] + k_1[CO]$ is equal to the CO (2–1) transition probability 0.03 $s^{-1}$. The CO mixing ratios at 46 km are compared in Figure 11.52 with those in the lowest scale height observed almost simultaneously. The results indicate a complicated dynamic exchange between the lower and middle atmospheres.

## 11.12 Nightglow, Polar Nightglow, and Aurora

The first attempt to detect nightglow on Mars was made in the visible range aboard the Mars 5 orbiter. The spectrometer had resolution of 2.5 nm, and no emissions were detected with an upper limit of 50 R on the nightside limb (Krasnopolsky and Krysko 1976). This limit refers to 77 km for the green line O 558 nm and to 60 km for OH (8–2) 590 nm.

### 11.12.1 NO Nightglow

Nightglow on Mars was discovered by Bertaux et al. (2005a) using the SPICAM/MEX observations. That was the NO nightglow, similar to that studied on Venus (Section 12.10.4) and excited by the two-body association of NO (Section 8.7). The averaged observed spectrum is compared in Figure 11.53 with a laboratory spectrum of the NO airglow of $\delta$- and $\gamma$-band systems. Cox et al. (2008) studied 42 limb crossings in the first Martian year of the Mars Express operation and had 30 detections of the NO nightglow with peak limb brightnesses of 0.2 to 10.5 kR. Solar activity was low with $F_{10.7} \approx 80$ at 1 AU. They fitted the observations by the Chapman layer (Section 2.4.1), and their mean data and standard deviations are in Table 11.5. Two nightglow limb profiles are shown in Figure 11.54.

The NO nightglow contributes to the SPICAM stellar occultations and can be extracted from those observations. Gagne et al. (2013) analyzed 2215 stellar occultations and detected the nightglow exceeding 0.5 kR in 128 events. Those detections had a mean peak brightness of 4.0 ± 3.5 kR and a peak altitude of 83 ± 24 km. Stiepen et al. (2015) continued this work and found that the detection probability strongly depends on solar activity, increasing from 2%–5% at $F_{10.7} < 45$ to 60% at $F_{10.7} = 58$. They studied ~5000 observations for 10 years of the Mars Express operation and had ≈200 detections in a range of 0.23 to 18.5 kR with a mean peak brightness of 5.0 ± 4.5 kR. The peak altitude was 72 ± 10 km and varied from 42 to 97 km.

Table 11.5 *Mean properties of the NO nightglow on Mars (Cox et al. 2008)*

| | |
|---|---|
| Limb peak intensity (kR) | 1.2 ± 1.5 |
| Peak altitude (km) | 73 ± 8 |
| Peak emission rate (R km$^{-1}$)[a] | 2.0 ± 2.8 |
| Peak altitude[a] (km) | 77 ± 9 |
| Scale height[a] (km) | 6.0 ± 1.7 |
| Vertical intensity[a] (R) | 36 ± 52 |

[a] Parameters of the Chapman layer.

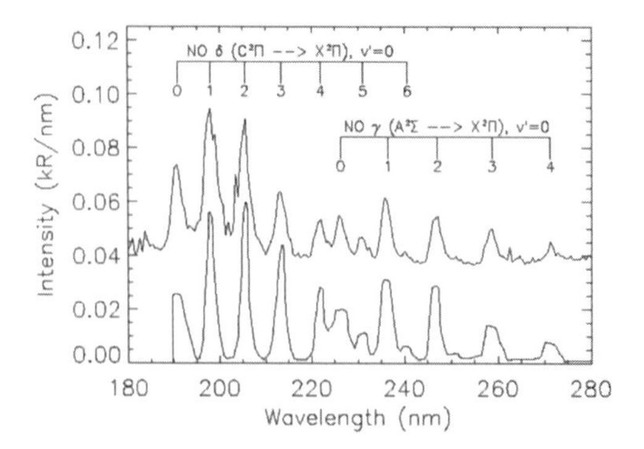


Figure 11.53 Averaged SPICAM/MEX and laboratory (bottom) spectra of the NO nightglow (Cox et al. 2008).

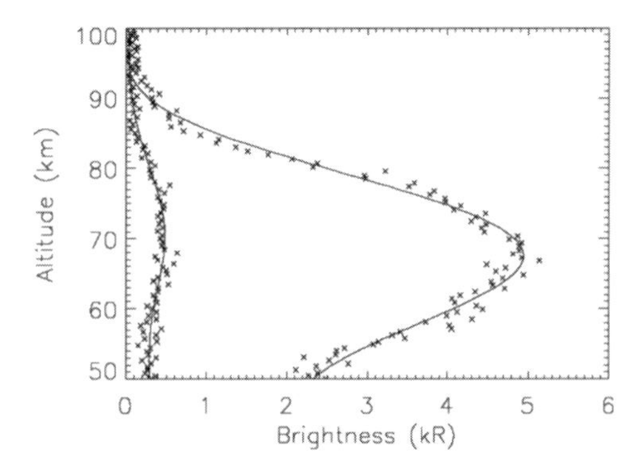


Figure 11.54 Two limb observations of the NO nightglow and their fitting by the Chapman layers (Cox et al. 2008).

The mean peak intensities in Gagne et al. (2013) and Stiepen et al. (2015) refer to the detections and are not corrected for the large numbers of the observations without detection. Furthermore, almost all observations were made at $F_{10.7} < 45$ with very low detection probability. Cox et al. (2008) corrected their results for the lack of detections in some observations and studied the nightglow vertical profiles. Therefore their data reflect probably better the mean nightglow properties (Table 11.5).

The NO nightglow properties from the LMD GCM that includes the Martian thermosphere and ionosphere (Gonzalez-Galindo et al. 2013) are discussed by Gagne et al. (2013). According to the model, the nightglow peaks up to 300 kR at the polar night and decreases to 0.5 kR at 30° of latitude in the winter hemisphere being immeasurable above this latitude and in the summer hemisphere. However, the nightglow is measurable at all latitudes near the equinoxes. The model predicts an increase in the nightglow intensity from evening to morning (LT = 19:00 to 06:00) by a factor of 6.

The model does not account for production of $N(^2D)$ in recombination of $NO^+$. Nitrogen chemistry is very sensitive to a fine balance between productions of N in the ground and metastable states. $N(^2D)$ is quenched by O, $N_2$, CO and forms NO in the reaction with $CO_2$. Then the reaction between N and NO is the major loss of both species in the thermosphere. The neglect of recombination of $NO^+$ as a source of $N(^2D)$ should significantly increase the flow of atomic nitrogen to the Martian nightside.

Significant seasonal variations of the NO nightglow at low and middle latitudes were observed by Stiepen et al. (2017) using IUVS at the MAVEN orbiter (Table 11.6). The observations were made at the Martian limb, and the observed mean peak intensities and altitudes are tabulated along with their standard deviations that reflect the observed variations. The results are compared with the updated LMD GCM, and the differences are very significant. The observations were made at medium solar activity ($F_{10.7} \approx 130$ at 1 AU), and the data for fall equinox and northern summer generally agree with the SPICAM observations (Table 11.5) that gave $1.2 \pm 1.5$ kR and $73 \pm 8$ km at $F_{10.7} \approx 80$ at 1 AU.

Table 11.6 *Mean peak limb intensities, altitudes, and their variabilities in IUVS observations and LMD GCM*

| Season | Northern winter ($L_S$ = 240°–300°) | | Fall equinox ($L_S$ = 13°–31°) | | Northern summer ($L_S$ = 75°–115°) | |
|---|---|---|---|---|---|---|
| IUVS/MAVEN | 4.7 ± 5.2 kR | 70 ± 8 km | 1.9 ± 1.0 kR | 71 ± 10 km | 1.7 ± 1.6 kR | 70 ± 9 km |
| LMD GCM | 0.5 ± 1.1 kR | 101 ± 20 km | 1.7 ± 1.4 kR | 95 ± 8 km | 0.3 ± 0.4 kR | 95 ± 11 km |

### *11.12.2 Polar Nightglow of $O_2$ at 1.27 μm*

An attempt of detection of the $O_2$ nightglow at 1.27 μm at low latitudes resulted in an upper limit of 40 kR (Krasnopolsky 2013c). The LMD GCM (Gagne et al. 2012) predicts a two-peak structure of this emission with peak altitudes at 55 and 80 km at the equator with total vertical intensities of 22 and 35 kR at $L_S$ = 0 and 180°, respectively. The predicted intensities of the Herzberg II band system are 280 and 400 R at these conditions. These data were calculated using the excitation and quenching parameters from Krasnopolsky (2011a; Table 12.8).

The polar nightglow of $O_2$ at 1.27 μm was discovered by Bertaux et al. (2012) using OMEGA/MEX, Fedorova et al. (2012) using SPICAM-IR/MEX, and Clancy et al. (2012b) using CRISM/MRO. The polar night on Mars is as long as the terrestrial year with a deep temperature minimum ≈145 K that is supported by the latent heat of the $CO_2$ condensation and influx of heat by atmospheric dynamics. These regions are sinks of rather strong flows from the middle latitudes that are enriched by atomic species and radicals. Atomic oxygen is the most abundant of those and produces a bright polar nightglow of $O_2$ at 1.27 μm. This nightglow is typically observed as a peak on the nightside limb near 50 km with intensity of ≈10 MR. The most detailed observations at six seasons were made by the CRISM team. Averaged inverted volume emission rates near the south pole close to the northern summer solstice are shown and compared with the LMD GCM simulations in Figure 11.55. The LMD GCM zonally mean vertical structure of the $O_2$ 1.27 μm airglow with two excitation mechanisms is shown in Figure 11.56. The nightglow was observed near the south pole at $L_S$ = 50°, 74°–96°, and 137°, near both poles at the fall equinox 166° and 193°, and near the north pole at 266°–301°. While LMD GCM properly simulates the phenomenon, there are some quantitative differences between the observations and the model.

### *11.12.3 Polar Nightglow of OH Meinel Bands*

Clancy et al. (2013) searched for the OH Meinel band emission in the Martian polar nightglow using CRISM/MRO. Ninety-six limb crossings were made in 2009–2012, and averaging of the data revealed the OH bands. The observed mean intensities are in Table 11.7. Mean profiles of volume emission rate of three detected OH bands are shown in Figure 11.57. Their shapes are rather similar to that of the $O_2$ emission at 1.27 μm, just like in the Venus nightglow.

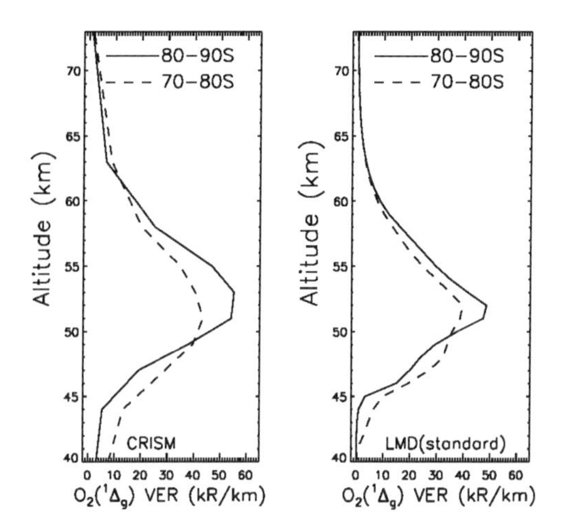


Figure 11.55 Averaged $O_2(a^1\Delta_g)$ volume emission rate in two southern polar zones at $L_S$ =74°–96°: CRISM observations and LMD GCM (Clancy et al. 2012b).

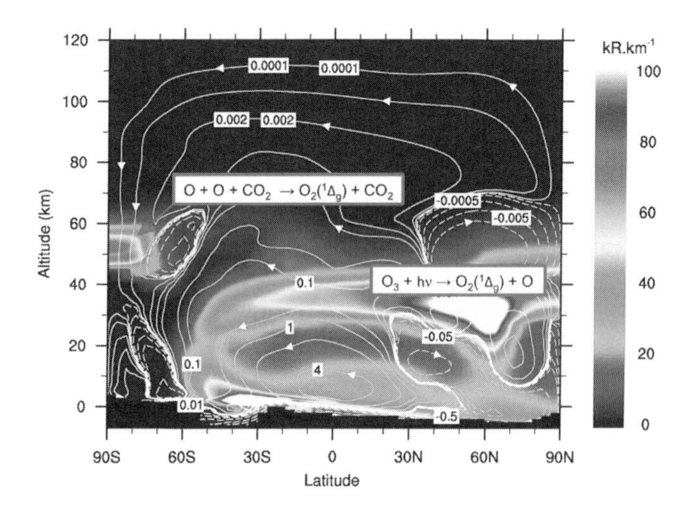


Figure 11.56 Two mechanisms of the $O_2$ 1.27 μm airglow excitation at $L_S$ = 95°–100° as simulated by LMD GCM. Labels give meridional stream functions in $10^9$ kg $s^{-1}$. From Clancy et al. (2012b). (A black and white version of this figure will appear in some formats. For the color version, please refer to the plate section.)

Ratios of the band intensities that originate from the same upper vibrational level should be equal to ratios of their transition probabilities. Therefore the expected (2–0)/(2–1) ratio is 0.41 while the observed ratio is 0.18 ± 0.12. The Venus OH nightglow demonstrates the similar problem, and the observed intensities of the Δv = 2 bands are smaller than those of the Δv = 1 bands scaled by ratios of the transition probabilities. The (2–1)/(1–0) ratio of 1.1 ± 0.6 is close to 0.88 observed on Venus and agrees with the collisional cascading model proposed by Krasnopolsky (2013a).

Table 11.7 *Mean limb intensities of the polar nightglow*

| Band | Wavelength (μm) | Intensity (kR) |
|---|---|---|
| OH (1–0) | 2.81 | 990 ± 280 |
| OH (2–1) | 2.94 | 1100 ± 480 |
| OH (3–2) | 3.10 | 590 ± 520 |
| OH (4–3) | 3.28 | 0 ± 300 |
| OH (2–0) | 1.42 | 200 ± 100 |
| OH (3–1) | 1.50 | 110 ± 140 |
| OH (4–2) | 1.58 | 31 ± 90 |
| OH (5–3) | 1.66 | 60 ± 190 |
| $O_2(^1\Delta_g)$ (0–0) | 1.27 | 9500 ± 46 |
| $O_2(^1\Delta_g)$ (0–1) | 1.58 | 200 ± 74 |

*Note.* From Clancy et al. (2013).

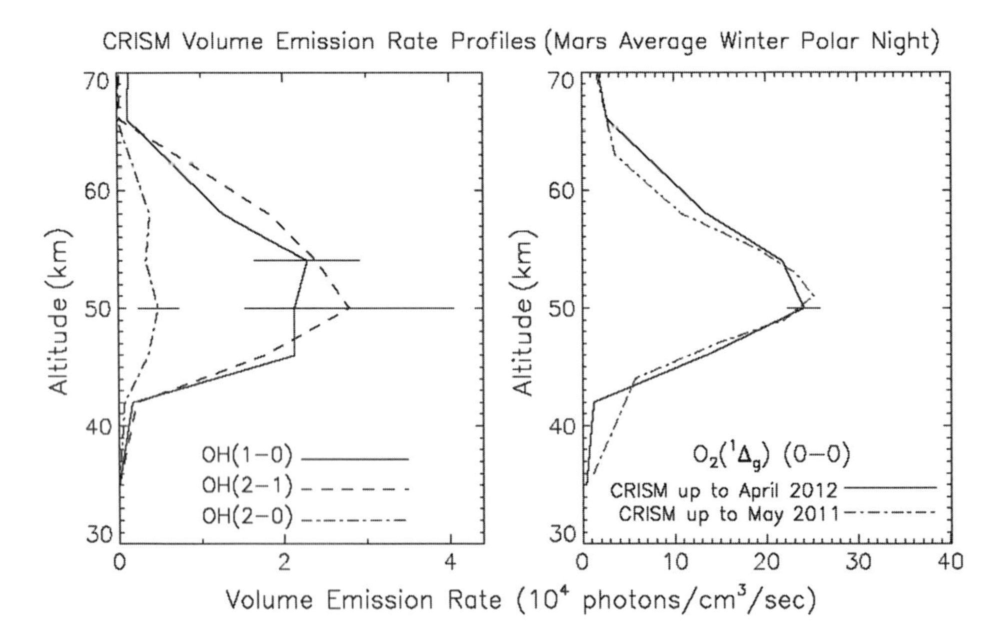


Figure 11.57 Mean vertical profiles of volume emission rate of three detected OH bands and the $O_2$ band at 1.27 μm in the Martian polar nightglow (Clancy et al. 2013).

### *11.12.4 Aurora*

While the NO $\delta$- and $\gamma$-band systems and H Lyman-alpha are typically present in the SPICAM spectra of the nighttime atmosphere, some observations demonstrate the CO Cameron and fourth positive bands at 190–270 nm and 135–170 nm, respectively, as well as the $CO_2^+$ band at 289 nm and O lines at 297 and 130 nm. These spectral features are similar to those in the Martian dayglow. They are excited on the nightside by electron impact and may therefore be classified as auroral emissions. Aurora was discovered on Mars by Bertaux et al. (2005b). A later study by Soret et al. (2016) determined a peak altitude of the strongest CO Cameron band emission at $137 \pm 27$ km. The observations were simulated by Monte-Carlo calculations of electron transport in the atmosphere for electron energies of 40 to 200 eV. Correlations between the SPICAM observations of aurorae and electron energy spectra measured by the ASPERA-3/ELS at Mars Express were studied. A complicated structure of the Martian magnetic field affects the phenomenon.

## 11.13 Upper Atmosphere and Ionosphere

### *11.13.1 Radio Occultations*

Measurements of the electron density profiles in the Martian ionosphere were made by radio occultations at all spacecrafts visited Mars since the first flyby missions. The dayside ionospheric peak fits fairly well to the Chapman layer (Section 2.4.2) with a peak altitude near 130 km at SZA $\approx 60°$ increasing to 170 km near the terminator (Zhang et al. 1990). Dust storms increase the peak altitude that generally corresponds to the slant $CO_2$ column abundance of $3\times10^{16}$ cm$^{-2}$. The peak electron density is $\approx 10^5$ cm$^{-3}$ at SZA $\approx 60°$.

The atmosphere is transparent for the solar EUV photons above 170 km; therefore

$$n \sum\nolimits_\lambda \sigma_\lambda I_\lambda = \alpha\, n_e^2.$$

Here $n$ and $n_e$ are the neutral and electron number densities, $\sigma_\lambda$ and $I_\lambda$ are the ionization cross sections and the EUV photon flux, and $\alpha$ is the recombination coefficient. This means that the plasma scale height is twice the neutral scale height, and this makes it possible to obtain the neutral temperature using the topside plasma scale height. However, the recombination coefficient depends on the electron temperature, the mean molecular mass varies significantly and is model-dependent above $\approx$170 km, and ambipolar diffusion may affect the ion balance in the above equation. Therefore neutral temperatures derived from the topside plasma scale heights are rather uncertain.

Nightside electron densities are very low for SZA $> 110°$ and mostly unmeasurable by radio occultations. The measured peak densities are $(3\text{–}5)\times10^3$ cm$^{-3}$ (Zhang et al. 1990). The ionization is caused by the plasma transport from the dayside and by suprathermal electron precipitations.

The basic ionospheric chemistry on Mars is simple and was understood after the flybys of Mariners 6 and 7:

$$\begin{aligned}
&CO_2 + h\nu \rightarrow CO_2{}^+ + e,\\
&O + h\nu \rightarrow O^+ + e,\\
&CO_2{}^+ + O \rightarrow O^+ + CO_2,\\
&\qquad\qquad\quad O_2{}^+ + CO\\
&O^+ + CO_2 \rightarrow O_2{}^+ + CO,\\
&O_2{}^+ + e \rightarrow O + O,\\
&CO_2{}^+ + e \rightarrow CO + O.
\end{aligned}$$

Therefore there are three major ions on Mars, and the dayside ionosphere is determined by solar activity, zenith angle, $O/CO_2$, and neutral and electron temperature. The latter affects rate coefficients of dissociative recombination that are proportional to $(300/T_e)^\beta$ with $\beta \approx 0.3$–$0.7$.

### *11.13.2 Viking Entry Measurements*

Mass spectrometers at the Viking 1 and 2 probes measured the atmospheric composition during the spacecraft entries at 200 down to 115 km (Figure 11.58). Altitude profiles of $CO_2$, $N_2$, Ar, CO, $O_2$, and NO (Nier and McElroy 1977) were measured by both probes as well as the isotope ratios that agree with those observed at the surface by the Viking landers.

Data on the ion composition on Mars were obtained using retarding potential analyzers (Hanson et al. 1977) at the Viking entries. The results from Viking 1 are shown in Figure 11.59. The Viking measurements remained the basic data on the thermospheric and ionospheric composition on Mars for a few decades.

### *11.13.3 Modeling*

Upper atmospheres at high altitudes become transparent to the solar EUV photons that dissociate and ionize the atmospheric species. Molecular diffusion dominates above the homopause and facilitates enrichment of light species at high altitudes. Chemical and mixing times are comparatively low in the upper atmospheres with significant variations of species abundances.

The first adequate models of the upper atmosphere and ionosphere appeared after the Mariner 6 and 7 flybys and were based on the exospheric temperature and the atomic oxygen abundance from the dayglow observations and the electron density profiles. The Mars Thermosphere General Circulation Model (MTGCM) simulates dynamics of the upper atmosphere along with some data on the composition. This model evolved with

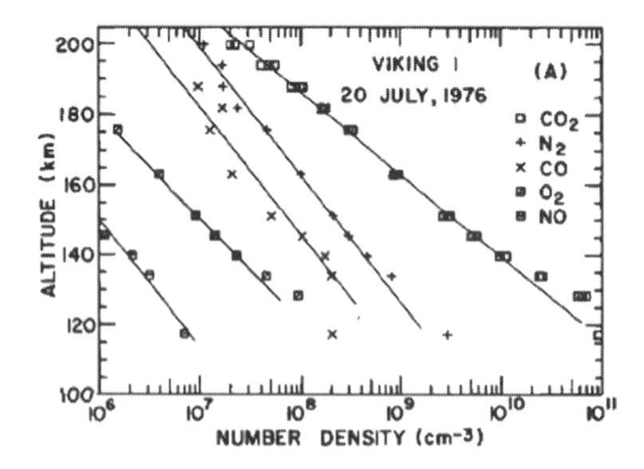


Figure 11.58 Composition of the Martian upper atmosphere observed by the Viking 1 mass spectrometer (Nier and McElroy 1977). The Viking 1 and 2 entries were at $F_{10.7} = 27$ and $L_S \approx 108°$.

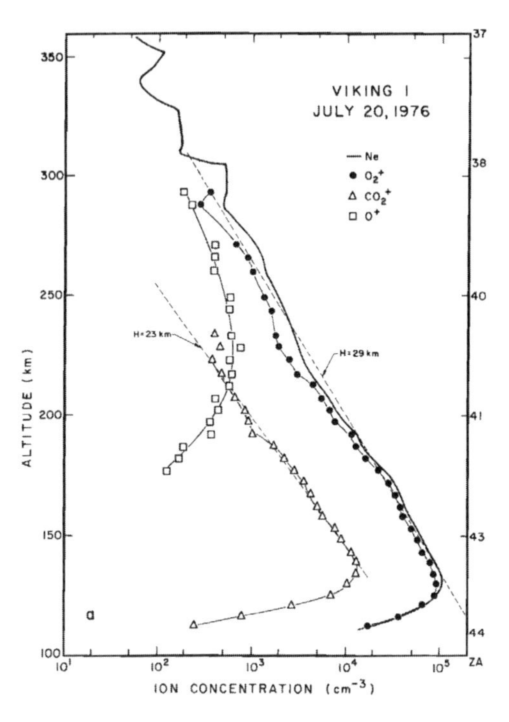


Figure 11.59 Martian ionosphere revealed by the Viking 1 retarding potential analyzer (Hanson et al. 1977).

evolution of our knowledge (Bougher et al. 2000, 2008, 2009, 2015a, 2017). The Mars Climate Database was extended up to the exobase and then included the ionosphere (Gonzalez-Galindo et al. 2009, 2013).

Thermal balance of thermosphere is discussed in Section 2.4.3. The upper part of thermosphere is isothermic with a so-called exospheric temperature $T_\infty$. Observed and modeled variations of the Martian exospheric temperature with solar activity at Mars position are shown in Figure 11.60. The observations are approximated by

$$T_\infty = 123 + 2.19F_{10.7}.$$

The recent MAVEN/NGIMS temperature appeared significantly higher than the model predictions, though rather close to the early model by Bougher et al. (2000) and the observational fit (Figure 11.60). The latest model (Bougher et al. 2015a, 2017) was adjusted to fit this value. We also applied relationship (2.36a)

$$T_\infty = \left(T_0^{s+1} + \beta F_{10.7}\right)^{\frac{1}{s+1}}$$

to the MAVEN value (solid line in Figure 11.60) using $s$ = 1.38 for $CO_2$ and $T_0 \approx 100$ K (Section 11.13.5).

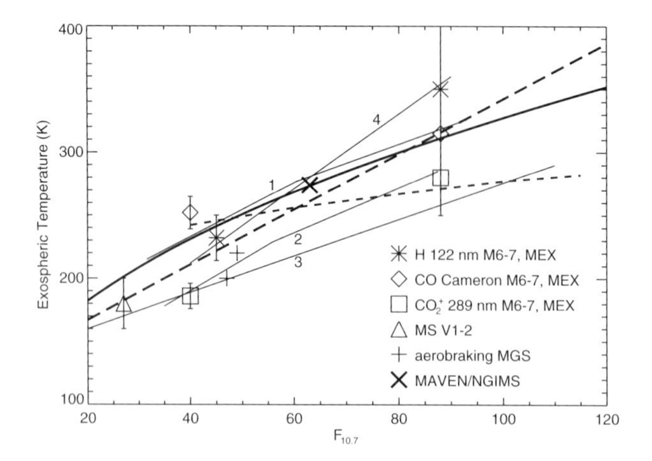


Figure 11.60 Exospheric temperature as a function of the solar activity index from observations and models. Linear fit to the observations is shown by dashed line. Lines 1, 2, 3, and 4 are models by Bougher et al. (2000, 2008, 2009, 2015a, respectively). Model by Gonzalez-Galindo et al. (2009) is similar to (3). All models are shown for fall equinox. Fit to the MGS densities at 390 km (short dashes) disagrees with the other data. Updated from Krasnopolsky (2010b).

The NO densities observed by the Vikings (Figure 11.58) reflects nitrogen chemistry in the upper atmosphere. Atomic nitrogen in the ground ($^4S$) and metastable ($^2D$) states is produced by predissociation and electron impact dissociation of $N_2$:

$$N_2 + h\nu(80 - 100\text{ nm}) \rightarrow N + N(^2D)$$

$$N_2 + e \rightarrow N + N(^2D),$$

and in some ion reactions, e.g.,

$$NO^+ + e \rightarrow N(^2D) + O$$

$$N_2{}^+ + O \rightarrow NO^+ + N(^2D).$$

$N(^2D)$ is quenched by O, CO, electrons and reacts with $CO_2$:

$$N(^2D) + CO_2 \rightarrow NO + CO.$$

NO partially recycles via photoionization by the solar Lyman-alpha and

$$O_2{}^+ + NO \rightarrow NO^+ + O_2.$$

Both reactions end by recombination of $NO^+$. Reaction of

$$N + NO \rightarrow N_2 + O$$

results in the irreversible sink of odd nitrogen (N, $N(^2D)$, and NO). These are basic ideas of odd nitrogen chemistry in the upper atmospheres of the terrestrial planets. ($O_2$ replaces $CO_2$ in this scheme for the Earth.) Fox et al. (1996) applied this chemistry and the MTGCM data at

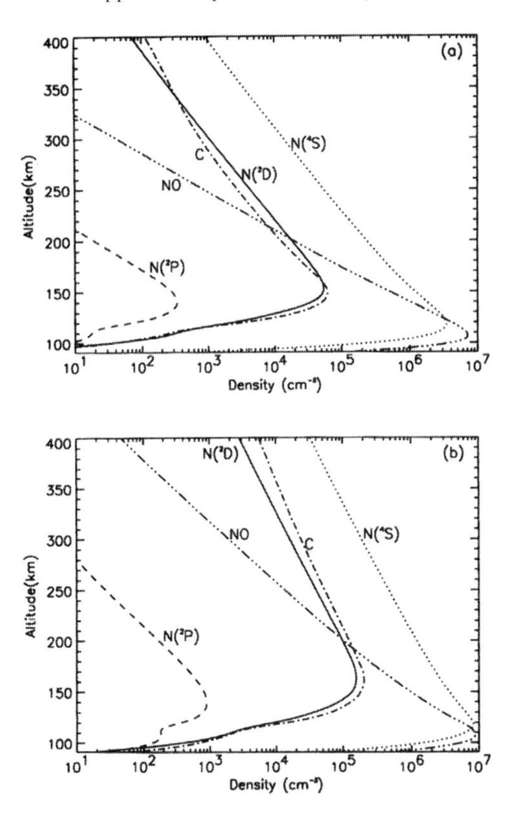


Figure 11.61 Vertical profiles of atomic nitrogen in the ground ($^4S$) and metastable states, NO, and C at low (a) and high (b) solar activity (Fox et al. 1996).

low and high solar activity (Figure 11.61). The calculated NO densities at low solar activity are slightly smaller (by a factor of 1.5) than those measured by Vikings (Figure 11.58).

Detections of the He, D, and $H_2$ emissions in the Martian dayglow (Section 11.11) required photochemical modeling for their interpretation. The problem was especially uncertain for D and $H_2$, whose observed abundances significantly disagreed with models created before the observations. Therefore a more detailed self-consistent model that included 86 reactions of 11 neutral and 18 ion species was developed at 80–300 km (Krasnopolsky 2002). The model involved both thermal and nonthermal escape of H, D, $H_2$, HD, and He (Table 10.3). The model was calculated for low, medium, and high solar activity to simulate the whole range of the variations and reflect various conditions of the observations of He, D, $H_2$, and also H (by Mariners 6, 7, and 9). Vertical transport by eddy, molecular, and ambipolar diffusions was involved in the model. Mixing ratios of He (10 ppm, Krasnopolsky and Gladstone (2005)), $H_2$ (17 ppm), and HD (12 ppb) were parameters to fit the EUVE, FUSE, and HST observations, and the solar activity index $F_{10.7}$ and proper exospheric temperatures were inputs. These temperatures were adopted at 200, 270, and 350 K for low, medium, and high $F_{10.7}$, respectively, in reasonable

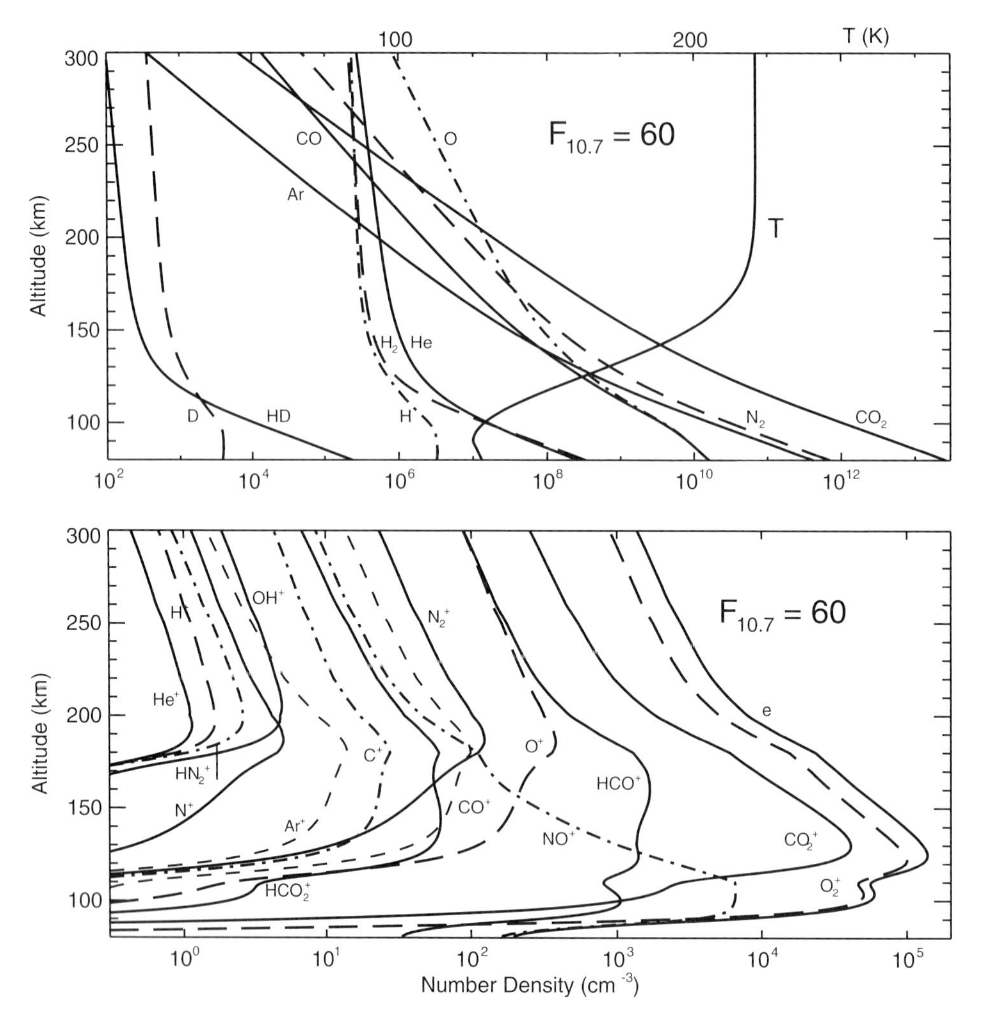


Figure 11.62 Photochemical model for the upper atmosphere (a) and ionosphere (b) at medium solar activity near perihelion with $H_2O$ = 30 ppm at 80 km (Krasnopolsky 2019a).

agreement with the recent model (Bougher et al. 2015a) and the MAVEN observations (Mahaffy et al. 2015). The model agrees with the Mariner observations of H, though there were no means to stimulate this agreement in the model. This proves that the derived $H_2$ mixing ratio is compatible with the Mariner observations of H Lyman-alpha. Figure 11.62 shows a recently updated model (Krasnopolsky 2019a) at medium solar activity near perihelion with $H_2O$ = 30 ppm at 80 km (see below).

### *11.13.4 Escape of H and $H_2$*

Escape of H varies weakly with solar activity in the model by Krasnopolsky (2002), from $1.8\times10^8$ to $2.1\times10^8$ cm$^{-2}$ s$^{-1}$, in accord to the Mariner 6, 7, and 9 observations (Anderson

1974). The escape is rather close to twice the column rate of $CO_2^+ + H_2 \rightarrow HCO_2^+ + H$, because the $HCO_2^+$ ions release H in their reactions. $CO_2^+$ is approximately proportional to square root of $F_{10.7}$ and varies by a factor of ~1.7, while $H_2$ near the ionospheric peak is smaller at high activity than that at low activity by a factor of ~1.4 because of this reaction. Though the model adopted a downward flow of H at the lower boundary of 80 km at the maximum diffusion rate $-K/H$ ($H$ is the scale height), almost all production of H in the above reaction moves up and escapes.

Escape of $H_2$ is mostly nonthermal at solar minimum with comparable nonthermal and thermal components near solar maximum (Table 10.3). It varies with solar activity by a factor of 5, being smaller than the escape of H by an order of magnitude. Total escape $\Phi_H + 2\ \Phi_{H2}$ is $1.9\times10^8$ and $2.4\times10^8$ cm$^{-2}$ s$^{-1}$ at $F_{10.7}$ = 25 and 90, respectively. It is smaller than the diffusion-limiting flow (Section 2.5.4) of $3.9\times10^8$ cm$^{-2}$ s$^{-1}$ for the $H_2$ mixing ratio of 17 ppm.

Variations of escape of H and $H_2$ with season and solar activity in the LMD GCM (Chaufray et al. 2015b, 2017) are shown in Figure 11.63. The model does not account for nonthermal escape; therefore the $H_2$ escape is underestimated at solar minimum. The seasonal component of the H escape varies by factors of 9 peak-to-peak and 4 smoothed. The variations of the H escape with solar activity are a factor of 2–3 and depend on season. The escape of H is $2.5\times10^8$ cm$^{-2}$ s$^{-1}$ at $L_S = 240°$ and solar maximum, being ~$6\times10^6$ cm$^{-2}$ s$^{-1}$ at $L_S = 50°$ and solar minimum.

The significant difference with the model by Krasnopolsky (2002) is unclear. In spite of the advantages of LMD GCM, Krasnopolsky (2002) is based on a more detailed photochemistry, includes nonthermal escape of the light species, and is rather transparent with all column reaction rates given in the paper, while LMD GCM is inevitably a black box to the reader.

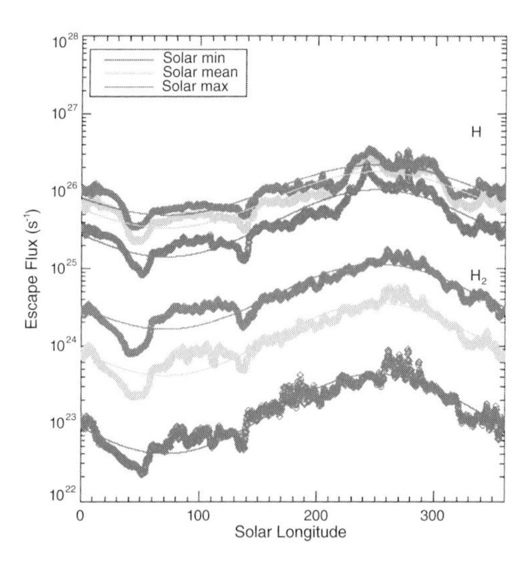


Figure 11.63 Total escape fluxes of H and $H_2$ versus solar longitude $L_S$ at various solar activity (Chaufray et al. 2015b). The lines are smoothed variations. These fluxes should be divided by the Mars surface area of $1.45\times10^{18}$ cm$^2$ to convert to cm$^{-2}$ s$^{-1}$.

Both models disagree with the very strong escape up to $10^9$ $cm^{-2}$ $s^{-1}$ observed near perihelion by MEX/SPICAM and HST (Section 11.11.2; Figure 11.44). This may be explained by significant water abundances in the lower thermosphere near perihelion that were observed by Maltagliati et al. (2013) and Fedorova et al. (2018) using the SPICAM IR solar occultations. A recent model (Figure 11.62) includes neutral and ion $H_2O$ chemistry and establishes quantitative dependences of hydrogen escape and thermospheric composition on water abundance and solar activity. Photolysis of water is the most effective at 160–190 nm, where the solar radiation varies within ~10%. This facilitates the low variations of hydrogen escape with solar activity.

### *11.13.5 Stellar Occultations*

SPICAM/MEX (Figure 11.64) and IUVS/MAVEN occultation spectra make it possible to retrieve limb abundances of $CO_2$ and $O_2$ and convert them into vertical profiles of $CO_2$, $T$, and $O_2$ mixing ratio. These data refer mostly to the nighttime atmosphere. The SPICAM data on $CO_2$ and $T$ were analyzed by Forget et al. (2009). They cover the altitudes from 50 to 130 km (Figure 11.65). Main features of the observed variations of $CO_2$ densities and temperatures are discussed in Forget et al. (2009). For example, seasonal variations of $CO_2$ at 85 km are shown in Figure 11.66.

The IUVS/MAVEN spectra are in the range of 115–170 nm, and the retrieved temperature profiles are from 100 to 145 km (Figure 11.67). Each profile is averaged of five stellar occultations observed at similar conditions indicated in the figure. The NGIMS/MAVEN profile measured during a deep dip down to 125 km at SZA $\approx 85°$ is shown in the figure and was used as an upper boundary condition for the retrievals. The MCD profiles for the observing conditions are shown as well.

The retrieved vertical profiles of $O_2$ mixing ratio extend from 90 to 130 km in the SPICAM observations (Sandel et al. 2015). The altitude-averaged values for six analyzed

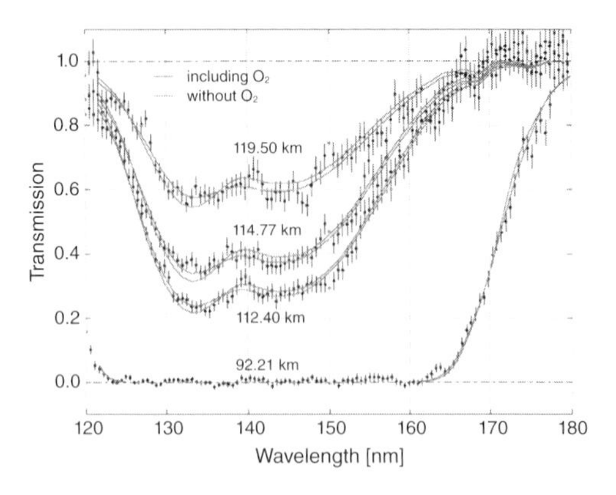


Figure 11.64 Fitting of the stellar occultation spectra by the $CO_2$ absorption without and including $O_2$ (gray and dark lines, respectively). From Sandel et al. (2015).

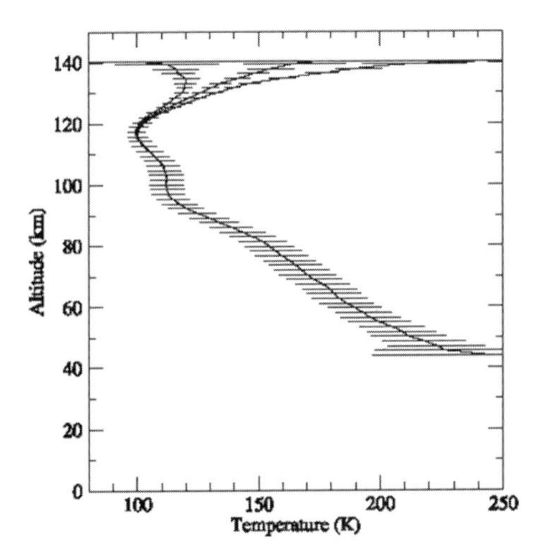


Figure 11.65 Temperature profile retrieved from SPICAM stellar occultation at $L_S$ = 247°, 4°S, LT = 02:00 for three adopted temperatures at 140 km (Forget et al. 2009).

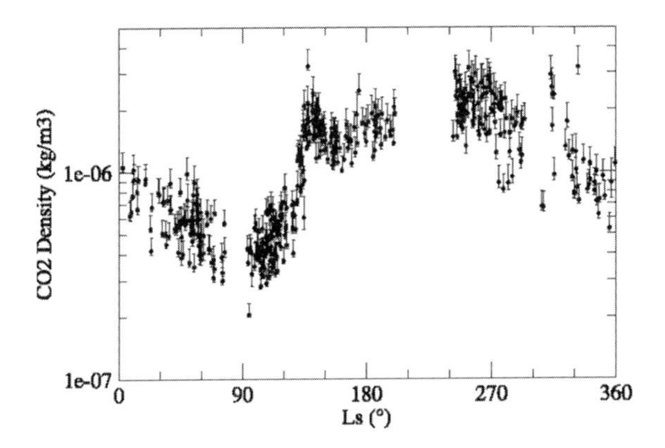


Figure 11.66 Seasonal variations of the $CO_2$ density at 85 km at low and middle latitudes observed by SPICAM stellar occultations (Forget et al. 2009). $10^{-6}$ kg m$^{-3}$ = $1.37\times10^{13}$ cm$^{-3}$ of $CO_2$.

occultations vary from $3.1\times10^{-3}$ to $5.8\times10^{-3}$, with a mean value of $4\times10^{-3}$. Sandel et al. (2015) compare their results with the Viking data (Figure 11.58) that were $1.7\times10^{-3}$ at 125–130 km increasing to $3\times10^{-3}$ at 145–150 km. The current $O_2$ annual-mean mixing ratio is $1.56\times10^{-3}$ in the lower atmosphere (Section 11.4.3). The increase to 150 km in the Viking measurements is due to diffusive separation above the homopause near 120 km with enrichment of light species including $O_2$. Therefore the Viking $O_2$ abundances agree with the MSL observations.

Groeller et al (2015) used a peak at 120 nm in the $O_2$ absorption spectrum for retrieval of $O_2$ from the IUVS/MAVEN stellar occultations. They mentioned that the observed $O_2$ mixing ratios are from $1.5\times10^{-3}$ to $5\times10^{-3}$ and agree with the SPICAM data.

### *11.13.6 Neutral Composition and Temperature from MAVEN/NGIMS*

The Mars atmosphere and volatile evolution (MAVEN) orbiter was inserted on the Martian orbit with periapsis of 150 km in September 2014. MAVEN has a few instruments for direct and remote studies of the thermosphere and ionosphere, including the neutral gas and ion mass spectrometer NGIMS. First results of the mission are available now, and vertical profiles of nine species between 155 and 300 km are shown in Figure 11.68. The observations were made in February to May 2015 during northern winter $L_S$ = 288°–326° at heliocentric distance of 1.41–1.47 AU, at low latitudes and SZA $\approx$ 45°. The mean $F_{10.7}$ was 128 on the Earth and 63 on Mars. Along with the regular orbits, the spacecraft had two deep dips down to 125 km. However, those data had not been completely analyzed in Mahaffy et al. (2015).

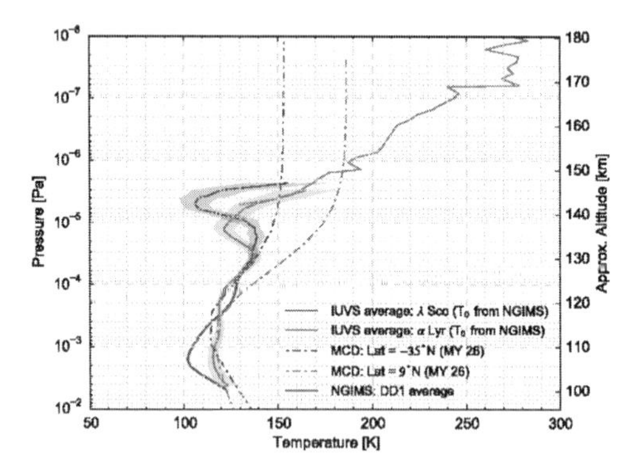


Figure 11.67 Two mean temperature profiles from five stellar occultations each, at LT = 22:00, SZA = 118°, 35°S, and LT = 06:00, SZA = 96°, 9°N. The NGIMS profile refers to dusk with SZA = 85°. From Groeller et al. (2015).

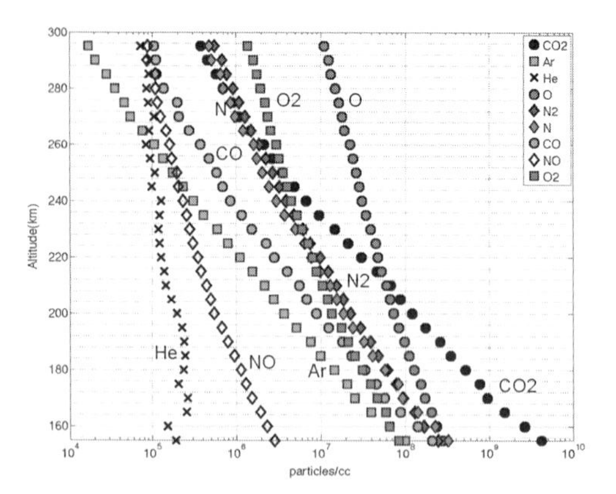


Figure 11.68 Vertical profiles of nine species measured by MAVEN/NGIMS in the upper atmosphere at $F_{10.7} \approx 63$, SZA $\approx$ 45°, $L_S$ = 288°–326°, latitudes 4°S to 46°N (Mahaffy et al. 2015). (A black and white version of this figure will appear in some formats. For the color version, please refer to the plate section.)

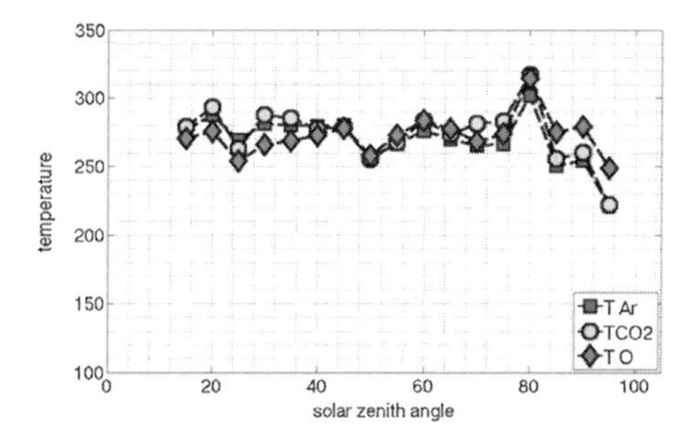


Figure 11.69 Exospheric temperatures derived from scale heights of Ar, $CO_2$, and O at 200–300 km at various SZA. The mean temperature is 274 ± 6 K at SZA = 15°–75° and $F_{10.7} \approx 63$. From Mahaffy et al. (2015).

The atmosphere at 200 to 300 km should be isothermal with diffusive equilibrium. Temperatures derived from the measured vertical profiles of three species at various SZA are shown in Figure 11.69. The temperatures are very similar for all species and do not vary with SZA from 15° to 75° with a mean value of 274 ± 6 K. These are the most accurate exospheric temperatures measured in the Martian atmosphere. They are compared with the other data in Section 11.13.3. Measurements during the deep dips showed an almost isothermal atmosphere with $T \approx 270$ K above 170 km steeply decreasing to 150 K at 125–130 km (at 145 km in Figure 11.67). The $Ar/CO_2$ and $N_2/CO_2$ ratios during the deep dips approach the values of 2% observed in the lower atmosphere at the Curiosity/MSL (Section 11.5.1).

There are some problems related to the data in Figure 11.68. $N_2/CO \approx 1$ at 155 km and 5 at 300 km, though this ratio is expected to be constant in this altitude range. $O_2/CO_2 = 0.02$ at 155 km in Figure 11.68 and 0.0036 in the Viking data (Figure 11.58). The Viking ratio at 127 km agrees with the Curiosity/MSL value for the lower atmosphere (Section 11.4.3). The difference is rather large to account for the solar activity. The scale height of $O_2$ is greater than that of $N_2$ in Figure 11.68, though the opposite is expected. NO is denser than that calculated by Fox et al. (1996; Figure 11.61) for high solar activity by a factor of ~5 at 155–200 km. Atomic nitrogen densities are similar to those of $N_2$ and exceed the values from Fox et al. (1996) by two orders of magnitude at 155–180 km.

Variations of temperature at 150–180 km with solar activity and season were analyzed by Bougher et al. (2017) using NGIMS and IUVS observations. The IUVS limb observations of scale height of the $CO_2^+$ doublet emission at 289 nm were studied along with scale heights of $CO_2$ and Ar from the MGIMS measurements. It looks like the seasonal behavior is presented fairly well by the $R^{-2}$ function, where $R$ is the heliocentric distance. The observed scale heights are just proportional to the solar Lyman-alpha radiation on Mars orbit measured by one of the MAVEN instruments. The NGIMS measurements near $L_S \approx 320$, 60, and 140° gave $T \approx 240°$, 190°, and 180°, respectively. These temperatures refer to 150–180 km, that is, should be slightly smaller than the exospheric temperatures.

### *11.13.7 Helium Bulge*

The He densities in the MAVEN data (Figure 11.68) are smaller than those fit the observed intensity of He 584 Å (Figure 11.62) by a factor of ≈3. Elrod et al. (2017) made a more detailed analysis of variations of helium with local time, season, and latitude. It is difficult to separate these types of the helium variability. However, the observations indicate a maximum at local solar time of 2–5 h and a secondary maximum at 22–24 h. The main maximum is of an order of magnitude relative to the daytime minimum.

Even a more complicated behavior of helium is predicted by the Mars Global Ionosphere-Thermosphere Model (MGITM; Bougher et al. 2015a). According to the model, peak He densities near the equinoxes are at ≈35°N and S and LT ≈ 5 h. Those at the solstices are at ≈60° in the winter hemisphere at LT ≈ 5 h. Formation of the morningside bulges of light species on Venus is discussed in Section 12.11.3, and similar mechanisms operate being weaker on Mars.

### *11.13.8 Ion Composition Measured by MAVEN/NGIMS*

NGIMS measurements in the ion observing mode revealed a very rich composition of the Martian ionosphere with detailed vertical profiles of 22 ions (Figure 11.70). Ion densities as low as 0.002 $cm^{-3}$ could be reliably measured. Comparing the major ions $O_2^+$, $CO_2^+$, and $O^+$ in the Viking observations (Figure 11.59) with the NGIMS data, one should bear in mind the significant difference in solar activity ($F_{10.7}$ = 27 and 63, respectively). $O_2^+/CO_2^+ \approx 7$ at 150–200 km at the Viking conditions and increasing from 5 to 10 for MAVEN, $O^+ = CO_2^+$ at 220 km and $O^+ = O_2^+$ at 300 km for both spacecraft.

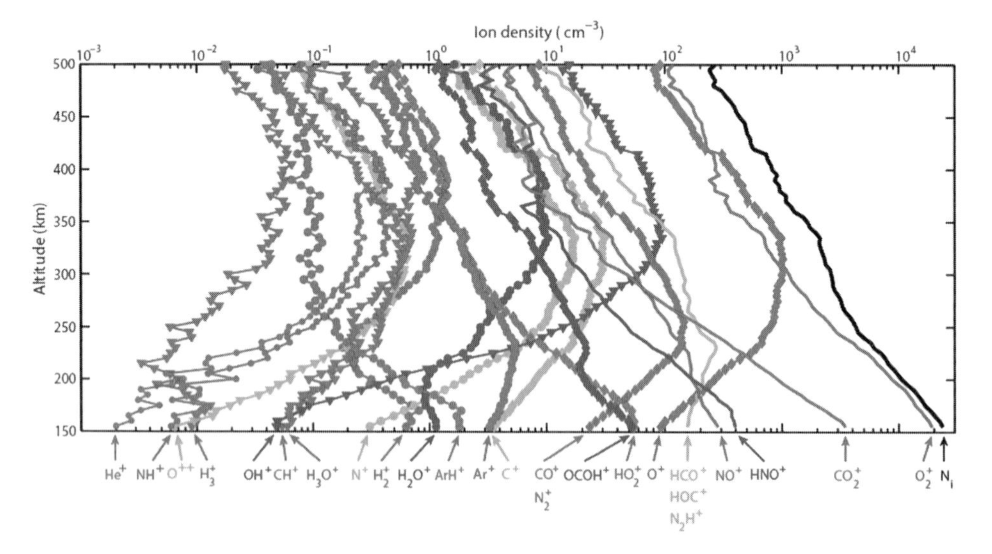


Figure 11.70 Vertical profiles of 22 ions measured by NGIMS at SZA = 60°. Measured isotopes $^{36}Ar^+$, $^{38}Ar^+$, and $^{18}OO^+$ are not shown. From Benna et al. (2015). (A black and white version of this figure will appear in some formats. For the color version, please refer to the plate section.)

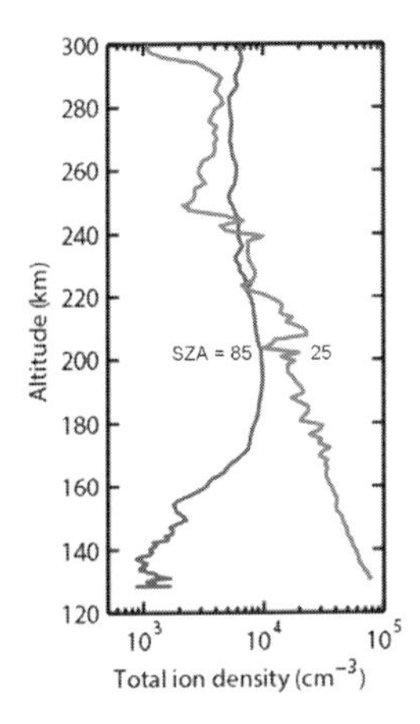


Figure 11.71 Vertical profiles of the total ion density for two deep dips at mean SZA = 85° and 25° (Benna et al. 2015).

Vertical profiles of total ion density observed at two deep dips are shown in Figure 11.71. These measurements were made near noon and terminator, and the data differ significantly.

Twelve of the measured 22 ions contain H. Fox et al. (2015) included 14 H-containing ions in her model of the Martian ionosphere. Actually that model is for low solar activity, and the new MAVEN data on the neutral composition have not been implemented into the model. Nevertheless the calculated ion densities reproduce fairly well the measured values (Figure 11.72). The required water abundance should be very low, 0.4 ppb at 80 km, otherwise $HCO^+$ is depleted via

$$HCO^+ + H_2O \rightarrow H_3O^+ + CO.$$

However, water stimulates production of $HCO^+$ via $H_2O^+ + CO \rightarrow HCO^+ + OH$ and weakly affects the $HCO^+$ densities (Krasnopolsky 2019) that are compatible with $H_2O$ up to 100 ppm at 80 km observed by the SPICAM IR occultations (Maltagliati et al. 2013; Fedorova et al. 2018; Section 11.3.2) and predicted by MCD. The model indicates that the mass-2 ion is $H_2^+$ with a minor contribution of $D^+$.

Median altitude profiles of the ion composition in the nighttime ionosphere at low and high suprathermal electron fluxes are shown in Figure 11.73 (Girazian et al. 2017). The low and high electron fluxes are those smaller and larger than $10^6$ eV $cm^{-2}$ $s^{-1}$ $sr^{-1}$.

Comparing the nighttime ion densities with those on the dayside in Figure 11.70, the difference in the electron densities is a factor of $\approx$10. Therefore the nighttime ion lifetimes are greater than those of the dayside by this factor, and the primary ions $CO_2^+$ and $O^+$ are significantly depleted on the nightside relative to the terminal ions $O_2^+$ and $NO^+$. For example, $O_2^+/CO_2^+ \approx 100$ and 10 are at 170 km on the nightside and dayside, respectively. The opposite behavior for $O_2^+/O^+ \approx 3$ and 10 at 220 km, respectively, may be explained, because $[CO_2] \approx [O]$ at this altitude on the dayside and much smaller than [O] on the nightside.

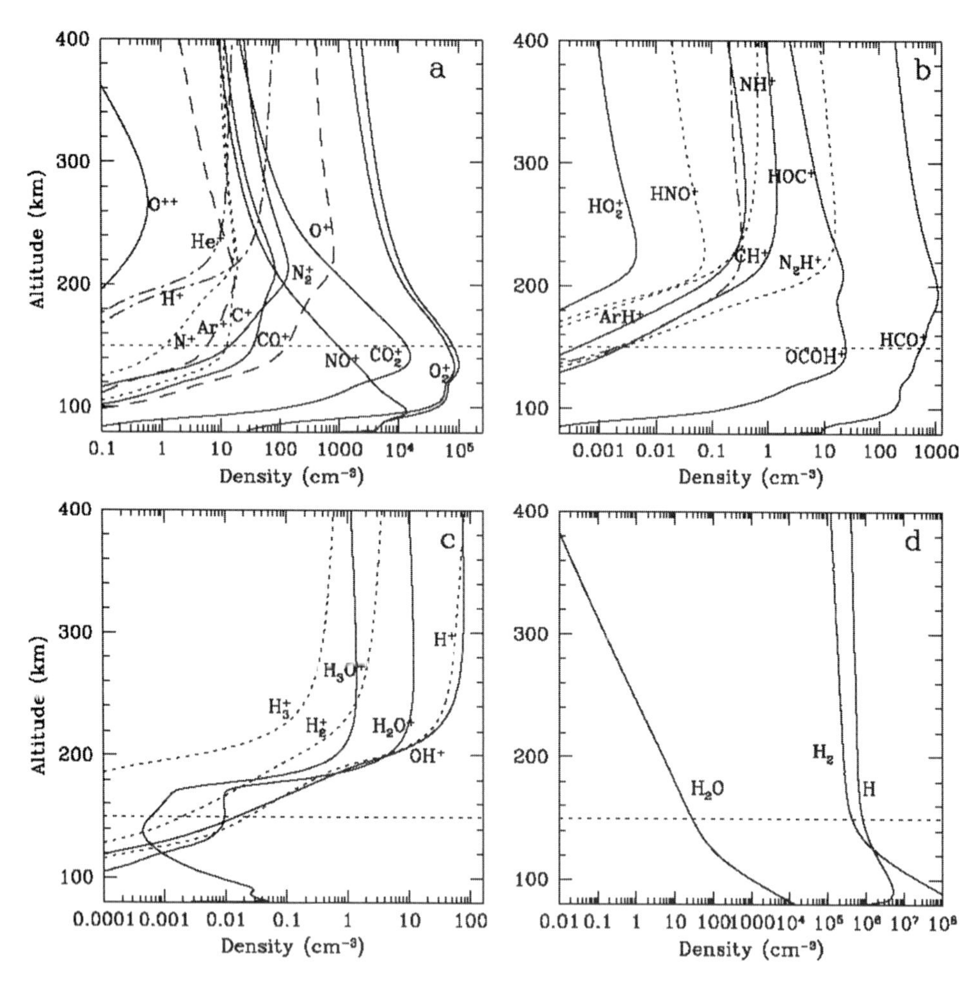


Figure 11.72 Calculated profiles of major ions (a), protonated ions (b, c), and hydrogen-bearing neutrals (d) in the model by Fox et al. (2015).

Even greater is the difference between $NO^+/CO_2^+ \approx 60$ and 0.1 at 170 km on the nightside and dayside, while $O_2^+/NO^+ \approx 5$ and 50, respectively. This is due to charge exchange

$$O_2^+ + NO \rightarrow NO^+ + O_2,$$

which is more effective on the nightside because of the longer chemical lifetimes.

## 11.14 Some Aspects of Evolution

### *11.14.1 Meteorite Impact Erosion*

Discussing evolution of Mars, we adopt a time scale with zero at the time when the Solar System formed and 4.6 Byr at the present. Melosh and Vickery (1989) argued that there was

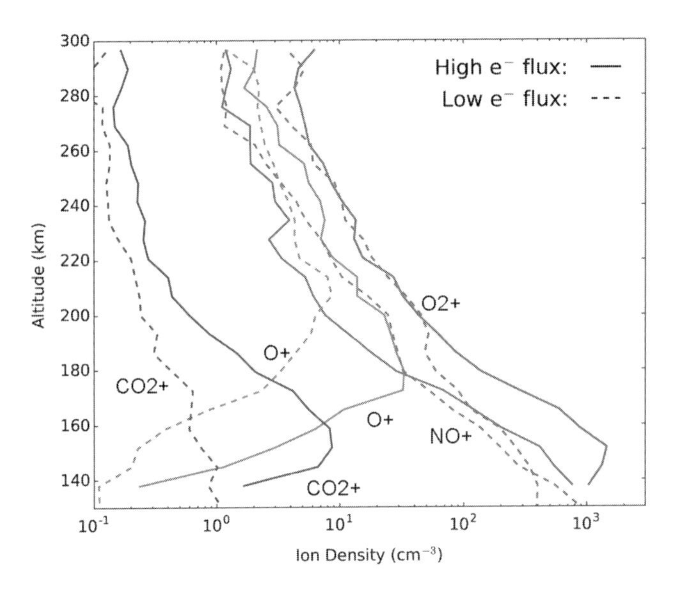


Figure 11.73 Median profiles of nightside ions at low and high electron fluxes (Girazian et al. 2017).

intense erosion of the Martian atmosphere by meteorite impacts in the first 0.8 Byr, and just a few percent of the initial atmosphere remained after $t = 0.8$ Byr.

Owen and Bar-Nun (1995) suggested using $^{84}$Kr to trace the history of $CO_2$ on Mars. This gas is concentrated mostly in the atmosphere, and its abundance was measured by the Viking mass spectrometers. Krypton is much heavier than $CO_2$ and therefore does not escape. C/$^{84}$Kr in the Martian atmosphere is smaller than the Earth's ratio of C in carbonates to $^{84}$Kr in the atmosphere by a factor of 10. Therefore the present abundance of 7.5 mbar of $CO_2$ in the atmosphere and the polar caps constitutes 10% of the total abundance. The rest escaped by sputtering and was transformed to carbonates. This total $CO_2$ abundance of 75 mbar refers to $t = 0.8$ Byr, because impact erosion affected krypton as well as all other gases. A reduction of $^{84}$Kr by a factor of 100 by impact erosion can be obtained comparing the abundances of this gas per gram on Mars and Earth. This factor should be the same for all atmospheric gases; therefore, the initial abundance of $CO_2$ was 7.5 bar. However, this is not the only evaluation of $CO_2$ before and after impact erosion in the literature.

### 11.14.2 Carbon and Oxygen Isotope Ratios

The measured $^{13}C/^{12}C = 1.046$ and $^{18}O/^{16}O = 1.048$ times the terrestrial standards (Mahaffy et al. 2013; Section 11.5.1), that is, $\delta^{13}C = 46 \pm 4$ ‰ and $\delta^{18}O = 48 \pm 5$ ‰, impose some constraints to evolution of volatiles on Mars.

The main process of carbon escape is atmospheric sputtering by $O^+$ ions. These ions are accelerated by the solar wind to energies exceeding 1 keV and reimpact the atmosphere due to magnetic field induced in the Martian ionosphere. The impacts dissociate $CO_2$ with a high yield of atomization and escape near or above the exobase. Heavy carbon is depleted

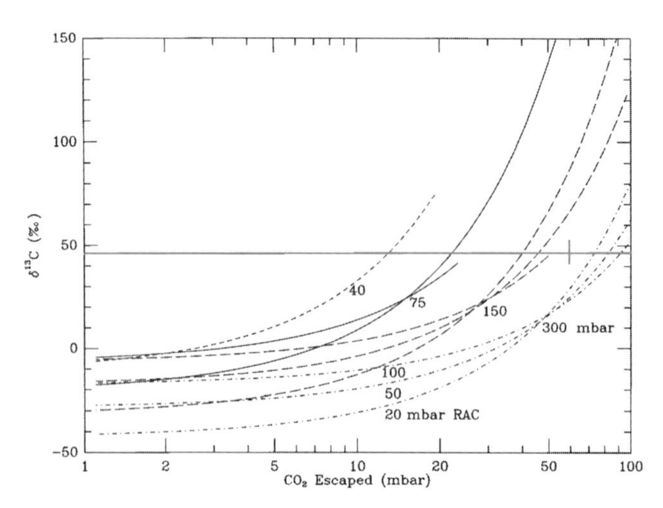


Figure 11.74 Carbon isotopes in the present atmosphere as a function of the $CO_2$ escape after the end of impact erosion at 0.8 Byr. The curves are calculated for various $CO_2$ abundances at 0.8 Byr (from 40 to 300 mbar) and the present regolith–atmosphere–polar caps reservoir of $CO_2$ (from 20 to 100 mbar). The horizontal line is the present carbon isotope ratio, and its intersection with the $CO_2$ escape of 60 mbar (Leblanc and Johnson 2002) is marked. Updated from Krasnopolsky et al. (1996).

in this escape due to diffusive separation, and the remaining heavy carbon is enriched by a factor of 1/0.891 (Krasnopolsky et al. 1996), that is, $\delta^{13}C = 0.109$ in this process. Formation of carbonates has the opposite effect with depletion of heavy carbon in $CO_2$ with $\delta^{13}C = -0.016$. The current carbon isotope ratio depends on the present abundance of $CO_2$ in the regolith, atmosphere, and polar caps (RAC), total escape, and total carbonates (Figure 11.74). Their sum is the $CO_2$ abundance at 0.8 Byr. Evidently the measured isotope ratio cannot uniquely determine all these values. Leblanc and Johnson (2002) calculated the total escape $CO_2$ at 60 mbar (Figure 11.74). Then the $CO_2$ abundance was ~200 mbar at 0.8 Byr and the total $CO_2$ in RAC and carbonates is ~200 – 60 = 140 mbar.

The oxygen isotope ratio involves fractionation between water and silicates, between water and $CO_2$, and escape of hydrogen (mostly thermal) and O (nonthermal and sputtering). The exchange with silicates and $CO_2$ was significant when water existed as a liquid ocean. Impact erosion of $CO_2$ is an additional parameter in the system.

### *11.14.3 Loss of Water from Mars*

According to Krasnopolsky (2015d), the $HDO/H_2O$ global-mean ratio is 4.6 times the terrestrial ratio and $\approx$7 in the ice of the north polar cap (NPC; Section 11.3.3). Photolysis of HDO is less effective than that of $H_2O$ by a factor of 2.5 on Mars (Section 10.5.3). $HDO/H_2O$ is reduced above the condensation level in the atmosphere because of the preferential condensation of HDO. Loss of HD in the reactions with OH and $O(^1D)$ is weaker than that of $H_2$. The measured $f_{H2}$ = 17 ppm and $f_{HD}$ =12 ppb (Section 11.13.3; Krasnopolsky 2002) result in $HD/H_2 \approx 2.25$ SMOW that is smaller than $HDO/H_2O \approx 7$ due to a combination of these factors. The calculated fractionation factor averaged over the solar activity cycle is

$$F = \frac{\Phi_D/\Phi_H}{D/H} = 0.105$$

(Krasnopolsky 2002). The water history on Mars may be divided in two periods before and after the first 0.3–0.5 Byr. $(D/H)_0$ was equal at that time to 1.9 in Leshin (2000) and 2.2–4 in Kurokawa et al. (2014); we adopt 2.2 to get enrichment $r = 7/2.2 = 3.2$ in the present NPC.

Here we apply a scheme with three reservoirs of water (Section 10.5.3): (1) the bulk water with $(D/H)_0$, (2) water in the permanent polar caps, and (3) water in the atmosphere and seasonal polar caps. Hydrogen escapes from (3) and is replenished from (2), and (2) is supplied from (1). Reservoirs (2) and (3) have D/H ratios that are determined by fractionation in the sublimation/condensation processes. A solution for this case (10.31) is

$$\alpha = \frac{1}{F} \ln \frac{1 - F}{1 - Fr} = 2.8,$$

using $F = 0.105$ and $r = 3.2$; $\alpha$ is the ratio of total loss of water to its abundance in reservoir (2). Amounts of water ice in the NPC and the south layered deposits are 10 m (Zuber et al. 1998) and 11 m (Plaut et al. 2007) of global equivalent layer, respectively. Then the lost water is $2.8\times21 \approx 60$ m for the last $\approx$4 Byr, and the total water was equal to 80 m $\approx$4 Byr ago.

There is some evidence, including abundances and isotope ratios of the noble gases, for hydrodynamic escape from the early Mars. Hydrodynamic escape of water is only possible for comparable abundances of $H_2O$ and $CO_2$ in the atmosphere (Kasting and Pollack 1983), and a liquid water ocean is insufficient to drive hydrodynamic escape on Mars. More probable is hydrodynamic escape of $H_2$ released by

$$Fe + H_2O \rightarrow FeO + H_2$$

from the hot Mars after the accretion. Fractionation factor for this escape is $F_h = 0.8$ (Zahnle et al. 1990), $(D/H)_0 = 1.275$ according to Kurokawa et al. (2014), the final D/H $\approx$ 2.2, and the final water abundance is 80 m. Then an initial water abundance is (10.25)

$$80 \times (2.2/1.275)^{1/(1-0.8)} = 1200 \text{ m}.$$

The terrestrial water ocean scaled to the Mars size and mass is 1 km deep, and the calculation confirms a conclusion by Dreibus and Waenke (1987) that Mars was initially even richer in water than Earth. Supply of water by cometary impacts is much smaller than the above estimates and cannot affect them (Krasnopolsky 2002). Escape of hydrogen is balanced by nonthermal escape and sputtering of oxygen and its loss in oxidation of the surface rocks.

Photochemical escape of hot oxygen was recently studied by Lillis et al. (2017) using observations by three MAVEN instruments for 1.5 years. The escape originates mostly from dissociative recombination of $O_2^+$ and varies from 1.2 to $5.5\times10^{25}$ s$^{-1}$ depending on season and solar activity. Its mean rate is $4.3\times10^{25}$ s$^{-1}$ = $3\times10^7$ cm$^{-2}$ s$^{-1}$ = 25 g cm$^{-2}$ Byr$^{-1}$ in the current epoch. The solar EUV was much brighter from the young Sun, and total loss of oxygen for 3.5 and 4 Byr was 480 and 2250 mbar, respectively (table 4 in

Lillis et al. 2017). The latter corresponds to 60 m of water and supports the above value deduced from the hydrogen isotope fractionation.

### 11.14.4 Evolution of Nitrogen

Compared with the terrestrial $^{14}N/^{15}N$ = 273, the Martian ratios 168 ± 17 and 173 ± 11 measured by Viking 1 and 2 and Curiosity/MSL correspond to $\delta^{15}N$ = 625 and 578‰, respectively, and agree within their uncertainties. The significant deviation from the terrestrial standard indicates effective loss of nitrogen with preferential escape of the light isotope. Evolution of nitrogen and its isotope ratio on Mars was analyzed for the first time by McElroy et al. (1976). Later Fox (1993) calculated escape of N that is equal to $4.8\times10^5$ cm$^{-2}$ s$^{-1}$ mostly by reactions

$$N_2^+ + e \rightarrow N + N \ (42\%)$$

$$N_2^+ + O \rightarrow NO^+ + N \ (20\%)$$

$$N_2 + h\nu \rightarrow N + N \ (21\%).$$

Fox (1993) took into account production of N($^2D$) by these reactions and vibrational excitation of $N_2^+$. Her results were used by Jakosky et al. (1994) to calculate evolution of nitrogen after the end of impact erosion at $t$ = 0.8 Byr (Figure 11.75). Their model adopted $CO_2$ = 750 mbar 3.8 Byr ago decreasing linearly in the logarithmic scale to the present 7 mbar because of carbonate formation, sputtering, and outgassing. The model applied available data on evolution of the solar EUV flux and some approximations to calculate sputtering and photochemical escape as functions of the solar flux, $CO_2$ and $N_2$ abundances.

The model [2] in Figure 11.75 for photochemical escape only gives too high enrichment by heavy nitrogen, while the model [3] for sputtering and outgassing of nitrogen gives too low enrichment. Evidently the isotope fractionation is much greater in photochemical escape than that in sputtering. Curve [1] that fits the present isotope ratio should involve outgassing along with photochemical escape and sputtering. Evolution of the $N_2$ abundance started at one-half of the present abundance 3.8 Byr ago and reached a maximum of

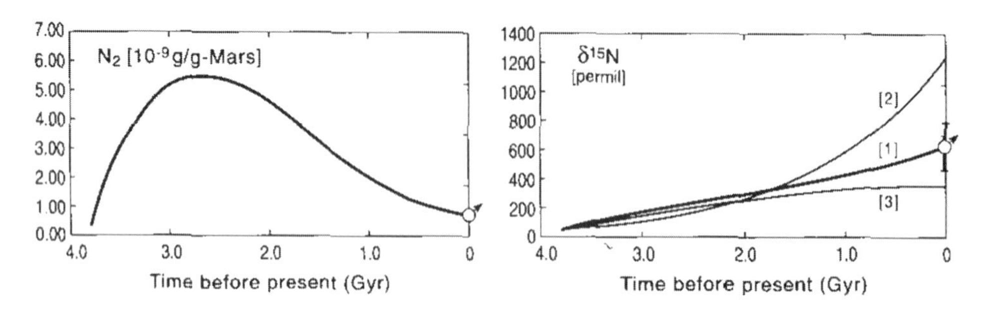


Figure 11.75 Evolution of atmospheric nitrogen and $\delta^{15}N$ on Mars. The present values are marked by Mars symbols. [1] is for sputtering, photochemical escape, and outgassing, [2] is for photochemical escape only, and [3] is for sputtering and outgassing, without photochemical escape. See details in text. From Jakosky et al. (1994).

Table 11.8 *Main natural radionuclides*

| Isotope | Ratio | Half-life[a] | Products[b] | Heating[c] |
|---|---|---|---|---|
| $^{238}$U | 0.993 | 4.51 | $^{4}$He ×8 | 0.976 |
| $^{235}$U | 7.2–3[d] | 0.71 | $^{4}$He ×7 | 5.71 |
| $^{232}$Th | 1 | 14.1 | $^{4}$He ×6 | 0.265 |
| $^{40}$K | 1.167–4 | 1.28 | $^{40}$Ar ×0.105[e] | 0.284 |

[a]In Byr. [b]Stable gaseous products and their yields in the total decay chains. [c]In erg $g^{-1}$ $s^{-1}$. [d]7.2–3 = $7.2\times10^{-3}$. [e]A dominant branch is formation of $^{40}$Ca.

7.5 times the present abundance 2.7 Byr ago (Figure 11.75). Unfortunately many numerical data on the model are missing from the paper.

### *11.14.5 Radiogenic Argon and Helium*

Uranium, thorium, and potassium are the most important natural radionuclides in the terrestrial planets. Their properties are given in Table 11.8. Waenke and Dreibus (1988) evaluated total element mass ratios in the Martian interior at 305 ppm for K, 16 ppb for U, and 56 ppb for Th.

Argon is lost by sputtering, and its current lifetime is 19 Byr and $\tau_{Ar} = 0.18\ e^{t\ (Byr)}$ Byr calculated by Krasnopolsky and Gladstone (1996) using Monte Carlo simulations by Jakosky et al. (1994). Lifetime of helium relative all types of escape (Table 10.3) was calculated at 0.13 Myr in the present atmosphere (Krasnopolsky and Gladstone 2005). Therefore $^{40}$Ar is a product of accumulation, outgassing, and loss throughout the history of Mars, while $^{4}$He reflects just the current conditions. Outgassing of $^{40}$Ar and $^{4}$He is equal to their abundances in the interior times outgassing coefficient. The outgassing should be very strong in the first 0.5 Byr to fit the present abundance of $^{36}$Ar to that was captured during Mars' formation. This strong outgassing weakly affects $^{40}$Ar which just began to accumulate. The adopted outgassing coefficient that fits the present abundance of $^{40}$Ar is $0.0065(t_0/t)^{1/2}$ Byr$^{-1}$, where $t_0$ = 4.6 Byr.

The model predicts that the losses of $^{40}$Ar by meteorite impact erosion and sputtering are 14% and 53% of its present abundance in the atmosphere, respectively. However, the calculated outgassing of helium is just 10% of its loss. Furthermore, the upper limit to outgassing of $SO_2$ (Section 11.8) and the lack of volcanism in the current epoch indicate very low outgassing.

An alternative source of helium is the solar wind alpha particles. Their mean flux is the solar wind flow of $2.5\times10^{8}$ cm$^{-2}$ s$^{-1}$ times 0.43 to account for the Martian heliocentric distance and the helium ratio of 0.06 in the solar wind. Then the required capture efficiency for the solar wind alpha particles is 0.3 referred to the solid disk of Mars. Therefore helium in the Martian atmosphere originates not from the radioactive decay of U and Th in the planet interior, but from the solar wind.

# 12

# Venus

## 12.1 General Properties and History of Studies

Except the Sun and the Moon, Venus is the brightest object on the Earth's sky and makes observable shadows in the night. The mean distance of Venus from the Sun is 0.723 AU (Table 12.1), and the maximal Venus elongation (Sun–Earth–object angle) is 47°. Therefore Venus can be observed in the nighttime for 2–3 hours before sunrise and after sunset even near the maximal elongations.

A challenging fact is that the rotation of Venus is retrograde, opposite to the counter-clockwise rotation of the Solar System, and very slow, so that its period is close to the Venus year. The obliquity of 177.4° reflects the retrograde rotation, and angle between the axes of the planet and its orbit is 2.6°. Coupled with the small eccentricity, this means no seasons on Venus.

The size, mass, and mean density of Venus are rather similar to those of the Earth. The atmosphere of Venus was discovered by Mikhail Lomonosov, who observed some refraction phenomena in the transit of Venus across the solar disk in 1761. Venus is covered by a thick and uniform cloud layer. Before the space era, these facts raised hypotheses that conditions at the Venus surface could be comfortable and even habitable, similar to those on the Earth in the dinosaur epoch.

However, microwave observations in 1950s revealed a strong signal from Venus indicating a hot and dense lower atmosphere. An alternative explanation was related to some ionospheric phenomena. The interpretations could be distinguished by observations of either limb darkening or limb brightening of the signal, respectively. Those observations were conducted by the first flyby of Venus by Mariner 2 in December 1962 that confirmed the hot and dense atmosphere of Venus.

The year 1967 was especially fruitful for the study of Venus because of three events. Venera 4, the first successful Russian descent probe, made direct observations of temperature, pressure, density, and abundances of $CO_2$ (~90%), $N_2$ (<7%), $O_2$ and $H_2O$ (the results are questionable) from 55 to 23 km. Photometer at H 122 nm and O 130 nm, magnetometer and two plasma instruments were installed at the flyby module.

The next significant event was the flyby of Mariner 5 just after the Venera 4 operation. The observations included radio occultation, a profile of H 122 nm, magnetic field and plasma measurements. Ground-based radar observations of Venus using large radio telescopes resulted in the Venus radius measurement of 6050–6056 km. With this value, the

Table 12.1 *Basic properties of Venus*

| Mean heliocentric distance (AU) | 0.723 |
|---|---|
| Eccentricity | 0.0068 |
| Obliquity | 177.4° |
| Radius (km) | 6052 |
| Mass | $4.87 \times 10^{27}$ g = 0.815 $M_E$ |
| Mean density (g cm$^{-3}$) | 5.24 |
| Surface gravity (equator, cm s$^{-2}$) | 887 |
| Sidereal year (Earth days) | 224.7 |
| Sidereal period of rotation (retrograde, Earth days) | −243.02 |
| Escape velocity at exobase (km s$^{-1}$) | 10.2 |

occultation covered the range measurement of 90–36 km, and the retrieved temperature profile had a tropopause near 67 km and a temperature gradient of −9 K km$^{-1}$ near 40 km. Extrapolation of the profile to the surface with this gradient gave the surface temperature of 737–792 K and pressure of 63–94 bar. The observed Lyman-alpha distribution showed a break near $h \approx 3000$ km that was later explained by thermal (cold) hydrogen below and hot hydrogen above the break.

Another important result was achieved by Pierre and Janine Connes and J. P. Maillard, who created a Fourier transform spectrometer (Section 6.4) for astronomical observations and measured high-resolution spectra of Venus, Mars, Jupiter, and Saturn at 1 to 2.5 μm. Spectra of Venus with resolving power of up to $10^5$ revealed measurable abundances of CO, HCl, and HF.

The next Veneras 5, 6, 7, and 8 were improved versions of Venera 4. Venera 8 (1972) reached the surface and continued operations for one hour after the landing. Along with the temperature and pressure sensors, it had an accelerometer to measure temperature and pressure profiles before the parachute phase of the descent below 60 km, a radar altimeter, gamma-ray spectrometer to measure abundances of K, U, and Th in the surface rocks to specify their type, and a photometer to study the cloud structure. The probe radio transmitter was designed to measure wind velocities by the Doppler shift of the signal. The bus contained two charged particle instruments and a UV photometer for H 122 nm and O 130 nm. The new data obtained resulted in significant progress in the study of Venus.

Mariner 10 was designed for a flyby of Venus (in 1974) and three flybys of Mercury. The new data on Venus included the first detailed and numerous images of the clouds, radio occultations of the middle atmosphere and the ionosphere, variations of temperature at the cloud tops observed by an infrared radiometer, and detection of He using its line at 58 nm.

Venera 9 and its duplicate Venera 10 delivered landers and the first orbiters to Venus in 1975. Each orbiter had a camera, NIR spectrometer, airglow spectrometer for 300–800 nm, IR radiometer, photopolarimeter, Lyman-alpha H/D detector, radio occultations, bistatic radar for mapping of the surface, and instruments to study the plasma environment of Venus.

Each lander included temperature and pressure sensors, accelerometer, photometer and nephelometer to study the clouds, anemometer and Doppler wind speed transmitter, mass spectrometer, gamma-ray spectrometer and densitometer, and panoramic cameras.

The first panoramic images of the surface of Venus were the most impressive results of the mission. The visible night airglow on Venus was discovered and studied as well, along with other new data.

A significant progress in the studies of Venus is related to the Pioneer Venus and Venera 11 and 12 missions that reached Venus in December 1978. Pioneer Venus involved two spacecraft for the multiprobe and orbiter missions. The multiprobe mission had a bus, a large probe, and three small probes. The bus included neutral and ion mass spectrometers that made measurements down to 110 km. The large probe had a gas chromatograph, mass spectrometer, particle size spectrometer, nephelometer, radiometers for the red and thermal infrared, temperature, pressure, and acceleration sensors. Each of three small probes had a nephelometer, net flux radiometer, temperature, pressure, and acceleration sensors. The radio transmitters on all probes were designed to observe Doppler shifts to measure wind and its fluctuations. The small probes were named North, Night, and Day probes and were deployed accordingly. The Day probe was in operation for one hour after the landing.

The Pioneer Venus orbiter had a surface radar mapper, cloud photopolarimeter, infrared radiometer, ultraviolet spectrometer, neutral and ion mass spectrometers, magnetometer and a few plasma instruments, and gamma-ray burst detector and could also make radio occultations and atmospheric drag observations. The orbiter operated up to 1992; however, the lowest periapsis of 150 km was supported only during the first months of the operation and later in 1992 for comparison with the Magellan radar data.

The Venera 11 and 12 missions included landers and flyby platforms. The instruments at each lander were color panoramic cameras, a gas chromatograph, mass spectrometer, nephelometer, X-ray fluorescent spectrometer, visible range spectrometer and scanning photometer, temperature, pressure, and acceleration sensors, penetrometer, soil analysis device, low-frequency radio sensor, and microphone/anemometer.

Each flyby platform had a UV spectrometer for 30–166 nm, magnetometer, plasma spectrometer and sensors, and two gamma-ray burst detectors.

Another important event at that time was detection of the night and day airglow of $O_2$ ($a^1\Delta_g \rightarrow X^3\Sigma_g^-$) at 1.27 μm by Connes et al. (1979) using ground-based high-resolution spectroscopy.

Veneras 13 and 14 were rather similar to Veneras 11 and 12 with some improvements in the instrument properties. For example, the scanning photometers were extended to the near UV range of 320–390 nm to study a vertical profile of the NUV absorber in the clouds. The spacecraft reached the planet in 1981.

The Venera 15 and 16 orbiters were aimed at the detailed mapping of the Venus northern hemisphere down to ~30°N using synthetic aperture radars (SAR). The spacecraft had polar orbits with a periapsis of ~1000 km near 62°N. The missions operated in 1983–1984 and included, except SARs, radar altimeters, infrared Fourier spectrometers, cosmic ray and solar plasma detectors.

Vega 1 and 2 were sophisticated missions, each of those deployed a lander and a balloon on Venus in 1985 and used the Venus gravity to be directed to and flyby through comet Halley in 1986. The mission name is a combination of Venus and Halley (Gallei in Russian). Most of the scientific instruments for the mission were designed in cooperation with a few European countries.

Payload of each lander included temperature, pressure, and acceleration sensors, a particle size spectrometer, aerosol analyzer, nephelometer, mass spectrometer, gas chromatograph, active UV spectrometer (with a light source), X-ray fluorescence spectrometer for clouds and soil, gamma-ray spectrometer, penetrometer, and drilling device. The landers operated on the surface for one hour.

Each balloon operated for 46 hours at low latitudes near the altitude of 54 km, that is, in the middle cloud layer, and in both nighttime and daytime. The balloon instruments were a thermometer, barometer, anemometer, nephelometer, and photometer. The main instruments to study the atmospheric dynamics were the balloons themselves that were tracked by 20 radio telescopes around the world.

Discovery of transparency windows in the Venus atmosphere at 1.7 and 2.3 μm (Allen and Crawford 1984) and the smaller windows at 1.1, 1.18, and 1.27 μm made it possible to observe thermal emission of the lower atmosphere at different altitudes on the nightside, where the reflected sunlight does not obscure the effect. This resulted in ground-based spectroscopic sounding of the chemical composition in the lower atmosphere. A summary of the first observations of this type, technique for their analysis, and retrieval of species abundances were made by Pollack et al. (1993).

The Magellan orbiter operated around Venus in 1990–1994. Its main objective was detailed radar mapping of the surface, and the radar system was used in three modes: as a synthetic aperture radar, altimeter, and radiometer. The orbiter periapsis was 295 km with a period of 3 hours for the most of the observations. Aerobraking was applied at the final phase to reduce the periapsis to 180 km that was advantageous for gravimetry and atmospheric drag observations. Magellan mapped ~95% of the surface of Venus, and most of the data have resolution of 100–250 m per pixel. The observed maps are of great value for geology of Venus.

Venus Express was the European Venus orbiter launched by a Russian rocket. It was in operation from 2006 to the end of 2014. The polar orbit had a periapsis of 250 km at 80°N and a period of 24 hours. The camera had near UV, visible, and two near-infrared narrow-band filters; the latter were in the transparency window at 1.01 μm and made it possible to map the surface of Venus.

The spectrograph SPICAV had a UV (120–320 nm), NIR (0.7–1.7 μm), and SOIR (Solar Occultation at InfraRed) channels. The UV and NIR channels (Sections 6.2 and 6.1; Figure 6.1) were used in the nadir, limb, solar, and stellar occultation modes. SOIR was an echelle spectrograph for the range of 2.3 to 4.2 μm with resolving power of $1.5\times10^4$ and could be used only at the solar occultation mode.

The imaging spectrograph VIRTIS had a visible (0.25–1.0 μm) and infrared (1–5 μm) low-resolution mapping channels (VIRTIS-M, resolution 0.01 μm in the infrared) and a medium-resolution ($\lambda/\delta\lambda$ = 1500) channel VIRTIS-H for 2 to 5 μm that was based on a combination of an echelle and a prism.

As in the previous missions, the atmosphere and ionosphere were sounded by radio occultations. The plasma environment was studied by a magnetometer and analyzer of space plasmas and energetic atoms (ASPERA-4).

The spacecraft studies of the Venus atmosphere were complemented in the last decade by ground-based high-resolution spatially resolved spectroscopy in the submillimeter and near-infrared regions.

## 12.2 $CO_2$, $N_2$, Model Atmosphere below 100 km, Atmospheric Dynamics, and Superrotation

### *12.2.1 $CO_2$ and $N_2$*

The $CO_2$ bands were discovered in the atmosphere of Venus by Adams and Dunham (1932), and the gas sensor at Venera 4 showed the $CO_2$ mixing ratio of ~90%. The Pioneer Venus and Venera 11–14 mass spectrometers and gas chromatographs established two major constituents in the Venus atmosphere, $CO_2$ and $N_2$, while all minor species are smaller than 0.1% altogether.

The observed abundances of $N_2$ were measured at $2.5 \pm 0.3\%$ by the V12 gas chromatograph (Gelman et al. 1979), $4.0 \pm 0.3\%$ by the V11–12 mass spectrometers (Istomin et al. 1983), $4.0 \pm 2.0\%$ by the PV Large Probe mass spectrometer (Hoffman et al. 1980), and $4.5 \pm 1.3\%$ by the PV bus mass spectrometer (von Zahn et al. 1980), so that the claimed uncertainties in the measurements by three mass spectrometers overlap. The smallest uncertainties were for the PV gas chromatograph (Oyama et al. 1980), and their values of $3.54 \pm 0.04\%$ at 42 km and $3.41 \pm 0.01\%$ at 22 km are close to the mean. However, the variation with altitude, especially for the first analysis at 52 km that gave $4.6 \pm 0.14\%$, significantly exceeds the uncertainty and is mysterious for such a long-living species. Lebonnois and Schubert (2017) considered the lower atmosphere as a supercritical fluid to explain these variations. Anyway, the value of 3.5% for the $N_2$ mixing ratio is accepted for the lower atmosphere by VIRA (1985). Then the mean molecular mass is 43.44 in the lower atmosphere.

### *12.2.2 Venus International Reference Atmosphere (VIRA)*

VIRA below 100 km (Seiff et al. 1985) is based on the PV and V8–14 direct measurements of temperature and pressure below 60 km, the accelerometer data above 60 km, the PV and V9–10 radio occultations at 40–95 km, and retrievals of temperature and pressure profiles using the multichannel infrared radiometer at 15 μm onboard the PV orbiter (Table 12.2). The data above 33 km are presented for five latitudes: less than 30° and at 45°, 60°, 75°, and 85°.

Zasova et al. (2006) updated the model by inclusion of the Vega 2 data that are superior to the V10 pressure and temperature measurements, the only source of the data below 12 km in VIRA. Other data available were the V15–16 and Magellan radio occultations and temperature–pressure profiles at 55–100 km from the V15 spectroscopy of the $CO_2$ band at 15 μm. The changes in the model by Zasova et al. (2006) refer mostly to high

Table 12.2 *Venus International Reference Atmosphere (VIRA) below 100 km at latitude of 45*

| h (km) | T (K) | p (bar) | n (cm$^{-3}$) | h (km) | T (K) | p (bar) | n (cm$^{-3}$) |
|---|---|---|---|---|---|---|---|
| 0 | 735.3 | 92.10 | 8.98 + 20 | 41 | 408.1 | 3.127 | 5.58 + 19 |
| 1 | 727.7 | 86.45 | 8.53 + 20 | 42 | 401.6 | 2.793 | 5.06 + 19 |
| 2 | 720.2 | 81.09 | 8.10 + 20 | 43 | 395.0 | 2.491 | 4.59 + 19 |
| 3 | 712.4 | 76.01 | 7.69 + 20 | 44 | 388.3 | 2.216 | 4.15 + 19 |
| 4 | 704.6 | 71.20 | 7.29 + 20 | 45 | 381.8 | 1.969 | 3.75 + 19 |
| 5 | 696.8 | 66.65 | 6.91 + 20 | 46 | 376.1 | 1.745 | 3.37 + 19 |
| 6 | 688.8 | 62.35 | 6.55 + 20 | 47 | 370.2 | 1.544 | 3.03 + 19 |
| 7 | 681.1 | 58.28 | 6.20 + 20 | 48 | 364.6 | 1.364 | 2.72 + 19 |
| 8 | 673.6 | 54.44 | 5.86 + 20 | 49 | 357.5 | 1.202 | 2.44 + 19 |
| 9 | 665.8 | 50.81 | 5.54 + 20 | 50 | 349.7 | 1.057 | 2.20 + 19 |
| 10 | 658.2 | 47.39 | 5.23 + 20 | 51 | 341.1 | 0.9258 | 1.97 + 19 |
| 11 | 650.6 | 44.16 | 4.93 + 20 | 52 | 332.5 | 0.8087 | 1.77 + 19 |
| 12 | 643.2 | 41.12 | 4.65 + 20 | 53 | 321.9 | 0.7036 | 1.59 + 19 |
| 13 | 635.5 | 38.26 | 4.38 + 20 | 54 | 312.3 | 0.6095 | 1.42 + 19 |
| 14 | 628.1 | 35.57 | 4.12 + 20 | 55 | 301.4 | 0.5255 | 1.27 + 19 |
| 15 | 620.8 | 33.04 | 3.87 + 20 | 56 | 290.2 | 0.4505 | 1.13 + 19 |
| 16 | 613.3 | 30.66 | 3.64 + 20 | 57 | 278.3 | 0.3839 | 1.00 + 19 |
| 17 | 605.2 | 28.43 | 3.42 + 20 | 58 | 267.4 | 0.3249 | 8.82 + 18 |
| 18 | 597.1 | 26.33 | 3.21 + 20 | 59 | 258.7 | 0.2733 | 7.67 + 18 |
| 19 | 589.3 | 24.36 | 3.01 + 20 | 60 | 253.3 | 0.2289 | 6.56 + 18 |
| 20 | 580.7 | 22.52 | 2.83 + 20 | 62 | 246.2 | 0.1591 | 4.69 + 18 |
| 21 | 572.4 | 20.79 | 2.65 + 20 | 64 | 240.7 | 0.1096 | 3.30 + 18 |
| 22 | 564.3 | 19.17 | 2.48 + 20 | 66 | 235.8 | 7.49–2 | 2.30 + 18 |
| 23 | 556.0 | 17.66 | 2.32 + 20 | 68 | 231.9 | 5.08–2 | 1.59 + 18 |
| 24 | 547.5 | 16.25 | 2.17 + 20 | 70 | 228.2 | 3.43–2 | 1.09 + 18 |
| 25 | 539.2 | 14.93 | 2.02 + 20 | 72 | 224.6 | 2.30–2 | 7.41 + 17 |
| 26 | 530.7 | 13.70 | 1.88 + 20 | 74 | 221.0 | 1.53–2 | 5.02 + 17 |
| 27 | 522.3 | 12.56 | 1.75 + 20 | 76 | 216.2 | 1.01–2 | 3.39 + 17 |
| 28 | 513.8 | 11.49 | 1.63 + 20 | 78 | 210.4 | 6.63–3 | 2.28 + 17 |
| 29 | 505.6 | 10.50 | 1.52 + 20 | 80 | 202.5 | 4.28–3 | 1.53 + 17 |
| 30 | 496.9 | 9.581 | 1.41 + 20 | 82 | 195.5 | 2.72–3 | 1.00 + 17 |
| 31 | 488.3 | 8.729 | 1.30 + 20 | 84 | 188.6 | 1.70–3 | 6.53 + 16 |
| 32 | 479.9 | 7.940 | 1.21 + 20 | 86 | 181.8 | 1.04–3 | 4.16 + 16 |
| 33 | 471.7 | 7.211 | 1.11 + 20 | 88 | 176.6 | 6.32–4 | 2.59 + 16 |
| 34 | 463.4 | 6.537 | 1.03 + 20 | 90 | 172.3 | 3.77–4 | 1.58 + 16 |
| 35 | 455.0 | 5,916 | 9.48 + 19 | 92 | 170.4 | 2.23–4 | 9.48 + 15 |
| 36 | 446.8 | 5.345 | 8.72 + 19 | 94 | 169.5 | 1.32–4 | 5.62 + 15 |
| 37 | 438.6 | 4.820 | 8.01 + 19 | 96 | 170.0 | 7.77–5 | 3.30 + 15 |
| 38 | 430.8 | 4.338 | 7.34 + 19 | 98 | 170.9 | 4.60–5 | 1.94 + 15 |
| 39 | 423.3 | 3.897 | 6.70 + 19 | 100 | 172.2 | 2.73–5 | 1.14 + 15 |
| 40 | 415.5 | 3.495 | 6.12 + 19 | | | | |

*Note.* 8.98 + 20 = $8.98 \times 10^{20}$. From Seiff et al. (1985).

latitudes. Furthermore, their model was made at four segments of the Venus sphere that corresponds to four local time intervals. However, the model variations of pressure with local time are rather moderate, ~10% even at 100 km.

The Venus Express radio occultations, solar and stellar occultations using the UV channel of SPICAV, the SOIR occultations and the nightside VIRTIS spectra of the $CO_2$ band at 4.3 μm may contribute to the model for Venus' atmosphere as well. We will consider those observations below.

VIRA and other models of this type are empirical. One should bear in mind that a local change in temperature at some level affects density and pressure throughout the atmosphere above this level. Furthermore, according to the thermal balance equation (2.28) (Section 2.3.4), heat sources and sinks are proportional to the second derivative of temperature, and a small change in temperature could mean a dramatic variation in the local thermal balance.

### 12.2.3 Atmospheric Dynamics and Superrotation

Details of atmospheric dynamics are beyond our scope. The lower and upper atmospheres present two limiting cases for atmospheric dynamics. The time of cooling by convection and radiative transfer is longer in the lower atmosphere than the solar day on Venus (117 Earth days). Therefore a Hadley cell with poleward winds in the upper part of the cell and the pole-to-equator winds in the lowest scale height is formed.

The cooling time is shorter than the solar day in the upper atmosphere, and this stimulates the subsolar-to-antisolar circulation in the upper atmosphere with the opposite return flow near the mesopause. Winds in the upper part are much stronger than those in the return flow because of the smaller atmospheric density. The atmosphere between these extreme conditions undergoes superrotation, that is, retrograde rotation with a period of 4 days at the cloud tops, much shorter than the solid body rotation period of 243 days. We have not found a simple and clear explanation of this phenomenon in the literature, though the current general circulation models successfully reproduce it.

Vertical profiles of the zonal wind were measured from all descent probes since V8 using the Doppler shift of the radio signal. The observed wind speed (Figure 12.1) shows a gradual increase from zero near the surface to ~80 m s$^{-1}$ at 60 km. The data were presented for VIRA as mean, maximum, and minimum profiles (Figure 12.2). The profiles were extrapolated to 100 km using the existing data on the thermal structure of the atmosphere and approximations of the cyclostrophic balance

$$u^2 \tan\varphi = -g\left(\frac{dz}{d\varphi}\right)_p \tag{12.1}$$

and thermal wind

$$\frac{du}{dp} = \frac{R}{2u\mu\tan\varphi}\left(\frac{dT}{d\varphi}\right)_p. \tag{12.2}$$

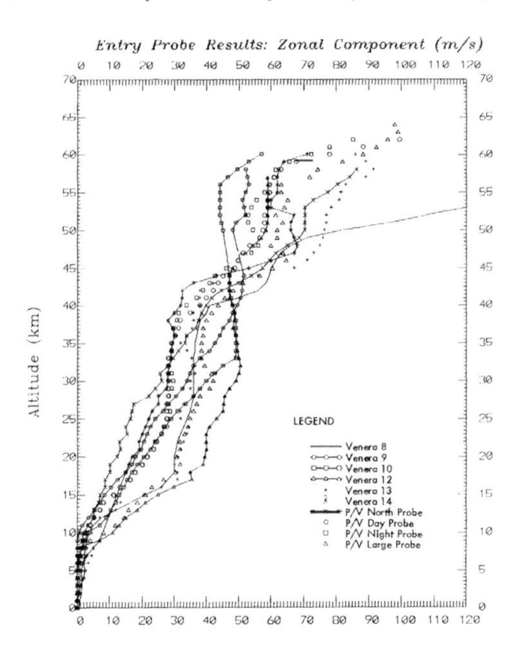


Figure 12.1 Zonal component of wind on Venus observed from the descent probes. From Kerzhanovich and Limaye (1985).

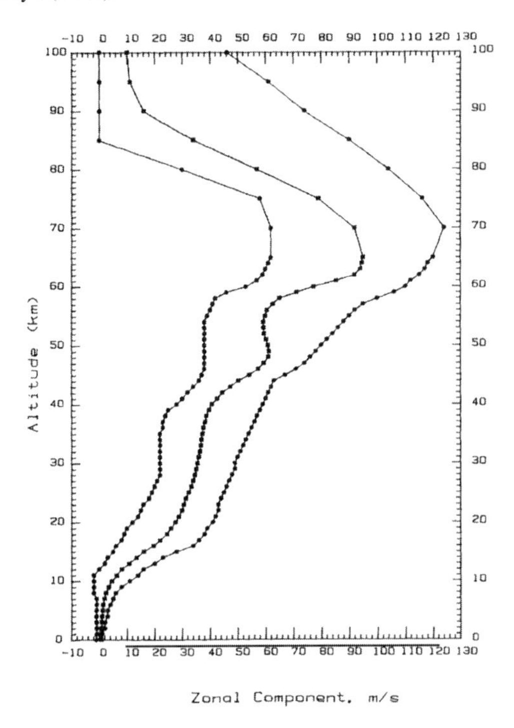


Figure 12.2 Mean, minimum, and maximum zonal component of wind on Venus recommended for VIRA. From Kerzhanovich and Limaye (1985).

Here $u$ is the zonal wind component, $\varphi$ is latitude, $g$ is gravity, $z$ is height of the pressure surface $p$, $R$ is the gas constant, and $\mu$ is the molecular mass. The cyclostrophic balance means that wind forms a centrifugal force that compensates the difference in pressure.

Wind speed at the cloud tops is traditionally measured by tracking of cloud features in images of Venus. These features are prominent in the NUV range at 320–400 nm. The most detailed and long-term study was made by Khatuntsev et al. (2013) and Patsaeva et al. (2015) using the data from the Venus Express monitoring camera. The results for 7 years obtained near 20°S are shown in Figure 12.3 and refer to the cloud tops near 67 km. While the mean values agree with the previous observations from Mariner 10, Pioneer Venus, and Galileo, a new and unexpected feature is the long-term increase from 85 m s$^{-1}$ in 2006 to 110 m s$^{-1}$ in 2012.

Latitudinal variations of both zonal and meridional components of the wind were measured as well (Figure 12.4). Despite the significant decrease in the zonal wind speed poleward of 50°, the rotation period does not remain constant and decreases from ~5 days near the equator to ~3 days near 75°S.

Images of Venus in the near-infrared filter at 965 nm were studied by Khatuntsev et al. (2017) as well. The contrasts in the NIR are much weaker than those in the NUV, and their analysis is more difficult. The data refer to 49–57 km, and the measured wind speeds are shown in Figure 12.4. Long-term variations of the wind at 49–57 km were studied at 20 ± 5°S. The retrograde zonal speed changed from 75 to 67 and then to 77 m s$^{-1}$ at orbit 220, 420, and 1300, respectively, while the meridional speed increased from 1.3 to 8.5 m s$^{-1}$ at orbits 220 and 1300, respectively.

The imaging mode of VIRTIS-M was used by Hueso et al. (2015) to study the wind field at 360–400, 570–680, and 900–955 nm. The results support the main conclusions by Khatuntsev et al. (2013). Generally, the lower cloud level near 50 km can be mapped on

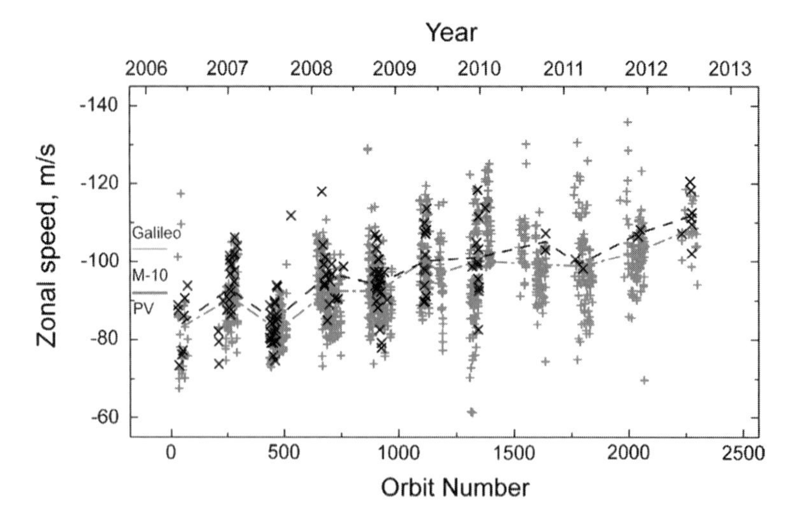


Figure 12.3 Long-term variations of the zonal wind speed near the cloud tops at 20° ± 2.5°S using the Venus Express monitoring camera. Two symbols and two mean curves refer to two methods of the data analysis. The results from Mariner 10 (92 m s$^{-1}$), Pioneer Venus (92 ± 3 m s$^{-1}$), and Galileo (103 m s$^{-1}$) are shown as well. From Khatuntsev et al. (2013).

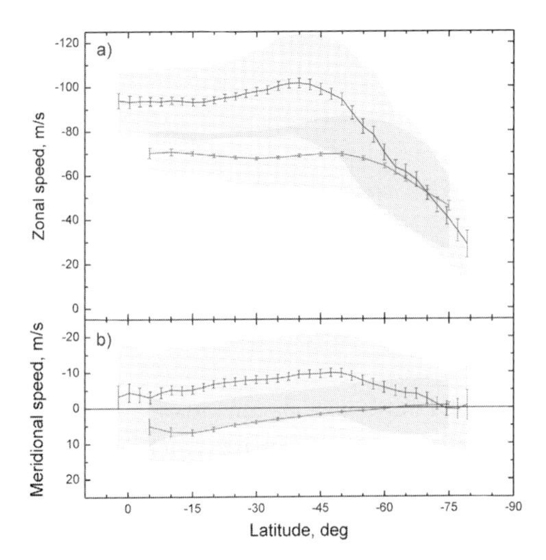


Figure 12.4 Latitudinal variations of the mean zonal and meridional wind speed at the cloud tops (67 km, 365 nm) and 49–57 km (965 nm) observed using the VEX monitoring camera for 10 Venusian years. From Khatuntsev et al. (2017).

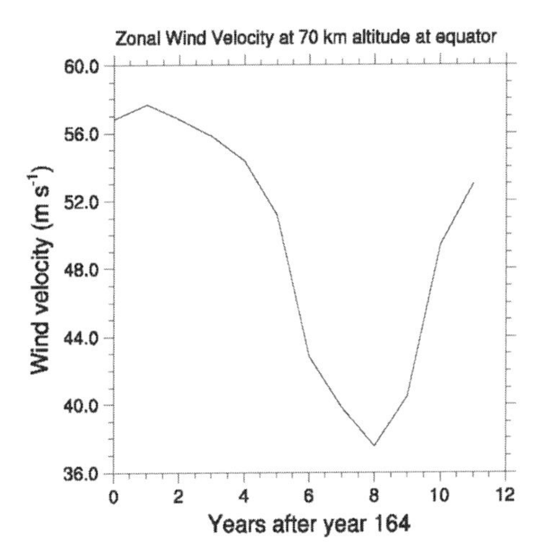


Figure 12.5 Long-term variations of the zonal wind speed in the general circulation model by Parish et al. (2011).

the nightside in the spectral windows at 1.7 and 2.3 μm. These maps require rather long exposures and have not been studied.

Dynamics of the Venus atmosphere was simulated using a GCM by Parish et al. (2011), who found variations of the zonal wind with a period of about 10 years (Figure 12.5). The computed peak-to-peak variations are a factor of 1.5 and cover the observed ratio of 1.3. However, the wind speed is smaller than the measured values by a factor of ~2 and the calculated vertical profile of the zonal wind does not match the spacecraft data in Figures 12.1 and 12.2.

There is some skepticism to both observed and modeled long-term variations of the wind speed. Bertaux et al. (2014) pointed out the effect of topography on the measured wind speed, and some instability might be suspected in the code of Parish et al. (2011).

## 12.3 Noble Gases and Isotopes

### *12.3.1 Helium*

Helium was measured on Venus by four teams using resonance scattering of its emission line at 584 Å and by two teams from in situ mass spectrometer data. The optical observations refer to a level where the mean slant optical depth for the $CO_2$ absorption at 584 Å is one, that is, the vertical optical depth for the sunlit hemisphere is 0.5. According to VIRA (Keating et al. 1985), this reference level is at 145 km on the dayside.

Helium on Venus was detected for the first time by the Mariner 10 UV spectrometer (Kumar and Broadfoot 1975). Their densities were retrieved using an exospheric temperature of 375 K, and a correction for the more realistic temperature of 275 K results in $[He]_{ref} = (1.8 \pm 0.9)\times10^6$ cm$^{-3}$. The V11–12 UV spectrometers were similar in design to that on Mariner 10, but had better rejection of stray light. The observed reference value is $[He]_{ref} = (2.6 \pm 1.2)\times10^6$ cm$^{-3}$ (Chassefiere et al. 1986). The EUVE observations of Venus (Figure 12.6) resulted in $[He]_{ref} = (4.3 \pm 1.9)\times10^6$ cm$^{-3}$ (Krasnopolsky and Gladstone 2005). The Cassini flyby of Venus gave $[He]_{ref} = 8\times10^6$ cm$^{-3}$ (Gerard et al. 2011). Interpretation of these spectroscopic data required model profiles of the major constituents ($CO_2$, $N_2$, O, CO, and He) in the upper atmosphere and radiative transfer calculation for the resonance line in the appropriate geometry of the observations.

Helium was also measured by mass spectrometers on board the PV bus (von Zahn et al. 1980) and orbiter (Niemann et al. 1980). The orbiter instrument revealed a helium bulge that peaks at 4:30 LT at a density exceeding the dayside value at 157 km by a factor of 14. VIRA for the neutral upper atmosphere (Keating et al. 1985) is based on the mass spectrometer data corrected for the atmospheric drag measurements. It gives $[He]_{ref} = 5.6\times10^6$ cm$^{-3}$.

The bus entry was near 8 LT beyond the bulge, and vertical profiles of $CO_2$, He, CO, and $N_2$ were measured down to 130 km. The measured data were fitted by a model with eddy diffusion $K = 1.4\times10^{13} n^{-1/2}$ cm$^2$ s$^{-1}$ ($n$ is the total number density in cm$^{-3}$) and helium mixing ratio of $\approx$12 ppm; $[He]_{ref} = 6.8\times10^6$ cm$^{-3}$.

The mean of the six measurements is $[He]_{ref} = (4.8 \pm 1.0)\times10^6$ cm$^{-3}$. Krasnopolsky and Gladstone (2005) argued in favor of some reduction of eddy diffusion to $K = 0.8\times10^{13}\, n^{-1/2}$ cm$^2$ s$^{-1}$ that results in the helium mixing ratio of $10 \pm 4$ ppm in the lower atmosphere. The given uncertainty reflects mostly the uncertainty of eddy diffusion. The retrieved mixing ratio agrees with that from the PV bus measurements.

Helium on Venus may be captured from the solar nebula during the planet formation, produced by the radioactive decay of uranium and thorium, and captured from the solar wind as alpha particles.

The PV Large Probe mass spectrometer showed a peak at mass 3, which, however, was mostly attributed to HD, and an upper limit for $^3He/^4He$ was established at $3\times10^{-4}$, much

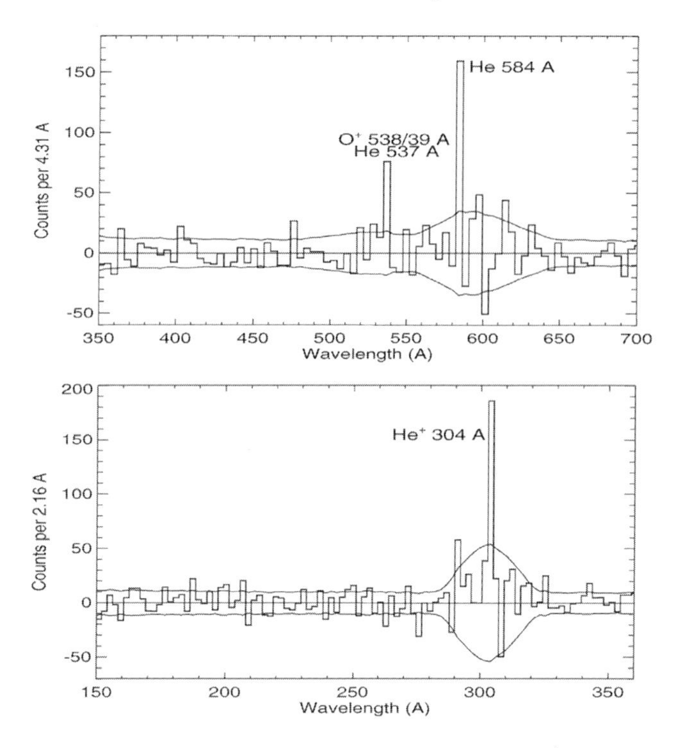


Figure 12.6 EUVE spectra of Venus. The detected emissions are He 584 Å (144 ± 32 R), He 537/$O^+$ 538/39 Å (34 ± 8 R), and $He^+$ 304 Å (69 ± 20 R). All brightnesses refer to the illuminated part of the disk which was 0.555 of the total disk area. The one-sigma envelops are shown as well. From Krasnopolsky and Gladstone (2005).

higher than the terrestrial value of $1.34\times10^{-6}$ and smaller than the solar wind ratio of $4.4\times10^{-4}$ (Bodmer et al. 1995).

### 12.3.2 Argon and Its Isotopes

Measurements of argon using the PV and V11–12 mass spectrometers and gas chromatographs were discussed by von Zahn et al. (1983). Here we can add improved data from the V13–14 mass spectrometers (Istomin et al. 1983) in the noble gas mode (Figure 12.7), when $CO_2$, $N_2$, and some chemically active species were removed by activated titanium. This resulted in an increase in the instrument sensitivity relative to V11–12 by a factor of 40 for the noble gases with the detection limit as low as ≈1 ppb.

The most reliable data on the argon mixing ratio were from the PV gas chromatograph (67 ± 2 ppm at 22 km; Oyama et al. 1980), the V12 gas chromatograph (40 ± 10 ppm below 42 km; Gelman et al. 1979), and the V13–14 mass spectrometers (100 ± 10 ppm below 25 km; Istomin et al. 1983). As for $N_2$, the differences exceed the claimed uncertainties. The mean value is 70 ± 20 ppm, similar to that recommended by von Zahn et al. (1983).

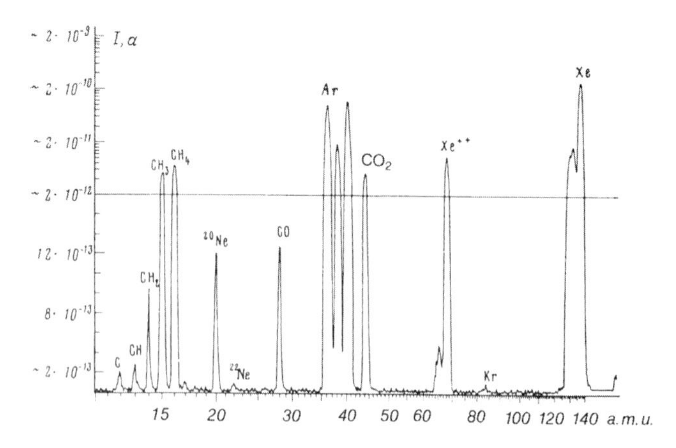


Figure 12.7 The mass spectrum measured by V13 at 25 km in the noble gas mode. Xenon is the calibration gas, and methane and its radicals are contaminations. From Istomin et al. (1983).

Argon has three stable isotopes with masses 36, 38, and 40. The light isotopes were captured during the planet formation, while $^{40}Ar$ is a product of the radioactive decay of $^{40}K$. The measured $^{40}Ar/^{36}Ar$ ratio is $1.03 \pm 0.04$ (Hoffman et al. 1980) and $1.11 \pm 0.02$ (Istomin et al. 1983). It is much smaller than the ratio of 300 in the terrestrial atmosphere. The $^{36}Ar/^{38}Ar$ ratio is 5.35, similar to that on the Earth.

Using the argon abundances and its isotope ratios, the $^{36}Ar$ and $^{38}Ar$ abundances scaled to the masses of the planets are greater on Venus than on Earth by a factor of 120, while radiogenic $^{40}Ar$ is depleted on Venus by a factor of 2.2. These differences reflect decreasing abundances of the noble gases with heliocentric distance in the solar nebula for $^{36}Ar$ and $^{38}Ar$ and more effective outgassing from the interior on the Earth for $^{40}Ar$.

### *12.3.3 Neon and Its Isotopes*

To evaluate the neon abundance, von Zahn et al. (1983) scaled the published values from the PV and V11–12 mass spectrometers and gas chromatographs to the recommended argon abundance of 70 ppm. Then the neon abundances were $10 \pm 7$ ppm (Hoffman et al. 1980), $8.6 \pm 3$ ppm (Istomin et al. 1983), and $4.3 \pm 4$ ppm (Oyama et al. 1980), and a value recommended by von Zahn et al. (1983) was $7 \pm 3$ ppm. The most accurate value of 8 ppm was from V13–14 (Istomin et al. 1983). Taking into account the scaling consideration, we support the previous recommendation with the reduced uncertainty: $7 \pm 1.5$ ppm.

The accurate V13–14 measurements of the neon isotopes (Figure 12.7) corrected for double ionized $CO_2$ and $^{40}Ar$ resulted in $^{20}Ne/^{22}Ne = 12.15 \pm 0.1$ (Istomin et al. 1983). Their measurements in the terrestrial air gave $10.07 \pm 0.35$, while the data in the literature were between 9.24 and 10.3 with the standard value of 9.78 (Table 10.2). The solar wind value is $13.7 \pm 0.3$, so that the neon isotope ratio on Venus is between those on the Earth and the solar wind. An upper limit of $^{22}Ne/^{21}Ne > 15$ gives $^{20}Ne/^{21}Ne > 180$; the terrestrial value is 335.

Table 12.3 *Isotope ratios in the Venus and Earth atmospheres and in the solar wind*

| Ratio | V11–14 | PV | Earth | SW[a] |
|---|---|---|---|---|
| $^{4}He/^{3}He$ | – | >3000 | $7.46\times10^{5}$ | 2500–3000 |
| $^{12}C/^{13}C$ | $91 \pm 2$ | $84 \pm 5$ | 89 | $84 \pm 5$ |
| $^{14}N/^{15}N$ | – | $274 \pm 70$ | 274 | $442 \pm 130$[b] |
| $^{16}O/^{18}O$ | $\approx500$ | $\approx500$ | 499 | $450 \pm 100$ |
| $^{20}Ne/^{22}Ne$ | $12.15 \pm 0.1$ | $14 \pm 5$ | 9.78 | $13.7 \pm 0.3$ |
| $^{22}Ne/^{21}Ne$ | >15 | – | 34.3 | $30 \pm 4$ |
| $^{35}Cl/^{37}Cl$ | $2.9 \pm 0.3$[c] | – | 3.13 | – |
| $^{40}Ar/^{36}Ar$ | $1.11 \pm 0.02$ | $1.03 \pm 0.04$ | 299 | – |
| $^{36}Ar/^{38}Ar$ | $\approx5.35$ | $\approx5.35$ | 5.35 | $5.5 \pm 0.3$ |

[a]Data from Wiens et al. (2004).
[b]From Marty et al. (2010).
[c]Deduced by Young (1972) from spectra of HCl.

Comparing the abundances of neon per the masses of Venus and Earth (18.2 ppm of Ne in the air), Venus is enriched in neon relative to the Earth by a factor of 45.

### *12.3.4 Krypton and Xenon*

A refined analysis of the PV mass spectrometer data resulted in an $^{84}$Kr mixing ratio of $25 \pm 3$ ppb (see Donahue and Pollack 1983). Using the terrestrial isotope ratios of krypton, its total abundance is $\approx$50 ppb. The V13–14 mass spectrometers showed 35 ppb for krypton (Figure 12.7) and an upper limit of 20 ppb for xenon. Using the V13–14 data and 1.14 ppm of Kr and 87 ppb of Xe in the Earth's atmosphere, the enrichment of krypton on Venus is a factor of 3.6, and that of xenon is smaller than a factor of 27. The significant decrease in the noble gas abundances from Venus to Earth and then Mars puts important constraints on processes of the planet formation (Donahue and Pollack 1983).

### *12.3.5 Isotopic Ratios of C, N, and O*

Isotopic ratios of C, N, and O were measured by the V11–14 and PV mass spectrometers being similar to the Earth's values. The observed isotope ratios are compared with those on the Earth and in the solar wind in Table 12.3.

## 12.4 Carbon Monoxide, Oxygen, and Ozone

### *12.4.1 Carbon Monoxide*

The first correct measurement of CO on Venus was made by Moroz (1964) using the CO (2–0) band at 2.35 μm; the derived CO abundance was 5 cm atm. Compared to the $CO_2$ abundance of 1 km atm above the clouds, this results in a CO mixing ratio of 50 ppm. Lines of the CO (2–0) band are clearly identified in the high-resolution spectra by Connes et al. (1968), who also retrieved the CO mixing ratio of 50 ppm above the cloud tops.

Observations of thermal emissions of the CO (1–0) band at 4.7 μm and the $CO_2$ band at 4.3 μm (Irwin et al. 2008) were conducted on the Venus nightside using the Venus Express VIRTIS-M spectrograph (see Section 12.1). The $CO_2$ band was used to retrieve temperature profiles in the atmosphere, while the CO band resulted in abundances of CO at 65–70 km that varied within 40 ± 10 ppm.

Ground-based spatially resolved high-resolution spectroscopy using IRTF/CSHELL (Figure 12.8) showed 52 ± 4 ppm in the morning and 40 ± 4 ppm in the afternoon at 68 km in the latitude range of ±60° (Krasnopolsky 2010c). The difference was attributed to some extent of the morningside bulge (see below). Later IRTF/CSHELL was applied by Marcq et al. (2015) to study the global distribution of CO near 70 km using the CO (1–0) R16 and R17 lines at 4.53 μm. The retrieved CO varied mostly in the range of 25–40 ppm on both dayside and nightside.

The SOIR spectrograph onboard Venus Express had resolving power of $2\times10^4$, and solar occultations of the lines of the CO (2–0) band revealed vertical profiles of the CO mixing ratio up to 120 km (Figure 12.9; Vandaele et al. 2015). The profiles show an order-of-magnitude increase from 84 km to ≈2000 ppm at 120 km and significant short-term

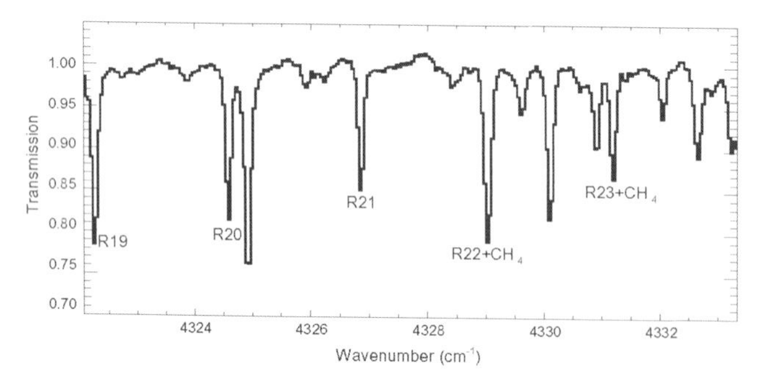


Figure 12.8 Spectrum of the CO (2–0) band near the R19–23 lines at the Venus equator. The CO and solar lines are Doppler-shifted by 0.183 $cm^{-1}$. Other features are telluric $H_2O$ and $CH_4$ and solar lines. From Krasnopolsky (2010c).

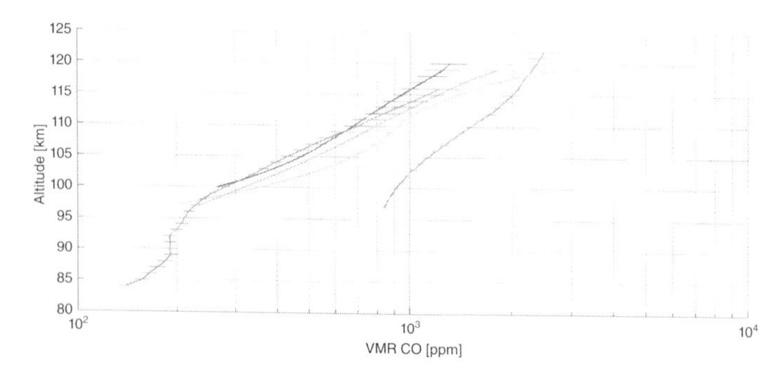


Figure 12.9 Vertical profiles of the CO mixing ratio observed by SOIR in July 2008 at 77°–87°N and LT = 06:00. From Vandaele et al. (2015).

variability. Most of the data refer to the polar and subpolar latitudes because of the orbit constraints and, evidently, to the morning or evening terminator.

It was discussed in Section 7.4 that simultaneous high-resolution observations of line shapes of the strong $^{12}CO$ ($J$ = 2→3) line at 346 GHz and the similar weak $^{13}CO$ line at 330 GHz make its possible to retrieve vertical profiles of temperature and CO mixing ratio. This technique was at first applied by Wilson et al. (1981) to the observations of the CO ($J$ = 0→1) lines at 2.6 mm and revealed a nighttime bulge of CO near 100 km.

The most detailed observations were conducted by Clancy et al. (2012a) using the James Clerk Maxwell Telescope (JCMT) on Mauna Kea, Hawaii. The telescope with diameter of 15 m was designed for the submillimeter spectral range. The observations covered a long period from 2000 to 2009 and were made at the inferior and superior conjunctions of Venus, when its nightside and dayside are seen with angular diameters of 58 and 10.5 arcs, respectively. The telescope beam size is 14 arcs, and ≈15 different positions were observed in each of five nightside sessions, while the beam covered the full disk in three dayside sessions. The retrieved vertical profiles of temperature and CO mixing ratio covered a pressure range from 10 to 0.01 mbar that corresponds to the altitudes from 76 to 103 km in VIRA. Disk averaged data for $T$ and CO, day and night, solar maximum (2000–2002) and minimum (2007–2009) are presented in Figure 12.10. Overall, the CO mixing ratio increases from ≈40 ppm at 76 km to ≈300 and ≈1200 ppm at 103 km on the dayside and nightside, respectively.

Using the interferometric technique, it is possible to improve spatial resolution in the microwave range. Gurwell et al. (1995) mapped CO at a few altitudes between 70 and 100 km using the CO line at 2.6 mm and the OVRO interferometer with six 10.4 m dishes and a beam size of 2.8 arcs. The maps show a morningside bulge of CO above 90 km (Figure 12.11).

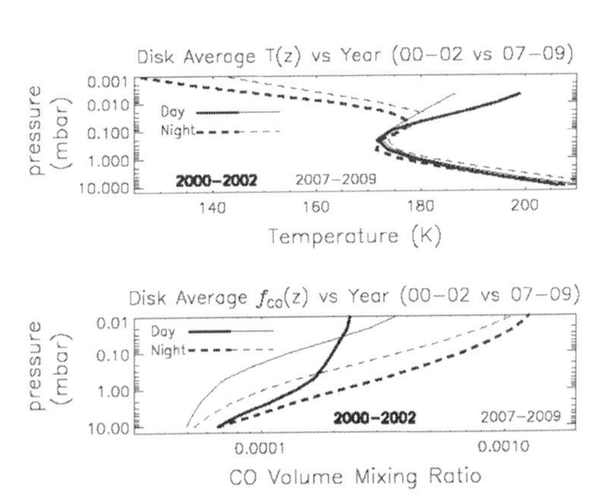


Figure 12.10 Disk average profiles of temperature and CO mixing ratio retrieved from the JCMT observations of the $^{12}CO$ and $^{13}CO$ line shapes at 346 and 330 GHz, respectively. The profiles reflect the day-night and solar minimum-maximum (2007–2009 and 2000–2002) variations. According to VIRA, presssures of 10, 1, 0.1, and 0.01 mbar correspond to 76, 86, 95, and 103 km, respectively. From Clancy et al. (2012a).

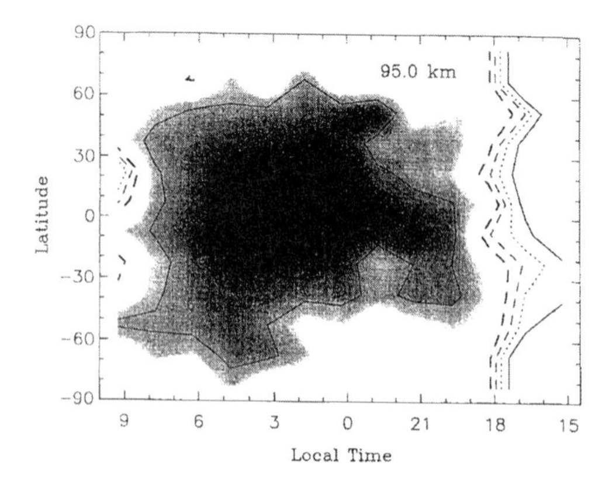


Figure 12.11 The CO bulge at 95 km observed with the OVRO millimeter interferometer (Gurwell et al. 1995).

Much better spatial resolution could be achieved with the Very Large Array (VLA) with 27 telescopes of 25 m each and resolution of 0.04 arcs, or the Atacama Large Millimeter/submillimeter Array (ALMA) with 66 telescopes of 12 m and 7 m each and expected resolution of 0.01 arcs. Recently Venus was observed (Encrenaz et al. 2015b) using ALMA in the compact configuration with resolution of 2 arcs; the Venus angular diameter was 10–12 arcs. The observations covered a range of 335–347 GHz with lines of CO, SO, $SO_2$, and HDO. The CO line was applied to retrieve temperature profiles with no new data on the CO distribution.

CO abundances and temperatures at 100–150 km were also retrieved from observations of the CO dayglow at 4.7 μm. CO in the upper atmosphere was observed by the PV mass spectrometers and a few UV spectrometers. Those data will be considered below.

The first measurements of CO in the lower atmosphere were made by the PV and V12 gas chromatographs (Oyama et al. 1980; Gelman et al. 1979). The measured mixing ratios were 20 ppm at 12 km and 22 km and 30 ppm at 40 km. Currently CO in the Venus lower atmosphere is measured using the spectral window at 2.3 μm (Figure 12.12). Spectra of the windows at 2.3 and 1.7 μm were measured with resolving power of $2\times10^4$ by Bezard et al. (1990). The first detailed analysis of the spectral windows was made by Pollack et al. (1993), who retrieved a CO abundance of $23 \pm 5$ ppm with a gradient of $1.20 \pm 0.45$ ppm km$^{-1}$ in the altitude range centered at 36 km. Later the observations were repeated by a few teams using spatially resolved ground-based spectroscopy and the Venus Express VIRTIS-M and VIRTIS-H spectrographs. Here we consider the latest results.

VIRTIS-H had resolving power of 1500 and did not cover the spectral windows at 1–1.3 μm and 1.7 μm. The observations in the window at 2.3 μm were analyzed by Marcq et al. (2008) and are shown in Figure 12.13a. VIRTIS-M with resolving power of $\approx$200 covered all spectral windows; the observed latitudinal variations of CO (Barstow et al. 2012) are presented in Figure 12.13b. CO retrieved from the ground-based observations by Arney et al. (2014) using the TripleSpec spectrograph with resolving power of 3500 at the Apache Point Telescope with diameter of 3.5 m are shown in Figure 12.13c. The similar

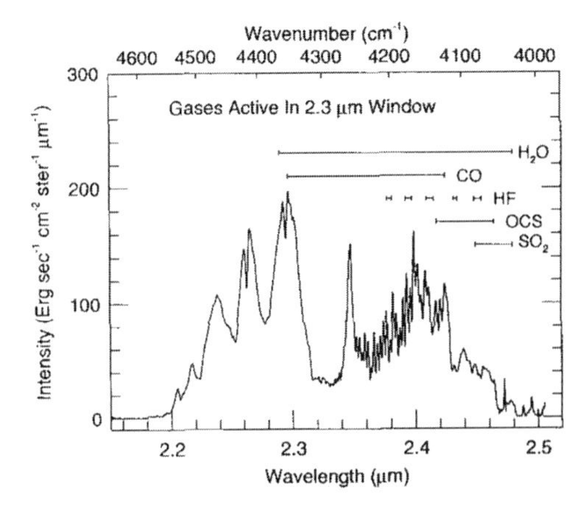


Figure 12.12 The spectral window at 2.3 μm observed at night with resolving power of 1800. The absorbing gases are identified. From Pollack et al. (1993).

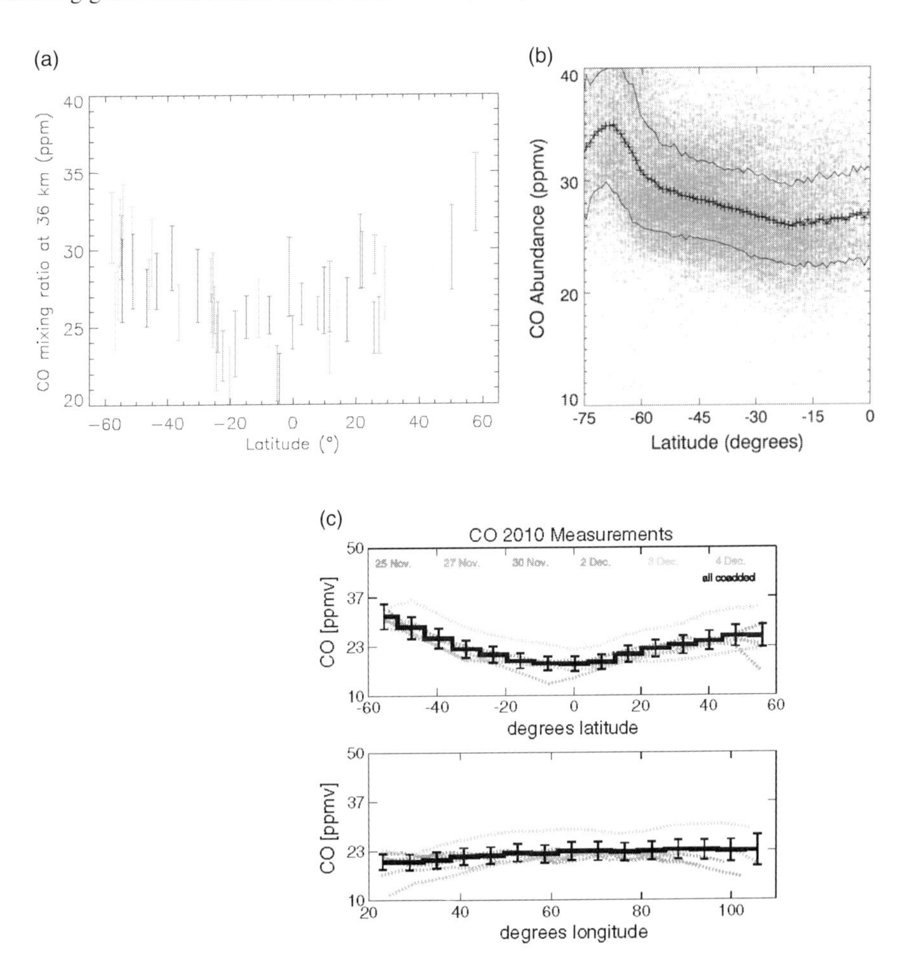


Figure 12.13 Variations of CO near 36 km retrieved from the VIRTIS-H (a; Marcq et al. 2008), VIRTIS-M observations onboard Venus Express (b; Barstow et al. 2012), and ground-based observations (c; Arney et al. 2014).

results were obtained by Cotton et al. (2012). Overall, the observed CO abundances are $\approx$25 ppm near 36 km at low latitudes with some increase to the higher latitudes.

### *12.4.2 Molecular Oxygen*

Each solar photon at $\lambda < 200$ nm and 200–218 nm impacting Venus is absorbed by $CO_2$ and $SO_2$, respectively. The solar photon fluxes are $2\times10^{13}$ and $8.5\times10^{13}$ ph cm$^{-2}$ s$^{-1}$ at the Venus orbit, respectively, and the global-mean column photolysis rates of $CO_2$ and $SO_2$ are quarters of these values. Almost all oxygen atoms formed are initially converted into $O_2$ by the three-body association and some catalytic reactions. The total column rate of the $O_2$ production is ~$10^{13}$ cm$^{-2}$ s$^{-1}$, and one might expect significant quantities of $O_2$ in the mesosphere of Venus.

However, $O_2$ has not been found on Venus, and the most restrictive observations were made by Trauger and Lunine (1983). ($O_2$ cannot exist in the hot reducing lower atmosphere of Venus, and some data on the presence of $O_2$ there are erroneous.) They used a triple pressure-scanned Fabry–Perot interferometer to search for a Doppler-shifted line of the $O_2$ band at 762 nm. The instrument had the resolving power $\lambda/\delta\lambda = 2.4\times10^5$. No absorption was found, and a 2-sigma upper limit to the equivalent width of the P9 line was $2\times10^{-6}$ nm, that is, $3.5\times10^{-5}$ cm$^{-1}$. Then Trauger and Lunine (1983) applied a radiative transfer code to this line assuming the uniformly mixed $O_2$ and derived an upper limit of 0.3 ppm to the $O_2$ mixing ratio.

This upper limit is actually weighted to a so-called level of line formation which is in the clouds at $2\tau\,(1-g) = 1$. Here $\tau$ is the cloud optical depth and $g \approx 0.75$ is the scattering asymmetry factor. This level is at 65 km in the atmosphere. All photochemical models predict variable vertical distribution of $O_2$ with the bulk oxygen above 65 km.

Therefore a simple reflection model is more adequate for the oxygen observation than the mixed oxygen. The solar light undergoes a two-way absorption in the atmosphere above a reflecting surface that is deeper than the level of line formation by a factor of 2, that is, at 62 km. The phase dependence of Venus near 762 nm (Irvine 1968) shows that the Lambert reflection is a reasonable approximation for Venus at this wavelength. (Brightness is proportional to cosine of solar zenith angle and does not depend on viewing angle for the Lambert reflection.) The observations of Trauger and Lunine (1983) were made for the light from the whole disk at phase (Sun–planet–Earth) angle of 90°. Then integration over the sunlit hemisphere of Venus results in a total airmass factor of $3\pi/2 \approx 4.7$. According to the spectroscopic database HITRAN 2012, the strength of the P9 line is $9\times10^{-24}$ cm at 250 K, and the 2-sigma upper limit of $3.5\times10^{-5}$ cm$^{-1}$ to the line equivalent width results in an upper limit to the $O_2$ column abundance of ~$10^{18}$ cm$^{-2}$. The current photochemical models cannot fit this limit.

### *12.4.3 Nighttime and Daytime Ozone*

Ozone $O_3$ was detected on Venus using stellar occultations by the SPICAV spectrograph (Montmessin et al. 2011). All detected limb abundances of the nighttime ozone are shown

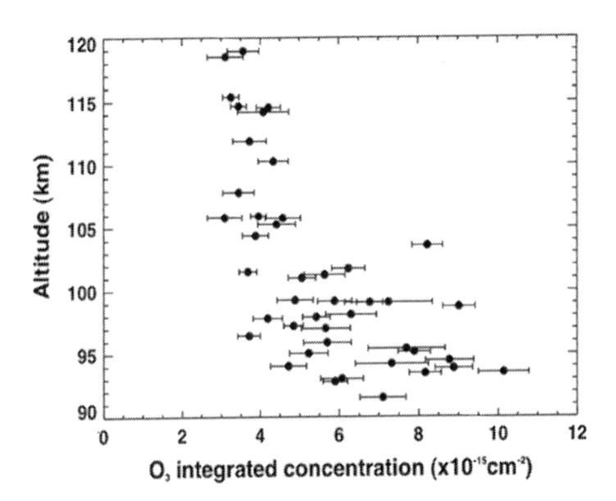


Figure 12.14 All positive detections of the nighttime ozone given as limb column abundances (Montmessin et al. 2011).

in Figure 12.14. Inversion to number densities for two observations resulted in peak number densities of $6\times10^7$ cm$^{-3}$ at 93 and 100 km.

The observed ozone peak densities are similar to those predicted by the global-mean photochemical models. However, the ozone photolysis is switched off at night, and the expected nighttime ozone densities exceeded very much the observed values. A significant flux of atomic chlorine from the dayside is required to fit the observed ozone densities (Krasnopolsky 2013a, 2019b).

Spectra of Venus in the middle UV range depend on scattering and absorption by the haze, $CO_2$, and $SO_2$. A thorough analysis of the SPICAV/VEX spectra by Marcq et al. (in preparation) revealed traces of ozone of 1–10 ppb near the cloud tops at high latitudes (>50°). Ozone at the middle and low latitudes (<50°) is below a detection limit of 1 ppb at the cloud tops.

## 12.5 Sulfur Species

Sulfur chemistry dominates in the lower atmosphere. Formation of the sulfuric acid clouds is the main feature of chemistry in the middle atmosphere as well.

### *12.5.1 Sulfur Dioxide and Monoxide*

Absorption spectrum of sulfur dioxide $SO_2$ is shown in Figure 12.15. $SO_2$ was discovered on Venus by Barker (1979) using a ground-based NUV spectrum that included the $SO_2$ bands at 305, 313, and 321 nm. Just after that $SO_2$ was detected by the PV UV spectrometer (Figure 12.16; Esposito et al. 1979). The observed spectra required a small scale height of $SO_2$ that was typically in the range of 1 to 4 km. Both $SO_2$ abundance and its scale height were extracted from the PV spectra.

Long-term PV observations (Esposito et al. 1988) were complemented by the V15 IR spectroscopy (Figure 7.2; Zasova et al. 1993) and observations from the International Ultraviolet Explorer, HST, and sounding rockets (Na et al. 1994). The gradual decrease in

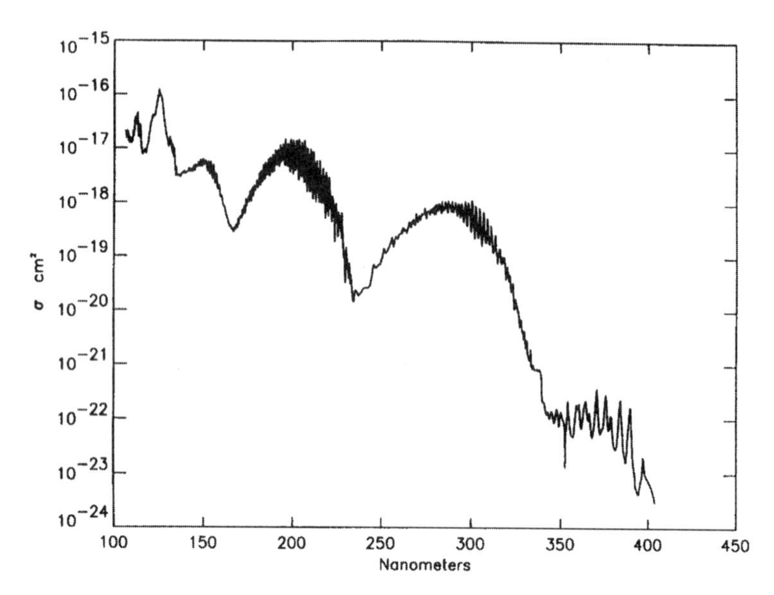


Figure 12.15 Absorption cross sections of $SO_2$ in the UV range. From Manat and Lane (1993).

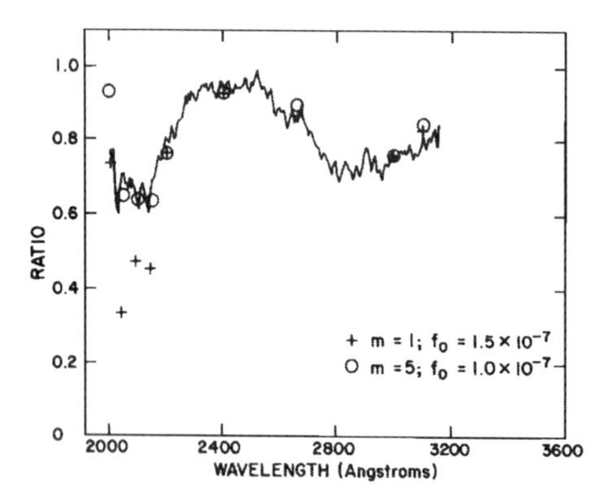


Figure 12.16 Ratio of the PV spectra of Venus observed at the dark and bright regions is compared with two models with different $SO_2$ abundance $f_0$ at 70 km and $m = H_{CO2}/H_{SO2}$ (Esposito et al. 1979).

the $SO_2$ abundances (Figure 12.17) stimulated a hypothesis of a huge volcanic release of $SO_2$ in 1978 or slightly earlier (Esposito et al. 1988). The situation was repeated in the SPICAV/VEX nadir observations of $SO_2$ (Marcq et al. 2013). However, according to the photochemical models (Krasnopolsky and Parshev 1981; Krasnopolsky 2012b, 2018a), very small variations in atmospheric dynamics (eddy diffusion) near 60 km induce strong variations of the $SO_2$ abundances near and above 70 km.

Vertical profiles of $SO_2$ were extracted from the SOIR/VEX solar occultations at the $SO_2$ (101–000) band near 4.05 μm (Belyaev et al. 2012, 2017; Mahieux et al. 2015a). The data (Figures 12.18, 12.19, and 12.20) extend from 66 to 82 km, and the mean values are ≈100 ppb at 70 km decreasing to ≈40 ppb at 80 km in Figure 12.18 being almost constant at ≈70 ppb in

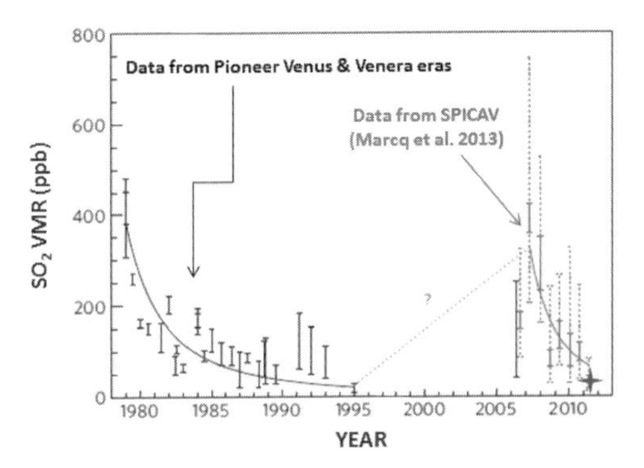


Figure 12.17 Long-term variations of the $SO_2$ abundance at 70 km. The data in 1979–1995 are from the PV, IUE, V15, and rocket observations, while those in 2006–2012 are the SPICAV/VEX nadir observations. Cross is the HST data by Jessup et al. (2015). From Vandaele et al. (2017).

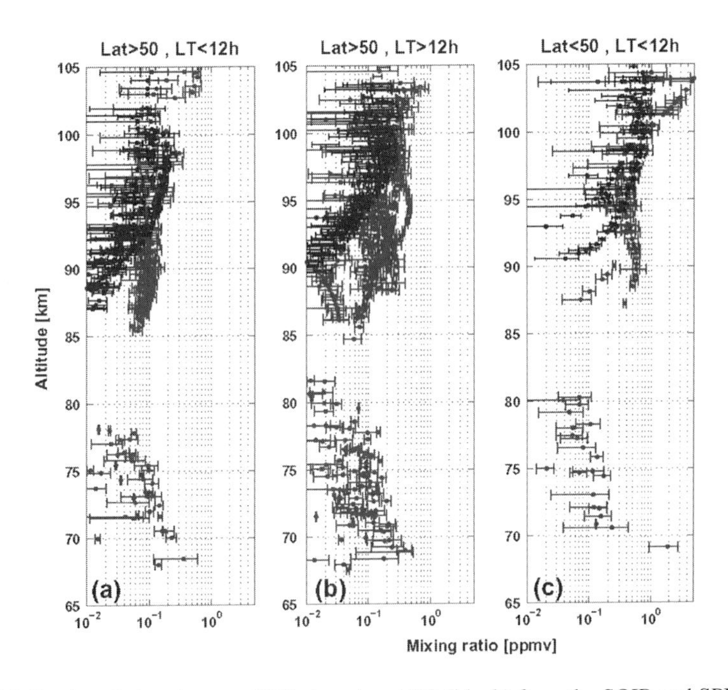


Figure 12.18 Retrieved abundances of $SO_2$ (gray) and SO (black) from the SOIR and SPICAV solar occultations: north polar latitudes, sunrise (a); south polar latitudes, sunset (b); middle latitudes, sunrise (c). From Belyaev et al. (2012).

Figure 12.19. The $SO_2$ abundance at 74 km (Figure 12.20) is variable but does not demonstrate long-term trends shown in Figure 12.17. The $SO_2$ abundances above 85 km were measured using the SPICAV solar occultations. The observations became impossible below 85 km because of the significant limb optical depth of the haze. The observed profiles end at 105 km; while the retrieved $SO_2$ mixing ratios are comparable at 85 and 105 km in Figure 12.18, the $SO_2$ number densities are smaller at 105 km than those at 85 km by two orders of magnitude.

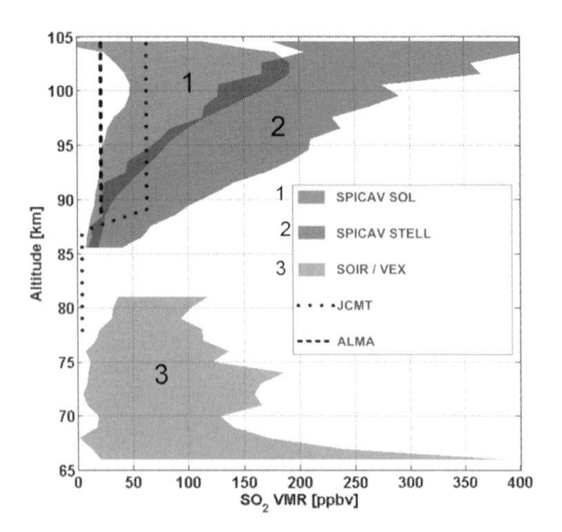


Figure 12.19 Retrieved abundances of $SO_2$ from SPICAV solar (1, sunset and sunrise, mostly high latitudes), stellar (2, night, low and middle latitudes), and SOIR solar occultations (3). The maximum $SO_2$ abundance observed at night using JCMT (Sandor et al. 2010; their mean abundance was 23 ppb) and the ALMA value (Encrenaz et al. 2015b) are shown for comparison. From Belyaev et al. (2017).

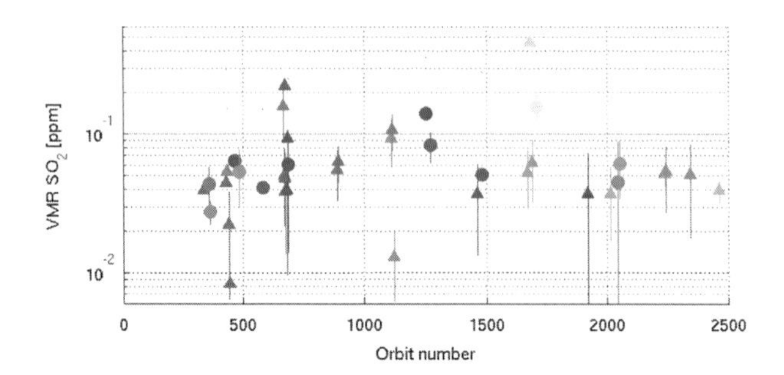


Figure 12.20 Long-term variations of $SO_2$ at a level of 16 mbar (~74 km). The orbit period is 24 hours, and the observations cover 2007–2014. From Mahieux et al. (2015a).

Analysis of a spectrum at 93 km is shown in Figure 12.21. There are four absorbers in the spectra (Figure 12.21a): $CO_2$, $SO_2$, SO, and haze. There is some similarity between the SO and $SO_2$ absorptions; however, the difference is significant at 201–212 nm and 219–226 nm. The haze is approximated by $\tau_\lambda = \tau_0\ (\lambda_0/\lambda)^\alpha$ with two fitting parameters $\tau_0$ and $\alpha$. Although the improvement from the inclusion of SO looks minor, it changes significantly $\chi^2$ and therefore is statistically plausible.

The retrieved vertical profiles of $SO_2$ and SO are shown in Figure 12.18. The data indicate a partial conversion of $SO_2$ into SO, and a sum of $SO_2$ and SO is greater at 105 km than, say, at 80 km. If this is not a local dynamic perturbation but a general feature, it requires a high-altitude source of $SO_X$. This source is highly questionable, and attempts to involve photolysis of the $H_2SO_4$ vapor or oxidation of sulfur aerosol suffer some drawbacks that will be discussed later.

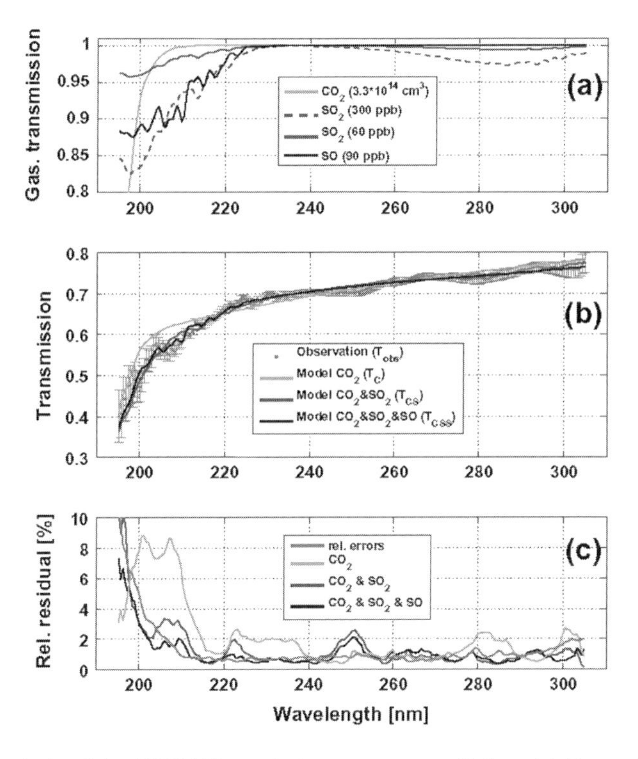


Figure 12.21 Analysis of the SPICAV occultation spectrum at 93 km. (a) Transmissions of three basic absorbers: $CO_2$, $SO_2$ (for two abundances), and SO. (b) Fitting of the observed spectrum by three models. (c) Residuals for three models; $\chi^2 = 3.5$ and 1.7 without and with SO, respectively. From Belyaev et al. (2012). (A black and white version of this figure will appear in some formats. For the color version, please refer to the plate section.)

Recent analysis of the SPICAV stellar occultations and some update of the SPICAV and SOIR solar occultations (Figure 12.19; Belyaev et al. 2017) favor even greater increase in $SO_2$ from 85 to 105 km.

Significant differences in measurements of the same object by different methods are not exceptional in physics. The useful part of the SPICAV occultation spectra is actually restricted to a narrow interval of 195–225 nm (Figure 12.21b). Krasnopolsky (2011c) pointed out that extinction by the haze particles with $r = 0.25$–$0.3$ μm varies considerably in this interval, and corrections for these variations can significantly affect the retrieved $SO_2$ and SO abundances. Furthermore, the $SO_2$ abundances retrieved from the SPICAV occultations disagree with submillimeter observations of $SO_2$ by an order of magnitude (see below). Sometimes data with too low and immeasurable values are removed from averaging, and this results in an overestimation of the mean values as well.

According to the submillimeter observations of the $SO_2$ and SO lines at 346.65 and 346.53 GHz, respectively (Sandor et al. 2010), their mixing ratios at 84–100 km vary within an order of magnitude with mean values of 23 and 7.5 ppb, respectively. However, the data require a deep decrease in $SO_2$ below 84 km and down to the cloud tops that is not supported by all other $SO_2$ observations. Variations of $SO_2$ by a few times within a few days present another problem in the submillimeter observations. The instrument observed the full disk of Venus or its large parts, and it is difficult to explain these variations.

Submillimeter observations using the largest interferometer ALMA resulted in disk-averaged mixing ratios of 12 ppb for $SO_2$ and 8 ppb for SO (Encrenaz et al. 2015b); those are constant above 88 km with steep cutoffs at this altitude, in accord with the observations by Sandor et al. (2010).

The V12 and PV gas chromatographs measured the $SO_2$ abundances of 130 $\pm$ 35 ppm (Gelman et al. 1979) and 180 $\pm$ 43 ppm (Oyama et al. 1980). Both values refer to the altitude of $\approx$20 km.

Observations in the filter of 320–390 nm at the V14 descent probe (Ekonomov et al. 1983b) revealed a strong absorption above 58 km with a minimum at 57 km and a gradual increase to the lower cloud boundary at 48 km. This increase may be caused by the $SO_2$ absorption with the effective mean cross section of $\approx 10^{-22}$ cm$^2$ (Figure 12.15). Then the $SO_2$ mixing ratio is 50, 20, and 10 ppm at 50, 54, and 57 km, respectively (Krasnopolsky 1986, p. 152).

The Vega 1 and 2 descent probes included UV spectrometers for a range of 220–400 nm that were designed for active spectroscopy with a xenon bulb as a light source and an absorption cell filled by the nearby air (Bertaux et al. 1996). Despite the excellent idea and reasonable quality of the data, the gas composition is poorly identified in the observed spectra. The results refer mostly to $SO_2$ and are rather uncertain: 150, 125, 38, and 25 ppm at 52, 42, 22, and 12 km, respectively.

The $SO_2$ bands contribute to the absorption in the nighttime spectral window at 2.3 μm (Figure 12.12); the derived $SO_2$ abundances refer to 35–45 km and are equal to 130 $\pm$ 40 ppm (Bezard et al. 1993), 130 $\pm$ 50 ppm (VIRTIS-H data; Marcq et al. 2008), 140 $\pm$ 37 and 126 $\pm$ 32 ppm in the observations in 2009 and 2010, respectively (Arney et al. 2014).

### *12.5.2 Carbonyl Sulfide OCS*

The first reliable detection of OCS in the lower atmosphere was made by Bezard et al. (1990) using nightside high-resolution spectroscopy at 2.3 μm. OCS abundances were also measured from the ground by Pollack et al. (1993), Marcq et al. (2006), Arney et al. (2014) and from Venus Express by Marcq et al. (2008) using medium resolution observations of the nightside emission feature at 2.3 μm (Figure 12.12). Retrievals from the ground-based observations by Pollack et al. (1993) and Marcq et al. (2006) adopted an OCS fraction decreasing linearly in the log scale near 33 km, and both the OCS fraction and its vertical gradient at 33 km were derived. However, actually vertical gradients of ln $f_{OCS}$ were obtained and may be applied to extend OCS from 30 to 36 km using a relationship

$$\frac{d \ln f_{OCS}}{dz} = \frac{1}{f_{OCS}} \frac{d f_{OCS}}{dz}.$$

Signal-to-noise ratios were insufficient in the VIRTIS-H/VEX observations of OCS to retrieve both the abundance and its gradient, and the measured mean abundance was 3.5 ppm at 33 km (Marcq et al. 2008). Summary of the OCS observations at 30–36 km is given in

Table 12.4 *OCS observations in the lower atmosphere of Venus*

| Paper | $d \ln f / dz$ | $f$(30 km) | $f$(33 km) | $f$(36 km) |
|---|---|---|---|---|
| Pollack et al. (1993) | −0.36±0.11 | 13±6 | **4.4±1.0** | 1.5±0.7 |
| Taylor et al. (1997) | −0.44±0.07 | **14±6** | 3.5± 1.0 | 0.9±0.4 |
| Marcq et al. (2006) | −0.47±0.13 | **12±7** | 2.6±1.0 | **0.55±0.15** |
| Marcq et al. (2008) | – | – | **3.5±0.5** | – |
| Arney et al. (2014) | – | – | – | **0.5±0.1** |

*Note.* Data from the papers are bold; the other values are calculated using the data from the papers.

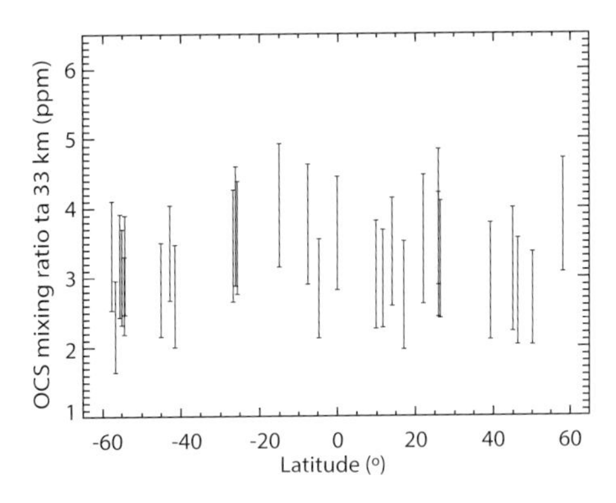


Figure 12.22 Latitudinal variations of OCS at 33 km from the VIRTIS-H observations (Marcq et al. 2008).

Table 12.4. Mean abundances in the observations are close to 13 ppm at 30 km and 3.5 ppm at 33 km. Latitudinal variations of OCS at 33 km observed using VIRTIS-H are shown in Figure 12.22. Differences between the observed OCS abundances become greater near 36 km.

To search for OCS in the upper cloud layer, Krasnopolsky (2010d) observed high-resolution spectra of Venus at the P-branch of the OCS band at 2.45 μm (Figure 12.23). The spectra were fitted by synthetic spectra without OCS, and the difference (observed minus synthetic) spectra in spectral intervals of $\pm 0.3$ cm$^{-1}$ centered at the expected Doppler-shifted positions of 16 OCS lines were summed to give OCS abundance. The results for three morningside and one afternoon observation are shown. The mean value is ≈3 ppb and refers to 65 km; the mean scale height of 2.5 km was measured using limb darkening. The SOIR observations resulted in an upper limit of $1.6 \pm 2$ ppb at 70 km (Vandaele et al. 2008); extrapolation of the mean values to 70 km gives ≈1 ppb and agrees with the upper limit.

### 12.5.3 $S_3$ and $S_4$

Spectra in the visible range observed at the V11–14 descent probes (Figure 7.14) revealed a so-called blue absorption with a significant decrease of the signal to the blue. This absorption is caused by the Rayleigh scattering in the dense lower atmosphere and a true absorption. Sanko (1980) calculated thermochemical equilibrium between the elemental

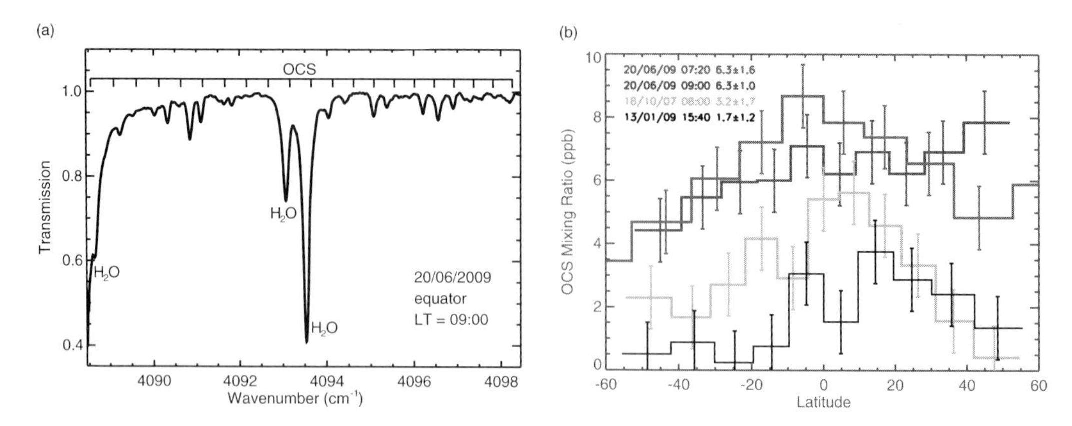


Figure 12.23 (a) Spectrum of Venus at the position of the P-branch of the OCS band at 2.45 μm. Strong telluric $H_2O$ lines are identified. Other features are weak telluric $H_2O$, $CH_4$, and solar lines. Expected (Doppler-shifted by $-0.185$ $cm^{-1}$) positions of the OCS lines are indicated. (b) Observed variations of OCS at 65 km. The lowest curve refers to the afternoon and may be more representative. From Krasnopolsky (2010d).

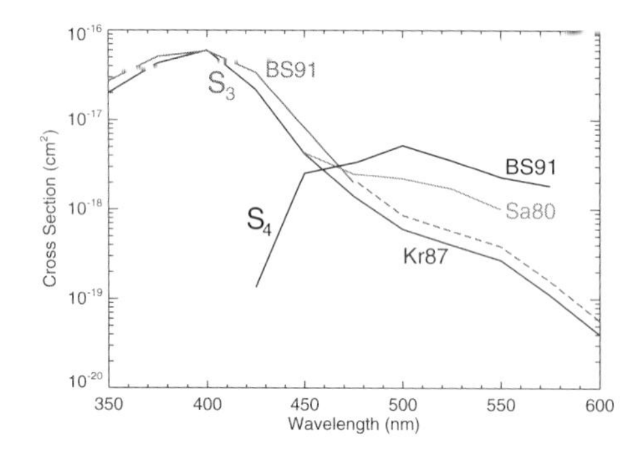


Figure 12.24 Absorption cross sections of $S_3$ and $S_4$ (Sanko 1980; Krasnopolsky 1987; Bilmers and Smith 1991). Extrapolation of the $S_3$ spectrum by Bilmers and Smith (1991) to 600 nm is shown by the dashed curve. From Krasnopolsky (2013b).

sulfur allotropes $S_X$ (X = 1–8) assuming a constant total sulfur mixing ratio. He also measured absorption spectrum of sulfur in the blue (Figure 12.24) and concluded that the observed true absorption can be produced by $S_3$ in equilibrium with 20–80 ppb of $S_2$. $S_2$ is a dominant allotrope of sulfur below 30 km on Venus.

More careful absorption spectra of $S_3$ and $S_4$ (Figure 12.24) were obtained by Krasnopolsky (1987) and Bilmers and Smith (1991). Maiorov et al. (2005) extracted spectra of the true absorption at 10–19 km and 3–10 km from the V11 observations (Figure 7.14), and Krasnopolsky (2013b) retrieved $S_3$ and $S_4$ abundances from those spectra using the $\chi^2$ fitting: $S_3 = 18 \pm 3$ ppt and $S_4 = 6 \pm 2$ ppt at 10–19 km, $S_3 = 11 \pm 3$ ppt and $S_4 = 4 \pm 4$ ppt at 3–10 km. (Part per trillion (ppt) means $10^{-12}$.) The chemical kinetic model (Krasnopolsky 2013b) does not support thermochemical equilibrium between $S_X$ below 20 km.

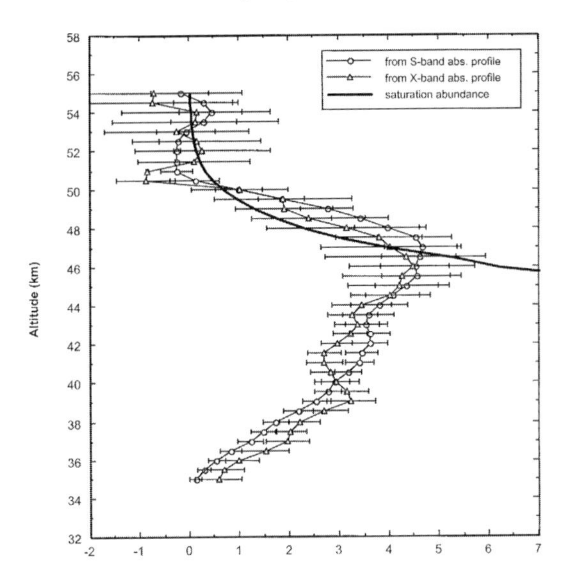


Figure 12.25 Vertical profile of $H_2SO_4$ vapor observed by the Magellan radio occultation at 67°N using the S and X-bands (13 and 3.6 cm, respectively). X axis is in ppm from Kolodner and Steffes (1998).

### 12.5.4 $H_2SO_4$

**Sulfuric acid vapor** absorbs microwave radiation and can be measured by radio occultations. A vertical profile observed by the Magellan orbiter at latitude of 67°N is shown in Figure 12.25. The $H_2SO_4$ vapor abundances are near zero above 50 km and reach a maximum of 4 ppm at 47 km, near the expected lower cloud boundary. Then the vapor mixing ratio gradually decreases to $\approx$0.5 ppm at 35 km.

The Venus Express radio occultations (Oschlisniok et al. 2012) at low latitudes on the dayside show an increase of the $H_2SO_4$ vapor from zero above 53 km to 3 ppm at 50 km.

The important upper limit of 3 ppb to $H_2SO_4$ vapor above 85 km was established by observations in the millimeter range by Sandor et al. (2012). This limit rules out photolysis of $H_2SO_4$ vapor as a significant source of $SO_X$ above 90 km.

### 12.5.5 Hydrogen Sulfide $H_2S$

$H_2S$ may originate in the reaction between pyrite $FeS_2$ and $H_2O$ and therefore was considered as one of the parent sulfur-bearing species. However, controversial detections by the PV mass spectrometer (3 $\pm$ 2 ppm in the lowest scale height; Hoffman et al. 1980) and the V13–14 gas chromatographs were not supported by thermochemical calculations by Krasnopolsky and Parshev (1979) and Fegley et al. (1997) and later by the chemical kinetic model (Krasnopolsky 2013b).

To search for $H_2S$ at the cloud tops, Krasnopolsky (2008) used the line at 2688.93 $cm^{-1}$ of the $H_2S$ (100–000) band. No absorption was found, and the observed upper limit of $3\times10^{16}$ $cm^{-2}$

was compared to the $CO_2$ abundance measured almost simultaneously at 4442 $cm^{-1}$ to give the $H_2S$ upper limit of 23 ppb. However, later observations of $CO_2$ at 2732 $cm^{-1}$ (Krasnopolsky 2010c) showed the $CO_2$ effective abundance that is smaller than that at 4442 $cm^{-1}$ by a factor of 3. Therefore the corrected upper limit to $H_2S$ is 70 ppb and refers to 74 km.

## 12.6 Hydrogen-Bearing Species: $H_2O$, HCl, HF, HBr, $NH_3$, and Their D/H Ratios

### *12.6.1 Water Vapor*

Water abundance above the Venus cloud tops is smaller than that on Mars by an order of magnitude, and geocentric velocities of Venus are smaller than those of Mars as well. Therefore $H_2O$ above the clouds is generally unmeasurable by the ground-based infrared spectroscopy. The first reliable detection of water was made by Fink et al. (1972) using the $H_2O$ bands at 1.4, 1.9, and 2.7 μm from an aircraft at 12 km with the overhead telluric water of $\approx$10 precipitable μm. The instrument resolving power was ~1000, and the retrieved water vapor mixing ratio was 0.6–1.0 ppm at 68 km. Later observations of the $H_2O$ band at 2.7 μm with resolving power of $1.3\times10^5$ from the Kuiper Airborne Observatory gave 2.1 $\pm$ 0.15 ppm of $H_2O$ at 72 km (Bjoraker et al. 1992).

A successful ground-based observation (Krasnopolsky et al. 2013) was occasionally made using CSHELL/IRTF at the exceptionally low telluric water of 0.3 pr. mm, much smaller than the mean water of 2 pr. mm above Mauna Kea with elevation of 4.2 km. The initial goal was to search for DF (Figure 12.26). The measured $H_2O$ fraction varied insignificantly between 55°S and 55°N with a mean value of 3.2 ppm at 74 km.

Water vapor was retrieved on Venus from the V15 FTS spectra (Figure 7.2). The spectra covered the $H_2O$ rotational band near 30 μm and the (010–000) band at 6.3 μm. Temperature profiles, aerosol mass loading, pressure at a level $\tau = 1$, and $H_2O$ mixing ratio at this level were

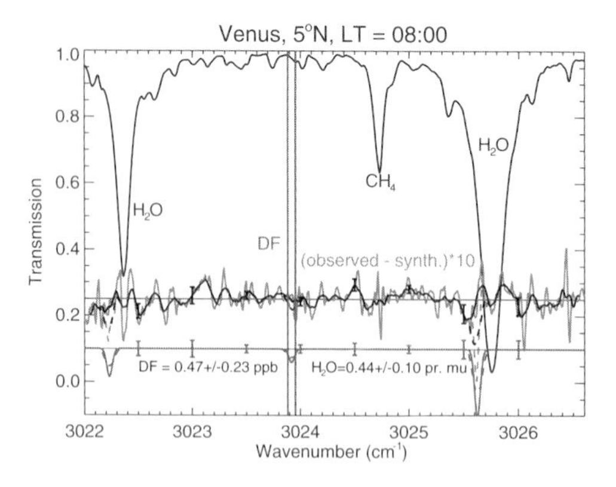


Figure 12.26 Spectrum of Venus near the equator at the expected Doppler-shifted position of the DF (1–0) R5 line. Difference between the observed spectrum and a synthetic spectrum for the terrestrial absorption is shown scaled by a factor of 10 unsmoothed and smoothed within 0.1 $cm^{-1}$. It reveals two Venusian $H_2O$ lines and the DF line. The derived species abundances are given. From Krasnopolsky et al. (2013).

Table 12.5 *Properties of the atmosphere extracted from the V15 FTS spectra*

| Latitude interval | $\varphi < 50°$ | $\varphi > 50°$ |
|---|---|---|
| Altitude $\tau = 1$ at 30 μm, km | $62.5 \pm 2.5$ | $56 \pm 2$ |
| $H_2O$ mixing ratio, ppm | $10 \pm 5$ | $8 \pm 6$ |
| $H_2O$ column, pr. μm | $8 \pm 5$ | $14 \pm 8$ |
| *T*, K | $250 \pm 10$ | $240 \pm 15$ |
| Pressure, mbar | $160 \pm 60$ | $370 \pm 100$ |
| Aerosol mass loading, g cm$^{-3}$ | $(2 \pm 1)\times10^{-9}$ | $(10 \pm 9)\times10^{-9}$ |

*Note.* Two-sigma deviations that reflect variabilities are given. From Ignatiev et al. (1999).

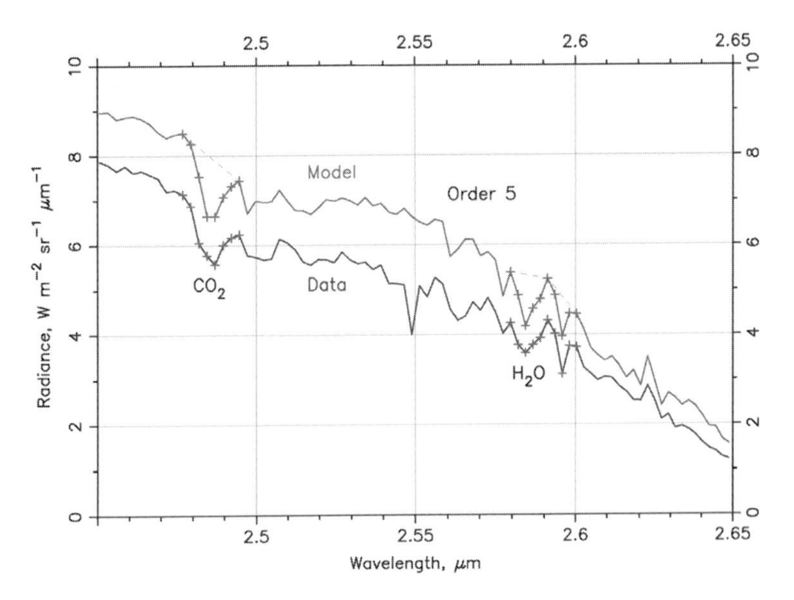


Figure 12.27 Part of the VIRTIS-H dayside spectrum that includes the $CO_2$ and $H_2O$ bands at 2.48 and 2.60 μm to determine the cloud top altitude and $H_2O$ abundance near this level. The model spectrum is shown as well. From Cottini et al. (2015).

extracted from the spectra self-consistently (Ignatiev et al. 1999). The retrieved atmospheric properties were rather different below and above latitude of 50° (Table 12.5). The data at high latitudes refer to the deeper atmosphere with the pressure increase by a factor of 2.3.

Water vapor was observed by Venus Express using the $H_2O$ band at 2.6 μm and the $CO_2$ band at 2.48 μm (Figure 12.27) in the VIRTIS-H spectra (Cottini et al. 2015). Fitting of the bands results in abundance of $H_2O$ and a cloud level of $\tau = 1$ it refers to. The cloud top level $\tau = 1$ is at 69 km at the low and middle latitudes and moves down to 64 km near 80°. The mean $H_2O$ mixing ratio is 3 ppm at the low latitudes, increases to 5 ppm at 70° and then drops to 4 ppm near the poles. Variations with local time are insignificant.

A similar analysis was made for the VEX/SPICAV-IR observations of the $H_2O$ band at 1.38 μm and the $CO_2$ bands at 1.4–1.6 μm (Fedorova et al. 2016). The retrieved effective altitude of the cloud level $\tau = 1$ is 62 km at these wavelengths and latitudes below 20°, and the $H_2O$ abundance is 6 ppm. This abundance has a shallow minimum of 5.4 ppm at the same altitude between 30° and 50°. The water content increases to 7 ppm at 60.6 km and high latitudes. Correlations with some geographic features have been found.

The SOIR occultations at the $H_2O$ band at 2.6 μm (Fedorova et al. 2008) resulted in vertical profiles of $H_2O$ mixing ratio from 70 to 110 km (Figure 7.12). The $H_2O$ abundances decrease from ≈2.5 ppm at 70 km to 1.5 ppm at 75–100 km and then to ≈0.7 ppm at 110 km. The profiles were observed at high latitudes of 63° to 90°; however, there is no latitude trend in the data.

Water vapor was detected in the millimeter range by Encrenaz et al. (1995) using the IRAM telescope with diameter of 30 m. The observed line at 183.3 GHz originates in the atmosphere at 75–100 km, and the observed $H_2O$ mixing ratio was $1^{+1}_{-0.5}$ ppm. Observations with the Submillimeter Wave Astronomy Satellite (SWAS) of the line at 556.9 GHz (Gurwell et al. 2007) showed highly variable water abundances from less than 0.03 ppm to 4.5 ppm.

Low-resolution spectrometers in the range of 0.4–1.2 μm at the V11–14 descent probes (Figure 7.14) covered the $H_2O$ bands at 0.7, 0.82, 0.95, and 1.13 μm. Initially analysis of the observed absorption in these bands resulted in the $H_2O$ mixing ratio decreasing from 200 ppm at 40–60 km to 20 ppm near the surface. Later progress in the spectroscopic data and radiative transfer using line-by-line discrete ordinate calculations showed a nearly constant $H_2O$ mixing ratio of ≈30 ppm from 60 km to the surface (Ignatiev et al. 1997).

Actually that was a confirmation of the $H_2O$ profile with the constant mixing ratio of $30 \pm 10$ ppm extracted by Pollack et al. (1993) from the ground-based nightside spectroscopy in the transparency windows at 1–1.3, 1.7, and 2.3 μm. Midpoints of the sensing regions were at 12, 23, and 33 km, respectively. Later ground-based (de Bergh et al. 1995; Chamberlain et al. 2013; Arney et al. 2014), VEX/SPICAV-IR (Bézard et al. 2011), and VEX/VIRTIS (Marcq et al. 2008; Haus and Arnold 2010; Tsang et al. 2010) observations confirmed a constant $H_2O$ abundance of 30 ppm in the lower atmosphere. Variations of water vapor are typically insignificant (Figure 12.28).

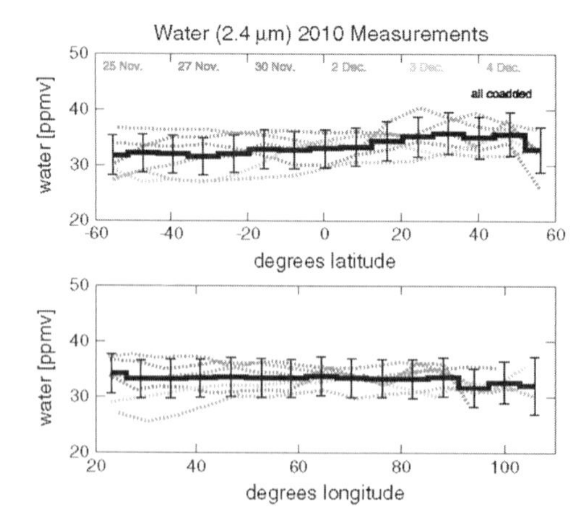


Figure 12.28 Variations of water vapor near 33 km (Arney et al. 2014).

### 12.6.2 HDO and $HDO/H_2O$ Ratio

The PV Large Probe mass spectrometer inlet was blocked by a sulfuric acid droplet in the altitude range of 50 to 27 km. That made it possible to study in detail the droplet composition and reveal a huge enrichment in HDO relative to the terrestrial value of $D/H = 1.56\times10^{-4}$. Comparison of the mass peak at 18 and 19 amu with corrections for $H^{18}O$, F, and double ionized $H^{37}Cl$ and $^{38}Ar$ resulted in a $HDO/H_2O$ ratio of $(3.2 \pm 0.4)\times 10^{-2}$, that is, D/H = 100 ± 13 times the terrestrial value (Donahue et al. 1982). This is an indication that Venus lost water more than its present abundance by a few orders of magnitude. Although that was unexpected, it looks natural when comparing the water column abundance of 1.3 cm on Venus with 2.7 km on the Earth.

De Bergh et al. (1991) analyzed a part of the high-resolution nightside spectrum of Venus in the window at 2.3 μm that was observed by Bezard et al. (1990). The range of 4140–4200 $cm^{-1}$ observed with resolution of 0.23 $cm^{-1}$ was sensitive to both $H_2O$ and HDO abundances at 32–42 km. The retrieved abundances were 34 ± 10 ppm and 1.3 ± 0.2 ppm, respectively, and resulted in D/H = 120 ± 40 times SMOW (standard mean ocean water).

High-resolution spectroscopy at 3770–3860 $cm^{-1}$ from the Kuiper Airborne Observatory (Bjoraker et al. 1992) gave D/H = 157 ± 15 times SMOW at 72 km. Donahue et al. (1997) reconsidered their D/H value from Donahue et al. (1982) and took into account all the D and H in HDO, $CH_3D$, HD, and $H_2O$ to get D/H = 157 ± 30 times SMOW.

The SOIR occultations revealed vertical profiles of both $H_2O$ and HDO (Figure 7.12) using the bands at 2.61 and 3.58 μm, respectively (Fedorova et al. 2008). The HDO profiles cover the altitudes of 70 to 95 km and give D/H = 240 ± 25 SMOW.

Finally, the ground-based detection of $H_2O$ (Figure 12.26) coupled with the simultaneous observations of HDO resulted in D/H = 95 ± 15 SMOW at 74 km (Krasnopolsky et al. 2013). We will discuss the problem of D/H on Venus later and just mention here that the differences between the observed values significantly exceed the claimed uncertainties.

Ground-based observations of HDO on Venus are more feasible than those of $H_2O$ because of the smaller telluric absorption. Those observations may be used to study variations of water vapor. However, they cannot be applied to get the D/H ratio by comparison with $H_2O$ abundances taken from other sources, because differences between the observed $H_2O$ are significant as well as its vertical gradient.

The most detailed observations in the millimeter range were made by Sandor and Clancy (2005) using the JCMT telescope. They conducted 17 observations in the period of 1998 to 2004, and no water was detected in five observations. Though the data refer to significant parts of the Venus disk, the observed variations were very strong even in short time intervals. It looks like this situation is typical of the microwave observations for almost all species and has not been explained. The D/H = 157 from Bjoraker et al. (1992) was applied for conversion of HDO to $H_2O$. The highest $H_2O$ was 3.5 ppm, and the mean and standard deviation for those 17 observations are 1.0 ± 1.0 ppm for 65–100 km. Six observations were vertically resolved and give 2.2 ± 1.1 ppm at 65–82 km and 1.8 ± 0.6 ppm at 82–100 km. While the mean values are rather similar in both altitude intervals, each of the six observations showed significant altitude gradients of both signs.

### *12.6.3 Hydrogen Chloride HCl and DCl*

HCl was discovered in the high-resolution spectra of the HCl (2–0) band at 1.76 μm by Connes et al. (1967) who evaluated its abundance at 600 ± 120 ppb. The estimated $CO_2$ abundance was 3.3 km-atm; corrected for the airmass, this gives a half-pressure level at 65 km. Young (1972) extracted 420 ± 70 ppb from the spectra by Connes et al. (1967).

Later Iwagami et al. (2008) observed the HCl (2–0) R3 line using CSHELL/IRTF and retrieved a mixing ratio of 740 ± 60 ppb at 60–66 km. Krasnopolsky (2010c) observed the HCl (1–0) P8 lines (Figure 12.29) with the same instrument. Using the $CO_2$ lines near 2734 $cm^{-1}$ for comparison, the measured HCl mixing ratio was 400 ± 30 ppb at 74 km in the latitude range of ±60°. Local variability is up to 10% in this range.

The SOIR occultation observations of HCl (Mahieux et al. 2015b) covered the period of 2006 to 2013 with 147 orbits presented in the cited paper. The HCl (1–0) R0 and R1 lines at 2900–2930 $cm^{-1}$ were used in that study. Summary of the observations is given in Figure 12.30. There are two unexpected features of the data. First, the measured HCl mixing ratio near 70 km is smaller than that from the ground-based infrared observations by almost an order of magnitude. Second, the increase in the mixing ratio from 70 to 110 km by a factor of 20 anticipates a strong source of HCl near 110 km and a strong sink near 70 km that look puzzling for a rather stable species of HCl. This does not support the element conservation (Section 2.2.5).

The JCMT observations of the HCl rotational line at 625.9 GHz (Sandor and Clancy 2012) revealed a complicated profile of the line (Figure 12.31). There were three nightside and one morningside observations. The best-fit profiles of HCl are shown in Figure 12.30, and the morningside profile is steeper. The retrieved HCl profiles are opposite to the SOIR profiles: they agree with the ground-based IR observations and show a decline with altitude. This decline is much steeper than the weak decrease to ≈300 ppb near 100 km predicted by the photochemical models (Yung and DeMore 1982; Krasnopolsky 2012b).

Mixing ratios of HCl near 24 km were extracted from the window at 1.7 μm in ground-based observations by Pollack et al. (1993; 480 ± 120 ppb), Taylor et al. (1997; 500 ± 150 ppb),

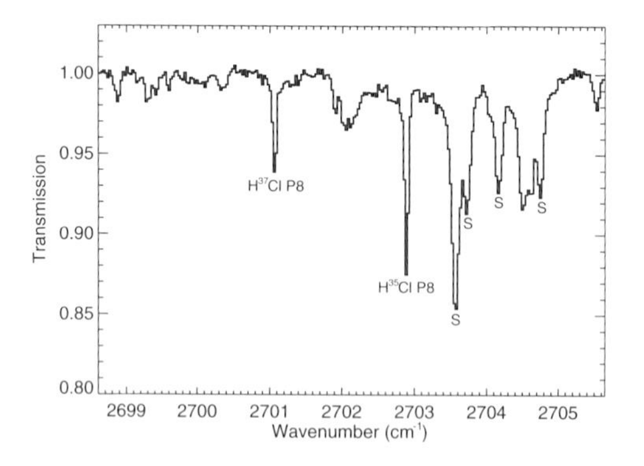


Figure 12.29 Spectrum of Venus near the HCl (1–0) P8 lines observed using CSHELL/IRTF. Strong solar lines are marked S. The wavenumber scale is Doppler-shifted by −0.122 $cm^{-1}$, so that the lines are at rest positions. From Krasnopolsky (2010c).

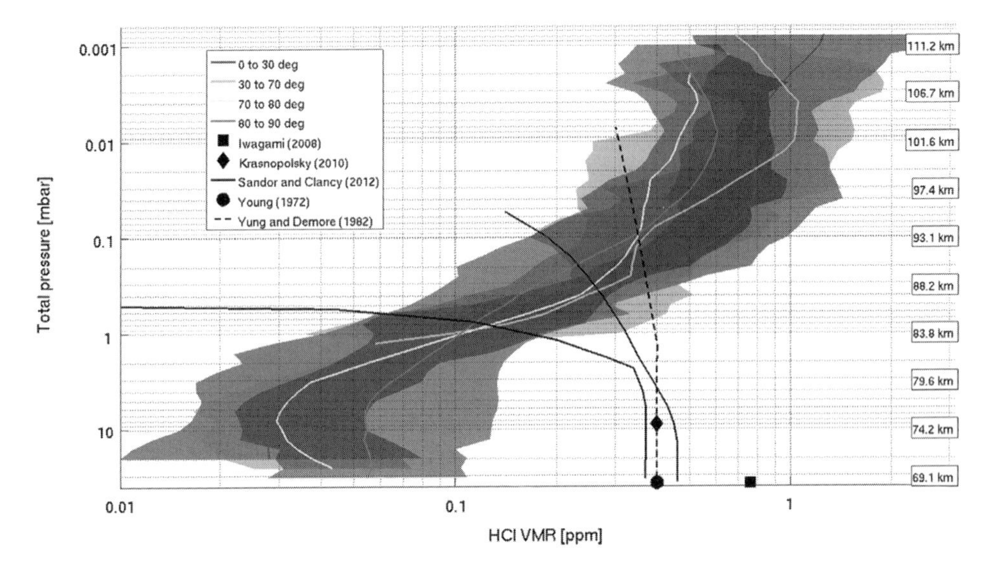


Figure 12.30 Vertical profiles of HCl from the SOIR occultations are compared wih the infrared and submillimeter observations and the model by Yung and DeMore (1982). From Mahieux et al. (2015b). (A black and white version of this figure will appear in some formats. For the color version, please refer to the plate section.)

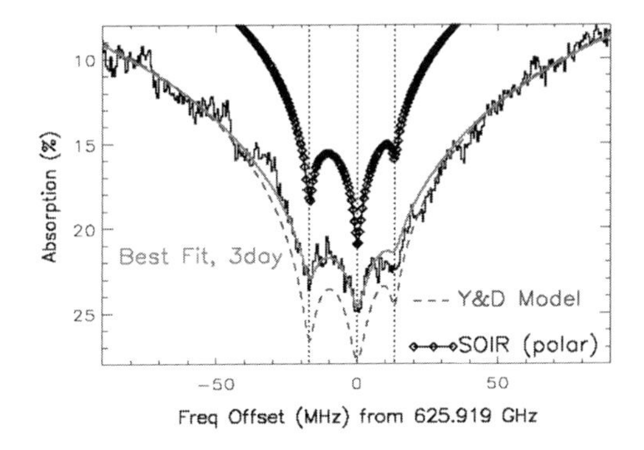


Figure 12.31 Observed central part of the HCl line at 625.92 GHz on the Venus nightside is compared with the best-fit profile of HCl shown in Figure 12.30 and the profiles from the model by Yung and DeMore (1982) and the initial version of the SOIR occultations. From Sandor and Clancy (2012).

Iwagami et al. (2008; 400 $\pm$ 50 ppb), and Arney et al. (2014; 410 $\pm$ 40 ppb). Spatially resolved observations by Arney et al. (2014) do not reveal any variations in HCl (Figure 12.32).

Search for DCl on Venus is hampered by contamination of lines of the (1–0) band at 2091 $cm^{-1}$ by telluric, solar, and the Venus $CO_2$ and CO lines. Krasnopolsky et al. (2013) chose the R4 line of $D^{35}Cl$ at 2141.54 $cm^{-1}$. A spectrum observed near the equator using CSHELL/IRTF is shown in Figure 12.33. Comparison with a synthetic spectrum that involved both DCl and $CO_2$ results in a $D^{35}Cl$ mixing ratio of 9.9 $\pm$ 3.6 ppb. Using the

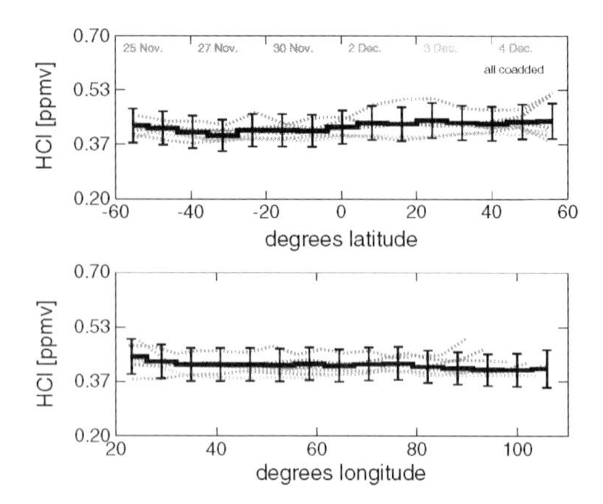


Figure 12.32 Variations of HCl near 24 km in observations by Arney et al. (2014).

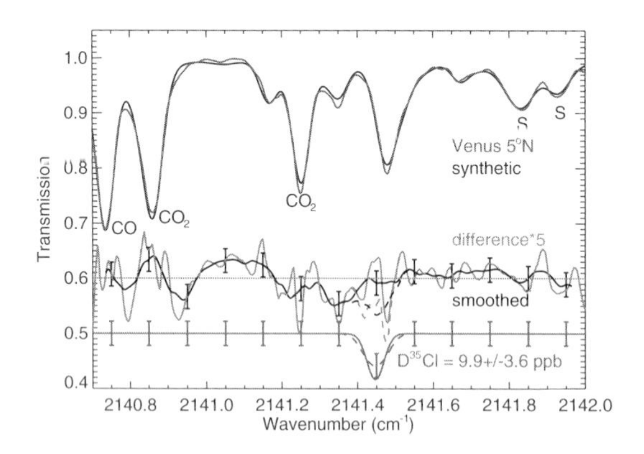


Figure 12.33 Detection of DCl on Venus. A spectrum observed near the equator is compared with synthetic spectra with and without the DCl line (solid and dashed curves, respectively). Their differences unsmoothed and smoother within 0.07 $cm^{-1}$ reveal the $D^{35}Cl$ line with the $D^{35}Cl$ abundance of 9.9 ± 3.6 ppb. From Krasnopolsky et al. (2013).

terrestrial $^{35}Cl/^{37}Cl$ ratio, the HCl mixing ratio of 400 ppb, averaging the data from all observed spectra, and taking into account possible systematic errors, the recommended D/H ratio in HCl is 190 ± 50 SMOW.

Isotopic shift in the DCl photolysis spectrum results in a reduction in the photolysis frequencies of DCl relative to those of HCl by a factor of 6 (Bahou et al. 2001; Section 10.5.3). A similar effect for $HDO/H_2O$ is a factor of 2.5, and this difference facilitates the greater D/H ratio in HCl.

### *12.6.4 Hydrogen Fluoride HF and DF*

Connes et al. (1967) discovered HF on Venus in their high-resolution spectra that covered both the (1–0) band at 3961 $cm^{-1}$ and the (2–0) band at 7750 $cm^{-1}$. The measured HF mixing ratio was 5 ± 2.5 ppb. The Kuiper Airborne Observatory gave 6.5 ± 0.3 ppb (Bjoraker et al. 1992). Observations of HF using CSHELL/IRTF are discussed in Section 5.4 and Figures 5.4–5.6; the retrieved mixing ratio is 3.5 ± 0.2 ppb at 68 km and does not vary within latitudes of ±60°.

The SOIR occultations of the HF (1–0) R1 line at 4038.96 $cm^{-1}$ (Mahieux et al. 2015b) result in vertical profiles of HF from 78 to 103 km (Figure 12.34). The mean value at 78 km is 4.4 ppb and agrees with the previous ground-based and aircraft observations. However, the observed increase to 50 ppb at 103 km raises the problems of a source and sink of HF and the element conservation, similar to those for HCl.

The window at 2.3 μm (Figure 12.12) includes absorption lines of HF. Pollack et al. (1993) evaluated its abundance at 1–5 ppb near 34 km. The similar value in Taylor et al. (1997) is 5 ± 2 ppb.

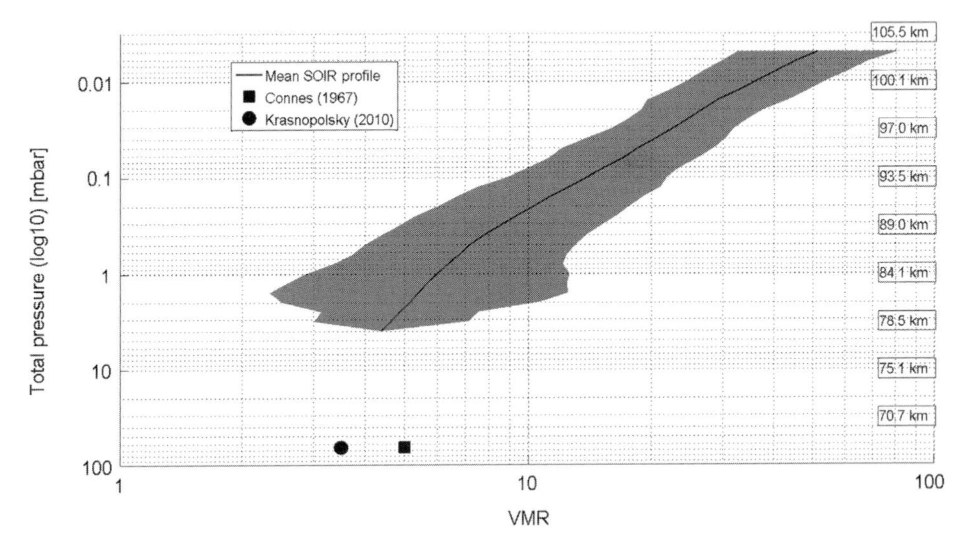


Figure 12.34 SOIR occultations of the HF (1–0) R1 line at 4038.96 $cm^{-1}$. The retrieved HF volume mixing ratios (VMR) are in ppb. From Mahieux et al. (2015b).

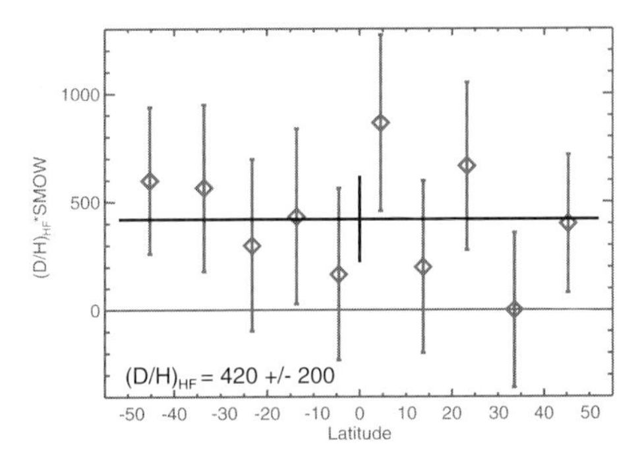


Figure 12.35 Observed D/H ratios in HF in 10 spectra at various latitudes. The bars show the mean ratio and its uncertainty. From Krasnopolsky et al. (2013).

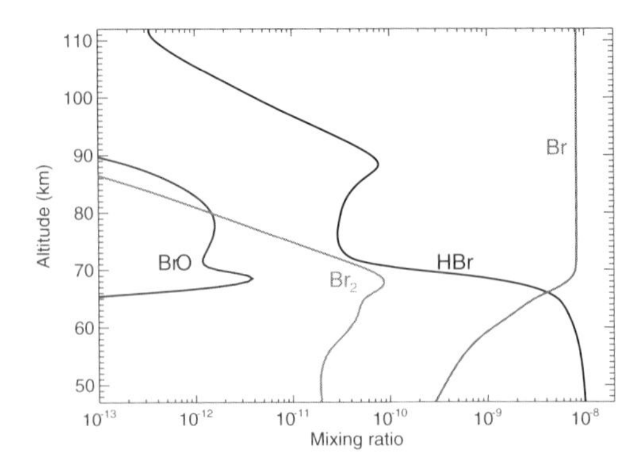


Figure 12.36 Bromine species on Venus calculated for HBr = 10 ppb below the clouds (Krasnopolsky and Belyaev 2017).

Search for DF on Venus was made using the DF (1–0) R5 line at 3024.0.54 $cm^{-1}$ (Figure 12.26). The retrieved DF abundances were scaled to HF = 3.5 ppb to get the D/H ratio. This ratio varied in 10 spectra observed at various latitudes (Figure 12.35), and the spectrum in Figure 12.26 gave the greatest value. Uncertainty of the mean value for 10 points is equal to the standard deviation reduced by a factor of $(n-1)^{1/2}=3$, and the mean D/H in HF is 420 ± 100 SMOW. The uncertainty was doubled to account for possible systematic errors.

### 12.6.5 HBr and $Br_2$

If assigned to $Br_2$, then a weak true absorption near 15 km retrieved from the spectrophotometric observations at the Venera 11–14 landing probes corresponds to an upper limit to $Br_2$ mixing ratio of 0.2 ppb (Moroz et al. 1981). Recently HBr was searched using its strongest line (1–0) R2 2605.8/6.2 $cm^{-1}$ with an upper limit of 1 ppb at 78 km (Krasnopolsky and Belyaev 2017). They developed a simplified version of bromine photochemistry that was implemented into the photochemical model by Krasnopolsky (2012b) and showed a steep decrease of HBr above 67 km because of the strong photolysis and reactions with O and H (Figure 12.36). A reanalysis of the observation using the calculated profile of HBr resulted in an upper limit to its mixing ratio of ≈30 ppb below 60 km. Thermochemical calculations prove that HBr is a dominant bromine species in the lower atmosphere, and this limit is therefore more restrictive than that of 0.2 ppb of $Br_2$. If $f_{HBr} \approx$ 10 ppb in the lower atmosphere (Figure 12.36), then its photochemical effect is smaller than but comparable to that of HCl.

### 12.6.6 Ammonia $NH_3$

Ammonia as a minor constituent of the Venus atmosphere was considered in some publications. To search for ammonia, Krasnopolsky (2012c) used the $NH_3$ line at 4484.11 $cm^{-1}$ in observations with CSHELL/IRTF. No absorption was detected in the spectra with an upper limit of 6 ppb to $NH_3$ at 68 km.

## 12.7 Nitric Oxide and Lightning

### *12.7.1 Nitric Oxide*

NO is formed in the upper atmospheres of the terrestrial planets, and the NO nightglow was studied in detail using the PV and VEX orbiters (Section 12.10.4). However, NO in the lower atmosphere of Venus may be formed only by lightning, and Yung and DeMore (1982) developed a version of the photochemical model with NO as a basic catalyst that controls the photochemistry. Therefore a search for NO was among the priority tasks for the Venus chemical composition. That was done using lines of the NO fundamental band at 5.3 μm that were observed with the TEXES spectrograph at NASA IRTF. The instrument covers a small interval in a range of 5 to 25 μm with resolving power of $8\times10^4$ (Lacy et al. 2002).

A part of the observed spectrum near the Doppler-shifted NO line at 1850.175 $cm^{-1}$ is shown in Figure 12.37. Analysis proved that the NO line is contaminated by a few very weak telluric lines that were calculated by scaling of similar lines observed in the spectrum. Subtraction of the continuum in the corrected spectrum revealed the NO absorption line.

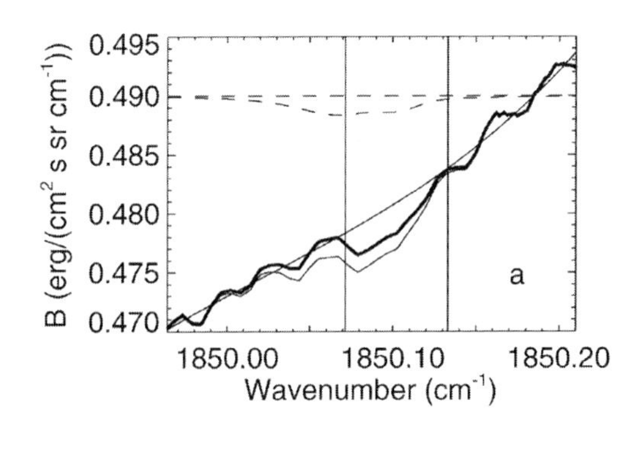


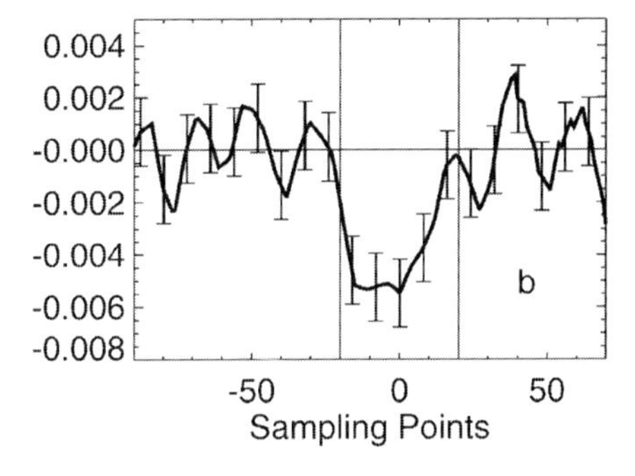


Figure 12.37 (a) Spectrum of Venus near the NO (1–0) line at 1850.175 $cm^{-1}$. The line is Doppler-shifted to the red by 0.072 $cm^{-1}$. The observed spectrum is the thin line, the calculated telluric spectrum is the dashed line, and the spectrum corrected for the telluric absorption is the solid line. The spectrum continuum is fitted by a third-degree polynomial. Vertical lines show an interval of two resolution elements centered at the NO line. (b) relative difference between the corrected spectrum and continuum. Sampling point is equal to 0.001544 $cm^{-1}$. From Krasnopolsky (2006a).

NO is not uniformly mixed in the middle atmosphere, and a simple photochemical model was applied to calculate the NO vertical profile. The model involved predissociation of NO at the bands at 191 and 183 nm, production of NO by two- and three-body associations, and loss in the reaction with N. The NO density at 50 km was a free parameter. Then extinction and radiation by the NO molecules and the mode 2 sulfuric acid aerosol were calculated to fit the observed equivalent widths of two NO and two $^{13}CO_2$ lines. That resulted in a NO mixing ratio of 5.5 $\pm$ 1.5 ppb at 50–60 km, and an upward NO flux of $(6 \pm 1.5)\times10^7$ cm$^{-2}$ s$^{-1}$ is required at 50–60 km to sustain this mixing ratio.

### *12.7.2 Effect of the Cosmic Rays*

Ionization of the Venus lower atmosphere by protons and muons from the cosmic rays was calculated by Upadhyay et al. (1994). Ionization by protons dominates, and the total column ionization rate is $2.7\times10^9$ cm$^{-2}$ s$^{-1}$. Using the data from Luna et al. (2003), Krasnopolsky (2006a) calculated yields of the ionization channels giving $N_2^+$, $N^+ + N$, $N^+ + N^+$, and $N^{++} + N$ at 0.585, 0.125, 0.125, and 0.02 per $N_2$ ionization event and assumed that one-half of all nitrogen atoms are formed in the doublet states, finally giving $N(^2D)$. $N_2^+$ and $N^{++}$ are lost in charge exchange with $CO_2$, $N(^2D)$ forms NO in the reaction with $CO_2$, and

$$N^+ + CO_2 \rightarrow CO_2^+ + N, \quad \gamma = 0.82$$
$$\qquad\qquad\quad CO^+ + NO, \quad \gamma = 0.18.$$

Taking into account the $N_2$ mixing ratio and its smaller mass, production of NO and N in the ground state by the cosmic rays are $7\times10^6$ and $2.3\times10^7$ cm$^{-2}$ s$^{-1}$, respectively. It is not clear if N can react with $CO_2$ at high temperatures and form NO. Using the reaction with $O_2$ as an analog, the total effect of the cosmic rays on the production of NO is negative at $(-0.7 \pm 0.9)\times10^7$ cm$^{-2}$ s$^{-1}$ because of the reaction between NO and N. Therefore the required production of NO is $\approx7\times10^7$ cm$^{-2}$ s$^{-1}$.

### *12.7.3 Lightning*

Despite the long history of studies of terrestrial lightning, there are significant uncertainties in some basic values. Orbiter observations established that the global flashing rate is $44 \pm 5$ s$^{-1}$ (Christian et al. 2003) instead of a previous value of 100 s$^{-1}$; 30% of this rate is the cloud-to-ground discharges the remaining being the intracloud flashes. Each flash involves a few strokes with a mean number of 4. A mean cloud-to-ground flash energy is estimated at $2\times10^8$ J by Hill (1979), $4\times10^8$ J by Borucki and Chameides (1984), $(1.2 \pm 0.3)\times10^9$ J by Liaw et al. (1990), $6.7\times10^9$ J by Price et al. (1997), and $10^9$ J by Cooray (1997). We adopt a value of $10^9$ J. The intracloud flashes are weaker by a factor of 2 (10 in the earlier estimate by Price et al. 1997). The above values correspond to the energy deposition of 0.06 erg cm$^{-2}$ s$^{-1}$.

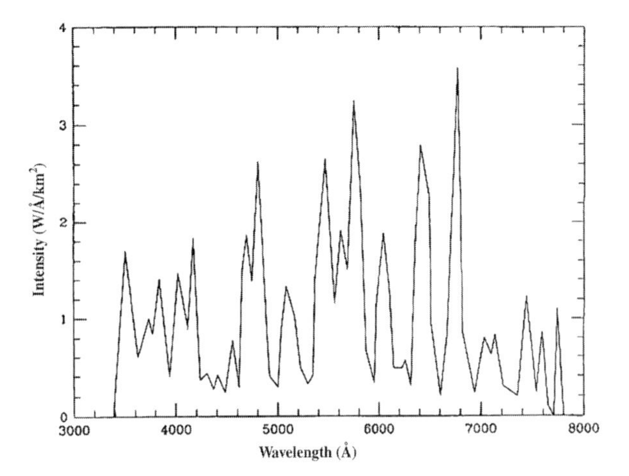


Figure 12.38 One of seven spectra of lightning observed by Venera 9; 480 points of the spectrum from 3000 to 8000 Å were consecutively measured in 10 s, therefore the spectrum also reflects a temporal behavior of the thunderstorm region. Mean duration of the flashes is 0.25 s and their light energy is $2\times10^7$ J (Krasnopolsky 1983b).

Electric signals detected by the Venera 11–14 landing probes, the Pioneer Venus orbiter, during the Galileo and Cassini flybys of Venus, and by Venus Express have been interpreted as originating from lightning (see Grebowsky et al. 1997; Russell et al. 2011 and references therein). However, lightning is a phenomenon which produces light; therefore optical observations are of the highest priority.

The optical spectrometers at the Venera 9 and 10 orbiters revealed one region with frequent flashes (Krasnopolsky 1983b). The lightning spectrum (Figure 12.38) is rather uniform from 340 to 800 nm. It is not significantly depleted by the Rayleigh extinction below 550 nm, and this rules out lightning in the lowest 20 km, e.g., from volcanic eruptions. The observed spectrum generally agrees with a spectrum simulated in laboratory for the atmosphere of Venus at 1 bar (Borucki et al. 1996). This pressure is typical of the lower cloud layer on Venus. The simulated spectrum consists of a strong continuum and a few lines. An increase in the simulated spectrum below 450 nm because of the CN bands is compensated in the observed spectrum by the blue absorption in the upper cloud layer. The O 777 nm line is not prominent in the observed spectrum because of the low sensitivity at this wavelength (~2% of the peak value) and random flashing. The mean light energy of a flash in the visible range is $2\times10^7$ J. Scaling the data to the total area observed by the spectrometers from both orbiters, a global flashing rate is 700 $s^{-1}$ or 45 $km^{-2}$ $y^{-1}$ and the lightning energy deposition is 1 erg $cm^{-2}$ $s^{-1}$ (Krasnopolsky 1983b). However, only one flashing region was observed, and these values are uncertain within $\pm100\%$. No near-surface lightning was detected with upper limits of 0.3 and 1 $km^{-2}$ $y^{-1}$ for the light energy exceeding $4\times10^5$ and $10^5$ J, respectively.

No flashes have been detected by the star sensor on the Pioneer Venus orbiter (Borucki et al. 1991) with an upper limit of 3 $km^{-2}$ $y^{-1}$. There were various and severe restrictions to those observations which resulted in a total observing time of just 83 s during two sessions in 1988 and 1990.

Each of the photometers on the Vega 1 and 2 balloons (Sagdeev et al. 1986) monitored illumination during the 45-h flights. The balloons were for ~30 h and covered ~7700 km at low latitudes in the nighttime conditions favorable for detection of lightning. The mean balloon altitude was 53 km. The photometers (silicon photodiodes) were sensitive at 400–1100 nm. The measured nighttime illumination was stable at 4 Lux. This is far above the nightglow and light of the stars and the Earth that are 0.002 Lux total (Krasnopolsky 1986). We attribute the observed illumination to radiation of the hot surface of Venus in a spectral window near 1 μm. No flashes had been detected. Unfortunately, a limiting illumination for the flash counter was not published; therefore it is impossible to get an upper limit to the flashing rate.

Ground-based observations of the Venus nightside revealed seven flashes per 3.7 h of the exposure (Hansell et al. 1995) with a rate of $8\times10^{-5}$ km$^{-2}$ y$^{-1}$. Six of those flashes were detected with a filter at 777 nm. The authors adopted in their analysis that the O 777 nm line constitutes 40% of the optical power of lightning on Venus. This assumption was based on photographic spectra of simulated lightning. Later photoelectric spectra (Borucki et al. 1996) showed a dominant contribution of the continuum which reduces the O 777 nm line to ~3%. Scaling the measured flash optical energies by a factor of 12 results in a minimum flash optical energy of $1.2\times10^9$ J which exceeds the mean flash optical energy on the Earth and Venus by factors of 250 and 60, respectively. Evidently it is difficult to compare the observed very low flashing rate with those from the Venera and Pioneer Venus observations.

Search for random flashes on the Venus nightside was made using the visible channel of VIRTIS/VEX (Moinelo et al. 2016). They concluded that the observed flashes originate from cosmic rays impacting the instrument detector and found no statistical evidence for lightning. The authors argued that this does not rule out the existence of lightning. The instrument sensitivity and the orbit conditions were not favorable for detection of lightning.

The Japanese orbiter Akatsuki has a special camera to search for lightning. However, its detection limit at the current orbit exceeds brightness of the mean terrestrial lightning by a factor of 2.

The detection of NO in the lower atmosphere of Venus is an independent confirmation of lightning on Venus. The NO flux of $(7\pm2)\times10^7$ cm$^{-2}$ s$^{-1}$ at 60 km, which is required to support the measured NO abundance, may be formed only by lightning. The NO mixing ratio at 60 km would be extremely low at 0.002 ppb without lightning in our model which is close to 0.003 ppb calculated for this case by Yung and DeMore (1982). A yield of NO from lightning on Venus is $(3.7 \pm 0.7)\times10^8$ molecules per erg (Levine et al. 1982), and the mean energy deposition by lightning on Venus is $0.19 \pm 0.06$ erg cm$^{-2}$ s$^{-1}$, that is, even higher than that on the Earth, 0.06 erg cm$^{-2}$ s$^{-1}$ (see above). If the total flash energy on Venus is equal to that on the Earth, $10^9$ J, then the derived energy deposition corresponds to a global flashing rate of $(87 \pm 25)$ s$^{-1}$ and $6 \pm 2$ km$^{-2}$ y$^{-1}$. This rate is close to the upper limit of 3 km$^{-2}$ y$^{-1}$ from the Pioneer Venus star sensor and comparable with the rate of $\approx 50\ s^{-1}$ estimated from the electric bursts observations (Russell et al. 2011; Delitsky and Baines 2015).

Delitsky and Baines (2015) discussed chemical production of species by lightning on Venus. It looks like that NO only has a global effect. While the production of CO is higher by two orders of magnitude, it is much smaller than the photochemical production. However, chemical perturbations by lightning may be significant in the thunderstorm areas.

Summary of chemical composition of the Venus atmosphere below 120 km is given in Table 12.6.

## 12.8 Thermosphere

Direct measurements in the neutral upper atmosphere of Venus were extensively made by the Pioneer Venus mission. Those measurements involved accelerometers at the four entry probes, bus and orbiter neutral mass spectrometers, and orbiter atmospheric drag measurements. The PV orbiter observations were mostly conducted in 1979–1980, when the orbiter periapsis was at 150–190 km and solar activity was high at $F_{10.7}$ = 180–200. The measurements were also made down to 130 km in 1992 before the orbiter destruction in the dense atmosphere. Those data refer to the nighttime thermosphere at mean solar activity $F_{10.7} \approx 130$.

Atmospheric drag measurements were also obtained in cycle 4 of the Magellan orbiter in 1992–1993 at 160–190 km and mean solar activity. Solar and stellar occultations by Venus Express gave data on the lower thermosphere near the morning and evening terminators and on the nightside, respectively.

### *12.8.1 PV Bus Neutral Mass Spectrometer*

The PV bus (von Zahn et al. 1980) entered the morningside atmosphere of Venus at the solar zenith angle of 60° and solar activity $F_{10.7}$ = 190. The mass spectra were measured down to 130 km, and the results are shown in Figure 12.39. Actually three peaks from $CO_2$, CO + $N_2$, and He were prominent in the spectra, and contributions of CO and $N_2$ were separated at 140–153 km. The profile of atomic oxygen was not extracted from that originated from decomposition of $CO_2$ in the instrument and is based on some model considerations. The measured total density agrees with the orbiter drag data and the extrapolated day probe accelerometer observations. The exospheric temperature is 275 K.

Detailed comparison of the He and $CO_2$ profiles resulted in eddy diffusion $K = 1.4\times10^{13}\ n^{-1/2}\ \mathrm{cm^2\ s^{-1}}$, and extrapolation to the middle atmosphere with this $K$ gives a helium mixing ratio of 12 ppm with the uncertainty within a factor of 2.

### *12.8.2 PV Orbiter Atmospheric Drag*

The PV orbiter drag (Keating et al. 1980) changes the orbiter period and is proportional to $\rho H^{1/2}$; here $\rho$ is the atmospheric density and $H$ is the scale height, and both values refer to

Table 12.6 *Chemical composition of Venus' atmosphere below 120 km*

| Gas | *h* (km) | Mixing ratio | References |
|---|---|---|---|
| $CO_2$ | – | 0.965 | Adams and Dunham (1932), von Zahn et al. (1983) |
| $N_2$ | – | 0.035 | Gelman et al. (1979), Istomin et al. (1983), Hoffman et al. (1980), Oyama et al. (1980) |
| He | – | 10 ppm | Kumar and Broadfoot (1975), von Zahn et al. (1980), Niemann et al. (1980), Krasnopolsky and Gladstone (2005) |
| Ne | – | 7 ppm | Hoffman et al. (1980), Istomin et al. (1980, 1983), Oyama et al. (1980) |
| Ar | – | 70 ppm | Gelman et al. (1979), Hoffman et al. (1980), Istomin et al. (1980), Oyama et al. (1980) |
| Kr | – | 35 ppb | Istomin et al. (1983), Donahue et al. (1981) |
| Xe | – | <20 ppb | Istomin et al. (1983) |
| CO | 120 | 2000 ppm | twilight, Vandaele et al. (2015) |
| | 100 | 1200 ppm | night, Clancy et al. (2012) |
| | | 300–1000 ppm | twilight, Vandaele et al. (2015) |
| | | 560 ppm | day, Krasnopolsky (2014c) |
| | | 300 ppm | day, Clancy et al. (2012) |
| | 85 | 400 ppm | twilight, Vandaele et al. (2015) |
| | 65–70 | 45 ppm | Moroz (1964), Connes et al. (1968), Irwin et al. (2008), Krasnopolsky (2010c, 2014c), Marcq et al. (2015) |
| | 36–42 | 30 ppm | Gelman et al. (1979), Oyama et al. (1980) |
| | 36 | 27 ppm | Pollack et al. (1993), Taylor et al. (1997), Marcq et al. (2006, 2008), Barstow et al. (2012), Cotton et al. (2012), Arney et al. (2014) |
| | 22 | 20 ppm | Oyama et al. (1980) |
| | 12 | 17 ppm | Gelman et al. (1979) |
| $O_2$ | 62 | $<10^{18}$ cm$^{-2}$ | Trauger and Lunine (1983) |
| $O_3$ | 100 | $5\times10^{7}$ cm$^{-3}$ | night, narrow layer, Montmessin et al. (2011) |
| | 70 | <1 ppb | $\varphi < 50°$, 1–10 ppb at $\varphi > 50°$, Marcq et al. (in preparation) |
| $H_2O$ | 110 | 0.7 ppm | Fedorova et al. (2008) |
| | 75–100 | 1.5 ppm | Sandor and Clancy (2005), Fedorova et al. (2008) |
| | 70 | 3 ppm | Fink et al. (1972), Bjoraker et al. (1992), Fedorova et al. (2008), Krasnopolsky et al. (2013), Cottini et al. (2015) |
| | 62 | 10 ppm | $\varphi < 50°$, Ignatiev et al. (1999) |
| | | 6 ppm | Fedorova et al. (2016) |
| | 56 | 8 ppm | $\varphi > 50°$, Ignatiev et al. (1999) |
| | 0–45 | 30 ppm | Pollack et al. (1993), de Bergh et al. (1995), Ignatiev et al. (1997), Marcq et al. (2008), Haus and Arnold (2010), Tsang et al. (2010), Bezard et al. (2011), Chamberlain et al. (2013), Arney et al. (2014) |
| HCl | 110 | 800 ppb | Mahieux et al. (2015b) |
| | 95 | <150 ppb | Sandor and Clancy (2012) |
| | 74 | 40 ppb | Mahieux et al. (2015b) |
| | 70 | 400 ppb | Connes et al. (1967), Young (1972), Krasnopolsky (2010b), Sandor and Clancy (2012) |

Table 12.6 (*cont.*)

| Gas | *h* (km) | Mixing ratio | References |
|---|---|---|---|
| | 15–30 | 400 ppb | Bezard et al. (1990), Pollack et al. (1993), Iwagami et al. (2008), Arney et al. (2014) |
| HF | 103 | 50 ppb | Mahieux et al. (2015b) |
| | 70 | 4 ppb | Connes et al. (1967), Bjoraker et al. (1992), Krasnopolsky (2010b), Mahieux et al. (2015b) |
| | 30–40 | 4 ppb | Bezard et al. (1990), Pollack et al. (1993) |
| $Br_2$ | 15 | <0.2 ppb | Moroz et al. (1981) |
| HBr | 78 | <1 ppb | Krasnopolsky and Belyaev (2017) |
| | 60 | <30 ppb | Krasnopolsky and Belyaev (2017) |
| NO | 65 | 5.5 ppb | Krasnopolsky (2006a) |
| $NH_3$ | 68 | <6 ppb | Krasnopolsky (2012c) |
| $SO_2$ | 90–105 | 100–700 ppb | Belyaev et al. (2012, 2017) |
| | 85–100 | 23 ppb | Sandor et al. (2010) |
| | 88–110 | 12 ppb | Encrenaz et al. (2015b) |
| | 80 | 10–70 ppb | Belyaev et al. (2012, 2017), Mahieux et al. (2015a) |
| | 70 | 20–1000 ppb | Barker (1979), Esposito et al. (1988, 1997), Marcq et al. (2013), Belyaev et al. (2012, 2017) |
| | 62 | 400 ppb | Zasova et al. (1993) |
| | 57 | 10 ppm | Krasnopolsky (1986) |
| | 54 | 20 ppm | Krasnopolsky (1986) |
| | 50 | 50 ppm | Krasnopolsky (1986) |
| | 52 | 150 ppm | Bertaux et al. (1996) |
| | 22 | 38 ppm | Bertaux et al. (1996) |
| | 12 | 25 ppm | Bertaux et al. (1996) |
| | 40 | 130 ppm | Gelman et al. (1979), Bezard et al. (1993), Bertaux et al. (1996), Marcq et al. (2008) |
| | 22 | 180 ppm | Oyama et al. (1980) |
| SO | 95–105 | 50–700 ppb | Belyaev et al. (2012) |
| | 85–100 | 8 ppb | Sandor et al. (2010), Encrenaz et al. (2015b) |
| OCS | 65 | 1–8 ppb | Krasnopolsky (2010c) |
| | 36 | 0.9 ppm | Pollack et al. (1993), Taylor et al. (1997), Marcq et al. (2006), Arney et al. (2014) |
| | 33 | 3.5 ppm | Pollack et al. (1993), Taylor et al. (1997), Marcq et al. (2006, 2008) |
| | 30 | 13 ppm | Pollack et al. (1993), Taylor et al. (1997), Marcq et al. (2006) |
| $H_2SO_4$ | 85–100 | <3 ppb | Sandor et al. (2012) |
| | 46 | 2–10 ppm | Kolodner and Steffes (1998), Butler et al. (2001), Jenkins et al. (2002) |
| $S_3$ | 15 | 18 ppt | Krasnopolsky (2013b) |
| | 6 | 11 ppt | Krasnopolsky (2013b) |
| $S_4$ | 15 | 6 ppt | Krasnopolsky (2013b) |
| | 6 | <8 ppt | Krasnopolsky (2013b) |
| $H_2S$ | 70 | <70 ppb | Krasnopolsky (2008) |

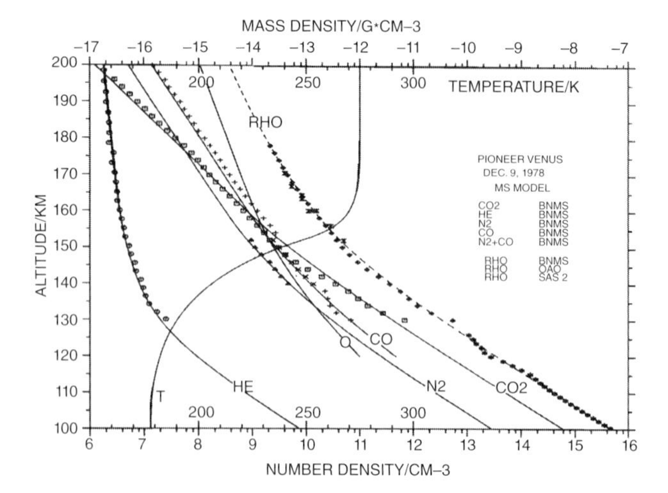


Figure 12.39 Model for the Venus upper atmosphere based on the PV bus neutral mass spectrometer data at high solar activity $F_{10.7}$ = 190 and solar zenith angle of ~60° (von Zahn et al. 1980). SAS 2 is the PV day probe accelerometer.

the altitude of $0.5H$ above the periapsis. A set of these values for various periapsides was fitted by a model that involved densities of $CO_2$, O, and CO + $N_2$ at a reference level and two parameters that specify the temperature profile. The retrieved parameters were used as scaling factors for the orbiter neutral mass spectrometer data to create a reference model for the Venus thermosphere (VIRA; Keating et al. 1985).

### *12.8.3 PV Orbiter Neutral Mass Spectrometer*

The PV orbiter neutral mass spectrometer (Niemann et al. 1980) is the main source of data on the composition and structure of the Venus thermosphere. Compositions of the atmosphere at 150–200 km at the subsolar and antisolar regions are shown in Figure 12.40. The subsolar atmosphere has temperature near 300 K at high solar activity, and atomic oxygen is a dominant species above 160 km. The midnight atmosphere is very cold with $T \approx 120$ K, and its density is smaller than the noon densities by factors of 25 at 150 km and 200 at 200 km.

A full set of the mass spectrometer measurements for the first years is presented by a model of Hedin et al. (1983). The observed densities are scaled by a factor of 1.63 in the model to account for the orbiter drag measurements. Diurnal variability of exospheric temperature over the equator is shown in Figure 12.41. The temperature is rather constant at ≈305 K over the most of the dayside centered at $13^h$. The nighttime minima are of 100 K and at $22^h$ and $3^h$, while the midnight maximum is 133 K.

Diurnal variations of the chemical composition at 150 and 100 km are depicted in Figure 12.42. While the $CO_2$ and O densities at 150 km are comparable in the daytime, O is more abundant than $CO_2$ at night by an order of magnitude. There is a significant asymmetry in the helium distribution that originates a morningside helium bulge. The curves at 150 km closely match the mass spectrometer observations.

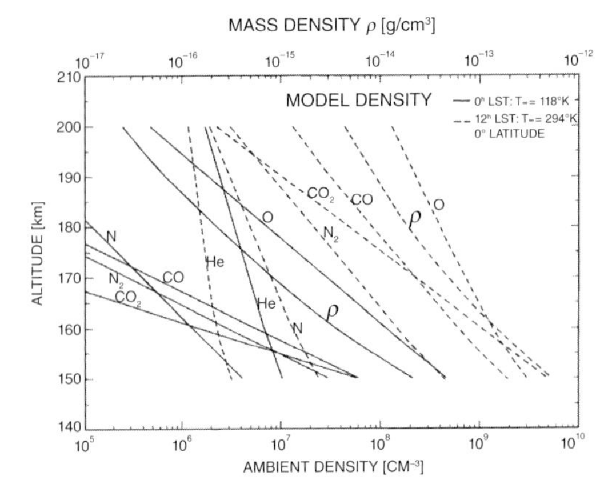


Figure 12.40 Neutral composition of the subsolar and antisolar thermosphere of Venus observed by the PV orbiter neutral mass spectrometer (Niemann et al. 1980). All densities should be scaled by a factor of 1.63 to account for the orbiter drag measurements (Hedin et al. 1983).

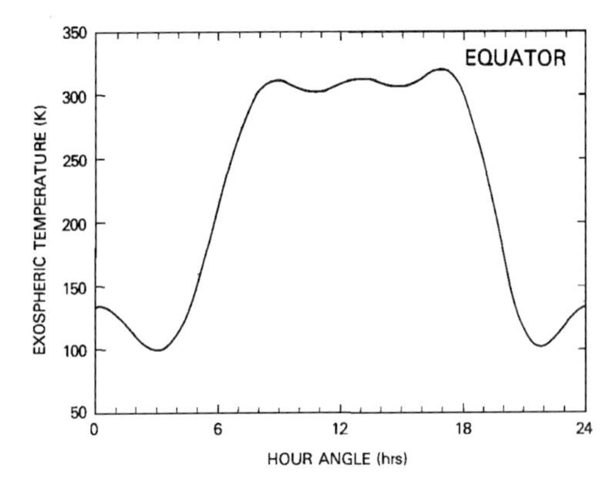


Figure 12.41 Variations of exospheric temperature at the Venus equator with local time based on the PV orbiter neutral mass spectrometer data at high solar activity (Hedin et al. 1983).

The data at 100 km are extrapolations of the model, because only total densities and relevant temperatures from the accelerometers at four probes were available here. For example, the CO observations in the millimeter range (Figures 12.10 and 12.11) demonstrate day-night variation of CO near 100 km by a factor of 4 and even higher, while the data in Figure 12.42 are rather constant.

The model for noon at the equator for $F_{10.7} = 200$ is shown in Table 12.7.

### *12.8.4 Venus Express Observations*

Periapsis of the Venus Express orbiter was at 460 km, that is, above the exobase on Venus, and this ruled out direct measurements of the thermosphere. However, solar and stellar

Table 12.7 *Venus model upper atmosphere for noon at the equator,* $F_{10.7}$ *= 200*

| $z$(km) | $\rho$(gcm$^{-3}$) | $T(k)$ | $N$(cm$^{-3}$) | $CO_2$(cm$^{-3}$) | O(cm$^{-3}$) | CO(cm$^{-3}$) | He(cm$^{-3}$) | $N_2$(cm$^{-3}$) | $\bar{M}$ |
|---|---|---|---|---|---|---|---|---|---|
| 100. | 5.50E–08 | 177. | 3.57E 03 | 7.33E 14 | 6.90E 10 | 1.16E 11 | 1.60E 09 | 3.10E 13 | 43.3 |
| 105. | 1.54E–08 | 179. | 4.40E 04 | 2.05E 14 | 4.99E 10 | 8.49E 10 | 6.41E 08 | 9.23E 12 | 43.3 |
| 110. | 4.37E–09 | 181. | 5.82E 05 | 5.79E 13 | 3.82E 10 | 6.39E 10 | 2.80E 08 | 2.82E 12 | 43.2 |
| 115. | 1.26E–09 | 185. | 8.25E 06 | 1.67E 13 | 3.07E 10 | 4.89E 10 | 1.34E 08 | 8.94E 11 | 43.1 |
| 120. | 3.74E–10 | 190. | 1.18E 08 | 4.90E 12 | 2.57E 10 | 3.75E 10 | 7.04E 07 | 2.95E 11 | 42.8 |
| 125. | 1.15E–10 | 197. | 6.78E 08 | 1.48E 12 | 2.17E 10 | 2.83E 10 | 4.03E 07 | 1.03E 11 | 42.3 |
| 130. | 3.72E–11 | 207. | 5.45E 08 | 4.65E 11 | 1.80E 10 | 2.04E 10 | 2.50E 07 | 3.76E 10 | 41.3 |
| 135. | 1.29E–11 | 221. | 3.16E 08 | 1.53E 11 | 1.43E 10 | 1.38E 10 | 1.67E 07 | 1.46E 10 | 39.6 |
| 140. | 4.85E–12 | 240. | 1.89E 08 | 5.31E 10 | 1.07E 10 | 8.73E 09 | 1.18E 07 | 6.09E 09 | 37.1 |
| 145. | 2.05E–12 | 259. | 1.19E 08 | 2.01E 10 | 7.70E 09 | 5.34E 09 | 8.91E 06 | 2.77E 09 | 34.3 |
| 150. | 9.68E–13 | 273. | 8.06E 07 | 8.27E 09 | 5.54E 09 | 3.25E 09 | 7.08E 06 | 1.37E 09 | 31.5 |
| 155. | 4.94E–13 | 283. | 5.68E 07 | 3.58E 09 | 3.98E 09 | 1.97E 09 | 5.84E 06 | 7.18E 08 | 28.8 |
| 160. | 2.69E–13 | 290. | 4.12E 07 | 1.61E 09 | 2.87E 09 | 1.20E 09 | 4.94E 06 | 3.92E 08 | 26.5 |
| 165. | 1.55E–13 | 295. | 3.05E 07 | 7.40E 08 | 2.09E 09 | 7.34E 08 | 4.26E 06 | 2.21E 08 | 24.4 |
| 170. | 9.34E–14 | 299. | 2.29E 07 | 3.47E 08 | 1.53E 09 | 4.52E 08 | 3.74E 06 | 1.27E 08 | 22.7 |
| 175. | 5.89E–14 | 301. | 1.75E 07 | 1.64E 08 | 1.13E 09 | 2.80E 08 | 3.31E 06 | 7.46E 07 | 21.3 |
| 180. | 3.86E–14 | 303. | 1.34E 07 | 7.85E 07 | 8.42E 08 | 1.74E 08 | 2.96E 06 | 4.43E 07 | 20.1 |
| 185. | 2.61E–14 | 305. | 1.04E 07 | 3.78E 07 | 6.31E 08 | 1.09E 08 | 2.67E 06 | 2.67E 07 | 19.2 |
| 190. | 1.81E–14 | 306. | 8.08E 06 | 1.83E 07 | 4.75E 08 | 6.83E 07 | 2.43E 06 | 1.62E 07 | 18.5 |
| 195. | 1.28E–14 | 306. | 6.32E 06 | 8.87E 06 | 3.60E 08 | 4.30E 07 | 2.21E 06 | 9.91E 06 | 18.0 |
| 200. | 9.25E–15 | 307. | 4.96E 06 | 4.32E 06 | 2.73E 08 | 2.71E 07 | 2.03E 06 | 6.11E 06 | 17.5 |
| 205. | 6.76E–15 | 307. | 3.90E 06 | 2.11E 06 | 2.08E 08 | 1.72E 07 | 1.86E 06 | 3.78E 06 | 17.2 |
| 210. | 4.99E–15 | 307. | 3.08E 06 | 1.03E 06 | 1.59E 08 | 1.09E 07 | 1.72E 06 | 2.35E 06 | 16.9 |
| 215. | 3.72E–15 | 308. | 2.43E 06 | 5.06E 05 | 1.22E 08 | 6.89E 06 | 1.59E 06 | 1.47E 06 | 16.7 |
| 220. | 2.80E–15 | 308. | 1.93E 06 | 2.49E 05 | 9.33E 07 | 4.38E 06 | 1.47E 06 | 9.22E 05 | 16.5 |
| 225. | 2.11E–15 | 308. | 1.53E 06 | 1.22E 05 | 7.17E 07 | 2.78E 06 | 1.37E 06 | 5.80E 05 | 16.3 |

| | | | | | | | | | |
|---|---|---|---|---|---|---|---|---|---|
| 230. | 1.60*E*–15 | 308. | 1.22*E* 06 | 6.02*E* 04 | 5.51*E* 07 | 1.77*E* 06 | 1.27*E* 06 | 3.66*E* 05 | 16.2 |
| 235. | 1.22*E*–15 | 308. | 9.66*E* 05 | 2.97*E* 04 | 4.24*E* 07 | 1.13*E* 06 | 1.19*E* 06 | 2.31*E* 05 | 16.0 |
| 240. | 9.36*E*–16 | 308. | 7.69*E* 05 | 1.46*E* 04 | 3.27*E* 07 | 7.20*E* 05 | 1.11*E* 06 | 1.46*E* 05 | 15.9 |
| 245. | 7.18*E*–16 | 308. | 6.13*E* 05 | 7.24*E* 03 | 2.53*E* 07 | 4.59*E* 05 | 1.03*E* 06 | 9.29*E* 04 | 15.8 |
| 250. | 5.52*E*–16 | 308. | 4.89*E* 05 | 3.58*E* 03 | 1.95*E* 07 | 2.93*E* 05 | 9.65*E* 05 | 5.91*E* 04 | 15.6 |

*Note*. From Hedin et al. (1983).

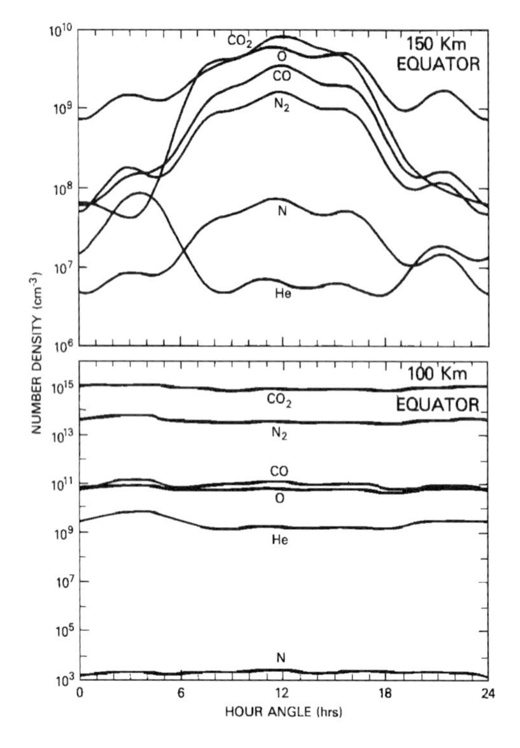


Figure 12.42 Diurnal variations of the atmospheric compositions at the equator at 150 and 100 km. The data at 150 km closely match the PV orbiter neutral mass spectrometer observations corrected by a factor of 1.63 to account for the PV orbiter drag observations. The data at 100 km are the model extrapolation. From Hedin et al. (1983).

occultations of the $CO_2$ absorption in the UV and IR ranges gave data on densities and temperatures at the altitudes of 70 to 150 km.

Detailed study of the nightside temperatures between 90 and 140 km was made by the SPICAV stellar occultations using the $CO_2$ absorption at 150–200 nm (Figure 12.43; Piccalli et al. 2015). The numerous retrieved temperature profiles are presented as temperature fields $T(z$, latitude) and $T(z$, local time), maps of $T$ at fixed altitudes, and mean temperature profiles for various combinations of latitude and local time. Despite the great variability, it looks like mean temperatures at the most of the nightside, say, at the solar zenith angle $\theta > 120^\circ$ are rather similar, and the density and temperature profiles in Figure 12.43 are typical.

Here the major problem is a great difference between the mean densities and temperatures at 90–100 km and those from VIRA. For example, $[CO_2] = 5.5\times10^{15}$ cm$^{-3}$ and $T = 215$ K at 90 km in Figure 12.43, being $1.53\times10^{16}$ cm$^{-3}$ and 172 K, respectively, in VIRA (Table 12.2). VIRA does not account for the local time variability. However, the data from the PV night probe accelerometer are similar to those from the other PV and V11–14 probes at 90–100 km. Temperature profiles from the V15 IR spectra extend up to 100 km and cover

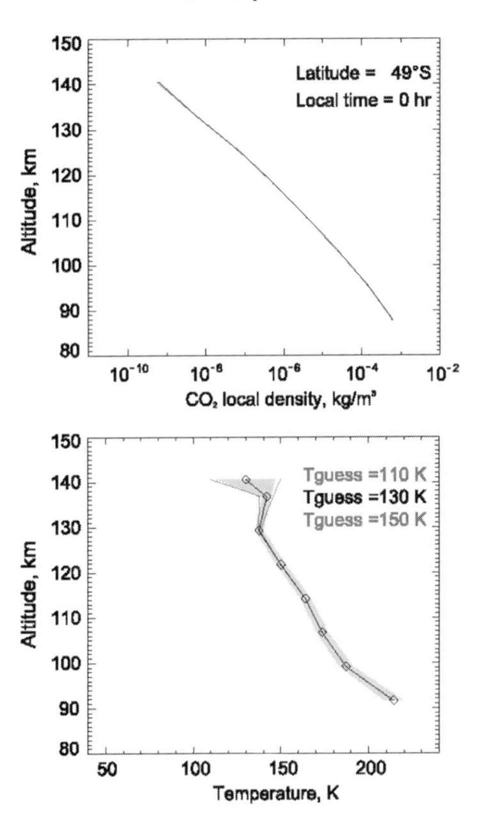


Figure 12.43 Typical density and temperature profiles in the nighttime atmosphere extracted from the SPICAV/VEX stellar occultations (Piccalli et al. 2015).

both dayside and nightside with differences of ~7% in the densities and ~4 K in the temperatures at 90 km (Zasova et al. 2006). Furthermore, rotational temperatures of the $O_2$ nightglow are ~185 K at 95 km (see below).

The UV absorption responsible for the retrieval near 90 km is at 190–200 nm, where the true $CO_2$ absorption is small and comparable with the Rayleigh extinction and absorption by other gases ($SO_2$, SO, $O_2$). The $CO_2$ absorption in this range may be a combination of continuum and bands, and the absorption by bands does not match the exponential law adopted for the occultation analysis.

SOIR/VEX observed 132 solar occultation events to retrieve $CO_2$ density and temperature profiles at the morning and evening terminator were discussed by Mahieux et al. (2015c) and Bougher et al. (2015b). Mean temperature profiles at low latitudes are shown in Figure 12.44, where they are compared with those predicted by the Venus thermosphere general circulation model (VTGCM). There are some similarities and differences between the profiles. Both the observations and the model show a wavelike structure with two minima near 85 and 125 km. A mean temperature of ~170 K at the whole range of 80 to

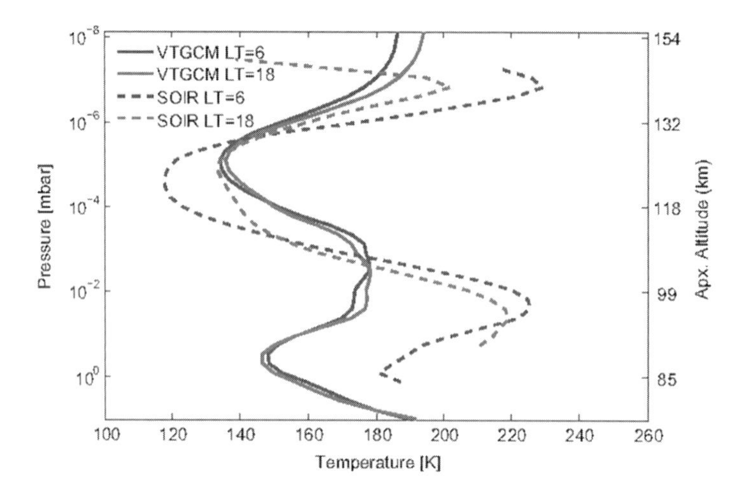


Figure 12.44 Mean SOIR sunrise and sunset temperature profiles at low latitudes ($\varphi < 30°$) are compared with the VTGCM profiles (Bougher et al. 2015b). (A black and white version of this figure will appear in some formats. For the color version, please refer to the plate section.)

150 km is similar for the observations and the model as well. However, the observed amplitude exceeds that in the model by a factor of 2–3, and differences in the temperatures reach 40 K.

Rotational temperatures from $CO_2$ line distributions observed in the SOIR/VEX occultations were derived by Mahieux et al. (2015d). However, uncertainties of the temperatures are typically 50 K and even higher, and differences between the rotational temperatures and those retrieved from the observed $CO_2$ density distributions are significant.

### *12.8.5 Venus Thermosphere General Circulation Model (VTGCM)*

This is a three-dimensional general circulation model for the Venus upper atmosphere. The model is supported by Bougher et al. for the last three decades with gradual improvements. The latest version of the model starts at 70 km and extends to the exobase. Some versions include restricted data on the chemical composition. Hereafter we will not consider physics of VTGCM but use its results for comparison with observational data and other models. The model involves two major components of the atmospheric circulation: subsolar to antisolar and the zonal superrotation.

Thermal structure of the atmosphere near the equator at solar minimum is shown in Figure 12.45. The main new feature is a warm region centered at the subsolar point at 115 km. It is caused by the near-IR heating, mostly in the $CO_2$ band at 4.3 μm. Temperature at 90–100 km is rather low and does not significantly increase at night, so that the model does not support high nighttime temperatures up to 250 K in some SPICAV stellar occultation data for these altitudes.

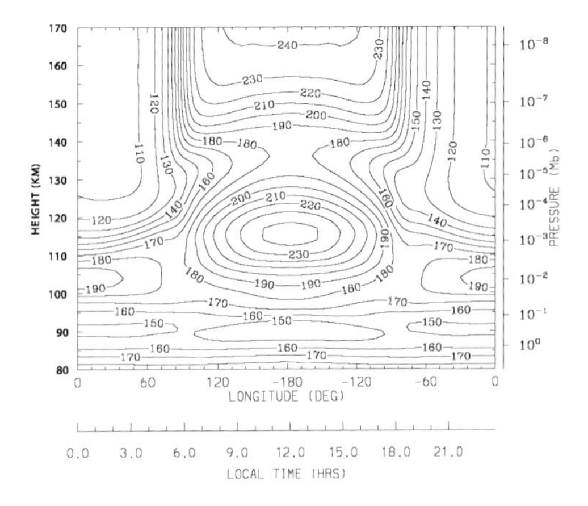


Figure 12.45 VTGCM: thermal structure of the atmosphere near the equator at solar minimum ($F_{10.7} = 70$). From Brecht and Bougher (2012).

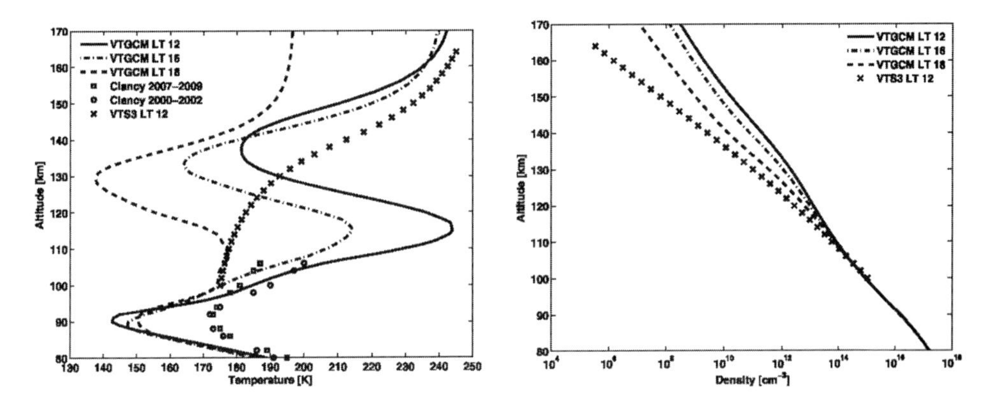


Figure 12.46 VTGCM: temperature (a) and $CO_2$ density (b) profiles at the equator at solar minimum ($F_{10.7} = 70$) at LT 12, 16, and $18^h$. Data from Clancy et al. (2012a) for low and high solar activity and from Hedin et al. (1983, VTS3 at solar minimum) are shown as well. From Brecht and Bougher (2012).

The model temperature profiles at noon, $16^h$, and sunset at the equator and solar minimum are compared with the profile from Hedin et al. (1983) for solar minimum and those from Clancy et al. (2012a; see Figure 12.10) in Figure 12.46a. $CO_2$ densities from these models are compared in Figure 12.46b. The $CO_2$ density at 160 km from VTGCM LT 12 exceeds that from Hedin et al. (1983) by a factor of 1000. The model by Hedin et al. (1983) reflects the PV orbiter neutral mass spectrometer data corrected for the PV orbiter drag observations that remain to be the most reliable source of data for the Venus thermosphere at 150–250 km. However, the PV data refer to high solar activity, while the data from Hedin et al. (1983) at low solar activity are based on combination of those for high activity and modeling.

### 12.8.6 Variations with Solar Activity

Both empirical models by Hedin et al. (1983) and Keating et al. (1985) give exospheric temperatures of 310 K and 245 K for $F_{10.7}$ = 200 and 70, respectively, at the subsolar region. VTGCM (Bougher et al. 1999) gave 310, 270, and 230 K for the subsolar region at $F_{10.7}$ = 200, 130, and 70, respectively.

Thermal balance in the thermosphere (Section 2.4.2) results in a temperature profile (2.35). If the thermospheric heating is proportional to the solar activity index $F_{10.7}$, then this relationship may be presented as

$$T_\infty^{s+1} = T_0^{s+1} + aF_{10.7}. \tag{12.3}$$

Using $T_\infty$ = 310 K at $F_{10.7}$ = 200, $T_0 \approx$ 180 K near 100 km, and thermal conductivity index $s$ = 1.38 for $CO_2$ (Table 2.3),

$$T_\infty = \left(2.3 \times 10^5 + 3100F_{10.7}\right)^{0.42}, \tag{12.4}$$

that is, $T_\infty$ = 274 and 237 K at $F_{10.7}$ = 130 and 70, respectively. Exospheric temperatures vary insignificantly over most of the dayside (Figures 12.41 and 12.46).

## 12.9 Ionosphere

Physics of the Venus ionosphere includes numerous problems related to interaction with the solar wind, configuration and variations of the magnetic field, thermal balance of plasma, etc., that are beyond the scope of this book. Here we will mostly consider observations and models of ion composition, sources of ionization, and their variations.

### 12.9.1 Radio Occultations

The ionosphere of Venus was studied by radio occultations from all flybys (Mariner 5 and 10) and orbiters (Venera 9 and 10, Pioneer Venus, Venera 15 and 16, Magellan, and Venus Express). Venera 9 and 10 were the first Venus orbiters; their observations were at solar minimum (Figure 12.47). The measured peak electron densities and their altitudes may be approximated by the Chapman layer, where a large set of the solar EUV emissions and the $CO_2$ ionization cross sections for those emissions are approximated by their sum $I$ and an effective mean cross section $\sigma$:

$$I\sigma n e^{-\sigma nH\sec\theta} = \alpha n_e^2; \quad n_e = \left(\frac{I\sigma n}{\alpha}\right)^{1/2} e^{-\frac{1}{2}\sigma nH\sec\theta}. \tag{12.5}$$

Here $\alpha$ is the effective mean recombination coefficient (Section 9.3.5) that is typically $\approx 3\times10^{-7}$ cm$^3$ s$^{-1}$ near the ionospheric peak, $n$ is the neutral number density, $n_e$ is the electron number density, and $\theta$ is the solar zenith angle (SZA). The left term is the ionization rate, the right term is the loss by recombination. The peak is at $\sigma nH\sec\theta = 1$. Comparing conditions at $\theta = 0°$ and 60°, the peak electron density should be smaller at 60° by a factor of $2^{1/2}$, and the peak altitude is higher by $H$ ln 2 = 0.7 $H$, in accord with the

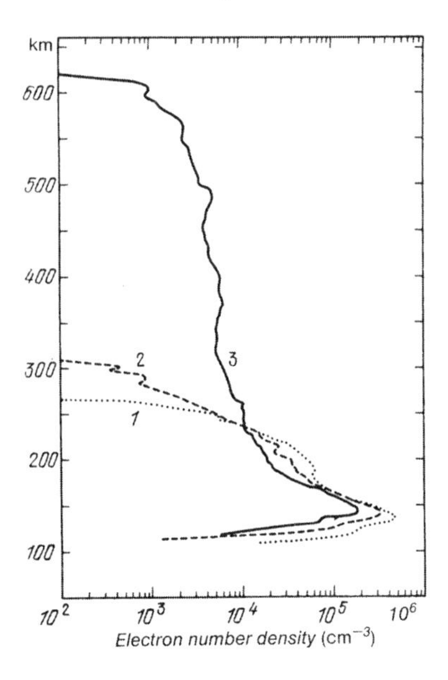


Figure 12.47 Dayside ionosphere of Venus measured by the V9–10 orbiters at solar minimum and SZA = 14°, 63°, and 83° (curves 1, 2, 3, respectively). From Ivanov-Kholodny et al. (1979).

profiles at 14° and 63° in Figure 12.47. The plane-parallel approximation, sec $\theta$, is inapplicable at $\theta > 75°$, and the Chapman function $Ch\ \theta$ replaces sec $\theta$. The effective absorption cross section of $CO_2$ is $\sigma \approx 3\times10^{-17}$ cm$^2$ in the EUV range (Figure 8.8), and the peak altitude is at a level where $N \sec \theta = 1/\sigma \approx 3\times10^{16}$ cm$^{-2}$. The shoulder below the main peak by ~20 km in Figure 12.47 is due to ionization by the solar soft X-rays.

The most impressive feature is a significant increase of the ionopause altitude with solar zenith angle. Ionopause is a level where the solar wind pressure is equal to the ion pressure, and the ion densities are strongly depleted above the ionopause. According to Figure 12.47, the ionopause at the solar minimum moves from ~250 km at the subsolar point to ~650 km near the terminator.

Mean electron density profiles observed by the PV radio occultations at the same SZA = 60°–70° in 1980 at solar maximum and in 1986 at solar minimum are compared in Figure 12.48. The profile for solar minimum is similar to that in the V9–10 observations at the same SZA (Figure 12.47). The difference in the peak electron densities in Figure 12.48 by a factor of 1.5 reflects the difference in the solar EUV by a factor of ~2.5. The increasing difference in the electron densities (by a factor of 20 at 400 km) is caused by the lower solar EUV and the atmospheric density at solar minimum.

### *12.9.2 V9–10 Data on the Nightside Ionosphere*

The V9–10 results remain of some interest because they refer to solar minimum, while the more detailed and extensive data from the PV orbiter reflect mostly the solar maximum conditions.

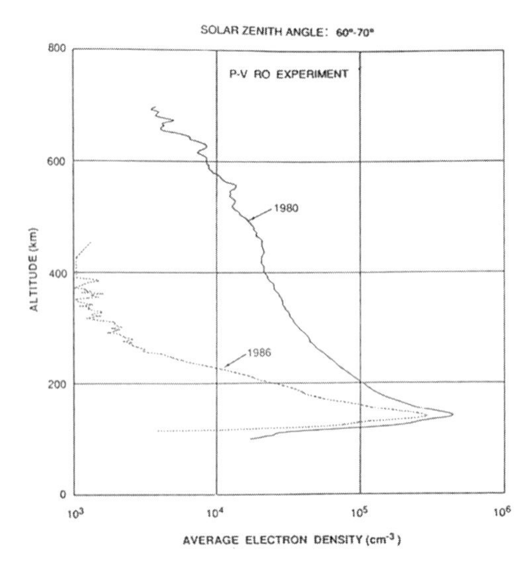


Figure 12.48 Average of 6 and 10 electron density profiles observed by the PV radio occultations at solar maximum in 1980 and minimum in 1986 at SZA = 60°–70°. From Kliore and Luhmann (1991).

The numerous observed nightside profiles of electron density are very variable. Sometimes they have a two-peak structure with the peaks near 140 and 115 km. The V9–10 observations in the Venus shadow revealed energetic electrons that may be described by a two-component Maxwellian distribution with $n_e$ = 1.5 cm$^{-3}$, $kT$ = 13 eV and $n_e$ = 0.2 cm$^{-3}$, $kT$ = 80 eV (Gringauz et al. 1979). These electrons might be responsible for the nighttime ionization peak near 140 km with a peak density of $\approx 10^4$ cm$^{-3}$. On the other hand, no nightglow had been detected at the ionospheric altitudes with an upper limit of 4 R (rayleighs) for the oxygen green line at 558 nm (Krasnopolsky 1979).

The $O(^1S–^1D)$ line at 558 nm is excited by electron impact of O and $CO_2$ and dissociative recombination of $O_2^+$. The upper limit agrees with the observed electron fluxes, if they entry the nighttime atmosphere at low magnetic field inclination $D < 18°$. (Inclination is the angle between the magnetic field and the horizontal plane, and ionization by the observed soft electrons is proportional to sin $D$.) Another restriction is to the $He^+$ flow from the dayside with an upper limit of $10^7$ cm$^{-2}$ s$^{-1}$ because of the process

$$He^+ + CO_2 \rightarrow CO^+ + O(^1S) + He.$$

This flow was considered earlier as a source of the nightside ionization.

The sporadic peak near 115 km was assigned to the meteorite flux with a mean total ionization rate of $2\times10^5$ cm$^{-2}$ s$^{-1}$ and a maximum at 115–120 km. The meteorite atoms Mg, Fe, Si, etc., have low ionization potentials and can be ionized by the middle and far UV photons and in charge exchange with other ions. Radiative recombination of the atomic ions is very slow with $k \approx 3\times10^{-12}$ cm$^3$ s$^{-1}$, and the ion attachment and detachment with $CO_2$ with the subsequent dissociative recombination explain the observed electron densities of $\approx 5\times10^3$ cm$^{-3}$ at the lower ionization maximum (Krasnopolsky 1979):

$$A^+ + CO_2 \rightarrow ACO_2{}^+ + h\nu; \qquad k_1 = 3\times10^{-17} cm^3\ s^{-1};$$
$$A^+ + CO_2 + CO_2 \rightarrow ACO_2{}^+ + CO_2; \qquad k_2 = 2\times10^{-29} cm^6\ s^{-1};$$
$$ACO_2{}^+ + CO_2 \rightarrow A^+ + CO_2 + CO_2; \qquad k_3 = 10^{-14} cm^3\ s^{-1};$$
$$ACO_2{}^+ + e \rightarrow A + CO_2; \qquad k_4 = 10^{-6} cm^3\ s^{-1}.$$

The rate coefficients are taken for A = Na. Sporadic layers at 105–115 km were observed by the Venus Express radio occultations even on the dayside (Paetzold et al. 2009).

Krasnopolsky (1979) estimated the He 584 Å interplanetary background at 10 R near Venus using the Mariner 10 and V9–10 EUV observations and concluded that a nighttime ionization peak from this emission is ~$3\times10^3$ cm$^{-3}$ at 140 km. Therefore this emission contributes to the nighttime ionosphere.

### *12.9.3 Dayside Ionosphere in the PV Observations*

The PV orbiter with a low periapsis down to 150 km had obvious advantages of the in situ studies of the ionosphere using the ion mass spectrometer, the retarding potential analyzer, and the electron temperature probe as well as radio occultations and measurements of the solar wind-magnetosphere interactions.

Composition of the ionosphere at the subsolar region from the PV ion mass spectrometer observations (Taylor et al. 1980) is shown in Figure 12.49. Vertical profiles of twelve ions are presented: $O_2{}^+$, $CO_2{}^+$, $NO^+$, $CO^+ + N_2{}^+$, $^{16}O^+$, $^{18}O^+$, $O^{++}$, $C^+$, $N^+$, $He^+$, $H^+$, and $D^+$.

A summary of the observations by the PV electron temperature probe is shown in Figure 12.50. Here the data reflect variations of vertical profiles of $T_e$ and $n_e$ with solar zenith angle. Variations of the ionopause are shown as well.

Mean neutral, ion, and electron temperatures measured by the PV instruments at SZA from 90° to 125° are depicted in Figure 12.51. The dayside temperatures are rather similar to those shown for 90°.

The daytime ionospheric chemistry reminds that on Mars. For example, $C^+$ is formed by dissociative photo- and photoelectron impact ionization of $CO_2$ and CO and lost in the reaction $C^+ + CO_2 \rightarrow CO^+ + CO$. $O_2{}^+$ is produced in the reactions of $CO_2{}^+ + O$ and $O^+ + CO_2$ and removed by dissociative recombination. Some $O_2{}^+$ ions react with NO and N to form $NO^+$ that then recombines. A model for the daytime ionospheric composition at solar maximum and SZA = 60° is shown in Figure 12.52. This SZA reflects the dayside mean conditions. The ion outflow velocity of $2\times10^4$ cm s$^{-1}$ at the upper boundary simulates the ion transport from the dayside to the nightside. Taking into account the difference in the solar zenith angles, there is a good agreement between the observed and calculated ionospheric compositions in Figures 12.49 and 12.52.

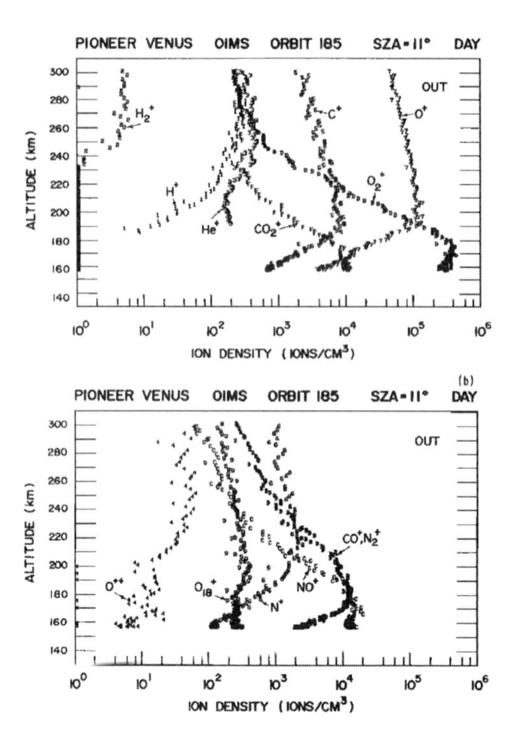


Figure 12.49 Composition of the Venus ionosphere in the subsolar region (SZA = 11°) at high solar activity observed using the PV orbiter ion mass spectrometer. Ion mass 2 was later ascribed to $D^+$. From Taylor et al. (1980).

### *12.9.4 Nighttime Ionosphere in the PV Observations and Models*

A variety of ionospheric phenomena was discovered on the Venus nightside. The ionosphere was "established" in some cases (Figure 12.53) with significant ion densities that remind the dayside ionosphere. In some periods the ionosphere is strongly depleted over the whole hemisphere (Figure 12.53), and there are transition cases between these extremes. The depleted ionosphere may be also localized in the so-called ionospheric holes.

The established nighttime ionosphere (Figure 12.54) demonstrates high densities of $H^+$, especially in the morningside sector. This is due to the morningside hydrogen bulge with densities of H exceeding the daytime values by two orders of magnitude. The PV radio occultations gave the mean nighttime peak electron density of $(1.7 \pm 0.7)\times 10^4$ cm$^{-3}$ and the peak altitude of $142 \pm 4$ km.

The most important source of ionization of the nighttime ionosphere is transport of the $O^+$ ions from the dayside. The ion flow velocities were measured using the retarding potential analyzer (Knudsen et al. 1980) with values $V \approx 2$ km s$^{-1}$ near the terminator. The product $[O^+]\, V$ is almost constant at $2\times 10^9$ cm$^{-2}$ s$^{-1}$ from 150 to 650 km at the terminator. Then the mean downward flux of $O^+$ on the nightside is

$$\Phi_{O+} = [O^+]V\frac{2\pi r\Delta r}{2\pi r^2} \approx 1.5 \times 10^8 \text{ cm}^{-2}\text{ s}^{-1}. \tag{12.6}$$

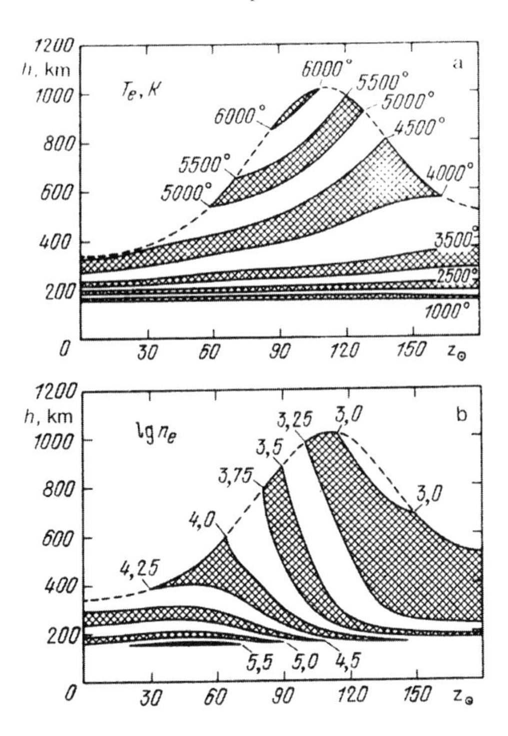


Figure 12.50 Empirical model for the observations of electron temperature and density using the PV electron temperature probe. Solar zenith angle is $Z_O$. From Theis et al. (1980).

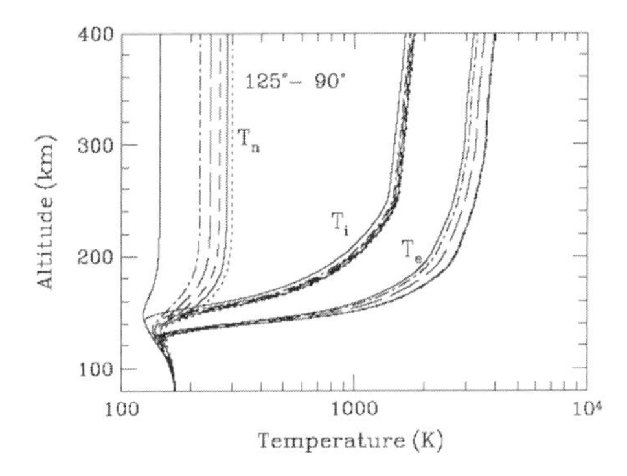


Figure 12.51 Neutral, ion, and electron temperatures at solar maximum, SZA = 125°–90° with intervals of 5° (Fox 2011).

Here $r$ is the planet radius and $\Delta r \approx 500$ km is the thickness of the ionosphere near the terminator. According to Fox (2011), column ionization rates from the ionospheric data are $1.4\times10^8$, $1.45\times10^8$, and $5.9\times10^7$ cm$^{-2}$ s$^{-1}$ for SZA = 105°, 115°, and 125°, respectively, with a mean value of $1.2\times10^8$ cm$^{-2}$ s$^{-1}$ for SZA = 90°–125°.

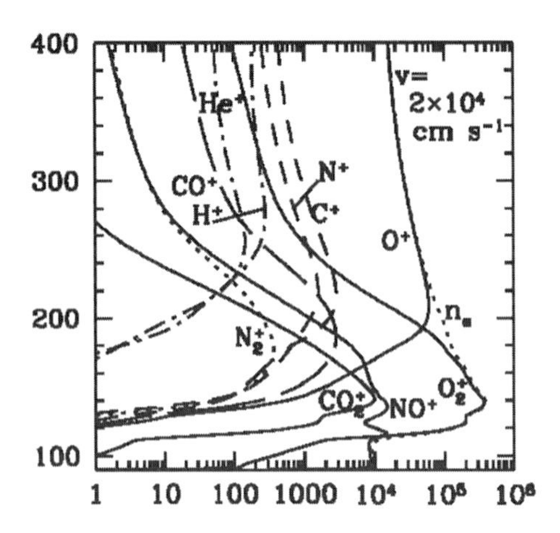


Figure 12.52 Calculated composition of the daytime ionosphere at solar maximum and SZA = 60°. Ion outflow velocity is $2\times10^4$ cm $s^{-1}$. The abscissa is number density in $cm^{-3}$, the ordinate is altitude in km. From Fox (2008).

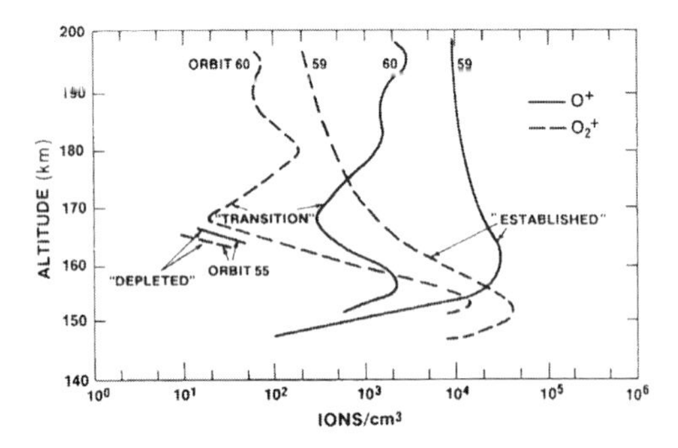


Figure 12.53 Altitude profiles of $O^+$ and $O_2^+$ (solid and dashed lines, respectively) for the established, transition, and depleted nightside ionospheres (orbits 59, 60, and 55, respectively). From Taylor et al. (1980).

The transport time for $O^+$ from the terminator to the nightside is ~1 h. The $O^+$ lifetime in the reaction with $CO_2$ is equal to the transport time at $[CO_2] = 3\times10^5$ $cm^{-3}$, that is, at 165 km. Therefore the loss of $O^+$ is small during its transportation above 165 km.

The V9–10 observations of the energetic electrons in the nighttime atmosphere were confirmed by the PV measurements, and relative roles of those electrons and the $O^+$ flows were studied by Spenner et al. (1996). They concluded that plasma transport from the dayside ionosphere is the main nighttime ionization source at high solar activity but becomes weaker at medium activity by a factor of 6 and comparable with the ionization by suprathermal electrons that dominate at low solar activity. Calculated profiles of the major ions for the plasma transport mechanism at solar maximum and SZA = 95° and 125° are shown in Figure 12.55 (Fox 2011). Both $O^+$ fluxes and the energetic electrons are

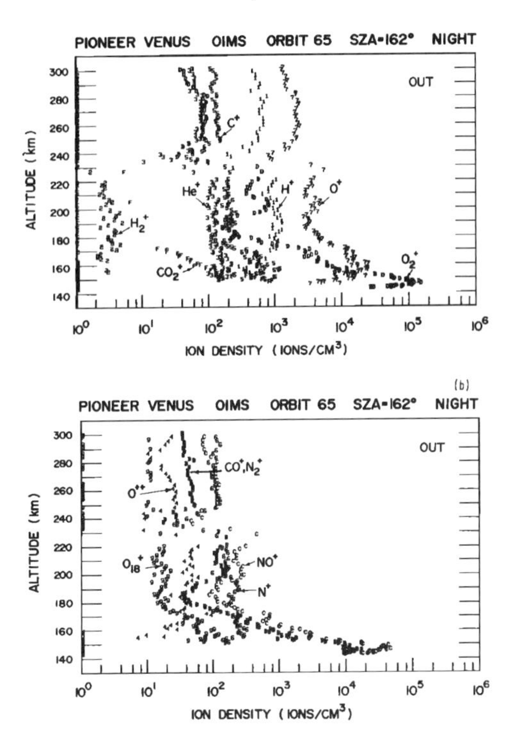


Figure 12.54 The "established" nightside ionosphere at SZA = 162° (Taylor et al. 1980). $H_2^+$ are actually $D^+$ ions.

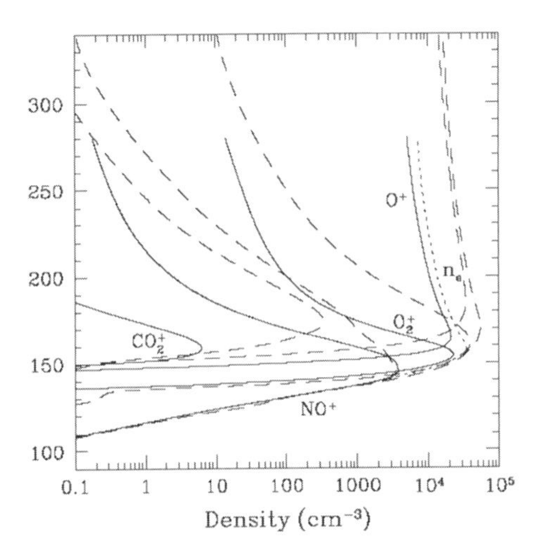


Figure 12.55 Calculated densities of the major ions in the Venus nighttime ionosphere induced by the transport of plasma at solar maximum, SZA = 125° and 95° (solid and dashed lines, respectively). From Fox (2011).

affected by some magnetospheric phenomena that result in high variability of the nighttime ionosphere. The calculated peak electron density altitude for the plasma transport mechanism is higher by ~10 km than the mean altitude of 142 km in the radio occultations.

## 12.10 Night Airglow

Night airglow is excited by chemiluminescent reactions of atoms and radicals (Section 8.7). These species may be formed in the nighttime, or remain in the atmosphere after sunset, or are transported from the dayside. Therefore nightglow studies are of both intrinsic interest and as diagnostics of the related species, atmospheric dynamics, and excitation processes.

### *12.10.1 Visible Nightglow*

The Venus nightglow was discovered in the visible range using the V9–10 orbiters (Krasnopolsky et al. 1976; Krasnopolsky 1983a) and studied in the laboratory by Lawrence et al. (1977), Slanger and Black (1978), and Kenner et al. (1979). The observed spectrum (Figure 12.56) consists of the prominent $O_2$ Herzberg II bands and traces of the Chamberlain and Herzberg I bands. Mean nightside intensities of the band systems are 2700, 200, and 140 R, respectively. All observed band systems are only presented by their (0–v″) progressions.

Later the HzII (0–10) band at 551 nm was observed using ground-based high-resolution spectrographs (Slanger et al. 2006), and the measured mean intensity of the HzII band system was evaluated at 3.1 kR. Finally, averaging of 420 hours of the nighttime limb observations using VIRTIS-M at Venus Express (García Muñoz et al. 2009a; Migliorini et al. 2013) resulted in a mean HzII limb peak intensity of 180 kR. Conversion to the vertical intensity gives 2.6 kR.

The ground-based observations by Slanger et al. (2006) revealed an event of significant nightglow of the green line O 558 nm with intensity of 170 R. Their other observations

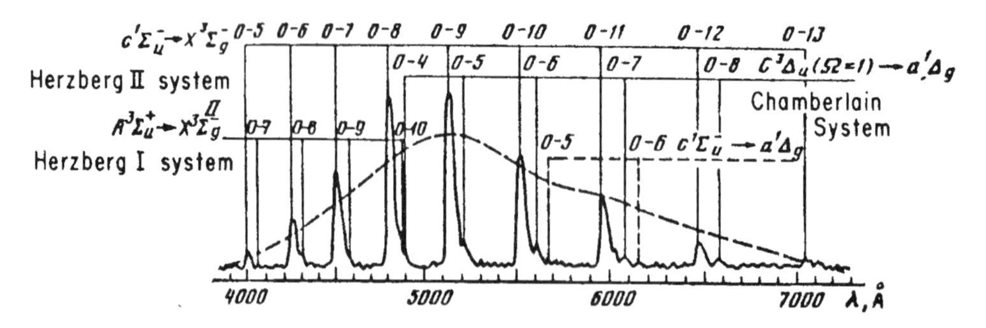


Figure 12.56 Spectrum of the Venus nightglow in the visible range. $O_2$ bands of the Herzberg II system are prominent, and those of the Chamberlain and Herzberg I systems are weak shoulders. All band systems are presented only by the (0–v″) progressions. The spectrometer sensitivity is shown by the dashed line. From Krasnopolsky (1983a).

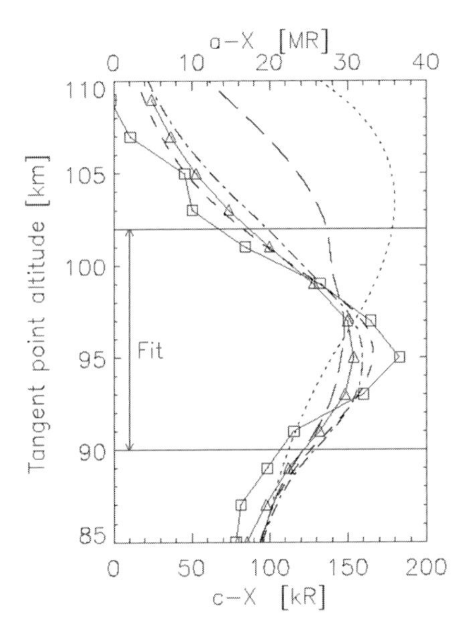


Figure 12.57 Mean limb intensities of the $O_2$ (c→X) Herzberg II band system (squares) and the (a→X) band at 1.27 μm (triangles). Other curves are fits of the observations by models. From García Muñoz et al. (2009a).

including that of 89 R at the limb correspond to the vertical intensities of ≤15 R. Furthermore, the observed intensities had not been corrected for the cloud reflection and the observing angle and therefore should be reduced by a factor of ≈2. Some detections of the visible nightglow were reported by Gray et al. (2014). No emission of the green line has been detected by the V9–10 and Venus Express orbiters. The V9–10 upper limits to the green and red O lines at 558 and 630/636 nm are 10–15 and 20–25 R, respectively.

A mean limb profile of the $O_2$ (c→X) system intensity is shown in Figure 12.57. The measured peak is at 95 km. According to Migliorini et al. (2013), a mean profile of the Chamberlain bands is higher than that of the HzII band by 4 km.

The observed nightglow band systems are similar to those on the Earth, because both nightglows are excited by the termolecular association of $O_2$. On the other hand, they are very different, because the Herzberg I and Chamberlain band systems dominate in the Earth's nightglow being rather weak on Venus. Furthermore, the terrestrial nightglow shows a developed vibrational structure, while the zeroth progressions are only seen on Venus.

### *12.10.2 $O_2$ Nightglow at 1.27 μm*

The Venus nightglow of the $O_2$ infrared atmospheric band system ($a^1\Delta_g \rightarrow X^3\Sigma_g^-$) was discovered by Connes et al. (1979) using the ground-based high-resolution spectroscopy. This nightglow is rather bright with a mean intensity of ~0.5 MR and almost completely

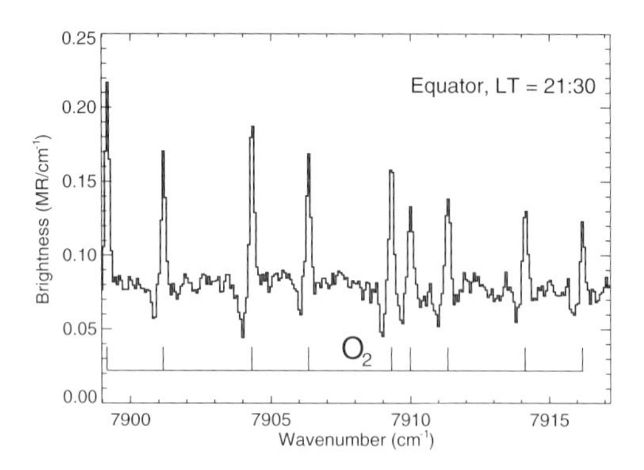


Figure 12.58 Spectrum of the $O_2$ nightglow at 1.27 μm observed at the equator using IRTF/CSHELL. Nine emission lines are prominent in the spectrum. From Krasnopolsky (2010a).

concentrated in the (0–0) band at 1.27 μm. Therefore this nightglow is a convenient tracer of transport of atomic oxygen from the dayside.

The $O_2$ nightglow at 1.27 μm (Figure 12.58) was observed from the ground by Crisp et al. (1996), Ohtsuki et al. (2008), Bailey et al. (2008), and Krasnopolsky (2010a). Detailed study of the nightglow distribution and variations was made in a number of papers using VIRTIS-M at Venus Express (see Piccioni et al. 2009; Soret et al. 2012a, 2015 and references therein).

High spectral resolution in the ground-based observations of the $O_2$ ($a \rightarrow$ X) nightglow made it possible to retrieve rotational temperatures of the nightglow. All five teams got temperatures of 185–193 K with a mean value of 187 K. These temperatures refer to ~95 km (Figure 12.57) and are lower than those retrieved from the SPICAV stellar occultations (Piccalli et al. 2015; Section 12.8.4).

Morphology of the $O_2$ ($a \rightarrow$ X) nightglow was studied using nadir and limb observations with VITRIS-M. The limb observations (Figure 12.57) show a mean maximum of 30 MR at 95 km, and the peak altitude varies typically within ±2 km. The retrieved peak altitudes of volume emission rate are 97.4 ± 2.5 km (Piccioni et al. 2009), and FWHM of the volume emission rate is 10 km at the low latitudes decreasing to 6 km at 50°–70°. Vertical profiles with two peaks were observed as well. Distribution of the nightglow intensity over the disk of Venus (Figure 12.59) demonstrates a bright region centered at the antisolar point with the intensity of ~1.2 MR. The mean nightside intensity is 0.5 MR. The high symmetry relative the antisolar point indicates a very low zonal component of the circulation near 100 km.

Using the termolecular association of $O_2$, Soret et al. (2012a) converted the observed volume emission rates into densities of atomic oxygen. Extraction from this three-dimensional field of atomic oxygen is shown in Figure 12.60. This figure depicts altitude profiles of atomic oxygen at the equator as a function of local time. An interesting feature is the midnight minimum. The retrieved O densities depend on the

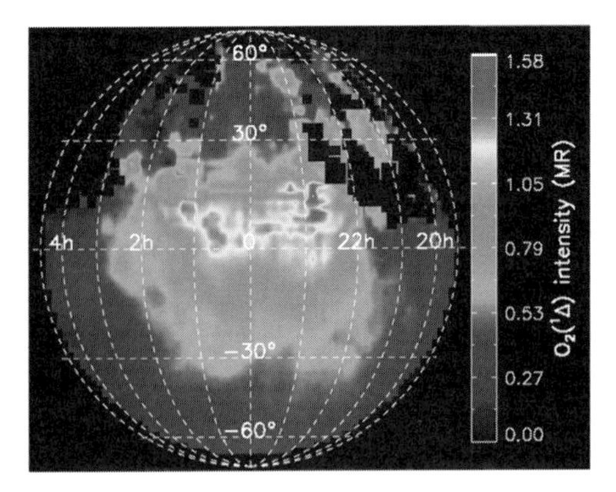


Figure 12.59 Distribution of the $O_2$ nightglow at 1.27 μm over the Venus nightside. From Soret et al. (2012a). (A black and white version of this figure will appear in some formats. For the color version, please refer to the plate section.)

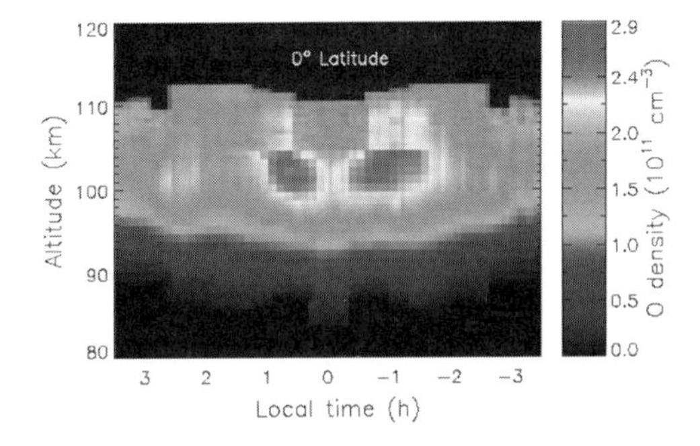


Figure 12.60 Vertical distribution of atomic oxygen at the equator derived from the $O_2$ ($a \rightarrow X$) nightglow observations (Soret et al. 2012a). (A black and white version of this figure will appear in some formats. For the color version, please refer to the plate section.)

adopted $CO_2$ densities that are taken in Figure 12.60 from Hedin et al. (1983). Significant correlation between the nightglow intensity and solar activity has not been found (Soret et al. 2015).

The $O_2$ ($a \rightarrow X$) (0–1) band at 1.58 μm was observed by VIRTIS as well. The measured (0–0)/(0–1) intensity ratio was 63 ± 8 (Piccioni et al. 2009).

### *12.10.3 Excitation of the Oxygen Nightglow on the Terrestrial Planets*

The observed $O_2$ nightglow on Venus is very different from that on the Earth, especially in the visible range, where the Herzberg II bands dominate on Venus and Herzberg I and the green line O 558 nm on the Earth. The $O_2$ nightglow on both planets is excited by the termolecular association of $O_2$, and yields of the $O_2$ electronic states (Table 8.4) should not depend on a third body ($CO_2$ on Venus and $N_2 + O_2$ on Earth). Therefore, quenching and excitation transfer processes are responsible for the difference. The problem was analyzed by Krasnopolsky (2011a), who combined the nightglow observations on both planets with the related laboratory data and some findings in theory. The $O_2$ electronic states are highly vibrationally excited on the Earth, while only v′ = 0 progressions are observed in the Venus nightglow. The latter is caused by rapid vibrational relaxation in $CO_2$. Quenching reactions depend on vibrational excitation, and this complicates the problem.

Radiative times, yields, and quenching coefficients in the suggested excitation scheme are given in Table 12.8. Three u-states are excited directly with yields calculated by Wraight (1982) (Table 8.4). Energy transfers from the upper states including $^5\Pi_g$ are essential for the b→X and a→X emissions at 762 nm and 1.27 μm, respectively. These transfers were estimated using the spin, sign, and statistical weight considerations. For example, quenching of $O_2(c^1\Sigma_u^-)$ by $CO_2$ to the ground state $O_2(X^3\Sigma_g^-)$ is spin-forbidden and to $O_2(b^1\Sigma_g^+)$ is sign-forbidden; therefore $O_2$(c) is quenched to $O_2(a^1\Delta_g)$. (Reactions $\Sigma^+ + \Sigma^- \leftrightarrow \Sigma^+ + \Sigma^+$ and $\Sigma^+ + \Sigma^- \leftrightarrow \Sigma^- + \Sigma^-$ cannot proceed via a planar collisional complex and are therefore less probable, that is, sign-forbidden.) Quenching of $O_2$(b) by $CO_2$ to $O_2$(X) is spin-forbidden; therefore both $O_2$(c) and $O_2$(b) populate $O_2$(a). If a yield of $O_2$(a) in quenching of $O_2(^5\Pi_g)$ is 0.9, then the effective yield of $O_2$(a) is 0.7 on Venus and Mars (see details in Krasnopolsky 2011a). The green line $O(^1S)$ 558 nm is excited by

$$O + O_2(A^3\Sigma_u^+, v \geq 6) \rightarrow O_2 + O(^1S),\ k = 1.5 \times 10^{-11}\ \text{cm}^3\ \text{s}^{-1}. \quad (12.7)$$

Evidently this reaction is insignificant on Venus because of the vibrational relaxation by $CO_2$.

Ionospheric source of the oxygen green and red lines from recombination of $O_2^+$ was calculated by Fox (2012) using the nighttime ionization by the $O^+$ fluxes from the dayside and yields of $O(^1S)$ and $O(^1D)$ for the vibrationally excited $O_2^+$ ions (Table 8.3). Calculated mean nightside intensities for solar maximum are 4.6 R and 26 R for the lines at 558 nm and 630/636 nm, respectively. However, the calculated nightside ionospheric peak altitude is significantly higher than those observed by radio occultations (155 and 142 km, respectively). These altitudes would be lower for ionization by energetic electrons (Brace et al. 1979). If the electrons observed by Gringauz et al. (1979; Section 12.9.2) could reach the atmosphere and enter vertically, then the green line emission from their impact of O and $CO_2$ would be 7 R by scaling the data from Krasnopolsky (1979).

### *12.10.4 Nitric Oxide Nightglow*

Venus' ultraviolet nightglow was detected by Feldman et al. (1979) using the International Ultraviolet Explorer (IUE) and studied by the Pioneer Venus orbiter (Stewart et al. 1980). The nightglow spectrum (Figure 12.61) consists of the (0–v″) progressions of the $\gamma$ (A→X)

Table 12.8 *Excitation, excitation transfer, and quenching processes in the $O_2$ nightglow*

| State | Bands | $\tau$ (s) | $\alpha$ | $\alpha_{TE}$ | $\alpha_{TV}$ | $k_O$ | $k_{O2}$ | $k_{N2}$ | $k_{CO2}$ |
|---|---|---|---|---|---|---|---|---|---|
| $A^3\Sigma_u^+$ | HzI | 0.14 | 0.04 | 0 | 0 | $1.3\times10^{-11}$ | $4.5\times10^{-12}$ | $3\times10^{-12}$ | $8\times10^{-12}$ |
| $A'^3\Delta_u$ | Chm | 2–4 | 0.12 | 0 | 0 | $1.3\times10^{-11}$ | $3.5\times10^{-12}$ | $2.3\times10^{-12}$ | $4.5\times10^{-13}$ |
| $c^1\Sigma_u^-$ | HzII | 5–7 | 0.03 | 0 | 0 | $8\times10^{-12}$ | $3\times10^{-14}/1.8\times10^{-11}$ | – | $1.2\times10^{-16}$ |
| $b^1\Sigma_g^+$ | 762 nm | 13 | 0.02 | 0.09 | 0.125 | $8\times10^{-14}$ | $4\times10^{-17}$ | $2.5\times10^{-15}$ | $3.4\times10^{-13}$ |
| $a^1\Delta_g$ | 1.27 μm | 4460 | 0.05 | 0.35 | 0.65 | – | $10^{-18}$ | $<10^{-20}$ | $10^{-20}$ |

*Note.* HzI, HzII, and Chm are the Herzberg I, II, and Chamberlain bands; $\tau$ is the radiative lifetime; if two values are given, then they refer to high and low vibrational excitation on the Earth and Venus, respectively; α is the direct excitation yield; $\alpha_{TE}$ and $\alpha_{TV}$ are excitation transfer yields for Earth and Venus from the upper states including $^5\Pi_g$ (see details in Krasnopolsky 2011a); and $k_X$ are quenching rate coefficients in $cm^3\ s^{-1}$. Two values of $k_{O2}$ are for $c^1\Sigma_g^+$ (v = 0 and 7–11). From Krasnopolsky (2011a).

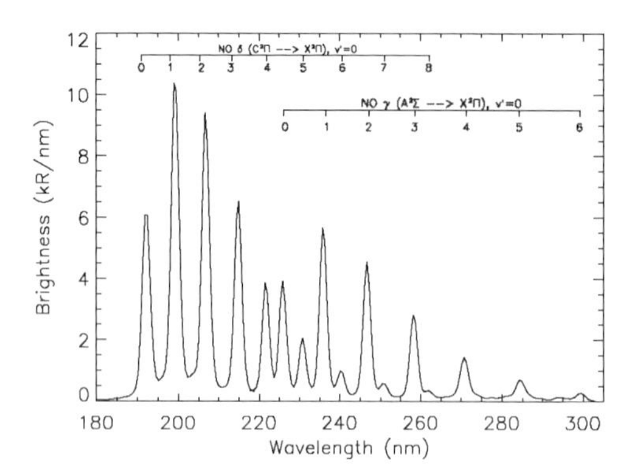


Figure 12.61 Spectrum of the NO nightglow observed using SPICAV at Venus Express. From Gerard et al. (2008).

and $\delta$ (C→X) band systems of NO that are excited by the two-body formation of NO (Section 8.7):

$$\mathrm{N} + \mathrm{O} \rightarrow \mathrm{NO}\left(\mathrm{C}^2\Pi, v = 0\right); \;\; k = 1.8 \times 10^{-17}(300/T)^{1/2}\ \mathrm{cm}^3\ \mathrm{s}^{-1}; \qquad (12.8)$$

$$\mathrm{NO}\left(\mathrm{C}^2\Pi, v = 0\right) \rightarrow \mathrm{NO}\left(\mathrm{A}^2\Sigma^+, v = 0\right) + hv(1.224\ \mu\mathrm{m}), \;\; \alpha = 1/3; \qquad (12.9)$$

$$\mathrm{NO}\left(\mathrm{C}^2\Pi, v = 0\right) \rightarrow \mathrm{NO}\left(\mathrm{X}^2\Pi\right) + hv(\delta - \mathrm{bands}), \;\; \alpha = 2/3; \qquad (12.10)$$

$$\mathrm{NO}\left(\mathrm{A}^2\Sigma^+, v = 0\right) \rightarrow \mathrm{NO}\left(\mathrm{X}^2\Pi\right) + hv(\gamma - \mathrm{bands}). \qquad (12.11)$$

Here $k$ is from Dalgarno et al. (1992) and the branching ratios $\alpha$ are from Garcia Muñoz et al. (2009b), who detected the emission at 1.224 μm using VIRTIS-M at Venus Express.

The $\delta$ (0–1) band at 198 nm is the strongest feature that accounts for 18% of the total emission. This band was used to study the NO nightglow distribution and variations from the Pioneer Venus orbiter. The latest data with some corrections are given by Bougher et al. (1990). The mean nightside airglow intensity of the band was $0.45 \pm 0.12$ kR, and its equatorial maximum was $1.9 \pm 0.6$ kR at LT = 02:00, that is, 2.5 kR and 10.6 kR, respectively, for the total NO nightglow. These values refer to the solar maximum conditions.

The NO nightglow was observed by Venus Express using the SPICAV UV spectrograph. The limb observations were analyzed by Gerard et al. (2008). The mean peak altitude on the limb is 113 km with variations from 95 to 132 km. The mean peak brightness of the total NO nightglow is 32 kR with variations from 5 to 440 kR. The Abel inversion (Section 7.7) of two limb profiles with limb brightnesses of 46 and 35 kR resulted in the vertical intensities of 0.9 and 0.8 kR. Then the mean limb brightness corresponds to the total vertical intensity of 0.7 kR.

Figure 11.1 Mars elevation map observed by MOLA/MGS. (A black and white version of this figure will appear in some formats.)

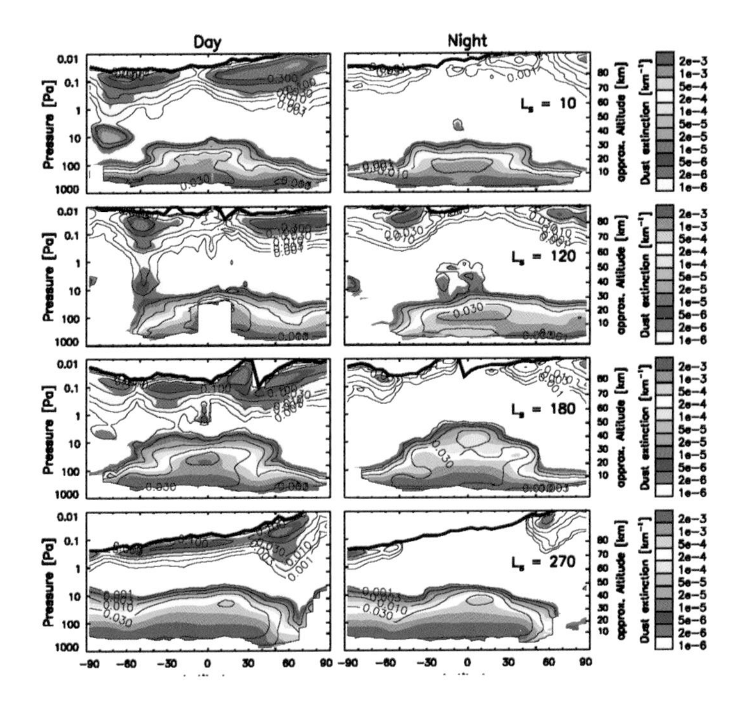


Figure 11.6 Vertical profiles of dust at various latitudes in four seasons observed by MRO/MCS (Kleinboehl et al. 2015). Solid lines are the upper boundaries of the observations. (A black and white version of this figure will appear in some formats.)

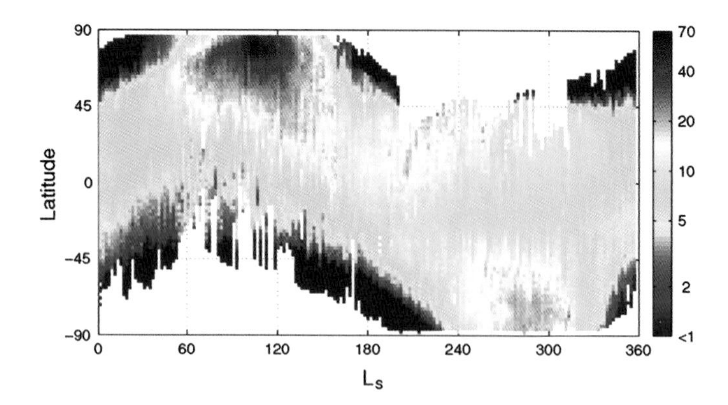


Figure 11.11 Seasonal and latitudinal variations of water vapor (in pr. $\mu$m) observed by MEX/SPICAM IR and averaged for 5 Martian years (Trokhimovsky et al. 2015). (A black and white version of this figure will appear in some formats.)

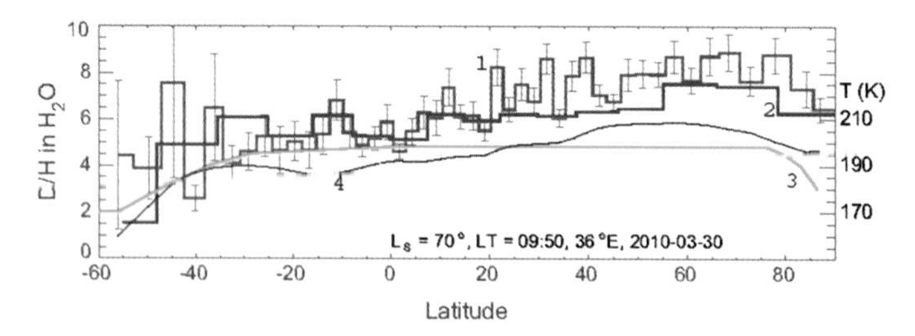


Figure 11.17 Variations of $HDO/H_2O$ (times the terrestrial ratio) retrieved using synthetic spectra (1) and spectra of the Moon (2) are compared with the model (3) by Montmessin et al. (2005). Temperature at 7 km is also shown (4) using the TES observations. From Krasnopolsky (2015d). (A black and white version of this figure will appear in some formats.)

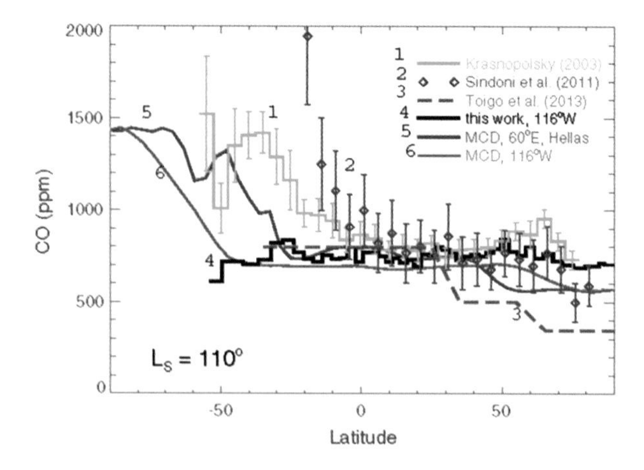


Figure 11.20 Comparison of the observational and model data on latitudinal variations of the CO mixing ratio at $L_S$ = 110°. The MCD predictions are shown for two longitudes. From Krasnopolsky (2015c). (A black and white version of this figure will appear in some formats.)

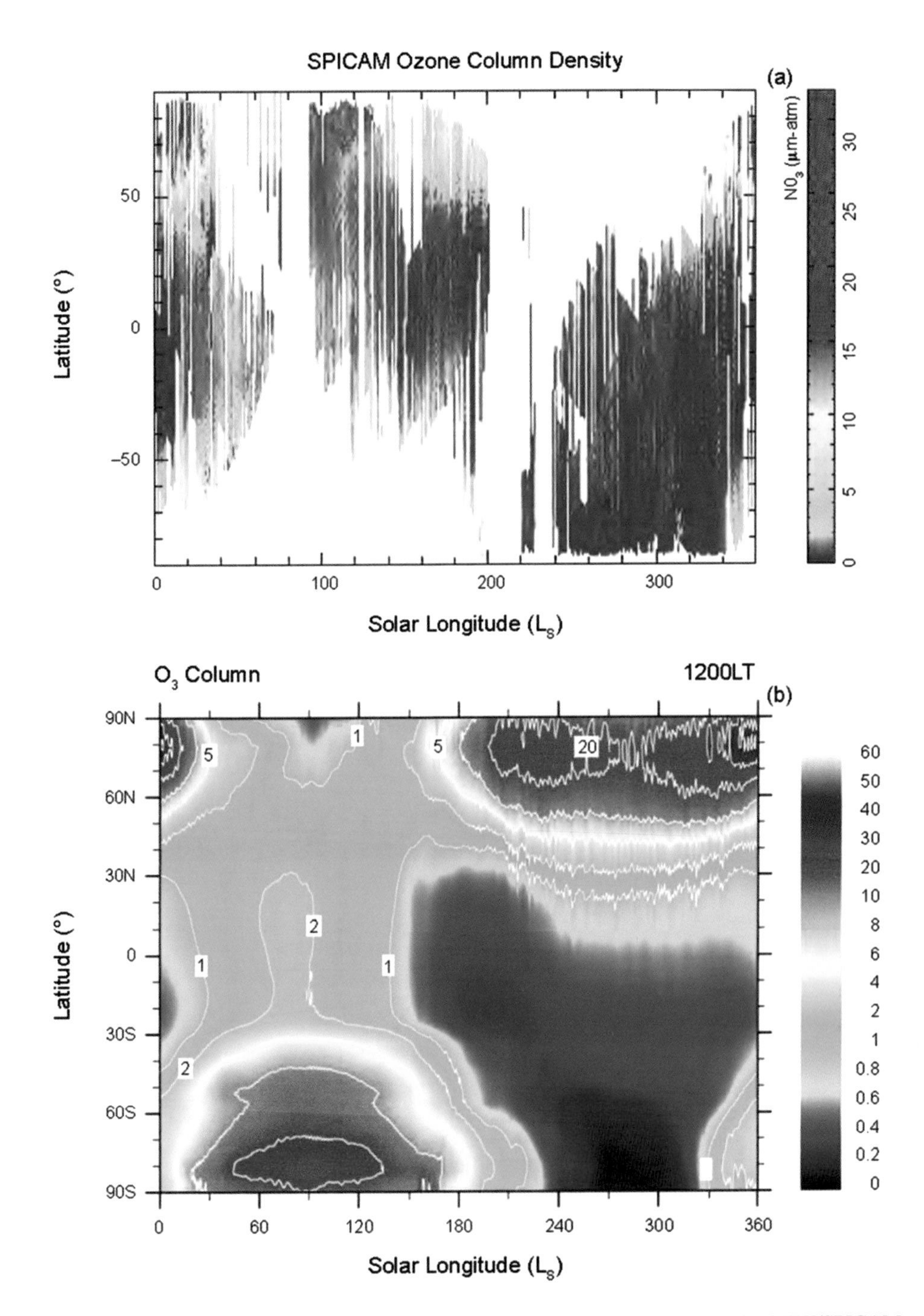


Figure 11.25 Seasonal variations of the daytime $O_3$ column ($\mu$m-atm) observed by MEX/SPICAM (upper panel; Perrier et al. 2006) and from the Mars Climate Database (lower panel; updated from Lefèvre et al. 2004). From Krasnopolsky and Lefèvre (2013). (A black and white version of this figure will appear in some formats.)

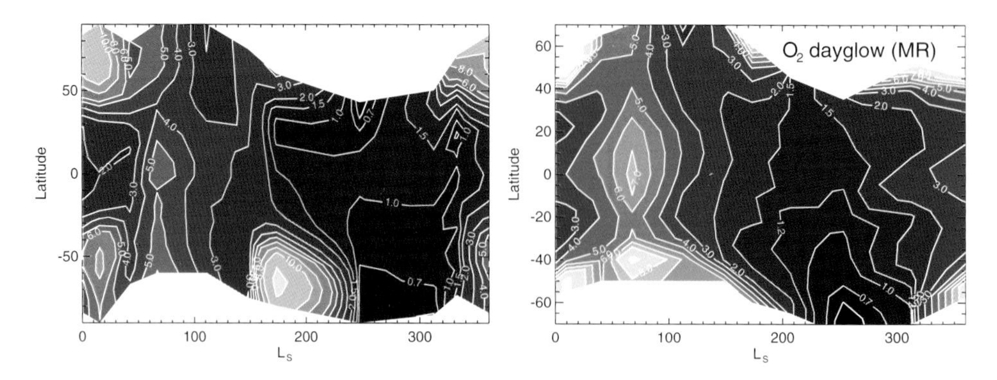


Figure 11.28 Seasonal-latitudinal variations of the $O_2$ dayglow at 1.27 $\mu$m: observations and model (upper and lower panels, respectively). From Krasnopolsky (2013c). (A black and white version of this figure will appear in some formats.)

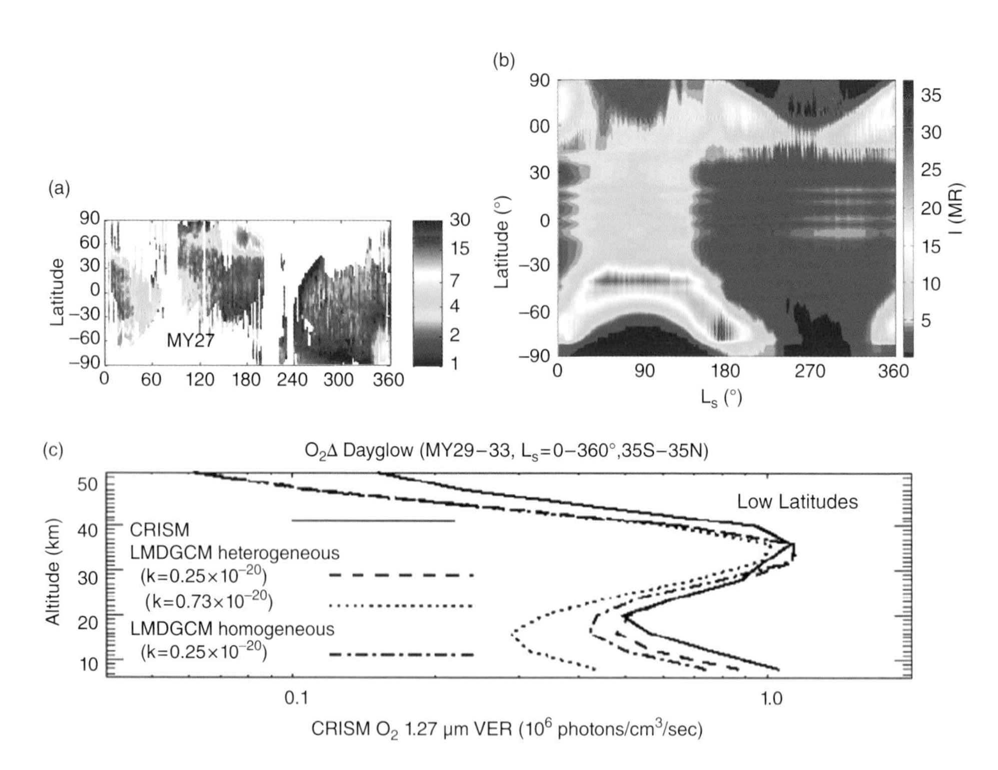


Figure 11.29 (*a*) SPICAM IR observations of the $O_2$ dayglow at 1.27 $\mu$m for 1 Martian year (Guslyakova et al. 2016). (*b*) Seasonal evolution of the dayglow at LT = 12 based on the MCD data (Gagne et al. 2012). (*c*) Mean vertical profile of the dayglow volume emission rate (VER) observed using CRISM/MRO at low latitudes compared with three model profiles (Clancy et al. 2017). (A black and white version of this figure will appear in some formats.)

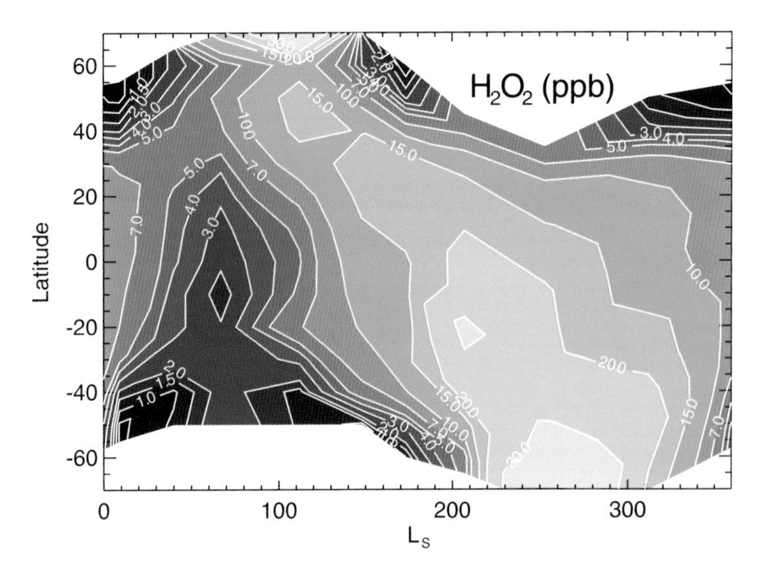


Figure 11.32 Seasonal and latitudinal variations of the $H_2O_2$ mixing ratio in the model by Krasnopolsky (2009b). (A black and white version of this figure will appear in some formats.)

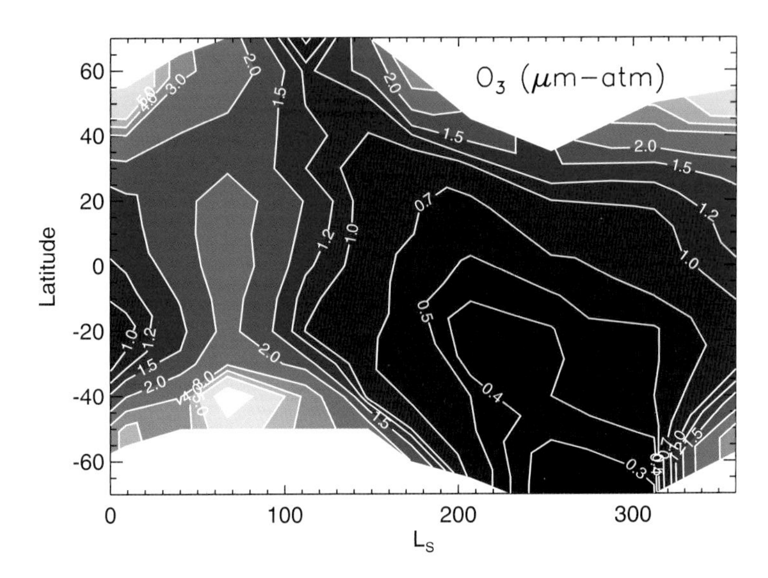


Figure 11.41 Seasonal and latitudinal variations of ozone from the model by Krasnopolsky (2009b). (A black and white version of this figure will appear in some formats.)

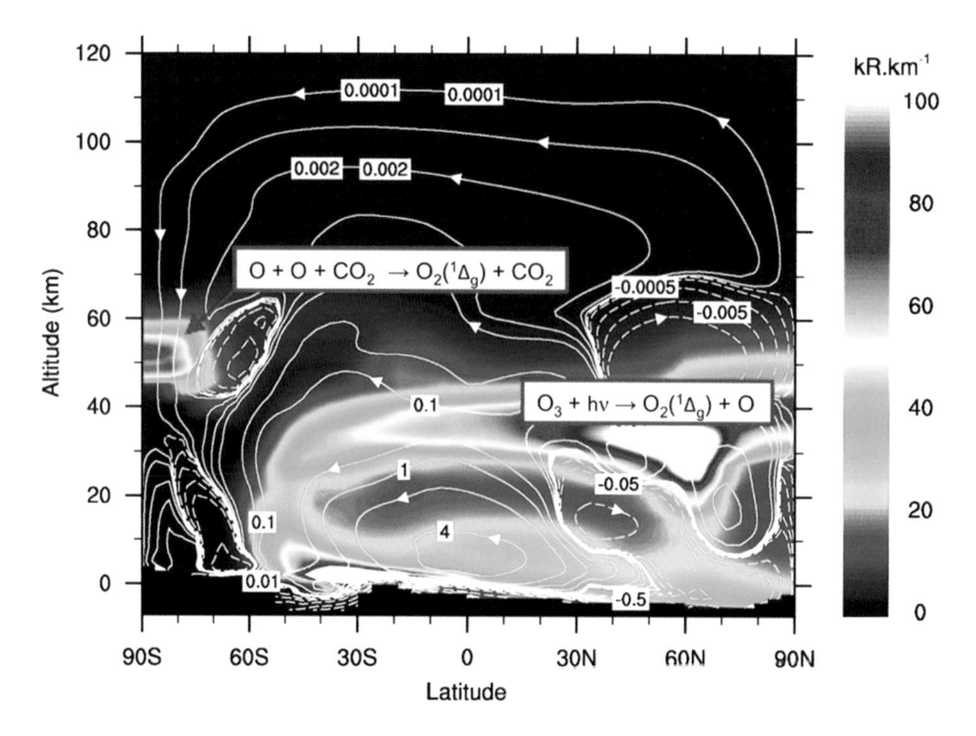


Figure 11.56 Two mechanisms of the $O_2$ 1.27 $\mu$m airglow excitation at $L_S = 95°–100°$ as simulated by LMD GCM. Labels give meridional stream functions in $10^9$ kg s$^{-1}$. From Clancy et al. (2012b). (A black and white version of this figure will appear in some formats.)

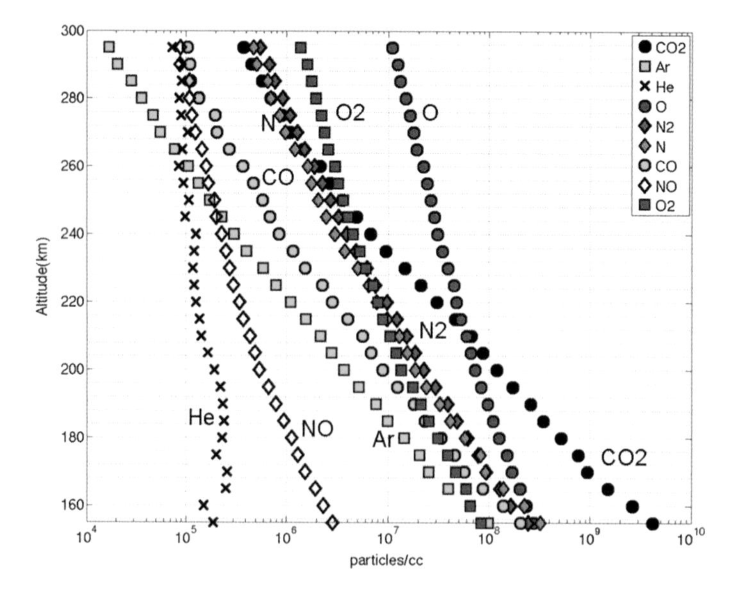


Figure 11.68 Vertical profiles of nine species measured by MAVEN/NGIMS in the upper atmosphere at $F_{10.7} \approx 63$, SZA $\approx 45°$, $L_S = 288°–326°$, latitudes 4°S to 46°N (Mahaffy et al. 2015). (A black and white version of this figure will appear in some formats.)

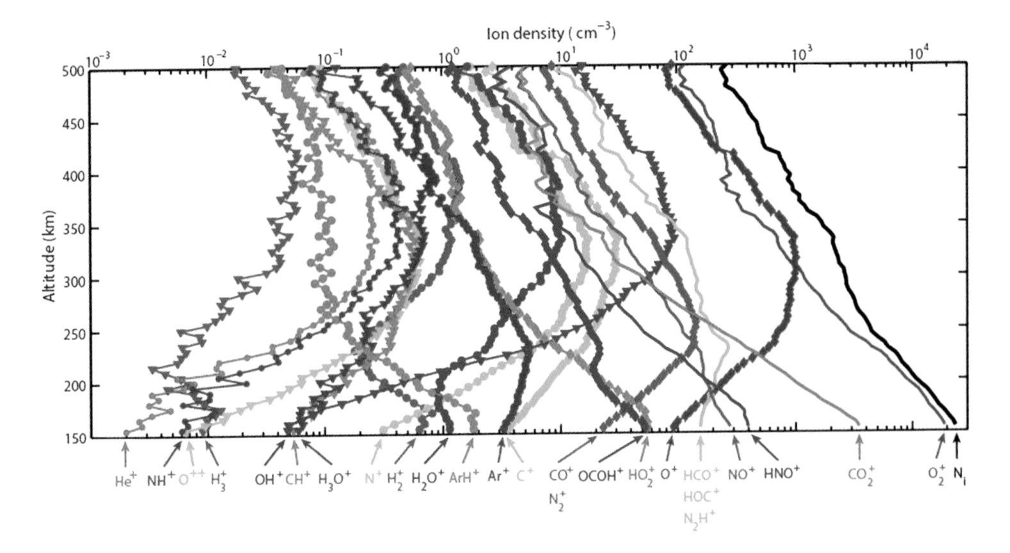


Figure 11.70 Vertical profiles of 22 ions measured by NGIMS at SZA = 60°. Measured isotopes $^{36}Ar^+$, $^{38}Ar^+$, and $^{18}OO^+$ are not shown. From Benna et al. (2015). (A black and white version of this figure will appear in some formats.)

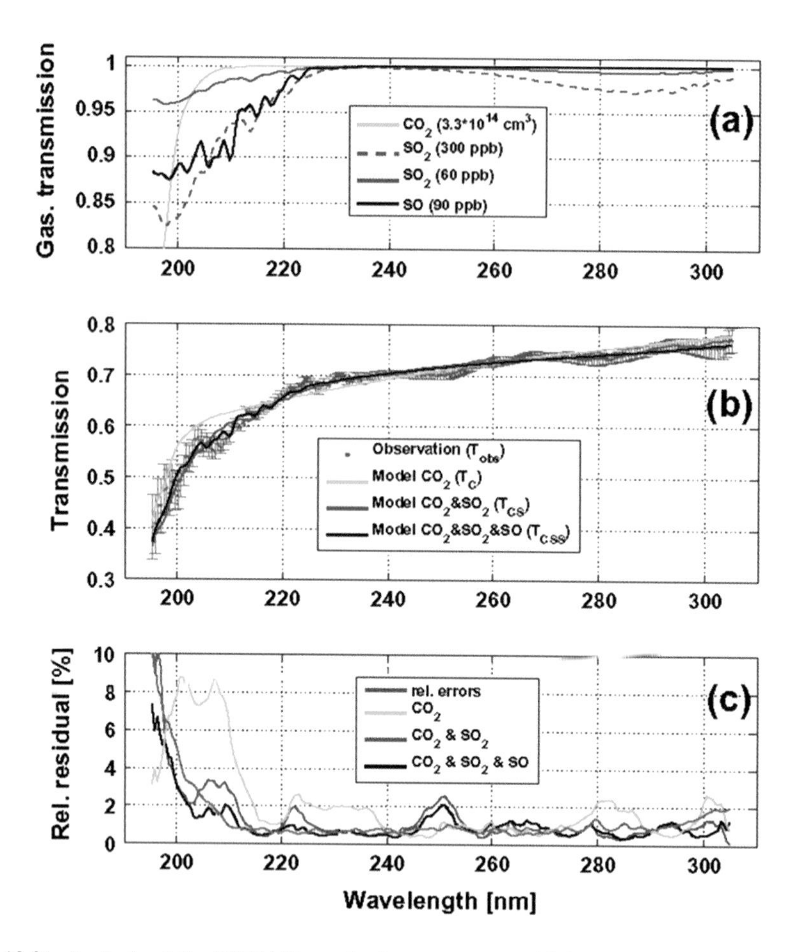


Figure 12.21 Analysis of the SPICAV occultation spectrum at 93 km. (*A*) Transmissions of three basic absorbers: $CO_2$, $SO_2$ (for two abundances), and SO. (*B*) Fitting of the observed spectrum by three models. *C*: residuals for three models; $\chi^2 = 3.5$ and 1.7 without and with SO, respectively. From Belyaev et al. (2012). (A black and white version of this figure will appear in some formats.)

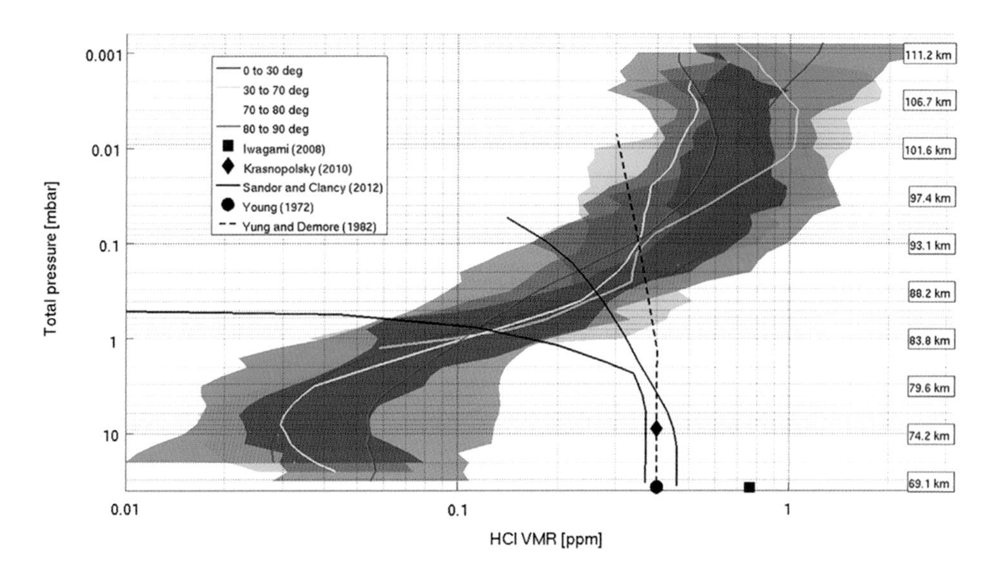


Figure 12.30 Vertical profiles of HCl from the SOIR occultations are compared wih the infrared and submillimeter observations and the model by Yung and DeMore (1982). From Mahieux et al. (2015b). (A black and white version of this figure will appear in some formats.)

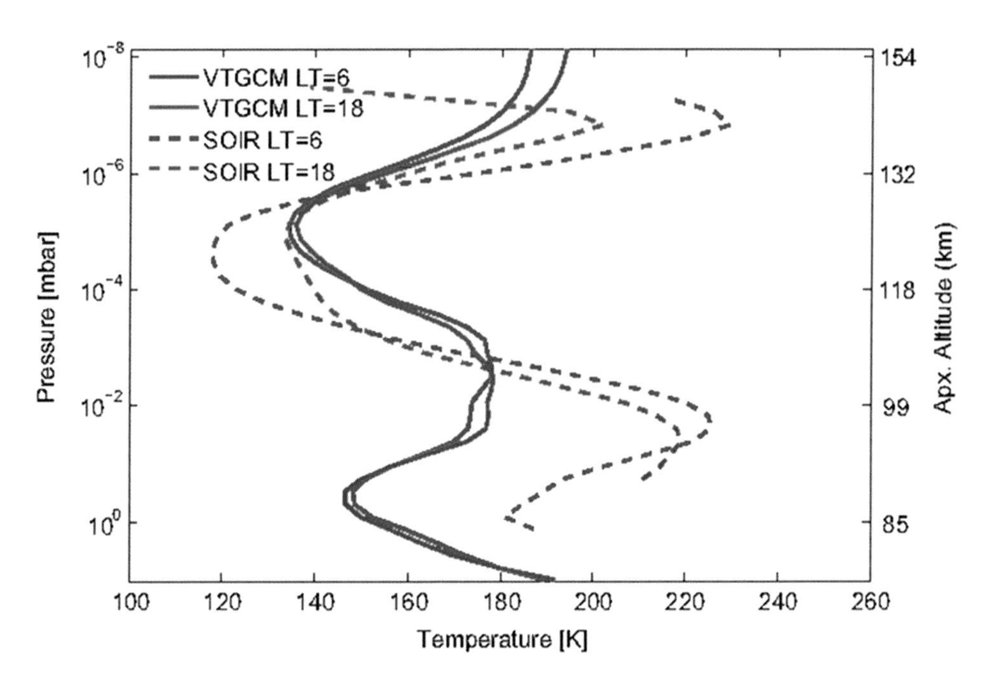


Figure 12.44 Mean SOIR sunrise and sunset temperature profiles at low latitudes ($\varphi < 30°$) are compared with the VTGCM profiles (Bougher et al. 2015b). (A black and white version of this figure will appear in some formats.)

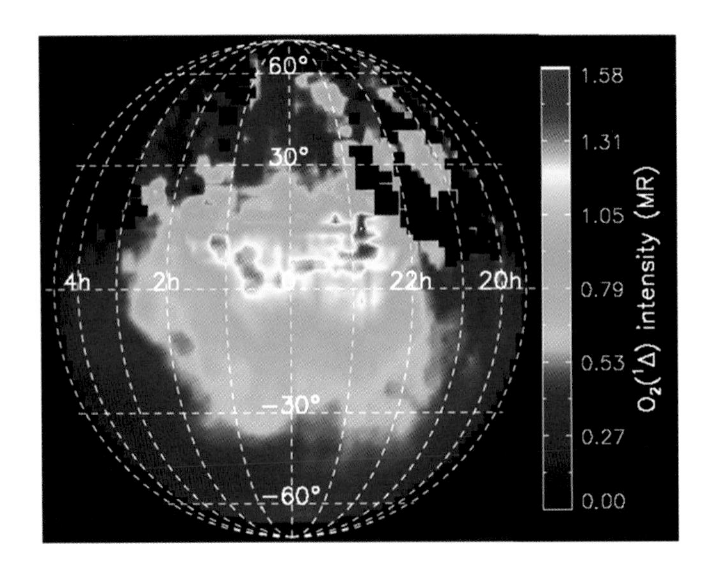


Figure 12.59 Distribution of the $O_2$ nightglow at 1.27 $\mu$m over the Venus nightside. From Soret et al. (2012a). (A black and white version of this figure will appear in some formats.)

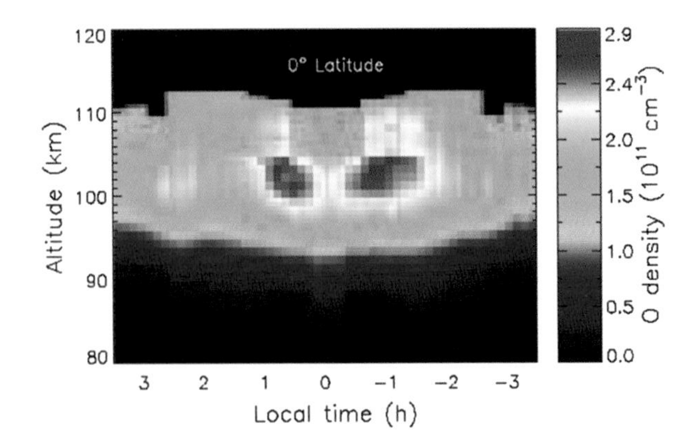


Figure 12.60 Vertical distribution of atomic oxygen at the equator derived from the $O_2$ ($a \rightarrow X$) nightglow observations (Soret et al. 2012a). (A black and white version of this figure will appear in some formats.)

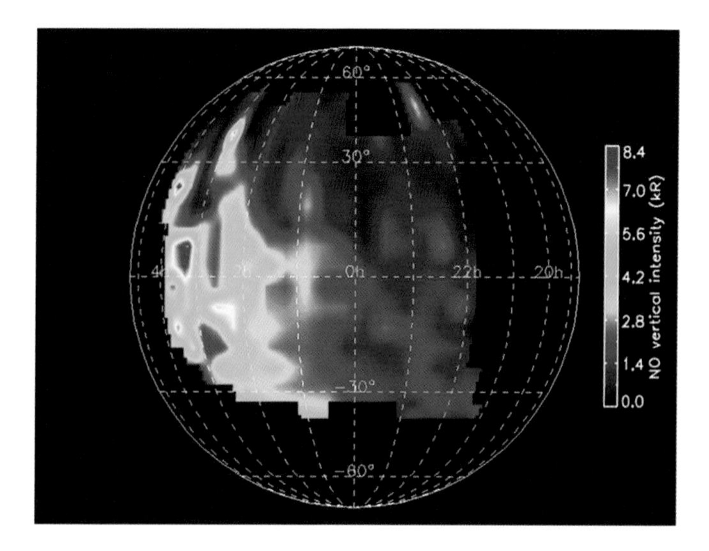


Figure 12.62 Distribution of the NO nightglow over the observed part of the Venus nightside. From Stiepen et al. (2013). (A black and white version of this figure will appear in some formats.)

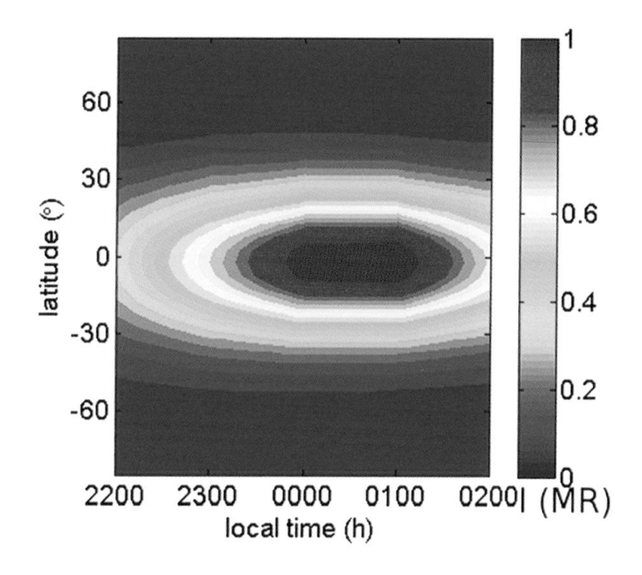


Figure 12.66 VTGCM brightness distribution of the $O_2$ nightglow at 1.27 $\mu$m, solar minimum (Gagne et al. 2012). (A black and white version of this figure will appear in some formats.)

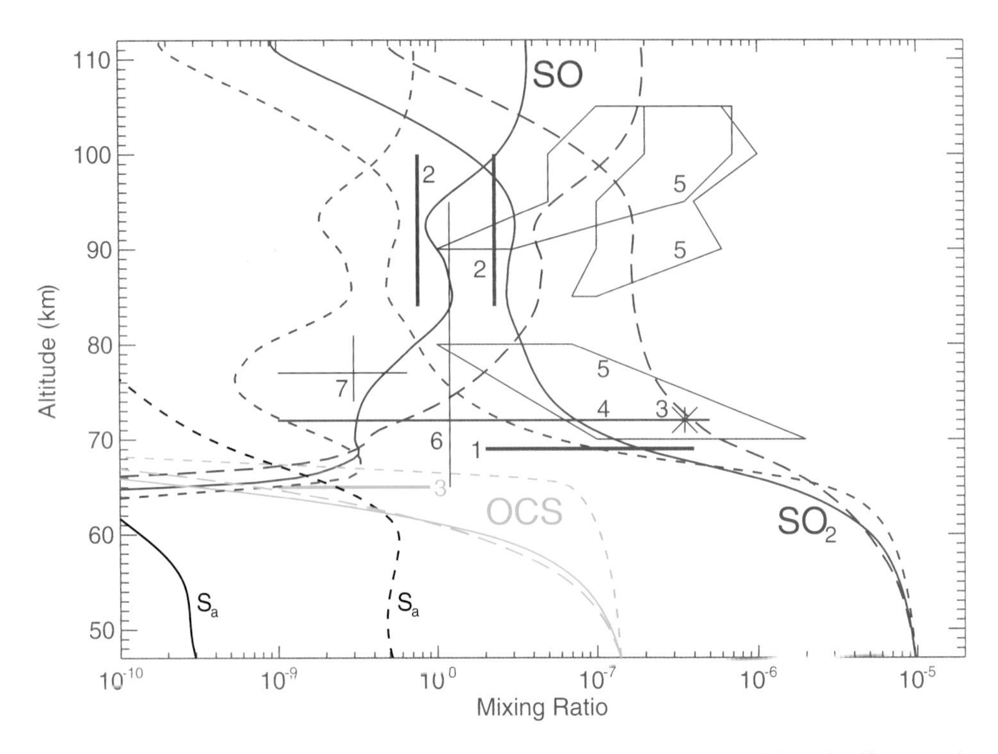


Figure 12.99 Basic sulfur species: model results and observations. $SO_2$, OCS, SO, and sulfur aerosol $S_a$ profiles are shown for the models with the eddy break at $h_e = 57$, 60, and 65 km (long dash, solid, and short dash curves, respectively). The $S_a$ mixing ratios refer to total number of sulfur atoms in the aerosol. Observations: (1) PV, Venera 15, HST and rocket data (Esposito et al. 1997); (2) mean results of the submillimeter measurements (Sandor et al. 2010); the observed $SO_2$ varies from 0 to 76 ppb and SO from 0 to 31 ppb; (3) IRTF/CSHELL (Krasnopolsky 2010d); (4) SPICAV_UV, nadir (Marcq et al. 2011); (5) VEX/SOIR and SPICAV-UV occultations (Belyaev et al. 2012); (6) rocket observation of SO (Na et al. 1994); (7) HST observations of SO (Jessup et al. 2015). From Krasnopolsky (2018a). (A black and white version of this figure will appear in some formats.)

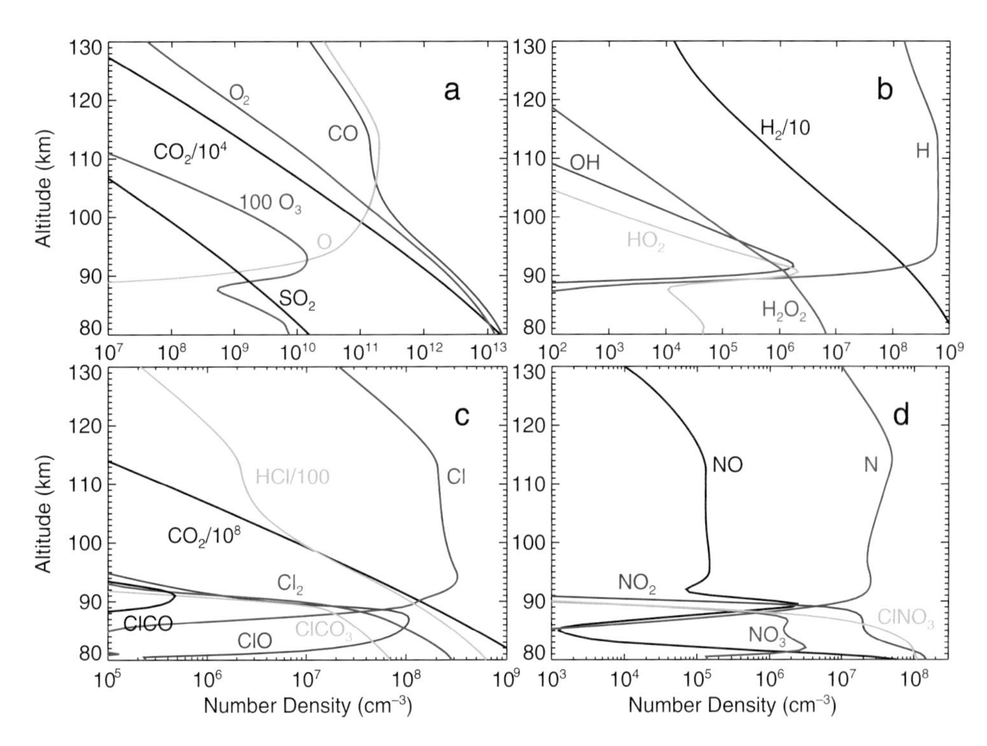


Figure 12.101 Vertical profiles of species in the model for the nighttime atmosphere (Krasnopolsky 2019b). (A black and white version of this figure will appear in some formats.)

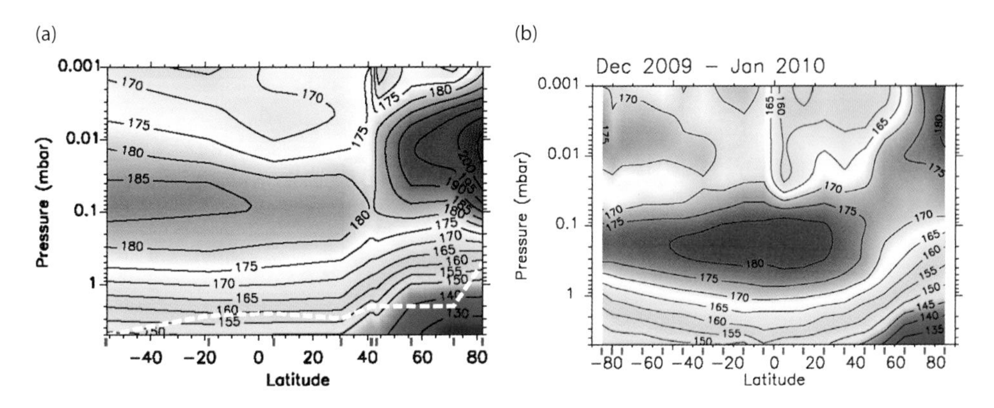


Figure 13.23 Thermal structure of the stratosphere observed using CIRS at the end of northern winter ($L_S$ = 305°–345°) and near vernal equinox ($L_S$ = 5°). 1, 0.1, 0.01, and 0.001 mbar correspond to ≈180, 300, 420, and 540 km, respectively. Temperatures are rather uncertain below the white dashed line. From Vinatier et al. (2010, 2015). (A black and white version of this figure will appear in some formats.)

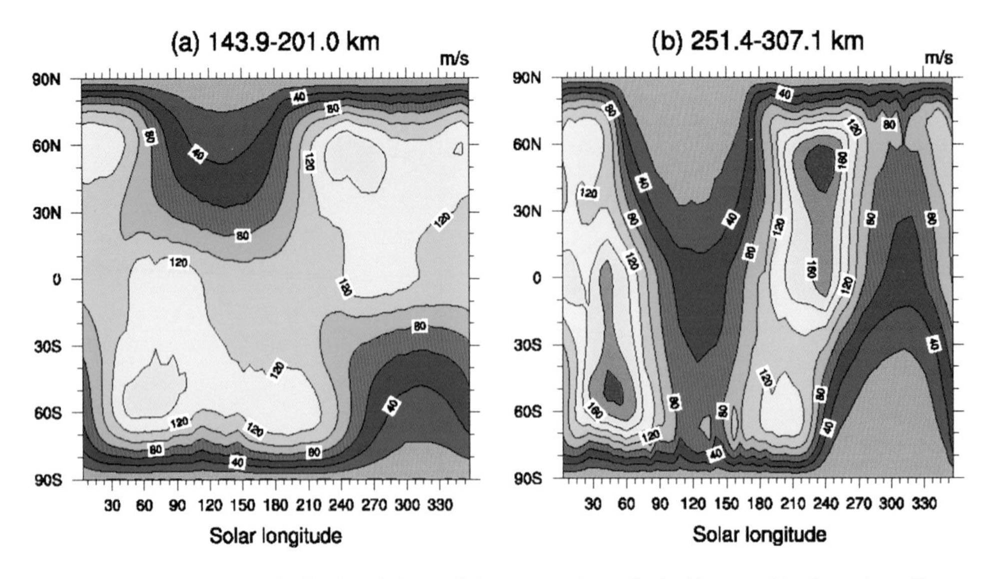


Figure 13.29 Seasonal-latitudinal variations of the eastward zonal wind in two altitude regions. From GCM by Lebonnois et al. (2012). (A black and white version of this figure will appear in some formats.)

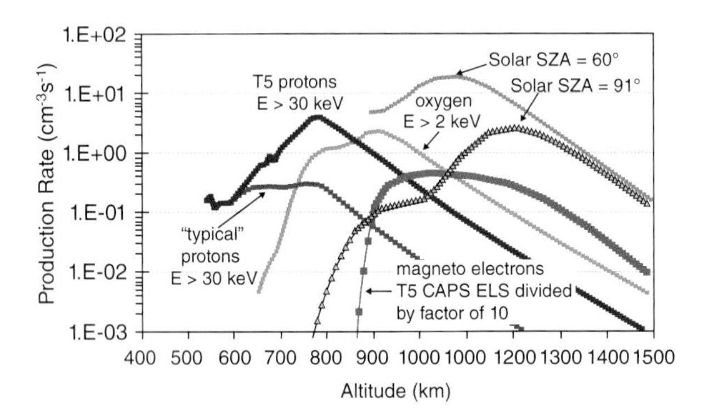


Figure 13.47 Ionization rates by magnetospheric electrons, protons, and oxygen ions at the strong precipitation event during flyby T5 compared to those by the solar EUV. From Cravens et al. (2008). (A black and white version of this figure will appear in some formats.)

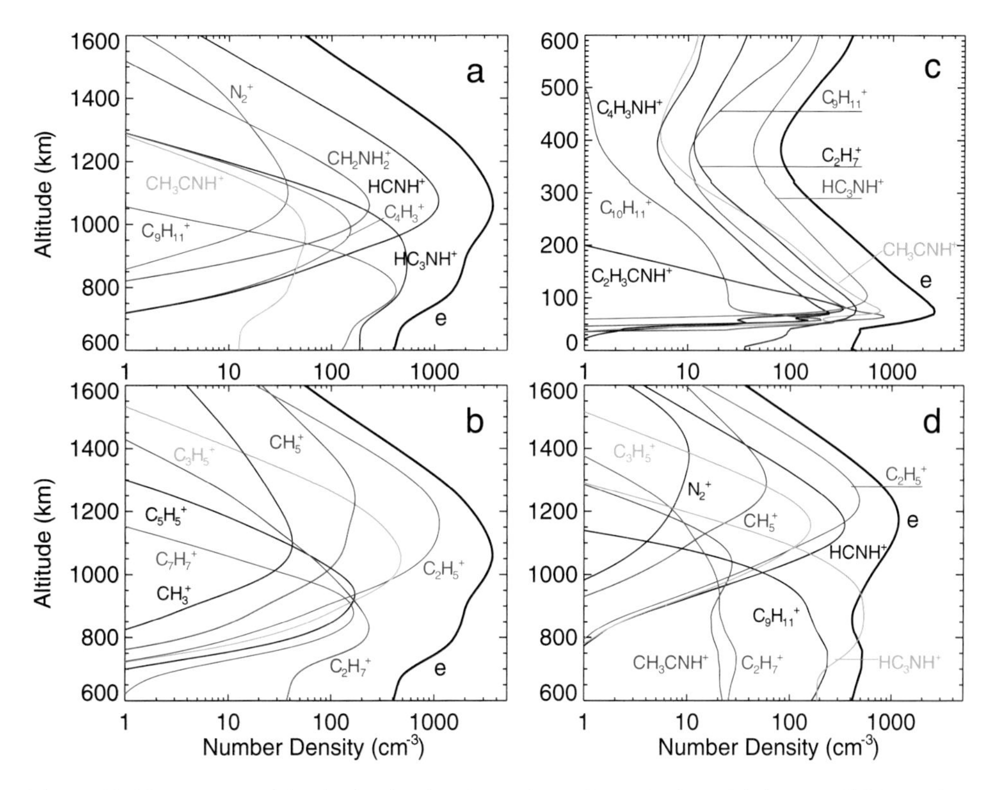


Figure 13.66 Ion composition in the daytime ionosphere above (*a, b*) and below (*c*) 600 km. The nighttime ionosphere (*d*) is calculated for the conditions of flyby T5 with strong precipitation of magnetospheric electrons. From Krasnopolsky (2012d). (A black and white version of this figure will appear in some formats.)

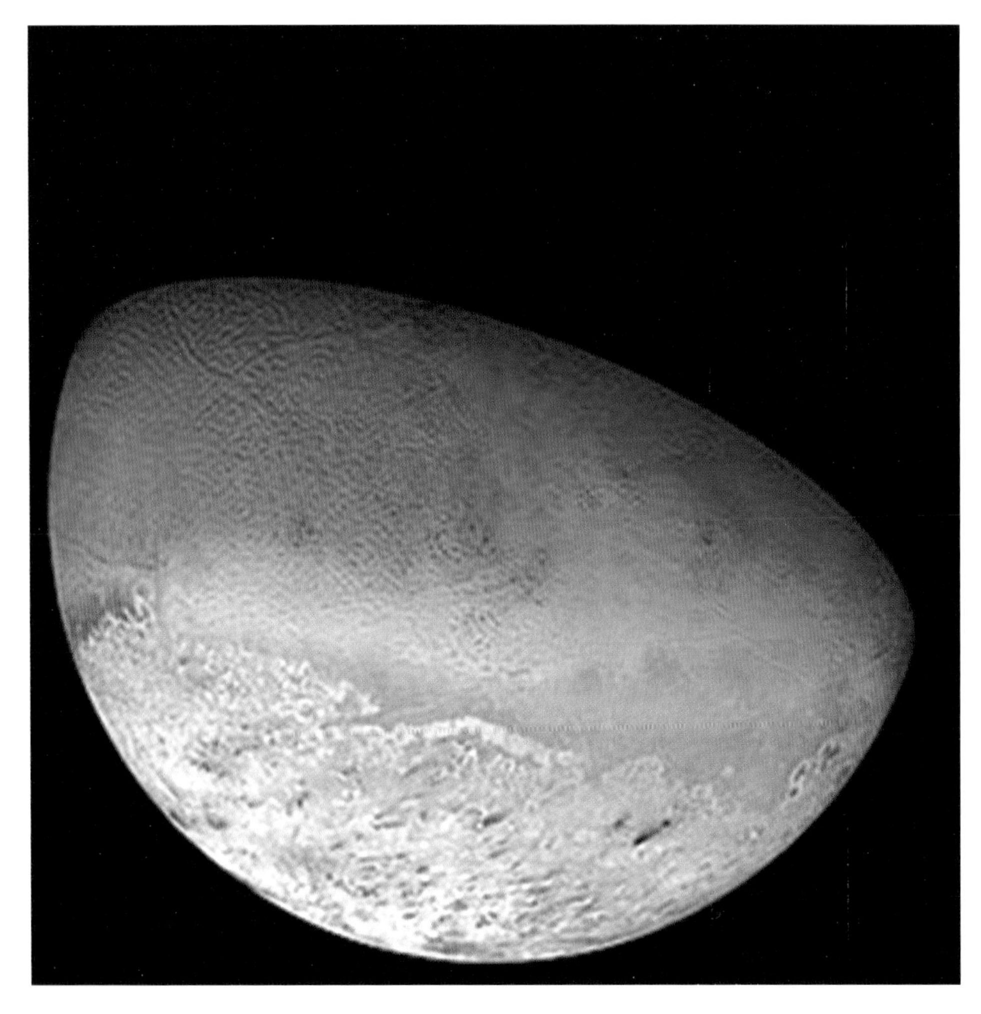

Figure 14.1 Photomosaic of Triton based on the Voyager 2 images. (A black and white version of this figure will appear in some formats.)

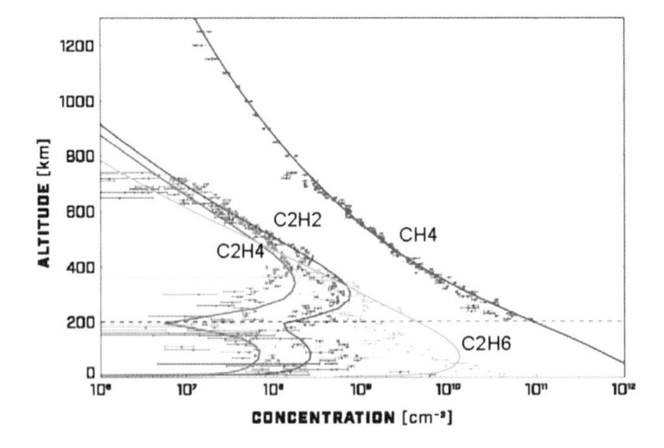


Figure 15.26 Number densities of $CH_4$, $C_2H_2$, $C_2H_4$, and $C_2H_6$ retrieved from the NH solar UV occultations (Gladstone et al. 2016) and calculated by Wong et al. (2017). (A black and white version of this figure will appear in some formats.)

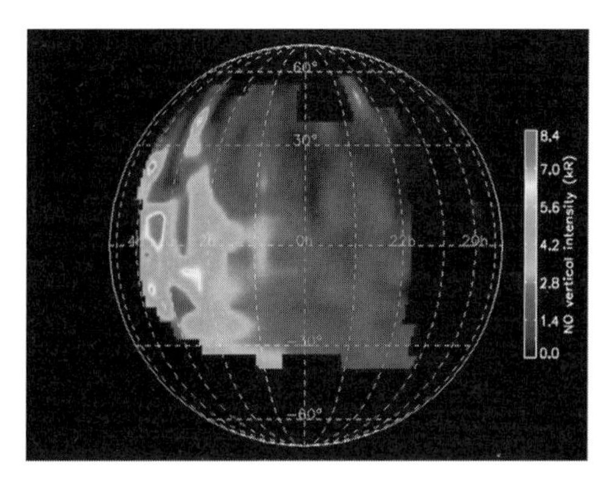


Figure 12.62 Distribution of the NO nightglow over the observed part of the Venus nightside. From Stiepen et al. (2013). (A black and white version of this figure will appear in some formats. For the color version, please refer to the plate section.)

The nadir observations of the NO nightglow by Venus Express were studied by Stiepen et al. (2013). A summary of the observations is the nightglow distribution over the Venus nightside (Figure 12.62). The observations covered a period of 2006 to 2011 at a low to medium solar activity. Stiepen et al. (2013) give a mean nightside intensity of 1.9 $\pm$ 1 kR that is close to the PV value at solar maximum but significantly greater than 0.7 kR based on Gerard et al. (2008).

The nightglow maximum is of 8 kR at LT = 03:20 in Figure 12.62. The authors consider a mean intensity at LT = 01:00 to 03:30 and from 25°S to 10°N for comparison with the PV data. This intensity is equal to 3.9 $\pm$ 1.5 kR, smaller than that at 03:20 by a factor of 2.

### *12.10.5 Hydroxyl Nightglow*

The VEX/VIRTIS observations revealed for the first time the OH rovibrational bands in the Venus nightglow (Piccioni et al. 2008; Gerard et al. 2012; Soret et al., 2012b). These bands are excited by the hydrogen–ozone reaction (Section 8.7). The observed nightglow involves the $\Delta v = 1$ and 2 band sequences at 2.9 and 1.4 μm, respectively (Figure 12.63). The mean limb intensity of the $\Delta v = 1$ sequence is $350^{+350}_{-210}$ kR (Soret et al. 2012b), that is, the vertical intensity is ~6 kR. The mean peak altitude is 96.4 $\pm$ 5 km on the limb. The sequence consists of the (1–0), (2–1), (3–2), and (4–3) bands with relative intensities of 45%, 39%, 9%, and 7%, respectively. The $\Delta v = 2$ sequence intensity is 0.38 $\pm$ 0.37 of that of $\Delta v = 1$. This sequence includes mostly the (2–0) band. There is a significant correlation between the OH and $O_2$ nightglows.

The data on the $\Delta v = 2$ sequence disagree with those on the $\Delta v = 1$ sequence (Krasnopolsky 2013a). Intensities of the $\Delta v = 2$ bands (2–0), (3–1), and (4–2) can be simply calculated using the known intensities of the (2–1), (3–2), and (4–3) bands of the $\Delta v = 1$ sequence and transition probabilities of the bands. Then the expected intensity ratio

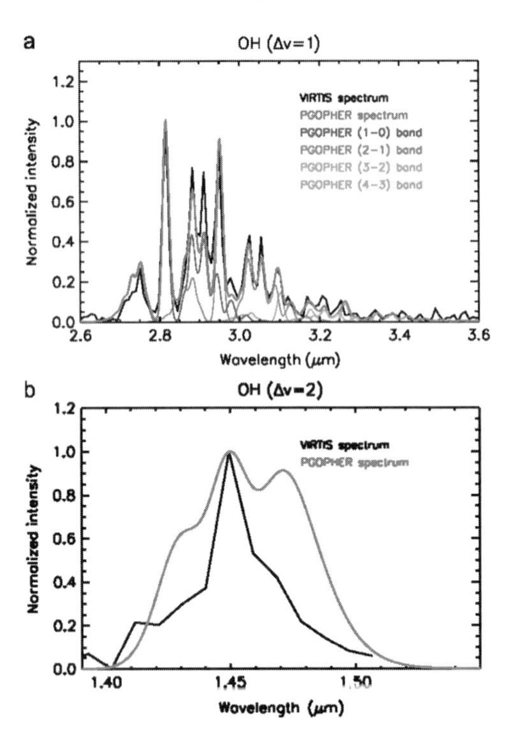


Figure 12.63 Mean normalized observed spectra of the OH $\Delta v$ = 1 and 2 sequences are compared with synthetic spectra of separate bands and their sums. From Soret et al. (2012b).

is 0.52, greater than that observed. If the only band in Δv = 2 is (2–0), then the intensity ratio is 0.16. Therefore the observational data on the Δv = 2 sequence and their interpretation are questionable.

Krasnopolsky (2010a) found a narrow window for ground-based observations of the OH nightglow on Venus and detected lines of the (1–0) and (2–1) bands that gave the OH nightglow vertical intensities of ~10 kR.

### 12.10.6 CO UV Nightglow

Stewart and Barth (1979) found traces of the $CO(a^3\Pi \rightarrow X^1\Sigma^+)$ Cameron band nightglow in the PV UV spectra. Only the (0–v″) progression was detected with the (0–1) 216 nm as the strongest band with intensity of ~100 R; then the progression intensity is ~250 R. Krasnopolsky (1982) calculated a profile of atomic C and estimated a mean downward flow of C on the nightside at $\sim 2\times10^9$ cm$^{-2}$ s$^{-1}$ corrected for the loss in the reaction with $O_2$ during the transport from the dayside. He suggested a reaction

$$C + O_2 \rightarrow CO\left(a^3\Pi, v = 0\right) + O - 0.019 \text{ eV}.$$

A current value of the total rate coefficient is $5.1 \times 10^{-11}$ $(300/T)^{0.3}$ $cm^3\ s^{-1}$, and a yield of the excited state is 0.7 $e^{-215/T}$ based on the statistical weights and the energy balance. The calculated nightglow was close to that observed.

### 12.10.7 VTGCM Simulations of the $O_2$ 1.27 μm and NO UV Nightglow

The model in Brecht et al. (2011) and Gagne et al. (2012) includes major ($CO_2$, CO, O, and $N_2$) and minor ($O_2$, N, $N(^2D)$, and NO) neutral species and ions ($CO_2^+$, $O_2^+$, $O^+$, and $NO^+$). The ion-neutral chemistry is based on Fox and Sung (2001). To account for the chlorine and hydrogen chemistry, density profiles of these species were copied from a nighttime model by Krasnopolsky (2010a). The model spans 70 to 300 km at the subsolar point and keeps this pressure range throughout the globe. The lower boundary conditions are the species global-mean densities from the model by Yung and DeMore (1982), and diffusive equilibrium (zero fluxes) is adopted at the upper boundary.

Calculated vertical profiles of O and N at the subsolar and antisolar points at solar minimum are shown in Figure 12.64. The midnight density of atomic oxygen peaks at 104 km and exceeds here the noon value by an order of magnitude. The similar difference is for N near 115 km.

The calculated volume emission rate of the $O_2$ airglow at 1.27 μm near the equator as a function of altitude and local time is shown in Figure 12.65. The adopted yield of $O_2(a^1\Delta_g)$ in the termolecular association is 0.75 that rules out quenching of $O_2(^5\Pi_g)$ by $CO_2$ to the

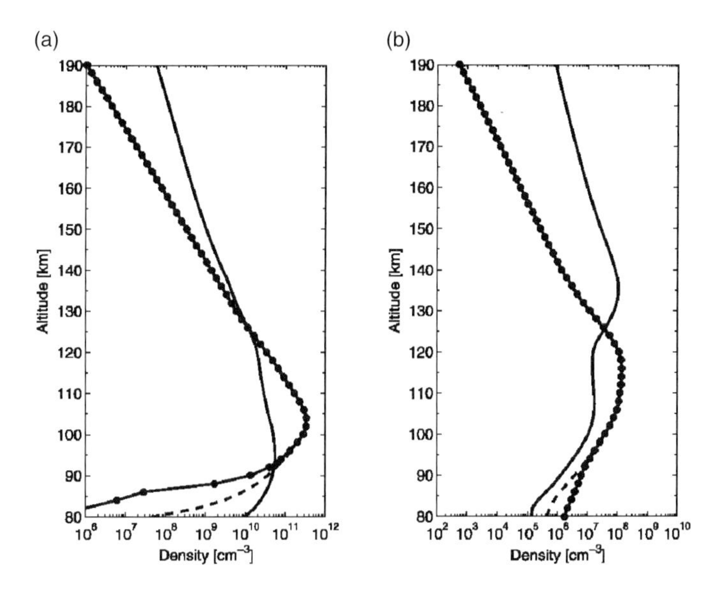


Figure 12.64 Vertical profiles of O (left) and N (right) at the subsolar and antisolar points (thin and dotted lines, respectively) at solar minimum. The profiles without chlorine and hydrogen chemistries are dashed. From Brecht et al. (2011).

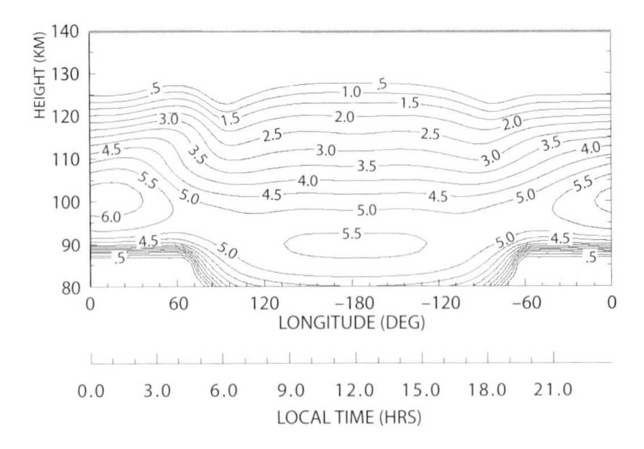


Figure 12.65 VTGCM altitude and local time distribution of the volume emission rate of the $O_2$ airglow at 1.27 μm near the Venus equator, solar minimum. The units are $\log_{10}$ (photons $cm^{-3}$ $s^{-1}$), and 6.0 means 0.1 MR $km^{-1}$. From Brecht et al. (2011).

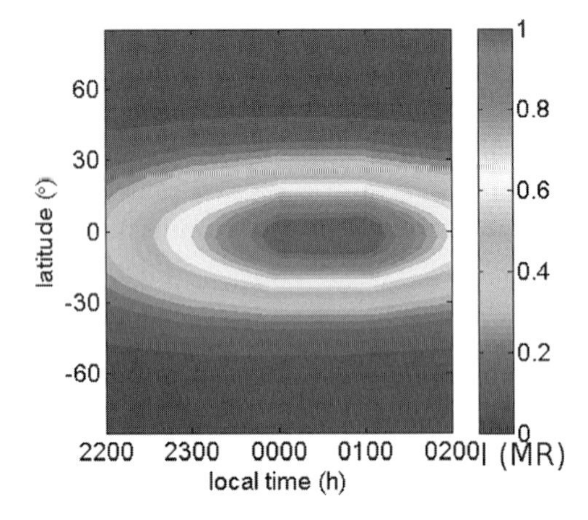


Figure 12.66 VTGCM brightness distribution of the $O_2$ nightglow at 1.27 μm, solar minimum (Gagne et al. 2012). (A black and white version of this figure will appear in some formats. For the color version, please refer to the plate section.)

$O_2$ states other than $O_2(a^1\Delta_g)$ (Section 12.10.3). Vertical intensities can be calculated by summing the layers along the altitude scale. The maximal vertical intensity is 1.76 MR, and a mean intensity in the box of ±60° in both latitude and longitude is 0.51 MR, in accord with the observations.

Latitudinal and local time distribution of the $O_2(a \rightarrow X)$ nightglow from a version of VTGCM in Gagne et al. (2012) is shown in Figure 12.66. The nightglow maximum is ~1 MR and shifted by 0.5 h to the morning. The isophotes look like ellipses with the axis ratio of 1 : 3; however, this is just the ratio of the $x$ and $y$ scales, and the true isophotes are circles.

Comparison of the model intensities averaged between LT = 22:00 and 02:00 with the observations is depicted in Figure 12.67. Two sets of the model parameters are

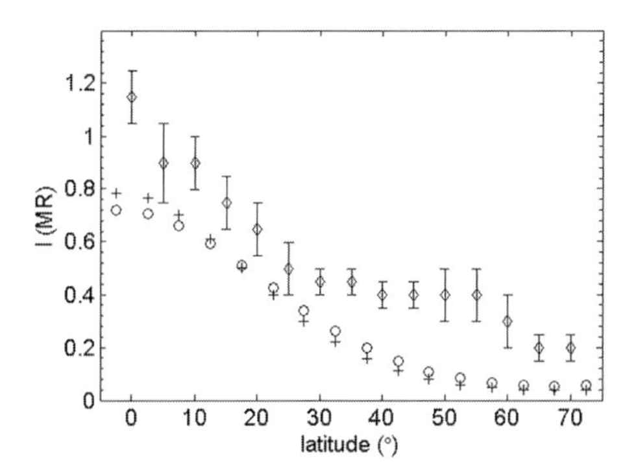


Figure 12.67 Latitudinal variations of the $O_2(a \rightarrow X)$ nightglow intensity averaged between LT = 22:00 and 02:00: the VEX observations (diamonds, from Piccioni et al. (2009)) and the VTGCM results from the first and second runs (circles and pluses, respectively). From Gagne et al. (2012).

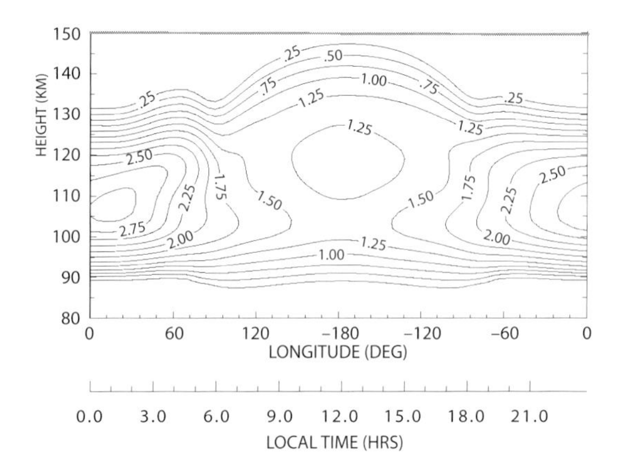


Figure 12.68 VTGCM altitude and local time distribution of the volume emission rate of the NO UV airglow near the Venus equator, solar minimum. The units are $\log_{10}$ (photons cm$^{-3}$ s$^{-1}$), and 2.0 means 10 R km$^{-1}$. From Brecht et al. (2011).

considered, and the results are rather similar. The difference between the model and the observations is moderate (~30%) at the low latitudes $\varphi < 30°$ and significant at the middle and high latitudes.

VTGCM results on the NO UV nightglow at solar minimum are illustrated in Figure 12.68 as volume emission rates. The peak value is 1200 photons cm$^{-3}$ s$^{-1}$ at 106 km near LT = 01:00, and the maximal vertical intensity is 1.83 kR, while the mean nightside value is 0.68 kR. The published results of the observations and the model are compared in Table 12.9.

The PV and VEX/nadir intensities look rather similar and are significantly greater than the VEX/limb and the model data. Both the morningside shift and the peak altitude are

Table 12.9 *The NO UV nightglow on Venus in observations and model*

| Source | Solar activity | $4\pi I$ (mean) | h (km) | $4\pi I$ (max) | LT (max) |
|---|---|---|---|---|---|
| PV (Bougher et al. 1990) | max | 2.2 kR | 115 | 10.6 kR | 02:00 |
| VEX limb (Gerard et al. 2008) | min | 0.7 | 113 | – | – |
| VEX nadir (Stiepen et al. 2013) | min and mean | 1.9 | – | 8 | 03:00 |
| VTGCM (Brecht et al. 2011) | min | 0.68 | 106–112 | 1.8 | 01:00 |

*Note.* All intensities are vertical.

smaller in the model than those observed. This shift reflects the retrograde superrotation near the peak altitude.

## 12.11 Day Airglow

Similar to the nightglow, dayglow and related phenomena present intrinsic interest and as a diagnostic tool for photochemistry and atmospheric dynamics.

### *12.11.1 Dayglow Spectra*

A spectrum of Venus observed by the Extreme Ultraviolet Explorer is shown in Figure 12.6. Two helium lines in the spectrum were discussed in Section 12.3.1. The $He^+$ 304 Å intensity of 69 ± 20 R is comparable to that observed at V11–12 (Bertaux et al. 1981) and too high for resonance scattering by the $He^+$ ions. Those ions were measured by the PV ion mass spectrometer with a peak density of 200 $cm^{-3}$ at 290 km. The line is excited by charge exchange between the solar wind alpha particles and the atmospheric species (Krasnopolsky and Gladstone 2005).

Spectra of Venus that covered a range of 80–120 nm were observed during the Galileo and Cassini flybys (Hord et al. 1991; Gerard et al. 2011b) and using the Hopkins Ultraviolet Telescope (HUT; Feldman et al. 2000) at the Space Shuttle. Both HUT and Cassini spectra (Figure 12.69) had resolution of 0.4 nm, and their observations were at solar minimum and maximum, $F_{10.7}$ = 82 and 214, respectively. Some lines in the HUT spectrum are significantly contaminated by the terrestrial dayglow. The observed spectral features are compared with calculations in Table 12.10. Overall, there is a fair agreement between the observations and calculations that are based on the model by Hedin et al. (1983; Table 12.7). Typically a few mechanisms contribute to a line excitation, multiple scattering of some lines complicates the problem, and densities of the emitting species have not been retrieved from the spectra.

Spectra in the range of 120–180 nm were observed by a few spacecraft including the PV and VEX orbiters. The most detailed Cassini/UVIS spectrum with all features identified is

Table 12.10 *Dayglow emissions observed using Cassini/UVIS and HUT at* $F_{10.7}$ = *214 and 82, respectively, compared with calculations for the UVIS conditions*

| λ (nm) | Emissions | UVIS (R) | HUT (R) | UVIS/HUT ratio | Model (R) |
|---|---|---|---|---|---|
| 83.4 | OII | 261 ± 4 | 91 ± 41 | 2.9 | 536 |
| 98.9 | OI | 110 ± 2 | 45 ± 33 | 2.4 | 94 |
| 102.5 | OI + Ly-ß | 180 ± 3 | 115 ± 23 | 1.6 | – |
| 104.0 | OI | 25 ± 1 | 25 ± 1 | 1 | – |
| 108.8 | CO C–X (0–0) + NII | 44 ± 6 | 63 ± 2 | 1.4 | 37[b] |
| 111.4 | CI | 14 ± 1 | – | – | – |
| 113.4 | NI | 27 ± 1 | 35 ± 11 | 0.8 | 18.1 |
| 115.2 | CO B–X (0–0) + OI | 211 ± 6 | 128 ± 10 | 1.6 | 177[b] |
| 115.8 | CI | 13 ± 3 | – | – | – |
| 120.0 | NI | 93 ± 4 | 77 ± 16 | 1.2 | 176 |
| 124.3 | NI | 23 ± 1 | – | – | 15.9 |
| 126.1 | CI | 15 ± 1 | – | – | – |
| 127.7 | CI | 175 ± 3 | – | – | – |
| 135.6 | OI | 776[a] ± 7 | 605[a] ± 28 | 1.3 | 840 |

*Note.* From Gerard et al. (2011b).
[a]Including blended CO fourth positive underlying bands. [b]Calculated for photoelectron impact on CO.

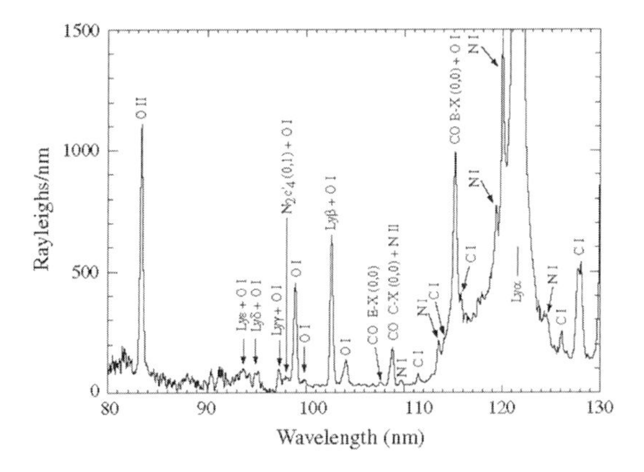


Figure 12.69 Average UVIS spectrum of Venus observed during the Cassini flyby (Gerard et al. 2011b).

shown in Figure 12.70. Main spectral features are the CO bands of the fourth positive system ($A^1\Pi \rightarrow X^1\Sigma^+$) and atomic lines of H Lyman-alpha 121.6 nm, O 130.4 and 135.6 nm, and C 156.1 and 165.7 nm.

The only published Venus spectrum that covers 180–300 nm is shown in Figure 12.71 (Chaufray et al. 2012a). It is similar to the SPICAM spectra on the Martian limb, properly

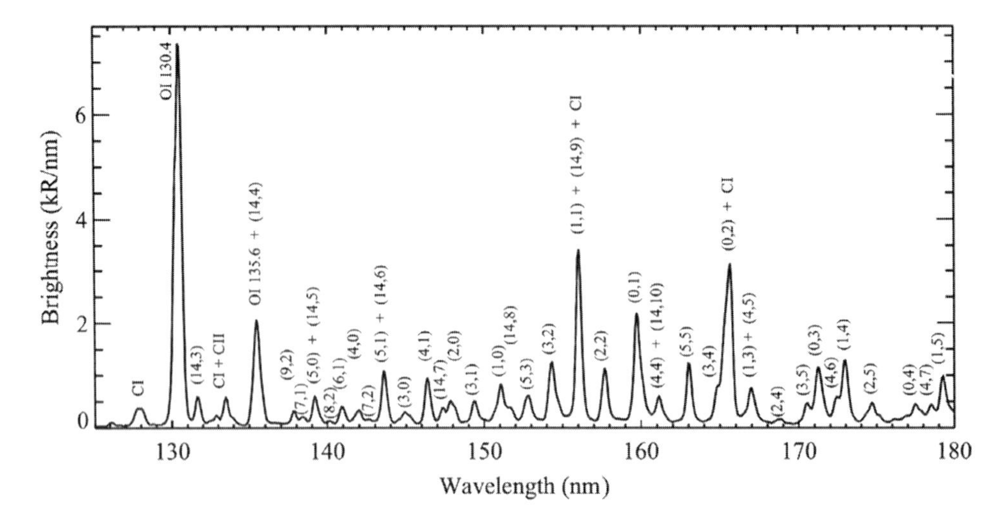


Figure 12.70 Average Cassini/UVIS spectrum of Venus at solar maximum with resolution of 0.4 nm. All features are identified, and, e.g., (2,2) means the CO (A → X) (2,2) band. From Hubert et al. (2010).

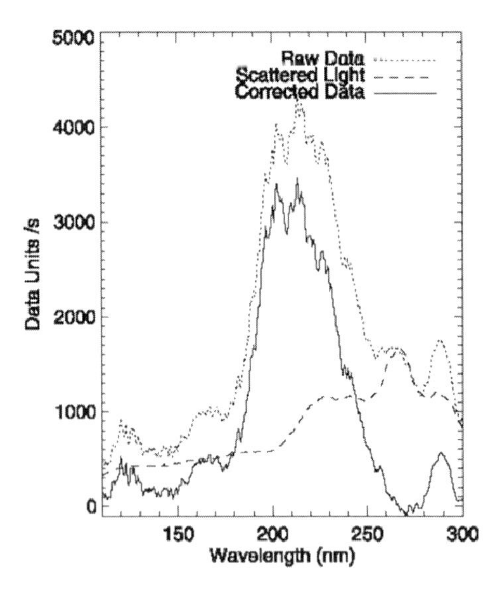


Figure 12.71 A SPICAV/VEX spectrum on the Venus limb with resolution of 10 nm. Raw data, scattered light, and the corrected spectrum are shown. From Chaufray et al. (2012a).

scaled and degraded to the resolution of 10 nm, except the H 122 nm line that is much weaker on Venus. Vertical profiles of the CO ($a$ → X) Cameron bands and the $CO_2^+$ (B → X) band at 289.6 nm were extracted from the observations. The Cameron bands are almost completely responsible for the broad feature at 180–270 nm in the spectrum in Figure 12.71.

The peak limb brightness of the Cameron bands is 2.0 ± 0.1 MR at 137 ± 1.5 km; $F_{10.7}$ = 140. It corresponds to a vertical brightness of 25 kR at the subsolar point. Scaling of the calculations by Gronoff et al. (2008) to the conditions of the observations results in 43 kR, higher by a factor of 1.7. (Here the dayglow is assumed proportional to $F_{10.7}$ and cos SZA. The $\cos^{1/2}$ law adopted by the authors is applicable to the peak electron density and invalid here.) The $CO_2^+$ (B → X) feature includes the O line at 297.2 nm with the expected contribution of 10%–15%. The peak limb brightness is 270 ± 20 kR at 135.5 ± 2.5 km, and the vertical brightness is 3.2 kR, smaller than the calculated and scaled intensities in Gronoff et al. (2008) by a factor of 5.

### *12.11.2 Atomic Hydrogen*

Vertical profiles of atomic hydrogen are retrieved from observations of the H Lyman-alpha emission at 121.6 nm. First measurements were made during the flybys of Mariners 5 and 10 and Veneras 4, 11, and 12. The flybys covered wide ranges of altitudes that are essential for the data analysis. The observations were made at the Venus oribters Veneras 9, 10, Pioneer Venus, and Venus Express as well.

The observed airglow profiles (Figure 12.72) look like sums of two exponents. There were attempts to explain this feature in the Mariner 5 data by significant abundances of $H_2$ and/or D in the upper atmosphere. The problem was solved by Anderson (1976), who suggested populations of thermal (cold) and hot hydrogen with excess energy from some exothermic processes in the exosphere. Table 12.11 gives a summary of the interpretation

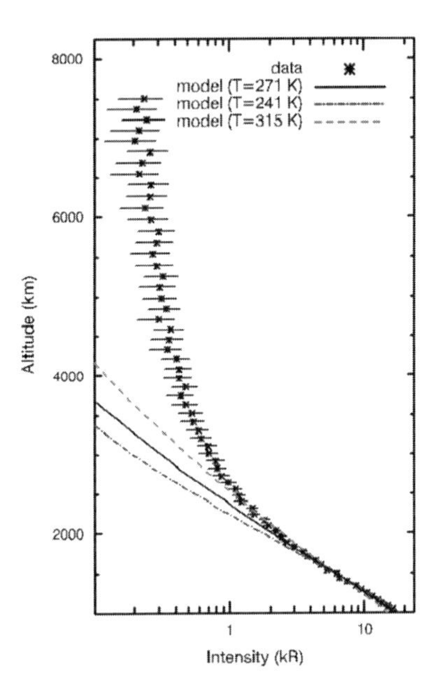


Figure 12.72 Observed profile of the H 122 nm dayglow near noon at solar minimum on Venus. The cold component with $T = 271^{+44}_{-30}$ K and $[H]_{ex} = 13.5^{+8.1}_{-5.6} \times 10^4$ cm$^{-3}$ dominates below 2000 km. From Chaufray et al. (2012b).

Table 12.11 *Summary of the flyby observations of the H 122 nm airglow*

| Mission, reference | $F_{10.7}$ | $n_{cd}$ | $T_{cd}$ | $n_{hd}$ | $T_{hd}$ | $n_{cn}$ | $T_{cn}$ | $n_{hn}$ | $T_{hn}$ |
|---|---|---|---|---|---|---|---|---|---|
| M5, Anderson (1976) | 120 | $(2\pm1)\times10^5$ | $275\pm50$ | 1300 | $1020\pm100$ | $(2\pm1)\times10^5$ | $150\pm50$ | 1000 | $1500\pm200$ |
| M10, Takacs et al. (1980) | 76 | $(1.5\pm1)\times10^5$ | $275\pm50$ | 500 | $1250\pm100$ | $(1\pm0.5)\times10^5$ | $150\pm25$ | – | – |
| V11–12, Bertaux et al. (1982) | 140 | $(4\pm2.5)\times10^4$ | $300\pm25$ | 1000 | 1000 | – | – | – | – |

Subscripts c, h, d, n mean cold, hot, day, night, respectively; number densities are in $cm^{-3}$ and refer to 200 km.

of the flyby observations. The data analysis by Anderson (1976) gave the first reasonable temperatures of both the dayside and nightside thermospheres.

Direct measurements of the neutral and ion composition using the mass spectrometers on board the Pioneer Venus orbiter revealed very strong diurnal variations of atomic hydrogen (Brinton et al. 1980). There are three basic processes that control its densities:

$$O^+ + H \rightarrow H^+ + O,\ k_1 = 5.7 \times 10^{-10}(T/300)^{0.36}e^{9/T}\ \text{cm}^3\ \text{s}^{-1};$$

$$H^+ + O \rightarrow O^+ + H,\ k_2 = 6.9 \times 10^{-10}(T_i/300)^{0.26}\ e^{-224/T_i}\ \text{cm}^3\ \text{s}^{-1};$$

$$H^+ + CO_2 \rightarrow HCO^+ + O,\ k_3 = 3.5 \times 10^{-9}\ \text{cm}^3\ \text{s}^{-1}.$$

Then the equilibrium results in

$$[H] = \frac{[H^+]}{[O^+]}\left([O]\frac{k_2}{k_1} + [CO_2]\frac{k_3}{k_1}\right). \tag{12.12}$$

The derived diurnal variations of atomic hydrogen are shown in Figure 12.73. Actually the data refer to 180 km on the dayside and 160 km on the nightside and have been corrected to 250 km to facilitate comparison with the Lyman-alpha observations. This technique revealed a morningside hydrogen bulge with the densities exceeding the daytime values by three orders of magnitude at solar maximum.

The above rate coefficients are from the UMIST 2012 database (Section 9.7) and slightly differ from those used by Brinton et al. (1980). Furthermore, photoionization and electron impact ionization of H and charge exchange $CO^+$ + H contribute to the production of $H^+$ as well, while recombination of $HCO^+$returns H. However, correction for all these sources is ≈30% and does not change the discovery of the hydrogen bulge.

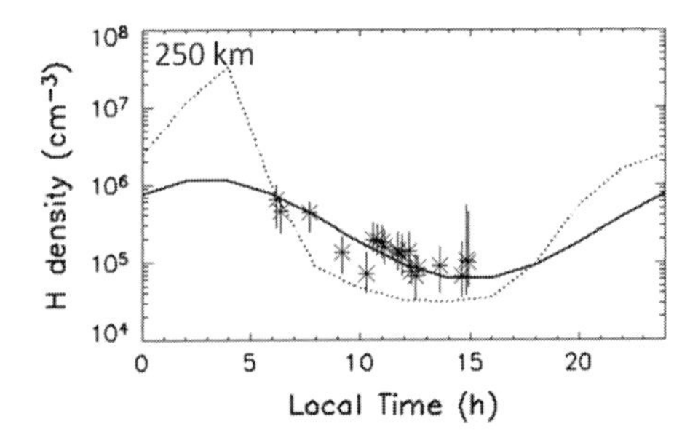


Figure 12.73 Diurnal variations of hydrogen density at 250 km extrapolated from the PV ion mass spectrometer data for 160–170 km at solar maximum (dots; Brinton et al. 1980) are compared with those retrieved from the SPICAV/VEX Lyman-alpha observations at solar minimum and extrapolated for the nighttime to fit the data at LT = 6, 12, and 18 h. From Chaufray et al. (2012b).

Observations of the H line at 122 nm from the PV orbiter (Paxton et al. 1985, 1988) were made in a spinning mode with scanning across the Venus disk from limb to limb. Limb brightening in this mode made it possible to retrieve profiles of H below the exobase that are determined by two parameters: density of H and upward flux at the exobase (relationship (2.43a)). The observed diurnal variations confirmed the H bulge, and the dayside data showed the exobase number density of $(6 \pm 1.5)\times10^4$ cm$^{-3}$, flux of $(7.5 \pm 1.5)\times10^7$ cm$^{-2}$ s$^{-1}$, and column abundance of $(3.6 \pm 1)\times10^{13}$ cm$^{-2}$ above 110 km.

There were two sets of the SPICAV observations of the H Lyman-alpha emission at the Venus Express orbiter. Dayside observations (Chaufray et al. 2012b) were made in the first eight months of the mission at solar minimum ($F_{10.7}$ = 60–90) and covered altitudes from 1000 to 8000 km (Figure 12.72). The signal was saturated below 1000 km; however, the cold component dominates and was well measured below 2000 km. Average data for the cold component from 17 orbits at LT = 9 to 17 h are $n = (12 \pm 4)\times10^4$ cm$^{-3}$ and $T = 270 \pm 15$ K at the exobase. Here the standard deviations reflect variabilities of the values.

The altitude extent was insufficient to measure and make correction for the interplanetary Ly-$\alpha$ background, and the data were taken from the SOHO/SWAN observations. This resulted in significant uncertainties in temperatures of the hot component, though its measured densities were rather accurate below 5000 km. Average cold and hot hydrogen densities at solar minimum between LT = 09:00 and 15.00 are shown in Figure 12.74. The horizontal bars show the full observed variabilities of the data. The results agree with a total hydrogen profile for LT = 15:00 from Hodges (1999).

The observed exobase densities of H are plotted versus local time in Figure 12.73, and the data are greater than those measured by PV at solar maximum. The SPICAV observations near terminators at LT = 6 and 18 h are sensitive to the hydrogen distributions beyond the terminators and require 3D modeling that involves the nighttime hydrogen. An attempt to fit the observations using the diurnal behavior from the PV curve in Figure 12.73 failed, and the best sinusoidal fit is shown in the figure. This fit indicates the morningside bulge of an order of magnitude at solar minimum, much smaller than that at solar maximum.

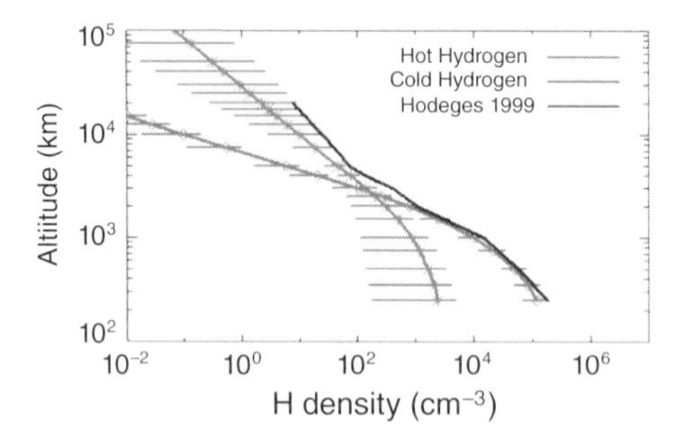


Figure 12.74 Average cold and hot hydrogen densities observed by SPICAV/VEX at solar minimum between LT = 9 to 15 h. The bars show the full observed variability of the data. A total hydrogen profile at LT = 15 h from Hodges (1999) is shown for comparison. From Chaufray et al. (2012b).

Hydrogen on the nightside near LT = 20 h was observed by SPICAV (Chaufray et al. 2015b) at four orbits in October 2011 at solar medium conditions ($F_{10.7}$ = 140). The observations involved the interplanetary background, emission of hydrogen from the illuminated part of the line of sight and absorption by hydrogen in the shadow. The retrieved exobase density was $3\times10^6$ cm$^{-3}$, greater than $2\times10^5$ cm$^{-3}$ at solar minimum and $7\times10^5$ cm$^{-3}$ at solar maximum (Figure 12.73). The derived hot hydrogen densities at the nightside exceed those on the dayside by a factor of 5.

### *12.11.3 Formation of Hydrogen Bulge*

The atmosphere above 150 km consists mostly of atomic oxygen, and its dayside density exceeds the nightside density by two orders of magnitude. Therefore the gas from the dayside expands to the nightside almost like to vacuum with a speed of $V \approx 0.5$ km s$^{-1}$, and all species are involved in this process. Then the mean flux of hydrogen from the day to nightside is

$$\Phi_H = n_0 V \frac{2\pi r H_H}{2\pi r^2} = n_0 V \frac{H_H}{r}. \tag{12.13}$$

Here $n_0$ is the H density at the lower boundary of the flow, $H_{\mathrm{H}}$ is the H scale height, and $r \approx 6200$ km is the radius at the flow boundary. This flux may be approximated by the diffusion limiting flux (Section 2.5.4)

$$\Phi_{\mathrm{H}}^{\max} = \frac{bf_H}{H} = \frac{bn_0}{Hn_{a0}}, \tag{12.14}$$

where $f_{\mathrm{H}}$ is the hydrogen element mixing ratio, $n_{a0}$ and $H$ are the total number density and scale height at the lower flow boundary, and $b = Dn_{a0}$ is the term in diffusion coefficient that is equal to $1.8\times10^{19}$ cm$^{-1}$ s$^{-1}$ for diffusion of H in $CO_2$ (Table 2.1) at 170 K typical of the mesopause. Using $\Phi_H = \Phi_{\mathrm{H}}^{\max}$ and the scale height for 170 K,

$$n_{a0} = \frac{br}{VHH_H} = 3.8 \times 10^{10}\ \mathrm{cm}^{-3}, \tag{12.15}$$

that is, the lower flow boundary is at 145 km according to Table 12.7. The hydrogen element mixing ratio near the mesopause may be evaluated as $f_H = f_{\mathrm{HCl}} + 2f_{\mathrm{H2O}} \approx (0.4 + 2)\times10^{-6}$ using $f_{\mathrm{H2O}} \approx 1$ ppm observed by the SOIR occultations at 100–110 km (Fedorova et al. 2008). Then $n_0 = n_{a0}f_H = 9\times10^4$ cm$^{-3}$, close to those observed on the dayside (Figure 12.73), and $\Phi_H = 1.2\times10^8$ cm$^{-2}$ s$^{-1}$ from (12.14).

The hydrogen mixing ratio on the dayside is almost constant up to the lower flow boundary, because the flow is diffusion-limiting. The hydrogen number density is almost constant at $n_0 = 9\times10^4$ cm$^{-3}$ above the flow boundary because of the large hydrogen scale height. The hydrogen density on the nightside for the known constant downward flux is almost constant down to a level where its mixing ratio is equal to 2.4 ppm. Then $dn/dz \approx 0$ and

$$\Phi_H = -\frac{nK}{H}, \quad n = n_a f_H, \quad K \approx 10^{13}\, n_a^{-1/2}\ \mathrm{cm^2\ s^{-1}}\ \text{(section 12.8.1)}; \tag{12.16}$$

therefore

$$n_a = \left(-\frac{\Phi_H H}{f_H \times 10^{13}}\right)^2 = 3.4 \times 10^{12}\ \mathrm{cm^{-3}}, \tag{12.17}$$

that is, down to 119 km on the nightside. The H mixing ratio is equal to 2.4 ppm at 145 km on the dayside and 119 km on the nightside, and the H number density night-day ratio is $n_a/n_{a0} \approx 100$. This hydrogen density ratio is almost constant up to the exobase and higher. The hydrogen morningside bulge was predicted by Kumar et al. (1978) using the similar consideration.

### *12.11.4 Atomic Deuterium*

The mass 2 ion in the PV ion mass spectrometer observations (Figures 12.49 and 12.54) was initially identified as $H_2^+$. McElroy et al. (1982) argued that the observed mass 2 ion densities agree with $D^+$ for the abundant HDO detected by the large probe mass spectrometer (Donahue et al. 1982; see Section 12.6.2). Hartle and Taylor (1983) studied altitude profiles of the ion mass 2 to 1 ratio to identify the mass 2 ion.

Densities of $H^+$, $D^+$, and $H_2^+$ are controlled by equilibriums:

$H^+ + O \leftrightarrow O^+ + H$, therefore $[H^+] = \alpha\,[H]\,[O^+]/[O]$;

$D^+ + O \leftrightarrow O^+ + D$, therefore $[D^+] = \alpha\,[D]\,[O^+]/[O]$; and $[D^+]/[H^+] = [D]/[H] \approx$ const;

$H_2 + h\nu, e \rightarrow H_2^+$, $H_2^+ + O \rightarrow OH^+ + H$, therefore $[H_2^+] = \beta\,[H_2]/[O]$.

Then

$$[H_2^+]/[H^+] = \frac{a}{[O^+]}\frac{[H_2]}{[H]} \approx \frac{\text{const}}{[O^+]}.$$

Here $\alpha$, $\beta$, and $a$ are ratios of the proper rate coefficients, and [D]/[H] and $[H_2]/[H]$ are almost constant because of the large scale heights of these species. The dependence of the ion ratio on $O^+$ is not supported by the observations that agree with the rather constant ratio expected for $[D^+]/[H^+]$. The observed D/H is equal to $(1.7 \pm 0.6)\times 10^{-2}$ at 160 km and $(2.2 \pm 0.6)\times 10^{-2} = 140 \pm 40$ SMOW below the homopause. This value agrees with the other measurements of D/H in water vapor (Section 12.6.2).

### *12.11.5 Atomic Oxygen*

The O ($2p^3 3s^3S \rightarrow {}^3P$) triplet at 130.4 nm is a standard tool to retrieve atomic oxygen abundances from UV dayglow of a planet. The PV observations (Figure 12.75) were analyzed by Meier et al. (1983), who adopted a partial frequency redistribution of photons after the scattering process. The emission is excited by resonance scattering and

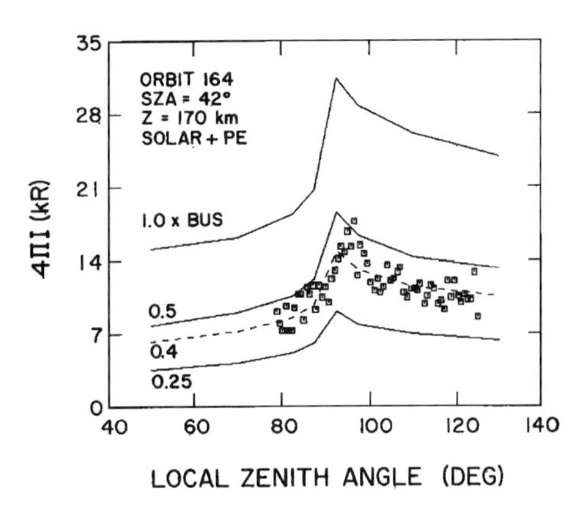


Figure 12.75 Observed brightness of the O 130.4 nm dayglow versus zenith angle of the optical axis of the PV UV spectrometer is compared with calculated distributions with the oxygen profile from von Zahn et al. (1980) scaled by a variable factor. From Meier et al. (1983).

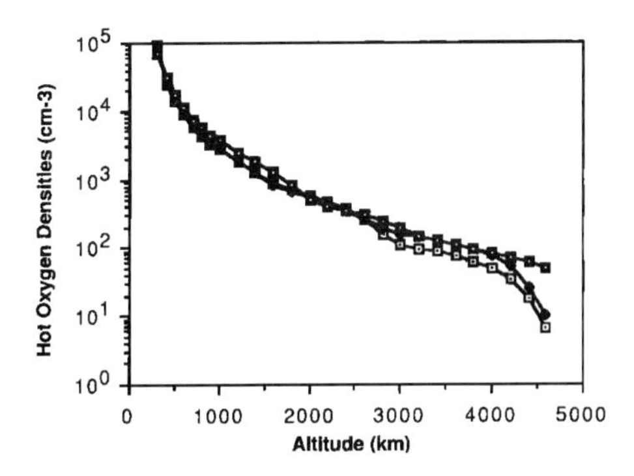


Figure 12.76 PV observations and models of hot oxygen on Venus (Nagy and Cravens 1988).

photoelectron excitation. The model atmosphere from the bus mass spectrometer measurements at LT = 8 h (von Zahn et al. 1980; Figure 12.39) was applied to calculate the O 130.4 nm emission in the geometry of the PV UV spectrometer observations (Figure 12.75). A scaling factor to the oxygen profile in the model was a fitting parameter. The best fit is for a factor of 0.4 that corresponds to the O/$CO_2$ ratio of 0.07 at the altitude of the ionospheric peak of 140 km. Paxton and Meier (1986) further improved the excitation and radiative transfer model to get the ratio of 0.1.

Later the O emissions at 130.4 and 135.6 nm in the high-quality Cassini/UVIS spectra (Figure 12.70) were analyzed by Hubert et al. (2010). Variations of the line intensities with solar zenith angle were reproduced by calculations based on the empirical thermosphere model by Hedin et al. (1983; Table 12.7), confirming the atomic oxygen densities in the model.

The O 130.4 nm emission extended in the PV observations up to 4500 km (Figure 12.76), indicating a hot oxygen component. Nagy and Cravens (1988) calculated density profiles of hot oxygen that is produced by dissociative recombination of $O_2^+$ (Section 8.7), and the results are in good agreement with the observations. Groeller et al. (2010) developed a sophisticated model for hot oxygen that was based on various assumptions on elastic, inelastic, and quenching collisions.

### *12.11.6 Atomic Carbon*

The C 166 nm line is the best of a few carbon lines in the Venus spectrum (Figure 12.70) to retrieve carbon abundances, because resonance scattering is the most effective for this line with emission rate factor $g = 4.4\times10^{-5}$ $s^{-1}$ at 1 AU (Table 8.2). The PV observations of this line at the limb (Paxton 1985) are shown in Figure 12.77.

To calculate densities of atomic carbon, Paxton (1985b) involved its production by photo- and electron impact dissociation of $CO_2$ and CO and recombination of $CO^+$. Atomic carbon is lost mostly in the reaction with $O_2$, which abundance is poorly known, and $O_2$/$CO_2 \approx 0.003$ at 135 km was adopted in calculations of C by Krasnopolsky (1982), Fox (1982), and Paxton (1985). (Fox (1982) calculated abundances of $C^+$ and compared them with the PV ion mass spectrometer data as well.) The calculated densities of C are rather similar in all three papers with a maximum of ~$5\times10^6$ $cm^{-3}$ at 140–150 km.

Next, prompt emissions from photo- and electron impact excitation of $CO_2$ and CO and then the CO(A $\rightarrow$ X) (0,2) and (3,4) bands that contaminate the emission at 166 nm were calculated (Figure 12.77). The prompt emissions originate from the exothermic reactions that give hot C atoms with high speed, and their emission is weakly absorbed by thermal C atoms. With all these sources taken into account, there is a significant tail in the measured profile above 250 km that is attributed to hot carbon atoms (Figure 12.77).

Fox and Paxton (2005) updated the model for C and $C^+$ on Venus using improved data on the CO photodissociation (Fox and Black 1989), production of C in dissociative recombination of $CO_2^+$ (Section 8.7), inclusion of charge exchange between $O^+$ and C, and some updated reaction rate coefficients. The model was compared with the PV observations of the

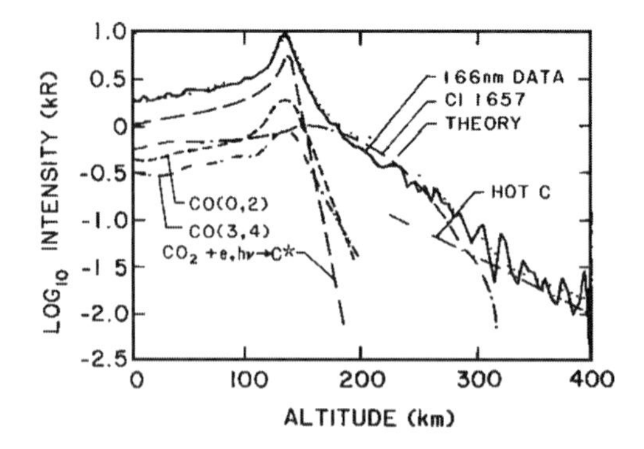


Figure 12.77 Observed and calculated profiles of the C 166 nm dayglow at the Venus limb. The emission involves the resonant scattering by atomic carbon, two CO (A $\rightarrow$ X) bands, electron and photon excitation of $CO_2$, and hot carbon. From Paxton (1985).

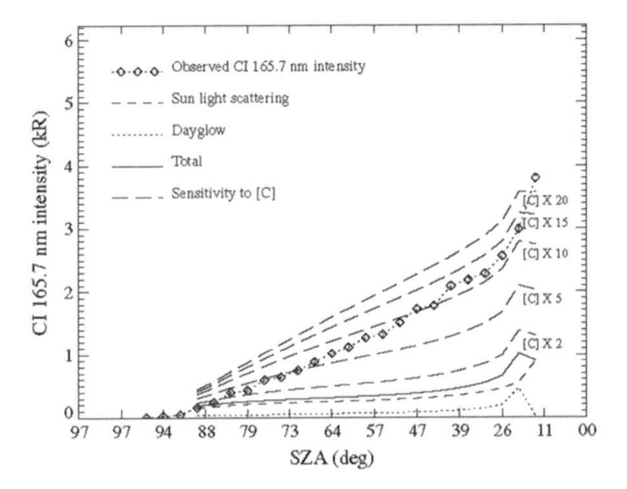


Figure 12.78 Observed intensities of the C 166 nm line in the Cassini/UVIS spectra. Viewing angle changed from 83° to 47° and 60° for SZA = 12°, 64°, and 90°, respectively. The C profile from Fox and Paxton (2005) was scaled by factors of 1, 2, 5, 10, 15, and 20 in the calculations. From Hubert et al. (2012).

C 166 nm line and the $C^+$ densities, and the best fit was for $O_2/CO_2 = 3\times10^{-4}$ in the lower thermosphere and the rate coefficient of $\approx10^{-10}$ $cm^3\ s^{-1}$ of the charge exchange reaction $O^+$ + C. The best-fit C profile did not change significantly from the earlier versions. The recommended $O_2/CO_2$ ratio is similar to that in the photochemical model by Krasnopolsky and Parshev (1981); other models give $10^{-3}$ (Yung and DeMore 1982), $1.5\times10^{-4}$ (Mills 1998), and $10^{-4}$ (Krasnopolsky 2012b), all at 112 km.

The carbon lines in the Cassini/UVIS spectrum of Venus (Figure 12.70) were analyzed by Hubert et al. (2012). High signal-to-noise ratio and resolution of 0.37 nm made it possible to separate the C 166 nm line from the nearby CO bands. Twenty-four spectra were measured along the track of the instrument line of sight from the morning terminator to the limb near the noon (Figure 12.78). Hubert et al. (2012) found a serious disagreement between the model and the observation, with a scaling factor for the C profile from Fox and Paxton (2005) varying from 1 at the terminator to 20 near the limb at noon.

We may indicate a source of this variability in true variations of C with SZA. The maximum of C is at 145 km with $[O_2] \approx 3\times10^{-4}[CO_2] = 6\times10^6$ $cm^{-3}$, the rate coefficient of C + $O_2$ is $k = 5.1\times10^{-11}(300/T)^{0.3}$ $cm^3\ s^{-1}$, and chemical lifetime of C is $\tau \approx (k\,[O_2])^{-1} \approx 1$ h. All sources of C are proportional to cos SZA, therefore densities of C should be proportional to cos SZA as well.

However, this explains only partly the disagreement between the calculations and observations in Hubert et al. (2012). The profile of C was calculated by Fox and Paxton (2005) for solar max at SZA = 25° and should be increased by a factor of 12 to fit the observation at this SZA (Figure 12.78). Some discontinuities in the excitation rates and comparison with the PV data (Figure 12.77) need explanation as well.

### *12.11.7 Carbon Monoxide*

Bands of the CO ($A^1\Pi \rightarrow X^1\Sigma^+$) fourth positive system are the main spectral features at 130–175 nm (Figure 12.70). These bands are excited by fluorescent scattering and electron impact of CO, photo- and electron impact dissociation of $CO_2$, and dissociative recombination of $CO_2^+$ (Section 8.7). Durrance (1981) calculated emission rate factors $g$ for the CO ($A \rightarrow X$) bands including the (14,v″) progression that is excited by the solar Lyman-alpha emission. This progression is rather strong, and its bands (14,3) 131.7 nm and (14,4) 135.4 nm contaminate the O 130.4 and 135.6 nm lines at low resolution. On the other hand, excitation of the (14,v″) progression by photoelectrons and photodissociation of $CO_2$ is negligible, and their band intensities are readily converted into CO column abundances. The best bands for this purpose are (14,3) 131.7 nm with $g = 3.5\times10^{-9}$ s$^{-1}$ and (14,5) 139.2 nm with $g = 2.9\times10^{-9}$ s$^{-1}$ at 1 AU (Table 8.2).

The CO (v′,0) bands are affected by self-absorption: their photons are absorbed by the CO molecules and reradiated in the (v′,v″) bands, so that the single scattering albedo is A(v′,0)/ΣA(v′,v″) < 1. The CO ($A^1\Pi$) molecules excited by photoelectrons and photodissociation of $CO_2$ have nonthermal rotational distribution, and their self-absorption is significantly weaker than that of the cold CO molecules.

Calculated spectra of the CO ($A \rightarrow X$) bands in Hubert et al. (2010) were based on the empirical model by Hedin et al. (1983; Table 12.7) and agreed with the observed spectra, thus confirming the model CO abundances. The authors mentioned that the measured intensity of the (14,3) band was 56 R at 11°S and LT = 16:15, and this resulted in a vertical CO column of $6.4\times10^{15}$ cm$^{-2}$ above the unspecified altitude of the unit optical depth, close to $8.7\times10^{15}$ cm$^{-2}$ at the same conditions in the model by Hedin et al. (1983).

### *12.11.8 CO Dayglow at 4.7 μm*

This dayglow was discovered by Crovisier et al. (2006) using ground-based high-resolution spectroscopy at the Canada–France–Hawaii Telescope. The dayglow presents emissions of the CO (1–0) and (2–1) vibrational bands. Later the dayglow was observed using CSHELL/IRTF (Figure 12.79). Each spectrum in the latitude range of ±60° included six (2–1) lines and three (1–0) lines. The CO (1–0) dayglow is produced by resonance scattering of the solar photons and the thermal emission at 4.7 μm. The dayglow is optically thick, weakly depends on the CO abundance, and its nadir observations are poorly accessible for diagnostics of the atmosphere.

The CO (2–1) dayglow is excited by absorption of the sunlight at the CO (2–0) band at 2.35 μm and photolysis of $CO_2$ by the solar Lyman-alpha emission. The dayglow is quenched by $CO_2$. The measured latitudinal variations of the dayglow rotational temperature are shown in Figure 12.80 along with temperature at 74 km derived from the CO absorption bands in the same spectra (Figure 12.79). Calculations show that the rotational temperatures reflect kinetic temperatures at 111 km. The measured mean temperatures are 203 K at 111 km (LT = 08:00) and 223 K at 74 km, in accord with VTGCM (Figure 12.45)

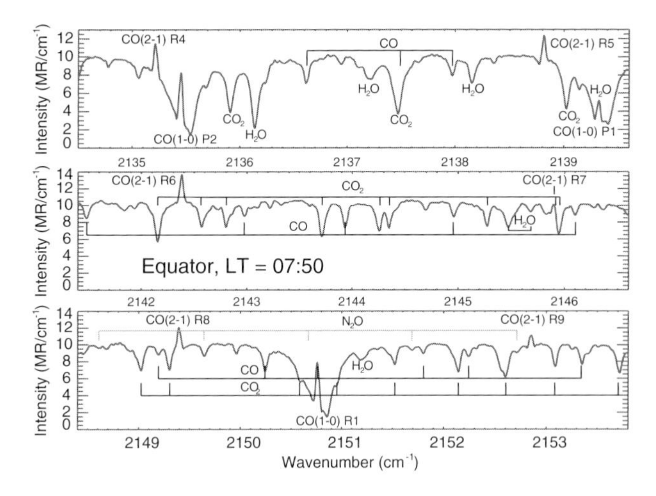


Figure 12.79 Spectra of Venus near 4.7 μm at the equator with resolving power of $4\times10^4$. Six CO (2–1) and three CO (1–0) dayglow lines are identified along with absorption lines of CO and $CO_2$ on Venus and telluric $H_2O$ and $N_2O$ lines. From Krasnopolsky (2014c).

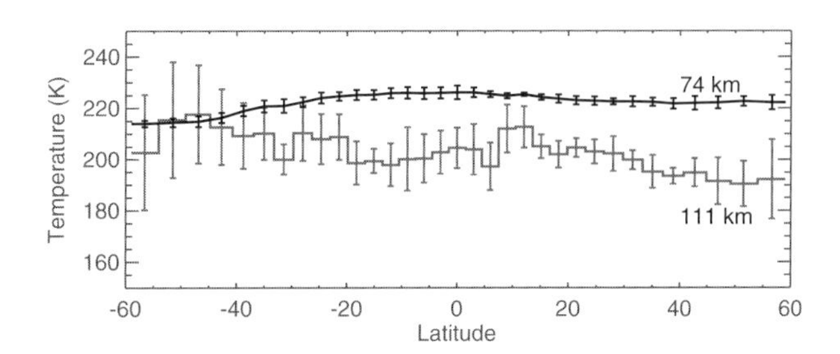


Figure 12.80 Retrieved latitudinal variations of temperature at 74 and 111 km (Krasnopolsky 2014c).

and VIRA (Table 12.2), respectively. The observations were conducted on August 17, 2012, at $F_{10.7} \approx 100$.

The observed (2–1) dayglow demonstrates significant limb brightening, while the vertical intensities are rather constant at 3.3 MR at SZA = 64° in the latitude range of ±50°. Using a model for the dayglow excitation and quenching, the observed intensities are converted into CO mixing ratios at 104 km (Figure 12.81) and compared with those at 74 km extracted from the CO absorption lines in the same spectra. The mean CO abundance is 560 ± 100 ppm at 104 km and ≈45 ppm at 74 km. The uncertainties is mostly systematic and originates from the model parameters.

Spectra of the CO dayglow at 4.7 μm were observed at the limb using VIRTIS-H at Venus Express (Gilli et al. 2015). The instrument resolving power was 1200 at 4.7 μm, and the field of view at the limb was a few km near the periapsis. The limb observations made it possible to retrieve vertical profiles of CO and temperature. An average spectrum from one

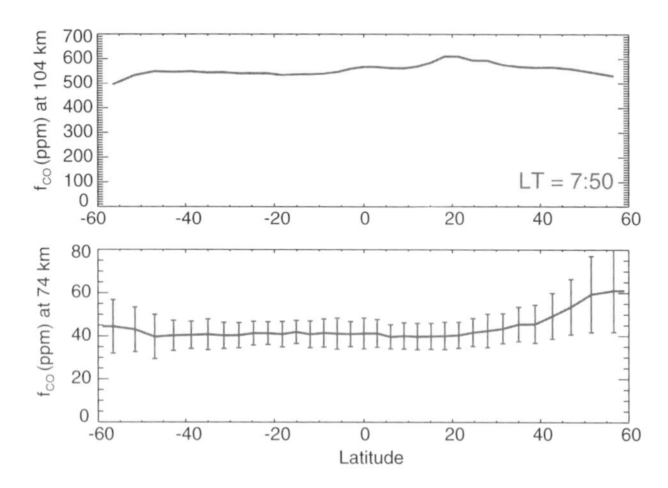


Figure 12.81 Retrieved latitudinal variations of the CO mixing ratio at 74 and 104 km (Krasnopolsky 2014c).

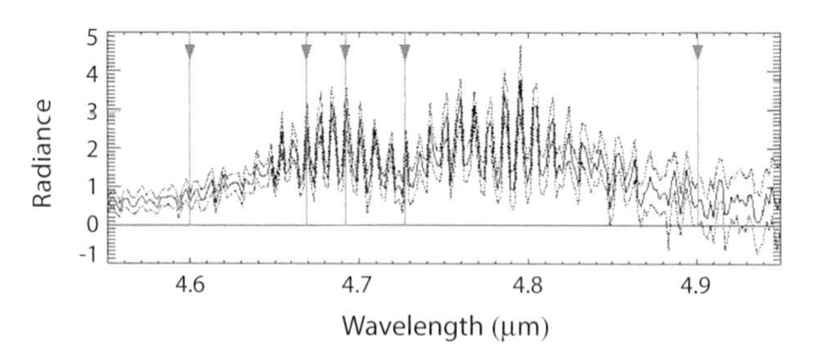


Figure 12.82 Average spectrum of the CO dayglow observed by VIRTIS-H at Venus Express. Arrows indicate spectral data used for analysis: 4.6 and 4.9 μm are the background, 4.67 and 4.73 μm are the CO (2–1) and (1–0) bands, and 4.69 μm is a sum of both bands. From Gilli et al. (2015).

orbit is shown in Figure 12.82. Five points in each spectrum were used for analysis: two background values, lines of the (2–1) and (1–0) bands, and their blend. Modeling of the (1–0) band required multiple scattering in a system with many rotational levels.

The derived CO number densities at low and middle latitudes near noon are shown in Figure 12.83. Excitation of the CO bands by photolysis of $CO_2$ was neglected in the analysis, though this process contributes ~50% of the dayglow at 115 km (Krasnopolsky 2014c). Therefore the data at 110–125 km are overestimated. The figure depicts the CO densities from Hedin et al. (1983) and VTGCM (Brecht and Bougher 2012) at the same conditions.

Retrieved temperatures at low latitudes are shown and compared with the models in Figure 12.84. The uncertainties are ~50 K, and the data are not restrictive. The observational data were collected for 5 years, and analogs of Figures 12.83 and 12.84 exist for other latitudes and local times.

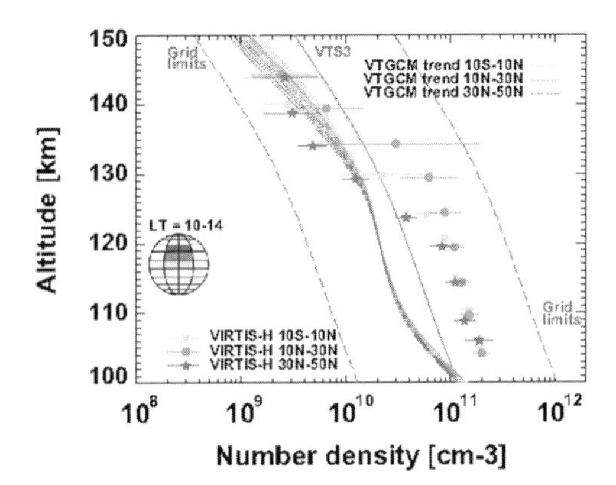


Figure 12.83 Retrieved CO number densities at 10°S to 50°N and LT = 10:00 to 14:00 are compared to those from VTS3 (Hedin et al. 1983) and VTGCM (Brecht and Bougher 2012). From Gilli et al. (2015).

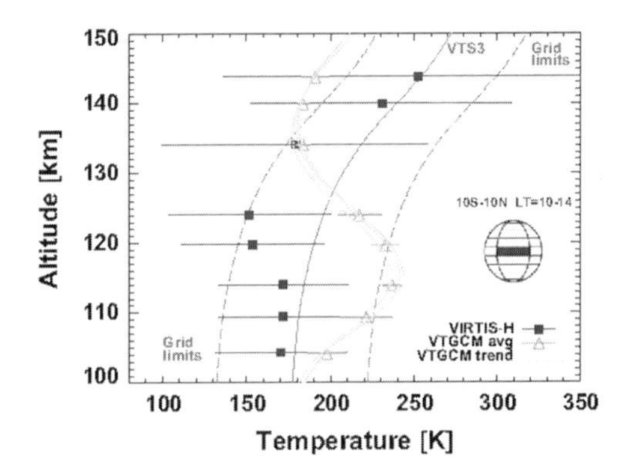


Figure 12.84 Retrieved temperatures at low latitude ±10° and LT = 10:00 to 14:00 are compared to those from VTS3 (Hedin et al. 1983) and VTGCM (Brecht and Bougher 2012). From Gilli et al. (2015).

## 12.12 Escape of H, O, and He and Evolution of Water

### *12.12.1 Hydrogen*

Thermal escape of hydrogen can be calculated using relationship (2.43) and the data of Table 12.11 on the cold and hot hydrogen. The Mariner 5 results are the most complete and refer to solar medium activity ($F_{10.7} = 120$). The calculated fluxes are 36 $cm^{-2}\ s^{-1}$ for cold dayside hydrogen, $2.3\times10^6\ cm^{-2}\ s^{-1}$ for hot dayside hydrogen, and $1.1\times10^7\ cm^{-2}\ s^{-1}$ for hot nightside hydrogen. Hence the mean escape rate is $6.6\times10^6\ cm^{-2}\ s^{-1}$.

There are two basic processes that produce hot hydrogen atoms: collisions of cold H with hot oxygen and charge exchange of $H^+$. Photo- and electron impact ionization of O above the exobase results in $O^+ + CO_2 \rightarrow O_2^+(v \leq 5) + CO$ and dissociative recombination of $O_2^+$. Vibrational distribution of $O_2^+$ in this process was calculated by Fox and Hac (2009), and yields of the O metastable states in four recombination channels for $v \leq 2$ are known (Section 8.7). The highest released energy is 6.98 eV and corresponds to velocity of the oxygen atoms of $V_0$ = 6.48 km $s^{-1}$. Hydrogen atoms collided by these nascent oxygen atoms have velocities

$$V = \frac{2V_0 \cos\varphi}{1 + m/m_0}. \tag{12.18}$$

Here $m/m_0$ = 1/16 is the mass ratio and $\varphi$ is the scattering angle of the H atom. The scattering phase function is poorly known, and we adopt a probability of 0.15 for the H atom to get $V$ exceeding the escape velocity of 10.2 km $s^{-1}$ in collision with the nascent O atom from McElroy et al. (1982). Here we follow that paper with some adjustments of numerical values.

According to the definition of the exobase, $\sigma N_0$ = 1, where $\sigma \approx 3\times10^{-15}$ $cm^2$ is the collisional cross section and $N_0 = 1/\sigma \approx 3\times10^{14}$ $cm^{-2}$ is the atomic oxygen column abundance at the exobase. Its ionization frequency is $J = 1.2\times10^{-6}$ $s^{-1}$ at solar maximum, and production rate of hot oxygen in recombination of $O_2^+$ is $2JN_0 = 7\times10^8$ $s^{-1}$ above the exobase. The exobase is at 205 km with [O] $\approx 2\times10^8$ $cm^{-3}$ (Table 12.7) and [H] $\approx 5\times10^4$ $cm^{-3}$ (Figure 12.73). Then the escape rate on the dayside is

$$\frac{0.5 \times 0.15 \times 16 \times 5 \times 10^4}{1.5 \times 2 \times 10^8} \times 7 \times 10^8 = 1.4 \times 10^5\ \text{cm}^{-2}\ \text{s}^{-1}.$$

Here 0.5 is half the total number of the hot hydrogen atoms moving up, 16 is the scale height ratio for H and O, and 1.5 is the adopted cross section ratio for O and H.

Escape of hydrogen on the nightside is concentrated in the bulge that extends from LT = 1 to 5 h with effective radius of 30°. This bulge covers 1 – cos 30° = 0.13 of the nightside. Extrapolation of the $O^+$ flow from the dayside to the bulge position using the data of Fox (2011; Section 12.9.4) gives the recombination rate ~$5\times10^7$ $cm^{-2}$ $s^{-1}$. [H] $\approx 2\times10^7$ $cm^{-3}$ is in the bulge at the exobase, and the escape rate is

$$\frac{0.5 \times 0.15 \times 16 \times 2 \times 10^7 \times 0.13}{1.5 \times 2 \times 10^8} \times 2 \times 5 \times 10^7 \approx 10^6\ \text{cm}^{-2}\ \text{s}^{-1}.$$

The global-mean escape rate from this mechanism is $6\times10^5$ $cm^{-2}$ $s^{-1}$, and averaging over the solar cycle would reduce this value to ~$3\times10^5$ $cm^{-2}$ $s^{-1}$.

Hydrogen atoms can get high energy and escape via charge exchange:

$$H^+ + H \rightarrow H^* + H^+ \qquad \text{and} \qquad H^+ + O \rightarrow H^* + O^+.$$

The ion temperature is a few thousand kelvins, and the neutral H atoms keep this temperature after charge exchange. Detailed calculations of this process by Hartle et al. (1996) involved variations of the Venus thermosphere and ionosphere with solar activity by comparing the PV observational data in the first 3 years at solar maximum and the final phase in 1992 at solar medium conditions. The calculated global-mean and solar cycle average escape by charge exchange is equal to $8\times10^6$ cm$^{-2}$ s$^{-1}$.

### *12.12.2 Escape of Ions*

Ions above the ionopause are swept out by the solar wind, and a significant part of them leaves the planet. Therefore ionization processes above the ionopause result in escape. These processes are photoionization, charge exchange with the solar wind ions, and electron impact. Some ionospheric ions move above the ionopause and form ion flow that escapes.

Escape of ions was measured by the ASPERA instrument onboard Venus Express. Total escape rates were measured at $2.7\times10^{24}$ s$^{-1}$ for $O^+$, $7.1\times10^{24}$ s$^{-1}$ for $H^+$, and $7.9\times10^{22}$ s$^{-1}$ for $He^+$ at solar minimum (Fedorov et al. 2011). These values are converted into cm$^{-2}$ s$^{-1}$ by dividing by the Venus total area $4\pi r^2 = 4.6\times10^{18}$ cm$^2$, so that $10^{24}$ s$^{-1}$ = $2.15\times10^5$ cm$^{-2}$ s$^{-1}$. Later Masunaga et al. (2013) doubled the value for $O^+$ to $5.4\times10^{24}$ s$^{-1}$, and it is not clear if the values for $H^+$ and $He^+$ should be doubled as well.

Curry et al. (2015) created a model for $O^+$ that gives the escape of $5.5\times10^{24}$ s$^{-1}$ = $1.2\times10^6$ cm$^{-2}$ s$^{-1}$ at solar minimum and $2.6\times10^{25}$ s$^{-1}$ = $5.6\times10^6$ cm$^{-2}$ s$^{-1}$ at solar maximum, so that the average value for the solar cycle is $3.4\times10^6$ cm$^{-2}$ s$^{-1}$. Neutral hot oxygen atoms do not escape, and this value is the total oxygen escape. If the scaling of the $O^+$ escape from Fedorov et al. (2011) to the final value is applicable to $H^+$ and $He^+$, then their mean escape rates are $9\times10^6$ and $10^5$ cm$^{-2}$ s$^{-1}$, respectively. Hence the total hydrogen (neutral plus ion) escape is $1.7\times10^7$ cm$^{-2}$ s$^{-1}$ and exceeds that of oxygen by a factor of 5.

Curry et al. (2015) calculated contributions of the four processes to the escape of $O^+$. Photoionization gives ≈60% and electron impact ~25%, the remaining is charge exchange with the solar wind ions and ion outflow.

### *12.12.3 Evolution of Water*

Thus there are three processes of the hydrogen escape on Venus: collisions with hot oxygen H + $O^*$, collisions with hot ions H + $H^+$, and sweeping out of $H^+$ above the ionopause. Their mean rates are $3\times10^5$, $8\times10^6$, and $9\times10^6$ cm$^{-2}$ s$^{-1}$, respectively. Dividing the $H_2O$ abundance of 30 ppm (Table 12.10) by the total H escape rate, the $H_2O$ lifetime is 0.16 Byr on Venus.

The mean D/H ratio from the six observations (Section 12.6.2) is 0.023 = 150 SMOW. To calculate escape of D, one needs fractionation factor

$$f = \frac{\Phi_D/\Phi_H}{\mathrm{D/H}} \tag{12.19}$$

in the escape processes. Using (12.18) and the escape velocity of 10.2 km s$^{-1}$, the lowest velocity of hot oxygen is 5.42 km s$^{-1}$ for H and 5.74 km s$^{-1}$ for D. Velocity distribution of the nascent hot oxygen may be found in Fox and Hac (1997) and results the D/H probability ratio of 0.82. The D/H scale height ratio is 0.5, and $f \approx 0.41$ for the escape from collisions with hot oxygen. The depletion of D relative to H from the homopause to the exobase is nearly compensated by the greater bulk flow velocity of H that reduces its densities. Hartle et al. (1996) adopted $f = 0.022$ for $H + H^+$ from Krasnopolsky (1985), and $f \approx 0.5/1.3 = 0.38$ may be adopted for the ion escape. Here $\approx 1.3$ is the depletion of D relative to H at the ionopause. Thus, the total escape of hydrogen is $1.7\times10^7$ cm$^{-2}$ s$^{-1}$ = $8\times10^{25}$ s$^{-1}$ with fractionation factor $f = \frac{0.41\times3\times10^5+0.022\times8\times10^6+0.38\times9\times10^6}{3\times10^5+8\times10^6+9\times10^6} = 0.22$.

Impacts of comets contribute to the balance of water on Venus. McKinnon et al. (1997) considered comets with mass up to $10^{18}$ g, that is, with size up to 13 km for the density of 1 g cm$^{-3}$, and their cumulative impact rate is

$$C(>m) = 9.3\times10^{-7}\left(\frac{m}{10^{15}\,\mathrm{g}}\right)^{-0.66}\mathrm{yr}^{-1}. \tag{12.20}$$

Water is half comet mass, and its delivery rate is

$$P_{H2O} = -0.5\int_0^{10^{18}} m\frac{dC}{dm}dm = 10^{10}\mathrm{g\ yr}^{-1}\ \text{and}\ P_{\mathrm{H}} = 4.5\times10^6\ \mathrm{cm}^{-2}\ \mathrm{s}^{-1}. \tag{12.21}$$

The production rate is comparable with the escape rate. Furthermore, it is proportional to $m_0^{0.34}$, where $m_0$ is the integration limit. If the integration is extended to comets with sizes up to 50 km, that is, the size of comet Hale–Bopp C/1995 O1, then the delivery of water by comets is equal to its loss by Venus. Mean D/H = 2.15 SMOW in four comets (Villanueva et al. 2009), and if the system is in steady state, then the present D/H = $2.15/f = 10$ (see (10.30)) and disagrees with the observed value. Grinspoon and Lewis (1988) applied $f = 0.022$ from Krasnopolsky (1985); then the present D/H = 100, in accord with the observations that give $150 \pm 50$. They pointed out that the D lifetime is longer than the H lifetime by a factor of $1/f$ and becomes comparable with the age of the planet for $f = 0.022$.

Using $P_{\mathrm{H}} = 4.5\times10^6$ cm$^{-2}$ s$^{-1}$, the net loss is $1.25\times10^7$ cm$^{-2}$ s$^{-1}$ and $f = \frac{0.22\times1.7\times10^7-4.5\times10^6\times2.15/150}{1.7\times10^7-4.5\times10^6} = 0.3$. Assuming this $f$ for the whole history of water and the initial D/H = SMOW, one gets the initial water amount exceeding the present value by a factor of $150^{1/(1-f)} = 1300$ (using (10.25)) and the initial amount a layer of 17 m deep, that is, 0.5% of the terrestrial ocean.

The comparable amounts of $N_2$ and $CO_2$ on Venus and Earth (including carbonates) indicate that the initial amounts of water could be comparable as well. Therefore the above consideration might be invalid in the initial period on Venus, when the $H_2O$ abundance was very large and the EUV emission of the young Sun could exceed the present value by an order of magnitude. Then H becomes the major thermospheric species, the scale height is very large and moves the heating level $r_1$ in (2.36) up while the temperature gradient changes insignificantly, and the solution (2.35) of thermal balance for a static thermosphere diverges indicating hydrodynamic flow of the atmosphere to space (Section 2.5.3).

Kasting and Pollack (1983) calculated properties of the atmosphere for various $H_2O$ abundances up to the full terrestrial ocean (fto, a layer of 2.7 km thick on the Earth) and the solar EUV. They concluded that hydrodynamic escape could be very effective to remove fto for ~0.4 Byr but ended when the remaining water was equal to ≈0.1 fto. Fractionation factor was $f \approx 0.6$, and the reduction of the $H_2O$ abundance by a factor of 10 increased D/H to $10^{1-f} = 2.5$ SMOW.

Kumar et al. (1983) calculated nonthermal escape of H and D for various abundances of elemental hydrogen at the homopause, from the present $2.5\times10^{-6}$ to 0.1. They concluded that the calculated D/H in the present atmosphere should be much greater than the observed value and the elapsed time to reach the present water abundance is ~20 Byr. They suggested that outgassing of water with the initial D/H could solve the problem. This possibility was studied by Krasnopolsky (1985) who found an outgassing history that could consent the current water with the initial fto. However, the derived outgassing was too artificial and sophisticated, and the initial abundance of ~1% fto looked more probable. Thus we return to the analysis we started that gives 0.5% fto.

### *12.12.4 Escape and Balance of Helium*

Krasnopolsky and Gladstone (2005) calculated four basic processes of the helium escape: photo- and electron impact ionization of He above the ionopause (2 and 12 times $10^{23}$ s$^{-1}$, respectively), ionization of He between the exobase and the ionopause ($8\times10^{23}$ s$^{-1}$), and collisions with hot oxygen ($2.2\times10^{24}$ s$^{-1}$).

Escape of $He^+$ is estimated above at $10^5$ cm$^{-2}$ s$^{-1}$ $\approx 5\times10^{23}$ s$^{-1}$ based on the Venus Express data. The contribution of photoionization above the ionopause agrees with this value, while that of the electron impact ionization requires a significant ion precipitation, in accord with Curry et al. (2015). Ionization of He between the exobase and the ionopause ends by exothermic reactions with $CO_2$ and $N_2$, and the released helium atoms have enough energy to escape.

The corrected escape rate of helium is $(2.2 + 0.5 + 0.8 = 3.5) \times 10^{24}$ s$^{-1}$ = $7.5 \times 10^5$ cm$^{-2}$ s$^{-1}$. Compared with the helium abundance of 10 ppm (Table 12.7), this results in a helium lifetime of 0.6 Byr that is very long but much shorter than the planet age of 4.6 Byr. Therefore the loss is compensated by production of He by the radioactive decay of U and Th with subsequent outgassing and by capture of the solar wind alpha particles. A model for outgassing of He in Krasnopolsky and Gladstone (2005) was based on equal relative abundances of K, U, and Th on Venus and Earth and eight episodes of strong resurfacing on Venus. Scaled to the radiogenic $^{40}$Ar, it gives 2.5 ppm of radiogenic helium. The remaining 7.5 ppm of He is from capture of the solar wind alpha particles with efficiency of 8% (10% in the paper).

The similar capture occurs on Mars with efficiency of 30% (Krasnopolsky and Gladstone 2005) that was later confirmed using the Mars Express data (Stenberg et al. 2011). Nevertheless, a recent evaluation of this capture on Venus resulted in a negligible value of $10^{21}$ $s^{-1}$ (Stenberg et al. 2015). However, the observed rather bright emission of $He^+$ 30.4 nm of 70 R (Figure 12.6) is excited by charge exchange between the solar wind $He^{++}$ and the atmospheric species O and H and indicates strong interaction between the alpha particles and the atmosphere, in favor of the significant capture efficiency.

## 12.13 Clouds and Haze

### *12.13.1 Cloud Properties*

Venus is covered by a thick and rather uniform cloud layer with reflectivity of 0.83 in the visible. Using relationship (4.31) and assuming conservative scattering in the clouds with asymmetry factor $g \approx 0.75$ and a low reflecting surface, this gives the cloud optical depth of 25, close to that observed by the Venera and Pioneer Venus landing probes.

Polarimetry has significant advantages to study the Venus cloud tops. The effect is mostly due to single scattering, because multiple scattering presents the light from numerous angles with low net polarization. The observed phase function of polarization at 365 nm is shown in Figure 12.85. It depends on properties of the aerosol medium, that is, refractive index and two parameters of the size distribution, and on pressure of the

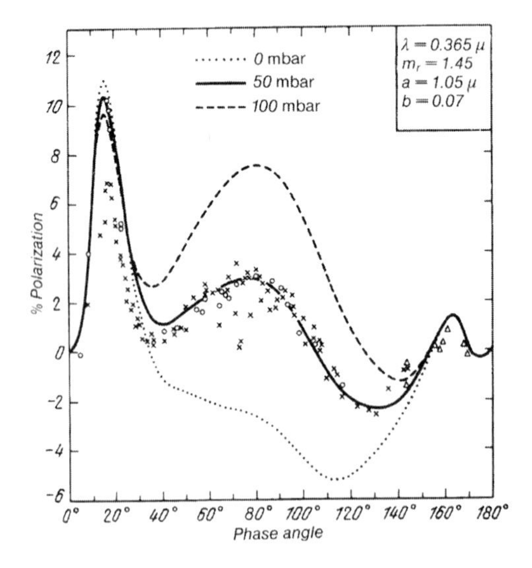


Figure 12.85 Phase function of the Venus polarization at 365 nm is fitted by aerosol with refractive index, modal radius, and width parameter of the gamma distribution shown in the inlet. The best-fit atmospheric pressure at a level of $\tau = 1$ is 50 mbar. From Hansen and Hovenier (1974).

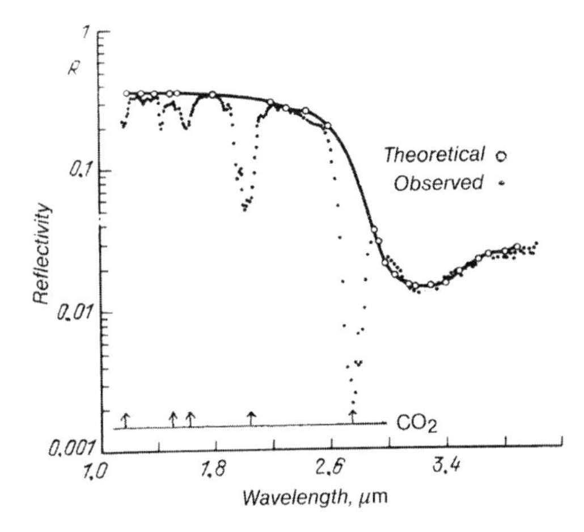


Figure 12.86 Reflectivity of Venus at a phase angle of 86° observed using the Kuiper Airborne Observatory. The light from Venus is scaled to that of the normally illuminated white Lambertian disk of the same size. The $CO_2$ bands are identified, and the sulfuric acid absorption with concentration of 85% is shown by the solid line. From Pollack et al. (1978).

gaseous Rayleigh-scattering atmosphere at a level of $\tau = 1$. The measured phase function has a complicated structure that makes it possible to retrieve all parameters with a reasonable accuracy (Hansen and Hovenier 1974). Similar observations and their analyses were done at 550 and 990 nm, and the derived spectral dependence of refractive index identified sulfuric acid with concentration of ≈80%. Sulfuric acid in the Venus clouds was first suggested by Sill (1972) and Young (1973).

Detailed study of the $H_2SO_4$ spectrum for various concentration was made by Palmer and Williams (1975), and strong infrared absorptions by sulfuric acid have been identified in the observed spectra of Venus (Figure 12.86).

Summary of the Venera and Pioneer Venus studies of the clouds is given in the aerosol chapter of VIRA (Ragent et al. 1985; see also Esposito et al. 1983). The most important results were obtained from the particle size spectrometer at the PV large probe (Figure 12.87). According to the data, there are three cloud layers: the upper, middle, and lower layers at 70–57, 57–51, and 51–48 km, respectively, and there are three particle modes. The upper cloud layer consists of the mode 2 particles with mean radius $r \approx 1.1$ μm and mode 1 with $r \approx 0.2$ μm. Mode 2 grows to 1.3 μm in the middle cloud layer, where mode 3 particles with $r \approx 3.8$ μm dominate in mass loading. The lower cloud layer involves three modes with radii of 0.2, 1.2, and 3.6 μm, respectively.

There were two hypotheses related to the mode 3 particles: (1) the minimum between mode 2 and mode 3 is instrumental, and mode 3 is just extension of mode 2, and (2) the mode 3 particles are crystals with a high aspect ratio. Then their optical depth and mass loading are significantly smaller than the evaluations for the spherical particles. Column mass loading at the PV Large Probe was 12 and 3.2 mg cm$^{-2}$, respectively.

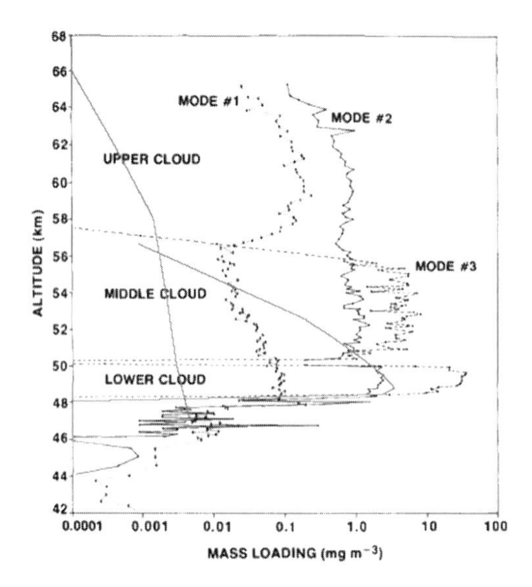


Figure 12.87 Three aerosol populations in three cloud layers observed by the PV particle size spectrometer (Knollenberg and Hunten 1980). Particles of modes 1, 2, and 3 have modal radii of 0.2, 1.2, and 3.6 μm, respectively. Model profiles of sulfur aerosol produced near 55–60 km and in the lower atmosphere are shown as well (Krasnopolsky 2016a). The adopted sulfur particle radii are 0.5 and 3.6 μm, respectively.

### *12.13.2 Cloud Variations*

Vertical profiles of volume extinction coefficient were measured by five V and four PV nephelometers, six V optical spectrometers and the PV particle size spectrometer and photometer. Mean optical depth of each layer is $\approx$ 10, and the values vary significantly in the lower cloud layer and rather stable in the middle cloud layer. The observed refractive indices in each layer agree within their uncertainties with those of sulfuric acid. The V11–14 optical spectrometer data were used to retrieve vertical profiles of water vapor (Ignatiev et al. 1997) and $S_3$ and $S_4$ (Krasnopolsky 2013b) in the lower atmosphere.

Variations of the cloud top altitude were retrieved in analyses of $H_2O$ abundances in the Venera 15 (Ignatiev et al. 1999) and Venus Express (Cottini et al. 2015; Fedorova et al. 2016) data. A level of $\tau = 1$ at 30 μm is at $\approx$62.5 km at the low and middle latitudes $\varphi < 50^\circ$ and at $\approx$56 km at high latitudes $\varphi > 50^\circ$ (Table 12.5; Ignatiev et al. 1999). A level of $\tau = 1$ at 2.5 μm is at 69 km at the low and middle latitudes lowering to 64 km at 80° (VIRTIS-H; Cottini et al. 2015). This level at 1.5 μm is at 70 km decreasing to 62–68 km at high latitudes (SPICAV-IR; Fedorova et al. 2016).

Backscatter nephelometers monitored the middle cloud layer at 54 km onboard the Vega balloons during 48 hours of their tracks of 11,000 km long in the atmosphere of Venus (Sagdeev et al. 1986). The cloud density was stable with just four events when deviations from the mean exceeded 30%. The balloon flights were at 7.5°N and S, and no conclusions can be made on latitudinal variability of the middle cloud layer from their data.

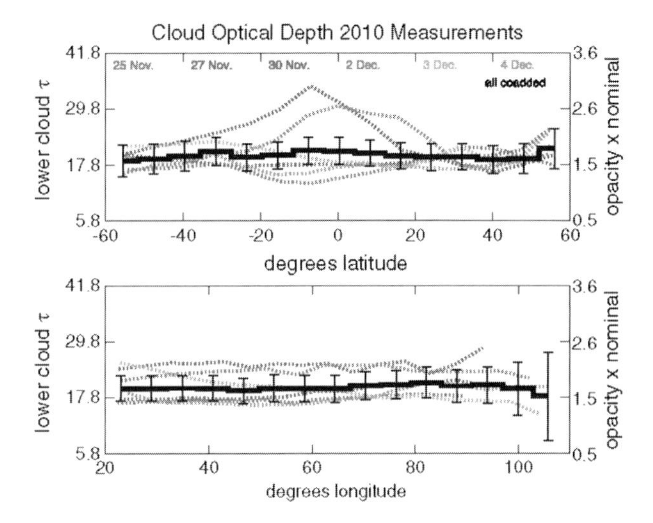


Figure 12.88 Variations of the lower cloud layer from the nightside observations at 2.3 μm by Arney et al. (2014).

Brightness of the spectral window at 2.3 μm is sensitive to the lower cloud optical depth that may be retrieved from the nightside observations. The mean data from the ground-based observations by Arney et al. (2014) in Figure 12.88 are constant at $\tau \approx 20$ at low and middle latitudes, while individual points vary up to a max/min ratio of 2.5. Data on the lower cloud layer from nine descent probes vary from $\tau = 2$ to 20 with a mean of $9 \pm 5$, where the standard deviation reflects the variability.

### *12.13.3 Cloud Composition: X-ray fluorescent spectroscopy*

The presence of sulfuric acid in the clouds of Venus was established by ground-based and aircraft spectroscopy, polarimetry, and mass spectrometry at the PV Large probe. Search for other species by direct analysis of the particulate matter at the descent probes is a difficult problem. The cloud particles are collected on and evaporate from the instrument filter while moving down to the hotter layers of the atmosphere, and only net effect of both processes is controlled. Evidently quantitative analyses of these data are rather uncertain.

X-ray fluorescent spectrometers at the V12, V14, and Vega 1 and 2 probes were designed to study elemental composition of the Venus clouds. The collected cloud substance was irradiated by X-rays from a radioactive source. The photons ionize K-shells of atoms; then those shells are restored by the K-capture of electrons from L-shells with emission of X-ray photons of certain energy for each atom. (K- and L-electrons are those with principal quantum number $n$ = 1 and 2, respectively.) However, peaks of different elements overlap in the spectra because of the low resolution of proportional gaseous counters in the spectrometers. For example, peaks of P, S, Cl, and Ar are at 2.02, 2.31, 2.62, and 2.96 keV.

The V12 spectrometer detected peaks of Cl and Fe (at 6.9 keV) in a proportion that corresponds to $FeCl_3$ (Petryanov et al. 1981). The mean mass loading for Cl was $4.3\times10^{-10}$ g cm$^{-3}$ and less than $10^{-10}$ g cm$^{-3}$ for S. These values may be compared with the mean total mass loading of $10^{-8}$ and $3\times10^{-9}$ g cm$^{-3}$ at the PV Large Probe assuming either spherical or crystal mode 3 particles. The visible spectrometers observed the greatest cloud optical depth at V12, and the low limit for sulfur looks puzzling. The V14 spectrometer measured mean mass loading of Cl and S at $1.6\times10^{-10}$ and $1.1\times10^{-9}$ g cm$^{-3}$, respectively (Surkov et al. 1983). The latter supports sulfuric acid as a major cloud constituent.

A few instruments onboard the Vega 1 and 2 probes were aimed to study the clouds. The mass spectrometers and gas chromatographs had pyrolytic devices that evaporated the collected aerosol for further analysis. The payload included X-ray spectrometers, two types of particle size spectrometers, nephelometers, and active UV spectrometers with light sources and cells filled in by the outside air (Section 12.5.1). Unfortunately, data from different instruments are sometimes contradictory or disagree with the previous studies of the cloud layer.

Two particle modes that are similar to mode 1 and 2 were observed, and mean mass loading was $\approx10^{-9}$ and $2\times10^{-9}$ g cm$^{-3}$ at 52–55 and 47–52 km, respectively, in accord with the previous V and PV data (Gnedykh et al. 1987). The authors found a ratio of 7 : 1 for mass loading of particles with refractive index of 1.4 (probably sulfuric acid) to those of 1.7 (probably sulfur). However, a distinct lower cloud boundary was lacking near 47 km, and the aerosol extended down to at least 33 km, similar to that observed at V8. The rather similar sulfuric acid to sulfur mass loading ratio of 10 : 1 was measured by the gas chromatographs (Porshnev et al. 1987). The mass spectrometer measured $\approx6\times10^{-9}$ g cm$^{-3}$ of sulfuric acid (Surkov et al. 1987).

The X-ray instruments (Andreichikov et al. 1987) were combinations of six radiometers, and sulfur filters on some detectors helped to resolve the sulfur and chlorine peaks. The instruments were designed to measure abundances of Cl, S, P, Fe, and those of elements lighter than phosphorus and heavier that iron (six parameters). However, errors and some instabilities restricted the data analysis, and abundances of S, Cl, and P were measured on the filter during the descent from 62 to 47 km. A mean abundance of iron of $(2 \pm 1)\times10^{-10}$ g cm$^{-3}$ at 52–47 km and an upper limit of $10^{-10}$ g cm$^{-3}$ to the heavy elements were obtained as well.

Two cubic meters of the air passed through the filter in the altitude range from 63 to 47 km. The deposit increased linearly to 4 mg down to 51.5 km with equal quantities of sulfur and chlorine and no phosphorus. Phosphorus dominates at 51.5 to 47 km with some traces of sulfur; chlorine is lacking there.

Krasnopolsky (1989) argued that if phosphorus exists in the Venus atmosphere, then dimer $P_4O_6$ is its major component. It can react with sulfuric acid and form phosphoric acid $H_3PO_4$ that precipitates and gradually loses water converting into phosphoric anhydride $P_2O_5$. It exists as aerosol down to 25 km and then evaporates. The required abundance of phosphorous anhydride $P_4O_6$ is 2 ppm below 25 km. Although this explanation is exotic,

Table 12.12 *Composition of Venus clouds based on the Vega probes*

| Species | $h$ (km) | Abundance |
|---|---|---|
| $H_2SO_4$ | 52–62 | ~5 mg/m$^3$ |
| Sulfur | 50–62 | $S_X/H_2SO_4 \approx 0.1$ by mass |
| $FeCl_3$ | 47–62 | ≈1% of the total column mass above 47 km |
| $H_3PO_4$ | ≤52 | ~5 mg/m$^3$ |
| $AlCl_3$ | 53–58 | ~3 mg/m$^3$ |

there are two facts in its favor: the observed aerosol extended down to 33 km where the nephelometer was switched off, and the similar aerosol altitude distribution was observed at V8. The aerosol is typically scarce below 45 km indicating the lack or low abundances of phosphorus.

An order-of-magnitude excess of Cl over Fe requires another chlorine aerosol species, in addition to $FeCl_3$. Aluminum chloride $AlCl_3$ is the most probable candidate that could condense in the middle cloud layer with the aerosol densities close to those measured. Composition of the clouds based on the Vega data and their analysis in Krasnopolsky (1989) is given in Table 12.12. These data look unusual and need further study and confirmation.

### *12.13.4 Near-UV Absorber and Iron Chloride*

The blue and NUV absorption in the Venus clouds (Figure 12.89) is known for a long time and responsible for the yellowish color of Venus. The absorption shortward of 320 nm is caused by $SO_2$, while that at 320–500 nm is here under discussion. The NUV features at the cloud tops observed by the orbiters are due to this absorber. They correlate with the observed $SO_2$, and this is explained by either dynamics (upward and downward flows if the absorber is under some haze) or chemistry (if the absorber is a product of the $SO_2$ chemistry).

The V14 photometer at 320–390 nm was designed to measure vertical distribution of the NUV absorber. The observed true absorption coefficient decreased from more than 0.05 km$^{-1}$ at 62–65 km to 0.02 km$^{-1}$ at 58–62 km and then to a minimum of 0.001 km$^{-1}$ at 55–58 km with an increase to 0.01 km$^{-1}$ at 48–52 km (Ekonomov et al. 1983b). Therefore the NUV absorber exists mostly near and above 60 km. The increase to the bottom of the cloud layer is caused by a very weak NUV absorption of $SO_2$ (Figure 12.15) and revealed its abundances at 57–50 km (Section 12.5.1).

Many candidates (Figure 12.89) were suggested for the absorber (e.g., Krasnopolsky 1986; Esposito et al. 1997). Some of them ($S_a$, $Cl_2$, $S_2O$, $SCl_2$) are photochemical products and should exist in the Venus atmosphere. However, these species (except $SCl_2$) poorly match the observed absorption, and the proposed abundances exceed the model predictions by two orders of magnitude. Problems with more exotic absorbers like ammonium pyrosulfite, nitrosylsulfuric acid, and perchloric acid are discussed in Krasnopolsky (1986).

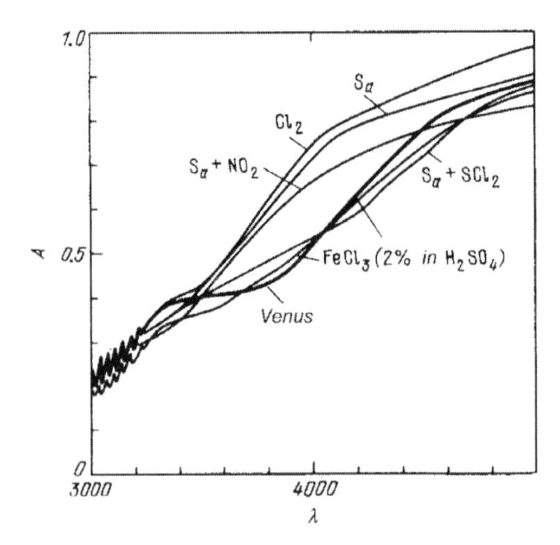


Figure 12.89 Albedo of Venus at 300–500 nm is compared with those calculated for various absorbers. From Krasnopolsky (1986, figure 115).

Solution of ~1% $FeCl_3$ in sulfuric acid (Zasova et al. 1981) is probably the best candidate (Krasnopolsky 2017a). Colors of the Venera landing sites favor the presence of ferric rocks (Pieters et al. 1986), though ferrous rocks are more stable on Venus. $FeCl_3$ was detected by the V12 probe (see Section 12.13.3). There is a significant confusion in the data on the iron chloride saturated vapor. Dimer $Fe_2Cl_6$ dominates in the gas phase, and

$$\ln P_{Fe_2Cl_6}(\text{atm}) = 29.69 \ln T - 0.06472T + 1.85 \times 10^{-5}T^2 - \frac{12453}{T} - 136.34 \quad (12.22)$$

according to Rustad and Gregory (1983) and

$$\lg P_{Fe_2Cl_6}(\text{atm}) = 13.19 - 7546/T \quad (12.23)$$

using thermodynamic data from Chase (1998). If the iron chloride mixing ratio is 17 ppbv in the atmosphere below the clouds, then it condenses at 47.6 km and its mass loading profile matches those of mode 1 in the lower and middle cloud layer for the upward flux of $Fe_2Cl_6$ of $1.2\times10^{-12}$ g cm$^{-2}$ s$^{-1}$ (Figure 12.90). The model in the figure accounts for condensation, vertical transport by eddy diffusion, the Stokes precipitation, and coagulation with the sulfuric acid droplets. Photochemical models by Mills (1998), Mills and Allen (2007), Zhang et al. (2012), and Krasnopolsky (2012b) predict the production of sulfuric acid at $6\times10^{11}$ cm$^{-2}$ s$^{-1}$ = $1.2\times10^{-10}$ g cm$^{-2}$ s$^{-1}$, so that the proportion between the flux of iron chloride and the production of sulfuric acid provides the concentration of ~1% required for the NUV absorption. Calculated $FeCl_3$ mixing ratio in the surface rocks is 19 ppb.

The $FeCl_3$ particles may be condensation centers for sulfuric acid in the upper cloud layer. The mode 2 to mode 1 particle mass ratio is ~100, and the expected concentration of

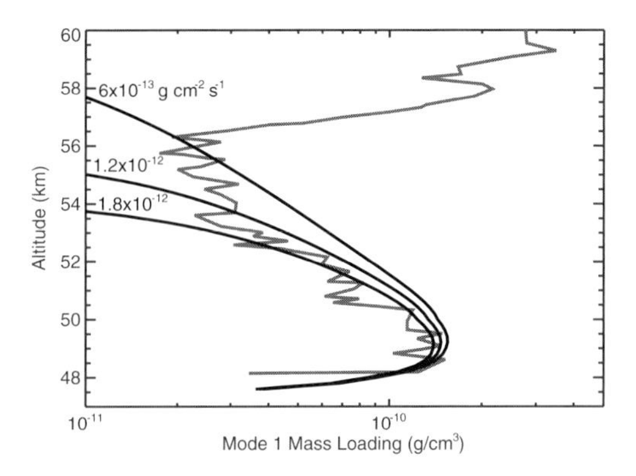


Figure 12.90 Observed mode 1 mass loading (in g $cm^{-3}$; from Knollenberg and Hunten 1980) is scaled by a factor of 1.5 to account for the higher $FeCl_3$ density and compared with the calculated vertical profiles of the iron chloride aerosol for three values of the upward flux. From Krasnopolsky (2017a).

~1% is just that fits to the observed absorption as well. Yet only a small part of the mode 1 particles in the upper cloud layer consists of $FeCl_3$ (Figure 12.90).

The reaction between $FeCl_3$ and concentrated $H_2SO_4$ is slow below 280 K typical of the clouds above 58 km, and the solution lifetime is ~ 1 month in the upper cloud layer. Colorless ferric sulfate replaces $FeCl_3$ near its bottom.

Absorption by the solid $FeCl_3$ mode 1 particles should be compatible with the V14 observations, and a critical point is the minimum of true absorption of 0.001 $km^{-1}$ at 55–58 km. The mode 1 extinction is equal to 0.05 $km^{-1}$ at this altitude in the PV particle size spectrometer data. Therefore the $FeCl_3$ mode 1 particles should have single scattering albedo exceeding 0.98 at 0.37 μm. Using the Mie formulas (Section 4.1) for the effective particle radius of 0.2 μm, the particle imaginary refractive index should be less than 0.005, that is, the absorption coefficient of solid $FeCl_3$ is less than 1700 $cm^{-1}$ at 0.37 μm. The only datum of 140 $cm^{-1}$ on this subject refers to the $FeCl_3$ 1% solution in 80% $H_2SO_4$ (figure 116 in Krasnopolsky 1986), and this consideration does not rule out $FeCl_3$ on Venus. The recent study of glory on Venus using the Venus Express camera favors $FeCl_3$ as the NUV absorber as well (Petrova 2018).

Frandsen et al. (2016) proposed disulfur dioxide $S_2O_2$ as the NUV absorber. They argue that $S_2O_2$ is formed as two OSSO isomers instead of the lowest energy isomer S=$SO_2$ adopted in the photochemical model by Krasnopolsky (2012b). They calculated absorption spectra of the OSSO isomers and their abundances assuming $f_{SO}$ = 12 ppb at 64–95 km from observations by Na et al. (1994). The improvements in the $S_2O_2$ chemistry by Frandsen et al. (2016) are important for Venus' photochemistry, because OSSO dissociates to SO + SO while S=$SO_2$ to $SO_2$ + S, and Krasnopolsky (2018a) updated his model using their results. However, the model abundance of OSSO is smaller than that required to explain the NUV absorption by two orders of magnitude.

### *12.13.5 Models for Sulfuric Acid in the Clouds*

A model for two-component diffusion and condensation of $H_2O$ and $H_2SO_4$ in the Venus clouds was developed by Krasnopolsky (2015b). Partial vapor pressures of these species in phase equilibrium with liquid or solid sulfuric acid are equal to

$$\ln p_i = \ln p_{si} + \frac{\mu_i - \mu_i^0}{RT}. \tag{12.24}$$

The subscripts $i = 0$ and 1 refer to $H_2O$ and $H_2SO_4$, $s$ means saturated vapor above a pure species, and $\mu_i - \mu_i^0$ is the difference of chemical potentials of a species in solution and a pure species. Chemical potentials of $H_2O$ and $H_2SO_4$ in sulfuric acid for various $m = H_2O/H_2SO_4$ were tabulated by Giauque et al. (1960).

Fluxes of the species in both gas and liquid phases are

$$\Phi_0 = -Kn\frac{df_0}{dz} - n_0 V\ , \quad \Phi_1 = -Kn\frac{df_1}{dz} - n_1 V. \tag{12.25}$$

Here $f$ is the gas mixing ratio (fraction), $n$ is the total atmospheric number density, $n_0$ and $n_1$ are the $H_2O$ and $H_2SO_4$ number densities in the aerosol, and $V$ is the precipitation velocity (equation (4.34)). $\Phi_0 + \Phi_1 = 0$ because of the element (hydrogen) conservation, $m = n_0/n_1$, and this results in

$$\Phi_1 = \frac{Kn}{1+m}\left(\frac{df_0}{dz} - m\frac{df_1}{dz}\right);\ \text{that is, } m = \frac{n\frac{df_0}{dz} - \frac{\Phi}{K}}{n\frac{df_1}{dz} + \frac{\Phi}{K}}. \tag{12.26}$$

Here $\Phi = \Phi_1$ is the column production of sulfuric acid that is very similar in the recent photochemical models and equal to $5.7\times10^{11}$ cm$^{-2}$ s$^{-1}$ in Krasnopolsky (2012b). The production is in a narrow layer at 66 km. A set of solutions of (12.26) for various eddy diffusion decreasing from $10^4$ cm$^2$ s$^{-1}$ above 54 km to $K$ = 2200–10,000 cm$^2$ s$^{-1}$ at 45 km is shown in Figures 12.91 and 12.92.

The current $H_2O$ observations indicate ~3 ppm at the cloud tops, and a preferable solution is that for $K_{45}$ = 3300 cm$^2$ s$^{-1}$. Mixing ratios of $H_2O$ decreases from 22.5 ppm at the lower cloud boundary of 47.5 km to 3 ppm above 68 km. Sulfuric acid vapor peaks at 7.5 ppm at the lower cloud boundary and is thermally decomposed below, increasing the $H_2O$ mixing ratio to 30 ppm. Concentration of the sulfuric acid aerosol decreases from 98% in the lower cloud layer to ~80% in the upper cloud layer. Atmospheric dynamics near the cloud bottom strongly affects water vapor at the cloud tops.

The steep decrease in the mixing ratio of $H_2SO_4$ in the lower cloud layer drives upward flux of this species that is returned as a strong downward flux of the aerosol (Figure 12.93) that forms the lower cloud layer (Krasnopolsky and Pollack 1994).

Particle size distributions of the sulfuric acid aerosol were calculated by Gao et al. (2014) and Parkinson et al. (2015). The models adopt vertical profiles of $H_2O$ that fix

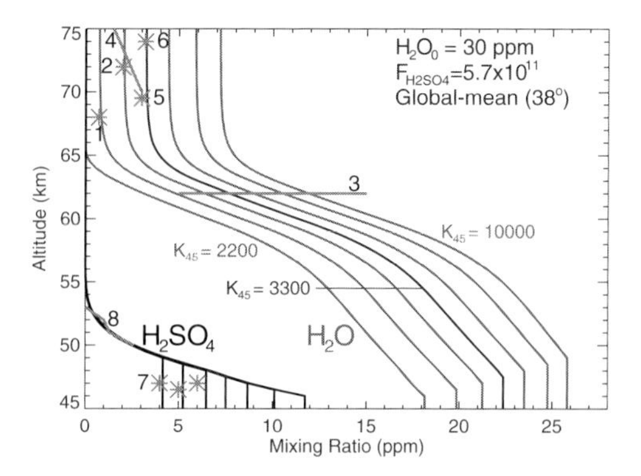


Figure 12.91 Vertical profiles of $H_2O$ and $H_2SO_4$ vapors for the global-mean flux of $H_2SO_4$ (middle latitudes). Eddy diffusion is adjusted below 54 km to fit LCB varying from 46 to 49 km. The preferable profile is for $K_{45}$ = 3300 cm$^2$ s$^{-1}$. Observations of $H_2O$: (1) Fink et al. (1972), (2) KAO (Bjoraker et al. 1992), (3) Venera 15 (Ignatiev et al. 1999), (4) VEX/SOIR occultations (Fedorova et al. 2008), (5) VEX/VIRTIS-H (Cottini et al. 2012), (6) IRTF/CSHELL (Krasnopolsky et al. 2013). Observations of $H_2SO_4$ vapor: (7) Magellan (Kolodner and Steffes 1998), (8) VEX/VeRa (Oschlisniok et al. 2012). From Krasnopolsky (2015b).

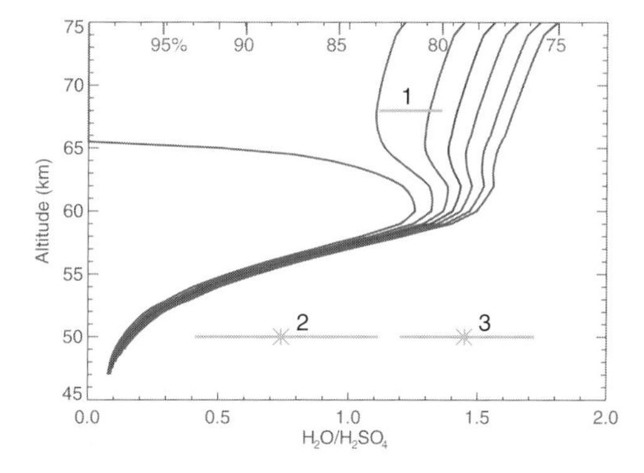


Figure 12.92 Variations of $m$ = $H_2O/H_2SO_4$ and concentration in the sulfuric acid droplets. The mean profile is preferable. Observations: (1) VEX/VIRTIS-H (Cottini et al. 2012), (2) VEX/VIRTIS-M (Barstow et al. 2012), (3) ground-based (Arney et al. 2014). From Krasnopolsky (2015b).

profiles of $H_2SO_4$ and its concentration in the aerosol as well. Other input data are productions of sulfuric acid and polysulfur in narrow layers centered at 61 km in Gao et al. (2014). Polysulfur is formed as condensation nuclei with radius of 10 nm, and meteoric dust particles are the condensation centers as well. Then coagulation and other processes that affect the particle size are calculated. The results (Figure 12.94) are in reasonable agreement with the observations. While Gao et al. (2014) adopted temperatures from VIRA, Parkinson et al. (2015) applied those from the VEX occultation data.

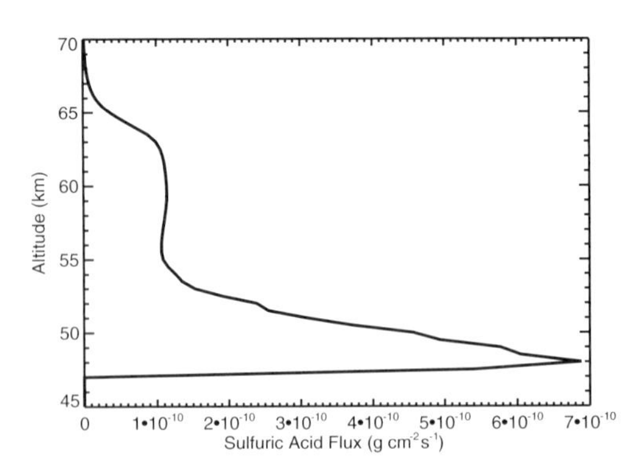


Figure 12.93 Downward flux of the sulfuric acid aerosol in the basic model in Figure 12.91. The increase down to 63 km reflects the photochemical production, while that with a peak at 48 km is caused by a strong gradient in $H_2SO_4$ vapor.

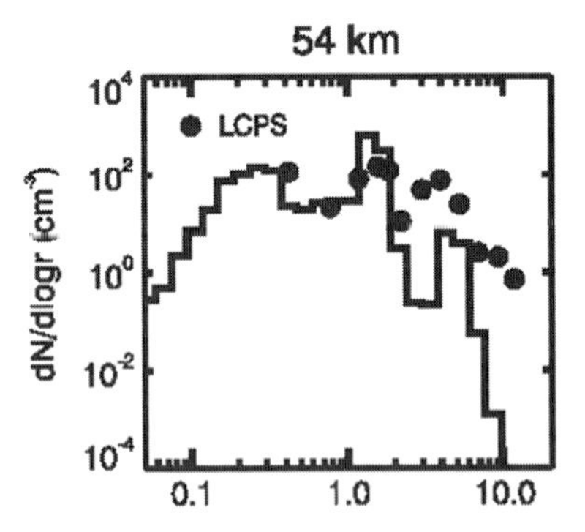


Figure 12.94 Particle size distribution at 54 km in the nominal model by Gao et al. (2014) is compared with the data of the PV particle size spectrometer.

### *12.13.6 Sulfur Aerosol*

Although sulfur aerosol has been discussed in the literature since the beginning of 1970s, it is lacking in the basic versions of photochemical models for the middle atmosphere by Yung and DeMore (1982), Mills (1998), and Mills and Allen (2007) and is negligible in Zhang et al. (2012). Krasnopolsky (2012b, 2016a) adjusted uncertainties in the sulfur chemistry to increase the sulfur production. According to the model, only 0.5% of sulfur atoms from the photolysis of OCS react with OCS to form $S_2$ and then $S_4$ that condenses. Column mass production of sulfur aerosol is smaller that that of sulfuric acid by a factor of 3000, and its mass loading, say, at 60 km is smaller than that of the mode 1 by a factor of 200 (Figure 12.87).

A chemical kinetic model (Section 12.14) predicts the $S_8$ mixing ratio of 2.5 ppm at 45–47 km. Based on this value, condensation and formation of an aerosol layer of sulfur is considered in Section 4.5.2 (Figure 12.87). The calculated sulfur mass loading is ~10% of the total in the lower cloud layer, similar to the Vega evaluations (Section 12.13.3). Sulfur cannot significantly contribute to the NUV absorption that peaks near or above 60 km.

### *12.13.7 Haze and Production of $SO_X$*

Haze extends from the main cloud deck at 70 km to ~100 km. It is optically thin and can be measured on the limb (Veneras 9 and 10) or by the solar occultation technique (Venus Express).

The Venera spectra at 270–700 nm were analyzed in Section 4.5.1 (Figures 4.6 and 12.95) using mean particle radius as a parameter and assuming refractive index of sulfuric acid. The SPICAV solar occultations at 220, 300, 757, and 1553 nm made it possible to retrieve a bimodal distribution of the haze particles with the mode radii of ~0.2 and ~0.7 μm (Wilquet et al. 2009). These particles are extensions of the modes 1 and 2 in the upper cloud layer (Figure 12.95). Using VIRTIS-M, thermal emission from the haze was observed at 4.5–5 μm on the nightside limb (de Kok et al. 2011). This emission is caused by the mode 2 particles with $r \approx 1$ μm, and their number densities extend to 85 km (Figure 12.95). Their data do not include the small particles, and the observed number densities are therefore smaller that those from the Venera and SPICAV observations. The aerosol models by Gao et al. (2014) and Parkinson et al. (2015) include the haze.

The most detailed study of the haze (Luginin et al. 2016) is based on 222 solar occultations observed using SPICAV-IR at 0.65 to 1.7 μm. The observed spectra were fitted by extinction of the spherical sulfuric acid particles with the proper refractive indices and effective radii as fitting parameters. Most of the observations are fitted by a single mode with $r = 0.54 \pm 0.25$ μm, and some orbits require two modes with $r = 0.12 \pm 0.03$ μm and $0.84 \pm 0.16$ μm. Here the standard deviations reflect variations,

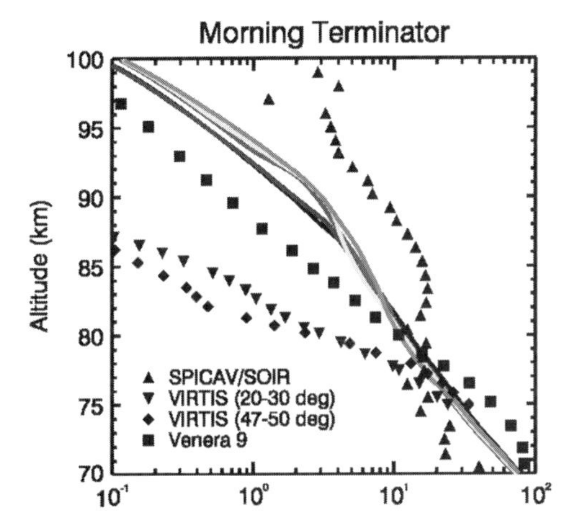


Figure 12.95 Model profiles of the haze particle number density (X-scale, in $cm^{-3}$) compared with those observed by Venera 9 (Krasnopolsky 1983a), VEX/SOIR (Wilquet et al. 2009), and VEX/VIRTIS (de Kok et al. 2011). From Parkinson et al. (2015).

while the mean values are constant from 74 to 88 km. Both mode 1 and 2 mean number densities decrease from 75 to 90 km by a factor of 10, so that their mean ratio is constant and equal to 500. Hence the haze mean scale height is 6.5 km, greater than the atmospheric scale height of 4.3 km (VIRA). This presumes a source of the haze above 90 km, while the photochemical models predict the aerosol production in the upper cloud layer and its delivery up to ≈100 km by eddy diffusion.

The single mode radius is greater than those of ≈0.2 μm in the V9–10 observations (Figure 4.6) and 0.23 μm at 80 km according to the PV polarimetry (Kawabata et al. 1980). However, only 2 of 12 occultation profiles in figure 13 of Luginin et al. (2016) show radii exceeding 0.5 μm, and most of the profiles have radii of 0.2–0.3 μm at 80–90 km.

There are attempts to explain the increase in the $SO_2$ + SO mixing ratio from 85 to 100 km in the SPICAV observations (Figure 12.18a; Belyaev et al. 2017) by evaporation of the haze and photolysis of $H_2SO_4$ vapor. Zhang et al. (2012) developed a photochemical model with the SOIR nighttime temperature profile that peaks at 250 K at 100 km. $H_2SO_4$ vapor was adopted at a quarter of the saturated vapor above the pure sulfuric acid. Its photolysis cross section at 195–330 nm was adopted at the upper limit value of $10^{-21}$ cm$^2$. The $H_2SO_4$ vapor mixing ratio peaked at 200 ppb, and its photolysis column rate was $10^9$ cm$^{-2}$ s$^{-1}$ above 90 km. The calculated $SO_2$ and SO profiles in Zhang et al. (2012) agreed with the SPICAV observations.

Krasnopolsky (2011c) calculated a vertical profile of $H_2SO_4$ vapor using the mean $H_2O$ profile from the SOIR occultations (Fedorova et al. 2008), a haze profile from the SPICAV occultations (Wilquet et al. 2009), and a temperature profile from the VEX radio occultations (Tellmann et al. 2009). This profile refers to 85°N and gives 195 K at 100 km instead of 166 K from VIRA at 85°. The model involved condensation and sublimation of $H_2SO_4$ and $H_2O$ with vertical transport by eddy diffusion. The calculated $H_2SO_4$ vapor number density is 500 cm$^{-3}$ at 100 km with a mixing ratio of $3\times10^{-13}$ and column photolysis rate of 23 cm$^{-2}$ s$^{-1}$, that is, absolutely negligible.

Temperature at 100 km is 170 K in VIRA and 175 K in VTGCM (Figures 12.44 and 12.45). Parkinson et al. (2015) adopt $T \approx 230$ K from the SOIR occultations. Saturated vapor of $H_2SO_4$ is denser at 230 K than that at 195 and 170 K by factors of 1500 and $1.8\times10^6$, respectively. (Temperatures in the lower thermosphere are discussed in Sections 12.8.4 and 12.8.5.) However, the haze mass loading is ~$5\times10^{-14}$ g cm$^{-3}$ at 90 km (Figure 4.6) and corresponds to an $H_2SO_4$ mixing ratio of 20 ppb. Therefore even complete evaporation of the haze above 90 km with immediate complete destruction by photolysis is insufficient to explain the increase above 85 km in the vertical profiles of $SO_2$ and SO in the SPICAV observations.

## 12.14 Chemical Kinetic Model for Lower Atmosphere (0–47 km)

The solar UV radiation does not reach the atmosphere below the clouds and cannot initiate photochemistry with highly reactive atomic species and radicals. The lower atmosphere consists of stable molecules; reactions between them are very slow and often have not been studied in the laboratory. However, significant efforts have been made to study the chemical composition of the lower atmosphere using the descent probes and the nightside spectroscopy (Table 12.6), and the results require a chemical kinetic model to be understood. Chemical kinetic modeling is similar to photochemical modeling, but applied to atmospheric regions where photolysis is relatively unimportant. This model (Krasnopolsky 2013b) will be considered below.

### *12.14.1 Sources of Chemistry and Boundary Conditions*

We will discuss in Section 12.15 that a photochemical effect of the middle atmosphere on the lower atmosphere may be described by a net reaction

$$SO_2 + CO_2 + H_2O \rightarrow H_2SO_4 + CO; \quad CR = 5.7 \times 10^{11}\ cm^{-2}\ s^{-1}. \qquad (12.27)$$

The column rate CR of this reaction is very similar in a few photochemical models and may be used with some confidence. The reaction transforms the stable reactants into the disequilibrium products that may drive some chemistry in the lower atmosphere.

The blue solar photons attain the lower atmosphere and can dissociate $S_3$ and $S_4$ stimulating chemical processes. Thermochemical reactions can proceed in the hot lowest scale height and tend to establish thermochemical equilibrium. Vertical transport by eddy diffusion moves some products to regions where they are in disequilibrium and drive chemistry. Assessment of the transport time is $\tau = H^2/K \approx 10^9$ s $\approx$ 30 years, and even very slow reactions may affect the atmospheric composition. The observed strong variations of the OCS and CO mixing ratios with altitude indicate a significant chemistry in the atmosphere.

Cosmic rays can be a source of chemistry as well. Calculations by Upadhyay et al. (1994) show that ionization by muons dominates in the lower atmosphere with a peak at 20 km and column rate of $1.2\times10^9$ cm$^{-2}$ s$^{-1}$. This rate is smaller than that of the photochemical transport from the middle atmosphere by a factor of 500 and is therefore neglected.

The model does not include $N_2$ and noble gases. HF is similar to HCl but less abundant by two orders of magnitude; furthermore, the HF bond energy is greater than that of HCl (Table 9.3), and HF is neglected in the model as well. Conditions at the lower boundary (surface) are the measured abundances of five parent species (Table 12.13). Conditions at the upper boundary (chosen at 47 km to avoid the aerosol problems) reflect exchange of

Table 12.13 *Boundary conditions for the chemical kinetic model at 0–47 km*

| | |
|---|---|
| 0 km: | $[CO_2] = 9.07\times10^{20}$ cm$^{-3}$, $f_{SO2} = 1.3\times10^{-4}$, $f_{H2O} = 3\times10^{-5}$, $f_{HCl} = 5\times10^{-7}$, $f_{NO} = 5.5\times10^{-9}$ |
| 47 km, fluxes (cm$^{-2}$ s$^{-1}$): | $H_2SO_4 = -5.7\times10^{11}$, $CO = -5.7\times10^{11} - 3v$ [OCS] $- v$ [$H_2S$], $H_2O = 5.7\times10^{11} - v$ [$H_2S$], $SO_2 = 5.7\times10^{11} - v$ [OCS], $S_X = v$ [$H_2S$] |
| 47 km, velocities (cm s$^{-1}$): | OCS = $v$, $H_2S = v$ |

*Note.* $v = K/2H$ =0.006 cm s$^{-1}$; all other boundary conditions are fluxes equal to zero (closed boundaries). From Krasnopolsky (2013b).

gases between the lower and middle atmosphere. Beside (12.27), it involves upward fluxes of OCS and $H_2S$ that dissociate in the middle atmosphere and return into the lower atmosphere via net reactions

$$OCS + 2\ CO_2 \rightarrow SO_2 + 3\ CO;\ v[OCS]_{47\ km} = 2.79 \times 10^{10}\ cm^{-2}\ s^{-1} \tag{12.28}$$

$$H_2S + CO_2 \rightarrow H_2O + CO + S;\ v[H_2S]_{47\ km} = 6.41 \times 10^{9}\ cm^{-2}\ s^{-1}. \tag{12.29}$$

The upward bulk velocity $v = K/2H = 0.006$ cm s$^{-1}$ is adopted at a half maximum diffusion velocity $K/H$, and the numerical values of (12.28–12.29) are calculated by the model. The exchange by OCS and $H_2S$ is weaker than that by $H_2SO_4$ + CO by a factor of 20. All other boundary conditions are fluxes equal to zero, that is, the closed boundaries.

### 12.14.2 Reactions and Rate Coefficients

Using Table 9.3, the weakest bond among the basic species in the lower atmosphere is OC=S with energy of 3.12 eV. Therefore chemistry is sulfur-driven in the Venus lower atmosphere. Krasnopolsky and Pollack (1994) explained the observed strong depletion in OCS from 13 ppm at 30 km to 1 ppm at 36 km (Section 12.5.2) by reactions

$$\begin{aligned} &SO_3 + OCS \rightarrow CO_2 + (SO)_2; && k = 10^{-11} e^{-10000/T}\ cm^3\ s^{-1},\\ &(SO)_2 + OCS \rightarrow CO + SO_2 + S_2; && k = 10^{-20}\ cm^3\ s^{-1},\\ &\text{Net}\quad SO_3 + 2\ OCS \rightarrow CO_2 + CO + SO_2 + S_2. && \end{aligned} \tag{12.30}$$

These reactions have not been studied in the laboratory, and their rate coefficients are adopted in the model. Yung et al. (2009) suggested another cycle to explain the reduction of OCS:

$$\begin{aligned} S_2 + S_2 + M &\rightarrow S_4 + M,\\ S_4 + h\nu &\rightarrow S_3 + S, \end{aligned} \tag{12.31}$$

$$S_3 + h\nu \rightarrow S_2 + S, \tag{12.32}$$

$$\begin{aligned} &2(S + OCS \rightarrow CO + S_2),\\ &\text{Net } 2\ OCS + 2h\nu \rightarrow 2\ CO + S_2. \end{aligned} \tag{12.33}$$

Yung et al. (2009) did not use this cycle in their model; the $S_3$ and $S_4$ photolysis yields were unknown, as was the expected effect of the cycle.

Limiting wavelengths of 438 and 453 nm for photolyses of $S_3$ and $S_4$ (12.31–12.32) can be derived from thermodynamic data. Their photolysis frequencies are calculated as functions of altitude using the V11 photon fluxes (Figure 7.14) and the species cross sections (Figure 12.24):

$$J_{S3} = \gamma \times 10^{-3}\left(4.4 + 1.36h + 0.063h^2\right)\ s^{-1} \tag{12.34}$$

$$J_{S4} = \gamma_0 \times 10^{-5}\left(8.5 + 2.4h + 0.15h^2\right)\ s^{-1}. \tag{12.35}$$

Here $\gamma$ and $\gamma_0$ are the photolysis yields, and their best-fit values are 0.017 and 1, respectively. Despite the high yield at $\lambda < 453$ nm, (12.31) is a small branch of the $S_4$ photolysis, and the main branch results in $S_2 + S_2$. Another model parameter is activation energy of decomposition of $S_4$ to $S_2 + S_2$, and the best-fit value is 7800 K.

The model includes 89 reactions of 28 species (Table 12.14). Chemistry of the sulfur allotropes $S_X$ (X = 1 to 8) is poorly known, and five allotropes are in the model: S, $S_2$, $S_3$, $S_4$, and $S_X = \Sigma n S_n$ for $n$ = 1–8, with local thermochemical equilibrium between $S_X$ and $S_2$. Rate coefficients of some reactions have not been studied and are evaluated using reactions of similar species. Data for inverse reactions are obtained using direct reactions and calculated constants of thermochemical equilibria. These problems are discussed in more detail in Krasnopolsky (2007).

Eddy diffusion decreases in the model from $10^4$ cm$^2$ s$^{-1}$ at 47 km to 1200 cm$^2$ s$^{-1}$ at and below 35 km. Our later study (Section 12.13.5; Figure 12.91) favors a decline in $K$ from $10^4$ at 54 km to 3300 at 45 km, that is, $K \approx 5000$ cm$^2$ s$^{-1}$ at 47 km. This change in eddy diffusion weakly affects the model results; the greatest variation is in $H_2SO_4$ vapor at 47 km that becomes 5 ppm instead of 3.8 ppm in the initial model.

### *12.14.3 Model Results*

Abundances of $SO_2$, $H_2O$, HCl, and NO are prescribed by the boundary conditions and cannot be used to test the model, while the observed chemical products OCS, CO, $S_3$, $S_4$, and $H_2SO_4$ vapor are appropriate for this purpose.

A test of the $S_4$ cycle (12.31–12.33) without the $SO_3$ + 2 OCS cycle (12.30) results in too high abundances of OCS and CO that exceed the observed values by an order of magnitude. Model results for the $SO_3$ + 2 OCS cycle with and without the $S_4$ cycle are shown in Figure 12.96. The OCS and CO profiles agree with the observations for both cases, while the $S_3$ profile does not fit the observed value at 3–10 km without the $S_4$ cycle.

Some reactions proceed with very high column rate. For example, the rate of R83 $S_2$ + $S_2$ + M exceeds that of the photolysis of $CO_2$ ($4\times10^{12}$ cm$^{-2}$ s$^{-1}$) in the middle atmosphere by a huge factor of ~$10^{11}$. However, the reaction is almost exactly compensated by the inverse reaction, their difference of $2.69\times10^{14}$ cm$^{-2}$ s$^{-1}$ is spent for photolysis, and only a small part of the photolysis results in the chemical transformation to $S_3$ + S. This part is equal to $2.48\times10^{12}$ cm$^{-2}$ s$^{-1}$, comparable to the photolysis of $CO_2$.

Sulfur is formed by

$$(CO + SO_2 \rightarrow CO_2 + SO) \times 2,$$

$$SO + SO \rightarrow SO_2 + S,$$

$$\text{Net } 2\,CO + SO_2 \rightarrow 2\,CO_2 + S;\ CR = 2.79 \times 10^{10}\ \text{cm}^{-2}\ \text{s}^{-1},\ h_r = 14\ \text{km}.$$

The OC=S bond energy is comparable with those in the sulfur allotropes, and exchange of sulfur between OCS and $S_X$ exceeds the source of sulfur by two orders of

Table 12.14 *Chemical reactions in Venus lower atmosphere, their rate coefficients, column rates, and mean altitudes*

| # | Reaction | Rate coefficient | Column rate | $h_r$ (km) |
|---|---|---|---|---|
| 1 | $SO_3 + H_2O + H_2O \rightarrow H_2SO_4 + H_2O$ | $2.3 \times 10^{-43} T\, e^{6540/T}$ | 2.57 + 16 | 40 |
| 2 | $H_2SO_4 + H_2O \rightarrow SO_3 + H_2O + H_2O$ | $7 \times 10^{-14}\, e^{-5170/T}$ | 2.57 + 16/5.70 + 11 | 36 |
| 3 | $SO_3 + CO \rightarrow CO_2 + SO_2$ | $10^{-11}\, e^{-13,000/T}$ | 2.70 + 10 | 37 |
| 4 | $SO_3 + OCS \rightarrow CO_2 + (SO)_2$ | $10^{-11}\, e^{-10,000/T}$ | 5.42 + 11 | 36 |
| 5 | $(SO)_2 + OCS \rightarrow CO + SO_2 + S_2$ | $10^{-20}$ | 5.42 + 11 | 36 |
| 6 | $SO + SO \rightarrow SO_2 + S$ | $10^{-12}\, e^{-1700/T}$ | 2.15 + 13/2.79 + 10 | 14 |
| 7 | $S + SO_2 \rightarrow SO + SO$ | $2.3 \times 10^{-11}\, e^{-5200/T}$ | 2.15 + 13 | 5 |
| 8 | $CO + SO_2 \rightarrow CO_2 + SO$ | $4.5 \times 10^{-12}\, e^{-24,300/T}$ | 5.30 + 12/5.71 + 10 | 15 |
| 9 | $SO + CO_2 \rightarrow SO_2 + CO$ | $1.5 \times 10^{-11}\, e^{-22,000/T}$ | 5.24 + 12 | 2 |
| 10 | $S_3 + h\nu \rightarrow S_2 + S$ | $\gamma * 10^{-3}(4.4 + 1.36h + 0.063h^2)$ | 7.54 + 12 | 14 |
| 11 | $S + S + M \rightarrow S_2 + M$ | $10^{-32}\, e^{200/T}$ | 117 | 9 |
| 12 | $S_2 + M \rightarrow S + S + M$ | $5 \times 10^{-8}\, e^{-50,800/T}$ | 33 | 2 |
| 13 | $S + S_2 + M \rightarrow S_3 + M$ | $10^{-32}\, e^{200/T}$ | 4.75 + 12 | 12 |
| 14 | $S_3 + M \rightarrow S + S_2 + M$ | $2.9 \times 10^{-7}\, e^{-31.200/T}$ | 9.09 + 9 | 2 |
| 15 | $S + CO + M \rightarrow OCS + M$ | $3 \times 10^{-33}\, e^{-1000/T}$ | 4.37 + 12 | 9 |
| 16 | $OCS + M \rightarrow CO + S + M$ | $2.2 \times 10^{-7}\, e^{-37,300/T}$ | 1.27 + 12 | 2 |
| 17 | $S + OCS \rightarrow CO + S_2$ | $2 \times 10^{-14}\, (T/300)^4\, e^{-580/T}$ | 5.55 + 13/1.95 + 12 | 28 |
| 18 | $S_2 + CO \rightarrow OCS + S$ | $10^{-12}\, e^{-17160/T}$ | 5.39 + 13 | 5 |
| 19 | $S + S_3 \rightarrow S_2 + S_2$ | $1.7 \times 10^{-10}\, e^{-2800/T}$ | 6.38 + 8 | 5 |
| 20 | $S_2 + S_2 \rightarrow S + S_3$ | $2.8 \times 10^{-11}\, e^{-23,000/T}$ | 1.65 + 10 | 5 |
| 21 | $CO + S_3 \rightarrow OCS + S_2$ | $10^{-11}\, e^{-20,000/T}$ | 7.55 + 8 | 2 |
| 22 | $S_2 + OCS \rightarrow CO + S_3$ | $2.4 \times 10^{-11}\, e^{-24,900/T}$ | 4.26 + 10 | 3 |
| 23 | $S + NO + M \rightarrow SNO + M$ | $3 \times 10^{-32}\, e^{940/T}$ | 2.65 + 11 | 9 |
| 24 | $S + SNO \rightarrow S_2 + NO$ | $5 \times 10^{-11}$ | 5.43 + 7 | 6 |
| 25 | $S_2 + SNO \rightarrow S_3 + NO$ | $10^{-17}$ | 2.65 + 11 | 9 |
| 26 | $SH + SH \rightarrow H_2S + S$ | $1.5 \times 10^{-11}$ | 1.68 + 12 | 6 |
| 27 | $S + H_2S \rightarrow SH + SH$ | $1.7 \times 10^{-10}\, e^{-3620/T}$ | 1.69 + 12/5.27 + 9 | 18 |
| 28 | $S + SH \rightarrow S_2 + H$ | $4.5 \times 10^{-11}$ | 2.51 + 8 | 7 |
| 29 | $H + S_2 \rightarrow SH + S$ | $5.3 \times 10^{-10}\, e^{-8830/T}$ | 1.69 + 8 | 5 |
| 30 | $H + SH \rightarrow H_2 + S$ | $2.5 \times 10^{-11}$ | 2.44 + 6 | 4 |
| 31 | $S + H_2 \rightarrow SH + H$ | $10^{-10}\, e^{-10,080/T}$ | 2.36 + 6 | 4 |
| 32 | $H + OCS \rightarrow CO + SH$ | $1.2 \times 10^{-11}\, e^{-1950/T}$ | 3.02 + 12/2.44 + 10 | 4 |
| 33 | $CO + SH \rightarrow OCS + H$ | $6.3 \times 10^{-14}\, e^{-7780/T}$ | 3.00 + 12 | 4 |
| 34 | $H + HCl \rightarrow H_2 + Cl$ | $1.7 \times 10^{-11}\, e^{-1770/T}$ | 1.46 + 11/9.46 + 8 | 13 |
| 35 | $Cl + H_2 \rightarrow HCl + H$ | $3 \times 10^{-11}\, e^{-2270/T}$ | 1.45 + 11 | 4 |
| 36 | $H_2S + Cl \rightarrow HCl + SH$ | $3.7 \times 10^{-11}\, e^{210/T}$ | 2.46 + 14 | 4 |
| 37 | $SH + HCl \rightarrow H_2S + Cl$ | $7.6 \times 10^{-12}\, e^{-5750/T}$ | 2.46 + 14/4.34 + 9 | 26 |
| 38 | $Cl + SH \rightarrow HCl + S$ | $8 \times 10^{-11}\, e^{210/T}$ | 1.59 + 9 | 4 |
| 39 | $S + HCl \rightarrow SH + Cl$ | $1.9 \times 10^{-10}\, e^{-9380/T}$ | 1.60 + 9/1.14 + 7 | 4 |
| 40 | $H + SNO \rightarrow NO + SH$ | $4 \times 10^{-10}\, e^{-340/T}$ | 5.11 + 6 | 3 |
| 41 | $H + SH + M \rightarrow H_2S + M$ | $10^{-30}\, (300/T)^2$ | 5.15 + 5 | 5 |
| 42 | $H_2S + M \rightarrow SH + H + M$ | $1.9 \times 10^{-7}\, e^{-44,750/T}$ | 2.74 + 5 | 2 |
| 43 | $H + S_3 \rightarrow SH + S_2$ | $1.2 \times 10^{-10}\, e^{-1950/T}$ | 2.91 + 7 | 4 |

Table 12.14 (*cont.*)

| # | Reaction | Rate coefficient | Column rate | $h_r$ (km) |
|---|---|---|---|---|
| 44 | $Cl + SO_2 + M \rightarrow ClSO_2 + M$ | $1.3 \times 10^{-34}\ e^{940/T}$ | 5.99 + 13/4.12 + 9 | 23 |
| 45 | $ClSO_2 + M \rightarrow Cl + SO_2 + M$ | $7 \times 10^{-16}\ e^{-10,540/T}$ | 5.99 + 13 | 5 |
| 46 | $ClSO_2 + ClSO_2 \rightarrow SO_2Cl_2 + SO_2$ | $10^{-12}$ | 1.36 + 14/3.44 + 9 | 24 |
| 47 | $SO_2Cl_2 + SO_2 \rightarrow ClSO_2 + ClSO_2$ | $10^{-12}\ e^{-11,000/T}$ | 1.36 + 14 | 10 |
| 48 | $ClSO_2 + Cl \rightarrow SO_2 + Cl_2$ | $10^{-12}$ | 4.44 + 8 | 4 |
| 49 | $ClSO_2 + S \rightarrow SO_2 + SCl$ | $10^{-12}$ | 2.03 + 8 | 9 |
| 50 | $ClSO_2 + H \rightarrow SO_2 + HCl$ | $10^{-11}$ | 3.03 + 7 | 5 |
| 51 | $SO_2Cl_2 + Cl \rightarrow ClSO_2 + Cl_2$ | $10^{-12}$ | 7.19 + 8 | 14 |
| 52 | $SO_2Cl_2 + S \rightarrow ClSO_2 + SCl$ | $10^{-12}$ | 2.64 + 9 | 23 |
| 53 | $SO_2Cl_2 + H \rightarrow ClSO_2 + HCl$ | $10^{-11}$ | 8.10 + 7 | 17 |
| 54 | $H + Cl_2 \rightarrow HCl + Cl$ | $8 \times 10^{-11}\ e^{-416/T}$ | 7014 | 7 |
| 55 | $S + Cl_2 \rightarrow SCl + Cl$ | $2.8 \times 10^{-11}\ e^{-300/T}$ | 3.05 + 5 | 15 |
| 56 | $Cl + SCl \rightarrow S + Cl_2$ | $2.8 \times 10^{-11}\ e^{-650/T}$ | 2.40 + 6 | 8 |
| 57 | $S + SCl \rightarrow S_2 + Cl$ | $10^{-12}$ | 4.77 + 5 | 22 |
| 58 | $H + SCl \rightarrow HCl + S$ | $10^{-11}$ | 2.03 + 4 | 12 |
| 59 | $SH + Cl_2 \rightarrow HSCl + Cl$ | $1.4 \times 10^{-11}\ e^{-690/T}$ | 1.17 + 9/2.65 + 4 | 11 |
| 60 | $HSCl + SH \rightarrow H_2S + SCl$ | $3 \times 10^{-12}\ e^{-500/T}$ | 1.17 + 9 | 11 |
| 61 | $HSCl + S \rightarrow SH + SCl$ | $1.7 \times 10^{-13}$ | 1.04 + 4 | 15 |
| 62 | $HSCl + H \rightarrow H_2 + SCl$ | $1.2 \times 10^{-13}\ e^{-2770/T}$ | 1.2 | 5 |
| 63 | $HSCl + Cl \rightarrow HCl + SCl$ | $2.5 \times 10^{-13}\ e^{-130/T}$ | 1.62 + 4 | 5 |
| 64 | $SH + SCl \rightarrow S_2 + HCl$ | $6 \times 10^{-13}\ e^{230/T}$ | 4.01 + 9 | 19 |
| 65 | $SH + OH \rightarrow H_2O + S$ | $2.5 \times 10^{-12}$ | 1.83 + 5 | 3 |
| 66 | $S + H_2O \rightarrow OH + SH$ | $4.7 \times 10^{-11}\ e^{-17,700/T}$ | 1.92 + 5 | 3 |
| 67 | $OH + H_2 \rightarrow H_2O + H$ | $2.8 \times 10^{-12}\ e^{-1800/T}$ | 1.22 + 8 | 3 |
| 68 | $H + H_2O \rightarrow OH + H_2$ | $1.3 \times 10^{-11}\ e^{-9420/T}$ | 1.32 + 8 | 3 |
| 69 | $OH + HCl \rightarrow H_2O + Cl$ | $2.6 \times 10^{-12}\ e^{-350/T}$ | 1.19 + 11 | 3 |
| 70 | $Cl + H_2O \rightarrow OH + HCl$ | $2 \times 10^{-11}\ e^{-8470/T}$ | 1.20 + 11/1.18 + 9 | 26 |
| 71 | $OH + H_2S \rightarrow H_2O + SH$ | $6.1 \times 10^{-12}\ e^{-75/T}$ | 1.26 + 11 | 3 |
| 72 | $SH + H_2O \rightarrow H_2S + OH$ | $10^{-11}\ e^{-14160/T}$ | 1.31 + 11/5.22 + 9 | 9 |
| 73 | $H + H_2S \rightarrow SH + H_2$ | $8.2 \times 10^{-11}\ e^{-1470/T}$ | 3.31 + 11 | 5 |
| 74 | $SH + H_2 \rightarrow H_2S + H$ | $3 \times 10^{-11}\ e^{-7930/T}$ | 3.32 + 11/9.56 + 8 | 4 |
| 75 | $CO + OH \rightarrow CO_2 + H$ | $2.4 \times 10^{-13}$ | 5.38 + 11/2.44 + 10 | 5 |
| 76 | $H + CO_2 \rightarrow CO + OH$ | $10^{-10}\ e^{-12,400/T}$ | 5.14 + 11 | 3 |
| 77 | $OH + OCS \rightarrow CO_2 + SH$ | $1.1 \times 10^{-13}\ e^{-1200/T}$ | 5.95 + 10 | 3 |
| 78 | $SH + CO_2 \rightarrow OH + OCS$ | $2.6 \times 10^{-13}\ e^{-19,360/T}$ | 7.76 + 10 | 3 |
| 79 | $SO + OH \rightarrow SO_2 + H$ | $2.7 \times 10^{-11}\ e^{335/T}$ | 1.48 + 8 | 3 |
| 80 | $SO_2 + H \rightarrow SO + OH$ | $3.7 \times 10^{-9}\ e^{-14,350/T}$ | 1.61 + 8 | 3 |
| 81 | $S + OH \rightarrow SO + H$ | $6.6 \times 10^{-11}$ | 217 | 3 |
| 82 | $SO + H \rightarrow S + OH$ | $4 \times 10^{-10}\ e^{-11,200/T}$ | 253 | 3 |
| 83 | $S_2 + S_2 + M \rightarrow S_4 + M$ | $10^{-32}\ e^{200/T}$ | 2.98 + 23/2.69 + 14 | 17 |
| 84 | $S_4 + M \rightarrow S_2 + S_2 + M$ | $1.2 \times 10^{-6}\ e^{-E/T}$ | 2.98 + 23 | 14 |
| 85 | $S_4 + h\nu \rightarrow S + S_3$ | $\gamma_0 * 10^{-5} * (8.5 + 2.4h + 0.15h^2)$ | 2.48 + 12 | 18 |
| 86 | $S_4 + h\nu \rightarrow S_2 + S_2$ | $0.01 * (1.4 + 0.535h - 0.0013h^2)$ | 2.67 + 14 | 17 |
| 87 | $S + S_3 + M \rightarrow S_4 + M$ | $10^{-32}\ e^{200/T}$ | 8.59 + 7 | 7 |

Table 12.14 (*cont.*)

| # | Reaction | Rate coefficient | Column rate | $h_r$ (km) |
|---|---|---|---|---|
| 88 | $S_4 + M \rightarrow S_3 + S + M$ | $3 \times 10^{-7} e^{-32,200/T}$ | 7.52 + 6 | 5 |
| 89 | $SO_3 + SO \rightarrow SO_2 + SO_2$ | $2 \times 10^{-15}$ | 1.31 + 9 | 34 |

*Note.* Rate coefficients are in $cm^3\ s^{-1}$ and $cm^6\ s^{-1}$ for two- and three-body reactions, respectively, and in $s^{-1}$ for sulfur photolyses. Column rates are in $cm^{-2}\ s^{-1}$, 2.57+16 = $2.57\times10^{16}$. Differences between column rates of direct and inverse reactions are shown for some reactions. References to and discussion of rate coefficients may be found in Krasnopolsky (2007); some changes are considered in the text. Activation energy is $E$ = 7800 K in R84 and quantum yields in the sulfur photolyses are $\gamma$ =0.017 (R10) and $\gamma_0$ = 1 (R85) in the basic model. From Krasnopolsky (2013b).

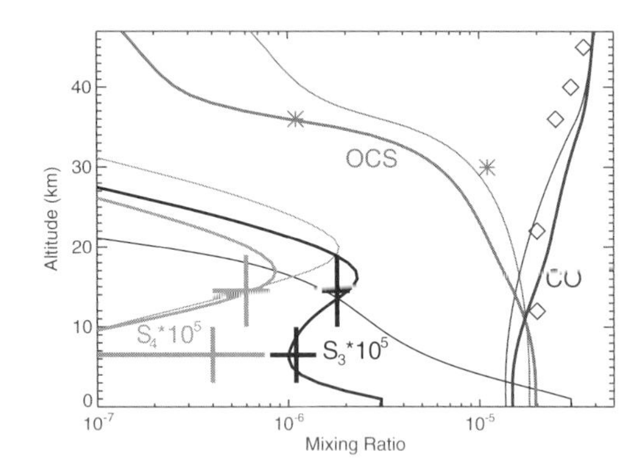


Figure 12.96 Calculated profiles of CO, OCS, S3, and S4 in the basic model (Krasnopolsky 2013b) and that without $S_4 + h\nu \rightarrow S_3 + S$ (solid and thin lines, respectively) compared with the observed abundances of these species.

magnitude. A sum of OCS + X $S_X$ is constant in this exchange, and its value is 20 ppm in the model. Thermolysis of $S_X$ and termolecular association of OCS are effective in the lowest scale height where OCS dominates in the sum (Figure 12.97), while the $SO_3$ + 2 OCS and $S_4$ cycles destroy OCS above 25 km. The only loss of the sum OCS + X $S_X$ is the upward flux of OCS at 47 km that exactly balances the sulfur production.

Sulfuric acid vapor peaks at the lower cloud boundary with a mixing ratio of 3.8 ppm (5 ppm for $K = 5000\ cm^2\ s^{-1}$ at 47 km). The layer half width at half maximum is 6 km, and both values agree with the observations (Section 12.5.4). $SO_3$ from thermolysis of $H_2SO_4$ is the only source of oxygen that is mostly removed in the $SO_3$ + 2 OCS cycle. Therefore CO + OCS is constant at 35 ppm up to 35 km and weakly increases to 40 ppm at 47 km because of the CO flux from the middle atmosphere. Oxidation of CO is via the $SO_3$ + 2 OCS cycle with minor contributions from the reactions of CO with $SO_2$, $SO_3$, and OH.

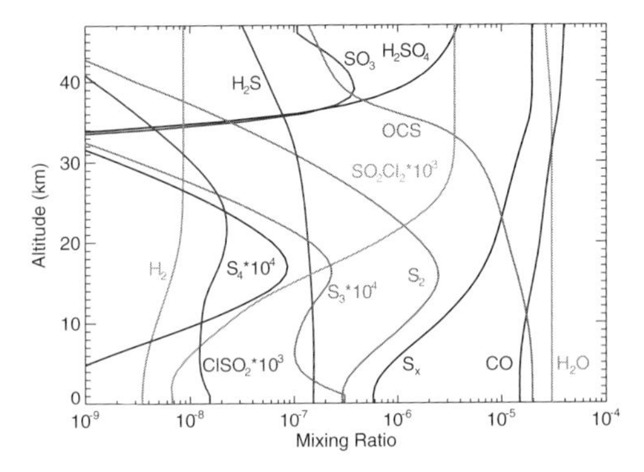


Figure 12.97 Calculated chemical composition of Venus' atmosphere below the clouds. Mixing ratios of $SO_2$ (130 ppm), HCl (0.5 ppm), and NO (5.5 ppb) are constant and not shown. $S_X = \Sigma\, nS_n$ for $n$ = 1–8. From Krasnopolsky (2013b).

A slow reaction R39 S + HCl → SH + Cl with a net rate of $1.14\times10^7$ cm$^{-2}$ s$^{-1}$ drives the SH, chlorine, and OH chemistries with rather abundant products $H_2S$, $H_2$, and $SO_2Cl_2$. For example, species recycle in

$$\text{R78} \quad SH + CO_2 \rightarrow OCS + OH,$$

$$\text{R75} \quad OH + CO \rightarrow CO_2 + H,$$

$$\text{R32} \quad H + OCS \rightarrow CO + SH$$

with no net result. These reactions are major sources and sinks of OH and H, and their column rates exceed that of R39 by three orders of magnitude. The most significant loss of radicals is R64 SH + SCl → HCl + $S_2$ with a column rate of $4.01\times10^9$ cm$^{-2}$ s$^{-1}$ and a mean altitude of 19 km. The interactions of radicals are complicated and can be analyzed using the data of Table 12.14 for each species.

Hydrogen sulfide $H_2S$ had some controversial detections and was even discussed as a parent sulfur species (Section 12.5.5). It is a product of sulfur chemistry and formed by reactions of SH with $H_2O$, HCl, HSCl, and $H_2$ with a total rate of $1.17\times10^{10}$ cm$^{-2}$ s$^{-1}$. This production is balanced by a loss in R27 S + $H_2S$ and the upward flux (12.29). The calculated profile of $H_2S$ varies from 150 ppb near the surface to 32 ppb at 47 km (Figure 12.97).

Production of $H_2$ in R34 H + HCl is balanced by its loss in R74 SH + $H_2$. The calculated profile of $H_2$ varies from 3.5 ppb near the surface to 8 ppb above 20 km (Figure 12.97).

The model predicts a constant mixing ratio of 3.5 ppb for $SO_2Cl_2$ above 30 km. However, chemistry of this species is poorly known, and this value is rather uncertain.

The model assumes the passive surface with no atmosphere-surface interaction. However, the model does not rule out this interaction, though a rate of this interaction looks smaller than those of the basic chemical processes in the lower atmosphere. The model presumes significant changes in the slow and fast atmospheric cycles of sulfur proposed by von Zahn et al. (1983).

## 12.15 Photochemistry of the Middle Atmosphere (47–112 km)

### *12.15.1 History of the Problem*

Photochemical modeling of the Venus middle atmosphere has a long history, and Prinn (1971) was the first who recognized the importance of $ClO_X$ chemistry before this chemistry was understood in the Earth's atmosphere. Sze and McElroy (1975) developed a model that included photochemistry of $CO_2$, $H_2O$, HCl and their products. Krasnopolsky and Parshev (1981, 1983) calculated a photochemical model with ClCO cycles and related chemistry that dominate in recombination of CO with O and $O_2$ on Venus. Yung and DeMore (1982) argued that the reaction

$$ClCO + O_2 \rightarrow CO_2 + ClO$$

from the model by Krasnopolsky and Parshev proceeds in two steps:

$$ClCO + O_2 + M \rightarrow ClCO_3 + M,$$

$$ClCO_3 + Cl \rightarrow CO_2 + Cl + ClO,$$

$$\text{Net } ClCO + O_2 \rightarrow CO_2 + ClO,$$

with the same net result. Yung and DeMore (1982) suggested three models based on abundant $H_2$ (model A), NO with a mixing ratio of 30 ppb produced by lightning in model B, and the ClCO chemistry in model B and C. Observations have not confirmed the assumptions of models A and B, and model C was the basic model that was further developed by Mills (1998) and later by Mills and Allen (2007).

Zhang et al. (2012) created a photochemical model (Section 12.13.7) that maximized photolysis of $H_2SO_4$ vapor to fit to an increase in the $SO_2$ mixing ratio from 85 to 100 km in the SPICAV observations (Figure 12.18a). Here we will consider in more detail a photochemical model by Krasnopolsky (2012b) with the recently updated $S_2O_2$ chemistry and modified lower boundary conditions according to the model for the lower atmosphere in the previous section (Krasnopolsky 2018a).

### *12.15.2 Input Data of the Model*

All models from Yung and DeMore (1982) to Zhang et al. (2012) were developed for the altitude range of 58–112 km with a step of 2 km. The upper boundary of 112 km allows neglecting molecular diffusion and ion chemistry, while effects of chemistry of the upper atmosphere and ionosphere may be simulated by fluxes of O, $O_2$, CO, and N at the upper boundary.

The lower boundary at 58 km is near the bottom of the upper cloud layer. The UV radiation is completely absorbed here. The boundary is shifted to 47 km in Krasnopolsky

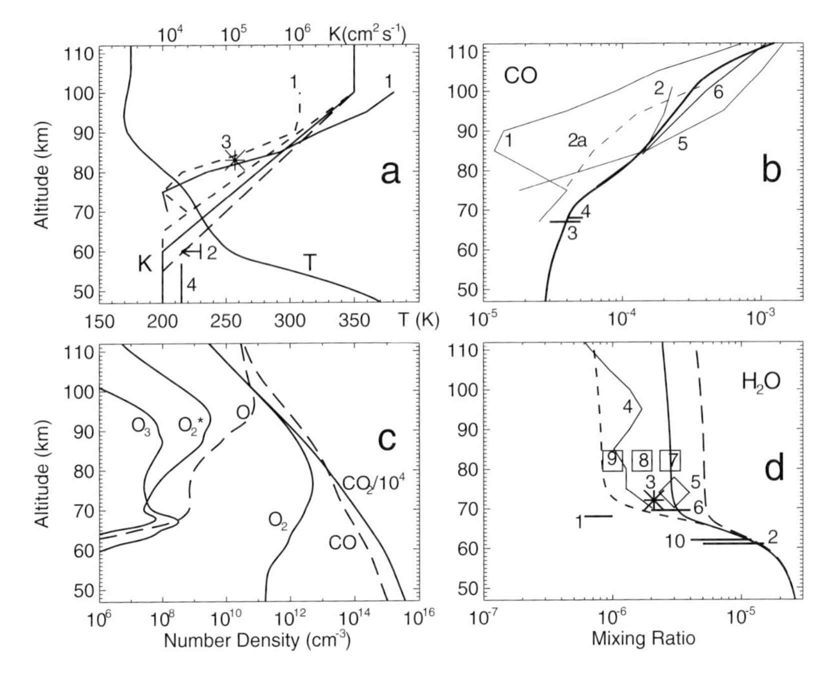


Figure 12.98 Some data from photochemical model for Venus atmosphere at 47–112 km (Krasnopolsky 2012b, 2018a). (a) Profiles of temperature and three versions of eddy diffusion in the model. Observations: (1) Krasnopolsky (1980, 1983a), (2) upper limit from Woo and Ishimaru (1981), (3) Lane and Opstbaum (1983), (4) Krasnopolsky (1985). (b) Calculated profile of the CO mixing ratio. Observations: (1) VEX/SOIR (Vandaele et al. 2008); (2) Clancy et al. (2012a), dayside observations in 2001–2002; (2a) Clancy et al. (2012a), dayside observations in 2007–2009; (3) VEX/VIRTIS (Irwin et al. 2008); (4) Connes et al. (1968), Krasnopolsky (2010c); (5) Lellouch et al. (1994), (6) Vandaele et al. (2015). (c) $CO_2$ and its photochemical products. $O_2$* is $O_2(^1\Delta_g)$ that emits the airglow at 1.27 μm. (d) Profiles of the $H_2O$ mixing ratio: models and observations. Three calculated profiles are for the models with the $SO_2$ mixing ratio of 10.7 (short dashes), 9.7 (solid), and 8.7 (long dashes) ppm at 47 km. Observations: (1) Fink et al. (1972); (2) Venera 15 (Ignatiev et al. 1999); (3) KAO (Bjoraker et al. 1992); (4) VEX/SOIR (Fedorova et al. 2008); (5) Krasnopolsky et al. (2013); (6) VEX/VIRTIS (Cottini et al. 2015), (10) SPICAV-IR (Fedorova et al 2016). Mean results of the microwave observations: (7) Encrenaz et al. (1995); (8) SWAS (Gurwell et al. 2007); (9) Sandor and Clancy (2005).

(2012b) to account for some results of the chemical kinetic model (Section 12.14) as the boundary conditions.

A vertical step was reduced from 2 km in the previous models to 0.5 km. This improves the numerical accuracy by an order of magnitude. Besides $CO_2$ and $SO_2$ as the basic UV absorbers and scattering in the clouds, the model includes a NUV absorber to simulate the absorption profile observed by V14 (Section 12.13.4). The temperature and eddy diffusion profiles (Figure 12.98a) are those from VIRA and based on the existing observational data, respectively.

Vertical profiles of water vapor were fixed in the previous models to avoid a complicated problem of phase equilibrium with the sulfuric acid aerosol. Here the $H_2O$ profile was calculated assuming that sulfuric acid is formed as $H_2SO_4$•1.36 $H_2O$, that is, with concentration of 80%.

Modeling by Krasnopolsky (2011c) and the observation by Sandor et al. (2012) show extremely low quantities of $H_2SO_4$ vapor in the middle atmosphere that cannot affect abundances of $SO_X$ (Section 12.13.7). This species is therefore neglected in the model. Elemental sulfur is restricted to S, $S_2$, $S_3$, and aerosol sulfur $S_a$. Chemistry of sulfur allotropes is poorly known, and some assumptions are inevitable here.

Conditions at the lower boundary (Table 12.15) for $H_2O$, OCS, and $H_2$ are based on the chemical kinetic model for the lower atmosphere (Krasnopolsky 2013b). The $SO_2$ abundance at 47 km is a model parameter that is slightly smaller than half the $H_2O$ abundance. Half maximum diffusion velocities $V = -K/2H = -0.0044$ cm s$^{-1}$ are the lower boundary conditions for the most of the species in the model.

The model is calculated for the global mean conditions, that is, solar zenith angle of 60° and the solar flux is halved to account for the nightside. The upper boundary conditions are fluxes (Table 12.16), and that of CO is equal to the number of the solar photons absorbed above 112 km. $O_2$ continues to dissociate above 112 km, and the condition reflects this fact. The flux of N is evaluated using the observed NO nightglow intensity, and the boundary is closed (flux is equal to zero) for all other species.

The model involves 153 reactions (Tables 12.17 and 12.18) of 44 species. The number of reactions is much smaller than those in the previous works because of removal of numerous insignificant processes. Even this list may be shortened as well, but some minor reactions do not look negligible a priori.

Table 12.15 *Conditions at the lower boundary (47 km)*

| $CO_2$ | $H_2O$ | $SO_2$ | HCl | OCS | NO | $H_2$ | CO | $O_2$ | other |
|---|---|---|---|---|---|---|---|---|---|
| $3.6\times10^{19}$ cm$^{-3}$ | 26 ppm | 9.7± 1 ppm | 400 ppb | 140 ppb | 5.5 ppb | 8.5 ppb | 0.1V | 2V | V |

V = $-K/2H = -0.0044$ cm s$^{-1}$.

Table 12.16 *Fluxes at the upper boundary (112 km)*

| CO | $O_2$ | O | N | other |
|---|---|---|---|---|
| $-6\times10^{11}$ | $[O_2]V$ | $-6\times10^{11}-2[O_2]V$ | $-6\times10^{8}$ | 0 |

Fluxes are in cm$^{-2}$ s$^{-1}$, V = $(J_4 + J_5)H$ = 0.3 cm s$^{-1}$.

Table 12.17 *Reactions of photodissociation*

| # | Reaction | $J$ | CR | $h$ (km) |
|---|---|---|---|---|
| 1 | $CO_2 + h\nu \rightarrow CO + O$ | – | 3.99 + 12 | 88 |
| 2 | $CO_2 + h\nu \rightarrow CO + O^*$ | $\lambda$<167 nm | 6.65 + 10 | 110 |
| 3 | $SO_2 + h\nu \rightarrow SO + O$ | – | 1.16 + 13 | 72 |
| 4 | $O_2 + h\nu \rightarrow O + O$ | – | 1.33 + 10 | 89 |
| 5 | $O_2 + h\nu \rightarrow O^* + O$ | $\lambda$<175 nm | 2.13 + 9 | 106 |
| 6 | $O_3 + h\nu \rightarrow O_2^* + O^*$ | 0.0085 | 1.45 + 12 | 81 |
| 7 | $SO_3 + h\nu \rightarrow SO_2 + O$ | – | 9.18 + 8 | 71 |
| 8 | $SO + h\nu \rightarrow S + O$ | – | 8.38 + 11 | 78 |
| 9 | $OCS + h\nu \rightarrow CO + S$ | – | 1.93 + 10 | 58 |
| 10 | $S_2O + h\nu \rightarrow SO + S$ | 0.04 | 2.62 + 10 | 68 |
| 11 | $S_2 + h\nu \rightarrow S + S$ | 0.006 | 7.18 + 6 | 60 |
| 12 | $S_3 + h\nu \rightarrow S_2 + S$ | 0.137 | 1.74 + 4 | 56 |
| 13 | $S_2O_2 + h\nu \rightarrow SO + SO$ | – | 4.00 + 13 | 68 |
| 14 | $H_2O + h\nu \rightarrow OH + H$ | – | 3.15 + 9 | 95 |
| 15 | $HCl + h\nu \rightarrow H + Cl$ | – | 9.71 + 10 | 74 |
| 16 | $ClO + h\nu \rightarrow Cl + O$ | 0.0065 | 4.78 + 11 | 76 |
| 17 | $ClO_2 + h\nu \rightarrow ClO + O$ | 0.009 | 7.52 + 7 | 79 |
| 18 | $ClCO_3 + h\nu \rightarrow CO_2 + ClO$ | 0.0032 | 8.13 + 10 | 76 |
| 19 | $Cl_2 + h\nu \rightarrow Cl + Cl$ | 0.0026 | 2.98 + 13 | 72 |
| 20 | $SCl + h\nu \rightarrow S + Cl$ | 0.05 | 3.48 + 9 | 70 |
| 21 | $SCl_2 + h\nu \rightarrow SCl + Cl$ | 0.005 | 8.15 + 8 | 71 |
| 22 | $COCl_2 + h\nu \rightarrow ClCO + Cl$ | $4\times10^{-5}$ | 2.44 + 8 | 72 |
| 23 | $NO + h\nu \rightarrow N + O$ | – | 6.43 + 6 | 83 |
| 24 | $NO_2 + h\nu \rightarrow NO + O$ | 0.01 | 5.96 + 11 | 71 |
| 25 | $NO_3 + h\nu \rightarrow NO_2 + O$ | 0.16 | 1.59 + 10 | 72 |
| 26 | $NO_3 + h\nu \rightarrow NO + O_2$ | 0.02 | 1.99 + 9 | 72 |
| 27 | $HO_2NO_2 + h\nu \rightarrow HO_2 + NO_2$ | $4.3\times10^{-4}$ | 6.63 + 7 | 66 |
| 28 | $ClNO + h\nu \rightarrow Cl + NO$ | 0.005 | 1.47 + 11 | 66 |
| 29 | $SNO + h\nu \rightarrow S + NO$ | $0.01 = J_{24}$ | 8.11 + 9 | 57 |
| 30 | $SO_2Cl_2 + h\nu \rightarrow ClSO_2 + Cl$ | $1.6\times10^{-4}$ | 2.82 + 12 | 66 |

*Note.* Numerical values of photolysis frequencies $J$ are in $s^{-1}$ for $\lambda > 200$ nm at 1 AU. Column rates CR are in $cm^{-2}\ s^{-1}$; $h$ is the weighted-mean altitude of a reaction; 3.99+12 = $3.99\times10^{12}$, $O_2^* = O_2(a^1\Delta_g)$, $O^* = O(^1D)$.

### *12.15.3 Model Results: Carbon Monoxide*

The calculated chemical composition is extremely sensitive to small variations of eddy diffusion with a breakpoint altitude $h_e = 60 \pm 5$ km (Figure 12.98a) and the $SO_2$ abundance of 9.7 ± 1 ppm at the lower boundary. Therefore five models have been calculated to account for these variations.

Table 12.18 *Reactions and their rate coefficients, column rates, and mean altitudes*

| # | Reaction | Rate coefficient | CR | $h$ (km) |
|---|---|---|---|---|
| 31 | $O^* + CO_2 \rightarrow O + CO_2$ | $7.5 \times 10^{-11}\ e^{115/T}$ | 1.52 + 12 | 83 |
| 32 | $O^* + H_2O \rightarrow OH + OH$ | $1.6 \times 10^{-10}\ e^{60/T}$ | 7.51 + 6 | 81 |
| 33 | $O^* + HCl \rightarrow OH + Cl$ | $1.5 \times 10^{-10}$ | 6.70 + 5 | 83 |
| 34 | $O_2^* \rightarrow O_2 + h\nu$ | $2.2 \times 10^{-4}$ | 7.97 + 11 | 92 |
| 35 | $O_2^* + CO_2 \rightarrow O_2 + CO_2$ | $10^{-20}$ | 1.08 + 12 | 79 |
| 36 | $O + CO + M \rightarrow CO_2 + M$ | $2.2 \times 10^{-33}\ e^{-1780/T}$ | 2.70 + 10 | 69 |
| 37 | $O + O + M \rightarrow O_2^* + M$ | $7.5 \times 10^{-33}\ (300/T)^{3.25}$ | 5.97 + 11 | 94 |
| 38 | $O + O_2 + M \rightarrow O_3 + M$ | $1.2 \times 10^{-27}\ T^{-2.4}$ | 7.82 + 12 | 76 |
| 39 | $O + O_3 \rightarrow O_2 + O_2$ | $8 \times 10^{-12}\ e^{-2060/T}$ | 1.41 + 8 | 85 |
| 40 | $H + O_2 + M \rightarrow HO_2 + M$ | $10^{-31}(300/T)^{1.3}$ | 5.80 + 11 | 85 |
| 41 | $H + O_3 \rightarrow OH + O_2$ | $1.4 \times 10^{-10}\ e^{-470/T}$ | 1.66 + 11 | 92 |
| 42 | $O + HO_2 \rightarrow OH + O_2$ | $3 \times 10^{-11}\ e^{200/T}$ | 4.15 + 11 | 90 |
| 43 | $O + OH \rightarrow O_2 + H$ | $1.1 \times 10^{-10}\ T^{-0.32}\ e^{177/T}$ | 4.63 + 11 | 92 |
| 44 | $CO + OH \rightarrow CO_2 + H$ | $2.8 \times 10^{-13}\ e^{-176/T}$ | 1.99 + 11 | 79 |
| 45 | $H + HO_2 \rightarrow OH + OH$ | $7.2 \times 10^{-11}$ | 3.76 + 9 | 91 |
| 46 | $H + HO_2 \rightarrow H_2 + O_2$ | $6.9 \times 10^{-12}$ | 3.61 + 8 | 91 |
| 47 | $H + HO_2 \rightarrow H_2O + O$ | $1.6 \times 10^{-12}$ | 8.28 + 7 | 91 |
| 48 | $OH + HO_2 \rightarrow H_2O + O_2$ | $4.8 \times 10^{-11}\ e^{250/T}$ | 2.75 + 7 | 88 |
| 49 | $OH + O_3 \rightarrow HO_2 + O_2$ | $1.7 \times 10^{-12}\ e^{-940/T}$ | 2.55 + 5 | 85 |
| 50 | $H + HCl \rightarrow H_2 + Cl$ | $6.6 \times 10^{-16}\ T^{1.44}\ e^{-1240/T}$ | 2.40 + 9 | 77 |
| 51 | $OH + HCl \rightarrow H_2O + Cl$ | $1.7 \times 10^{-12}\ e^{-230/T}$ | 4.85 + 9 | 73 |
| 52 | $O + HCl \rightarrow OH + Cl$ | $10^{-11}\ e^{-3300/T}$ | 1.05 + 9 | 70 |
| 53 | $Cl + H_2 \rightarrow HCl + H$ | $3.9 \times 10^{-11}\ e^{-2310/T}$ | 2.97 + 9 | 73 |
| 54 | $Cl + HO_2 \rightarrow HCl + O_2$ | $1.8 \times 10^{-11}\ e^{170/T}$ | 7.14 + 10 | 76 |
| 55 | $Cl + HO_2 \rightarrow OH + ClO$ | $6.3 \times 10^{-11}\ e^{-570/T}$ | 7.76 + 9 | 74 |
| 56 | $OH + ClO \rightarrow HCl + O_2$ | $10^{-13}\ e^{600/T}$ | 1.49 + 6 | 83 |
| 57 | $OH + ClO \rightarrow HO_2 + Cl$ | $1.9 \times 10^{-11}$ | 1.29 + 7 | 81 |
| 58 | $H + Cl_2 \rightarrow HCl + Cl$ | $8 \times 10^{-11}\ e^{-416/T}$ | 2.42 + 10 | 72 |
| 59 | $Cl + Cl + M \rightarrow Cl_2 + M$ | $5 \times 10^{-26}\ T^{-2.4}$ | 1.68 + 13 | 73 |
| 60 | $Cl + O_3 \rightarrow ClO + O_2$ | $2.8 \times 10^{-11}\ e^{-250/T}$ | 6.18 + 12 | 75 |
| 61 | $ClO + O \rightarrow Cl + O_2$ | $2.5 \times 10^{-11}\ e^{110/T}$ | 3.06 + 12 | 78 |
| 62 | $Cl + CO + M \rightarrow ClCO + M$ | $4 \times 10^{-33}\ (300/T)^{3.8}$ | 4.46 + 15/4.24 + 12[a] | 72 |
| 63 | $ClCO + M \rightarrow Cl + CO + M$ | $1.2 \times 10^{-9}\ e^{-2960/T}$ | 4.46 + 15 | 72 |
| 64 | $ClCO + Cl \rightarrow CO + Cl_2$ | $1.1 \times 10^{-10}\ e^{-706/T}$ | 2.54 + 11 | 75 |
| 65 | $ClCO + NO_2 \rightarrow CO_2 + NO + Cl$ | $6 \times 10^{-13}\ e^{600/T}$ | 1.39 + 9 | 73 |
| 66 | $ClCO + ClNO \rightarrow COCl_2 + NO$ | $8 \times 10^{-11}\ e^{-573/T}$ | 1.03 + 8 | 71 |
| 67 | $ClCO + Cl_2 \rightarrow COCl_2 + Cl$ | $4.2 \times 10^{-12}\ e^{-1490/T}$ | 1.94 + 8 | 72 |
| 68 | $ClCO + O \rightarrow CO_2 + Cl$ | $3 \times 10^{-11}$ | 3.51 + 11 | 85 |
| 69 | $ClCO + S \rightarrow OCS + Cl$ | $3 \times 10^{-11}$ | 8.45 + 6 | 77 |
| 70 | $ClCO + O_2 + M \rightarrow ClCO_3 + M$ | $\frac{5.7 \times 10^{-32} e^{500/T}}{1 + M/2 \times 10^{18}}$ | 3.64 + 12 | 75 |
| 71 | $ClCO_3 + Cl \rightarrow CO_2 + Cl + ClO$ | $10^{-11}$ | 3.21 + 12 | 74 |
| 72 | $ClCO_3 + O \rightarrow CO_2 + Cl + O_2$ | $10^{-11}$ | 2.62 + 11 | 78 |
| 73 | $ClCO_3 + H \rightarrow CO_2 + Cl + OH$ | $10^{-11}$ | 1.79 + 8 | 83 |
| 74 | $COCl_2 + O \rightarrow ClCO + ClO$ | $10^{-14}$ | 5.25 + 7 | 72 |

Table 12.18 (*cont.*)

| # | Reaction | Rate coefficient | CR | $h$ (km) |
|---|---|---|---|---|
| 75 | $N + O \rightarrow NO + h\nu$ | $3.3 \times 10^{-16}\ T^{-0.5}$ | 7.66 + 7 | 99 |
| 76 | $N + O + M \rightarrow NO + M$ | $3.5 \times 10^{-31}\ T^{-0.5}$ | 1.94 + 8 | 94 |
| 77 | $N + NO \rightarrow N_2 + O$ | $2.2 \times 10^{-11}\ e^{160/T}$ | 3.44 + 8 | 93 |
| 78 | $NO + O + M \rightarrow NO_2 + M$ | $10^{-27}\ T^{-1.5}$ | 5.91 + 11 | 69 |
| 79 | $NO + O_3 \rightarrow NO_2 + O_2$ | $3 \times 10^{-12}\ e^{-1500/T}$ | 4.97 + 9 | 66 |
| 80 | $NO + HO_2 \rightarrow NO_2 + OH$ | $3.5 \times 10^{-12}\ e^{250/T}$ | 1.06 + 11 | 65 |
| 81 | $NO_2 + O \rightarrow NO + O_2$ | $5.1 \times 10^{-12}\ e^{210/T}$ | 4.76 + 11 | 73 |
| 82 | $NO_2 + O + M \rightarrow NO_3 + M$ | $1.7 \times 10^{-26}\ T^{-1.8}$ | 2.78 + 10 | 71 |
| 83 | $NO_3 + NO \rightarrow 2\ NO_2$ | $1.5 \times 10^{-11}\ e^{170/T}$ | 9.22 + 9 | 70 |
| 84 | $NO_3 + O \rightarrow NO_2 + O_2$ | $10^{-11}$ | 6.99 + 8 | 72 |
| 85 | $N + O_2 \rightarrow NO + O$ | $2 \times 10^{-18}\ T^{2.15}\ e^{-2560/T}$ | 1.03 + 6 | 90 |
| 86 | $NO + Cl + M \rightarrow ClNO + M$ | $2 \times 10^{-31}(300/T)^{1.8}$ | 7.28 + 12 | 71 |
| 87 | $ClNO + Cl \rightarrow Cl_2 + NO$ | $5.8 \times 10^{-11}\ e^{100/T}$ | 7.13 + 12 | 71 |
| 88 | $ClNO + O \rightarrow NO + ClO$ | $8.3 \times 10^{-12}\ e^{-1520/T}$ | 7.51 + 7 | 69 |
| 89 | $NO + ClO \rightarrow NO_2 + Cl$ | $6.2 \times 10^{-12}\ e^{295/T}$ | 2.53 + 12 | 72 |
| 90 | $NO + ClO_2 \rightarrow ClNO + O_2$ | $4.5 \times 10^{-11}$ | 2.73 + 8 | 73 |
| 91 | $NO_2 + SCl \rightarrow NO + OSCl$ | $2.3 \times 10^{-11}$ | 8.86 + 7 | 71 |
| 92 | $SO_2 + O + M \rightarrow SO_3 + M$ | $5 \times 10^{-22}\ T^{-3}\ e^{-2400/T}$ | 5.06 + 11 | 66 |
| 93 | $SO_2 + OH + M \rightarrow HSO_3 + M$ | $3.5 \times 10^{-20}\ T^{-4.3}$ | 8.71 + 10 | 65 |
| 94 | $SO_2 + Cl + M \rightarrow ClSO_2 + M$ | $1.3 \times 10^{-34}\ e^{940/T}$ | 8.67 + 12 | 69 |
| 95 | $ClSO_2 + H \rightarrow SO_2 + HCl$ | $10^{-11}$ | 3.61 + 9 | 67 |
| 96 | $ClSO_2 + Cl \rightarrow SO_2 + Cl_2$ | $10^{-12}$ | 5.40 + 12 | 71 |
| 97 | $ClSO_2 + O \rightarrow SO_2 + ClO$ | $10^{-12}$ | 4.34 + 11 | 70 |
| 98 | $ClSO_2 + SCl \rightarrow SO_2 + SCl_2$ | $10^{-13}$ | 8.02 + 6 | 69 |
| 99 | $HSO_3 + O_2 \rightarrow SO_3 + HO_2$ | $1.3 \times 10^{-12}\ e^{-330/T}$ | 8.71 + 10 | 65 |
| 100 | $S + O_2 \rightarrow SO + O$ | $2.3 \times 10^{-12}$ | 9.21 + 11 | 77 |
| 101 | $S + O_3 \rightarrow SO + O_2$ | $1.2 \times 10^{-11}$ | 1.15 + 8 | 76 |
| 102 | $S + NO + M \rightarrow SNO + M$ | $3 \times 10^{-32}\ e^{940/T}$ | 8.27 + 9 | 57 |
| 103 | $SNO + O \rightarrow NO + SO$ | $10^{-11}$ | 1.57 + 8 | 69 |
| 104 | $SO + O + M \rightarrow SO_2 + M$ | $1.3 \times 10^{-30}\ (300/T)^{2.2}$ | 4.82 + 12 | 71 |
| 105 | $SO + O_2 \rightarrow SO_2 + O$ | $1.25 \times 10^{-13}\ e^{-2190/T}$ | 9.75 + 10 | 70 |
| 106 | $SO + O_3 \rightarrow SO_2 + O_2$ | $3.4 \times 10^{-12}\ e^{-1100/T}$ | 1.40 + 10 | 68 |
| 107 | $SO + OH \rightarrow SO_2 + H$ | $2.7 \times 10^{-11}\ e^{335/T}$ | 1.56 + 10 | 84 |
| 108 | $SO + ClO \rightarrow SO_2 + Cl$ | $2.8 \times 10^{-11}$ | 4.26 + 12 | 73 |
| 109 | $SO + ClO_2 \rightarrow SO_2 + ClO$ | $1.9 \times 10^{-12}$ | 2.22 + 7 | 76 |
| 110 | $SO + HO_2 \rightarrow SO_2 + OH$ | $2.8 \times 10^{-11}$ | 6.21 + 10 | 68 |
| 111 | $SO + SO \rightarrow SO_2 + S$ | $3.5 \times 10^{-15}$ | 4.59 + 10 | 70 |
| 112 | $SO + NO_2 \rightarrow SO_2 + NO$ | $1.4 \times 10^{-11}$ | 2.16 + 12 | 71 |
| 113 | $SO + SO_3 \rightarrow SO_2 + SO_2$ | $2 \times 10^{-15}$ | 5.69 + 8 | 69 |
| 114 | $SO_3 + O \rightarrow SO_2 + O_2$ | $2.3 \times 10^{-16}\ e^{-487/T}$ | 8.96 + 5 | 70 |
| 115 | $SO_3 + H_2O + H_2O \rightarrow H_2SO_4 + H_2O$ | $2.3 \times 10^{-43} T\ e^{6540/T}$ | 4.15 + 11 | 66 |
| 116 | $Cl + O_2 + M \rightarrow ClO_2 + M$ | $5 \times 10^{-33}(300/T)^{3.1}$ | 4.79 + 14/1.83 + 10[a] | 73 |
| 117 | $ClO_2 + M \rightarrow Cl + O_2 + M$ | $7 \times 10^{-10}\ e^{-1820/T}$ | 4.79 + 14 | 73 |
| 118 | $O + Cl_2 \rightarrow ClO + Cl$ | $7.4 \times 10^{-12}\ e^{-1650/T}$ | 3.87 + 10 | 72 |
| 119 | $Cl + O + M \rightarrow ClO + M$ | $5 \times 10^{-32}$ | 3.54 + 11 | 73 |

Table 12.18 (*cont.*)

| # | Reaction | Rate coefficient | CR | $h$ (km) |
|---|---|---|---|---|
| 120 | $S + S + M \rightarrow S_2 + M$ | $10^{-32}\ e^{206/T}$ | 147 | 66 |
| 121 | $S + S_2 + M \rightarrow S_3 + M$ | $10^{-32}\ e^{206/T}$ | 1.74 + 4 | 56 |
| 122 | $S_2 + S_2 + M \rightarrow 4\ S_a + M$ | $10^{-32}\ e^{206/T}$ | 1.30 + 7 | 54 |
| 123 | $S + S_3 \rightarrow S_2 + S_2$ | $1.7 \times 10^{-10}\ e^{-2800/T}$ | 5e-5 | 55 |
| 124 | $S_2 + S_3 + M \rightarrow 5S_a + M$ | $10^{-32}\ e^{206/T}$ | 0.6 | 55 |
| 125 | $O + S_2 \rightarrow SO + S$ | $2.2 \times 10^{-11}\ e^{-84/T}$ | 3.08 + 5 | 58 |
| 126 | $O + S_3 \rightarrow SO + S_2$ | $5 \times 10^{-11}$ | 0.06 | 58 |
| 127 | $O + S_a \rightarrow SO$ | $\gamma = 0.01$ | 3.07 + 6 | 71 |
| 128 | $SO + SO + M \rightarrow S_2O_2 + M$ | $10^{-30}$ | 4.01 + 13 | 68 |
| 129 | $S_2O_2 + O \rightarrow S_2O + O_2$ | $3 \times 10^{-14}$ | 2.39 + 9 | 70 |
| 130 | $S_2O_2 + O \rightarrow SO_2 + SO$ | $3 \times 10^{-15}$ | 2.39 + 8 | 70 |
| 131 | $S_2O_2 + SO \rightarrow S_2O + SO_2$ | $3.3 \times 10^{-14}$ | 2.77 + 10 | 68 |
| 132 | $S_2O + O \rightarrow SO + SO$ | $1.5 \times 10^{-12}$ | 7.11 + 8 | 68 |
| 133 | $S_2O + Cl \rightarrow SCl + SO$ | $10^{-12}$ | 3.19 + 9 | 69 |
| 134 | $ClCO_3 + SO \rightarrow Cl + SO_2 + CO_2$ | $10^{-12}$ | 5.68 + 10 | 73 |
| 135 | $ClCO_3 + SO_2 \rightarrow Cl + SO_3 + CO_2$ | $10^{-14}$ | 2.33 + 10 | 69 |
| 136 | $O + OCS \rightarrow SO + CO$ | $1.6 \times 10^{-11}\ e^{-2150/T}$ | 3.76 + 7 | 62 |
| 137 | $S + OCS \rightarrow S_2 + CO$ | $2.4 \times 10^{-24}\ T^4\ e^{-580/T}$ | 2.71 + 7 | 54 |
| 138 | $S + CO + M \rightarrow OCS + M$ | $3 \times 10^{-33}\ e^{-1000/T}$ | 6.32 + 9 | 55 |
| 139 | $S + Cl_2 \rightarrow SCl + Cl$ | $2.8 \times 10^{-11}\ e^{-290/T}$ | 5.68 + 9 | 72 |
| 140 | $O + SCl \rightarrow SO + Cl$ | $1.2 \times 10^{-10}$ | 5.29 + 9 | 71 |
| 141 | $Cl + SCl + M \rightarrow SCl_2 + M$ | $10^{-30}$ | 7.36 + 8 | 71 |
| 142 | $Cl_2 + SCl \rightarrow SCl_2 + Cl$ | $7 \times 10^{-14}$ | 7.10 + 7 | 71 |
| 143 | $ClO_2 + O \rightarrow ClO + O_2$ | $2.4 \times 10^{-12}\ e^{-960/T}$ | 3.13 + 5 | 84 |
| 144 | $ClO_2 + Cl \rightarrow Cl_2 + O_2$ | $10^{-10}$ | 8.86 + 9 | 76 |
| 145 | $ClO_2 + Cl \rightarrow ClO + ClO$ | $10^{-10}$ | 8.86 + 9 | 76 |
| 146 | $SO + Cl + M \rightarrow OSCl + M$ | $5 \times 10^{-33}(300/T)^{3.1}$ | 3.04 + 11 | 72 |
| 147 | $OSCl + O \rightarrow SO_2 + Cl$ | $10^{-12}$ | 1.99 + 10 | 71 |
| 148 | $OSCl + Cl \rightarrow SO + Cl_2$ | $10^{-12}$ | 2.84 + 11 | 72 |
| 149 | $ClSO_2 + ClSO_2 \rightarrow SO_2Cl_2 + SO_2$ | $10^{-12}$ | 2.83 + 12 | 66 |
| 150 | $ClSO_2 + SCl \rightarrow SO_2Cl_2 + S$ | $10^{-13}$ | 8.02 + 6 | 69 |
| 151 | $NO_2 + HO_2 + M \rightarrow HO_2NO_2 + M$ | $5 \times 10^{-31}(300/T)^{3.1}$ | 8.70 + 7 | 65 |
| 152 | $HO_2NO_2 + M \rightarrow HO_2 + NO_2 + M$ | $10^{-4}\ e^{-10{,}600/T}$ | 2.07 + 7 | 64 |
| 153 | $HO_2NO_2 + OH \rightarrow NO_2 + O_2 + H_2O$ | $1.3 \times 10^{-12}\ e^{380/T}$ | 3.72 + 4 | 66 |

*Note.* Rate coefficients are in $cm^3\ s^{-1}$ and $cm^6\ s^{-1}$ for two- and three-body reactions, respectively. Aerosol uptake coefficients are dimensionless. Column rates CR are for the basic model, in $cm^{-2}\ s^{-1}$; 1.58+12 = $1.58 \times 10^{12}$. $O^*$ is $O(^1D)$, $O_2^*$ is $O_2(^1\Delta_g)$. Yield of $O_2(^1\Delta_g)$ is 0.7 in r37.

[a] Differences between the rates: r62 – r63 and r116 – r117, respectively.

CO (Figure 12.98b) is formed by photolysis of $CO_2$ with a column rate of $4.66 \times 10^{12}$ $cm^{-2}\ s^{-1}$ including the flux from the upper atmosphere. Photolysis of OCS is weaker by two orders of magnitude and partly compensated by the reverse termolecular reaction 138. The ClCO cycles are the main processes of loss of CO:

$$CO + Cl + M \rightarrow ClCO + M, \quad (12.36)$$

$$ClCO + O_2 + M \rightarrow ClCO_3 + M, \quad (12.37)$$

$$ClCO_3 + X \rightarrow CO_2 + Cl + XO, \quad (12.38)$$

$$\text{Net } CO + O_2 + X \rightarrow CO_2 + XO\left(3.55 \times 10^{12}\ \text{cm}^{-2}\ \text{s}^{-1}\right), \quad (12.39)$$

where X = Cl, O, SO, $SO_2$, and H. Photolysis of $ClCO_3$ has a similar effect, as well as

$$CO + Cl + M \rightarrow ClCO + M, \quad (12.40)$$

$$ClCO + O \rightarrow CO_2 + Cl, \quad (12.41)$$

$$\text{Net } CO + O \rightarrow CO_2\left(3.51 \times 10^{11}\ \text{cm}^{-2}\ \text{s}^{-1}\right). \quad (12.42)$$

Reactions of CO with OH and O are much weaker than the ClCO cycles that give $3.98\times10^{12}$ cm$^{-2}$ s$^{-1}$, and the excess of the production over the loss is equal to $6.8\times10^{11}$ cm$^{-2}$ s$^{-1}$ and moves down into the lower atmosphere.

The CO model profile is compared with the observations in Figure 12.98b. Later analysis of the VEX/SOIR occultations (Vandaele et al. 2015; Figure 12.9) show a decrease in CO from ~1000 ppm at 110 km to ~140 ppm at 84 km, in accord with the model.

### 12.15.4 Sulfur Species

The observed high variability of sulfur species $SO_2$, SO, and OCS needs explanation, and sulfur chemistry in the middle atmosphere is driven by photolysis of $SO_2$. Though $SO_2$ is a minor species, its column photolysis rate exceeds that of $CO_2$ by a factor of 2.4. Formation of the dense clouds of sulfuric acid is among main features of Venus' photochemistry. Sulfuric acid mostly forms in

$$SO_2 + O + M \rightarrow SO_3 + M \quad (12.43)$$

$$SO_3 + H_2O + H_2O \rightarrow H_2SO_4 + H_2O, \quad (12.44)$$

with a column rate of $4.15\times10^{11}$ cm$^{-2}$ s$^{-1}$. The production is in a narrow layer at 66 km with a width of 4 km at half maximum and a yield of 0.036 relative to the $SO_2$ photolysis. The peak altitude is higher than those in the previous models (~62 km) because of the inclusion of the NUV absorption.

The formation of sulfuric acid reduces the $SO_2$ abundance at 70 km by two orders of magnitude (Figure 12.99). This reduction is very sensitive to minor variations of atmospheric dynamics (eddy diffusion) near the cloud tops and the $SO_2/H_2O$ ratio near the lower cloud boundary. Therefore the observed variations of $SO_2$ (Figure 12.99; Section 12.5.1) do not require volcanism for explanation. SO produced by the $SO_2$ photolysis above

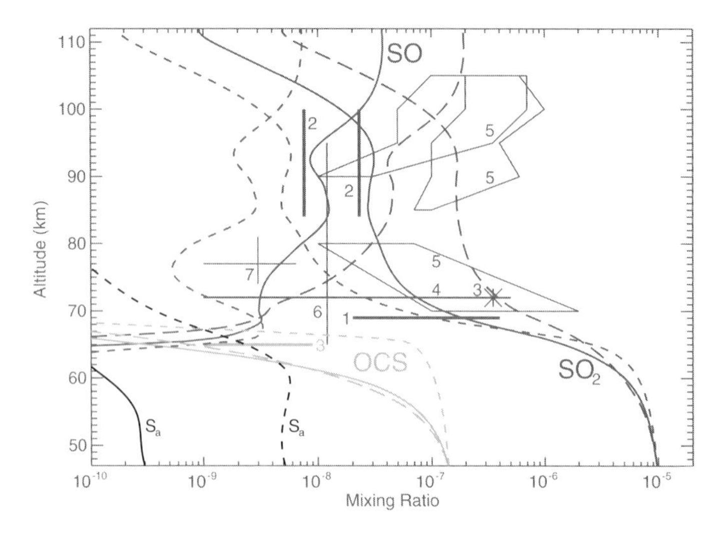


Figure 12.99 Basic sulfur species: model results and observations. $SO_2$, OCS, SO, and sulfur aerosol $S_a$ profiles are shown for the models with the eddy break at $h_e$ = 57, 60, and 65 km (long dash, solid, and short dash curves, respectively). The $S_a$ mixing ratios refer to total number of sulfur atoms in the aerosol. Observations: (1) PV, Venera 15, HST and rocket data (Esposito et al. 1997); (2) mean results of the submillimeter measurements (Sandor et al. 2010); the observed $SO_2$ varies from 0 to 76 ppb and SO from 0 to 31 ppb; (3) IRTF/CSHELL (Krasnopolsky 2010d); (4) SPICAV_UV, nadir (Marcq et al. 2011); (5) VEX/SOIR and SPICAV-UV occultations (Belyaev et al. 2012); (6) rocket observation of SO (Na et al. 1994); (7) HST observations of SO (Jessup et al. 2015). From Krasnopolsky (2018a). (A black and white version of this figure will appear in some formats. For the color version, please refer to the plate section.)

70 km regenerates $SO_2$ in the reactions with ClO, $NO_2$, and SO + O + M. The regeneration becomes ineffective above 100 km, and a sum of $SO_2$ and SO is nearly constant above 80 km and is equal to ~40 ppb in the basic model.

Photolysis of OCS is strongly affected by absorption by $SO_2$ and the NUV absorber. One-third of the photolysis is compensated by the reverse association S + CO + M, and the remaining two-thirds result in a steep decline in the OCS mixing ratio above 60 km, in accord with the observations (Figure 12.99).

Sulfur atoms are formed by photolyses of SO (90%), OCS, $S_2O$, and the reaction SO +SO. Almost all sulfur is lost in

$$S + O_2 \rightarrow SO + O\ \left(9.2 \times 10^{11}\ \text{cm}^{-2}\ \text{s}^{-1}\right), \tag{12.45}$$

which helps to remove $O_2$. $O_2$ (Figure 12.98c) is delivered below 60 km by eddy diffusion and controls abundances of atomic sulfur in the upper cloud layer. Only 0.14% of S from the OCS photolysis reacts with OCS to form $S_2$ and then sulfur aerosol (Section 12.13.6). Its abundance is 0.3 ppb in terms of the sulfur atoms in the basic model (Figure 12.99) and very sensitive to atmospheric dynamics near the cloud tops. The calculated profiles of the sulfur species are in reasonable agreement with the observations (Figure 12.99).

### 12.15.5 Water and Oxygen Allotropes

Water vapor (Figure 12.98d) is also greatly affected by the formation of sulfuric acid. The $H_2O$ mixing ratio is very sensitive to small variations of the $SO_2/H_2O$ ratio at the lower boundary. The model $H_2O$ profile agrees with the observations.

The calculated vertical profiles of $O_2$, O, $O_2(^1\Delta_g)$, and $O_3$ are shown in Figure 12.98c. Reactions O + (ClO, O + M, $NO_2$, and OH) contribute 67, 13, 10, and 10% to the production of $O_2$, respectively. The $O_2$ total production is $4.6\times10^{12}$ cm$^{-2}$ s$^{-1}$ and removes 56% of oxygen atoms released by photolyses of $CO_2$ and $SO_2$. $O_2$ is lost in the ClCO cycle as a difference between the reactions

$$ClCO + O_2 + M \rightarrow ClCO_3 + M \tag{12.46}$$

$$ClCO_3 + O \rightarrow CO_2 + O_2 + Cl. \tag{12.47}$$

This difference accounts for 74% of the $O_2$ loss, and 20% is lost by S + $O_2$. The calculated $O_2$ column abundance is $8.3\times10^{18}$ cm$^{-2}$ exceeding the observed upper limit (Section 12.4.2) by a factor of 8. This is a traditional problem of the $O_2$ chemistry on Venus that remains unsolved. Attempts to solve it by a significant increase of a ClCO equilibrium constant (Mills 1998; Mills and Allen 2007) are not justified by the laboratory data and result in insufficient improvements for $O_2$ and significant problems for CO.

Recent analysis of the SPICAV nadir spectra revealed 1–10 ppb of ozone near the cloud tops at subpolar latitudes $\varphi > 50^\circ$ with an upper limit of 1 ppb at the low and middle latitudes $\varphi < 50^\circ$ (Marcq et al., in preparation). Our model gives 0.15 ppb, in accord with this upper limit.

### 12.15.6 Chlorine and Hydrogen Chemistries

Chlorine and hydrogen chemistries in the middle atmosphere of Venus (Figure 12.100a) originate from photolysis of HCl that forms equal quantities of H and Cl. Odd hydrogen is

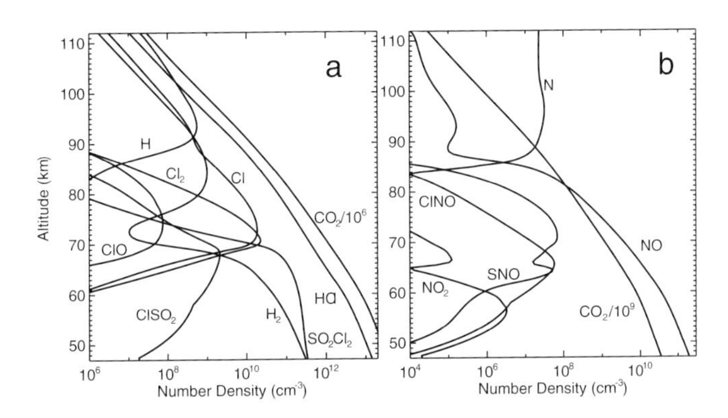


Figure 12.100 Some chlorine species and $H_2$ (a); some nitrogen species (b). Updated from Krasnopolsky (2012b).

also formed by photolysis of $H_2O$. Its column rate is smaller than that of HCl by a factor of 30. However, the $H_2O$ photolysis exceeds that of HCl above 90 km, and the H densities are greater than those of Cl above 90 km.

Molecular chlorine is formed in termolecular association Cl + Cl + M and via

$$\mathrm{Cl + X + M \rightarrow ClX + M} \tag{12.48}$$

$$\mathrm{ClX + Cl \rightarrow Cl_2 + X} \tag{12.49}$$

$$\mathrm{Net\ Cl + Cl \rightarrow Cl_2.} \tag{12.50}$$

Here X = NO, $SO_2$, O, CO, SO, and $O_2$, in the decreasing order, and $Cl_2$ immediately dissociates with a column rate of $2.98 \times 10^{13}$ cm$^{-2}$ s$^{-1}$ that significantly exceeds the rates of the $CO_2$ and $SO_2$ photolyses.

A balance between odd hydrogen and chlorine species is established via reactions

$$\mathrm{H + HCl \rightarrow H_2 + Cl}\ \left(2.40 \times 10^9\right), \tag{12.51}$$

$$\mathrm{Cl + H_2 \rightarrow HCl + H}\ \left(2.97 \times 10^9\right), \tag{12.52}$$

$$\mathrm{OH + HCl \rightarrow H_2O + Cl}\ \left(4.85 \times 10^9\right). \tag{12.53}$$

Their column rates in parentheses are in cm$^{-2}$ s$^{-1}$. The first and second reactions are opposite with almost equal rates. The last reaction converts odd hydrogen into odd chlorine, and this explains why the former is less abundant than the latter below 90 km. The strong depletion of $H_2O$ by the formation of sulfuric acid and the depletion of odd hydrogen in the last reaction diminish odd hydrogen chemistry on Venus.

The first and second reactions forms and removes $H_2$, respectively. Their altitude distributions are different, and this difference results in a deep minimum of $H_2$ at 72 km (Figure 12.100a). The $H_2$ mixing ratio is 8.5 ppb below 65 km and 60 ppb at 100 km.

Loss of HCl in reactions

$$\mathrm{HCl} + h\nu \rightarrow \mathrm{H + Cl}\ \left(9.71 \times 10^{10}\right) \tag{12.54}$$

$$\mathrm{HCl + OH \rightarrow H_2O + Cl}\ \left(4.85 \times 10^9\right) \tag{12.55}$$

$$\mathrm{HCl + O \rightarrow OH + Cl}\ \left(1.05 \times 10^9\right) \tag{12.56}$$

$$\mathrm{Total}\qquad 1.030 \times 10^{11}\ \mathrm{cm^{-2}\ s^{-1}} \tag{12.57}$$

is almost balanced by its production:

$$Cl + HO_2 \rightarrow HCl + O_2 \left(7.14 \times 10^{10}\right) \quad (12.58)$$

$$H + Cl_2 \rightarrow HCl + Cl \left(2.42 \times 10^{10}\right) \quad (12.59)$$

$$\underline{H + ClSO_2 \rightarrow HCl + SO_2 \left(3.61 \times 10^{9}\right)} \quad (12.60)$$

$$\text{Total} \qquad 9.92 \times 10^{10}\ \text{cm}^{-2}\ \text{s}^{-1}. \quad (12.61)$$

The difference of $3.8\times10^{9}$ cm$^{-2}$ s$^{-1}$ is released as a weak flow of $SO_2Cl_2$ into the lower atmosphere. This flow exceeds by a factor of 2 the difference between the production of $H_2O$ in R51 HCl + OH and its loss by photolysis. (The factor of 2 is because two odd hydrogen species are formed by photolysis of $H_2O$.) The calculated $SO_2Cl_2$ mixing ratio is 36 ppb at 68 km and 10 ppb at 47 km; the $Cl_2$ mixing ratio is 24 ppb at 72 km.

The total chlorine mixing ratio is constant at 420 ppb in the model, while HCl varies from 400 ppb at 47 km to 330 ppb at 66 km, then to 370 ppb at 110 km. However, the model does not support a steep decrease from ~400 ppb above 80 km in the submillimeter observations by Sandor and Clancy (2012). The SOIR observations (Vandaele et al. 2015; Figure 12.30) show HCl $\approx$ 50 ppb at 70–80 km increasing to ~700 ppb at 95–110 km. They disagree with both the model and other observations and do not conform to the element conservation (Section 2.2.5).

### 12.15.7 Odd Nitrogen Chemistry

Odd nitrogen chemistry was calculated by Yung and DeMore (1982) in their model B for 30 ppb of NO and 500 ppb of $H_2$ with a major cycle

$$H + O_2 + M \rightarrow HO_2 + M \quad (12.62)$$

$$NO + HO_2 \rightarrow NO_2 + OH \quad (12.63)$$

$$CO + OH \rightarrow CO_2 + H \quad (12.64)$$

$$\underline{NO_2 + h\nu \rightarrow NO + O} \quad (12.65)$$

$$\text{Net } CO + O_2 \rightarrow CO_2 + O. \quad (12.66)$$

This cycle is weak at 5.5 ppb of NO and 8.5 ppb of $H_2$ in our model, and more important are the cycles

$$NO + O + M \rightarrow NO_2 + M \quad (12.67)$$

$$\underline{NO_2 + O \rightarrow NO + O_2} \quad (12.68)$$

$$\text{Net } O + O \rightarrow O_2, \quad (12.69)$$

which provides 10% of the $O_2$ production,

$$NO + Cl + M \rightarrow ClNO + M \quad (12.70)$$

$$\underline{ClNO + Cl \rightarrow NO + Cl_2} \quad (12.71)$$

$$\text{Net } Cl + Cl \rightarrow Cl_2, \quad (12.72)$$

responsible for a quarter of the $Cl_2$ production, and

$$NO + O + M \rightarrow NO_2 + M \quad (12.73)$$

$$\underline{SO + NO_2 \rightarrow SO_2 + NO} \quad (12.74)$$

$$\text{Net } SO + O \rightarrow SO_2, \quad (12.75)$$

which compensates one-fifth of the $SO_2$ photolysis. The calculated profiles of some odd nitrogen species are shown in Figure 12.100b. Overall, NO below 80 km on Venus is important as a convincing proof of lightning (Section 12.7.3) and an effective catalyst in the above cycles.

## 12.16 Nightglow and Nighttime Chemistry at 80–130 km

The extensive observations of the $O_2$, NO, and OH nightglow (Section 12.10) require two types of modeling for their simulation. The nightglow is excited by fluxes of atomic O, N, and H that are transported from the dayside, and the VTGCM is a tool to reproduce these phenomena (Section 12.10.7). However, photochemistry is restricted in VTGCM, and another approach to solve the problem is to model a photochemical response of the nighttime atmosphere to input fluxes of O, N, and H. Here we will consider a one-dimensional model for nighttime chemistry and nightglow on Venus by Krasnopolsky (2019b).

The observed nighttime ozone (Montmessin et al. 2011; Section 12.4.3) is rather similar to that predicted by the global-mean models (Figure 12.98c). However, the UV photolysis of ozone is switched off on the nightside, and an initial version of the nighttime model (Krasnopolsky 2010a) predicted ozone abundances exceeding those observed by two orders of magnitude. A significant flux of atomic chlorine from the dayside is required to fit the observed nighttime ozone. The recent detection of the nighttime ClO abundance of 2.6 +/– 0.6 ppb in a 10-km layer at 90 km (Sandor and Clancy 2019b) supports this idea.

The model involves 86 reactions of 29 species. The lower boundary is at 80 km to avoid the detailed sulfur chemistry. Densities of $CO_2$, HCl, $SO_2$, and a flux of NO are the lower boundary conditions, while $V = -K/2H = -0.116$ cm s$^{-1}$ is adopted for the remaining species. Fluxes $\Phi$ of O, N, H, and Cl at 130 km are the model parameters. These downward fluxes are equal to $3\times10^{12}$, $1.2\times10^{9}$, $9\times10^{9}$, and $3\times10^{9}$ cm$^{-2}$ s$^{-1}$, respectively, in the basic nightside-mean version of the model.

The calculated species altitude profiles are shown in Figure 12.101. The excess of the odd hydrogen flux over that of chlorine is partly transformed into chlorine:

$$OH + HCl \rightarrow H_2O + Cl, \qquad 1.53 \times 10^9 \text{ cm}^{-2} \text{ s}^{-1} \quad (12.76)$$

$$H + HCl \rightarrow H_2 + Cl, \qquad 8.34 \times 10^8 \text{ cm}^{-2} \text{ s}^{-1} \quad (12.77)$$

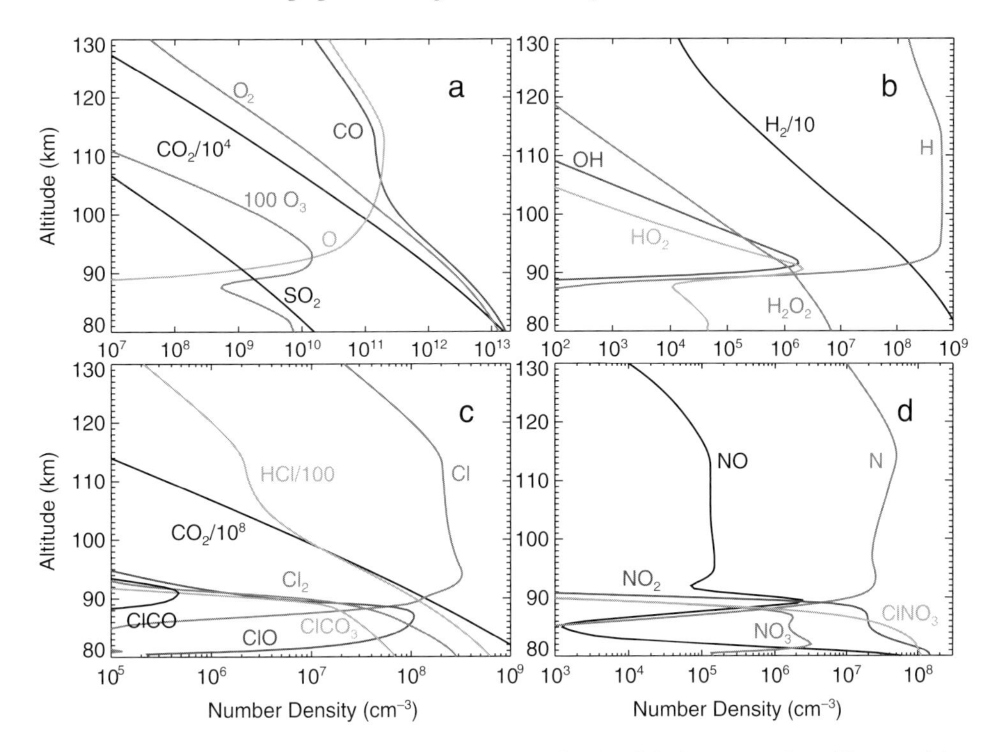


Figure 12.101 Vertical profiles of species in the model for the nighttime atmosphere (Krasnopolsky 2019b). (A black and white version of this figure will appear in some formats. For the color version, please refer to the plate section.)

Major losses of odd hydrogen and odd chlorine are via reactions

$$Cl + HO_2 \rightarrow HCl + O_2, \qquad 5.08 \times 10^9 \text{ cm}^{-2} \text{ s}^{-1} \tag{12.78}$$

$$H + Cl_2 \rightarrow HCl + Cl, \qquad 2.97 \times 10^8 \text{ cm}^{-2} \text{ s}^{-1} \tag{12.79}$$

Therefore the Cl production in (12.76), (12.77) plus the Cl flux of $3\times10^9$ cm$^{-2}$ s$^{-1}$ are balanced by the Cl loss in (12.78), (12.79). The excess of the hydrogen flux over its loss in (12.76)–(12.79) is spent on production of $H_2$ and $H_2O$ in the reactions H + $HO_2$ and OH + $HO_2$.

Major odd hydrogen and odd chlorine cycles are

$$X + O_3 \rightarrow XO + O_2,$$

$$O + XO \rightarrow X + O_2,$$

$$\text{Net } O + O_3 \rightarrow 2\, O_2.$$

Here X = H and Cl with the $O_2$ net production of $5.67\times10^{11}$ and $1.7\times10^{11}$ cm$^{-2}$ s$^{-1}$, respectively. These cycles remove almost half the O flux at 130 km, and the remaining half is lost in the termolecular association of $O_2$. Oxidation of CO is weak above 80 km, and all O atoms form $O_2$ with a mixing ratio of $\Phi_O/(2n_{80}V)$ = 82 ppm at 80 km increasing to 256 ppm at 100 km (Figure 12.101a).

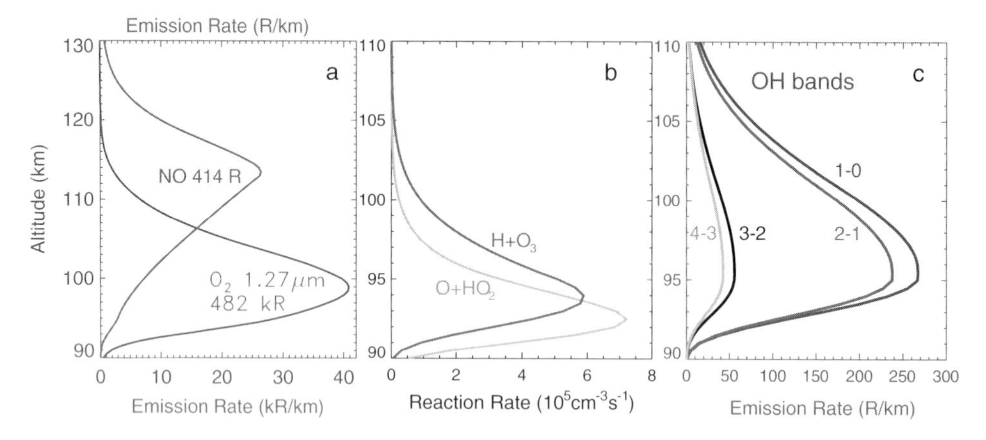


Figure 12.102 Calculated vertical profiles of the $O_2$ 1.27 μm and NO UV total ($\gamma$ + $\delta$ bands) nightglow (a), two reactions of the OH excitation (b), and four bands of the OH nightglow in the $\Delta$v = 1 sequence (c). From Krasnopolsky (2019b).

The calculated intensity and profile of the $O_2$ nightglow at 1.27 μm (Figure 12.102a) agrees with the VEX observations, and the nightglow intensity may be approximated by

$$4\pi I_{O_2} = 127\left(\Phi_O/10^{12}\right)^{1.22}\text{kR}; \tag{12.80}$$

that is, the $O_2$ nightglow quantum yield is 0.16 per O atom for $\Phi_O = 3\times10^{12}$ cm$^{-2}$ s$^{-1}$.

The above H and Cl cycles balance the production of ozone in the termolecular association O + $O_2$ + M. The calculated $O_3$ profile (Figure 12.101a) agrees with those observed by the SPICAV stellar occultations (Montmessin et al. 2011). Variations of the peak $O_3$ limb column abundance may be approximated by

$$[O_3]_{\text{limb}} = 1.37\times10^{15}\frac{\left(\Phi_O/10^{12}\right)^{1.2}}{x^{0.74+0.28\,lnx}y^{0.29}}\text{cm}^{-2};\quad x=\frac{\Phi_H}{10^{10}},\ y=\frac{\Phi_{Cl}}{10^{10}}. \tag{12.81}$$

The model profile of the NO nightglow (Figure 12.102a) agrees with the observations as well, and its vertical intensity is equal to

$$4\pi I_{NO} = 225\left(\Phi_N/10^9\right)\left(\Phi_O/10^{12}\right)^{0.35}\ R. \tag{12.82}$$

The observed vibrational excitation of the OH nightglow is v $\leq$ 4 on Venus and $\leq$ 9 on the Earth. There are two possible processes for the OH vibrational excitation on the terrestrial planets:

$$H + O_3 \rightarrow OH(v \leq 9) + O_2 \tag{12.83}$$

$$O + HO_2 \rightarrow OH(v \leq 6) + O_2. \tag{12.84}$$

Kaye (1988) argued that the O–H bond is formed in (12.83) and large v values are probable. However, OH is the remaining part of $HO_2$ in (12.84) with a low vibrational excitation v $\leq 3$. Krasnopolsky (2019a) adopted the OH transition probabilities, excitation yields in (12.83), and quenching rate coefficients by $CO_2$ from García Muñoz et al. (2005) and proposed collision cascade yields that fit the observed band distribution in the $\Delta v = 1$ sequence. The calculated vertical intensity of the $\Delta v = 1$ sequence at 2.9 μm (Figure 12.63a) is approximated by

$$4\pi I_{OH(\Delta v=1)} = 0.41\left(\Phi_O/10^{12}\right)^{1.45}\frac{x^{0.17-0.38\,\ln x}}{y^{1.58+0.49\,\ln y}}. \tag{12.85}$$

Reaction O + $HO_2$ (Figure 12.28b) peaks below the observed maximum of the OH nightglow, and its effect in the nightglow excitation is neglected in the basic model. The calculated profile of CO (Figure 12.27a) agrees with the nighttime microwave observations (Clancy et al. 2012a), and the model fits all observational constraints.

The ClO abundance in a layer of 10 km at 90 km is approximated by

$$f_{ClO} = 16.6\left(\Phi_O/10^{12}\right)^{0.47}\frac{y^{1.61-0.48\,\ln y}}{x^{2.23+0.41\,\ln x}}\text{ ppb at 90 km} \tag{12.86}$$

Major reactions that control ClO are

$$Cl + O_3 \rightarrow ClO + O_2$$

$$ClO + O \rightarrow Cl + O_2.$$

Using their rate coefficients, $[ClO] \approx 0.8\,e^{-285/T}\,[Cl][O_3]/[O]$. All these species are highly variable near 90 km (Figure 12.101), presuming high variability of ClO.

The calculated relationships (12.80–12.82, 12.85–86) make it possible to predict the nightglow intensities and ozone abundances for known fluxes of the atomic species and to convert observed nightglow intensities and ozone and ClO abundances into fluxes of the atomic species.

## 12.17 Some Unsolved Problems

The variety of conditions and the complicated chemical composition, structure, and dynamics of the Venus atmosphere suggest many problems to study. Here we will mention only some critical points in an arbitrary order.

1 Mutual conversions between $SO_3$, OCS, CO, and $S_x$ are the key problem for chemistry in the lower atmosphere. The current model is based on the reactions

$$SO_3 + OCS \rightarrow (SO)_2 + CO_2$$
$$(SO)_2 + OCS \rightarrow CO + SO_2 + S_2$$

from Krasnopolsky and Pollack (1994) that have not been studied in the laboratory and may substitute a more complicated chemistry.

2 The observed $SO_2$ abundance of 130 ppm in the lower atmosphere is incompatible with the photochemical models of the middle atmosphere that require ~10 ppm at the lower cloud boundary.

3 The observed increase in $SO_2$ above 80 km is highly questionable, and an alternative interpretation of the SPICAV spectral data should be studied. The Mie resonances in the haze scattering may affect the current interpretation.
4 There are significant disagreements between the vertical profiles of HCl and HF in the middle atmosphere from the SOIR occultations and the ground-based submillimeter and infrared observations.
5 There are significant disagreements between the temperature profiles from the SPICAV and SOIR occultations and those from the other observations and models.
6 The last VTGCM versions poorly agree with the Pioneer Venus data on the thermospheric structure and composition.
7 Strong variations of species in the microwave observations that cover significant parts of the planet are puzzling, as well as some data on their vertical distributions.
8 The upper limit to $O_2$ above the clouds observed three decades ago is smaller than the model predictions by an order of magnitude.
9 The models predict ~30 ppb of $SO_2Cl_2$ at 30 to 70 km. Chemistry of this species is poorly known, and the prediction is uncertain. However, it is worth to be checked by observations.
10 There are indirect indications of lightning on Venus. However, optical observations of lightning are scarce and uncertain.
11 Some data on the chemical composition of the clouds are uncertain and need further study.

This list is incomplete and may be extended.

# 13

# Titan

## 13.1 General Properties and Pre-Voyager Studies

### *13.1.1 General Properties*

Titan is the largest satellite of Saturn with radius of 2575 km. It is just slightly smaller than Ganymede, the largest moon of Jupiter with $r$ = 2631 km, and greater than the Moon (by a factor of ~1.5, $r$ =1738 km) and even Mercury ($r$ = 2439 km).

Titan's orbit is in the equatorial plane of Saturn and tidally locked, that is, Titan shows only one side to Saturn, likewise to the Moon to the Earth. Then the duration of day is the orbit period of 16 days (Table 13.1). The solar illumination conditions are determined by the Saturn orbit: the sunlight is near 1% of that on the Earth, the significant obliquity and eccentricity result in seasons on Titan with the recent spring equinox on August 11, 2009, and the annual cycle of 29.5 years. The heliocentric distance varies from 9.03 to 10.07 AU with a change in the insolation by a factor of 1.24. The recent perihelion was in July 2003 and the aphelion is in March 2018. Using the terrestrial calendar, the perihelion is on January 8 and aphelion on July 8, that is, the southern summer is warmer and the southern winter is colder on Titan, similar to those on Mars. Like on Mars, seasons on Saturn and Titan are specified by the solar longitude $L_S$ that is 0° at the vernal equinox, 90° at the summer solstice, 180° at the fall equinox, and 270° at the winter solstice.

The mean density of 1.88 g cm$^{-3}$ indicates that Titan consists of almost equal quantities of water ice and rocks. Rocks dominate in a central core with $r \approx$ 1700 km and water ice in the outer shell. A low abundance of Ar and a lack of Kr and Xe (see below) favor ammonia as a precursor of the atmospheric nitrogen, and some ammonia is believed to be kept in the interior. A mixture of water and ammonia can remain liquid down to 176 K. This originates a hypothesis of a subsurface liquid ocean. Reflection of low-frequency radio waves and observed variations of some surface features support this hypothesis. The interior keeps some heat from the period of accretion; Saturn's tides and decay of the natural radionuclides K, U, and Th heat the interior as well.

Lakes of probably methane with some ethane and HCN were observed. Those are the only liquid features observed in the Solar System except the Earth. However, the amount of methane in the lakes is insufficient for long-term support of the irreversible photochemical loss of methane.

Table 13.1 *Basic properties of Titan*

| | |
|---|---|
| Mean heliocentric distance of Saturn (AU) | 9.539 |
| Eccentricity of Saturn orbit | 0.056 |
| Sidereal length of year (years) | 29.46 |
| Saturn obliquity | 26.7° |
| Last vernal equinox | August 2009 |
| Distance from Saturn (km) | $1.22\times10^6$ |
| Period of rotation (days) | 15.95 |
| Radius (km) | 2575 |
| Mass (g) | $1.345\times10^{26}$ |
| Mean density (g cm$^{-3}$) | 1.88 |
| Surface gravity (equator, cm s$^{-2}$) | 135.2 |
| Escape velocity at exobase (km s$^{-1}$) | 2.64 |

### *13.1.2 Pre-Voyager Studies*

Titan was discovered in 1655 by the Dutch astronomer Christiaan Huygens. Much later in 1908 Comas Sola from Spain claimed detection of limb darkening that was explained as an evidence for a substantial atmosphere. This interpretation is generally questionable, because, e.g., the Lambertian hard sphere shows limb darkening as well at small phase angles accessible for Titan from the Earth. A much more reliable identification of the atmosphere was made by Kuiper (1944), who detected bands of gaseous methane in the red. The $CH_4$ bands in the red and near-infrared were observed on Titan from the ground till the end of 1970s and used to retrieve the methane abundance and atmospheric pressure; the results were generally uncertain. Elliot et al. (1975) observed a lunar occultation of the Saturn system, and the measured radius of Titan with the atmosphere was 2900 km, indicating a very extended atmosphere. Gillett (1975) detected all $C_2H_X$ hydrocarbons and deuterated methane in a low-resolution spectrum at 7.8–13.3 μm (Figure 13.1). Khare and Sagan (1973) started laboratory experiments in mixtures of methane with nitrogen and other gases irradiated by UV photons or electrons. Brownish products in those experiments were called tholins and expected to be analogs of Titan's haze. Later, Khare et al. (1984) studied in detail optical constants of tholins from X-rays to millimeter wavelengths (Figure 13.2), and these data are currently used for analyses of optical observations of the haze. It became clear that methane chemistry produces numerous species that form the haze and are responsible for the low UV albedo of Titan observed by that time.

Analysis of Titan's phase variations and spectral data to retrieve properties of the haze by Rages and Pollack (1980) is an example of how reasonable results can be achieved from rather restricted observational data. Phase (Sun–object–observer) angle of Titan varies at the Earth's orbit from 0° to 6.4°, and the phase variations were observed by Noland et al. (1974) at 8% at 350 nm decreasing to ~1% at 750 nm. The effect depends on both refractive index and particle radius, and the derived values are $n_r \approx 1.7$ and $r \approx 0.25$ μm.

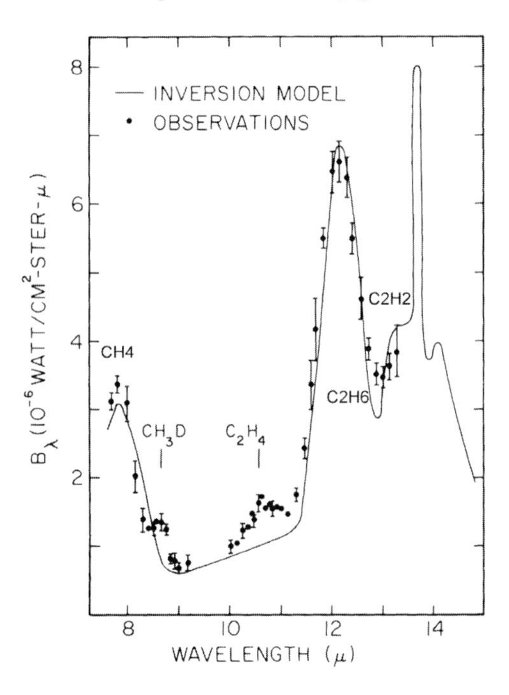


Figure 13.1 Low-resolution ($\lambda/\delta\lambda \approx 50$) spectrum of Titan with identified thermal emissions of $CH_4$, $CH_3D$, $C_2H_4$, $C_2H_6$, and $C_2H_2$ (Gillett 1975). The model does not include $CH_3D$ and $C_2H_4$.

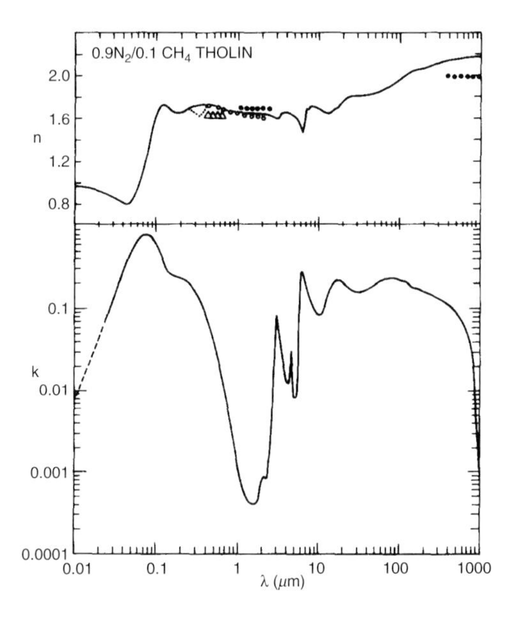


Figure 13.2 Real and imaginary refraction indices of analogs to Titan's haze in the range from X-rays to millimeter (Khare et al. 1984).

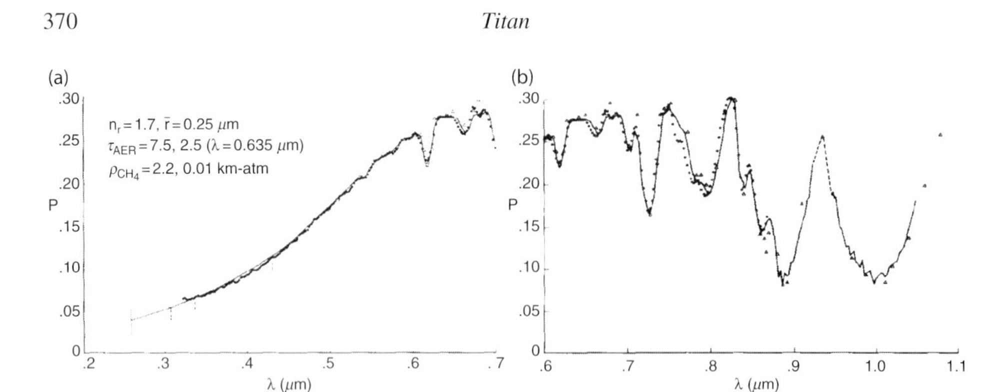


Figure 13.3 Fitting of the observed IUE (International Ultraviolet Explorer) and ground-based spectra of Titan's geometric albedo by a two layer model of the atmosphere. The model parameters are given in the text and in (a). From Rages and Pollack (1980).

Weak and strong methane bands in the observed spectra of Titan (Figure 13.3) indicate different methane abundances. Therefore the authors adopted a two-layer model of the atmosphere with abundance of methane and haze optical depth in each layer as parameters to fit the observed spectra of Titan. The best-fit data are 0.01 km-atm of $CH_4$ and $\tau = 2.5$ at $\lambda - 635$ nm for the upper layer and $CH_4 = 2.2$ km-atm and $\tau = 7.5$ in the lower layer. According to the Huygens data (Niemann et al. 2010), the boundary between the layers at $CH_4 = 0.01$ km-atm corresponds to ~100 km, while $CH_4 = 2.2$ km-atm is close to the total amount of methane in the atmosphere; the haze optical depth at 634 nm is 2 at 100 km and 6.5 near the surface (Tomasko et al. 2008a).

To fit the observed geometric albedo (Figure 13.3), Rages and Pollack (1980) varied the haze imaginary index. The derived index was 0.1 at 0.25 μm and decreased to 0.007 at 0.8 μm. Both real and imaginary indices are rather close to those measured for tholins by Khare et al. (1984; Figure 13.2).

The atmosphere appeared to be too complicated and deep to be studied in detail using the $CH_4$ bands and other means of the ground-based infrared spectroscopy. The observations showed that methane is not a major constituent of the atmosphere, and $N_2$, Ar, and Ne were plausible candidates. Finally, Titan's effective temperature was measured at $87 \pm 9$ K using the Very Large Array (VLA) at 6 cm (Jaffe et al. 1980), just prior to the Voyager flyby. Then a surface temperature was expected at ~90 K, and a model of the pressure-induced opacity of $N_2$ on Titan (Hunten 1978) required a surface pressure of 1–2 bar for this temperature.

## 13.2 Voyager 1 Observations

Pioneer 11 was the first spacecraft that visited the Saturn system in 1979. It was a small probe, and its flyby at 363,000 km from Titan gave a few distant images of the satellite. A breakthrough in the studies of Titan's atmosphere was related to the Voyager 1 flyby in November 1980. The Voyager 1 and 2 spacecraft were the most fruitful mission to study the Solar System. They explored the complicated systems of Jupiter, Saturn, Uranus, and Neptune, their satellites and environments, and space plasma up to the heliopause and beyond. The spacecraft were launched in 1977, and later it was decided to direct Voyager 1 to Titan with a flyby at 6970 km from Titan's center. The flyby was in November 1980, a few months after the vernal equinox.

Titan was observed using the narrow- and wide-angle cameras, the infrared interferometer spectrometer (IRIS), the ultraviolet spectrometer (UVS), the photopolarimeter-radiometer (PPR), and radio occultations.

### 13.2.1 Haze

Titan was rather uniform in the images with the southern hemisphere brighter (by ~25% in the blue) and less red than the northern hemisphere. This is caused by seasonal effects with a significant time lag. A more complicated structure was observed at high latitudes near the north pole. The main haze extends up to 200 km with a detached layer with $\tau \approx 0.01$ near 330 km (Rages and Pollack 1983).

The haze shows very high polarization at phase angles of ~90° that requires very small particles $r \approx 0.05$ μm, while the observed scattering at various phase angles implies particle radii of ~0.3 μm, in accord with the pre-Voyager analysis by Rages and Pollack (1980). This was explained by fractal haze particles that consist of numerous small monomers. Rannou et al. (1999) developed and published a code to calculate extinction cross section, single scattering albedo, and scattering asymmetry parameter of a fractal particle that contains $N$ monomers with radius $r$ and known real and imaginary refractive indices. The calculated data make it possible a further radiative transfer analysis.

### 13.2.2 Lower Atmosphere from the Radio Occultations

Interpretation of the radio occultation data depends on adopted mean molecular mass, and a reasonable agreement with the VLA data and thermal balance calculations required $N_2$ as a major atmospheric constituent. Assuming $\mu = 28$, the radio occultations gave a surface pressure of $1.50 \pm 0.02$ bar, temperature of $94 \pm 0.7$ K, and radius of $2575 \pm 0.5$ km (Lindal et al. 1983). The troposphere extends up to 42 with tropopause at 130 mbar and $71.4 \pm 0.5$ K. The stratosphere above this level was observed up to 200 km where $p = 0.75$ mbar and $T \approx 170$ K. Later these data were slightly corrected for a poorly known abundance of $CH_4$ and possible presence of Ar in the lower atmosphere.

### 13.2.3 Composition of the Stratosphere from the IRIS Observations

IRIS was a Fourier transform spectrometer that covered a range of 180 to 2500 $cm^{-1}$ (4–55 μm) with an apodized resolution of 4.3 $cm^{-1}$. The spectral continuum reflects thermal radiation of the surface affected by pressure-induced opacities in the $N_2$–$CH_4$–$H_2$ mixture and the haze. The observed continuum opacity at 200–600 $cm^{-1}$ revealed the surface temperature of 94–97 K, the

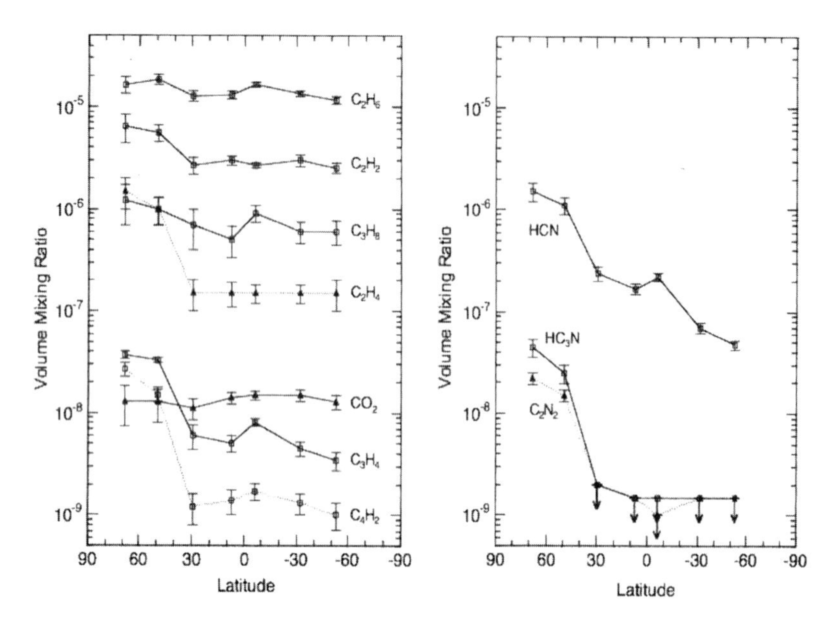


Figure 13.4 Latitudinal distribution of minor constituents in Titan's stratosphere observed by Voyager/IRIS during vernal equinox. All hydrocarbons and especially nitriles are enriched in the northern hemisphere. From Coustenis and Bezard (1995).

mean molecular mass of 28.3–29.2, and the $H_2$ mixing ratio of $0.002 \pm 0.001$ (Samuelson et al. 1981). Later, Courtin et al. (2005) revised this value to $(1.0 \pm 0.4)\times 10^{-3}$.

Spectral bands of species in the warmer stratosphere look like thermal emissions. Their identification and radiative transfer analysis resulted in species mixing ratios that were assumed constant at 70–300 km (Figure 13.4). The flyby was at vernal equinox, and the retrieved temperatures at 1 mbar (~170 km) were rather constant at ~168 K from 60°S to 20°N decreasing to 152 K at 70°N (Coustenis and Bezard 1995), and the colder northern hemisphere at the equinox reflects thermal inertia of the atmosphere. The latitudinal distributions of species observed by IRIS include six hydrocarbons, three nitriles, and $CO_2$. All hydrocarbons and especially nitriles show greater abundances in the northern hemisphere. Variations of the spectra at 50°–90°N near the limb (limb darkening/ brightening) made it possible to get some data on vertical profiles of the species. The retrieved abundances were assigned to 70°N and referred to 160 and 270 km (Coustenis et al. 1991). While the bands of $C_2H_4$, $C_3H_8$, and $CO_2$ were rather weak for the retrieval at 270 km, all other species showed increase in mixing ratio with altitude. Comparison of the observed $CH_3D$ and $CH_4$ resulted in a D/H ratio of $1.5^{+1.4}_{-0.5} \times 10^{-4}$, similar to that on the Earth but uncertain within a factor of 3.

### *13.2.4 UVS Solar Occultations*

UVS covered a range of 530 to 1700 Å with resolution of 30 Å. The observations were aimed to study the composition and structure of the upper atmosphere by solar occultations and airglow on the disk and limb of Titan. The instrument had a mechanical collimator, the Sun moved within the field of view, scattered light from the grating affected the observed spectra, and corrections for these effects required significant efforts.

The UVS solar occultations were the basic data on Titan's upper atmosphere up to the beginning of the Cassini observations in 2005. The altitude range of 500–900 km is currently accessible only to UV occultations, and the UVS occultation data are still important and will be considered here. The UVS occultations were analyzed by Smith et al. (1982), Strobel et al. (1992), Vervack (1997), and Vervack et al. (2004); we will discuss the latter.

The ingress and egress occultations referred to evening and morning, tangential latitudes of 4°N and 16°S, and tangential distances of 6760 and 18,100 km, respectively. Photoabsorption spectra of hydrocarbons and nitriles that may contribute to the observations are shown in Figure 13.5. $N_2$ is almost completely responsible for the absorption at

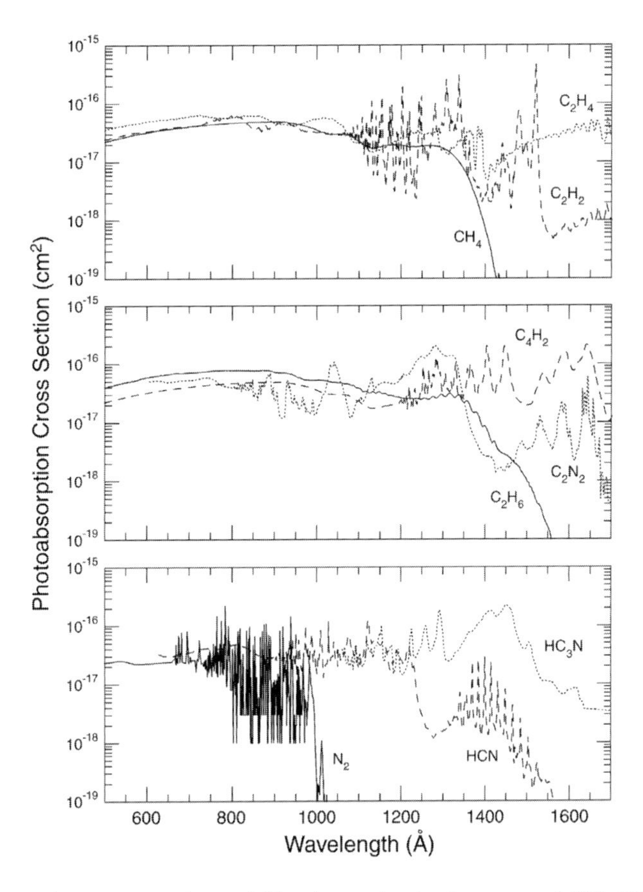


Figure 13.5 Absorption cross sections of Titan's species in the range of 500–1700 Å (Vervack et al. 2004).

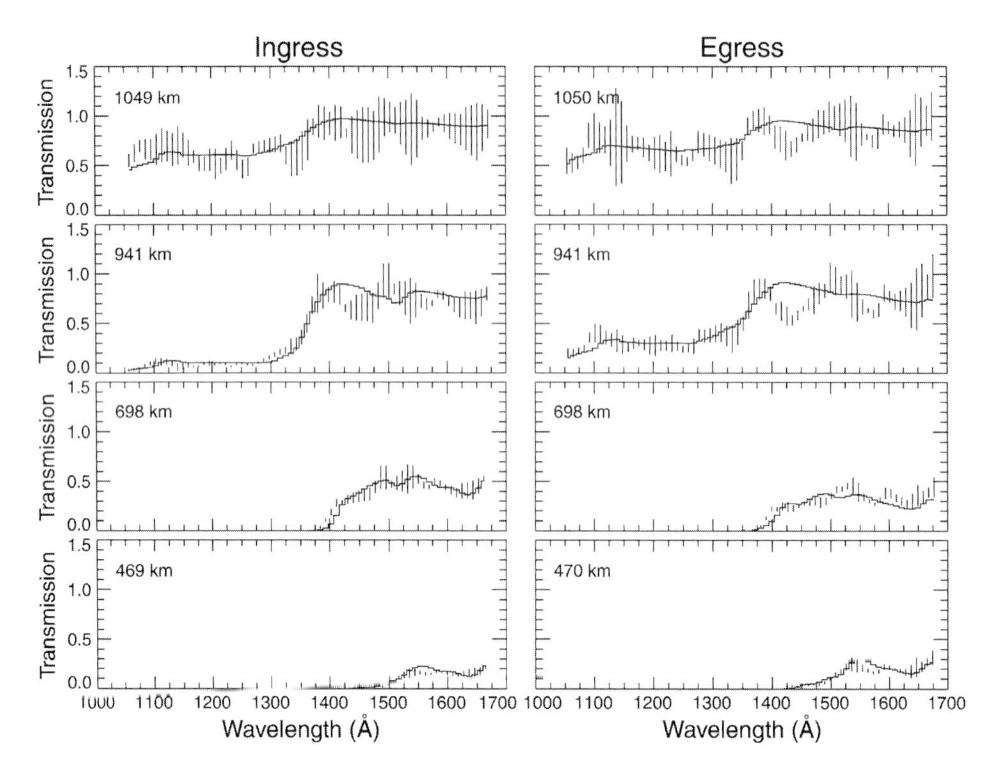


Figure 13.6 Transmission spectra of Titan's atmosphere at four tangent altitudes observed using UVS/Voyager in the solar occultation mode are compared with the model fittings (Vervack et al. 2004).

$\lambda < 1000$ Å and $CH_4$ at 1000–1400 Å, while various combinations of species are generally possible at 1400–1700 Å.

Some of the observed transmission spectra are depicted in Figure 13.6. The absorption by methane is similar in the ingress and egress spectra at 1050 km and very different at 941 km. Only minor constituents are measurable below 700 km. Retrieved vertical profiles of $N_2$ and $CH_4$ are shown in Figure 13.7. The $N_2$ profiles result in temperatures of 153 and 158 K. The mean $[N_2] = 8\times10^8$ cm$^{-3}$ at 1200 km with $T = 156$ K may be compared to 0.75 mbar at 200 km from the radio occultations to get a mean temperature at 200–1200 km. Using the barometric formula $n = n_0 \exp(\lambda - \lambda_0)$ (2.38) for the isothermal extended atmosphere and a methane mixing ratio of 0.017 for the mean molecular mass, the calculation gives 163 K.

The solid lines in Figure 13.7 are for $CH_4$ mixing ratios of 0.024 and 0.011 at the ingress and egress, respectively, with the mean of 0.017 used above. There is a statistically significant deviation from the constant mixing ratio above 1000 km in the egress data, indicating a homopause at 1050 km.

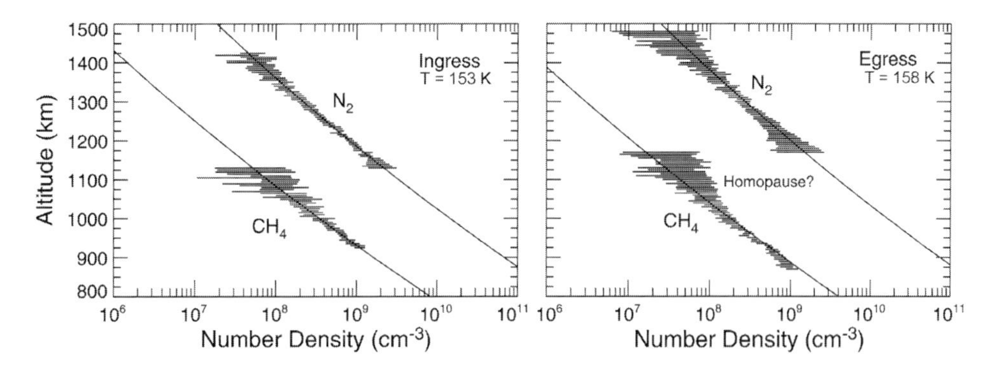


Figure 13.7 Retrieved number density profiles of $N_2$ give thermospheric temperatures of 153 and 158 K. The observed $CH_4$ profiles are compared with those for methane mixing ratio of 0.024 and 0.011, respectively. The deviation above 1050 km in the egress data may indicate homopause. From Vervack et al. (2004).

The retrieved profiles of acetylene $C_2H_2$ are compared in Figure 13.8 with the previous retrievals and predictions of photochemical models. The results are smaller than those in Smith et al. (1982) by a factor of ~40, agree with the models at 850–1100 km and smaller than the model predictions below 800 km. The observed profiles of ethylene $C_2H_4$ (Figure 13.8) are in reasonable agreements with the models. The measured profiles of hydrogen cyanide HCN and cyanoacetylene $HC_3N$ are shown in Figure 13.9 and compared with the model data.

The observed light curves for 1500–1700 Å are compared in Figure 13.10 with those calculated for the photochemical models. It is clear that the models overestimate abundances of photochemical products below 750 km. The light curves are flat at 380–330 km with steep cutoffs below that indicate a haze layer.

### *13.2.5 Dayglow and Ionosphere*

The UVS dayglow observations (Strobel et al. 1992) revealed emissions of nitrogen excited by the solar photons and photoelectrons (Table 13.2). The $N_2$ ($a^1\Pi_g \rightarrow X^1\Sigma_g^+$) Lyman–Birge–Hopfield band system is the brightest in the dayglow. The observed intensities were fitted by a model that involved the solar UV photons and photoelectrons and magnetospheric electrons with a total power of $2.3\times10^8$ W, i.e., 10% of the solar EUV power input at $\lambda < 800$ Å. The model included methane that slightly reduces the $N_2$ excitation and absorbs the emissions. The $N_2$ emissions at 958 and 981 Å are poorly resolved at the instrument resolution of 30 Å, and their sum agrees with the model with no $CH_4$, while $CH_4 \approx 3\%$ is preferable for the $N^+$ 1085 Å emission.

The observed Titan's ionosphere was near a detection limit of the Voyager radio occultations. The ionospheric maximum was at $h_{max}$ = 1180 ± 150 km with a peak electron density $n_{emax}$ = 2400 ± 1100 cm$^{-3}$ (Bird et al. 1997).

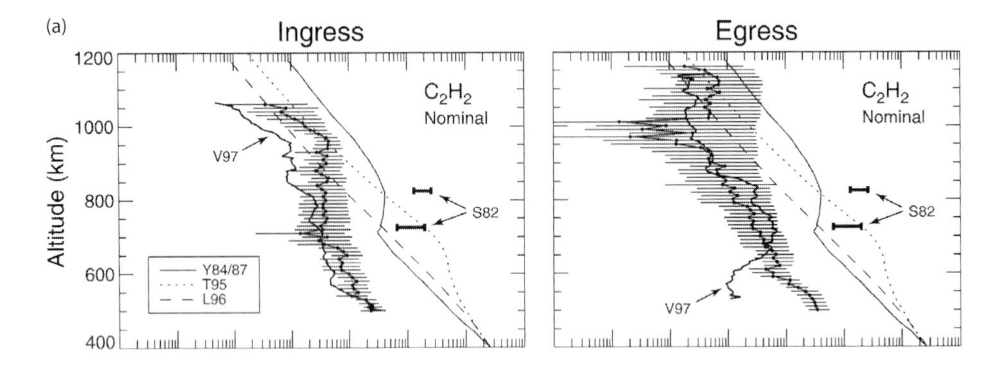


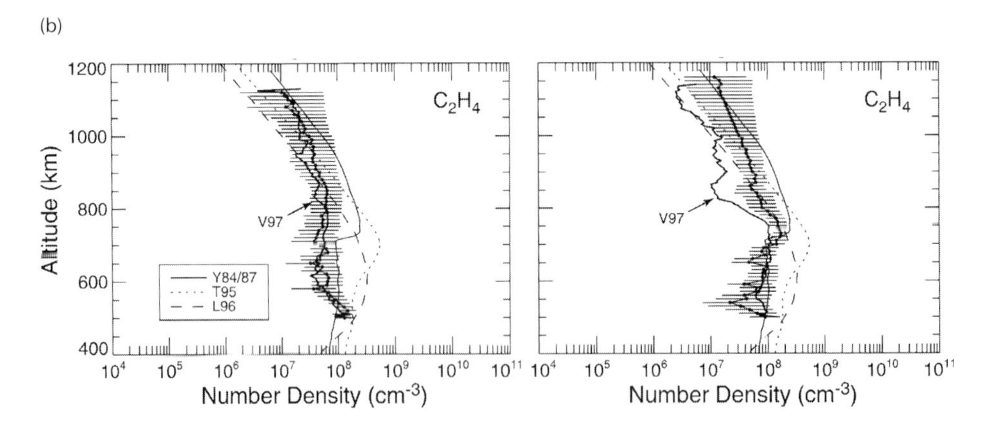


Figure 13.8 Number density profiles of $C_2H_2$ and $C_2H_4$ retrieved from the UVS/Voyager solar occultations are compared with the earlier analyses by Smith et al. (1982) and Vervack (1997) and with predictions of photochemical models by Yung et al. (1984), Yung (1987), Toublanc et al. (1995), and Lara et al. (1996). From Vervack et al. (2004).

## 13.3 Ground-Based and Earth-Orbiting Observations

Diameter of Titan seen from the Earth at the opposition is 0.83 arcs and equal to four resolution elements of the Hubble Space Telescope (HST) that provides the best spatial resolution. Therefore spatially resolved ground-based spectroscopy is very restricted for Titan and the observations refer mostly to the whole disk.

### *13.3.1 Some Near-Infrared Observations*

High-resolution spectroscopy revealed the CO (3–0) band at 1.57 μm with a CO mixing ratio of 60 ppm (Lutz et al. 1983). After that detection CO was observed by a few teams using the fundamental (1–0) band at 4.7 μm and lines in the millimeter range. The measured mean CO mixing ratio is 50 ppm.

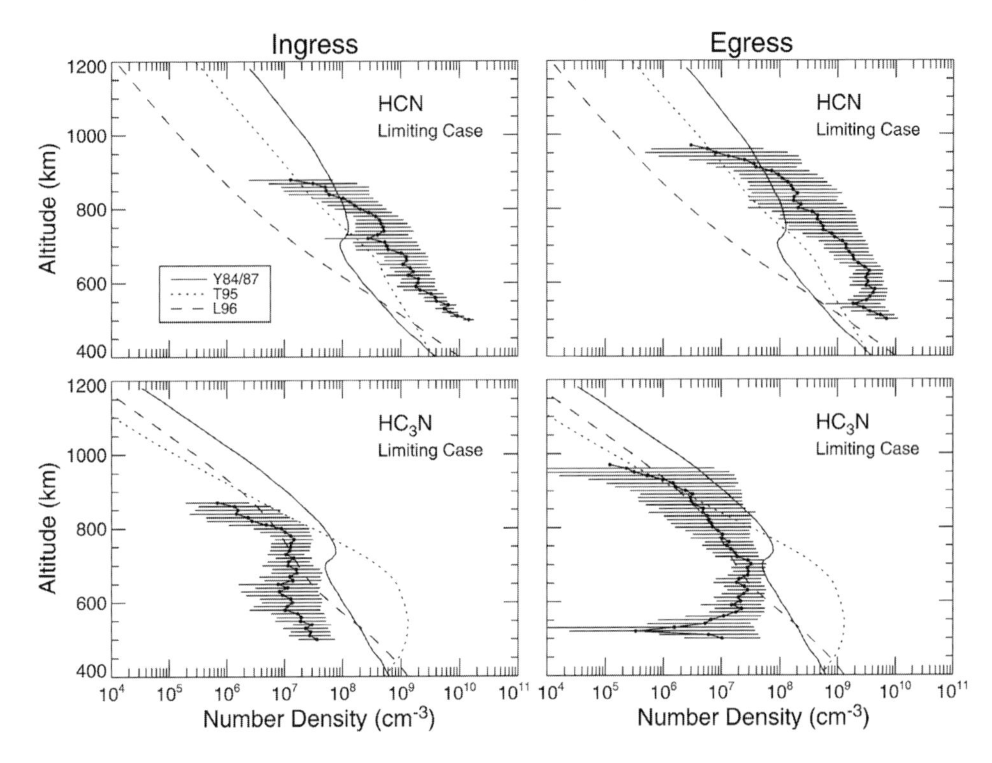


Figure 13.9 Profiles of HCN and $HC_3N$ from the UVS/Voyager solar occultations are compared with the data of photochemical models (Vervack et al. 2004).

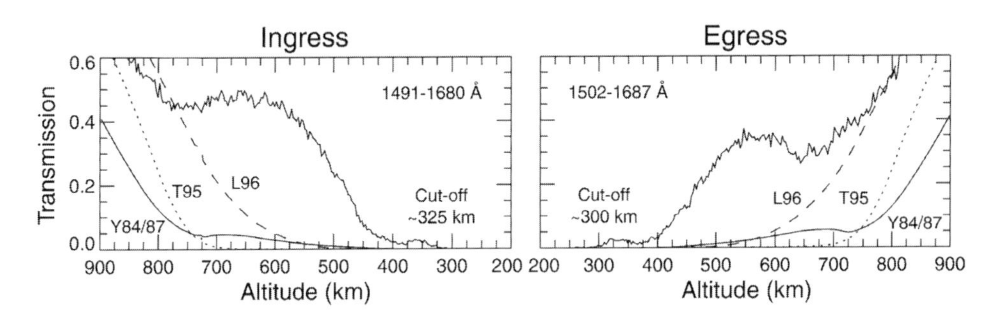


Figure 13.10 Observed light curves for 1500–1700 Å are compared with those calculated using the data of photochemical models. The models overestimate abundances of absorbing species below 750 km. The cutoff at ~300 km is caused by a haze layer. From Vervack et al. (2004).

True absorption by the haze is low at 0.7–2.5 μm (Figure 13.2), the haze size parameter $2\pi r/\lambda$ is ~2.5 at $\lambda = 0.63$ μm and ~0.8 at 2 μm for the effective particle radius $r \approx 0.25$ μm (Rages and Pollack 1980; Section 13.1.2). This reduction of the size parameter corresponds to reduction in the optical depth by an order of magnitude (Figure 4.1). Therefore a spectral

Table 13.2 *Observed and calculated dayglow emission intensities (in R) on Titan*

| | $N^+$ 916 Å | $N_2$ 958 Å | $N_2$ 981 Å | $N^+$ 1085 Å | $N_2$ LBH |
|---|---|---|---|---|---|
| Observed | 0.6 | 8 | 6 | 8 | 96 |
| CH4 = 0% | 2.3 | 11 | 1.8 | 11 | 103 |
| CH4 = 2.6% | 1.6 | 7.3 | 1.3 | 8.8 | 100 |
| CH4 = 3.4% | 1.4 | 6.4 | 1.0 | 7.5 | 85 |

*Note*. From Strobel et al. (1992).

window between the $CH_4$ bands at 2 μm may be used to sound the deep atmosphere and surface of Titan. For example, Griffith et al. (2000) observed variations in spectra of the CH4 band wing at 2.0–2.25 μm during the 16-day rotation of Titan. The spectra vary within a factor of 1.3 at 2.0–2.05 μm. Here the atmospheric opacity is minimal, and the variations reflect the surface features. Both signal and its variation are low at 2.16–2.25 μm, where the methane absorption is strong and the sounding is in the stratosphere. The signal at 2.1–2.15 μm sounds the troposphere and does not correlate with that at 2.0–2.05 μm indicating variable tropospheric clouds.

Measurements of the D/H ratio in methane by comparison of the $CH_4$ and $CH_3D$ bands were started by Gillet (1975), and the latest results were obtained by Penteado et al. (2005) using IRTF/TEXES and de Bergh et al. (2012) using KPNO/FTS. The measured ratios are $(1.25 \pm 0.25)\times10^{-4}$ and $(1.13 \pm 0.25)\times10^{-4}$, respectively.

### *13.3.2 Microwave Observations*

The millimeter range appeared fruitful to study nitriles on Titan. The heterodyne technique provides very high spectral resolution that makes it possible to retrieve vertical profiles of the observed species (Section 7.4). The first observation and analysis of the HCN line was made by Tanguy et al. (1990) using the IRAM 30 m radio telescope for the millimeter range in Spain, and the later results for hydrogen cyanide (HCN), cyanoacelylene ($HC_3N$), and acetonitrile ($CH_3CN$; Figure 13.11) are shown in Figure 13.12. Comparing the observed lines of $HC^{14}N$ and $HC^{15}N$, Marten et al. (2002) found $HC^{14}N/HC^{15}N$ = 60–70, smaller than the terrestrial $^{14}N/^{15}N$ = 272 (Table 10.2) by a factor of ~4. Their search for $HC_5N$ resulted in an upper limit to its mixing ratio of $4\times10^{-10}$ in the lower stratosphere.

Later Titan was observed by Gurwell (2004) using the Submillimeter Array (SMA) that consists of eight 6-m telescopes. A few temperature profiles (Figure 13.13a) were adopted for analysis of the data including profile A from Coustenis and Bezard (1995) that was used by Marten et al. (2002). CO was measured at 51 ± 4 ppm, the retrieved HCN abundances (Figure 13.13b) are greater than those from Marten et al. (2002) above 250 km, while those for $HC_3N$ agree with Marten et al. (2002). The measured $HC^{14}N/HC^{15}N$ = 94 ± 13 for profile A and 72 ± 9 for profile D, and the uncertainties of the latter overlap with those

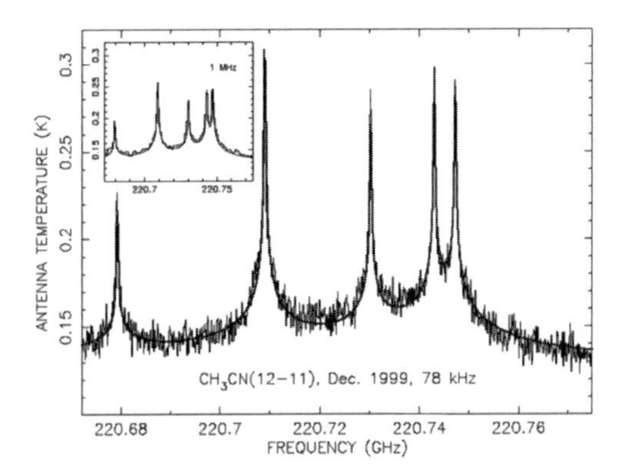


Figure 13.11 IRAM spectrum of five lines of acetonitrile on Titan with resolution of 78 kHz and 1 MHz fitted by a model in Figure 13.12. From Marten et al. (2002).

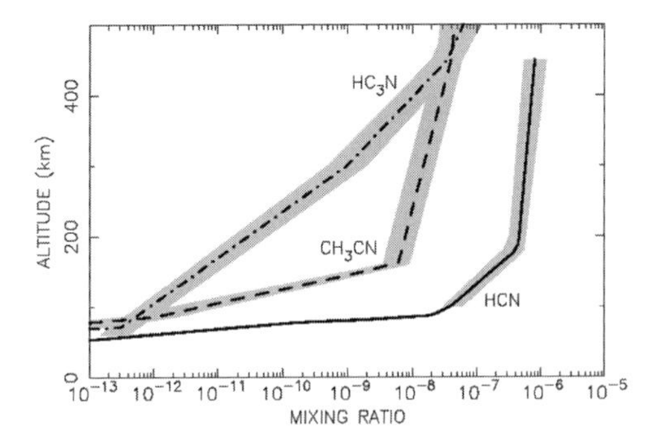


Figure 13.12 Vertical profiles of three nitriles retrieved from the IRAM observations (Marten et al. 2002).

from Marten et al. (2002). $H^{12}CN/H^{13}CN$ was measured as well with values of 132 ± 25 and 108 ± 20 for profiles A and D, respectively. These ratios are above the terrestrial ratio $^{12}C/^{13}C$ = 89 (Table 10.2).

Recently Titan was observed using the Atacama Large Millimeter Array (ALMA) in Chile. This array is at the altitude of 5 km and consists of 66 telescopes of 12 m or 7 m each. The observations of nitriles (Cordiner et al. 2014, 2015) were made in July 2012 with the spatial resolution of ~0.4 arcs that was comparable with Titan's radius. HNC was detected with a column abundance of $1.9\times10^{13}$ cm$^{-2}$ mostly above 400 km. $HC_3N$ was observed with a mean abundance of $2.3\times10^{14}$ cm$^{-2}$ distributed over 70 to 600 km and enhanced over the poles. Ethyl cyanide $C_2H_5CN$ was detected for the first time with 19 emission lines and blends exceeding the $3\sigma$ level. It is distributed mostly above 200 km and in the southern hemisphere with a mean column abundance of (1–5)$\times10^{14}$ cm$^{-2}$.

The ALMA observations of a few CO lines in the millimeter range (Serigano et al. 2016) confirmed its mixing ratio of 50 ± 2 ppm. CO is uniformly mixed on Titan, and the

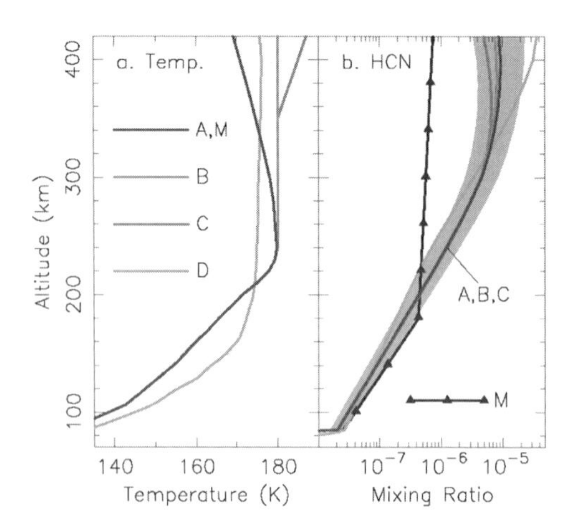


Figure 13.13 Temperature profiles adopted for analysis of the SMA observations and retrieved profiles of HCN are compared with that from Marten et al. (2002, designated M). From Gurwell (2004).

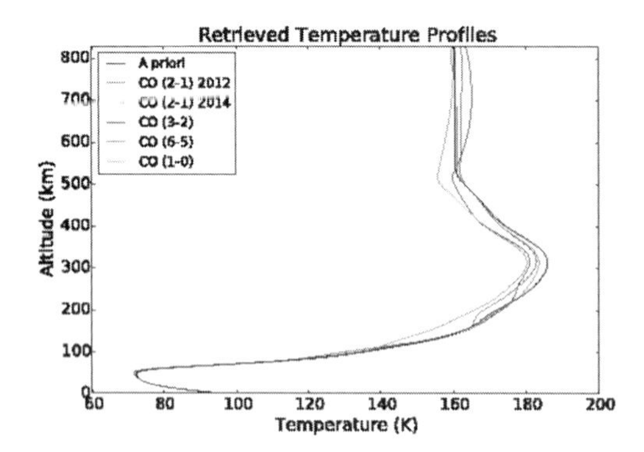


Figure 13.14 Temperature profiles of Titan's atmosphere retrieved from the ALMA observations of a few CO lines in the millimeter range. From Serigano et al. (2016).

observed line shapes may be used to derive temperature profiles (Figure 13.14). The profiles from five lines are in excellent mutual agreement. The CO isotopic lines are used to get isotopic ratios in CO: $^{12}C/^{13}C = 90 \pm 3$, $^{16}O/^{18}O = 486 \pm 22$, and $^{16}O/^{17}O = 2917 \pm 359$. These ratios are the most accurate for CO on Titan and agree within the uncertainties with the terrestrial ratios (Table 10.2).

### *13.3.3 Infrared Space Observatory*

ISO was an Earth-orbiting observatory operated by the European Space Agency (ESA) in 1996–1998. Titan was observed using ISO in January, June, and December 1997. Basic results for Titan were obtained using the Short-Wavelength Spectrograph (SWS) that

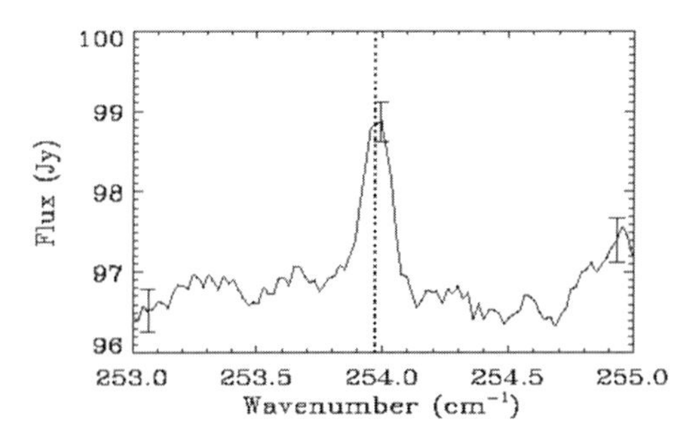


Figure 13.15 Detection of the $H_2O$ line in the ISO spectrum of Titan. Resolving power is 1900, and two lines were detected at a level of 8σ. From Coustenis et al. (1998).

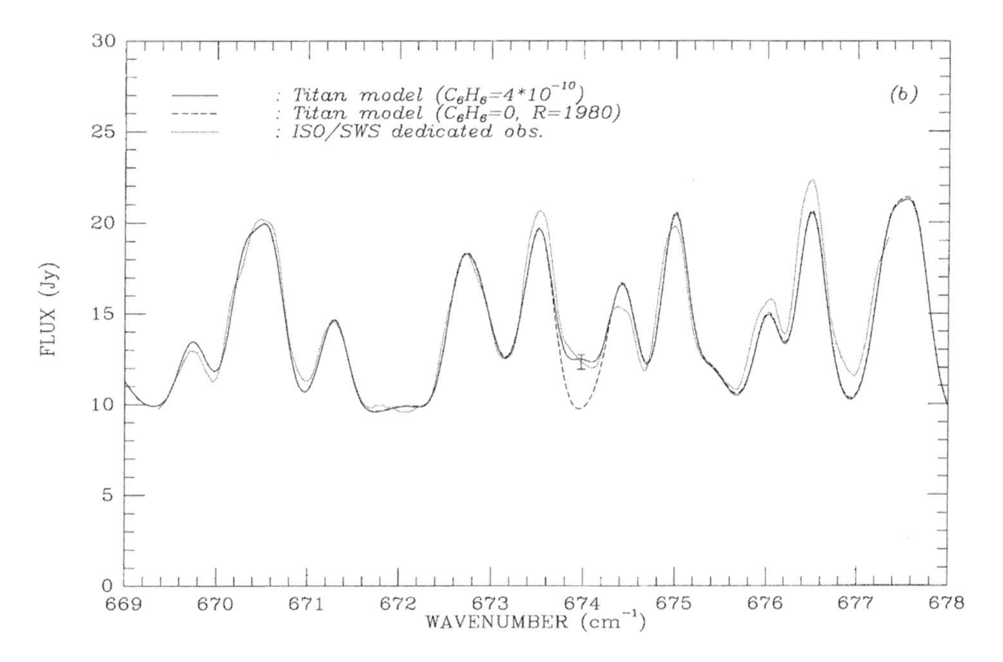


Figure 13.16 Detection of benzene $C_6H_6$ in the ISO/SWS spectrum (Coustenis et al. 2003).

covered a range of 2.4 to 45 μm. The ISO observations have better spectral resolution than the Voyager/IRIS spectra but refer to the full disk of Titan.

Two $H_2O$ lines were detected in the spectra of Titan near 40 μm (Figure 13.15). That was the first detection of water on Titan. The observed column density was $(2.6 \pm 1.7)\times 10^{14}$ cm$^{-2}$ above the surface and a mixing ratio of $(8 \pm 5)\times 10^{-9}$ at 400 km (Coustenis et al. 1998).

The observed $C_2H_2$ spectrum at 670–780 cm$^{-1}$ was fitted with a significant increase in the acetylene mixing ratio from $2\times 10^{-6}$ near 70 km to $4\times 10^{-5}$ at 450–500 km (Coustenis et al. 2003). Benzene $C_6H_6$ was detected with a mixing ratio of $(4 \pm 3)\times 10^{-10}$ at 100–200 km using the line at 674 cm$^{-1}$ (Figure 13.16). Upper limits to allene, the symmetric isomer of methylacetylene, of $2\times10^{-9}$, $CH_3CN$ of $5\times10^{-10}$, and $C_2H_5CN$ of $10^{-10}$ were established from the lack of the bands at 353, 1041, and 1075 $cm^{-1}$, respectively. The observed $CH_3D$ corresponds to $D/H = 8.7^{+3.2}_{-1.9}\times10^{-5}$ and is smaller than the terrestrial ratio.

Except the above results, all species detected by the Voyager/IRIS at low latitudes were observed by the ISO/SWS as well and typically with better uncertainties. The Voyager data refer to vernal equinox, while the ISO observations were conducted in the beginning of northern fall. However, it is difficult to make any conclusions on the seasonal variations from their comparison, because the ISO results are disk-average.

The ISO and ground-based infrared observations of Titan's stratospheric composition are tabulated in Coustenis et al. (2013).

### *13.3.4 Herschel Space Observatory*

Herschel Space Observatory was another ESA mission that was designed for spectroscopy and imaging in the far infrared and submillimeter ranges. The mission was active in the period of 2009–2013 and had three instruments. PACS was a camera and spectrometer for 55 to 210 μm with resolving power varying from 1000 to 5000. SPIRE had the similar functions in the range of 194–672 μm with resolving power up to 1000, and HIFI was a heterodyne spectrometer for a range of 157–625 μm with resolving power up to $10^7$.

Observations of Titan by HSO in 2010–2011 revealed $H_2O$ rotational lines at 66.4, 75.4, and 108 μm in the PACS range and at 557 and 1097 GHz in the HIFI range (Moreno et al. 2012). Weighting functions (Sections 7.1 and 7.4) of the $H_2O$ lines measured by PACS are maximal near 100 km, while those observed by HIFI are near 300 km. This made it possible to derive three parameters of the $H_2O$ vertical profile adopted as $f = f_0\ (p_0/p)^n$. Fitting of the observed lines resulted in the $H_2O$ fraction $f_0 = (2.3 \pm 0.6)\times10^{-11}$ at $p_0 = 12.1$ mbar (~100 km) and a power index $n = 0.49 \pm 0.07$ that corresponds to $f = 5\times10^{-10}$ at 400 km.

Search for ammonia $NH_3$ in the SPIRE spectra at 32–50 $cm^{-1}$ (Teanby et al. 2013) resulted in an upper limit to the $NH_3$ column abundance of $1.2\times10^{15}$ $cm^{-2}$ at 65–110 km, that is, the mixing ratio of less than 0.19 ppb.

The PACS spectra of Titan contain lines of $CH_4$, CO, and HCN, and the observed $CH_4$ mixing ratio is constant at $1.29 \pm 0.03\%$, the CO abundance is $50 \pm 2$ ppm, and HCN shows an increase from 40 ppb at 100 km to 400 ppb at 200 km (Rengel et al. 2014). (The claimed 4 ppm at 200 km in their abstract is wrong.) The $^{12}C/^{13}C$ isotopic ratios in CO and HCN were measured at $124 \pm 58$ and $66 \pm 35$, respectively.

## 13.4 Observations from the Huygens Landing Probe

### *13.4.1 Cassini–Huygens Mission*

This was a complicated NASA–ESA mission to study Saturn, Titan, and their environments. It was launched in October 1997 and then approached Venus two times with flybys in April 1998 and June 1999. Those flybys were mostly aimed to get momentum from Venus' gravity; however, some important observations were conducted as well (Section

12.11.1; Figures 12.69 and 12.70). Flybys of the Earth in August 1999 and Jupiter in December 2000 also helped to get momentum to overcome the solar gravity and reach Saturn. The flyby of the Earth and Moon was used for some calibrations, while that of Jupiter gave some valuable scientific data.

The spacecraft entered an orbit around Saturn in July 2004. The first close flyby through the upper atmosphere of Titan at an altitude of 1200 km was in October 2004. The Huygens probe was released on December 25, 2004, and entered Titan's atmosphere after some tests and operations on January 14, 2005. The probe descent in the atmosphere was for 2.5 hours, it continued to send data for 1.5 hours after landing. The Huygens data were transmitted to the Cassini orbiter and then to Earth. The Huygens probe was designed and operated by ESA. The probe could land on either solid or liquid surface. The landing site at 10°S and 192°W was chosen using data of the Cassini flyby of Titan in October 2004. The Cassini orbiter studied Saturn and Titan for 13 years and was finally directed into the atmosphere of Saturn in September 2017.

Instrumentation of the Huygens probe included the Gas Chromatograph Mass Spectrometer (GCMS), the Descent Imager/Spectral Radiometer (DISR), the Huygens Atmospheric Structure Instrument (HASI), the Surface Science Package (SSP), the Doppler Wind Experiment (DWE), and the Aerosol Collector and Pyrolyser (ACP).

### *13.4.2 GCMS Measurements (Niemann et al. 2010)*

The instrument was switched on at 146 km and measured the atmospheric composition down to the surface for 148 min. It survived the impact of landing and operated for 72 min after the landing.

The observed methane mixing ratio was constant at $(1.48 \pm 0.09)\%$ in the lower stratosphere at 140 to 75 km. It increased in the troposphere to $(5.65 \pm 0.18)\%$ in the lowest 7 km. This means that methane is subsaturated near the surface at a relative humidity of $\approx$50%. Its abundance fits the saturation conditions near 7 km and matches these conditions up to the tropopause being constant in the stratosphere because of mixing. The methane aerosol should be liquid at 7 to 14 km and solid above 14 km. $N_2$ is dissolved in methane and slightly affects the phase equilibrium.

The measured $H_2$ fraction was constant at $(1.0 \pm 0.16)\times10^{-3}$ during the descent and near the surface. The value agrees with $(1.0 \pm 0.4)\times10^{-3}$ from the Voyager/IRIS (Courtin et al. 2005) and $(9.6 \pm 2.4)\times10^{-4}$ from the Cassini/CIRS (Courtin et al. 2012) observations.

Precise measurements of some isotope ratios were made and resulted in $^{14}N/^{15}N$ = $167.6 \pm 0.6$ in $N_2$, $^{12}C/^{13}C = 91.1 \pm 1.4$ in methane, and $D/H = (1.35 \pm 0.30)\times10^{-4}$ in $H_2$. The measured fractions of the argon isotopes are $^{40}Ar = (3.39 \pm 0.12)\times10^{-5}$, $^{36}Ar = (2.1 \pm 0.8)\times10^{-7}$, and an upper limit of $5\times10^{-8}$ to $^{38}Ar$. Only $^{22}Ne$ was detected at $(2.8 \pm 2.1)\times10^{-7}$, while the signal of the main isotope $^{20}Ne$ was contaminated by the double ionized $^{40}Ar$; its upper limit was $2\times10^{-5}$. Krypton and xenon were below a detection limit of $10^{-8}$. Isotopes of the major atmospheric constituents and abundances of noble gases are keys to the origin and evolution of the atmosphere. Isotopic ratios for the major constituents are less affected by isotope fractionations in photochemical reactions that are typically poorly known.

The probe was warmer than the surface and resulted in some evaporation from the surface. $CH_4$ increased immediately from 5.7% to 7.7% and remained constant at this value. $C_2H_6$ became measurable in 20 min after the landing and reached a constant abundance of 150 ppm in 40 min. $C_2H_2$ and $CO_2$ grew continuously up to 80 ppm and 25 ppm, respectively, and $C_2N_2$ was at ~1 ppm after the landing. These data are relevant to composition of the surface ice.

All data from the gas chromatographic subsystem of GCMS were below unspecified detection limits.

### *13.4.3 Descent Imager/Spectral Radiometer (DISR)*

Descent Imager/Spectral Radiometer (DISR) included upward- and downward-looking visible and infrared spectrometers and violet photometers and a camera, that is, seven instruments total. The spectrometers covered ranges of 480–960 nm and 850–1600 nm, respectively, and the photometer bandpass was at 350–470 nm.

Images of the surface (Figure 13.17) and spectra of the surface reflectivity were observed (Tomasko et al. 2005). The reflectivity is maximal of 0.17 near 800 nm decreasing to 0.1 at 500 nm and 0.07 at 1600 nm. Wind speed increasing from ~0 at 8 km to 28 m $s^{-1}$ eastward at 55 km was derived from the observed images.

The observed spectra show continua of the haze extinction and the methane absorption bands. The retrieved methane abundance was 5% ± 1% near the surface (Tomasko et al. 2005). Analysis of the upward looking infrared spectrometer observations of the $CH_4$ band at 1.4 μm gave a stratospheric fraction of methane at $1.44^{+0.27}_{-0.11}$% (Bezard et al. 2014). Both values are in accord with the GCMS data.

Figure 13.17 Image of Titan's surface at the Huygens landing site. Globules of probably water ice are of 10–15 cm in size.

Analysis of the haze extinction from the observations at various altitudes, zenith and azimuthal angles was made by Tomasko et al. (2008a). The observed polarization reached 60% at phase angles of ~90° and required small monomers with $r \leq 0.05$ μm, while the phase functions were typical of particles with radii greater by an order of magnitude. This problem was met in the Voyager observations of the haze and solved using the fractal particles. The best fit to the DISR data is for $r \approx 0.05$ μm and a mean number of monomers of $N \approx 3000$ in the particle. Radius of the equal volume sphere is 0.72 μm and of the sphere with equal projected area is 2.03 μm.

The particle number density is ~5 cm$^{-3}$ at 80 km and diminishing with a scale height of 65 km above 80 km. Condensation of the photochemical products occurs in spaces between the monomers at 30–80 km, increasing the particle single-scattering albedo, while the number density changes insignificantly. A weak wavelength dependence of the extinction below 30 km indicates appearance of large particles with $r \approx$ 3–10 μm with much smaller number densities due to condensation of methane in the troposphere.

The retrieved vertical profiles of the haze extinction at various wavelengths are shown in Figure 13.18. The total optical depth at 1583 nm is smaller than that at 355 nm by an order of magnitude, and the difference is even greater at the higher altitudes. Phase functions at various wavelengths above and below 80 km are shown in Figure 13.19. Backscattering is very much weaker than the forward scattering for these particles, while it would be almost equal for very small particles of the monomer radii. Single scattering albedos as functions of wavelength are

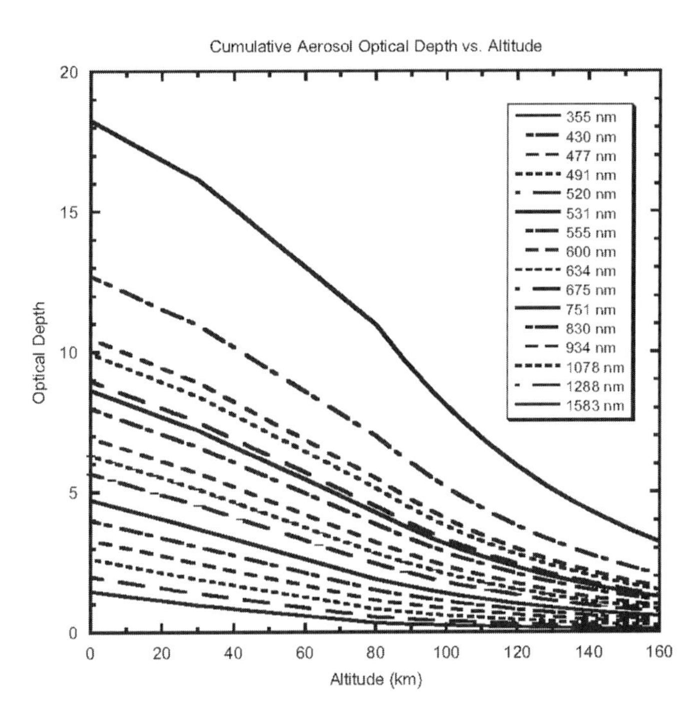


Figure 13.18 Haze extinction optical depth for different wavelengths observed by Huygens/DISR. The total depth decreases from 18 at 355 nm to 1.4 at 1583 nm. From Tomasko et al. (2008a).

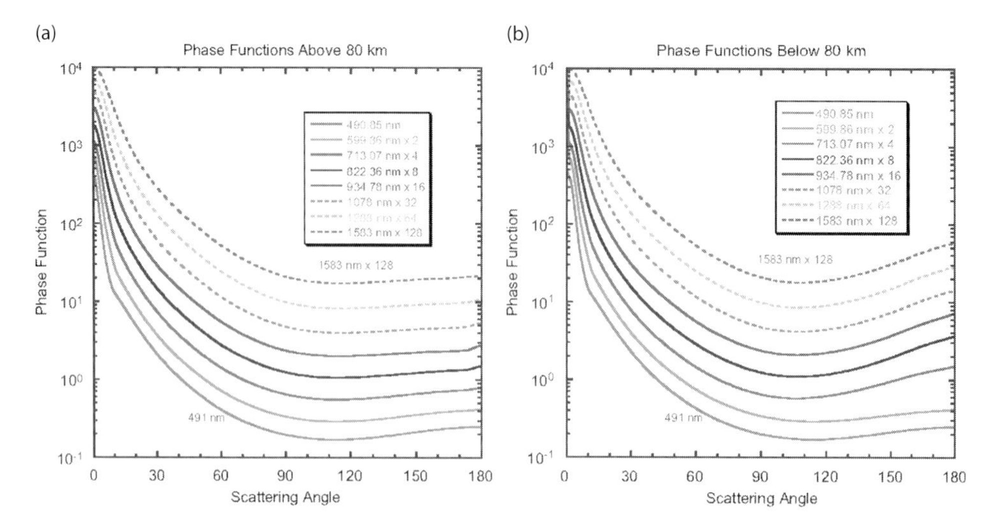


Figure 13.19 Haze scattering phase functions retrieved from the Huygens/DISR observations above and below 80 km (Tomasko et al. 2008a).

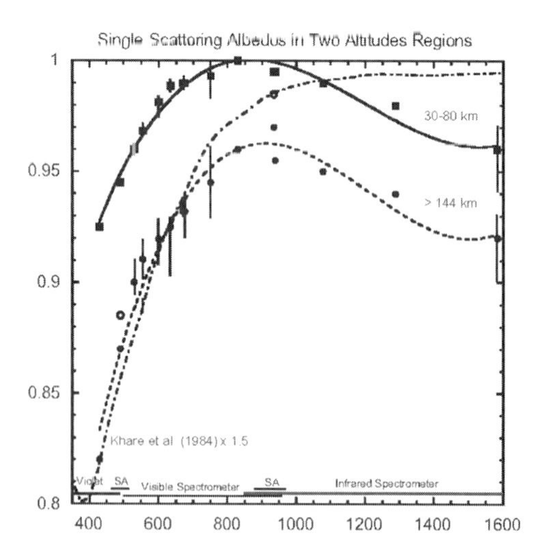


Figure 13.20 Single scattering albedos of the haze above 144 km and at 30–80 km compared with those above 144 km for the imaginary indices from Khare et al. (1984, figure 201) scaled by a factor of 1.5. The abscissa is wavelength in nm. From Tomasko et al. (2008a).

shown for two altitude regions in Figure 13.20 and compared with those calculated for the imaginary indices from Khare et al. (1984) scaled by a factor of 1.5. The values agree at $\lambda < 800$ nm and are very different at the longer wavelengths.

Doose et al. (2016) studied thin (<1 km) aerosol layers at 21, 11, and 7 km and a thicker layer at 55 km. They found a decrease in the haze scale height from 65 km below 140 km to 45 km at higher altitudes. Applications of the observed haze properties to the thermal balance of the atmosphere were considered in Tomasko et al. (2008b).

### *13.4.4 Other Huygens Measurements*

The Huygens Atmospheric Structure Instrument (HASI; Fulchignoni et al. 2005) measured temperature and density of the atmosphere using an accelerometer below 1400 km and direct sensors below 150 km. The results below 200 km agree with the Voyager radio occultations and Cassini data, while the accelerometer values above 200 km are greater than those expected from the Voyager observation. The temperature profile shows a wavelike structure around a constant value of 170 K at 200 to 1400 km. The atmospheric densities above 600 km are typically larger than those inferred from the Voyager data by a factor of ~2.5.

Two sensors of HASI, mutual impedance and relaxation probes, could discriminate the ionospheric conductivity caused by positive and negative charges. The electron mobility is very much greater than that of negative ions, and the negative charge conductivity is completely due to electrons. The measured conductivity has a broad peak at 140 to 40 km with a maximum near 65 km. The peak density of positive ions is $\approx 2000$ $cm^{-3}$ and of electrons is $\approx 450$ $cm^{-3}$ (Lopez-Moreno et al. 2008). Evidently the difference refers to negative ions. The observed values are close to those expected for the cosmic ray ionization (Molina-Cuberos et al. 1999).

The Doppler Wind Experiment (Bird et al. 2005) measured Doppler shift of the signal from the Huygens probe to extract a zonal wind profile during the descent. The signal was tracked at the Cassini orbiter and the ground-based Green Bank Telescope (Folkner et al. 2006). The observed zonal wind speed was 100 m $s^{-1}$ at 150–120 km and steeply reduced to ~5 m $s^{-1}$ at 70 km, increased to 30–35 m $s^{-1}$ at 65–55 km and gradually diminished to ~0 near the surface.

The Aerosol Collector and Pyrolyser (Israel et al. 2005) made analyses of the aerosol collected at 130–35 km and 25–20 km. The collected matter was pyrolized at 600°C and analyzed by the mass spectrometer within GCMS. The mass spectra included strong features at $m/z = 2$ ($H_2$), 12–17 ($CH_X$ and $NH_3$), 27–31 ($N_2$, $C_2H_X$, HCN), 20 and 40 (background Ar), and 44 (background $CO_2$). $NH_3$ and HCN were identified as the main pyrolysis products.

## 13.5 Cassini Orbiter Observations below 500 km

### *13.5.1 Cassini Instruments*

The Cassini–Huygens mission was briefly discussed in Section 13.4.1. The Cassini part of the mission was an orbiter of Saturn that studied the planet, its satellites, and their environments. The orbiter had ~100 flybys of Titan through its upper atmosphere at altitudes down to 880 km, facilitating in situ and close remote observations. The orbiter was in operation for 13 years, covering the 11-year solar cycle and the significant part of the Saturn and Titan 30-year annual cycle.

The orbiter was powered by three radioisotope thermoelectric generators that transformed heat from the radioactive decay of 33 kg of $^{238}$Pt to electricity. Radio Science

Subsystem, designed mostly for transmission of the orbiter data to Earth, was also used for radio occultations to study the lower atmosphere and ionosphere of Titan.

The radar mapped the surface of Titan and searched for its radio emission. Nine percent of Titan's surface was studied by the radar, and a full topographic map of Titan has been made by interpolation of the data (Corlies et al. 2017). The topographic features are typically within $\pm 0.5$ km, and deviations from sphericity are $\approx 0.3$ km.

The Imaging Science Subsystem included wide- and narrow-angle cameras with sets of filters. The Visual and Infrared Mapping Spectrometer (VIMS) was a camera that took pictures in 352 different wavelengths between 0.3 and 5.1 μm.

The Composite Infrared Spectrometer (CIRS) had three spectral channels for ranges of 10 to 600 $cm^{-1}$, 600 to 1100 $cm^{-1}$, and 1100 to 1400 $cm^{-1}$. It covered the middle and far infrared spectral ranges and the submillimeter range from 7 μm to 1 mm total. The instrument was designed to study the structure and chemical composition of Titan's atmosphere at 100–500 km and for some tasks related to Saturn, its rings and satellites. The instrument spectral resolution varied from 0.5 to 15.5 $cm^{-1}$.

The Ultraviolet Imaging Spectrograph (UVIS) had two channels (EUV and FUV) for spectral ranges of 56–118 nm and 110–190 nm, a third channel for stellar occultations, and a D/H absorption cell to measure this ratio from D and H Lyman-alpha emissions.

There were a few instruments for direct sensing of the upper atmosphere and ionosphere during the close Cassini flybys. The Ion Neutral Mass Spectrometer (INMS) was a quadrupole sensor that covered an $m/z$ range from 0.5 to 100 and measured mass spectra of neutral molecules ionized inside the instrument by electrons with energies of 70 eV. The instrument in the ion mode measured positive ions in the ionosphere.

Cassini Plasma Spectrometer (CAPS) was in operation from 2004 to 2012 and included three sensors. The electron spectrometer ELS measured electron energy in a range of 0.7 eV to 30 keV and their directions. The ion beam spectrometer IBS made the same for positive ions in a range of 1 eV to 50 keV, and the ion mass spectrometer IMS analyzed composition of positive ions. It had much lower mass resolution than INMS but much broader mass range.

The plasma environment was studied by Magnetometer (MAG), Magnetospheric Imaging Instrument (MIMI), and Radio and Plasma Wave Science (RPWS).

### *13.5.2 VIMS Observations*

VIMS observations revealed albedos of various regions on Titan in a few spectral windows between the methane bands up to 5 μm and wavelength variations of the haze single scattering albedo from 0.92 at 1.6 μm to 0.7 at 2.5 μm with a steep decrease to 0.5 at 2.6 μm. The VIMS solar occultations at the $CH_4$ bands at 1.4 and 1.7 μm resulted in a methane stratospheric abundance of $1.28 \pm 0.08\%$ (Maltagliati et al. 2015), smaller than measured by GCMS.

Dinelli et al. (2013) found an excess in the R-branch of the $CH_4$ band at 3.3 μm (Figure 13.21a) that is not identified to any observed gas on Titan and may refer to

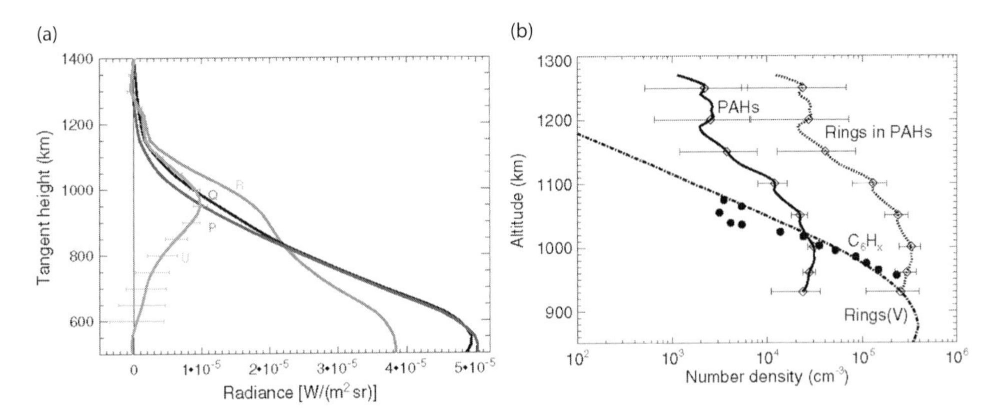


Figure 13.21 (a) Limb intensities of the R, Q, and P-branches of the $CH_4$ band at 3.3 μm observed by VIMS. The excess U of the observed R-branch over that calculated using the measured Q and P branches is shown as well and not identified with the known gases in Titan's atmosphere. It may be caused by polycyclic aromatic hydrocarbons (PAH). From Dinelli et al. (2013). (b) Retrieved densities of PAHs in numbers of molecules and rings. Solid dots are sums of benzene and $C_6H_X$ observed by INMS. The dash-dotted line is the ring number density calculated by Vuitton et al. (2008). From Lopez-Puertas et al. (2013).

polycyclic aromatic hydrocarbons (PAH). This feature was analyzed by Lopez-Puertas et al. (2013) using the NASA Ames PAH IR Spectroscopic Database. Electronic states of PAHs are excited by absorption of the visible and UV photons and result in vibrational excitation after internal rearrangement of the excitation energy. Finally this energy is partly emitted at 3.3 and ≈10 μm. The observed emission at 3.3 μm corresponds to the PAH number densities shown in Figure 13.21b. The mean number of rings in PAHs is ≈10, mean number of carbons is 34, mean mass is 430, and mean surface area is 0.5 $nm^2$. About one-third of PAHs have nitrogen atoms.

### 13.5.3 CIRS Observations during the Nominal Mission

CIRS observations during the nominal mission were summarized by Vinatier et al. (2010). The nominal mission covered 2004–2008 and referred to the end of northern winter ($L_S$ = 305°–345°). The CIRS limb observations during the close flybys of Titan gave detailed information of the thermal and chemical structure of the atmosphere at 100–500 km. We will use the term "stratosphere" for this region, though the upper boundary of the stratosphere is below 500 km. Spatial resolution of those observations is typically 40 km of altitude. The observations were made in high- and low-resolution modes of 0.5 and 15.5 $cm^{-1}$, respectively. Parts of the spectra at 610–930 $cm^{-1}$ measured in both modes at identical conditions are shown in Figure 13.22. Bands of seven hydrocarbons, two nitriles and $CO_2$ are identified in the high-resolution spectrum. Evidently the high-resolution observations are preferable but take much longer time.

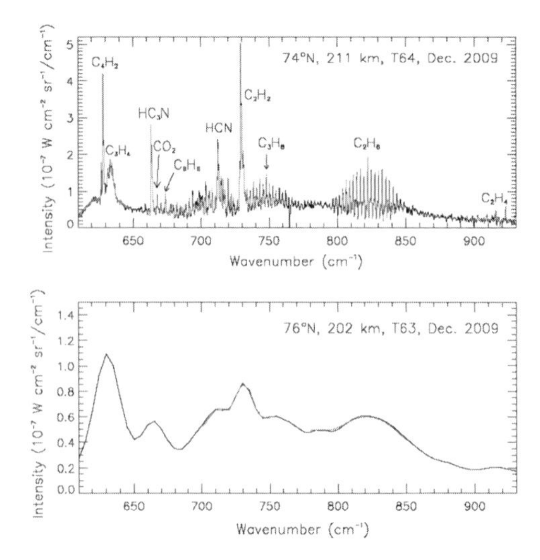


Figure 13.22 CIRS limb spectra at 610–930 $cm^{-1}$ observed at the same conditions with resolution of 0.5 and 15.5 $cm^{-1}$ (Vinatier et al. 2015).

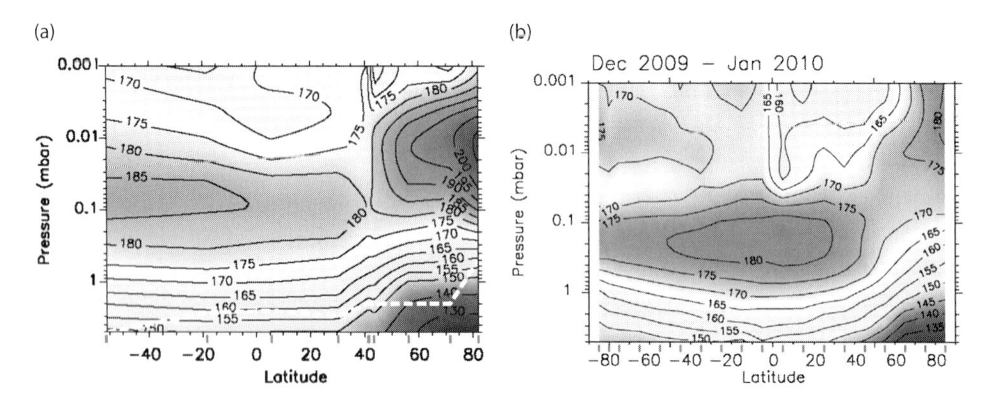


Figure 13.23 Thermal structure of the stratosphere observed using CIRS at the end of northern winter (a; $L_S$ = 305°–345°) and near vernal equinox (b; $L_S$ = 5°). 1, 0.1, 0.01, and 0.001 mbar correspond to ≈180, 300, 420, and 540 km, respectively. Temperatures are rather uncertain below the white dashed line. From Vinatier et al. (2010, 2015). (A black and white version of this figure will appear in some formats. For the color version, please refer to the plate section.)

Limb intensities in the thermal infrared range are determined by vertical profiles of temperature and emitting species. The best way to retrieve a temperature profile is to fit the observed shape of the $CH_4$ band at 1305 $cm^{-1}$ using a constant mixing ratio of methane in the stratosphere that was observed using GCMS and DIRS at 1.48% and 1.44%, respectively. Some initial profile $T(\ln p)$ is adopted and then adjusted to minimize the $\chi^2$ between observed and calculated spectra. The observed thermal structure of the stratosphere at $L_S$ = 305°–345° is shown in Figure 13.23a. The stratosphere at the end of northern winter is

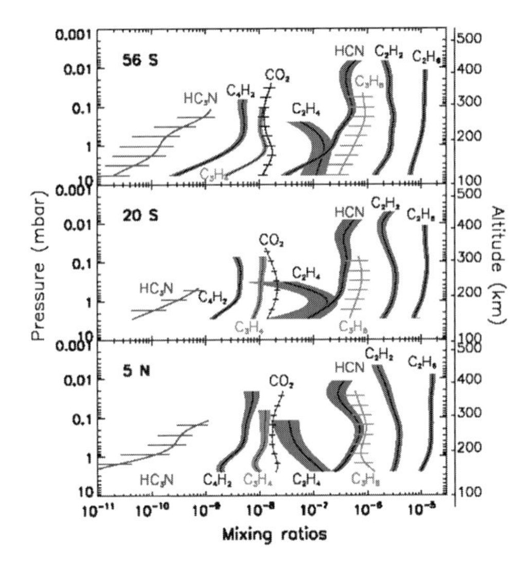


Figure 13.24 Vertical profiles of nine species from CIRS limb observations in the southern hemisphere at the end of northern winter ($L_S$ = 305°–345°). From Vinatier et al. (2010).

warmer near the north pole than that at the low and middle latitudes above ~300 km and colder below this level.

The next step is to fit the observed limb profile of a species band intensity using the derived temperature profile and an initial profile of the species mixing ratio. The initial profile is adjusted to get the best $\chi^2$ fit. The retrieved vertical profiles of nine species at various latitudes in both hemispheres are shown on Figures 13.24 and 13.25.

The analysis of the CIRS observations is based on the assumption of local thermodynamic equilibrium, when populations of vibrational and rotational levels of species are determined by collisions with negligible effects of photon absorption and emission. This assumption is questionable above ~400 km, where the collision rate becomes low and excitation by the solar photons may result significant deviations from local thermodynamic equilibrium.

All observed minor constituents and especially nitriles are more abundant near the north pole than at the middle latitudes. The polar enhancement is greater above ~300 km than below this altitude.

Nixon et al. (2013) averaged the CIRS limb high-resolution spectra at 880–940 cm$^{-1}$ measured between July 2004 and July 2010 in the altitude range of 100–250 km and at latitudes of 30°S to 10°N. They detected the band of propene $C_3H_6$ at 912.5 cm$^{-1}$ and extracted vertical profiles of this and five other species ($C_2H_2$, $C_2H_4$, $C_2H_6$, $C_3H_4$, and $C_3H_8$). Teanby et al. (2009) detected cyanogens $C_2N_2$ with a mixing ratio of $5.5\times10^{-11}$ at southern and equatorial latitudes. The Voyager/IRIS values were ~$1.5\times10^{-8}$ at 45°–70°N with an upper limit of 1.5 $\times10^{-9}$ at 30°N–50°S (Figure 13.4). Jolly et al.

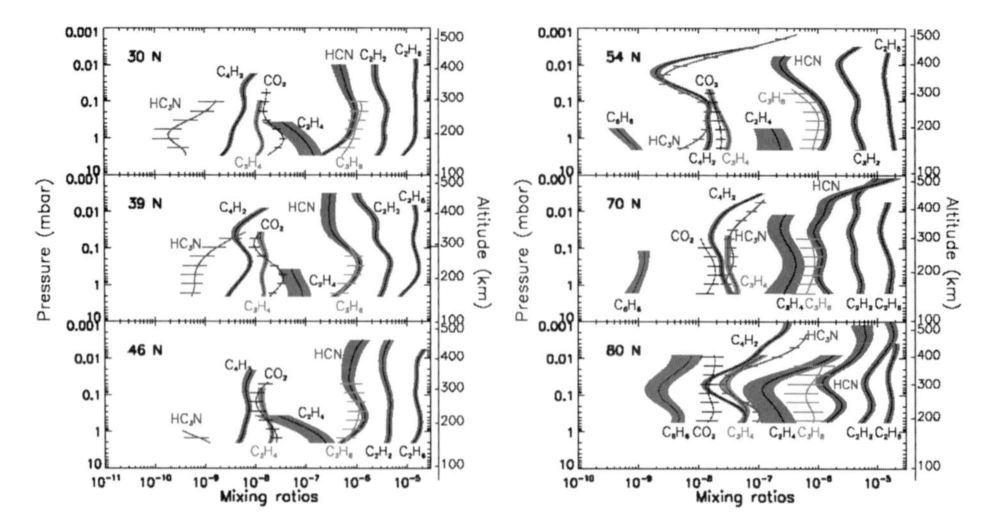


Figure 13.25 Vertical profiles of nine species from CIRS limb observations in the northern hemisphere at the end of northern winter ($L_S$ = 305°–345°). From Vinatier et al. (2010).

(2015) searched for $C_4N_2$ in the CIRS nadir and limb spectra and established an upper limit of $1.5 \times 10^{-9}$ at 150 km.

Nixon et al. (2012) studied the bands of $^{13}CH_4$, $CH_3D$, and $^{13}CH_3D$ in the CIRS spectra and derived $^{12}C/^{13}C = 86.5 \pm 8.2$ and D/H = $(1.59 \pm 0.33) \times 10^{-4}$ in methane on Titan. Combining two ground-based observations (see Section 13.3.1) and four CIRS-based values, they obtained a weighted-mean ratio D/H = $(1.36 \pm 0.08) \times 10^{-4}$ in $CH_4$, which coincides with that in $H_2$ measured using GCMS and is slightly smaller than the terrestrial ratio of $1.56 \times 10^{-4}$.

Combining the $^{12}C/^{13}C$ ratios in methane measured by GCMS, INMS (Mandt et al. 2012a), and two CIRS values, the weighted-mean ratio is 89.7 ± 1.0 that is similar to the terrestrial ratio of 89.4.

### *13.5.4 Haze Properties from the CIRS Observations*

Haze properties from the CIRS observations were analyzed by Anderson and Samuelson (2011) and Vinatier et al. (2012). The observations result in spectra of volume extinction coefficient of the haze near 190 km. Using the DISR data on the haze number densities, these data are converted to extinction cross sections of the haze particles (Figures 13.26 and 13.27). These cross sections are compared with those extracted from the DISR observations and calculated for the fractal particles of 3000 monomers with the monomer radius of 0.05 μm. According to the DISR observations, the haze consists of these particles. The refractive indices from Khare et al. (1984; Figure 13.2) are adopted for this curve.

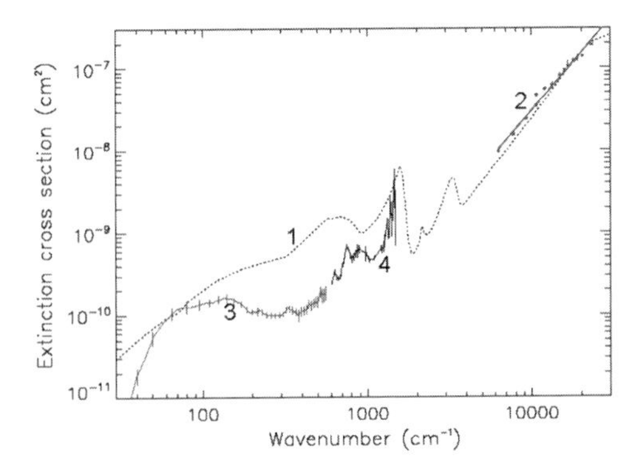


Figure 13.26 Extinction cross sections of the haze particles near 190 km extracted from the CIRS observations by Anderson and Samuelson (2011; curve 3) and Vinatier et al. (2012; curve 4) are compared to those from the DISR observations (Tomasko et al. 2008a; curve 2) and for the fractal particles (curve 1) of 3000 monomers with $r$ = 0.05 μm having refractive indices from Khare et al. (1984, figure 246). From Vinatier et al. (2012).

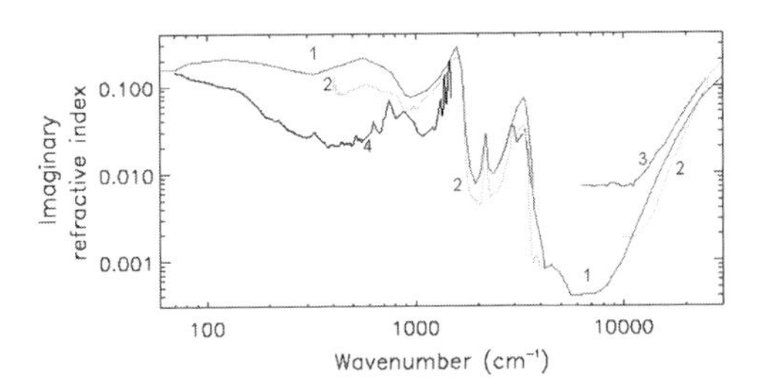


Figure 13.27 Imaginary refractive indices of the haze from the CIRS (Vinatier et al. 2012, curve 4) and VIMS (Rannou et al. 2010, curve 3) observations compared with the laboratory tholins (Khare et al. 1984, curve 1; Imanaka et al. 2012, curve 2). From Vinatier et al. (2012).

Spectra of both real and imaginary refractive indices of the haze are extracted from the spectra of volume extinction coefficient. The derived imaginary refractive index is compared to those of tholins and obtained from the VIMS observations by Rannou et al. (2010).

### *13.5.5 Main Features of the Atmospheric Dynamics*

Likewise Venus, Titan has the dense atmosphere around the slowly rotating solid body. Similar to that on Venus, the atmosphere of Titan is superrotating with strong eastward winds of ~100 m $s^{-1}$. The atmospheric dynamics is even more complicated that that on Venus because of the significant seasonal effects that are lacking on Venus.

The radiative times in the troposphere are longer than the year on Titan, and seasonal variations are expected to be comparatively weak in the troposphere. These times in the

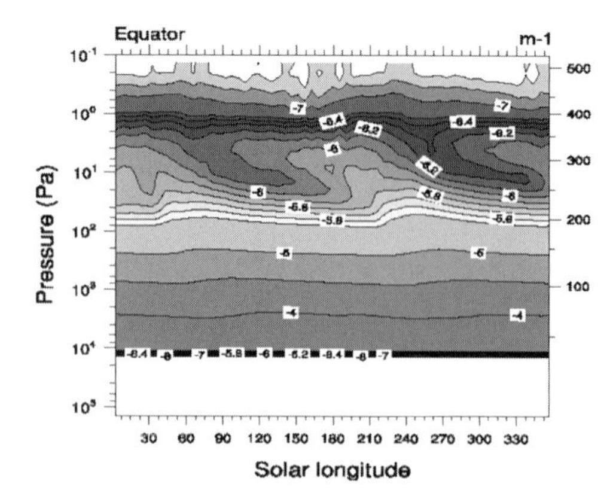


Figure 13.28 Seasonal variations of the haze opacity at the equator. Labels give log of the haze opacity in $m^{-1}$ at 700 nm, so that $-6$ corresponds to $10^{-6}$ $m^{-1}$. From GCM by Lebonnois et al. (2012).

stratosphere become comparable to and even shorter than the year, and significant seasonal effects are therefore expected. Meridional transport is presented by the only circulation cell with a northward wind in the upper atmosphere and the return flow in the lower atmosphere during the northern winter with the opposite wind directions during the northern summer. These dynamics switch near the equinoxes and form two Hadley cells with circulations from the equator to the poles. This transient mode exists for a comparatively short time. All these processes occur with some time lag because of the significant thermal and dynamical intertia of the atmosphere.

Seasonal behavior of the atmosphere is reproduced by general circulation models. For example, seasonal variations of the haze vertical profile at the equator from a GCM by Lebonnois et al. (2012) are shown in Figure 13.28. The predicted haze variations are of a factor of 2.5 near 300 km. The calculated seasonal-latitudinal variations of the zonal wind speed (Figure 13.29) look rather complicated and different at ~170 and ~280 km.

Numerous data on seasonal and latitudinal variations of temperature and chemical composition of the stratosphere are presented by Vinatier et al. (2015) based on the CIRS limb observations. The downward gas flow near the north pole heats the atmosphere above 300 km during the northern winter (Figure 13.23a), and this effect is disappearing soon after the equinox (Figure 13.23b).

Mixing ratios of photochemical products increase typically with altitude; therefore downward flows result in enrichments of the products in the atmosphere, while the upward flows have opposite effects. These trends are the most significant for nitriles that form in the upper atmosphere where $N_2$ can dissociate. Variations of vertical profiles of HCN and $C_2H_2$ near the south pole are shown in Figure 13.30. The HCN profiles are rather stable below 350 km and vary at ~470 km from $L_S$ = 5° to 30° by three orders of magnitude. The variations of $C_2H_2$ are similar but smaller, by two orders

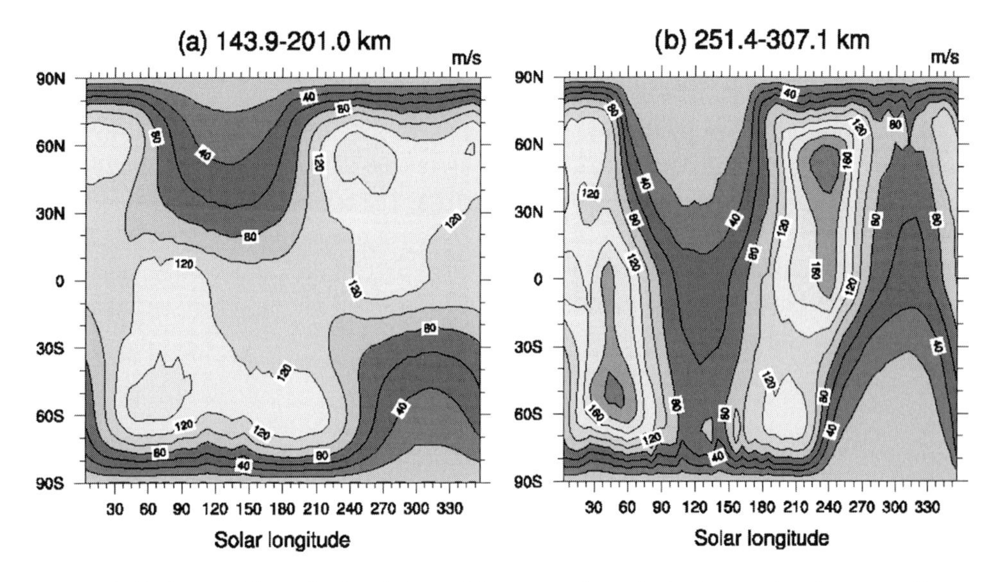


Figure 13.29 Seasonal-latitudinal variations of the eastward zonal wind in two altitude regions. From GCM by Lebonnois et al. (2012). (A black and white version of this figure will appear in some formats. For the color version, please refer to the plate section.)

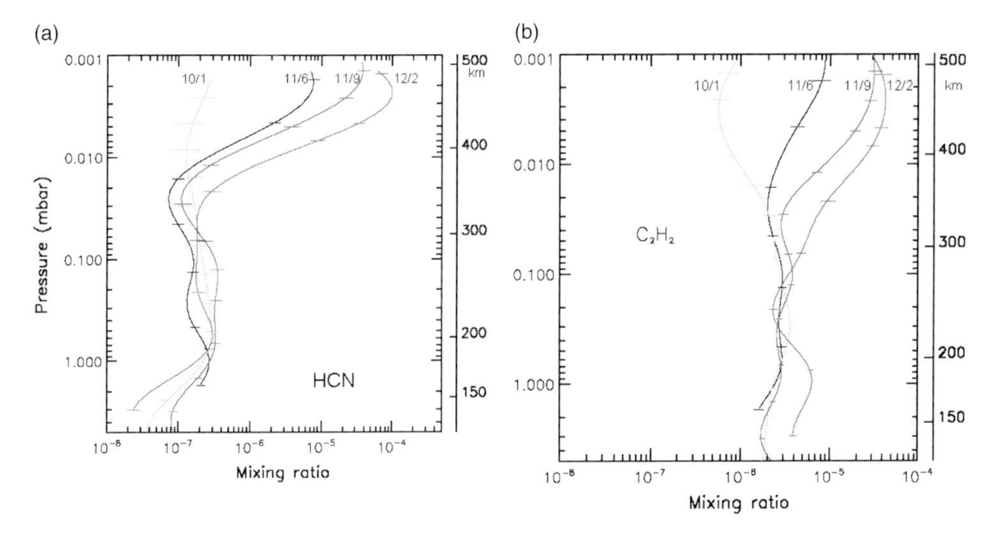


Figure 13.30 Seasonal variations of the CIRS vertical profiles of HCN (a) and $C_2H_2$ (b) near the south pole (84°S); 10/1 means January 2010. From Vinatier et al. (2015).

of magnitude. Seasonal evolution of the stratosphere in the CIRS observations near the poles was considered by Coustenis et al. (2018). Sylvetre et al. (2018) analyzed the CIRS observations of seasonal variations of $C_2N_2$, $C_3H_4$, and $C_4H_2$ in the lower stratosphere.

## 13.6 Cassini/UVIS Occultations and Airglow Observations

### *13.6.1 Occultations*

Titan's atmosphere above 500 km was studied by the Cassini orbiter using the UVIS stellar and solar occultations. Koskinen et al. (2011) analyzed the stellar occultations observed by the FUV channel that covers a range of 1120–1910 Å. The measured spectra of stars agree with previous observations of those spectra by other instruments.

The observed occultation spectra are presented as tangent optical depths versus wavelength (Figure 13.31). These spectra are fitted by sums of eight absorbers including the haze that was approximated by the power ($\lambda^{-\alpha}$) function.

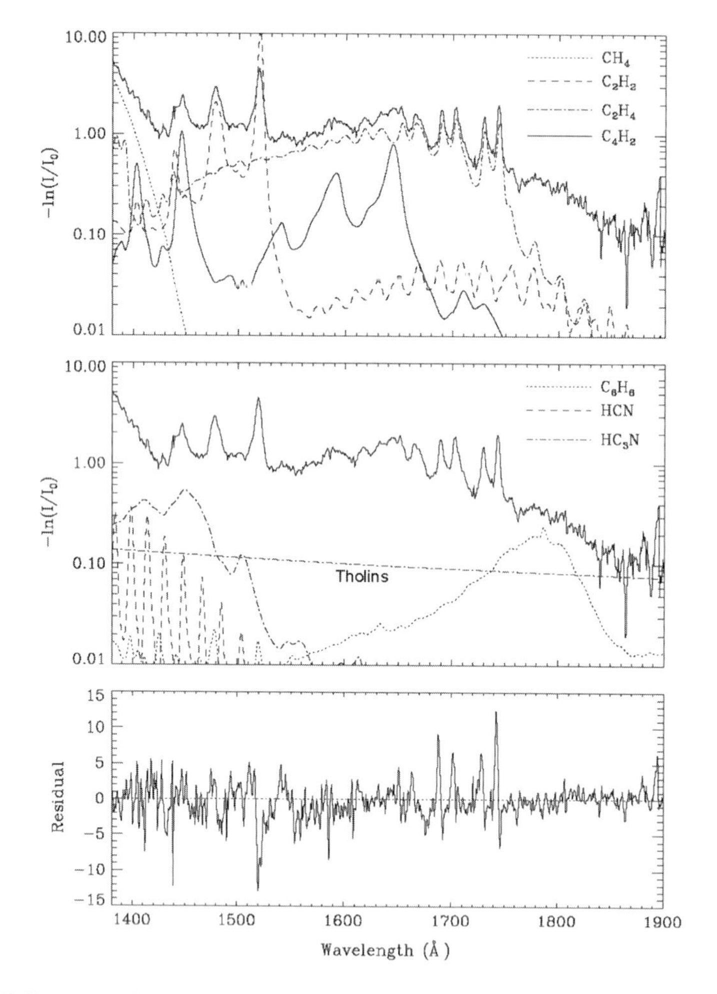


Figure 13.31 Spectrum of tangential optical depth of the atmosphere at 700–750 km observed by the UVIS stellar occultation T41 is fitted by absorption spectra of seven species and aerosol (tholins). Absorption by $C_2H_6$ is very low at this altitude and not shown. The lower panel shows differences between the observed and model optical depth divided by uncertainties of the measurements in terms of optical depth. From Koskinen et al. (2011).

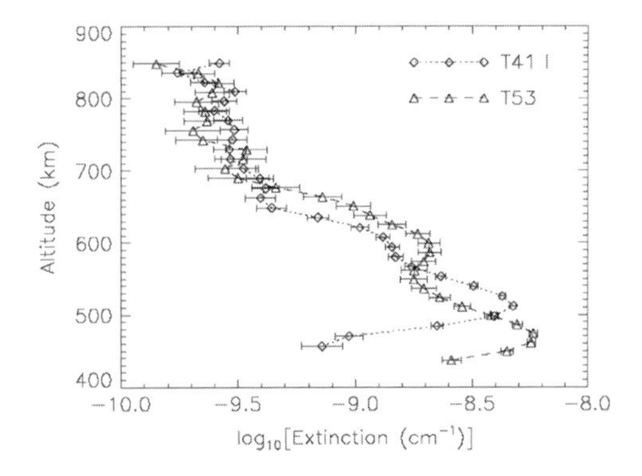


Figure 13.32 Vertical profiles of haze extinction at 1850–1900 Å in two UVIS stellar occultations (Koskinen et al. 2011). T53 was observed in April 2009 and refers to LT = 14 h and 39°N.

Koskinen et al. (2011) analyzed data of twelve stellar occultations. Most of their figures refer to flyby T41 in February 2008 with local times of 2.4 and 13.3 h, latitudes of 6°S and 27°S for ingress and egress, respectively. The measured vertical profiles of the haze are shown in Figure 13.32. The haze forms a layer at ≈500 km with a peak extinction of $\sigma \approx 5\times10^{-9}$ cm$^{-1}$ that diminishes to ≈$2\times10^{-10}$ cm$^{-1}$ at 850 km. Therefore the haze scale height is $H \approx 110$ km and the limb optical depth at 500 km is

$$\tau = \sigma\sqrt{2\pi(R+h)H} = 0.7,$$

that is, a well measured value. (Here $R$ is Titan's radius and $h$ is the altitude.) The vertical optical depth is 0.055.

Absorption by ethane was found insignificant in the spectrum in Figure 13.31 at 700–750 km and is not shown. The $\chi^2$ test gives 5.9 for the set of the data in Figure 13.31 and 8.4 if $C_6H_6$ is lacking. Therefore the detection of benzene is statistically significant. The retrieved tangent column abundances as functions of altitude are presented in Figure 13.33. They are converted into vertical profiles of number densities (Section 7.7) and then into the species mixing ratios (Figure 13.34) using the HASI density profile. The results are compared with the CIRS retrievals below 450 km and the INMS measurements near 1000 km. Currently these values along with the Voyager/UVS solar occultations (Vervack et al. 2004) are the only observational data on the chemical composition of Titan's atmosphere between 500 and 900 km.

Kammer et al. (2013) analyzed stellar occultations observed using the EUV channel of UVIS. This channel covers a range of 563–1182 Å. However, the ionization continuum of atomic hydrogen begins at 911 Å, and all stellar spectra are completely absorbed by the interstellar hydrogen below this wavelength. The authors studied the spectra at 920–1100 Å to get abundances of $N_2$, using the absorption bands below 1000 Å, and $CH_4$ that has dissociative and ionization continua in this range.

Stability of the spacecraft pointing to a chosen star restricts significantly quality of the data, and four occultations of 20 were taken for the analysis. Retrieved vertical profiles of $N_2$ and $CH_4$ for one of those events are shown in Figure 13.35. The measured profiles result in the following mean values: the $CH_4$ mixing ratio is (0.83 ± 0.15)% at 1000 km

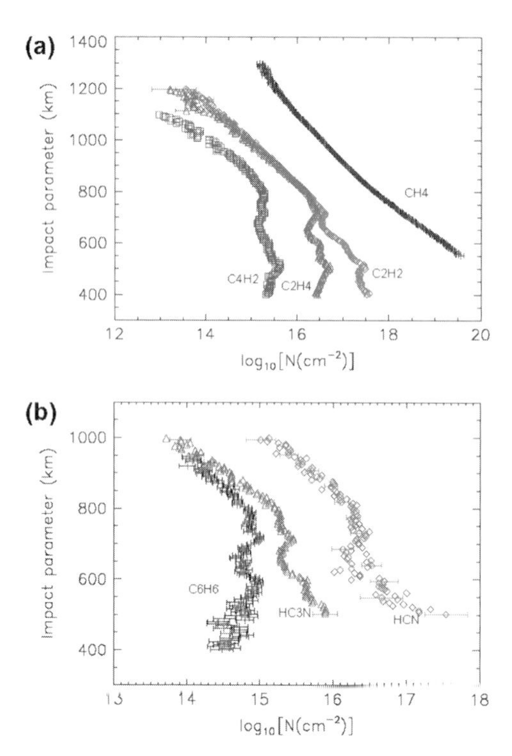


Figure 13.33 Vertical profiles of tangential column densities of seven species retrieved from the UVIS stellar occultation T41 (Koskinen et al. 2011).

and (4.2 ± 1.6)% at 1400 km, the mean temperature at 1000–1400 km is 159 ± 26 K, and the mean $N_2$ density at 1000 km is $(2.7 \pm 0.7)\times10^{10}$ cm$^{-3}$.

The UVIS/EUV solar occultations were analyzed by Capalbo et al. (2015). Eight occultations were studied to get $N_2$ and $CH_4$ density profiles and mean temperatures. Strong solar lines down to He 584 Å were used to improve uncertainties of the retrieved data.

The measured temperatures varied from 113 ± 4 K to 179 ± 9 K. Evidently the true variations of temperature are greater than the claimed observational errors $\sigma$. Therefore weights of all measured temperatures should be the same to get the mean temperature that is equal to 150 K. The standard deviation is 24 K, and the uncertainty of the mean value is $24/(8-1)^{1/2} = 9$, that is, 150 ± 9 K. If the statistical weights were adopted proportional to $1/\sigma^2$, then the weighted-mean value would be 155 ± 1 K. However, this assumption is invalid, and the uncertainty of the mean temperature of 150 ± 1 K claimed by the authors is incorrect. Generally, standard deviation of the true variations and that of the observational errors are statistically independent and should be added as squares.

The retrieved $N_2$ and $CH_4$ density profiles in two occultations are compared in Figure 13.36 with the INMS measurements in those flybys, results of the stellar occultation T21 (Kammer et al. 2013), the Voyager/UVS solar occultations, and the HASI density profile. All methane mole fractions from the UVIS solar occultations are shown in Figure 13.37.

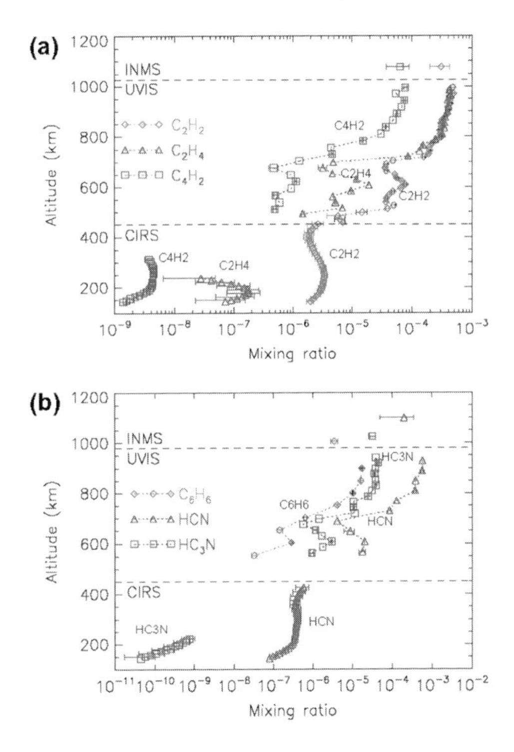


Figure 13.34 Vertical profiles of mixing ratios of six species retrieved from the UVIS stellar occultation T41 relative to the HASI density profile. The results are compared with those measured by CIRS and INMS. From Koskinen et al. (2011).

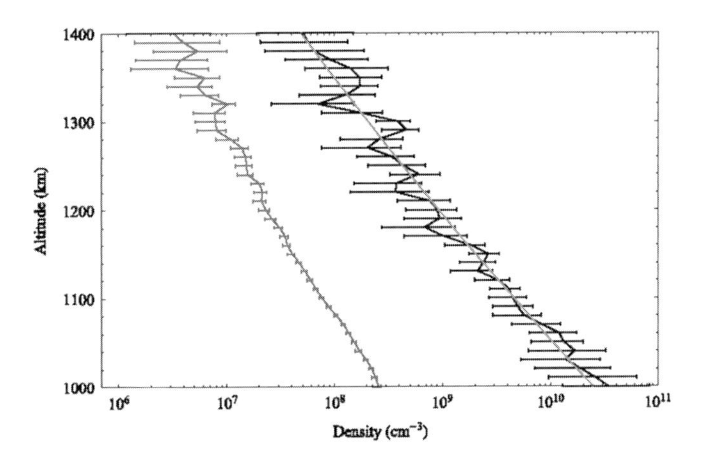


Figure 13.35 Number density profiles of $N_2$ and $CH_4$ from the UVIS stellar occultation T41. The smooth line is for $T$ = 137 K. From Kammer et al. (2013).

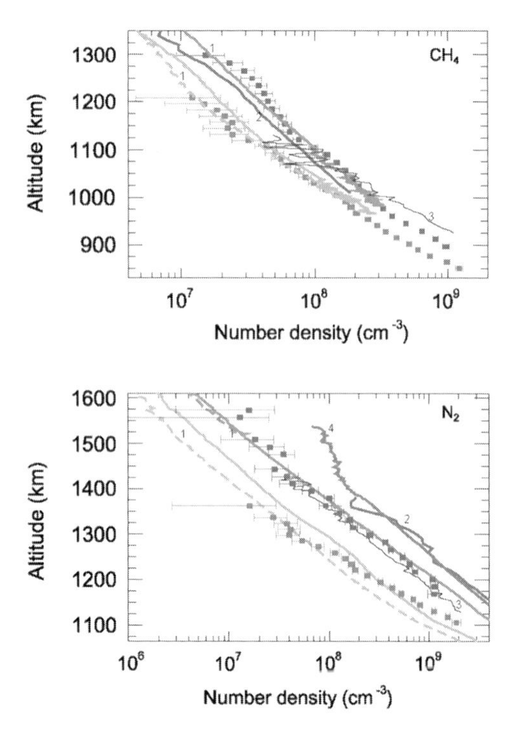


Figure 13.36 Number density profiles of $N_2$ and $CH_4$ from (1) two UVIS solar occultations are compared with the INMS data at the same flybys, (2) UVIS stellar occultation T21 (Kammer et al. 2013), (3) Voyager/UVS (Vervack et al. 2004), and (4) Huygens/HASI (Fulchignoni et al. 2005). From Capalbo et al. (2015). (According to Teolis et al. (2015), the INMS data should be scaled by a factor of 2.2.)

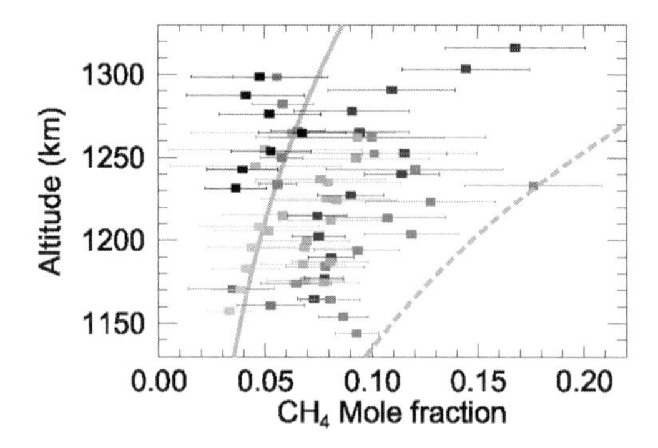


Figure 13.37 Methane mole fractions from all UVIS solar occultations. Two profiles are calculated by Cui et al. (2012) neglecting photochemical loss for the $CH_4$ escape rate of $3.8\times10^{27}$ s$^{-1}$ ($4.6\times10^{9}$ cm$^{-2}$ s$^{-1}$) and diffusive equilibrium at $K = 2\times10^{7}$ cm$^{2}$ s$^{-1}$. From Capalbo et al. (2015).

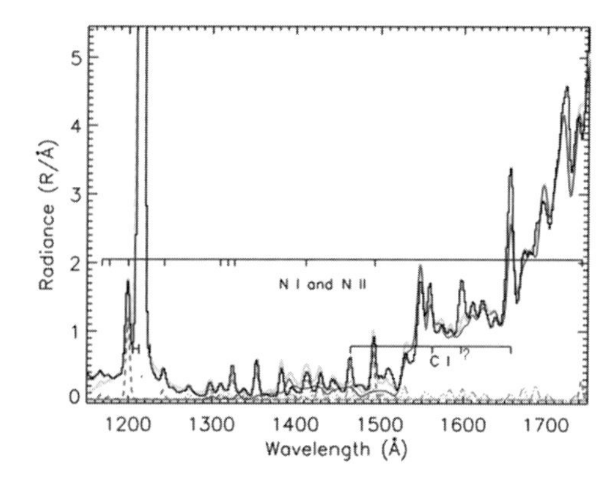


Figure 13.38 UVIS spectrum of Titan's dayside disk. The reflected sunlight is significant at 1350–1520 Å and dominate above 1520 Å. The LBH bands of $N_2$, N, and $N^+$ lines excited by photodissociative ionization of $N_2$ are present in the spectrum. H Lyman-alpha, three carbon lines, and an unidentified line at 1597 Å are shown as well. From Ajello et al. (2008).

### 13.6.2 Airglow

Ajello et al. (2007) analyzed observations of Titan's disk using the UVIS/EUV channel. The observations were conducted on December 13, 2004, at low solar activity. While the EUV channel covers 561–1182 Å, most of the data were collected at 900–1140 Å. Sixteen lines and bands are identified, and their intensities are tabulated. Those include three $N_2$ band systems, N and $N^+$ multiplets, and H Lyman-$\beta$ at 1026 Å. The total observed dayglow intensity at 900–1140 Å is 16.6 R. All $N_2$, N, and $N^+$ emissions are caused by photo- and photoelectron excitation, dissociation, ionization, and dissociative ionization. The dayglow is weaker than that observed by the Voyager/UVS at solar maximum.

The UVIS/FUV spectrum (Figure 13.38) observed simultaneously was analyzed by Ajello et al. (2008). Its main feature is a steep increase longward of 1500 Å caused by the reflected sunlight. This process is modeled using Rayleigh scattering by $N_2$ and $CH_4$, absorption by $C_2H_2$, $C_2H_4$, $C_4H_2$, and scattering and absorption by tholins. The adopted particle radius is 0.0125 μm with the refractive indices from Khare et al. (1984; see Figure 13.2).

The authors calculate single scattering albedo for a semi-infinite atmosphere and then the atmospheric reflectivity. They conclude that the bulk of the observed sunlight is scattered near 250 km. The DISR observations are sensitive to this region, and the DISR aerosol data (Section 13.4.3) look incompatible with the adopted radius. The best-fit mixing ratios are $2.4\times10^{-6}$ for $C_2H_2$, $5.7\times10^{-8}$ for $C_2H_4$, $1.2\times10^{-8}$ for $C_4H_2$, and $2.4\times10^{-11}$ for the tholin particles with $r$ = 0.0125 μm. The hydrocarbon abundances are in reasonable agreement with the CIRS data (within a factor of 2).The authors indicate that the chosen particle radius is not critical for the model, and a similar fit was obtained for $r$ = 25 Å. The observed geometric albedo is 0.017 at 1850 Å, in reasonable agreement with the HST observation by McGrath et al. (1998), who measured the albedo increasing from 0.02 at 1800 Å to 0.044 at 3300 Å.

The observed dayglow emissions are H Lyman-$\alpha$ (208 R), the LBH bands of $N_2$ (43 R total), the N multiplets (1200 Å at 6.8 R, 1493 Å at 3.0 R, 16 R total), C 1657 Å and

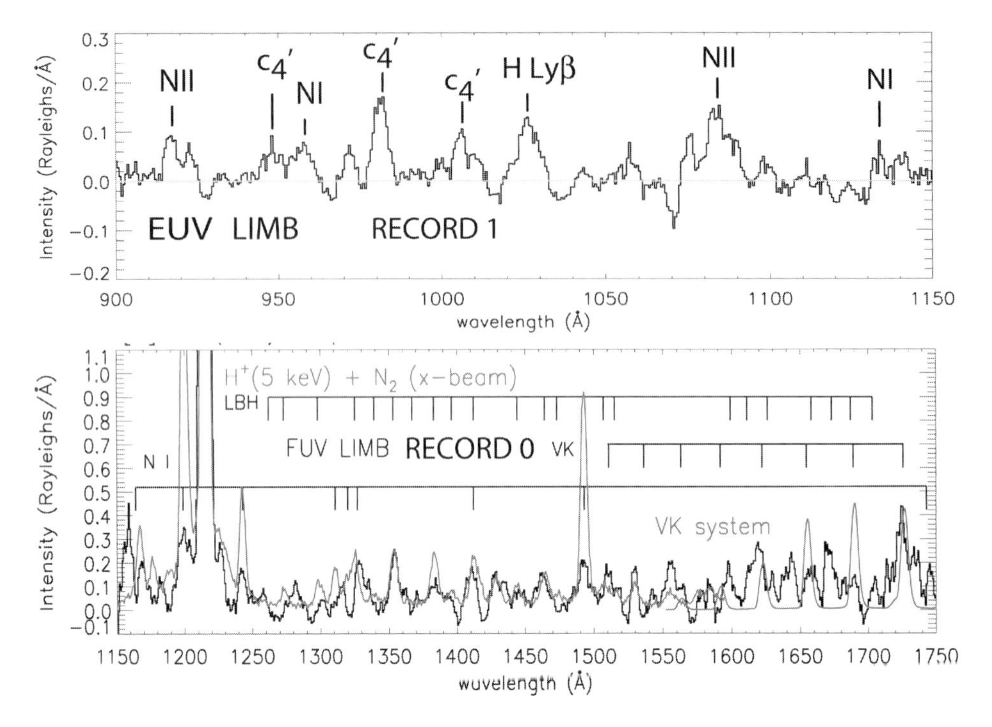


Figure 13.39 UVIS spectra of the dark limb. Positions of the LBH bands of $N_2$ and a laboratory spectrum excited by protons with energy of 5 keV are shown. Synthetic spectrum of the Vergard-Kaplan bands of $N_2$ is overplotted. From Ajello et al. (2012).

1561 Å, and an unidentified emission at 1597 Å. Below 1350 Å, the model includes absorption by $CH_4$ with a slant column abundance of $3\times10^{16}$ cm$^{-2}$ and places an upper limit of $3\times10^{14}$ cm$^{-2}$ to the the slant column of $C_2H_2$. The Lyman–Birge–Hopfield band system of $N_2$ ($a^1\Pi_g \rightarrow X^1\Sigma_g^+$) is parity-forbidden ($g \leftrightarrow u$ is allowed) and excited by electron or proton impacts.

Nighttime spectra on the dark limb are shown in Figure 13.39 (Ajello et al. 2012). Comparison with laboratory and synthetic spectra favors excitation by both magnetospheric protons and electrons. The Vegard–Kaplan $N_2$ ($A^3\Sigma_u^+ \rightarrow X^1\Sigma_g^+$) band system was observed at 1500–1750 Å. This system is spin-forbidden and excited by electron impacts. It is hidden by the much stronger reflected sunlight in the dayglow spectra (Figure 13.38).

Dayglow limb observations of the N 1493 Å and $N^+$ 1085 Å multiplets at 800–1300 km made it possible to retrieve vertical profiles of $N_2$ and $CH_4$ (Stevens et al. 2015). Both multiplets are excited by photofragmentation of $N_2$, and $N^+$ 1085 Å is absorbed by $CH_4$. Despite significant uncertainties, the data are in reasonable agreement with the corrected (Teolis et al. 2015) INMS observations.

Dependence of the dayglow intensity corrected for the solar activity on solar zenith angle may be fitted by a Chapman layer (Royer et al. 2016). This means that the UV and EUV dayglow on Titan is excited mostly by the solar photons, while magnetospheric precipitations dominate in the nightglow excitation.

## 13.7 Ion/Neutral Mass Spectrometer Measurements

### *13.7.1 Neutral Upper Atmosphere from INMS Measurements*

The Cassini flybys through the upper atmosphere at altitudes down to 880 km made it possible to directly study the thermospheric structure and chemical composition using INMS. Results from the measurements in the so-called closed source neutral mode for the nominal mission (2004–2008) were analyzed by two independent teams (Cui et al. 2009; Magee et al. 2009). This period was at low solar activity with a mean $F_{10.7} \approx 75$.

A sample of an atmospheric gas should be ionized to be studied by a mass spectrometer, and this is done in INMS by the electron beam with energy of 70 eV. These electrons ionize and sometimes double ionize the species and dissociate and excite them. The products react in the instrument. Collision rates are low in the upper atmospheres, and the instrument walls stimulate rather intense heterogeneous chemistry as well. All these effects change significantly the intrinsic atmospheric composition. Therefore complicated laboratory simulations and some modeling are required to analyze the INMS data on the neutral composition.

One of the measured mass spectra is shown in Figure 13.40. The count rate varies within the spectrum by a factor of ~$10^7$. A spectral feature at $m/z$ = 13–17 is depicted in Figure 13.41. It involves methane, its heavy isotope, and dissociation products $CH_3$, $CH_2$, and CH. $N_2$ dissociates by electron impact, and $^{14}N$ and $^{15}N$ contribute to the feature as well. Double ionized $C_2H_2$ and $C_2H_4$ also add to the feature. Analysis of the mass spectra should account for all these effects.

The latest analysis of the INMS sensitivity was made by Teolis et al. (2015). They concluded that all previously published INMS number densities of neutral species should be scaled by a factor of $2.2 \pm 0.5$, while those of ions by a factor of $1.55 \pm 0.33$. Evidently the measured mixing ratios remain unchanged under these conditions. The species mixing ratios retrieved by Cui et al. (2009) and Magee et al. (2009) are presented in Table 13.3 along with the GCMS measurements and the CIRS data at low latitudes from the nominal mission.

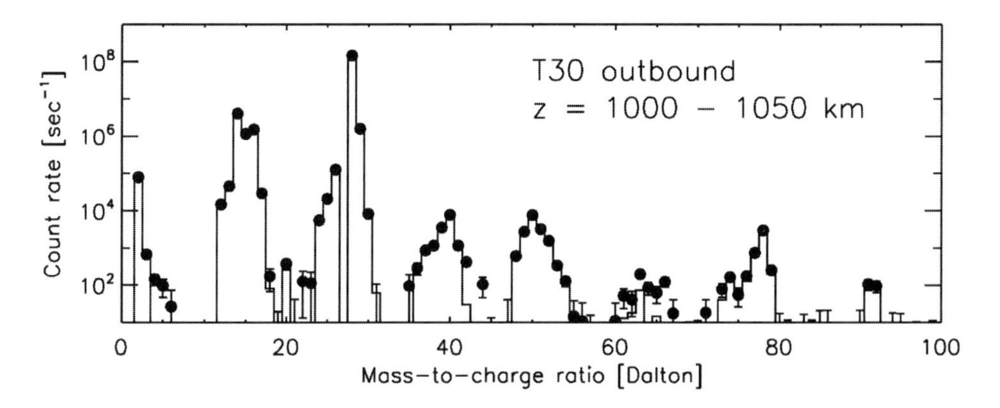


Figure 13.40 INMS mass spectrum measured at 1000–1050 km on the nightside (Cui et al. 2009).

Table 13.3 *Cassini/Huygens (GCMS, CIRS, UVIS, and INMS) data on Titan's chemical composition (mixing ratios) at low and middle latitudes*

| *h* (km) | 0[a] | 100–200[b] | 150[c] | 300[c] | 400[c] | 550[h] | 700[h] | 900[h] | 1050[d] | 1077[e] | 1100[f] |
|---|---|---|---|---|---|---|---|---|---|---|---|
| CH4 | 0.057 | 0.015[a] | – | – | – | – | – | – | 0.022 | 0.023 | – |
| Ar | 3.4–5[g] | – | – | – | – | – | – | – | 1.3–5 | 1.1–5 | – |
| $H_2$ | 1.0–3 | 9.6–4[i] | – | – | – | – | – | – | 3.4–3 | 3.9–3 | – |
| $C_2H_2$ | – | 3–6 | 3–6 | 3–6 | 2–6 | 4–5 | 4–5 | 4–4 | 3.4–4 | 2.3–4 | – |
| $C_2H_4$ | – | 1.3–7 | 1.5–7 | 3–8 | – | 5–6 | 5–6 | 4–4 | 3.9–4 | 7.5–5 | 1–3 |
| $C_2H_6$ | – | 8–6 | 1–5 | 1.4–5 | 1.7–5 | – | – | – | 4.6–5 | 7.3–5 | – |
| $C_3H_4$ | – | 5–9 | 1–8 | 1–8 | – | – | – | – | 9.2–6 | 1.4–4 | – |
| $C_3H_6$ | – | 2.6–9[j] | – | – | – | – | – | – | 2.3–6 | – | – |
| $C_3H_8$ | – | 5–7 | 1.3–6 | 1–6 | 5–7 | – | – | – | 2.9–6 | <5–5 | – |
| $C_4H_2$ | – | 1.3–9 | 2–9 | 6–9 | 8–9 | 5–7 | 1–6 | 6–5 | 5.6–6 | 6.4–5 | 1–5 |
| $C_6H_2$ | – | – | – | – | – | – | – | – | – | – | 8–7 |
| $C_6H_6$ | – | 2–10 | – | – | – | 3–8 | 6–7 | 2–5 | 2.5–6 | 9–7 | 3–6 |
| $C_7H_4$ | – | – | – | – | – | – | – | – | – | – | 3–7 |
| $C_7H_8$ | – | – | – | – | – | – | – | – | 2.5–8 | <1.3–7 | 2–7 |
| $NH_3$ | – | *<2–10*[n] | – | – | – | – | – | – | – | 3–5 | 7–6 |
| $CH_2NH$ | – | – | – | – | – | – | – | – | – | – | 1–5 |
| HCN | – | 1–7 | 2.5–7 | 5–7 | 3–7 | 2–5 | 4–6 | 6–4 | 2.5–4 | – | 2–4 |
| $CH_3CN$ | – | – | *1–9*[o] | *1.5–8*[o] | *3–8*[o] | – | – | – | – | 3.1–5 | 3–6 |
| $C_2H_3CN$ | – | – | – | – | – | – | – | – | 3.5–7 | <1.8–5 | 1–5 |
| $C_2H_5CN$ | – | *<1–10*[p] | – | – | – | – | – | – | 1.5–7 | – | 5–7 |
| $HC_3N$ | – | 4–10 | 1–11 | 1–9 | – | 1–6 | 1–6 | 3–5 | 1.5–6 | 3.2–5 | 4–5 |
| $C_4H_3N$ | – | – | – | – | – | – | – | – | – | – | 4–6 |
| $HC_5N$ | – | – | – | – | – | – | – | – | – | – | 1–6 |
| $C_5H_5N$ | – | – | – | – | – | – | – | – | – | – | 4–7 |
| $C_6H_3N$ | – | – | – | – | – | – | – | – | – | – | 3–7 |

| | | | | | | | | | | | |
|---|---|---|---|---|---|---|---|---|---|---|---|
| $C_6H_7N$ | – | – | – | – | – | – | – | – | – | – | 1–7 |
| $C_2N_2$ | – | 5.5–11[k] | – | – | – | – | – | – | 2.1–6 | 4.8–5 | – |
| $C_4N_2$ | – | <1.5–9[l] | – | – | – | – | – | – | – | – | – |
| $H_2O$ | – | – | 2–10[m] | – | – | – | – | – | – | – | <3–7 |
| CO | – | 4.5–5 | – | – | – | – | – | – | – | – | – |
| $CO_2$ | – | 1.2–8 | 2–8 | 1.8–8 | 3–8 | – | – | – | – | – | – |

*Note.* Other observations are in italics.

[a]GCMS (Niemann et al., 2010). [b]CIRS nadir (Coustenis et al., 2010). [c]CIRS limb (Vinatier et al., 2010). [d]INMS (Magee et al., 2009). [e]INMS (Cui et al., 2009, tables 3 and 4). [f]Derived from INMS ion spectra (Vuitton et al., 2007). [g]3.4–5 = $3.4\times10^{-5}$. [h]UVIS stellar occultations (Koskinen et al. 2011). [i]CIRS (Courtin et al. 2012). [j]CIRS (Nixon et al. 2013). [k]CIRS (Teanby et al. 2009). [l]CIRS (Jolly et al. 2015). [m]CIRS (Cottini et al. 2012). [n]Herschel (Teanby et al. 2013). [o]IRAM (Marten et al. 2002). [p]ISO (Coustenis et al. 2003).

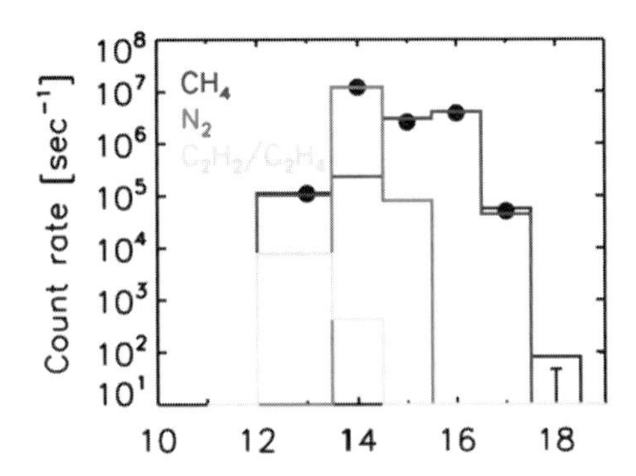


Figure 13.41 Fragment of the mass spectrum at $m/z$ = 10 to 18 that includes $CH_4$ and its derivatives ($^{13}CH_4$, $CH_4$, $CH_3$, $CH_2$, CH), those of $N_2$ ($^{15}N$ and $^{14}N$), and double ionized $C_2H_4$ and $C_2H_2$. From Cui et al. (2009).

### 13.7.2 Escape of $CH_4$ and $H_2$

Mean density profiles of $N_2$, $CH_4$, $H_2$, and $^{40}Ar$ are shown in Figure 13.42. The inbound and outbound densities are almost equal. Therefore these species are weakly affected by heterogeneous chemistry inside the instrument. The INMS observations of these species were analyzed in a few papers (see Cuie et al. 2012 and references therein) to get escape rates of methane and molecular hydrogen and establish a type of the escape (thermal, nonthermal, hydrodynamic; see Section 2.5).

Cui et al. (2012) compared the INMS measurements of $^{40}Ar$ above 1000 km with those by GCMS below 140 km using a constant eddy diffusion coefficient above 600 km as a fitting parameter. This results in $K = 2\times10^7$ cm$^2$ s$^{-1}$. With this eddy diffusion and the $CH_4$ mixing ratio of 1.48% observed by GCMS, Cui et al. (2012) calculated profiles of methane in the upper atmosphere for diffusive equilibrium (no escape) and for escape rate as a parameter to fit the INMS measurements of methane. The required escape of methane is $3.8\times10^{27}$ s$^{-1}$, that is, $4.6\times10^9$ cm$^{-2}$ s$^{-1}$. The model does not account for the chemical loss of methane that is comparable to the claimed escape value above ~1000 km. Both methane profiles are shown in Figure 13.37.

We will consider here how the results of Cui et al. (2012) would change from the scaling of all INMS neutral densities by the factor of 2.2 (Teolis et al. 2015). The profile of the $^{40}Ar$ mixing ratio (Cui et al. 2012, figure 5) remains unchanged as well as the argon homopause. The diffusion coefficient is $D_i = b_i/n$ (2.13); its value at the homopause is smaller by the factor of 2.2 because of the higher total density $n$, and $K = 2\times10^7/2.2 = 9\times10^6$ cm$^2$ s$^{-1}$. The profile of the $CH_4$ mixing ratio for diffusive equilibrium (Figure 13.37; Cui et al. 2012, figure 6) also does not change, because the methane homopause is at the same altitude. The eddy component of flux (2.16) is

$$\Phi_i = -Kn\frac{\partial f_i}{\partial z}.$$

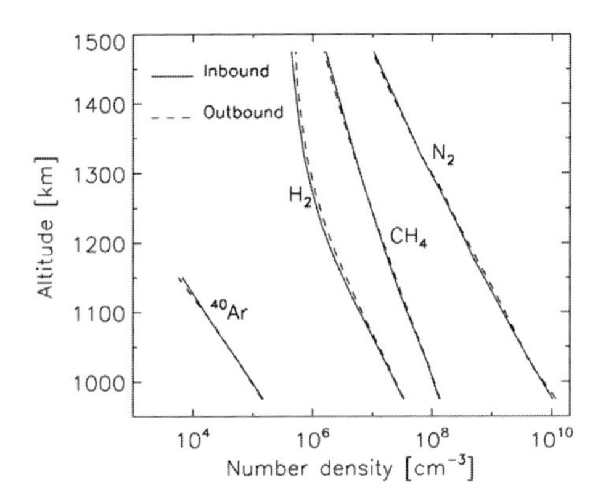


Figure 13.42 Density profiles of the major components and $^{40}Ar$ averaged over 15 flybys. The inbound and outbound profiles are identical, indicating that the wall effects are negligible for these species. From Cui et al. (2009). (According to Teolis et al. (2015), the data should be scaled by a factor of 2.2.)

*Kn* does not change, and the eddy flux is the same. The molecular diffusion component of flux (2.12a) is

$$\Phi_i = -b_i\left[\frac{\partial f_i}{\partial z} + f_i\left(\frac{1}{H_i} - \frac{1}{H} - \frac{\alpha_i}{T}\frac{\partial T}{\partial z}\right)\right]$$

and is not affected by the scaling factor as well. Therefore we conclude that the analyses in Cui et al. (2012) and in the other papers on this subject are not sensitive to the scaling factor to the INMS number densities.

The high escape rate of methane in Cui et al. (2012) and the preceding papers favors hydrodynamic escape of methane from Titan. However, a similar analysis by Bell et al (2014), which involved some chemistry and chemical losses of methane, resulted in much smaller escape of $CH_4$, $1.4\times10^7$ cm$^{-2}$ s$^{-1}$, and $H_2$, $1.1\times10^{10}$ cm$^{-2}$ s$^{-1}$. The value for $H_2$ agrees with thermal escape in photochemical models (Krasnopolsky 2014b), and the value for $CH_4$ requires a rather moderate nonthermal escape.

Direct Monte Carlo simulations, which are equivalent to solving the Boltzmann kinetic equation, result in no hydrodynamic escape from Titan (Johnson et al. 2009). Johnson et al. (2009) argued that thermal conduction in the equations for hydrodynamic escape ceases at the exobase, where the atmosphere becomes collisionless, and the equations become invalid. Carbon ions are scarce in the Saturn magnetospheric plasma, and this does not favor hydrodynamic escape of $CH_4$ as well. Krasnopolsky (2010e) applied tests for existence of exobase and static thermosphere to Titan that gave positive results and ruled out hydrodynamic escape from Titan.

The photochemical loss of methane in the thermosphere is greater at solar maximum, and its thermospheric abundances anticorrelate with solar activity (Westlake et al. 2014b). This anticorrelation is revealed by comparison of the INMS data in the periods of the minimum and maximum in the 11-yr solar cycle.

### *13.7.3 Nighttime Ion Mass Spectra*

The most interesting results were obtained during flyby T5 in April 2005 at the closest approach of 1027 km with local time of 23:15 and latitude of 74°N. That flyby coincided with a strong magnetospheric precipitation, and the ion nighttime densities were comparatively high. The observed mass spectra between 1027 and 1200 km were summed to give an average spectrum that may be referred to 1100 km (Figure 13.43). Ions of 63 different masses are present in the spectrum, and the ion densities as low as 0.1 $cm^{-3}$ are measurable. That was a great progress relative to Voyager 1, where direct measurements in the ionosphere were impossible.

Vuitton et al. (2007) developed a model to fit the observed spectrum. The model adopted photochemical equilibrium neglecting transport of ions, and this is a reasonable approximation at 1100 km. The best-fit ionization rate was just 1 $cm^{-3}$ $s^{-1}$, and only $N_2$ and $CH_4$ were initially ionized, while other ions were formed as products of ion-neutral reactions. The total number of reactions was 1250. The most abundant neutral species were adopted in the model from the INMS observations, and abundances of about 20 minor species were fitting parameters of the model. Therefore the model had a few goals: to give a reliable identification to the observed mass spectrum, to reveal main processes of the ion formation and loss, and to improve the data on the chemical composition of the upper atmosphere.

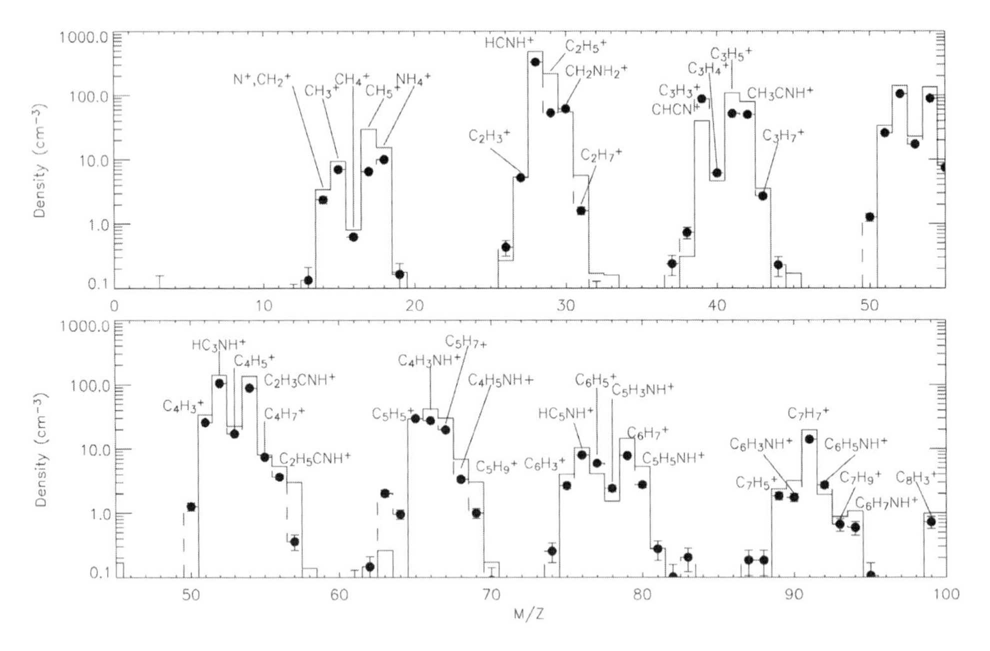


Figure 13.43 The INMS mass spectrum (dots) of the nighttime ionosphere near 1100 km during a strong precipitation of magnetospheric electrons in flyby T5. Model fitting to the observed spectrum is shown by histogram with identification of major ions. From Vuitton et al. (2007).

The model results are compared with the observed mass spectrum in Figure 13.43, where major ions for each mass are identified. Best-fit abundances of neutral species are given in Table 13.3. While these data are not directly measured, they originate from the measured spectrum and present a synthesis of the observation and modeling.

The scaling factors of 1.55 and 2.2 for ions and neutrals (Teolis et al. 2015) affect the results of the model. For example, the ionization rate would probably change to $1.55^2 = 2.4\ cm^{-3}\ s^{-1}$. However, if production of an ion is greater by a factor of 2.4 and density of a neutral that reacts with this ion is greater by a factor of 2.2, then the ion density is almost unchanged.

## 13.8 Ionosphere

### *13.8.1 Radio Occultations and RPWS Data*

Thirteen profiles of electron density were measured by the Cassini radio occultations (Kliore et al. 2008, 2011). Seven of those were observed at low latitudes, and the other profiles at subpolar regions. Mean dusk and dawn profiles are shown in Figure 13.44. The difference between the profiles is partly caused by different solar zenith angles that are equal to 87° and 95°, respectively. The profiles are compared with that from the Voyager radio occultation. The considerable electron densities below 800 km cannot be explained by the solar EUV ionization and require other sources of ionization.

Three observed profiles show electron densities that significantly exceed those in the other profiles. The measured peak electron densities are compared with those observed by the RPWS Langmuir Probe data in Figure 13.45a. The peak densities vary from $\approx 3000\ cm^{-3}$ in the subsolar region to $\approx 700\ cm^{-3}$ at night.

The electron peak altitude as a function of solar zenith angle from the RPWS observations (Agren et al. 2009) is shown in Figure 13.45b. The peak altitude increases from 1050 km near noon to 1120 km near terminator and 1170 km on the nightside. The measured electron peak temperature varies from 250 to 700 K with a mean value of 500 K. However, it does not show any dependence on the solar zenith angle.

The RPWS observations during flyby T18 are depicted in Figure 13.46. The closest approach for this flyby was near the terminator at SZA = 88°, and the inbound and outbound electron density profiles refer to the nightside and dayside, respectively.

### *13.8.2 Magnetospheric Precipitations and Other Ionization Processes*

Water ice from the rings of Saturn is released into Saturn's magnetosphere, where protons and $O^+$ are the most abundant ions. The Voyager PLS measurements and the Cassini LEMMS and MIMI data from the T5 flyby give energetic spectra of protons

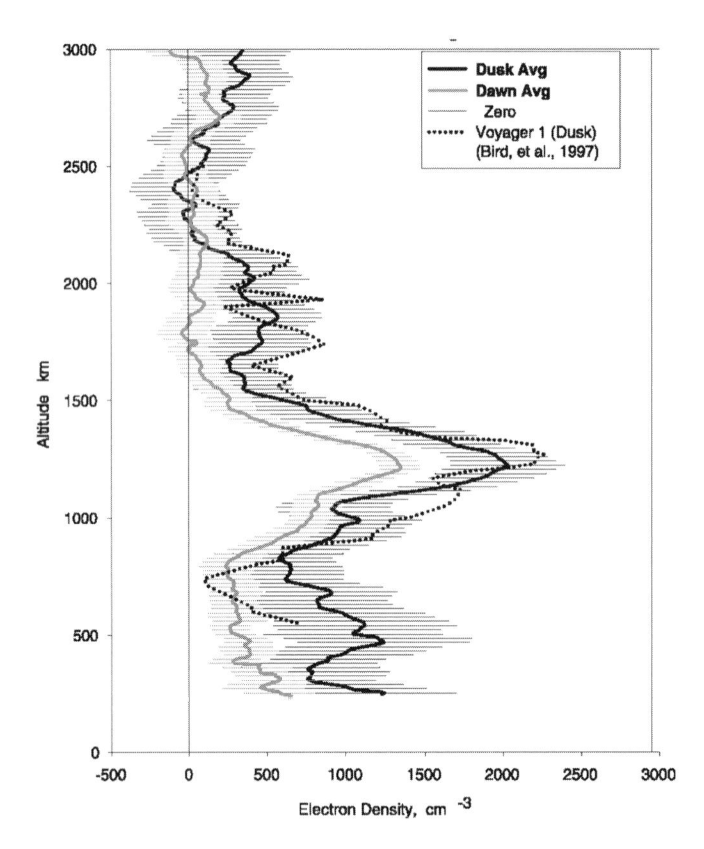


Figure 13.44 Dusk and dawn average electron density profiles observed by the Cassini radio occultations. Mean solar zenith angles are 87° and 95°, respectively, and the observations refer to low latitudes. The Voyager radio occultation is shown by dotted line. From Kliore et al. (2008).

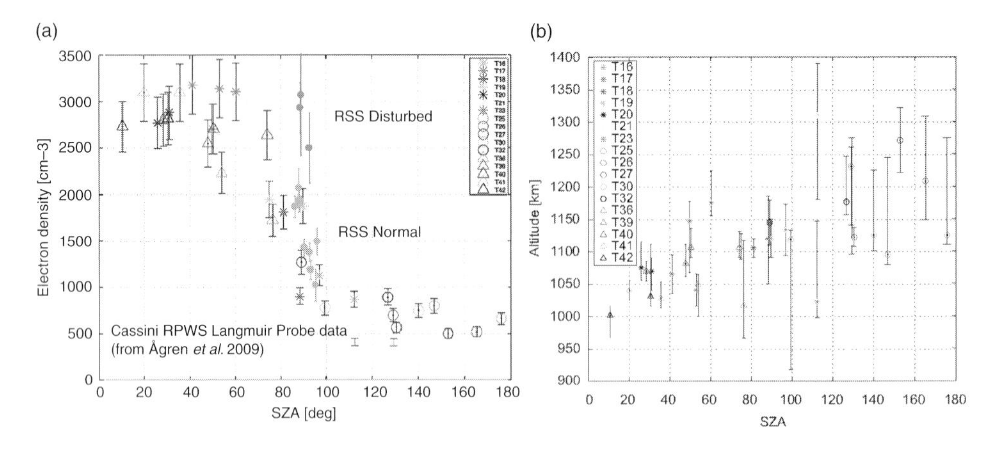


Figure 13.45 (a) Peak electron density versus solar zenith angle from the RPWS Langmuir Probe data (Agren et al. 2009) and radio occultations (filled circles, RSS, Kliore et al. (2011)). Three upper circles refer to disturbed conditions. (b) Electron peak altitude as a function of solar zenith angle (Agren et al. 2009).

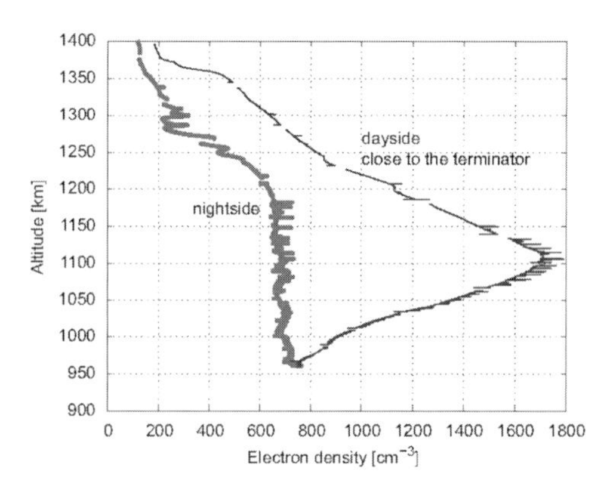


Figure 13.46 Electron density profiles from the inbound and outbound orbits of flyby T18. The closest approach was at SZA = 88°. From Agren et al. (2009).

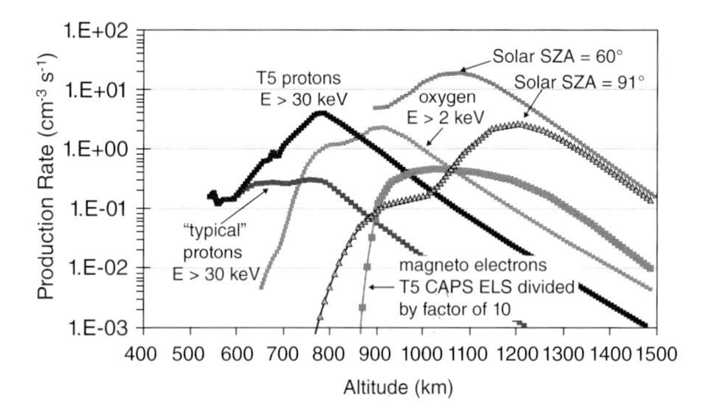


Figure 13.47 Ionization rates by magnetospheric electrons, protons, and oxygen ions at the strong precipitation event during flyby T5 and typical values compared to those by the solar EUV. From Cravens et al. (2008). (A black and white version of this figure will appear in some formats. For the color version, please refer to the plate section.)

and oxygen ions up to 4 MeV (Cravens et al. 2008). This makes it possible to calculate vertical profiles of ionization rate for these sources of ionization. The data from Cravens et al. (2008) for the strong precipitation event T5 are shown in Figure 13.47. Ionization by the solar EUV and magnetospheric electrons extends down to 800 km, while those by oxygen ions and protons can reach 700 and 500 km, respectively. The ionization rates at typical conditions are recommended to be reduced by an order of magnitude relative to the T5 values.

Prior to the Cassini/Huygens mission, Molina-Cuberos et al. (2001) calculated ionization of Titan's atmosphere by the meteor ablation. Atomic ions are formed in this process, and $Si^+$, $Fe^+$, and $Mg^+$ are the most abundant. Their radiative recombination is slow, and termolecular association

$$Mg^+ + N_2 + N_2 \rightarrow Mg^+N_2 + N_2,\ k \approx 10^{-30}\ cm^6\ s^{-1}$$

with subsequent dissociative recombination of $Mg^+N_2$ is more effective. Their model that involves transport by eddy diffusion predicts a layer with a peak ion/electron density of $10^4$ $cm^{-3}$ at 750 km. The layer width is ≈100 km at half maximum. The predicted electron density significantly exceeds those observed (Figure 13.44). Termolecular associations may be faster than adopted in the model, and recombination between atomic ions and negative ions was ignored in the model and may significantly reduce the ion densities. On the other hand, ionization potentials of Si, Fe, and Mg are ≈8 eV, the solar photons $\lambda \leq$ 150 nm reach the altitudes near 750 km and significantly increase the production of the meteoric ions. (This process was ignored in the model.) Therefore the problem remains uncertain.

Ionization rate by the galactic cosmic rays (Molina-Cuberos et al. 1999) is comparable with that of the solar EUV photons but peaks in the deep stratosphere at 70 km (Figure 13.48) and steeply decreases above the peak being proportional to the atmospheric density. Using this mechanism and the photochemical model by Lara et al. (1996), Molina-Cuberos et al. (1999) calculated ion composition of the atmosphere below 400 km. Electron density peaks at 2000 $cm^{-3}$ near 90 km, and the most abundant ion is $C_7H_X{}^+$; x is not specified.

Measurements by the Huygens/HASI mutual impedance and relaxation probes (Section 13.4.4) are converted into number densities of positive ions and electrons (Figure 13.49). The total number density of positive ions is similar to that calculated by Molina-Cuberos et al. (1999) at 70–80 km but decreases with altitude steeper than that calculated. Mishra et al. (2014) calculated ionization by the middle UV photons of the haze particles, very small embryos of ≈7 Å, and photodetachment of negative ions (Figure 13.48).

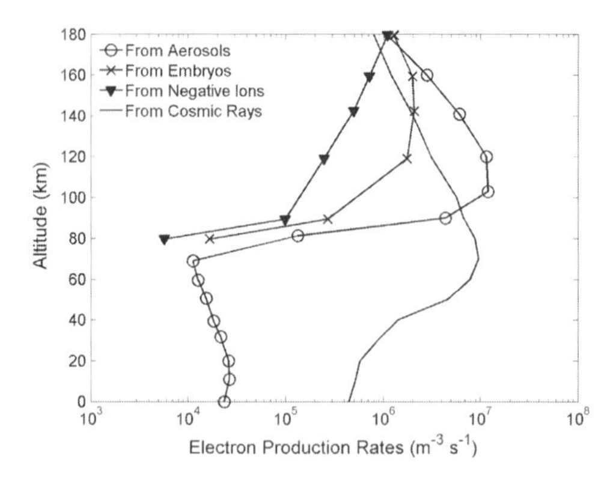


Figure 13.48 Ionization of the lower atmosphere by the galactic cosmic rays (Molina-Cuberos et al. 1999) and photoionization of the haze particles, small embryos, and photodetachment of negative ions (Mishra et al. 2014).

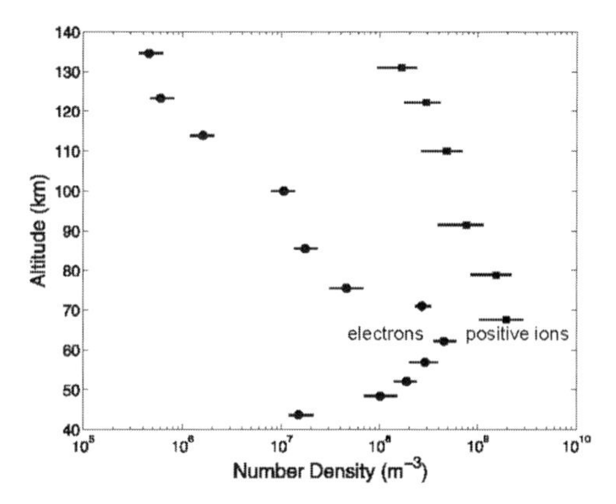


Figure 13.49 Number density profiles of electrons and positive ions from the Huygens/HASI mutual impedance and relaxation probes (Lopez-Moreno et al. 2008).

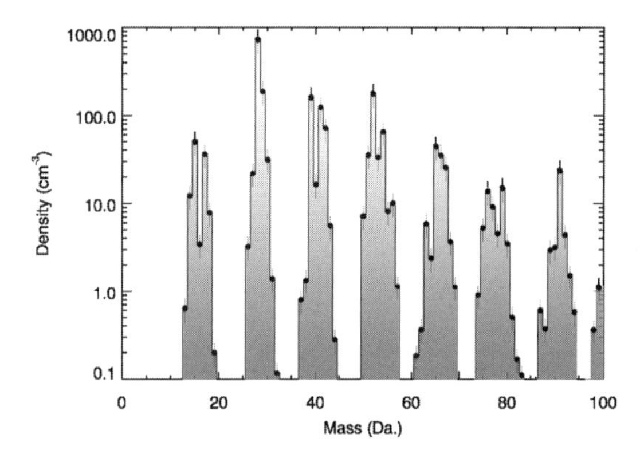


Figure 13.50 The ion mass spectrum measured using INMS during flyby T40 at 1080 km and SZA $\approx 40^\circ$ (Westlake et al. 2012). The data are not scaled by the factor of 1.55 (Teolis et al. 2015).

### *13.8.3 Composition of the Daytime Ionosphere*

The INMS spectrum at 1080 km and SZA $\approx 40^\circ$ is shown in Figure 13.50. The spectrum resolution is reduced to 1 mass unit. The spectrum consists of eight broad features with the number of heavy atoms C and N increasing from 1 to 8. The difference between the masses of C and N and different numbers of hydrogen atoms determine structure of each feature.

Vertical profiles of number densities for some ion masses may be found in Westlake et al. (2012) for the data from the INMS measurements in flyby T40 in January 2008. The closest approach was at 1010 km, 12°S, SZA = 38°, local solar time 13:00. Mandt et al. (2012b) averaged the INMS ion measurements that referred to the dayside, morning and evening terminators, and nightside with some corrections for the data of RPWS Langmuir

Table 13.4 *Dayside and nightside mean ion densities at 1100 and 1300 km from INMS data*

| | | Day | | Night | |
|---|---|---|---|---|---|
| Mass | Species | 1100 km | 1300 km | 1100 km | 1300 km |
| 28 | $HCNH^+$, $N_2^+$, $C_2H_4^+$ | 824 ± 173 | 392 ± 148 | 164 ± 106 | 173 ± 128 |
| 29 | $C_2H_5^+$, $N_2H^+$, $H_3CN^+$ | 206 ± 144 | 176 ± 55 | 27 ± 20 | 71 ± 44 |
| 39 | $C_3H_3^+$, $CHCN^+$ | 100 ± 20 | 18 ± 11 | 34 ± 15 | 8.5 ± 8.5 |
| 41 | $C_3H_5^+$, $CH_3CN^+$ | 108 ± 18 | 32 ± 16 | 20 ± 12 | 18 ± 15 |
| 15 | $CH_3^+$, $NH^+$ | 39 ± 7 | 23 ± 10 | – | – |
| 17 | $CH_5^+$, $NH_3^+$ | 36 ± 7 | 54 ± 10 | – | – |
| 27 | $C_2H_3^+$, $HCN^+$ | 16 ± 3 | 8 ± 5 | – | – |
| 30 | $C_2H_6^+$, $NO^+$, $CH_3NH^+$, $CH_2NH_2^+$ | 30 ± 5 | 9.4 ± 5.5 | 22 ± 9 | 6 ± 5 |
| 40 | $CH_2CN^+$, $C_3H_4^+$ | 12 ± 2 | 3.2 ± 1.9 | – | – |
| 42 | $CH_3CNH^+$, $C_3H_6^+$ | 51 ± 11 | 8.6 ± 4.6 | 19 ± 9 | 5.2 ± 5.3 |
| 51 | $C_4H_3^+$, $HC_3N^+$ | 23 ± 6 | 1.3 ± 0.8 | 6.8 ± 5.9 | 1.7 ± 2.7 |
| 52 | $HC_3NH^+$, $C_4H_4^+$, $C_2N_2^+$ | 96 ± 33 | 3.1 ± 2.5 | 34 ± 24 | 4.2 ± 6.8 |
| 53 | $C_4H_5^+$, $HC_2N_2^+$, $C_2H_3CN^+$ | 24 ± 4 | 4.8 ± 2.4 | 6.5 ± 4 | 3.7 ± 3.5 |
| 54 | $C_2H_3CNH^+$, $C_4H_6^+$ | 48 ± 12 | 2.5 ± 2.0 | 27 ± 21 | 2.9 ± 4.5 |
| 65 | $C_5H_5^+$, $C_4H_3N^+$ | 27 ± 7 | 1.0 ± 0.6 | 11 ± 7 | 1.9 ± 2.7 |
| 66 | $C_4H_3NH^+$ | 19 ± 7 | 0.28 ± 0.22 | 9.4 ± 7.5 | 0.56 ± 0.98 |
| 67 | $C_5H_7^+$, $C_4H_5N^+$ | 15 ± 4 | 0.59 ± 0.36 | 7.6 ± 5.1 | 0.9 ± 1.3 |
| 91 | $C_7H_7^+$ | 12 ± 4 | 0.27 ± 0.22 | 5.3 ± 1.7 | 0.11 ± 0.12 |
| 14 | $N^+$, $CH_2^+$ | 7.6 ± 2.1 | 3.5 ± 1.8 | – | – |
| 18 | $NH_4^+$ | 7.1 ± 0.9 | 3.0 ± 0.9 | – | – |
| 55 | $C_4H_7^+$ | 5.9 ± 1.4 | 0.71 ± 0.45 | – | – |
| 76 | $HC_5NH^+$, $C_6H_4^+$ | 6.2 ± 3.6 | 0.04 ± 0.04 | – | – |
| 79 | $C_6H_7$+, $C_5H_5N$+ | 6.6 ± 2.8 | 0.09 ± 0.08 | – | – |
| Total | | 1720 | 746 | 393 | 297 |

*Note.* Number densities are given in $cm^{-3}$. From Mandt et al. (2012b). According to Teolis et al. (2015), the data should be scaled by the factor of 1.55.

Probe (Figure 13.45) and CAPS Ion Beam Spectrometer. The dayside and nightside mean data from Mandt et al. (2012b) are given in Table 13.4. These data may be used for comparison with models of Titan's ionosphere.

Scaling the total ion densities at 1100 km by the factor of 1.55 (Teolis et al. 2015), the mean dayside and nightside ion densities at 1100 km are 2670 $cm^{-3}$ and 610 $cm^{-3}$, respectively. These values are in excellent agreement with the RPWS Langmuir Probe data (Figure 13.45).

INMS does not cover ions with mass exceeding 100, while CAPS IBS is extended to very large ion masses with a lower spectral resolution. The INMS and CAPS IBS ion mass spectra are compared in Figure 13.51. The measurements were made at 955 km on the

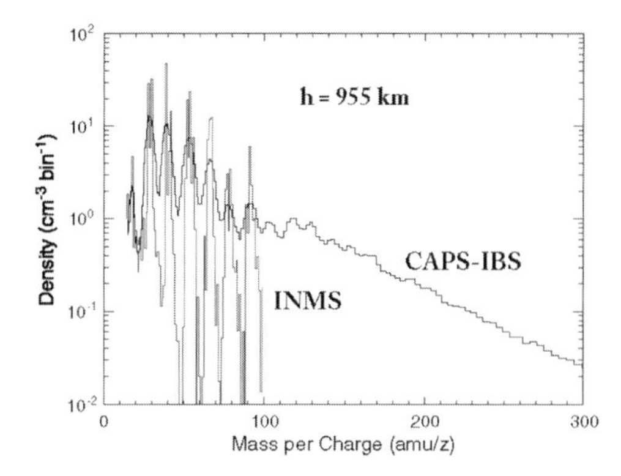


Figure 13.51 Comparison of the CAPS-IBS and INMS nighttime ion mass spectra at 955 km, flyby T57 (Westlake et al. 2014a).

nightside. Heavy ions decrease steeply with altitude because of diffusive separation of heavy neutrals they originate from.

Edberg et al. (2013) analyzed peak electron densities measured by the RPWS LP for SZA $< 120°$. They concluded that the observed peak electron densities are proportional to the solar EUV at 1 to 80 nm with a power index of $0.54 \pm 0.18$, in accord with the index of 0.5 for the Chapman layer.

### *13.8.4 Negative Ions*

Discovery of the abundant negative ions of large mass is among unexpected results of the Cassini mission. The negative ions are detected using the RPWS Langmuir Probe (Shebanits et al. 2013) and the CAPS electron spectrometer (Coates et al. 2009). A summary of the RPWS LP measurements on the dayside and nightside and near the terminators is shown in Figure 13.52. The negative ion densities are $\approx 1000$ cm$^{-3}$ near 950 km decreasing with altitude exponentially on the nightside and gradually on the dayside. While the expected behavior near the terminators is intermediate between those on the dayside and nightside, negative ions are lacking near the terminators above 1000 km, steeply increasing to $\approx 1000$ cm$^{-3}$ near 900 km.

The CAPS ELS measurements of negative ions were presented by Coates et al. (2009) as maximum ion mass as function of altitude, latitude, and SZA. For example, the maximum ion mass is smaller than 300 above 1200 km increasing to $\approx 10^4$ near 950 km. This correlates with the VIMS detection of polycyclic aromatic hydrocarbons (PAH) with a mean mass of 430 (Section 13.5.2; Dinelli et al. 2013; Lopez-Puertas et al. 2013). Further coagulation of these large molecules may form aerosol particles with masses $\approx 10^4$.

A mass spectrum of negative ions measured by the CAPS ELS at 1015 km and SZA $\approx$ 40° is shown in Figure 13.53. The instrument resolving power is 6, and three features at $m = 22 \pm 4$, $44 \pm 8$, and $82 \pm 14$ are identified along with a broad distribution with a maximum at $m \approx 500$, close to the mean PAH mass of 430.

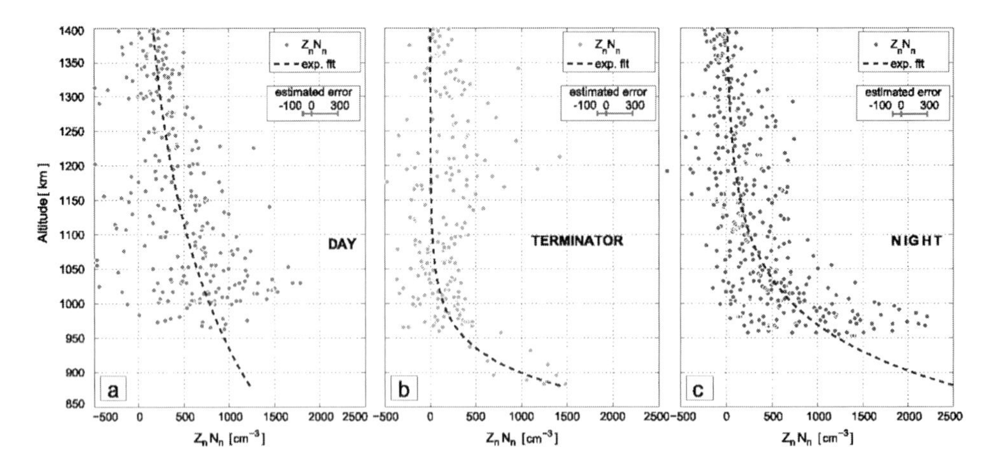


Figure 13.52 Cassini RPWS LP observations of negative ion densities (Shebanits et al. 2013). $Z_n$ and $N_n$ are the charge and number density of negative ions.

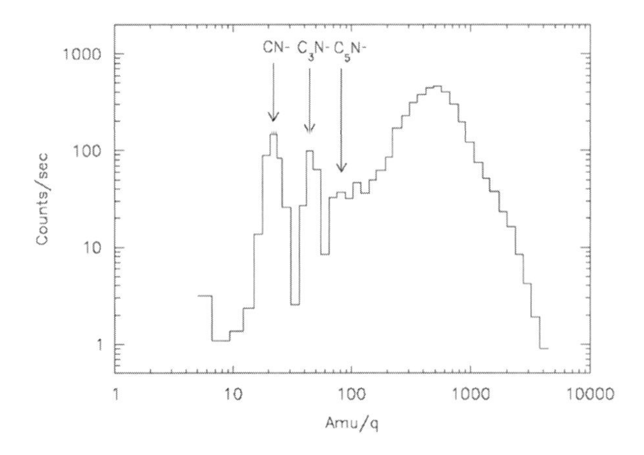


Figure 13.53 CAPS ELS negative ion mass spectrum at 1015 km during flyby T40, SZA $\approx 40°$ (Vuitton et al. 2009).

Vuitton et al. (2009) developed a model for negative ions on Titan. Electron affinity and gas phase acidity of the atmospheric species are important input data of the model. Both values are the bond energies between species and electron in a negative ion and between species and proton, that is, in $AH \rightarrow A^- + H^+$. The data from Vuitton et al. (2009) are given in Table 13.5.

The model involved 19 neutrals and 11 anions. Neutral densities were based on the INMS observations in both neutral and positive ion modes (Section 13.7.3).

The excess of energy in electron attachment to species with electron affinity is either radiated (if the molecule is simple) or relaxed in vibrational and rotational excitation (if the molecule is complex and have numerous vibrational modes). Rate coefficient of the process varies therefore from $3\times10^{-16}$ cm$^3$ s$^{-1}$ for H to $6\times10^{-8}$ cm$^3$ s$^{-1}$ for $C_6H$. Electron

Table 13.5 *Electron affinity and gas phase acidity of some molecules (in eV)*

| Species | H | $CH_2$ | $CH_3$ | $C_2H$ | $C_4H$ | $C_6H$ | $C_8H$ | $C_6H_5$ | $CH_2C_6H_5$ | CN | $CH_2CN$ | $C_3N$ | $C_5N$ | $NH_2$ | $CH_2N$ | O | OH |
|---|---|---|---|---|---|---|---|---|---|---|---|---|---|---|---|---|---|
| Affinity | 0.75 | 0.65 | 0.08 | 3.0 | 3.6 | 3.8 | 4.0 | 1.1 | 0.9 | 3.8 | 1.5 | 4.6 | 4.5 | 0.8 | 0.5 | 1.5 | 1.8 |
| Species | $H_2$ | $CH_4$ | $C_2H_2$ | $C_4H_2$ | $C_6H_2$ | $C_8H_2$ | $C_6H_6$ | $CH_3C_6H_5$ | HCN | | $CH_3CN$ | | $HC_3N$ | $NH_3$ | $CH_2NH$ | $H_2O$ | |
| Acidity | 17.1 | 17.7 | 16.1 | 15.3 | 15.2 | 15.1 | 17.1 | 16.2 | 14.9 | | 15.9 | | 14.9 | 17.2 | 16.5 | 16.7 | |

*Note.* From Vuitton et al. (2009).

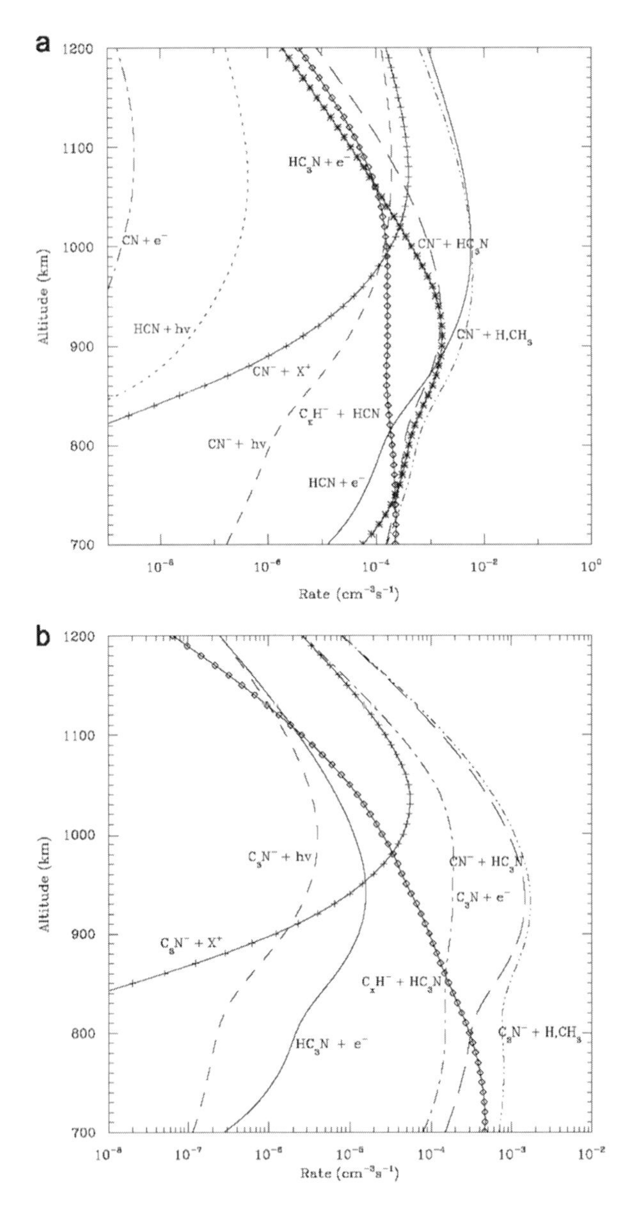


Figure 13.54 Production and loss of $CN^-$ (a) and $C_3N^-$ (b; Vuitton et al. 2009).

attachment is negligible in the production of $CN^-$ and the main process of the $C_3N^-$ formation (Figure 13.54).

Dissociation energy usually exceeds electron affinity, and dissociative electron attachment requires suprathermal electrons with energies of a few eV. Cross sections of this

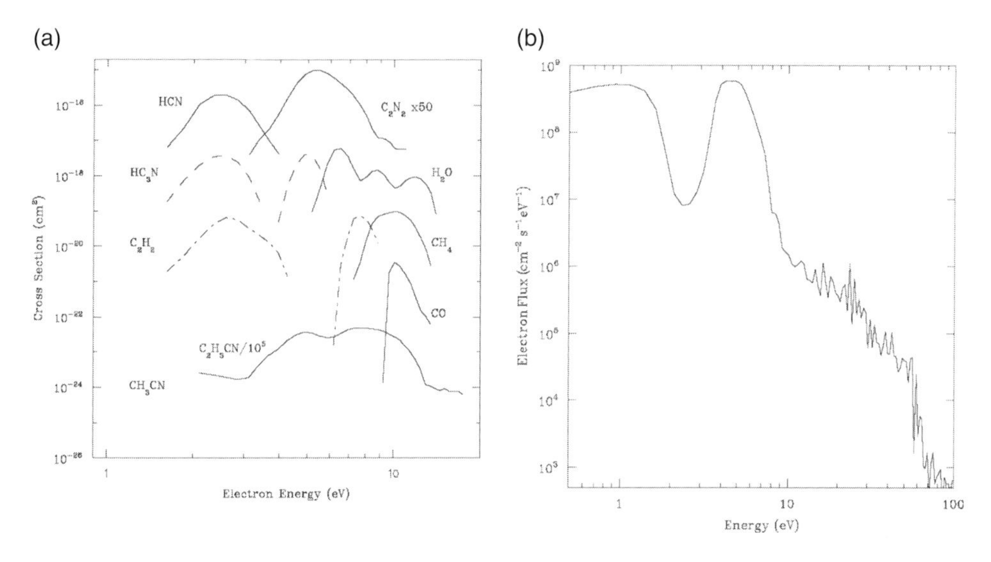


Figure 13.55 Cross sections of dissociative electron attachment (a) and energetic electron flux at 1015 km (b). From Vuitton et al. (2009).

process for some species of interest are shown in Figure 13.55a. Calculation of the low-energy electron spectrum is a more sophisticated problem than that for photoelectrons with $E > 10$ eV (Section 8.6). This task was solved, and the calculated spectrum is shown in Figure 13.55b. Dissociative electron attachment to HCN and $HC_3N$ dominate in the production of $CN^-$ and is negligible for $C_3N^-$ (Figure 13.54).

Ion-pair photoionization is similar to but much less effective than dissociative photoionization. Its contribution is minor to $CN^-$ and negligible to $C_3N^-$.

Ion-neutral reactions may be significant in both production and loss of negative ions. For example, $CN^-$ is formed by $C_XH^-$ + HCN and lost in the reactions with H, $CH_3$, and $HC_3N$ (Figure 13.54a). The similar reactions of $C_3N^-$ are shown in Figure 13.54b. Rate coefficients of the ion-neutral reactions of negative ions are close to those of positive ions, that is, $\approx 10^{-9}$ cm$^3$ s$^{-1}$ (9.22).

Rate coefficients of the ion-ion recombination $X^+ + Y^- \rightarrow$ products are $\approx 10^{-7}$ cm$^3$ s$^{-1}$. It is significant in the loss of both ions. Another loss mechanism is photodetachment, and its cross section was assumed at

$$\sigma = \sigma_0(1 - EA/h\nu)^{1/2}, \ h\nu > EA, \sigma_0 = 10^{-17}\ \text{cm}^2.$$

This loss is shown in Figure 13.54 as well.

Calculated vertical profiles of negative ions are depicted in Figure 13.56. The most abundant ions are $CN^-$, $C_3N^-$, and $C_5N^-$, in accord with the observed mass spectrum in Figure 13.53. However, the calculated total number density of negative ions peaks at 1.5 cm$^{-3}$

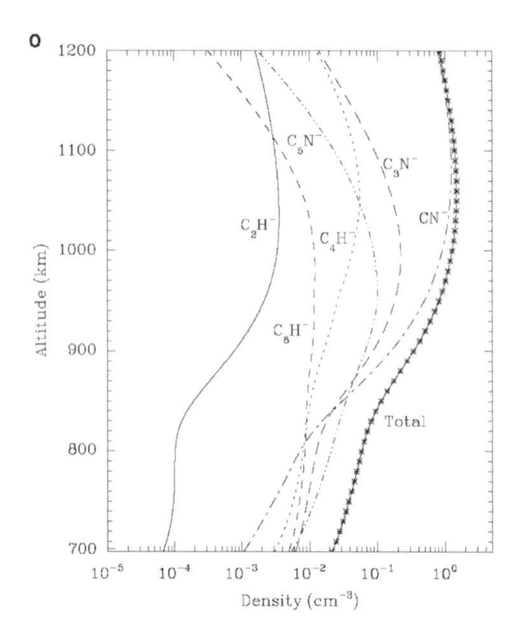


Figure 13.56 The most abundant negative ions (Vuitton et al. 2009).

near 1000 km, smaller than the measured values (Figure 13.52) by three orders of magnitude. Thus the observed high densities of negative ions near 1000 km remain unexplained.

## 13.9 Isotope Ratios

### *13.9.1 Observations*

Isotope ratios in Titan's atmospheric species were measured using the GCMS and INMS mass spectrometers, CIRS, and the microwave instruments (ALMA, SMA, and IRAM). The first detection of $CH_3D$ was made by Gillet (1975) using a spectrometer with a circular variable filter. A summary of the observed isotope ratios is in Table 13.6.

### *13.9.2 Nitrogen Isotopes*

$N_2$ is very abundant on Titan, while its photochemical loss and nonthermal escape are comparatively low, $6.4\times10^8$ atoms cm$^{-2}$ s$^{-1}$ based on Krasnopolsky (2009a). Therefore the lifetime of $N_2$ is 25 Byr, significantly exceeding the age of the Solar System.

It was believed prior to the Cassini/Huygens mission that $N_2$ on the terrestrial planets, Titan, and comets originated from ammonia $NH_3$ in the protosolar nebula, and their $^{14}N/^{15}N$ ratios may be similar. However, heavy nitrogen is enriched on Titan relative to that on Earth by a factor of 1.6, and recent detections of $^{15}NH_2$ in comets (Rousselot et al. 2014; Shinnaka et al. 2014) result in $^{14}N/^{15}N = 127 \pm 32$ in the cometary ammonia. ($NH_2$ is a product of photolysis of $NH_3$ in comets.) Furthermore, the

Table 13.6 *Isotope ratios in Titan's atmosphere*

| Ratio | Species | Value | Standard | Instrument, reference |
|---|---|---|---|---|
| $^{14}N/^{15}N$ | $N_2$ | $167.6 \pm 0.6$ | 274 | GCMS, Niemann et al. (2010) |
| | HCN | $60 \pm 6$ | | IRAM, Marten et al. (2002), SMA, Gurwell (2004) CIRS, Vinatier et al. (2007) |
| $^{40}Ar/^{36}Ar$ | Ar | $160 \pm 60$ | 299 | GCMS, Niemann et al. (2010) |
| D/H | $CH_4$ | $(1.36 \pm 0.09)\times 10^{-4}$ | $1.56\times 10^{-4}$ | CIRS, GB,[a] Nixon et al. (2012) |
| | $H_2$ | $(1.35 \pm 0.30)\times 10^{-4}$ | | GCMS, Niemann et al. (2010) |
| $^{12}C/^{13}C$ | $CH_4$ | $89.7 \pm 1.0$ | 89.4 | GCMS, INMS, CIRS, Nixon et al. (2012) |
| | CO | $89.9 \pm 3.4$ | | ALMA, Serigano et al. (2016) |
| | $C_2H_2$ | $84.8 \pm 3.2$ | | CIRS, Nixon et al. (2008a) |
| | $C_2H_6$ | $89.8 \pm 7.3$ | | CIRS, Nixon et al. (2008a) |
| | $C_4H_2$ | $90 \pm 8$ | | CIRS, Jolly et al. (2010) |
| | HCN | $89.8 \pm 2.8$ | | ALMA, Molter et al. (2016) |
| | $HC_3N$ | $79 \pm 17$ | | CIRS, Jennings et al. (2008) |
| | $CO_2$ | $84 \pm 17$ | | CIRS, Nixon et al. (2008b) |
| | WM[b] | $88.3 \pm 1.8$ | | |
| $^{16}O/^{18}O$ | CO | $486 \pm 22$ | 499 | ALMA, Serigano et al. (2016) |
| | $CO_2$ | $380 \pm 142$ | | CIRS, Nixon et al. (2009) |
| $^{16}O/^{17}O$ | CO | $2917 \pm 359$ | 2680 | ALMA, Serigano et al. (2016) |

[a]Ground-based observatories (NASA IRTF and KPNO). [b]Weighted mean for all photochemical products.

Table 13.7 *$^{14}N/^{15}N$ in the Solar System*

| Object | Species | $^{14}N/^{15}N$ |
|---|---|---|
| Solar wind | N | $442 \pm 130$ |
| Jupiter | $NH_3$ | $435 \pm 65$ |
| Chondrites | organics | $259 \pm 15$ |
| Mars meteorite | N-compounds | $280 \pm 5$ |
| Mars atmosphere | $N_2$ | $168 \pm 17$ |
| Earth | $N_2$ | $272 \pm 0.3$ |
| Venus | $N_2$ | $273 \pm 56$ |
| Titan | $N_2$ | $168 \pm 0.7$ |
| Comets | $NH_3$ | $127 \pm 32$ |

*Note.* Based on Mandt et al. (2014) and references therein.

recent study by Marty et al. (2010) established the solar wind ratio similar to that in Jupiter (Table 13.7).

Therefore $N_2$ was the primary nitrogen species in the protosolar nebula with $^{14}N/^{15}N \approx 440$, while $NH_3$ and HCN were less abundant as ices. The Sun and

Jupiter got their nitrogen as $N_2$. Nitrogen in the inner Solar System (Earth, Venus, Mars, and chondrites) was probably a mixture of the primordial $N_2$ and $NH_3$ in proportion 3 : 1. The smaller ratio on Mars is explained by preferential nonthermal escape of the light isotope.

### 13.9.3 Photochemical Enrichment of Heavy Nitrogen

Photochemistry of nitrogen is initiated on Titan in dissociation, ionization, and dissociative ionization of $N_2$ by the solar EUV photons, photoelectrons, magnetospheric electrons, protons, and cosmic rays. The ionization has the highest rate; however, almost all $N_2^+$ ions return $N_2$ in charge exchange with $CH_4$ and other species. Furthermore, laboratory measurements show a weak isotope fractionation in photoionization of $N_2$ averaged over the solar spectrum at $\lambda < 80$ nm (Croteau et al. 2011). Total production of N and $N^+$ is $6.4\times10^8$ cm$^{-2}$ s$^{-1}$ (Krasnopolsky 2016b), and predissociation at 80–100 nm is 28% of this production.

The $N_2$ absorption bands in the 80–100 nm range form spectral windows between the lines, and small isotopic shifts move the isotopic lines to those windows increasing the absorption and predissociation by $^{15}N^{14}N$. Liang et al. (2007) calculated quantum mechanically the absorption spectra of $N_2$ and $^{15}N^{14}N$ and then predissociation of both isotopologues in Titan's atmosphere (Figure 13.57). According to Liang et al. (2007), if the predissociation is the only source of atomic nitrogen that forms HCN, then $HC^{14}N/HC^{15}N = 23$ for the ratio of 183 in $N_2$ that was initially claimed by the GCMS team.

Correction for $^{14}N/^{15}N = 168$ results 21, and this source is 28% of the total production of N. Then the expected ratio

$$HC^{14}N/HC^{15}N = \frac{1}{\frac{0.28}{21} + \frac{0.72}{168}} = 57.$$

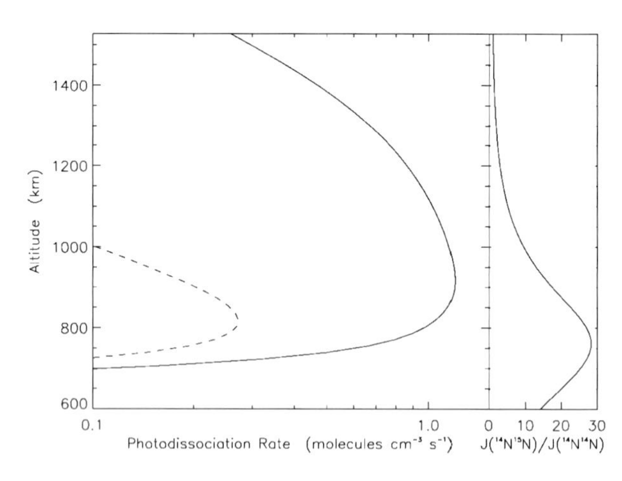


Figure 13.57 Calculated predissociation rates for $N_2$ and $^{15}N^{14}N$ (solid and dashed curves, respectively). Absorption by $CH_4$ has been included in the calculations. Ratio of the isotopic predissociation coefficients is shown as well. From Liang et al. (2007).

Therefore the quantum mechanical calculations by Liang et al. (2007) combined with the data of the photochemical model by Krasnopolsky (2009a) are in excellent agreement with the observed nitrogen isotope ratio in HCN (Table 13.6). The $^{14}N/^{15}N$ ratio in other nitriles should be similar as well.

The C≡N triple bonds are strong and cannot be broken in Titan's atmosphere. Therefore all nitriles are irreversibly lost by condensation and polymerization with a total rate of 392 g cm$^{-2}$ Byr$^{-1}$ of nitrogen (Krasnopolsky 2014b, 2016b). Fractionation factor (Section 10.5.4) is 168/60 = 2.8 for this process. Another loss is ejection of nitrogen by magnetospheric and pick-up ions (sputtering) that was estimated at 60 g cm$^{-2}$ Byr$^{-1}$ (De la Haye et al. 2007) with fractionation factor of 0.73 (Mandt et al. 2014).

According to Penz et al. (2005), the young Sun was brighter in the EUV (by a factor of 6 at $t$ = 1 Byr since the Solar System formation) and ejected a stronger wind (by a factor of 20 at $t$ = 1 Byr). These variations may be approximated by functions $(t_0/t)^\beta$. Assuming hydrodynamic escape with low fractionation in the first billion years (Mandt et al. 2014) and the linear response of the atmosphere to the solar EUV and wind, the calculated initial $^{14}N/^{15}N$ = 108. If the response is significantly nonlinear and, say, proportional to square root of the values, then the initial $^{14}N/^{15}N$ = 125. Therefore the current data agree with the initial $^{14}N/^{15}N$ ratio on Titan similar to that of ammonia in comets (Krasnopolsky 2016b), and nitrogen appeared on Titan as ammonia ice.

### *13.9.4 Argon Isotopes*

The origins of two major isotopes of argon, $^{36}Ar$ and $^{40}Ar$, are very different, and it is better to discuss their mixing ratios rather than the isotope ratio. The $^{36}Ar$ mixing ratio measured using GCMS is $(2.1 \pm 0.8)\times10^{-7}$ (Niemann et al. 2010). It reflects the conditions in the protosolar nebula near Saturn. The authors compare $^{14}N/^{36}Ar \approx 10^7$ in Titan's atmosphere with those on Mars and Earth, $4\times10^4$, and 30 in the Sun.

$^{40}Ar$ is much more abundant with a mixing ratio of $(3.39 \pm 0.12)\times10^{-5}$ (Niemann et al. 2010). It is a product of radioactive decay of $^{40}K$. If the rock mass fraction of Titan is 0.55 and the potassium abundance is similar to that of CI-chondrites, then the atmospheric $^{40}Ar$ is ≈8% of the total radiogenic production (McKinnon 2010). This value is 2.6% on Mars; however, two-thirds of the atmospheric argon was lost by the intense meteorite erosion before 0.8 Byr and by sputtering (Krasnopolsky and Gladstone 1996, 2005). Therefore the total $^{40}Ar$ released into the Martian atmosphere is ≈8% of the radigenic production as well.

Hodyss et al. (2013) studied solubility of argon in liquid methane and ethane. They conclude that "large subsurface reservoirs of liquid methane and ethane could be sufficient to trap much of the argon outgassing from the interior."

### *13.9.5 Carbon Isotope Ratios*

Mandt et al. (2012a) analyzed the INMS measurements in 30 flybys of the Cassini orbiter and derived $^{12}C/^{13}C = 88.5 \pm 1.4$ in methane extrapolated to the surface and fractionation factor of $0.736 \pm 0.045$ for carbon escape. Nixon et al. (2012) deduced $^{12}C/^{13}C = 86.5 \pm 8.2$

in methane using the CIRS spectra of methane bands. These ratios combined with $^{12}C/^{13}C$ = 91.1 ± 1.4 from the GCMS measurements in methane (Niemann et al. 2010) give weighted-mean $^{12}C/^{13}C$ = 89.7 ± 1.0, very close to the terrestrial value of 89.4.

Krasnopolsky (2018c) scaled the measured $^{12}C/^{13}C$ in the photochemical products (Table 13.6) by their column rates in the production of the haze from the photochemical model (Krasnopolsky 2014b). Then the weighted-mean ratio in the haze is $^{12}C/^{13}C$ = 88.3 ± 3.0. The most significant production of the haze is by condensation of $C_2H_6$ and polymerization in the reactions $C_6H + C_4H_2$, $C_3N + C_4H_2$, and $HC_3N + C_4H$. Carbon isotope ratios in $C_3N$ and $C_4H$ were adopted equal to those in $HC_3N$ and $C_4H_2$, respectively. Total loss of methane is 7 kg cm$^{-2}$ Byr$^{-1}$, and correction for sputtering and ion escape of 23 g cm$^{-2}$ Byr$^{-1}$ (De la Haye et al. 2007) results in $^{12}C/^{13}C$ = 88.5 ± 3.0 in the total loss of carbon.

Lifetime of methane in Titan's atmosphere is a ratio of its column abundance to the total loss rate. It is equal to 32 Myr, much smaller than the age of the Solar System. Most probably, the loss of methane is compensated by its outgassing and cryovolcanism from the interior, though direct evidence of those have not been found. Water ice is near one-half of Titan's mass, and methane may exist as methane clathrate hydrate $CH_4 \times 5.75H_2O$. Then methane in the interior should have $^{12}C/^{13}C$ = 88.5 ± 3.0 as well. It is smaller but within uncertainties of $^{12}C/^{13}C$ = 92.4 ± 5.4 in the outer Solar System (Jupiter and Saturn) and 89.4 on the Earth. The difference is even greater for a scenario with injection of methane into Titan's atmosphere and its gradual depletion to the present abundance.

Nair et al. (2005) calculated that fractionation of carbon isotopes is negligible in photolysis of $CH_4$. Nixon et al. (2012) applied the transition state theory to calculate kinetic isotope effect in $C_2H + CH_4 \rightarrow CH_3 + C_2H_2$ that is equal to $k_{12}/k_{13}$ = 1.019 at 175 K.

### *13.9.6 Other Isotope Ratios*

D/H = $(1.6 \pm 0.2)\times10^{-5}$ on Saturn is smaller than that on Titan by an order of magnitude. Nixon et al. (2012) concluded that "the original D/H value is too uncertain to provide any constraint."

Oxygen precipitates into Titan's atmosphere as the meteorite water and $O^+$ ions in the Saturn environment that originate from $H_2O$ on the rings and Enceladus. The most accurate measurements of $^{16}O/^{18}O$ and $^{16}O/^{17}O$ in CO using ALMA (Table 13.6) agree with the standard terrestrial ratios. Loison et al. (2017) found that the photochemical fractionation of oxygen isotopes is very low and well within uncertainties of the observed isotope ratios (Table 13.6).

## 13.10 Photochemical Modeling of Titan's Atmosphere and Ionosphere

### *13.10.1 Photochemical Models for Titan*

Photochemical processes in the dense $N_2/CH_4$ atmosphere initiate formation of numerous hydrocarbons, nitriles, their ions, and haze particles. This makes photochemical modeling for Titan a challenging problem. Self-consistent models for Titan's neutral atmosphere were developed by Yung et al. (1984), Toublanc et al. (1995), Lara et al. (1996), Hebrard et al. (2007), and Lavvas et al. (2008a, 2008b).

Pure ionospheric models adopt a background neutral atmosphere and neglect the effects of the ion chemistry on the neutral composition. However, these effects are significant, for example, benzene $C_6H_6$ is mostly formed by the ion reactions on Titan. Therefore coupled models of Titan's atmosphere and ionosphere have significant advantages. These models were made by Banaszkiewicz et al. (2000), Wilson and Atreya (2004), Krasnopolsky (2009a, 2010e, 2012d, 2014b), and Dobrijevic et al. (2016).

Banaszkiewicz et al. (2000) created the first model of this type. However, they calculated hydrocarbons neglecting effects of nitriles and ions on their abundances, then calculated nitriles using the calculated hydrocarbons but neglecting effects of ions, and finally calculated ions.

The model by Wilson and Atreya (2004) involves interactions between all types of species. This is a rather detailed and accurate model. It was made before the Cassini/Huygens mission and obviously does not account for the mission results.

Lavvas et al. (2008a, 2008b) is a detailed model of Titan's neutral chemistry. It is the only model that couples photochemistry with thermal balance, radiative transfer calculations, and the haze microphysics. The model involves 520 reactions of 68 neutral species.

Hebrard et al. (2007) initiated numerous publications with incremental improvements of their model for the neutral atmosphere. Main feature of their model is that it includes uncertainties of the reaction rate coefficients and calculates both species densities and their uncertainties. The latest version of the model was developed by Dobrijevic et al. (2016).

Willacy et al. (2016) calculated a model that accounts for sublimation of some species in the troposphere. Their model is mostly focused on nitriles in the stratosphere and troposphere.

### 13.10.2 Titan's Atmosphere and Ionosphere in the Model by Dobrijevic et al. (2016)

The model involves neutral species and positive and negative ions. The authors intended to reduce the numbers of species and their reactions. The model included 74 neutral species and 310 neutral reactions. Hydrocarbons were extended up to $C_4H_{10}$, and four hydrocarbons (including benzene), argon, and five nitriles from the list of the observed species in Table 13.3 are missing in the model. There are 47 positive ions with 376 reactions in the model, including ionization, ion-neutral reactions, and dissociative recombination. Six negative ions with 70 reactions of a few types complement the model. The upper boundary is at 1500 km; ion diffusion is neglected, thermal escape is the boundary condition for H and $H_2$ with no escape for the other species. O and OH influxes from meteorite water and the Saturn environment are fitted at $1.6\times10^6$ and $5.2\times10^5$ cm$^{-2}$ s$^{-1}$, respectively. The model is calculated for the global-mean conditions (SZA = 60° and the solar EUV is halved to account for the nightside).

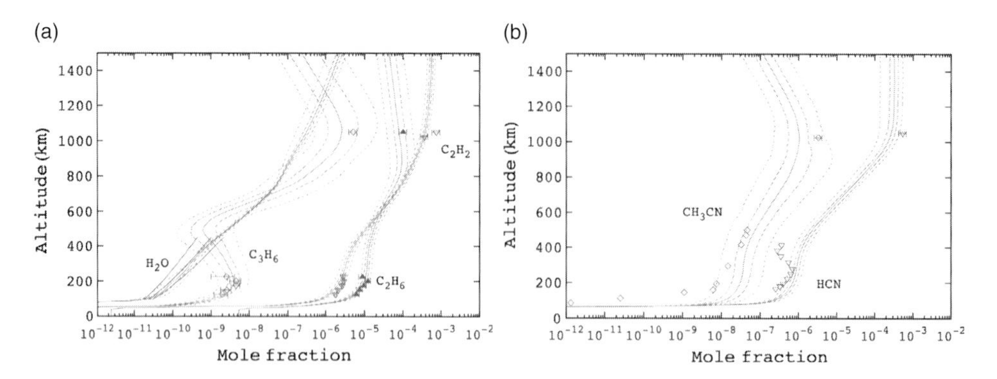


Figure 13.58 Vertical profiles of three hydrocarbons, two nitriles, and $H_2O$ from the model by Dobrijevic et al. (2016) compared with the CIRS, INMS, IRAM ($CH_3CN$), and Hershel ($H_2O$) observations. The INMS data on the figure should be reduced by the factor of 2.2 (see text). Long and short dashed curves indicate $1\sigma$ and $2\sigma$ uncertainties induced by uncertainties of the reaction rate coefficients.

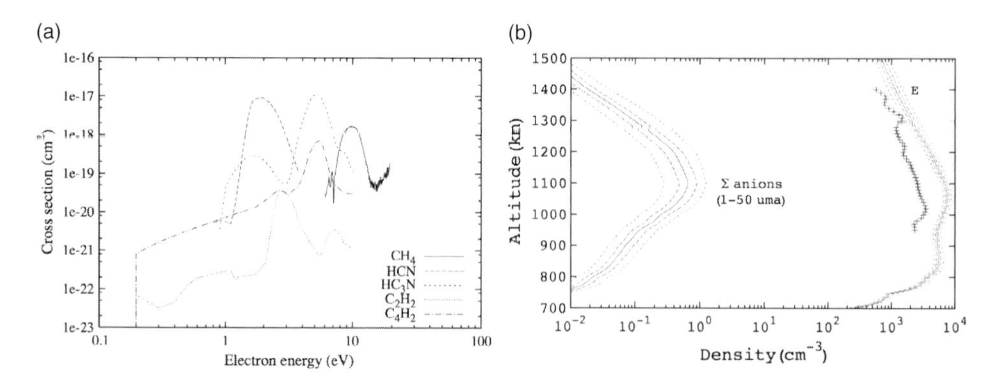


Figure 13.59 Cross sections of dissociative electron attachment (a) and (b) vertical profiles of electrons and negative ions in the model by Dobrijevic et al. (2016). The RPWS-LP electron density profile measured at SZA = 60° is shown for comparison (crosses).

$N_2$, $CH_4$, $H_2 = 0.003$, and $CO = 5.1\times10^{-5}$ are specified at the surface. Actually $H_2$ and CO are the photochemical products and should be given by effusion velocities at the lower boundary, e.g., $v = 0$ at the chemically passive surface. Conditions at the lower boundary for the other species are lacking in the paper. All INMS mixing ratios in Dobrijevic et al. (2016) are erroneously scaled by the factor of 2.2 (Teolis et al. 2015). Actually this factor refers to number densities and does not affect mixing ratios.

Calculated vertical profiles of three hydrocarbons, two nitriles, and one oxygen species (water) are shown in Figure 13.58. The calculated species in Figure 13.58 are in good agreement with the observations.

The electron density profile in the model (Figure 13.59) shows a peak with $n_e$ = 7000 cm$^{-3}$ at 1070 km and a significant downward continuation to $n_e$ = 5000 cm$^{-3}$ at

850 km. The model electron (and hence positive ion) densities exceed the observed values by a factor of $\approx 2$. The calculated profile of densities of negative ions is similar to that in Vuitton et al. (2009; Figure 13.56), but smaller by a factor of 2. Some initial data for negative ions were updated in Dobrijevic et al. (2016; compare Figures 13.59a and 13.55). Detailed data on the reactions, their rate coefficients, uncertainties and column rates may be found in the model supplements.

### *13.10.3 Photochemical Model for Titan's Atmosphere and Ionosphere by Krasnopolsky (2009–2014)*

There are four versions of the model. Two versions in Krasnopolsky (2009a) adopt hydrodynamic escape of light species ($m < 20$) and either "standard" eddy diffusion or that proposed by Hörst et al. (2008). They suggested an increase in $K \sim n^{-2}$ from 75 to 300 km by five orders of magnitude to fit better the CIRS limb observations. Our model with this $K$ has both advantages and shortcomings relative to the first version.

It is discussed in Section 13.7.2 that the direct Monte Carlo simulations of the escape processes and the low abundance of the carbon bearing ions in Saturn's magnetosphere do not support hydrodynamic escape of $CH_4$. Tests in Krasnopolsky (2010e) are not favorable for hydrodynamic escape on Titan either, and versions 3 and 4 were calculated without hydrodynamic escape. Yelle et al. (2006) considered $K = (4 \pm 3)\times 10^9$ cm$^2$ s$^{-1}$ as an alternative to the hydrodynamic escape, and $K = 10^9$ and $6\times 10^8$ cm$^2$ s$^{-1}$ were adopted above 700 km in versions 3 (Krasnopolsky 2010e) and 4 (Krasnopolsky 2012d, 2014b), respectively. Two reactions that stimulate formation of $NH_3$ were included in these versions. Finally, version 4 is further improved using the Troe approximation for the termolecular reactions (see Section 9.5; formula (9.37)), some changes in eddy diffusion, and two radiative association reactions. Here we will consider the latest version from Krasnopolsky (2012d, 2014b).

### *13.10.4 Initial Data*

Basic initial data are the $N_2$ and $CH_4$ densities near the surface and profiles of $T$ and $K$ (Figure 13.60a). The observed temperature profile is slightly adjusted to bind the GCMS and INMS observations of $N_2$. The adopted $K$ results in a homopause for $CH_4$ at 1000 km. The model upper boundary is at 1600 km, near the top of the INMS observations. The calculated exobase is at 1360 km.

The UV absorption by the haze significantly affects Titan's photochemistry and was calculated using the Huygens haze data (Tomasko et al. 2008a): the haze particles are aggregates of ~3000 monomers with the monomer radius of ~0.05 μm, the particle number density is 5 cm$^{-3}$ below 80 km decreasing with a scale height of 65 km above 80 km. We calculated scattering properties of the particles using a code for aggregate particles by Rannou et al. (1999) and refractive indices from Khare et al. (1984; Figure 13.2). Radiative

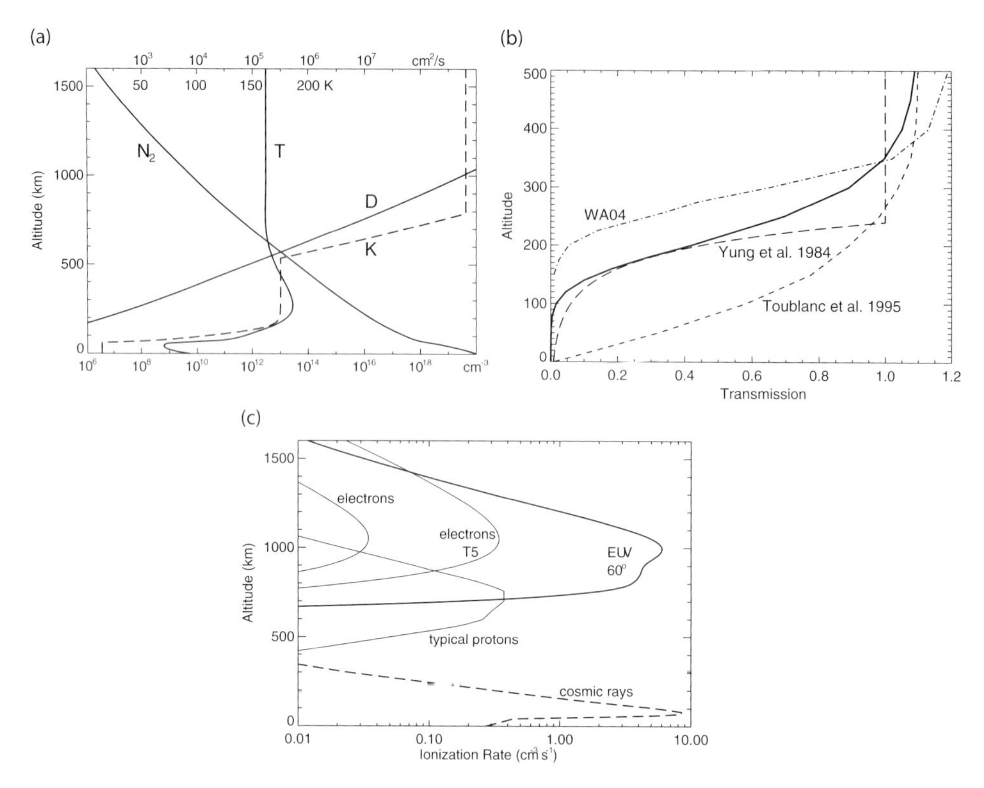


Figure 13.60 A: Vertical profiles of temperature, $N_2$ density, eddy diffusion, and diffusion coefficient of $CH_4$ in $N_2$. B: Calculated haze transmission plus reflection based on the DISR data compared with those from the previous models. C: Ionization rates by the solar EUV photons, magnetospheric electrons and protons, and cosmic rays. From Krasnopolsky (2009a).

transfer with multiple scattering in the optically thick haze was calculated (Figure 13.60b) using some approximations from van de Hulst (1980).

Dissociation and ionization of the atmospheric species by the solar photons (Figure 13.60c) were calculated self-consistently using the known species cross sections. Precipitations of magnetospheric electrons for a typical case and a strong event were taken from the Cassini encounters T21 and T5, respectively (Cravens et al. 2009; Vuitton et al. 2007). The ion densities measured by INMS during T5 were especially high and accurate. Productions of $N_2^+$, $N^+$, N, and $N(^2D)$ by electron impact were obtained using the data of Fox and Victor (1988) and the Born approximation. Ionization by magnetospheric protons and the cosmic rays (Figure 13.60c) were taken from Cravens et al. (2008) and Molina-Cuberos et al. (1999), respectively.

With a general trend to remove unimportant species and reactions, the model involves 419 reactions of 83 neutrals and 33 positive ions, and column rates are given for all reactions in Krasnopolsky (2009a). Hydrocarbon chemistry is extended to $C_{12}H_{10}$ for neutrals and $C_{10}H_{11}^+$ for ions. The model involves ambipolar diffusion (Section 2.2.3) and escape of ions and calculates the $H_2$ and CO densities near the surface that were assigned in some previous models. Formation of haze by polymerization processes and recombination of heavy ions is considered as well.

The lower boundary is at the surface and closed (i.e., chemically passive with a zero flux) for all incondensable species except $N_2$, $CH_4$, and Ar that are given by their densities. Twenty-four species in the model condense near the tropopause. Their saturated vapor pressures were found in the literature and approximated by $\log p_s = a - b/T - c \ln T$ in Table 13.8. The loss term for condensation is $L_{ic} = A \ln(n_i/n_{is})$, where $n_{is}$ is the saturation vapor density and $A = 10^{-9}$ s$^{-1}$ is taken from Yung et al. (1984). The model had a vertical step increasing from 2.5 km near the surface to 10 km at the upper boundary.

### 13.10.5 Hydrocarbons, Ar, and $H_2$

Argon does not react and escape, and diffusive separation is the only process that affects the Ar mixing ratio (Figure 13.61). The calculated profile demonstrates two homopauses near 550 and 1000 km, in accord with the adopted eddy diffusion in Figure 13.60a. The second homopause is an exotic feature that is lacking in the other atmospheres. The region near 550 km is poorly covered by the observations, and direct observational evidences for the second homopause are missing. However, this helps to connect the GCMS and INMS data for argon in the lower and upper atmospheres, respectively.

The calculated $CH_4$ profile (Figure 13.61) perfectly fit the GCMS observations in the lower atmosphere. The effect of the second homopause is smoothed because of significant chemical loss above 550 km. Photolysis of $CH_4$ results in a quarter of the total loss of $CH_4$. The photolysis products $CH_3$ and $CH_2$ radicals are quickly converted to CH that reacts with $CH_4$ to form $C_2H_4$. These reactions almost double the loss of methane by photolysis.

Photolyses of $CH_4$, $C_2H_2$, and $C_4H_2$ begin at 140, 200, and 260 nm, respectively. Radicals $C_2H$ and $C_2$ released by photolyses of $C_2H_2$ and $C_4H_2$ react with methane. This indirect photolysis results in 30% of the methane loss; the remaining loss is in reactions with other radicals and ions. Chemical production of methane is much smaller than its loss, and the difference is spent to precipitation of the haze, condensation of photochemical products in the troposphere, and escape of $H_2$ and H. The methane column abundance is 230 g cm$^{-2}$, and its lifetime is 30 Myr. Water ice is near one-half of Titan's mass, and methane clathrate hydrate $CH_4 \times 5.75H_2O$ may be a source of methane in the interior and compensate the atmospheric loss by outgassing and cryovolcanism.

The C:H:N ratio in the total aerosol flux is 1:1.2:0.06 in the numbers of atoms. The ratio in methane is 1:4; this means that 70% of hydrogen released from methane escapes to space. The models with and without hydrodynamic escape of $\mathbf{H_2}$ give very similar escape rates that are diffusion-limited in both cases. The H–H bond is slightly stronger than the H bond in some radical, and much of the H atoms are converted into $H_2$ by reactions like

Table 13.8 *Saturated vapor pressures used in the model*

| | | |
|---|---|---|
| $CH_4$ | 67–91 K | $9.43507 - 453.92414/T - 4055.6016/T^2 + 115{,}352.19/T^3 - 1{,}165{,}560.7/T^4$ |
| $C_2H_2$ | 80–145 K | $30.493 - 1644.1/T - 3.224 \ln T$ |
| $C_2H_4$ | 89–104 K | $10.85 - 901.6/(T - 2.555)$ |
| $C_2H_6$ | 30–90 K | $12.135 - 1085/(T - 0.561)$ |
| $C_3H_4$ | 160–200 K | $7.645 - 1375/T + 0.55 \ln T$ |
| $C_3H_8$ | 115–143 K | $46.309 - 2047/T - 6.035 \ln T$ |
| $C_4H_2$ | 127–249 K | $57.409 - 3300.5/T - 7.224 \ln T$ |
| $C_4H_4$ | 177–200 K | $33.013 - 2227/T - 3.561 \ln T$ |
| $C_4H_6$ | 141–175 K | $21.04 - 1732/T - 1.771 \ln T$ |
| $C_4H_8$ | 134–165 K | $23.105 - 1731/T - 2.082 \ln T$ |
| $C_4H_{10}$ | 139–170 K | $44.113 - 2382/T - 5.464 \ln T$ |
| $C_6H_6$ | 230–293 K | $12.36 - 2413/T$ |
| $C_7H_8$ | 195–242 K | $5.243 - 1843/T + 0.798 \ln T$ |
| HCN | 132–168 K | $13.53 - 2318/T$ |
| $HC_3N$ | – | $8.347 - 1913/T$ |
| $CH_3CN$ | 226–257 K | $10.583 - 1912/T$ |
| $C_2N_2$ | 148–175 K | $21.381 - 2031/T - 1.502 \ln T$ |
| $C_4N_2$ | 147–162 K | $10.394 - 2155/T$ |
| $C_2H_3CN$ | 200–250 K | $21.058 - 2371/T - 1.560 \ln T$ |
| $C_2H_5CN$ | 100–140 K | $9.715 - 2597/T + 0.836 \ln T$ |
| $C_4H_5N$ | 208–250 K | $65.036 - 4240/T - 8.345 \ln T$ |
| $C_5H_5N$ | 250–280 K | $11.06 - 2266/T$ |
| $CO_2$ | 114–137 K | $11.983 - 1385/T + 0.0343 \ln T$ |
| $H_2O$ | 110–273 K | $4.1477 - 2485.582/T + 1.533 \ln T - 0.003163\, T$ |

*Note.* These relationships give $\log_{10}P_S$ (Pa). Temperature ranges of measurements are given for each species.

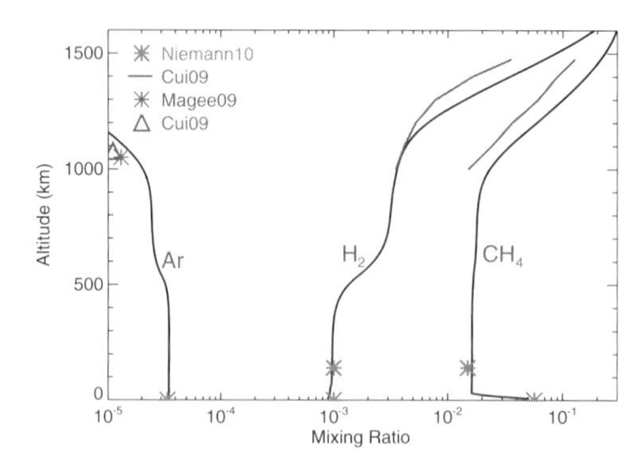


Figure 13.61 Vertical profiles of Ar, $CH_4$, and $H_2$: model and observations (Krasnopolsky 2014b).

$$\begin{aligned} &H + C_4H_2 + M \rightarrow C_4H_3 + M, \\ &H + C_4H_3 \rightarrow C_4H_2 + H_2, \\ &\text{Net } H + H \rightarrow H_2. \end{aligned}$$

$H_2$ is not fixed at the lower boundary in our model; therefore the agreement of the model with the observations (Figure 13.61) may be considered as very good. The second homopause is well seen in the $H_2$ vertical profile.

Photolysis of methane peaks near 750 km; photolyses of the photochemical products are maximal typically at 300–500 km. Therefore basic reactions that determine species balances occur usually around 500 km. Species mixing ratios are typically constant above 500 km up to the homopause at 1000 km and decrease, increase, or stable above 1000 km depending on its molecular mass that may be greater, smaller, or close to that of $N_2$, respectively.

Most of the photochemical products condense in the lowest ~100 km, and here their mixing ratios are much smaller than those near ~500 km. Therefore there is an inevitable steep decrease in the product mixing ratios from ~500 to ~100 km. The calculated profiles usually intersect the CIRS nadir values that refer to 100–200 km. However, these profiles are steeper than those from the CIRS limb observation. The chosen steep increase in eddy diffusion at 70–180 km in our model (Figure 13.60a) partly compensates for this difference. Even steeper increase in eddy diffusion was suggested by Hörst et al. (2008); however, our test shows that the eddy diffusion profile in Figure 13.60a is optimal.

We mentioned above that almost all $CH_x$ products of the methane photolysis form CH and then $CH + CH_4 \rightarrow C_2H_4 + H$. This reaction is the major source of ethylene that is mostly lost by photolysis. The calculated densities of $\mathbf{C_2H_4}$ (Figure 13.62) are below the condensation level in the troposphere. $C_2H_4$ near 300 and 1100 km is within the uncertainties of the CIRS and INMS values, respectively.

Acetylene $\mathbf{C_2H_2}$ is produced by photolyses of $C_2H_4$ and $C_4H_{2x}$ ($x = 1, 2, 3$), the reactions of $C_2H$ with $CH_4$ and $C_2H_6$, and by

$$\begin{aligned} &C_2H_3 + H \rightarrow C_2H_2 + H_2 \\ &C_3H_4 + H \rightarrow C_2H_2 + CH_3. \end{aligned}$$

It is lost by photolysis (42%), reactions with $C_2H$ and CN (32%) that form diacetylene $C_4H_2$ and cyanoacetylene $HC_3N$, and by condensation (Table 13.9). The calculated mixing ratios at 100–300 and 1100 km agree with the CIRS and INMS values.

Ethane $\mathbf{C_2H_6}$ is produced by the reaction

$$CH_3 + CH_3 + M \rightarrow C_2H_6 + M$$

and lost by condensation and reactions with radicals (two-thirds and one-third, respectively). The calculated densities (Figure 13.62) are compared with the CIRS and INMS observations.

The $\mathbf{C_3H_{2x}}$ hydrocarbons (propyne, propylene, and propane) are shown in Figure 13.63. Production and loss of propyne are mostly by the reactions

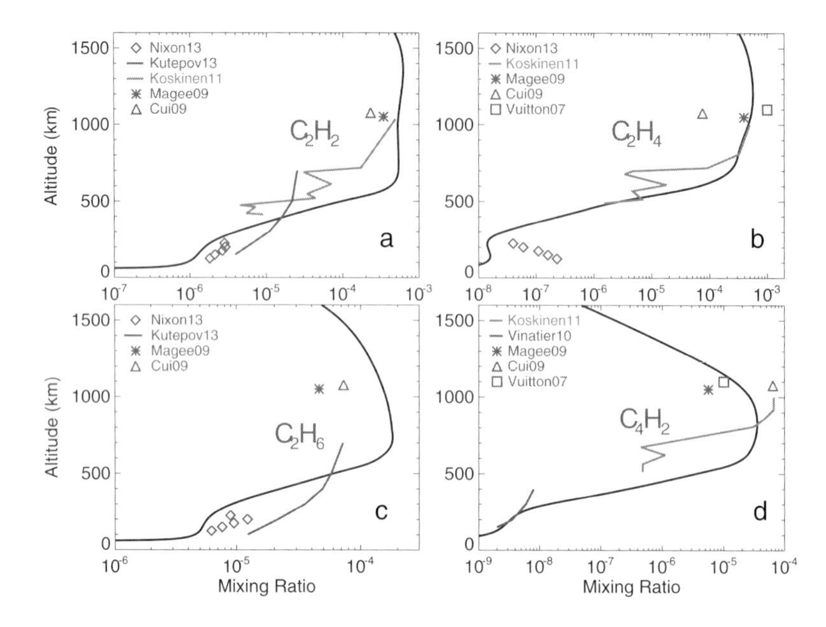


Figure 13.62 Vertical profiles of $C_2H_2$, $C_2H_4$, $C_2H_6$, and $C_4H_2$: model and observations (Krasnopolsky 2014b).

$$C_3H_3 + H + M \rightarrow C_3H_4 + M$$
$$C_3H_4 + H \rightarrow C_2H_2 + CH_3,$$

which proceed near 300–350 km, while its high-altitude source is by

$$C_2H_4 + CH \rightarrow C_3H_4 + H.$$

Propylene is formed by a few radical reactions and lost by photolysis. Its high-altitude source is

$$C_2H_6 + CH \rightarrow C_3H_6 + H.$$

Propane is produced by

$$C_2H_5 + CH_3 + M \rightarrow C_3H_8 + M$$

and lost mostly by condensation.

Photolysis of $C_2H_2$ forms $C_2H$ and $C_2$ that react with acetylene and ethylene and make the $C_4H_x$ hydrocarbons. Diacetylene $\mathbf{C_4H_2}$ (Figure 13.62) is the most abundant of those. It is a catalyst in indirect recombination of $H_2$ and $CH_4$:

$$C_4H_2 + H \rightarrow C_4H_3 + h\nu$$
$$C_4H_3 + H \rightarrow C_4H_2 + H_2$$
$$\underline{C_4H_3 + CH_3 \rightarrow C_4H_2 + CH4}$$
$$\text{Net } H + H \rightarrow H_2, \ H + CH_3 \rightarrow CH4.$$

Photolysis and polymerization with $C_6H$ and $C_3N$ are the major loss of $C_4H_2$.

There are three sources of benzene $\mathbf{C_6H_6}$ (Figure 13.63), which is the simplest polycyclic aromatic hydrocarbon (PAH):

Table 13.9 *Column production and loss rates, escape/precipitation flows, and mean chemical lifetimes of some species*

| Species | H | $H_2$ | $CH_4$ | $C_2H_2$ | $C_2H_4$ | $C_2H_6$ | $C_3H_4$ | $C_3H_6$ | $C_3H_8$ |
|---|---|---|---|---|---|---|---|---|---|
| Production ($cm^{-2}\ s^{-1}$) | 1.55+10 | 1.17+10 | 1.13+9 | 5.83+9 | 2.66+9 | 1.54+9 | 4.97+8 | 2.51+8 | 2.29+8 |
| Loss ($cm^{-2}\ s^{-1}$) | 1.36+10 | 1.94+8 | 9.60+9 | 5.67+9 | 2.66+9 | 5.17+8 | 4.97+8 | 2.51+8 | 5.02+7 |
| Flow ($g\ cm^{-2}\ Byr^{-1}$) | 101 | 1210 | 7132 | −212 | – | −1610 | – | – | −415 |
| Lifetime (yr) | 0.003 | 5.95+5 | 2.85+7 | 17.5 | 0.61 | 562 | 0.44 | 0.72 | 264 |
| **Species** | $C_4H_2$ | $C_6H_6$ | HCN | $HC_3N$ | $CH_3CN$ | $C_2N_2$ | $H_2O$ | CO | $CO_2$ |
| Production ($cm^{-2}\ s^{-1}$) | 5.60+9 | 2.15+8 | 1.20+9 | 1.59+9 | 1.58+7 | 6.17+6 | 5.71+6 | 4.35+6 | 4.35+6 |
| Loss ($cm^{-2}\ s^{-1}$) | 5.60+9 | 1.92+8 | 1.05+9 | 1.52+9 | 7.88+6 | 4.22+6 | 1.04+7 | 4.35+6 | 2.56+6 |
| Flow ($g\ cm^{-2}\ Byr^{-1}$) | – | −9.44 | −207 | −169 | −17.4 | −5.3 | – | – | −4.1 |
| Lifetime (yr) | 0.03 | 0.73 | 13.2 | 2.66 | 27.7 | 53.7 | 2.18 | 7.57+7 | 443 |

*Note.* Positive flows are escape or sublimation from the surface; negative flows are condensation. Precipitation of haze is not included (see Table 13.11). All values are reduced to the surface; 1.55 + 10 = $1.55\times10^{10}$.

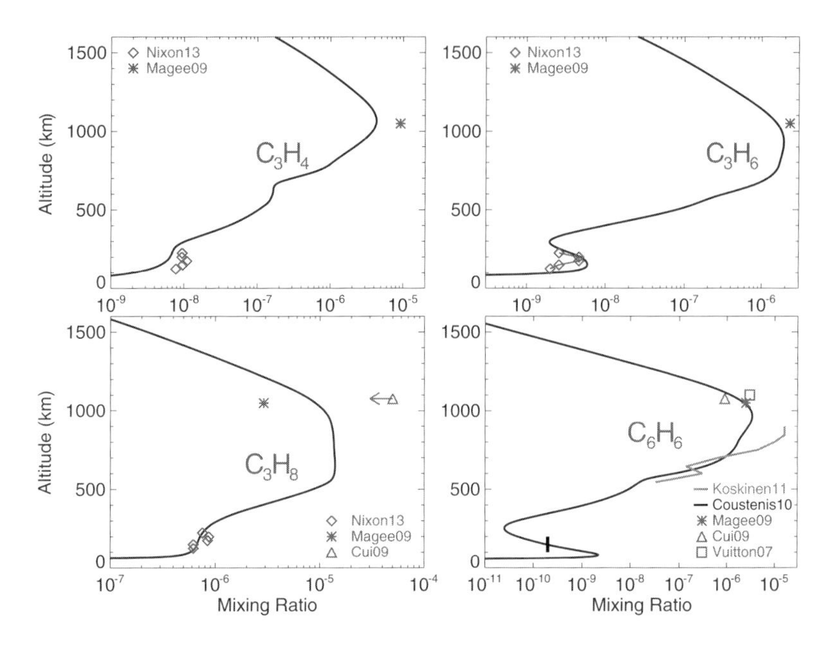


Figure 13.63 Vertical profiles of $C_3H_4$, $C_3H_6$, $C_3H_8$, and $C_6H_6$: model and observations (Krasnopolsky 2014b).

$$C_4H_5 + C_2H_2 \rightarrow C_6H_6 + H,$$
$$C_3H_3 + C_3H_3 + M \rightarrow C_6H_6 + M,$$

and dissociative recombination of $C_6H_7^+$ and $C_7H_7^+$. The first reaction is effective in the lower atmosphere at ≈100 km, the second in the middle atmosphere near 350 km, while the dissociative recombination forms benzene near 800–900 km. Benzene recycles in the major photolysis branch and is lost in the second branch $C_6H_6 + h\nu \rightarrow C_6H_4 + H_2$ as well as in condensation (Table 13.9).

Vertical profiles of other hydrocarbons and their reactions may be found in Krasnopolsky (2009a, 2012d). There is a reasonable agreement between the model and the observations, taking into account their uncertainties.

### *13.10.6 Nitriles*

The N≡N triple bond in the $N_2$ molecule is very strong with dissociation energy of 9.76 eV. Ionization of $N_2$ by the solar photons $\lambda < 80$ nm, photoelectrons, magnetospheric electrons, protons, and cosmic rays (Figure 13.60c) forms mostly $N_2^+$ ions that return $N_2$ after charge exchange with hydrocarbons. Predissociation of $N_2$ at $\lambda = 80-100$ nm, dissociative ionization at $\lambda < 51$ nm, and electron, proton and cosmic ray impact dissociations originate N and $N(^2D)$ atoms and $N^+$ ions with a total column rate of $7\times10^8$ cm$^{-2}$ s$^{-1}$. Most of them (72%) form nitriles $C_xH_yCN$. The basic reaction of the nitrile formation is $N + CH_3 \rightarrow H_2CN + H$, which consumes one-half of the atomic nitrogen production. The C≡N bonds are strong (7.85 eV in CN and 9.58 eV in HCN); therefore nitriles do not recombine to $N_2$ in Titan's atmosphere, and condensation and polymerization with precipitation to the surface are the ultimate fate of nitriles on Titan.

$H_2CN$ is converted to **HCN** in the reaction with H. Hydrogen cyanide HCN (Figure 13.64) is the most abundant nitrile on Titan. Beside $H_2CN$ (24%), reactions of CN with $CH_4$, $C_2H_4$, $C_2H_6$ (56%), and ion reactions (20%) contribute to the production of HCN. HCN is lost in reactions with CH, $C_2H_3$, $C_3N$ (68%), ion reactions (14%), condensation (12%), and photolysis (6%). The calculated profile (Figure 13.64) is in excellent agreement with the observations.

Cyanoacetylene **$HC_3N$** is another abundant nitrile that is formed by reactions of CN + $C_2H_2$ (60%) and $C_3N + CH_4$ (40%) and lost by photolysis (86%), condensation (4%), polymerization (1%), reactions with $C_2H$ and CN (1%), and ion reactions (8%). The calculated $HC_3N$ abundance agrees with the INMS data by Cui et al. (2009) and those extracted from the INMS ion spectrum by Vuitton et al. (2007). However, the difference between the model and the CIRS and IRAM (Marten et al 2002) observations below 500 km is significant.

The reaction $C_3N + CH_4 \rightarrow HC_3N + CH_3$ has not been studied in the laboratory and was suggested by Yung (1987), who adopted its rate coefficient at $5\times10^{-14}$ cm$^3$ s$^{-1}$. The effective mean altitude of this reaction is 370 km in the model; maybe, its rate is significantly overestimated. There are some indications that the $HC_3N$ loss to

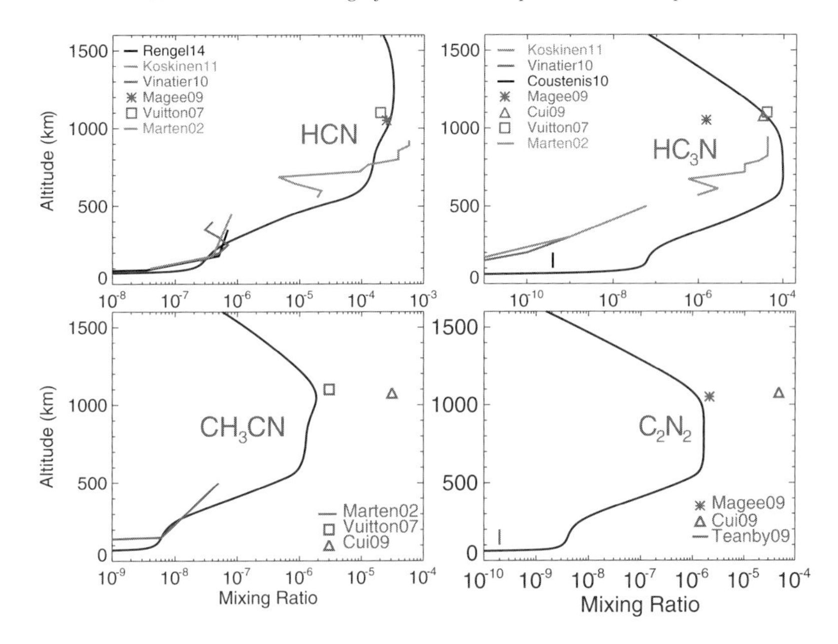


Figure 13.64 Vertical profiles of HCN, $HC_3N$, $CH_3CN$, and $C_2N_2$: model and observations (Krasnopolsky 2014b).

polymerization may be underestimated in the model that is based on the data from Lavvas et al. (2008a).

Acetonitrile $\mathbf{CH_3CN}$ (Figure 13.64) is formed by

$$CH_2CN + H + M \rightarrow CH_3CN + M\ (53\%, 142\ \text{km}),$$
$$N(^2D) + C_2H_4 \rightarrow CH_3CN + H\ (25\%, 981\ \text{km}),$$
$$N + C_2H_3 + M \rightarrow CH_3CN + M\ (20\%, 62\ \text{km}).$$

Condensation (50%) and ion reactions dominate in the loss of $CH_3CN$.

Formation of cyanogens $\mathbf{C_2N_2}$ is by reactions

$$CN + CH_2CN \rightarrow C_2N_2 + CH_2 (85\%, 660\ \text{km})$$
$$N + CHCN \rightarrow C_2N_2 + CH\ (15\%, 246\ \text{km}).$$

It is lost by photolysis (68%) and condensation (32%). Data on other nitriles and their reactions may be found in Krasnopolsky (2009a, 2012d).

The model involves nine species with **N–H** bonds. These bonds are ~3.5 eV and comparable with the C–H bonds. Therefore species with the N–H bonds may return N and recombine to $N_2$. That is why they are less abundant than nitriles. The calculated profile of $CH_2NH$ agrees with the INMS value while that of $NH_3$ is smaller than that observed by INMS.

Some species in the model that are not discussed above are compared with the INMS observations in Table 13.10.

Table 13.10 *Mixing ratios of some species retrieved from the INMS observations and in the model*

| Reference | h (km) | $C_6H_2$ | $C_7H_4$ | $C_7H_8$ | $NH_3$ | $CH_2NH$ | $C_2H_3CN$ | $C_2H_5CN$ | $C_4H_3N$ | $HC_5N$ | $C_5H_5N$ |
|---|---|---|---|---|---|---|---|---|---|---|---|
| Magee et al. (2009) | 1050 | – | – | 2.5–8 | – | – | 3.5–7 | 1.5–7 | – | – | – |
| Cui et al. (2009) | 1077 | – | – | <1.3–7 | 3–5 | – | <1.8–5 | – | – | – | – |
| Vuitton et al. (2007) | 1100 | 8–7 | 3–7 | 2–7 | 7–6 | 1–5 | 1–5 | 5–7 | 4–6 | 1–6 | 4–7 |
| This model | 1075 | 2.8–6 | 1.5–9 | 3.8–9 | 1–6 | 1.8–5 | 1.1–5 | 1.5–7 | 4.5–6 | 5–6 | 5–7 |

*Note.* $2.5–8 = 2.5\times10^{-8}$.

### *13.10.7 Oxygen Species*

Ground-based, Earth-orbiting, and CIRS observations of CO, $CO_2$, and $H_2O$ are described in Sections 13.3 and 13.5.

Oxygen is delivered into Titan's atmosphere as the meteoritic water and ions $O^+$. Using the study by Pereira et al. (1997), Wong et al. (2002) argued that a basic reaction of the CO production in the prior models, $OH + CH_3 \rightarrow CO + 2H_2$, proceeds with products $H_2O + {}^1CH_2$ and results in recycling of $H_2O$. They adopted the $H_2O$ influx of $1.5\times10^6$ cm$^{-2}$ s$^{-1}$ and release of CO at $1.1\times10^6$ cm$^{-2}$ s$^{-1}$ at the surface to fit the observations.

Hartle et al. (2006) analyzed the Voyager and Cassini/CAPS data and found a magnetospheric influx of $O^+$ on Titan with a rate of ~$10^6$ cm$^{-2}$ s$^{-1}$. These ions may originate from water in Saturn's ring and Enceladus. Hörst et al. (2008) calculated a model for the oxygen species on Titan using the influxes of OH and O as fitting parameters. Their results agreed with the observations available at that time but exceed the later measurements of $H_2O$ by an order of magnitude.

Our model for oxygen species (Figure 13.65) applied basic ideas of Wong et al. (2002) and Hörst et al. (2008) and is calculated for the $H_2O$ and $O^+$ influxes of $3\times10^6$ and $1.7\times10^6$ cm$^{-2}$ s$^{-1}$. $O^+$ is neutralized by the charge exchange with $CH_4$ and then forms the C=O bonds in reactions with $CH_3$ and $C_2H_4$. OH from photolysis of $H_2O$ reacts with either $CH_3$ and returns $H_2O$ or with CO to form $CO_2$. Photolysis of $CO_2$ returns two CO, because O is converted to CO as well. There is no production of CO from the interior in the model, and CO is not fixed at the surface. Production of formaldehyde $H_2CO$ is a significant branch of the reaction of O + $CH_3$. Photolysis quickly converts $H_2CO$ to CO. However, the formaldehyde abundance reaches 1 ppm near 1100 km. The model agrees with the observed oxygen species.

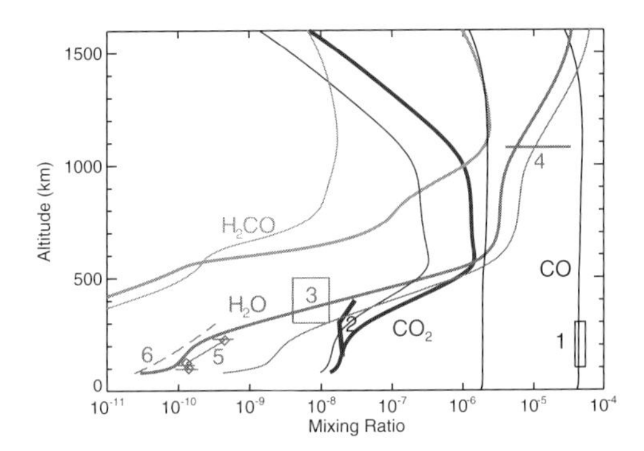


Figure 13.65 Oxygen species in the model (Krasnopolsky 2014b). Observations: (1) De Kok et al. (2007), (2) Vinatier et al. (2010), (3) Coustenis et al. (1998), (4) Cui et al. (2009), (5) Cottini et al., 2012), (6) Moreno et al. (2012). The model without flux of O+ is shown by thin lines.

### *13.10.8 Ionosphere*

Sources of ionization in Titan's atmosphere are discussed in Section 13.8.2 and shown in Figure 13.60. The profiles of ionization by magnetospheric electrons are approximated for the typical conditions and a strong precipitation event T5 using data from Vuitton et al. (2007) and Cravens et al. (2008, 2009). Ionization by protons is from Cravens et al. (2008) and by the cosmic rays from Molina-Cuberos et al. (1999). Ionization by the solar EUV photons is calculated using the standard cross sections and the solar EUV fluxes for the solar zenith angle $z = 60°$ and reduced by a factor of 2 to account for the nightside (global mean conditions).

Except 308 neutral reactions, our model involves 111 reactions of 33 ions, much smaller than those in the ionospheric models by Vuitton et al. (2007) and Cravens et al. (2009). Rate coefficients of ion-neutral reactions and dissociative recombinations were taken from McEwan and Anicich (2007) and the UMIST Database for Astrochemistry (Section 9.7). The model includes ambipolar diffusion and escape of ions at half maximum diffusion rate.

Calculated profiles of the most abundant ions are shown in Figure 13.66. The calculations were made for the global-mean conditions (Figure 13.66a, 66b, and 66c), typical nighttime conditions, and for the T5 event (Figure 13.66d). The recent correction of the INMS neutral densities by a factor of 2.2 (Teolis et al. 2015) means that the altitude scale in

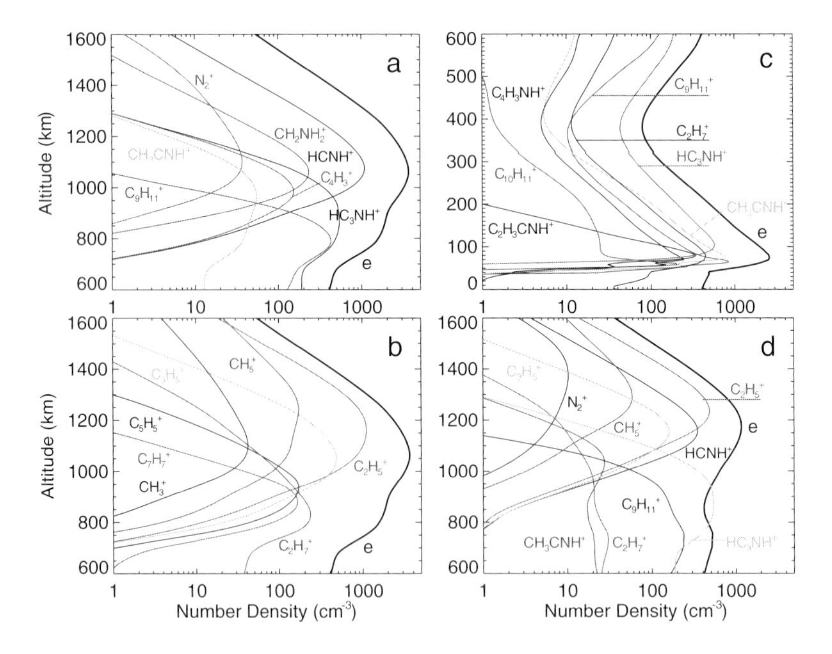


Figure 13.66 Ion composition in the daytime ionosphere above (*a*, *b*) and below (*c*) 600 km. The nighttime ionosphere (*d*) is calculated for the conditions of flyby T5 with strong precipitation of magnetospheric electrons. From Krasnopolsky (2012d). (A black and white version of this figure will appear in some formats. For the color version, please refer to the plate section.)

Figure 13.66 should be shifted upward by 55 km above ≈800 km, so that, e.g., 1000 km in Figure 13.66 corresponds to 1055 km.

Many species and radicals on Titan have significant proton affinity, hydrogen is the abundant element in the atmosphere, and the most abundant ions are protonated species. For example, $HCNH^+$, $HC_3NH^+$, $C_2H_5^+$ are HCN, $HC_3N$, $C_2H_4$ plus proton, respectively.

Production and loss of each ion in the model can be easily understood using the table of ion reactions with their column rates in Krasnopolsky (2009a). For example, ionization of $N_2$ by all sources results in production of $N_2^+$ with column rate (CR) of $2.83 \times 10^8$ cm$^{-2}$ s$^{-1}$. The major loss is

$$N_2^+ + CH_4 \rightarrow CH_3^+ + H + N_2, \; CR = 1.94 \times 10^8 \text{ cm}^{-2} \text{ s}^{-1}.$$

$CH_3^+$ is also formed by photoionization of methane and lost mostly via

$$CH_3^+ + CH_4 \rightarrow C_2H_5^+ + H_2, \; CR = 2.69 \times 10^8 \text{ cm}^{-2} \text{ s}^{-1}.$$

Proton exchange is very effective between $C_2H_5^+$ and nitriles that have greater proton affinity:

$$C_2H_5^+ + HCN \rightarrow HCNH^+ + C_2H_4, \; CR = 1.69 \times 10^8 \text{ cm}^{-2} \text{ s}^{-1}$$
$$C_2H_5^+ + HC_3N \rightarrow HC_3NH^+ + C_2H_4, \; CR = 2.59 \times 10^7 \text{ cm}^{-2} \text{ s}^{-1}.$$

Proton exchange proceeds between nitriles as well:

$$HCNH^+ + HC_3N \rightarrow HC_3NH^+ + HCN, \; CR = 6.68 \times 10^7 \text{ cm}^{-2} \text{ s}^{-1}.$$

Dissociative recombination removes ions:

$$HCNH^+ + e \rightarrow HCN + H, \; CR = 9.08 \times 10^7 \text{ cm}^{-2} \text{ s}^{-1}$$
$$HC_3NH^+ + e \rightarrow C_3N + H_2, \; CR = 9.27 \times 10^7 \text{ cm}^{-2} \text{ s}^{-1}.$$

These reactions are main production and loss processes for $N_2^+$, $CH_3^+$, $C_2H_5^+$, $HCNH^+$, and $HC_3NH^+$. More detailed data for all processes and all ions may be found in the table in Krasnopolsky (2009a). The above reactions demonstrate transitions from the primary to secondary, tertiary, and terminal ions on Titan.

While modeling of Titan's ionosphere has a long history, the first models to fit the Cassini INMS ion spectra were made Vuitton et al. (2007) and Cravens et al. (2009) for the T5 event of a strong nighttime precipitation of magnetospheric electrons (Section 13.7.3 and Figure 13.43).

To make a quantitative assessment to fitting of observations by models, we suggested a difference factor that is equal to a mean ratio of observed-to-model values or vise versa, so that the biggest value is always in the numerator (Krasnopolsky 2009a). Therefore the difference factor for a perfect fitting is 1. This factor is 2.42 for the T5 model in Figure 13.66d, while it is 1.74 for Vuitton et al. (2007) and 5.83 for Cravens et al. (2009). Our model is self-consistent and does not include densities of neutral species as fitting parameters; therefore the agreement with the observed INMS ion spectrum is very good.

Mean INMS results on the daytime and nighttime ionospheres are discussed in Section 13.8.3 based on the data from Mandt et al. (2012) in Table 13.4. The ion densities in Table 13.4 should be scaled by a factor of 1.55 (Teolis et al. 2015). The daytime data may be compared with those in Figures 13.66a and 13.66b. This comparison for 1100 km results in a difference factor of 2.45. A model by Robertson et al. (2009) for flyby T17 with SZA = 34° was published with complete sets of the observed and calculated ion densities at 1156–1206 km that makes it possible to calculate a difference factor of 4.68. Published papers on other models do not contain full data set for calculation of difference factor.

A model for the daytime ion composition was developed by Westlake et al. (2012). De la Haye et al. (2008) constructed a model of diurnal variations of the atmosphere and ionosphere above 600 km at 39°N and 74°N. These latitudes correspond to the conditions of the $T_A$ and T5 flybys; $z = 62°$ at noon for 39°N and the polar night conditions at 74°N. The model involved 35 neutral and 44 ion species. Neutrals with lifetimes exceeding $1.5\times10^4$ s were fixed at 600 km to fit the mean INMS data near 1000 km or taken from Lebonnois (2005). All ions were calculated neglecting ambipolar diffusion and escape. Wahlund et al. (2009) analyzed data of INMS, two CAPS channels, and two RPWS channels during three flybys of Titan with the ionospheric peak densities of 900–3000 cm$^{-3}$. Heavy (>100 amu) positive ions constitute about one-half of the total ions near 1000 km, and negative ion densities are <100 cm$^{-3}$. Heavy ions may dominate below 950 km, and recombination of heavy ions may contribute to the haze production.

Our model predicts $C_9H_{11}^+$ (119 amu) as a major ion at 500–900 km with densities of ~200 cm$^{-3}$ (Figure 13.66). $C_{12}H_{10}$ and $C_{10}H_{11}^+$ with 154 and 131 amu are the heaviest neutral and ion species, respectively, in our model.

### *13.10.9 Production of Haze*

Haze is formed on Titan by polymerization and condensation of hydrocarbons and nitriles and recombination of heavy ions. Condensation rates for photochemical products are given in Table 13.9; haze production by polymerization and recombination is shown in Figure 13.67.

Polymerization of $C_4H_2$ in reactions with $C_6H$ and $C_3N$ produce 75% and 95% of hydrocarbon and nitrile polymers. Both processes peak near 570 km, and polymerization of nitriles becomes significant even at 1000 km. Recombination of heavy ions is another source of haze. The heaviest ions in our model are $C_{10}H_{11}^+$, $C_{10}H_9^+$, and $C_9H_{11}^+$ with masses of 131, 129, and 119 amu, and their recombination peaks at 770 km. Strong secondary peaks of all three sources of haze are near 100 km and due to chemistry initiated by the cosmic rays.

Summary of deposition rates to the surface from all photochemical products is given in Table 13.11. Polymerizations of hydrocarbons and nitriles have equal rates while their condensation rates are very different. The deposition by condensation does not account for possible partial sublimation near the surface, where the atmosphere is warmer than that at

Table 13.11 *Precipitation of photochemical products (in g cm$^{-2}$ Byr$^{-1}$)*

| | |
|---|---|
| $C_XH_Y$ polymerization | 1645 |
| $C_XH_YN$ polymerization | 1650 |
| Recombination of heavy ions | 26 |
| $C_XH_Y$ condensation | 2246 |
| $C_XH_YN$ condensation | 400 |
| Total | 5967 |

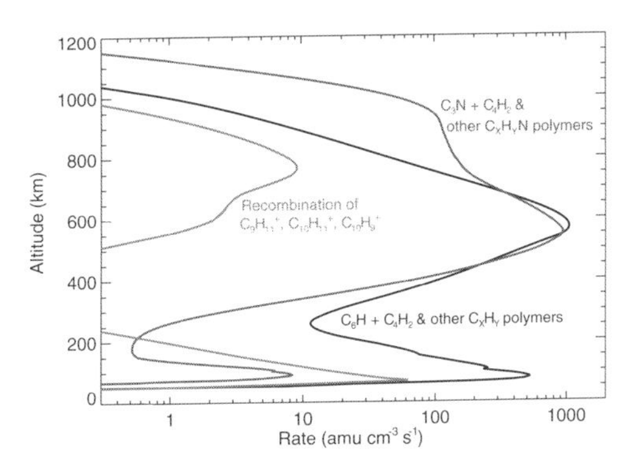


Figure 13.67 Production of haze on Titan by polymerization of hydrocarbons and nitriles and recombination of heavy ions (Krasnopolsky 2014b).

the tropopause (94 and 70 K, respectively). The deposition of nitrogen is 390 g cm$^{-2}$ Byr$^{-1}$ in total, that is, 7% of the total deposition rate of 6 kg cm$^{-2}$ Byr$^{-1}$. The model implies a surface deposit thickness of ~60 m per billion years. The total flow of $6\times10^{-6}$ g cm$^{-2}$ yr$^{-1}$ looks negligible compared with precipitation of ~100 g cm$^{-2}$ yr$^{-1}$ of water on the Earth. It is not clear if this deposit could be reprocessed and in some part returned into the atmosphere.

## 13.11 Unsolved Problems

Though the Huygens/Cassini mission was very successful, some unsolved problems related to the chemical composition of Titan's atmosphere and ionosphere remain. First of all, the atmosphere between 350 and 950 km has not been studied in detail, and this is the region where the basic photochemical processes proceed.

Photochemical products on Titan are typically formed near 500 km, and their mixing ratios are rather constant up to ~1000 km due to the strong eddy diffusion. The product mixing ratios decrease to ~100 km, where the species condense. Therefore the expected gradients of the mixing ratios at 100–500 km are the differences between the INMS observations and the CIRS nadir values divided by ~350 km. However, some CIRS limb observations demonstrate smaller gradients. This is another general problem that may be partly related to non-LTE effects. Limb emissions of species observed by CIRS are analyzed assuming local thermodynamic equilibrium. However, deviations from LTE may be significant above ≈300 km and affect the retrieved species abundances.

INMS is the major source of the data on Titan's chemical and ion composition. However, retrieval of species densities from the INMS observations is a difficult process that was done by two independent teams and results in significant differences and uncertainties.

The problem of heavy ions and negative ions remain uncertain both in observations and modeling. There are significant differences between the existing observations and models.

The composition of Titan's atmosphere and ionosphere is very complicated and involves many species. Maybe, there are some species that contribute to the UVIS occultations at 140–190 nm apart from nine compounds used in the analysis. There are problems related to some of the observed species that we have not discussed here.

# 14

# Triton

## 14.1 General Properties and Pre-Voyager Studies

Triton (Figure 14.1) is the only big satellite of Neptune, and its radius of 1353 km is close to those of Pluto (1190 km, 30–50 AU) and Eris (1163 km, 96.3 AU). Conditions on Triton are significantly determined by Neptune's orbit that is almost circular with a mean radius of 30.1 AU (Table 14.1) and a period of 164.8 years.

The angle between Neptune's equatorial and orbital planes (obliquity) is 28°. Triton's orbit is retrograde and tilted relative Neptune's equator by −23°, that is, 156.9°. (Minus reflects the retrograde rotation.) Therefore Triton's total inclination to Neptune's orbit is −51°, that is, 129.6°. The annual period combined with a comparatively short precession period of 688 years results in a complicated seasonal behavior on Triton (Figure 14.2).

The retrograde rotation indicates that Triton could not originate from the Neptune subnebula and is a captured body. The capture requires a significant loss of the initial energy of Triton. It is suggested that Triton might have had a satellite before the capture, similar to Pluto and Charon, that got the excess energy and expelled. Triton's mass is currently 99.7% of the entire Neptune's satellite system.

Triton was discovered by British astronomer Lassell in 1846. Basic data on Triton were obtained during the Voyager 2 flyby at 39,800 km from Triton's center in August 1989. The pre-Voyager studies of Triton were rather scarce, and their review may be found in Atreya (1986). Triton's radius was evaluated at 1750 km, so that the disk surface was overestimated by a factor of 1.7, while the albedo was underestimated by the same factor and even more to a value of 0.4. Then the simple thermal balance equation (2.27) with an adopted emissivity of 0.8 resulted in the subsolar and disk average temperatures of 67 and 57 K, respectively, much higher than that measured by Voyager 2. Cruikshank et al. (1984) and Cruikshank and Apt (1984) observed the $N_2$ band at 2.16 μm and the $CH_4$ band at 2.3 μm. Densities of both gases should be very significant at the above temperatures, and the atmosphere of Triton was expected to be rather dense.

Table 14.1 *Basic properties of Triton*

| | |
|---|---|
| Semi-major axis of Neptune (AU) | 30.11 |
| Eccentricity | 0.0095 |
| Orbital period of Neptune (years) | 164.8 |
| Neptune obliquity | 28.3° |
| Distance from Neptune (km) | $3.548 \times 10^5$ |
| Period of rotation (days) | 5.877 |
| Inclination to Neptune's equator | 156.9° |
| Inclination to Neptune's orbit | 129.6° |
| Radius (km) | 1353 |
| Mass (g) | $2.14 \times 10^{25}$ |
| Mean density (g cm$^{-3}$) | 2.06 |
| Surface gravity (cm s$^{-2}$) | 78 |
| Escape velocity (km s$^{-1}$) | 1.455 |

Figure 14.1 Photomosaic of Triton based on the Voyager 2 images. (A black and white version of this figure will appear in some formats. For the color version, please refer to the plate section.)

## 14.2 Interior and Surface

### *14.2.1 General Features*

The accurate Voyager measurements of Triton's radius and mass (Table 14.1) revealed its mean density of 2.06 g cm$^{-3}$. This means that Triton has a water-ice crust of ≈400 km thick and a silicate and metal core. Tidal dissipation and radiogenic heating (≈$10^{11}$ W) are essential in thermal balance of the interior. Some versions of the problem presume a thin liquid ocean of water with some ammonia at the bottom of the crust.

Voyager 2 observed Triton (Figure 14.1) near a peak of the warmest southern summer (Figure 14.2). The images revealed a rather uniform surface at the low latitudes and an extended polar cap of nitrogen ice. The surface is very young with a small number of

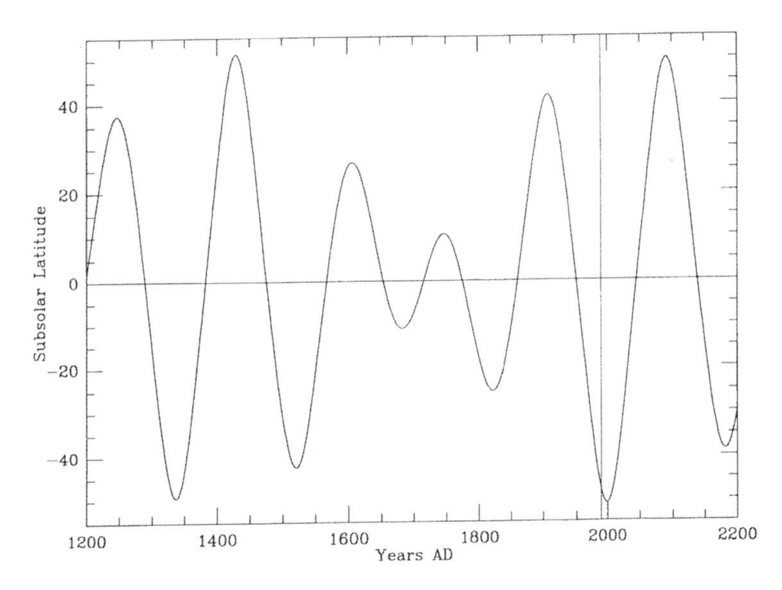


Figure 14.2 Seasonal variations of the subsolar latitude on Triton. Vertical line corresponds to the Voyager 2 encounter.

impact craters and the estimated age of 5 to 50 Myr. The observed topographic features are typically smaller than 200 m in altitude.

There are many geyser-like eruptions (plumes) on the south polar cap. Gas and dust in the plumes move vertically up to 8 km and then make long horizontal tails up to 150 km. This is a rare case of cryovolcanism in the Solar System. All plumes were observed near the subsolar latitude at 50°–57°S. It looks like the nitrogen ice covers dusty subsurfaces at the plume locations. Absorption of the sunlight by dust heats and evaporates the ice. Major findings in Neptune and Triton by the Voyager 2 encounter are summarized in "Neptune and Triton" (Cruikshank 1995).

### *14.2.2 Surface Temperature*

An average spectrum of Triton at 20–50 μm observed using Voyager/IRIS (Conrath et al. 1989) showed some signal at 40–50 μm that resulted in the value of Triton's temperature of $38^{+3}_{-4}$ K.

Nitrogen ice is in phase $\beta$ above 35.6 K and in phase $\alpha$ below this temperature. Structure of the $N_2$ ice band at 2.16 μm is sensitive to temperature near 35.6 K. Ground-based spectroscopy of this band made it possible to determine the ice temperature of $38^{+2}_{-1}$ K (Tryka et al. 1993).

The Voyager radio occultations of Triton revealed the surface atmospheric pressure of 14 ± 1 μbar and an equivalent isothermal atmospheric temperature of 42 ± 4 K (Gurrola 1995). Assuming equilibrium between the atmospheric nitrogen and the surface ice and taking into account the IRIS data, a recommended surface temperature is 37.5 ± 0.5 K (Gurrola 1995) for the surface emissivity of 0.5. This agrees with the IRIS data that give 41 ± 5 K for the emissivity of 0.5 (Yelle et al. 1995).

Condensation and sublimation of nitrogen result in heating and cooling of the ice and the atmosphere. Topographic features are low, and the atmosphere is almost isobaric near the surface. The ice is in equilibrium with the $N_2$ gas; therefore the nitrogen ice is isothermal throughout Triton's surface.

Based on the Voyager 2 images, Triton's bolometric hemispherical albedo is 0.82 (Hiller et al. 1994). Using the thermal balance equation (2.27), Triton's temperature is 39.4, 37.6, and 36.2 K for emissivity of 0.5, 0.6, and 0.7, respectively. Overall 38 K is a rather reliable value for the surface temperature of Triton at the Voyager 2 encounter.

### *14.2.3 Composition of the Surface Ice*

This problem is solved by means of the near-infrared spectroscopy. The first results were obtained before the Voyager 2 flyby (Cruikshank et al. 1984; Cruikshank and Apt 1984), and the first quantitative assessment of the ice composition was made later (Cruikshank et al. 1993). These studies continue now (Holler et al. 2016), while the most detailed spectroscopic analysis was made by Quirico et al. (1999). Their results will be briefly discussed here.

Spectra of Triton observed in 1995 using the cooled grating spectrograph CGS4 at the UK Infrared Telescope (UKIRT, Hawaii, elevation 4.2 km) are shown in Figure 14.3. Bands of five ices ($N_2$, $CH_4$, CO, $H_2O$, and $CO_2$) are identified in Triton's spectra, and laboratory spectra of these ices are depicted in Figure 14.3 as well.

Hapke (1993) adjusted the Mie theory to calculate reflectance of solid surfaces that consist of particles and grains with given properties. Quirico et al. (1999) made some changes in Hapke's method to fit better to their problem. They conclude that Triton's ice in some regions is a solid solution of $CH_4$ and CO in $N_2$. Other regions are either mixtures of $H_2O$ and $CO_2$ ices or covered by pure $H_2O$ and $CO_2$ ices. Retrieved quantities of species and their grain sizes are given in Table 14.2.

According to the models, 55% of Triton's surface is covered by $N_2$ ice with 0.1% of $CH_4$ and 0.05% of CO diluted in the ice. The ice pieces are large, $\approx$10 cm. The remaining 45% are covered by $CO_2$ and $H_2O$ (either mixture or separate regions) with an approximative proportion 1:2. Their grain sizes are $\approx$0.2 mm.

A very weak spectral feature at 2.406 μm (Figure 14.3) may be caused by either $^{13}CO$ or $C_2H_6$ ices diluted in $N_2$ ice. DeMeo et al. (2010) argued that both ices contribute to the observed band. Holler et al. (2016) measured spectra of Triton at various rotational phases. They found no correlation between the band at 2.406 μm and the CO bands at 1.58 and 2.35 μm. This indicates solid ethane on Triton.

Triton's spectra observed by two instruments onboard the Japanese orbiter AKARI detected a band at 4.76 μm (Figure 14.4) that was tentatively identified as that of HCN ice (Burgdorf et al. 2010).

Thus seven species have been observed in Triton's ice. $N_2$, $CH_4$, and CO have significant vapor densities and should be present in the atmosphere. $H_2O$ and $CO_2$ ices are inert at Triton's temperatures, while $C_2H_6$ and HCN are products of photochemistry.

Table 14.2 *Composition and grain sizes of Triton's ice*

| Model a | | Model b | | Model c | |
|---|---|---|---|---|---|
| $N_2 : CH_4 : CO$ | $H_2O + CO_2$ | $N_2 : CH_4 : CO$ | $H_2O + CO_2$ | $N_2 : CH_4 : CO$ | $H_2O + CO_2$ |
| $G_{N2:CH4:CO} = 55\%$ | $G_{CO2+H2O} = 45\%$ | $G_{N2:CH4:CO} = 55\%$ | $G_{CO2+H2O} = 45\%$ | $G_{N2:CH4:CO} = 55\%$ | $G_{CO2+H2O} = 45\%$ |
| $D_{N2} = 1.1\ 10^5$ μm | $D_{CO2} = 300$ μm | $D_{N2} = 1.1\ 10^5$ μm | $D_{CO2} = 600$ μm | $D_{N2} = 1.1\ 10^5$ μm | $D_{CO2} = 130$ μm |
| $C_{CH4} = 0.08\%$ | $D_{H2O} = 160$ μm | $C_{CH4} = 0.08\%$ | $D_{H2O} = 160$ μm | $C_{CH4} = 0.11\%$ | $D_{H2O} = 160$ μm |
| $C_{CO} = 0.05\%$ | $A_{CO2} = 20\%$ | $C_{CO} = 0.05\%$ | $A_{CO2} = 13\%$ | $C_{CO} = 0.05\%$ | $A_{CO2} = 74\%$ |
| $g = -0.4$ | $A_{H2O} = 80\%$ | $g = -0.4$ | $A_{H2O} = 87\%$ | $g = -0.4$ | $A_{H2O} = 26\%$ |
| | $g = -0.4$ | | $g = -0.4$ | | $g = -0.4$ |

*Note.* From Quirico et al. (1999).

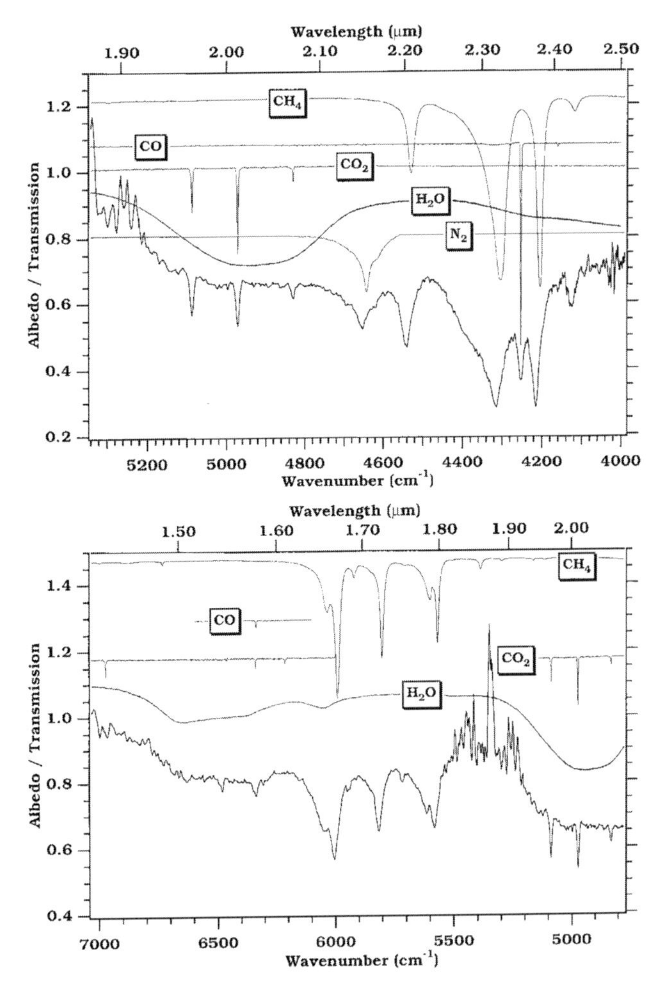


Figure 14.3 Observed spectra of Triton and laboratory spectra of five ices contributing to Triton's spectra. From Quirico et al. (1999).

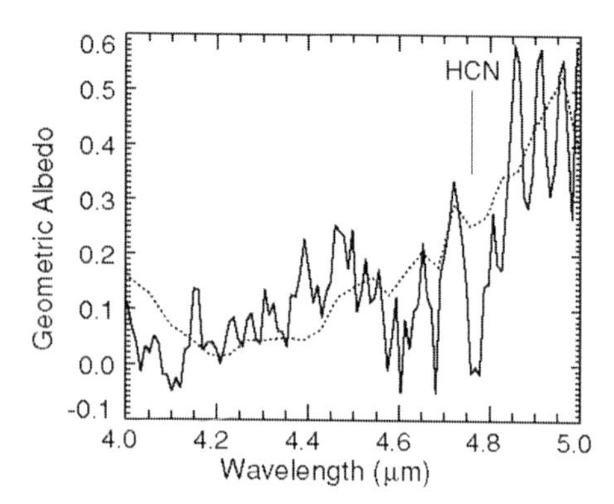


Figure 14.4 Spectra of Triton observed by two instruments onboard the AKARI orbiter indicate the absorption band of HCN (Burgdorf et al. 2010).

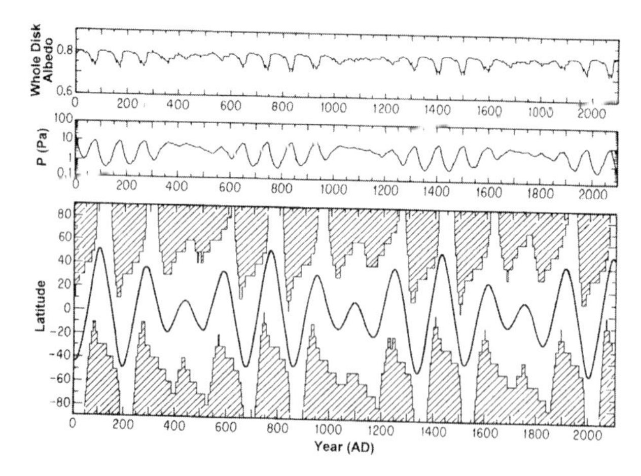


Figure 14.5 Model of seasonal variations of the whole disk albedo, atmospheric pressure, the subsolar latitude, and $N_2$ ice coverage (shaded areas) on Triton (Hansen and Paige 1992).

### *14.2.4 Seasonal Variations*

Triton's year is long, and the subsolar latitude varies up to $\pm 50°$. This presumes strong seasonal variations in the atmosphere and ice. A model of these variations by Hansen and Paige (1992) is shown in Figure 14.5. The model was calculated for the background ice ($H_2O$ and $CO_2$) with albedo of 0.8 and the $N_2$ frost with other admixtures having albedo of 0.62. The calculated whole disk albedo varies from 0.73 to 0.8. Atmospheric pressure varies from 5 to 100 μbar, while the frost boundary varies from the poles to the equator. The model predicts a decrease in the atmospheric pressure after the Voyager flyby that disagrees with the stellar occultations (Section 14.5).

Modeling of seasonal variations on Triton by Spencer and Moore (1992) involved 15 versions for various assumptions. Three versions with all permanent frost in the southern hemisphere show a pressure increase to $\approx 22$ μbar in 2000.

### 14.2.5 Spatial Distribution of Ices

Grundy et al. (2010) conducted long-term observations of Triton's spectra of ices using the IRTF/SpeX spectrograph in a period of 2000–2009. This made it possible to study variations of the ice band equivalent widths as functions of sub-Earth longitude. Peak-to-peak variations are by a factor of 1.7 for the $N_2$ ice band at 2.15 μm and that of CO at 2.35 μm. The bands are maximal near the sub-Neptune region and shifted to the east at $31 \pm 3°$ and $56 \pm 4°$, respectively. Variations of the $CH_4$ ice bands at 1.73 μm and 0.89 μm are smaller and shifted to the west at ≈40 and ≈90°W, respectively. Ices of $CO_2$ and $H_2O$ demonstrate very low variability. These are disk-average observations, and the true variations are significantly smoothed by this averaging.

## 14.3 Atmosphere

### 14.3.1 Voyager Radio Occultations

It was mentioned in 14.2.2 that the radio occultations resulted in the atmospheric pressure of $14 \pm 1$ μbar, the surface temperature of $37.5 \pm 0.5$ K, and the equivalent isothermal temperature of $42 \pm 4$ K below 60 km (Gurrola 1995).

Radio occultations of Triton's ionosphere (Tyler et al. 1989) revealed peak electron densities of $2.3 \times 10^4$ cm$^{-3}$ at the ingress (dawn) and $4.6 \times 10^4$ cm$^{-3}$ at the egress (dusk). The peak altitude is 340 km, and the topside plasma scale height is $128 \pm 25$ km.

### 14.3.2 Haze

The Voyager images at the limb show a thin haze up to 20–30 km and clouds below 5 km that cover 37% of the limb poleward of 30°S (Pollack et al. 1990). Absorption by the haze at 140–168 nm (Herbert and Sandel 1991) was measured by solar occultations using the ultraviolet spectrometer (UVS). The retrieved vertical profile of volume extinction coefficient is shown in Figure 14.6. The haze vertical optical depth was 0.022 at 150 nm. The initial interpretation involved Rayleigh-scattering particles with $r \approx 0.015$ μm. However, extrapolation of Rayleigh scattering to the visible yields much smaller brightness of the haze than that observed and shown in Figure 14.6.

A model to fit the observed vertical profiles of the haze at 150 and 470 nm (Krasnopolsky et al. 1992, 1993) assumed production and condensation of ethylene $C_2H_4$ by photolysis of methane with a quantum yield of ≈1. The condensation begins at 30 km, and the effective particle radius at 30 km is a model parameter. The particles precipitate with the Stokes–Davis velocity (4.34) and grow via condensation. The atmosphere is denser at low altitudes, the particle velocity becomes smaller, and therefore the particle number density increases. The model agrees with the observations (Figure 14.6), and the observed and calculated properties of the haze are in Table 14.3.

Rages and Pollack (1992) studied radial limb scans at radii of 1200–1400 km and large phases angles ≈150°. Their best-fit models for latitude of 15°S (close to that in Figure 14.6) are for the particle radii either 0.17 μm or 0.025 μm.

Table 14.3 *Measured and model data for haze on Triton*

| h. | Measured | Values | Model | | | | | |
|---|---|---|---|---|---|---|---|---|
| km | $\sigma_{UV}$, $cm^{-1}$ | $\sigma_B$, $cm^{-1}$ | $\sigma_{UV}$, $cm^{-1}$ | $\sigma_B$, $cm^{-1}$ | $\sigma_{vis}$, $cm^{-1}$ | n, $cm^{-3}$ | $r_c$, μm | $r_{UV}$, μm |
| 30 | 1.8(−9) | 6.4(−11) | 1.63(−9) | 3.70(−11) | 1.86(−10) | 2.42 | 0.098 | 0.095 |
| 20 | 3.8(−9) | 1.6(−10) | 3.98(−9) | 1.47(−11) | 8.09(−10) | 3.08 | 0.124 | 0.119 |
| 10 | 7.6(−9) | 3.6(−10) | 7.55(−9) | 3.66(−10) | 2.14(−9) | 5.37 | 0.135 | 0.132 |
| 0 | 1.56(−8) | 7.5(−10) | 1.47(−8) | 7.22(−10) | 4.22(−9) | 10.4 | 0.135 | 0.133 |

*Note.* $\sigma_{UV}$ and $\sigma_{vis}$ are volume extinction coefficients at 150 and 470 nm, $\sigma_B$ is the volume brightness coefficient at 470 nm, $n$ is the particle number density, $r_c$ is the geometrical cross section average radius, $r_{UV}$ is the cross section weighted-mean radius at 150 nm, $1.8(-9) = 1.8 \times 10^{-9}$. From Krasnopolsky (1993a).

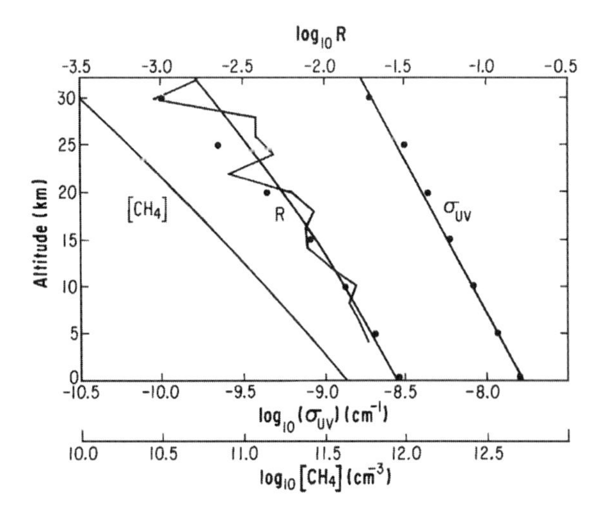


Figure 14.6 Mean profiles of methane and volume extinction coefficient ($\sigma_{UV}$) of the haze measured by the UVS solar occultations at 120–140 and 140–165 nm, respectively (Herbert and Sandel 1991). The broken line is the haze brightness coefficient $R = \pi I/I_0$ in the limb images at 13.6°S and at 470 nm (Pollack et al. 1990). Its least square fit by a second-order polynomial is shown as well. Dots are results of the haze model. From Krasnopolsky et al. (1992).

### 14.3.3 $CH_4$, N, and $N_2$ Densities from the UVS Solar Occultations

Herbert and Sandel (1991) retrieved methane density profiles using the observed absorption at 119–140 nm in the solar occultation spectra (Figure 14.7). Both profiles are near saturation at the surface and steeply decrease with altitude because of photolysis by the solar Lyman-alpha and the interplanetary background at this line. The irradiation is stronger in summer with a steeper decrease in the observed methane. Fitting the observed profiles by models with various eddy diffusion coefficients is shown.

Transmission curves at four altitudes in the range of 54–91 nm are depicted in Figure 14.8. These curves demonstrate a strong absorption by the $N_2$ ionization continuum at $\lambda < 80$ nm and a step-like feature at 85 nm that corresponds to the ionization continuum of atomic nitrogen. Retrieved slant column abundances of both species were converted to number densities that are shown in Figures 14.9 and 14.10. The observed profiles of atomic nitrogen are fitted by diffusive equilibrium with [N] = $(1.0 \pm 0.25)\times 10^8$ cm$^{-3}$ at 400 km and $T = 100 \pm 7$ K, and the $N_2$ profiles give [$N_2$] = $(4 \pm 0.4)\times 10^8$ cm$^{-3}$ at 575 km and

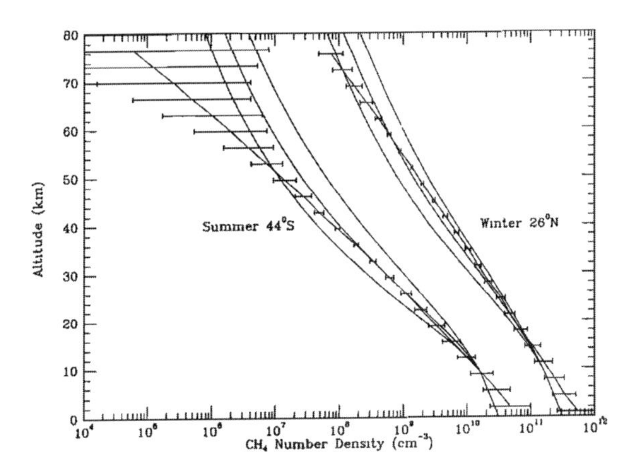


Figure 14.7 Methane profiles extracted from the UVS solar occultations (Herbert and Sandel 1991). The summer profile is displaced to the left by a factor of 10. The model curves are for eddy diffusion coefficient $K$ = 2000, 4000, and 8000 cm$^2$ s$^{-1}$ and $K = 10^5$ cm$^2$ s$^{-1}$ below 10 km. From Krasnopolsky and Cruikshank (1995).

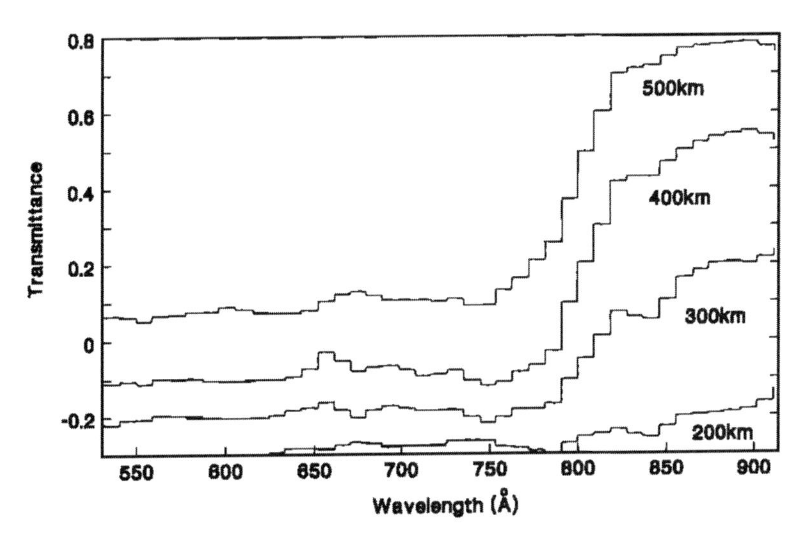


Figure 14.8 Transmission spectra from the UVS solar occultation (entrance, 26°N) show a steplike feature at 85 nm caused by the ionization continuum of atomic nitrogen. Spectra at 400, 300, and 200 km are displaced down by 0.1, 0.2, and 0.3, respectively. From Krasnopolsky et al. (1993).

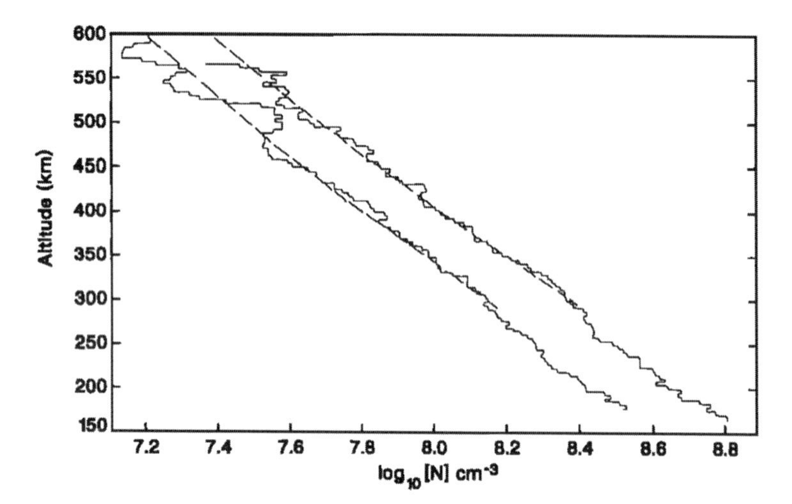


Figure 14.9 Atomic nitrogen number densities derived from the UVS solar occultations (exit at 44°S is left and displaced by 0.2). Dashed curves are for diffusive equilibrium at $T$ = 100 K and [N] = $1.06\times10^8$ cm$^{-3}$ at 400 km. From Krasnopolsky et al. (1993).

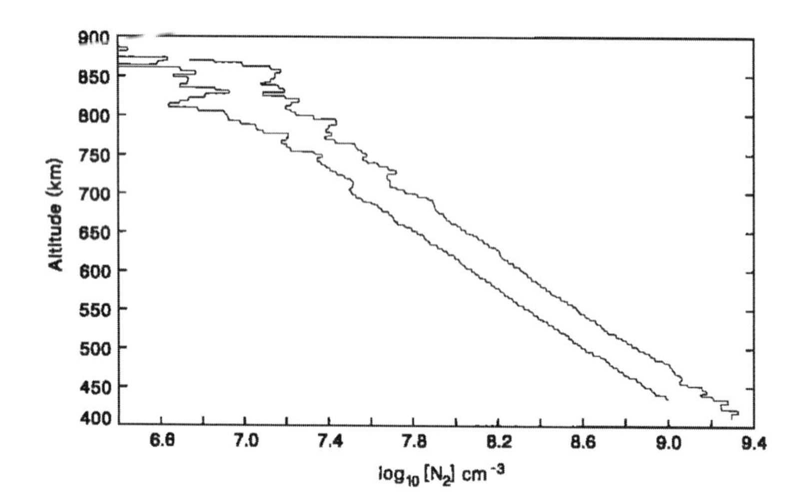


Figure 14.10 $N_2$ density profiles from the UVS solar occultation. Exit at 44°S is left and displaced by 0.2 (Krasnopolsky et al. 1993).

$T$ = 102 ± 3 K. The N and $N_2$ profiles are similar at the entrance and exit occultations within the claimed uncertainties, despite the very different conditions. This may be explained by strong winds that globally mix the atmosphere. Using the observed N and $N_2$ profiles, the exobase altitude is at 870 km and thermal escape of N is $9\times10^{24}$ s$^{-1}$.

Coupling the occultation data at 80–100 nm with the total atmospheric pressure of 14 μbar and a calculated temperature profile (see below) made it possible to deduce relationships (9.31, 9.32) for absorption of $N_2$ in this range, which is difficult to calculate by the standard methods.

### 14.3.4 Thermal Balance and Temperature Profile

Yelle et al. (1991) argued that the observed altitude of 8 km of the geysers (plumes) indicates a tropopause. The tropospheric temperature gradient follows the $N_2$ saturation and is equal to $-0.13$ K km$^{-1}$ (Section 2.1.4, relationship (2.11)) with $T = 37$ K at the tropopause. Thermal radiation of the atmospheric species is smaller than a heat flux from the thermosphere; therefore mesosphere is lacking and thermosphere begins at the tropopause.

Stevens et al. (1992) and Krasnopolsky et al. (1993) calculated thermal balance of Triton's thermosphere. Their models are different in details but rather similar in conclusions. Here the model by Krasnopolsky et al. (1993) will be briefly discussed. It includes calculations of heating efficiencies for some key processes that will be omitted here. Sources of energy are the solar EUV at high solar activity ($F_{10.7} \approx 200$) and magnetospheric electrons, whose ionization rate profile was calculated by Strobel et al. (1990) for crossing of Neptune's equator. A similar profile was shifted upward by two scale heights in Summers and Strobel (1991). The model versions for these two profiles are labeled $e_{low}$ and $e_{high}$. The effect is maximal at the crossing by Neptune's equator, while a mean effect is scaled by a factor of $\beta$.

The most abundant molecules $N_2$, $CH_4$, and $H_2$ have no dipole moment, and rotational transitions are forbidden for them. Furthermore, even multiplet splitting is lacking for the atomic species N and H in the ground states. Therefore CO is the only molecule that radiates in its rotational lines. The lines are narrow and thermal-broadened with strong saturation effects, and the calculated total CO cooling varies by a factor of 4 when the adopted CO mixing ratio varies from $10^{-4}$ to $10^{-2}$.

Ionization of $N_2$ has the highest rate in both the solar EUV absorption and magnetospheric electrons deceleration. It is mostly followed by

$$N_2^+ + H_2 \rightarrow N_2H^+ + H,$$
$$N_2H^+ + e \rightarrow N_2 + H,$$
$$\text{Net } N_2^+ + H_2 + e \rightarrow N_2 + H + H + 11.1 \text{ eV}.$$

The same refers to $CO^+$. Heating efficiencies of these processes are unknown, and two values, $\gamma = 1/3$ and $2/3$, were used in the model. Sixteen versions of the model were calculated (Table 14.4) using relationship (2.35), and $\beta$ was a parameter to fit the observed $[N_2] = (4 \pm 0.4)\times10^8$ cm$^{-3}$ and $T = 102 \pm 3$ K at 575 km. Nine versions are within the claimed uncertainties of the above values, and the thermal balance modeling does not give clear preferences to three model parameters. However, the differences between the parameters are moderate, and heating efficiencies of both solar EUV photons and magnetospheric electrons are $\approx$0.25, while the mean effect of the magnetospheric electrons is $\approx$0.2 of that at the crossing of Neptune's equator.

Table 14.4 *Values of normalized electron flux ($\beta$), temperature, and $[N_2]/10^8$ $cm^{-3}$ at 575 km in 16 versions of thermal balance in Triton's thermosphere*

| Model | $f_{CO} = 0$ | $f_{CO} = 10^{-4}$ | $f_{CO} = 10^{-3}$ | $f_{CO} = 10^{-2}$ | $\epsilon_{hv}$ | $\epsilon_e$ |
|---|---|---|---|---|---|---|
| $e_{low}$, $\gamma = 1/3$ | 0.265/99.5/4.89 | 0.270/99.5/4.83 | 0.298/99.5/4.34 | 0.399/100.8/4.00 | 0.235 | 0.199 |
| $e_{low}$, $\gamma = 2/3$ | 0.176/99.0/4.63 | 0.184/99.0/4.60 | 0.204/100.0/4.50 | 0.272/101.0/4.10 | 0.252 | 0.290 |
| $e_{high}$, $\gamma = 1/3$ | 0.190/102.5/ 4.02 | 0.195/102.8/ 3.97 | 0.214/103.5/3.73 | 0.275/105.0/3.22 | 0.235 | 0.215 |
| $e_{high}$, $\gamma = 2/3$ | 0.144/102.2/ 3.95 | 0.148/102.9/ 4.01 | 0.162/102.7/3.56 | 0.213/106.2/3.46 | 0.252 | 0.277 |

*Note.* Sixteen versions of the thermal balance model are given which cover two profiles of electron precipitation (low, from Strobel et al. 1990, and high, from Summers and Strobel 1991), two heating efficiencies $\gamma = 1/3$ and $2/3$ of the reactions $N_2^+$, $CO^+ + H_2$ with subsequent recombination, and four values of the CO mixing ratio, $f_{CO}$. Each version is described by the ratio $\beta$ of the mean electron flux to the maximum flux, which was chosen to provide the best fit to the measured $T = 102 \pm 3$ K and $[N_2] = (4 \pm 0.4) \times 10^8$ $cm^{-3}$ at 575 km, and the calculated values of $T$ and $[N_2]$ in the format: $\beta/T/[N_2]_{575}$ km (in $10^8$ $cm^{-3}$). The mean photon ($\lambda < 800$ Å) and electron heating efficiencies, $\epsilon_{hv}$ and $\epsilon_e$, are also shown. From Krasnopolsky et al. (1993).

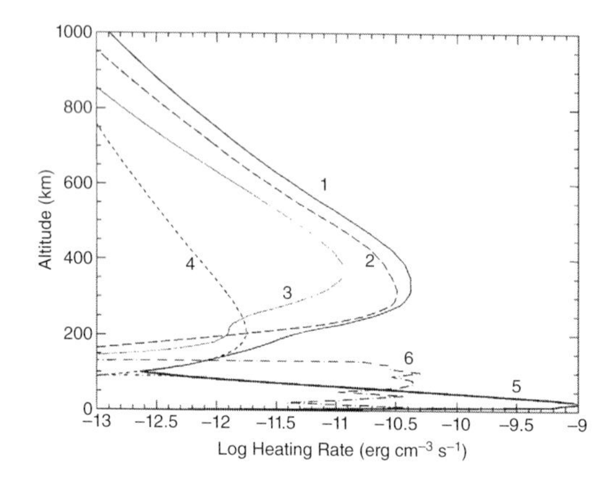


Figure 14.11 Heating rates in Triton's atmosphere: total rate (1), by magnetospheric electrons (2), the solar EUV ($\lambda < 80$ nm (3) and 80–100 nm (4)), and by methane absorption (5). Cooling by the CO rotational lines is (6). The chosen model is with $e_{high}$, $\gamma = 2/3$, $f_{CO} = 10^{-3}$, and $\beta = 0.162$ (Krasnopolsky et al. 1993).

Heating and cooling rates in the atmosphere are shown in Figure 14.11. The former exceeds the latter, and the excess heats the troposphere. Calculated temperature profile for one of the models is presented in Figure 14.12. Energy input from magnetospheric electrons is $\approx 10^8$ W and exceeds that from the solar EUV at $\lambda < 80$ nm by a factor of $\approx 2$.

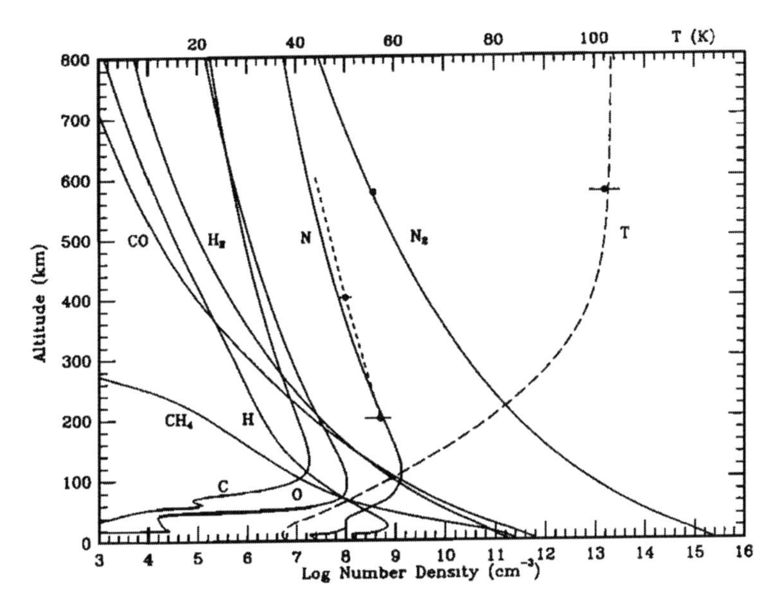


Figure 14.12 Composition of Triton's atmosphere: the most abundant species (Krasnopolsky and Cruikshank 1995). The $N_2$ and $T$ profiles and the measured N profile (short dashes), $N_2$, and $T$ at 575 km are from Krasnopolsky et al. (1993).

Stevens et al. (1992) calculated five models, and their preference is for a model with (in our terms) $e_{\text{high}}$, $\beta = 0.15$, and mean heating efficiency $\varepsilon = 0.24$. It is close to the model in Figures 14.11 and 14.12 that has $\varepsilon = 0.25$ for photons with $\lambda \leq 80$ nm and $\varepsilon = 0.22$ for magnetospheric electrons. The calculated $T = 96$ K in the preferable model by Stevens et al. (1992) was claimed to fit the UVS solar occultations but uncorrected for the atomic nitrogen absorption. Some significant differences between the models are outlined in Krasnopolsky et al. (1993).

Using the chosen thermal balance model and some chemistry of N, $H_2$, and H, Krasnopolsky et al. (1993, table 9) developed a reference model of Titan's atmosphere during the Voyager encounter. The model includes profiles of $T$, $N_2$, $H_2$, N, H, and $CH_4$. Pressure is 13.5 μbar near the surface and 0.67 μbar at a reference level $r = 1400$ km ($h = 47$ km) for stellar occultations (Section 14.5.1); temperatures are 38.0 and 45.4 K, respectively. A mean temperature at 0–50 km may be calculated using the pressures at these altitudes. It is equal to 41.3 K, while the radio occultations give $42 \pm 4$ K (Gurrola 1995).

### 14.3.5 UV Dayglow

The UVS spectra of dayglow on Triton and Titan are shown in Figure 14.13. The observed emissions of $N_2$ and $N^+$ are compared with calculated emissions in Table 14.5 (Strobel and Summers 1995). The measured H Lyman-alpha is 86 R on the dayside and 72 R on the nightside. No emissions of Ar, Ne, and CO have been detected.

Table 14.5 *Nitrogen dayglow on Triton*

| Emission | $N^+$ 916 Å | $N_2$ ($c_4$' (0,0)) 958 Å | $N_2$ ($c_4$' (0,1)) 981 Å | $N^+$ 1085 Å | $N_2$ LBH |
|---|---|---|---|---|---|
| Observed intensity (R) | ? | ≈0 | 2–3 | 1.2–5 | ? |
| Calculated intensity (R) | 0.3 | 8.3 | 1.3 | 2.3 | 33 |

*Note.* Calculations are for magnetospheric electron energy input of $5\times10^8$ W. From Strobel and Summers (1995).

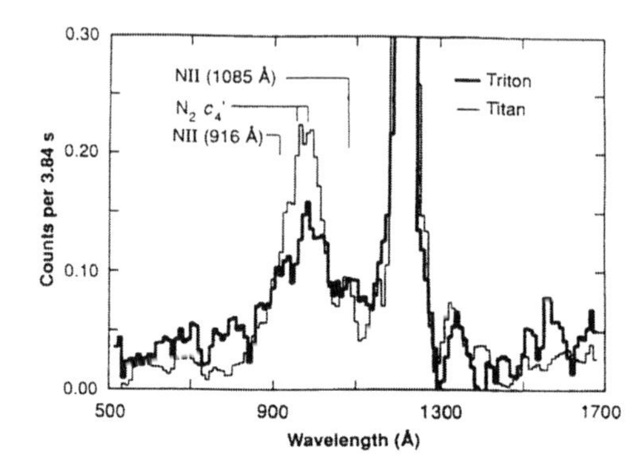


Figure 14.13 Dayside disk-averaged UVS spectra of Triton compared with a dayglow spectrum of Titan measured by the same instrument. The spectrum of Titan is normalized to Triton near 1085 Å. From Broadfoot et al. (1989).

## 14.4 Photochemistry

Triton's ionosphere is denser than that on Titan by a factor of 20, though the solar radiation is weaker on Triton by an order of magnitude. This means that Triton's ionosphere consists of atomic ions, whose radiative recombination is slower than dissociative recombination of molecular ions by five orders of magnitude. Magnetospheric electrons on Triton are an additional factor that favors the observed difference.

The first ionospheric models (Yung and Lyons 1990; Majeed et al. 1990) implied a significant production of $N^+$ by dissociative ionization of $N_2$, ionization of N formed by recombination of $N_2^+$, and charge exchange between $N_2^+$ and N. $N^+$ reacts with $H_2$ that is a product of the $CH_4$ photolysis. This photolysis removes methane to negligible abundances at the ionospheric altitudes. If the production of $N^+$ is smaller than that of $H_2$, then the ion and electron densities are low. If they are comparable, then some part of $N^+$ recombines radiatively and explains the observed ionospheric densities. $N^+$ is a dominant ion in those models.

Summers and Strobel (1991) reduced the requirement to the ion production rate by formation of $H^+$ in charge exchange of $N_2^+$ with H. Similar to $N^+$, $H^+$ recombines radiatively or is lost in reactions with hydrocarbons and nitriles, whose abundances are low in the ionosphere. Later laboratory measurements gave a low upper limit to the rate coefficient of $N_2^+$ + H. Furthermore, $H^+$ densities are strongly depleted in the presence of CO.

The next approximation was suggested by Lyons et al. (1992), who developed a solar only model (without magnetospheric electrons) and $C^+$ as the most abundant ion. Atomic carbon originated from

$$CH_4 + h\nu \rightarrow C + \text{products} \ (\gamma \approx 0.004),$$
$$CH + H \rightarrow C + H_2,$$
$$CN + N \rightarrow C + N_2.$$

Photoionization of C and charge exchange of C with $N_2^+$ are sources of $C^+$ that is lost in reaction with $CH_4$. The model agrees with the observed ionosphere for a rate coefficient of $10^{-9}$ cm$^3$ s$^{-1}$ for $N_2^+ + C \rightarrow C^+ + N_2$. That rate coefficient had not been studied in the lab, and its current value is $1.1\times10^{-10}$ cm$^3$ s$^{-1}$, too low to fit the observed ionosphere.

Further progress in the problem is related to the detection of CO in Triton's ice (Cruikshank et al. 1993) that presumes some CO in the atmosphere. Evidently CO facilitates production of C and therefore $C^+$ and involves oxygen chemistry. Photochemical models including CO were developed by Lyons et al. (unpublished) and Krasnopolsky and Cruikshank (1995). The latter will be discussed below.

### 14.4.1 General Features of the Model

The model is for the global mean conditions (solar zenith angle of 60°, the solar flux is halved to account for the nightside) at solar maximum ($F_{10.7}$ = 200) with the Lyman-alpha interstellar background of 340 R. Magnetospheric electron deposition during crossing Neptune's equator is scaled by $\beta$ = 0.162 and equals $10^8$ W. It is shifted upward by two scale heights (Section 14.3.4; Table 14.4). The proper temperature profile was calculated and discussed above (Section 14.3.4; Figure 14.12). Thermal escape of atomic species and $H_2$ and escape of ions with effusion velocity of 150 cm s$^{-1}$ are the upper boundary conditions. The lower boundary conditions are densities of the parent species ($N_2$, $CH_4$, CO), the closed boundary for $H_2$, either closed boundary or scavenging for radicals, and condensation on the haze for hydrocarbons and HCN. The model involves 139 reactions of 32 neutral and 21 ion species. Calculated reaction column rates are reduced to the surface and given in the model. This facilitates quantitative analysis of the photochemistry.

### 14.4.2 Methane and Eddy Diffusion

Balance of methane is determined by photolysis and two reactions:

$$CH_4 + h\nu \rightarrow \text{products} \ (CH_3, CH, H, H_2) \qquad 2.25 + 8,$$
$$CH_4 + CH \rightarrow C_2H_4 + H \qquad 9.46 + 7,$$
$$CH_3 + H + M \rightarrow CH_4 + M \qquad 7.18 + 7.$$

The numbers to the right are column rates, so that 2.25 + 8 = $2.25 \times 10^8$ $cm^{-2}$ $s^{-1}$. Effective quantum loss of methane is therefore 1.1. The loss is compensated by sublimation from the surface ice and eddy diffusion. To fit the observed vertical profiles, eddy diffusion is $K = 4000$ $cm^2$ $s^{-1}$ (Figure 14.7) and the homopause is at 32 km. Eddy diffusion is adopted at $10^5$ $cm^2$ $s^{-1}$ in the troposphere below 8 km (Yelle et al. 1991), and a preferable $CH_4$ density near the surface is $3.1 \times 10^{11}$ $cm^{-3}$, within the uncertainties of the observed methane.

### 14.4.3 Hydrocarbons, HCN, and $H_2$

The conventional chemistry of hydrocarbons is concentrated in the lowest 50 km on Triton and strongly affected by condensation at temperatures of $\approx$40 K. The haze properties in Table 14.3 below 30 km and extrapolation of those above 30 km are used in the model. Loss of the condensable species is adopted with sticking coefficient of one. The calculated hydrocarbon gas number densities are very low, reaching $4 \times 10^6$ $cm^{-3}$ for $C_2H_4$ at 20–50 km.

The authors suggested a cycle of indirect photolysis of $N_2$:

$$\begin{aligned}
&CH_4 + h\nu \rightarrow CH + H + H_2,\\
&CH + N_2 + M \rightarrow HCN_2 + M,\\
&HCN_2 + H \rightarrow HCN + NH,\\
&NH + H \rightarrow N + H_2,\\
&\text{Net } CH_4 + N_2 + H + h\nu \rightarrow HCN + N + 2\,H_2.
\end{aligned}$$

This cycle is initiated by the Lyman-alpha photon that cannot dissociate $N_2$ directly. It is effective only at low methane mixing ratios, otherwise CH reacts mostly with $CH_4$. The cycle efficiency is also lowered by other possible branches of $HCN_2$ + H and by photolysis of $HCN_2$. The cycle is a major source of HCN on Triton, and the detection of HCN in Triton's ice (Section 14.2.3) supports the proposed cycle.

Two hydrogen atoms are released in the photochemical transformation of $CH_4$ into the $C_2H_{2x}$ hydrocarbons that form the haze. Therefore the total production of H is $\Phi$ = $2.25 \times 10^8 \times 2 \times 1.1$ = $5 \times 10^8$ $cm^{-2}$ $s^{-1}$ = $1.15 \times 10^{26}$ $s^{-1}$; here 1.1 is the effective quantum loss of methane. There are mutual transformations between $H_2$ and H, and $H_2$ dominates throughout the atmosphere (Figure 14.12). Finally both $H_2$ and H escape, the escape is diffusion-limited, so the ($2H_2$ + H) mixing ratio $f_H$ should be rather constant and equal to

$$f_H = \Phi H/b = 2 \times 10^{-4}\ ;\ b = Dn = 1.88 \times 10^{17} T^{0.82}\ cm^{-1}\ s^{-1}.$$

Here $H \approx 15$ km is the scale height and $D$ is the diffusion coefficient of $H_2$ in $N_2$. The model confirms this simple evaluation (Figure 14.12).

### 14.4.4 CO, Atomic Species, and Ionosphere

Magnetospheric electrons result in two-thirds of the ionization events on Triton, and total productions of N, $N_2^+$, and $N^+$ are 15, 8.6, and 1.7 times $10^7$ $cm^{-2}$ $s^{-1}$, respectively. Therefore, the column ionization rate is $10^8$ $cm^{-2}$ $s^{-1}$. $N^+$ becomes neutral after charge exchange; dissociative recombination of $N_2^+$, its reaction with O with subsequent recombination of $NO^+$, and the CH + $N_2$ cycle add to a total atomic nitrogen production that is equal to $2.77\times10^8$ cm$^{-2}$ s$^{-1}$. Loss of atomic nitrogen is $2.43\times10^8$ cm$^{-2}$ s$^{-1}$ in formation of NO, CN, and NH, their reactions with N, and termolecular association of $N_2$. The remaining $3.4\times10^7$ cm$^{-2}$ s$^{-1}$ = $7.8\times10^{24}$ s$^{-1}$ escape. The model data for N are in excellent agreement with the observations (Figure 14.12).

A CO mixing ratio near the surface is adopted at $3\times10^{-4}$ in the basic model (Figure 14.12). The $CH_4$ mixing ratio of $\approx10^{-4}$ and the CO fraction of $3\times10^{-4}$ may be compared with those of $\approx2\times10^{-4}$ and $\approx7\times10^{-4}$ observed in 2009 by Lellouch et al. (2010; see below). Conditions on Triton in 2009 significantly differ from those in 1989; however, the overall comparison favors the model values.

One may expect that the CO fraction is constant throughout the atmosphere because of the similarity with $N_2$ in mass and chemical behavior (both species are very inactive). However, CO is ionized in the charge exchange with $N^+$ that depletes its mixing ratio above 400 km by an order of magnitude.

Atomic oxygen is produced mostly by recombination of $CO^+$ and lost in reactions with CH, $CH_2$, $CH_3$, $C_2H_4$, CN, and CNN below 200 km. Recombination of $CO^+$ gives one-half of the atomic carbon production. The other half is supplied mostly by reactions CH + H, CN + N, $C_2$ + N. Reaction of C + $N_2$ + M that is followed by reactions of CNN + (H, O) is the major loss of atomic carbon. Charge exchange with $N_2^+$ and $CO^+$ is $1.83\times10^6$ cm$^{-2}$ s$^{-1}$ (10% of the total loss), and escape of C is $4.68\times10^6$ cm$^{-2}$ s$^{-1}$ = $1.1\times10^{24}$ s$^{-1}$.

The calculated composition and structure of Triton's ionosphere is shown in Figure 14.14 (basic model). Though the production of $C^+$ in charge exchange of C with $N_2^+$ and $CO^+$ is 2% of the total ionization rate, $C^+$ is the major ion at 250–500 km that covers the ionospheric peak. $C^+$ is lost by radiative recombination and in reactions with $CH_4$ and HCN.

Charge exchange $N_2^+$ + N doubles the primary production of $N^+$. The $N^+$ densities are comparable to those of $C^+$ above 500 km, where loss of $N^+$ is small due to the low densities of CO and $H_2$.

$N_2^+$ produces $N_2H^+$ in the reaction with $H_2$, and $N_2H^+$ reacts with CO and forms $HCO^+$. This ion is also formed by the reaction of $CO^+$ and $H_2$. A total production of $HCO^+$ is $10^7$ cm$^{-2}$ s$^{-1}$, and it is the most abundant ion below 250 km (Figure 14.14). Using the terrestrial terms, Triton's ionosphere is of E-type below 250 km and of F-type above this altitude. The model agrees with the electron density profiles observed by the Voyager 2 radio occultations.

### *14.4.5 Versions of the Model*

There are some significant uncertainties in the input data of the model. The CO mixing ratio in the lower atmosphere was unknown, and its estimates varied in the literature form $10^{-4}$ to $10^{-3}$. Interactions between the radicals and the haze are unknown as well, and the limiting cases are either no interaction (the basic model) or that with sticking coefficient of one (scavenging). Reaction R60 $HCN_2$ + H $\rightarrow$ HCN +NH with $k = 10^{-14}$ cm$^3$ s$^{-1}$ is a key

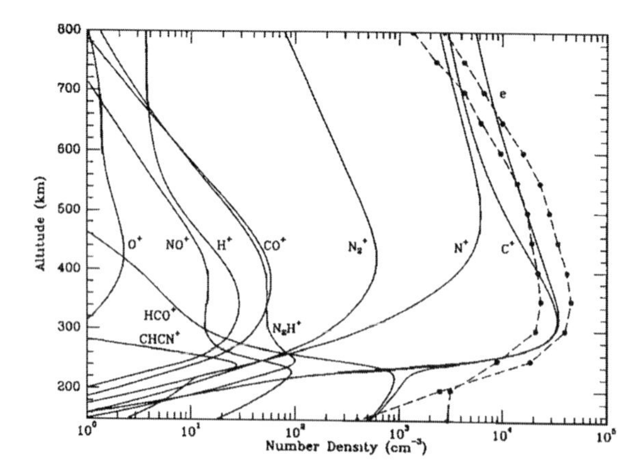


Figure 14.14 Composition of Triton's ionosphere (Krasnopolsky and Cruikshank 1995). Dash-dotted lines are the electron densities observed by the Voyager radio occultations (Tyler et al. 1989).

reaction in the formation of HCN and indirect photolysis of $N_2$, but has not been studied in the laboratory. Reactions of CNN with N, H, C, and O are hypothetical as well and may be switched off by neglect of R25 C + $N_2$ + M. Effusion velocity of the ion escape is another uncertain parameter. Combination of these uncertainties originated eight versions of the model that are presented in Table 14.6.

Four versions do not completely fit the observational constraints. Models 1, 6, and 7 with the neglect of R60 show a strong reduction in the production and precipitation of HCN (by a factor of 50 for model 1), and it would be hardly possible to detect such low productions of HCN in the ice spectra (Figure 14.4). Model 3 with scavenging of the radicals results in a peak electron density exceeding the measured values. The solar only model 6 results in atomic nitrogen densities below the observed values.

The parent species $N_2$, $CH_4$, and CO sublime from the ice with rates of 40, 208, and 0.3 g $cm^{-2}$ $Byr^{-1}$, respectively, in the basic model. Photochemistry produces hydrocarbons $C_2H_4$, $C_2H_6$, $C_2H_2$, and nitrile HCN that precipitate to the ice with rates of 135, 28, 1.3, and 29 g $cm^{-2}$ $Byr^{-1}$, respectively. $H_2$ and atomic species N, H, O, and C are formed in the atmosphere, and their escape is significant (Table 14.5). This is a basis for evolution of the atmosphere–ice system on Triton.

## 14.5 Triton's Atmosphere after the Voyager Encounter

### *14.5.1 Stellar Occultations*

Broadband stellar occultations in the visible and near-infrared ranges (Section 7.8) are a tool to monitor atmospheric pressure and temperature at 20–100 km on Triton.

Triton's angular diameter is 0.13 arcs that is smaller than resolution elements of the best ground-based and on-orbit telescopes. Therefore Triton is a point source for the ground-based astronomy. However, angular diameters of stars are very much smaller than that of

Table 14.6 *Photochemical models of Triton's atmosphere and ionosphere*

| Value | Measured | Basic | 1[a] | 2[b] | 3[c] | 4 | 5 | 6[d] | 7[e] |
|---|---|---|---|---|---|---|---|---|---|
| $f_{CO}, \times 10^{-4}$ | – | 3 | 3 | 3 | 3 | 1 | 10 | 1 | 2.5 |
| $k(N_2^+, CO^+ + C), \times 10^{-11}$ | – | 4 | 4 | 4 | 4 | 8 | 2 | 10 | 1 |
| $v_i$, cm s$^{-1}$ | – | 150 | 150 | 150 | 150 | 150 | 150 | 10 | 10 |
| $h_{max}$, km | 340 | 320 | 320 | 300 | 300 | 290 | 330 | 280 | 290 |
| $e_{max} \times 10^4$cm$^{-3}$ | 3.5 ± 1 | 3.5 | 3.6 | 4.25 | 5.1 | 3.6 | 3.0 | 3.3 | 3.4 |
| $e_{700\ km}, \times 10^4$ cm$^{-3}$ | 0.55 ± 0.1 | 0.76 | 0.75 | 0.78 | 1.01 | 0.77 | 0.56 | 1.00 | 1.52 |
| $[N^+]_{700\ km}, \times 10^4$ cm$^{-3}$ | – | 0.34 | 0.34 | 0.33 | 0.57 | 0.45 | 0.13 | 0.09 | 0.71 |
| $[N]_{200\ km}, \times 10^8$ cm$^{-3}$ | 5 ± 2.5 | 5.7 | 5.7 | 5.9 | 5.9 | 6.5 | 4.9 | 2.5 | 7.5 |
| $[N]_{400\ km}, \times 10^7$ cm$^{-3}$ | 10 ± 3 | 7.2 | 7.2 | 7.5 | 7.5 | 7.7 | 6.5 | 3.1 | 9.6 |
| $[CO]_{300\ km}, \times 10^6$ cm$^{-3}$ | 1.1 | 1.1 | 1.0 | 0.7 | 0.3 | 6.5 | 0.6 | 0.7 | |
| $[C]_{300\ km}, \times 10^6$ cm$^{-3}$ | – | 3.9 | 3.8 | 4.3 | 4.4 | 1.4 | 7.1 | 0.9 | 4.0 |
| $\Phi_N, \times 10^{24}$ s$^{-1}$ | 10 ± 3 | 7.7 | 7.7 | 8.0 | 8.2 | 8.6 | 6.8 | 3.2 | 10.0 |
| $\Phi_{H2}, \times 10^{25}$ s$^{-1}$ | – | 4.5 | 4.6 | 4.8 | 2.1 | 4.4 | 4.8 | 5.4 | 4.2 |
| $\Phi_H, \times 10^{25}$ s$^{-1}$ | – | 2.4 | 2.4 | 2.2 | 1.6 | 2.7 | 1.8 | 0.7 | 3.3 |
| $\Phi_O, \times 10^{22}$ s$^{-1}$ | – | 4.4 | 4.4 | 5.1 | 4.8 | 1.7 | 8.4 | 0.8 | 4.3 |
| $\Phi_C, \times 10^{24}$ s$^{-1}$ | – | 1.1 | 1.1 | 1.2 | 1.2 | 0.4 | 2.0 | 0.24 | 1.1 |
| $\Phi_{C+}, \times 10^{23}$ s$^{-1}$ | – | 2.6 | 2.6 | 2.8 | 2.8 | 2.0 | 2.7 | 0.4 | 0.4 |
| $S_{C2H2}$, g cm$^{-2}$ Byr$^{-1}$ | – | 1.3 | 1.5 | 3.3 | 0.7 | 1.2 | 0.8 | 1.3 | 1.4 |
| $S_{C2H4}$, g cm$^{-2}$ Byr$^{-1}$ | – | 135 | 164 | 133 | 155 | 137 | 134 | 166 | 166 |
| $S_{C2H6}$, g cm$^{-2}$ Byr$^{-1}$ | – | 28 | 22 | 28 | 81 | 29 | 27 | 22 | 22 |
| $S_{HCN}$, g cm$^{-2}$ Byr$^{-1}$ | – | 29 | 0.57 | 36 | 10 | 32 | 28 | 3 | 3 |

*Note.* Here, $f_{CO}$ is the CO mixing ratio, $v_i$ is the ion escape velocity, $h_{max}$ and $e_{max}$ are the altitude and electron density at the ionospheric peak, and Φ and *S*, are the total escape and sedimentation fluxes of *i* species. From Krasnopolsky and Cruikshank (1995).
[a]Basic model without (R60) $HCN_2$ + H. [b]Basic model without (R25) C + $N_2$ + M. [c]Basic model with condensation and scavenging of all species except $N_2$, $CH_4$, CO, and $H_2$. [d]Analog of the solar model of LYA: (R60) $HCN_2$ + H is neglected, radiative recombination coefficients are $5 \times 10^{-12}$ cm$^3$ s$^{-1}$.
[e]Analog to Strobel and Summers (1995) with $P = 1.4 \times 10^8$ W and without (R60).

Triton. Orbiting Neptune while Neptune is orbiting the Sun, Triton covers stars from time to time, and those occultations are used to study its atmosphere. Stars are point sources in these events, and Triton is an extended body. Very careful astrometric calculations are required to get a position of the occultation chord on Triton's disk that depends on geographic coordinates of the observatory. A few observatories are typically involved in observations of an occultation event, and the data obtained are compared and discussed.

If light from a star and a planet are *S* and *P* before or after the occultation, while the light of their combination during the occultation is *SP*, then transmission of the starlight (occultation light curve) is (*SP* – *P*)/*S*. The most accurate stellar occultation by Triton was observed in November 1997 using Fine Guidance Sensor (FGS) #3 aboard the Hubble Space Telescope (Elliot et al. 2000a). The measured light curve is shown in Figure 14.15.

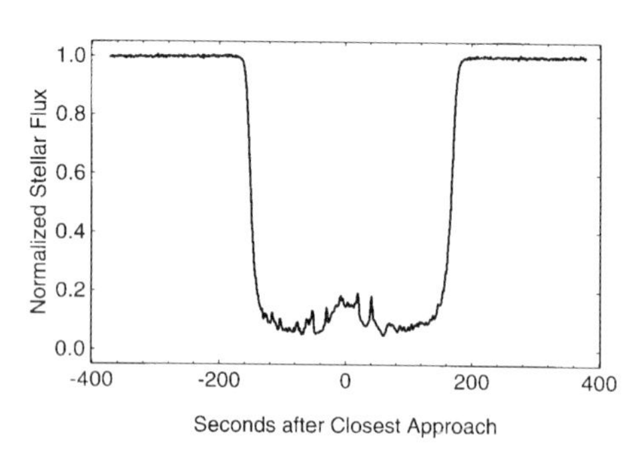


Figure 14.15 Light curve of the stellar occultation of Triton observed using the HST/FGS#3 in November 1997 (Elliot et al. 2000a).

Evidently transmission is unity before and after the occultation. The starlight drops down when the star is crossing the limb. The shape is determined by refraction in the atmosphere that is proportional to gas column abundance along the ray. The refraction keeps some signal after the star is occulted. There is focusing of the refracted light at the center of the curve that makes a peak. The observed structure near the central peak reflects turbulence and winds in the atmosphere.

At first radii of the immersion and emersion half-light points are determined from the light curve. These radii vary typically within 1450 ± 30 km. Then the observed slope is fitted by an isothermal model with density and temperature as fitting parameters. Finally the density is extrapolated to a reference radius of 1400 km, and weighted-mean temperature and pressure at 1400 km from all observations of the occultation event are given.

Four stellar occultations of Triton have been observed. The first event in 1993 had significant uncertainties, and results of the other three events are compared with the model for the Voyager data (Figure 14.12) in Table 14.7.

A temperature profile for the Voyager conditions in Strobel et al. (1996) has exospheric temperature of 100 K (within the uncertainty of the UVS solar occultations, $T = 102 \pm 3$ K) and a warmer atmosphere at 10–100 km. It is not clear how it fits the observed density $[N_2] = (4 \pm 0.4)\times 10^8$ cm$^{-3}$ at 575 km. The temperatures from Tr148 and Tr176 agree with that from Krasnopolsky et al. (1993), while those from Tr180 agree with Strobel et al. (1996). The overall observed increase in pressure near 50 km for 8 years since the Voyager encounter is a factor of 2–3.

The HST observations of Tr180 resulted in the data with high signal-to-noise ratio that made it possible to retrieve the temperature profiles without the assumption of isothermal atmosphere. The data for both cases are compared in Table 14.6. The error of the isothermal approximation is ≈20% in pressure and 2 K in temperature.

The retrieved temperature profiles are compared with a few models in Figure 14.16 (Elliot et al. 2000a). The major problem is the isothermal atmosphere at 52 K from 50 to 20 km. The "conduction" model (dash-dots) is for ionospheric heating above 145 km and heat conduction to the surface. The "add CO" (thin line) involves CO rotational line cooling with a mixing ratio of $2\times 10^{-4}$. The "add $CH_4$" (dots) combines the conduction

Table 14.7 *Pressure (μbar) and temperature (K) at 1400 km* (h = *47 km) from the Voyager-based models (Figure 14.12) and stellar occultations*

| Event | Date | Pressure | Temperature | Reference |
|---|---|---|---|---|
| Voyager | 1989-08-25 | 0.67 ± 0.1 | 45.4 ± 1 | Krasnopolsky et al. (1993) |
| | | 0.8 ± 0.1 | 50.3 ± 0.3 | Strobel et al. (1996) |
| Tr148 | 1995-08-14 | 1.49 ± 0.14 | 46.7 ± 1.4 | Olkin et al. (1997) |
| Tr176 | 1997-07-18 | 2.23 ± 0.28 | 43.6 ± 3.7 | Elliot et al. (2000b) |
| Tr180 | 1997-11-04 | 2.15 ± 0.02 | 49.4 ± 0.2 | Elliot et al. (2000b) |
| Tr180* | 1997-11-04 | 1.8 ± 0.1 | 51.6 | Elliot et al. (2000a) |

*Note.* The last line is for inversion of the temperature profile, the other three lines are for isothermal atmosphere above $r$ = 1400 km.

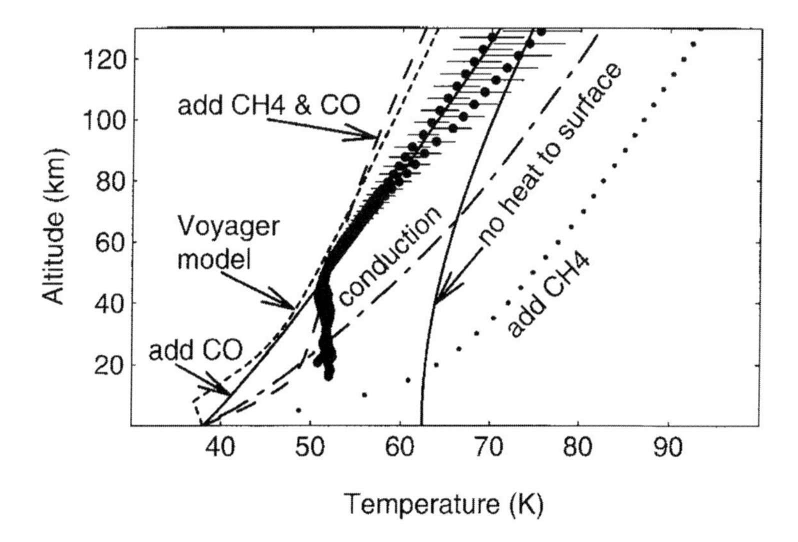


Figure 14.16 Retrieved temperature profiles from the Tr180 occultation (dots with error bars) are compared with a few models (see text). From Elliot et al. (2000a).

model with UV heating by photolysis of $CH_4$, near-IR heating by the methane bands and cooling by the $CH_4$ band at 8 μm. The "add $CH_4$ & CO" model (dashes) is for CO fraction of 0.002 and methane is half that from the UVS solar occultations. "No heat to surface" is similar to "add CO" but without heat conduction to the surface. The "Voyager model" (short dashes) is from Strobel et al. (1996).

The "add CO" model fits the observations above 50 km but disagrees below 50 km. The "add $CH_4$ and CO" model reproduces the isothermal region at 20–50 km better than the other models but has a smaller temperature gradient above 50 km than those in the observed profiles.

Extrapolating to the surface, the authors got a surface pressure of 19 ± 1.5 μbar, exceeding the Voyager value by a factor of 1.35, smaller than the increase at 47 km by the factor of 2–3.

Comparing the surface pressure with 1.8 μbar at 47 km, a mean temperature at 0–47 km is 52.4 K. The authors argued that this may occur if ≈10% of the surface is dark with albedo of 0.2. However, no dark regions have been observed by Voyager. The stellar occultations Tr180 were at 5°S and 8°S, far from the subsolar latitude of ≈50°S and with low insolation. Triton's crust is of water ice to the depth of 400 km, and an abundant dark material (seen in the plumes and required for the albedo of 0.2) needs explanation. On the other hand, dark regions exist on Pluto. Therefore the problem remains unclear and needs further study.

### *14.5.2 High-Resolution Spectroscopy*

The combination of the Very Large Telescope with diameter of 8.2 m and the CRIRES echelle spectrograph (Section 6.3) is perfect to search for minor species in Triton's atmosphere. This study was conducted by Lellouch et al. (2010) in July 2009. They observed spectra near 2.35 μm that cover lines of the CO (2–0) band and the $CH_4$ $\nu_3 + \nu_4$ band. A total exposure was four hours. They applied adaptive optics mode with a slit of 0.4 arcs and resolution $\lambda/\delta\lambda = 6\times10^4$.

A small part of the observed spectrum that includes four methane lines is compared to four synthetic spectra with different methane abundances in Figure 14.17. Using all observed methane lines, the measured $CH_4$ column abundance is 0.08 ± 0.03 cm-atm. If the $CH_4$ vertical profile is similar to that in the UVS solar occultation (ingress), then a

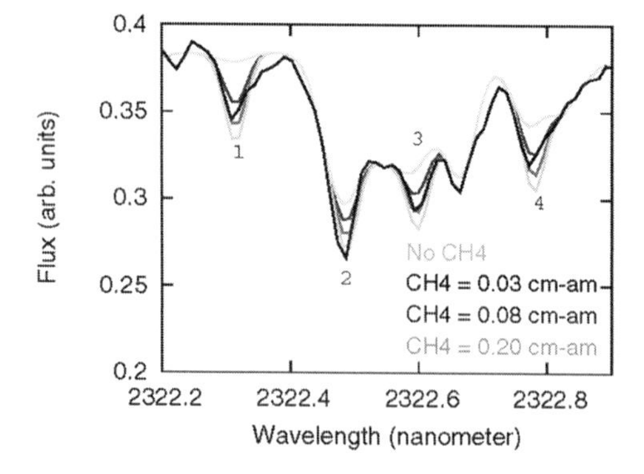


Figure 14.17 A part of the VLT/CRIRES spectrum of Triton (the brightest curve) is compared to four synthetic spectra with $CH_4$ abundances of 0, 0.03, 0.08, and 0.20 cm-atm. The plot covers four $CH_4$ lines. The observed spectrum is between 0.03 and 0.08 at line 1 and 4, at 0.20 for line 2, and at 0.08 for line 3. Using all observed $CH_4$ lines, detected methane abundance is 0.08 ± 0.03 cm-atm. From Lellouch et al. (2010).

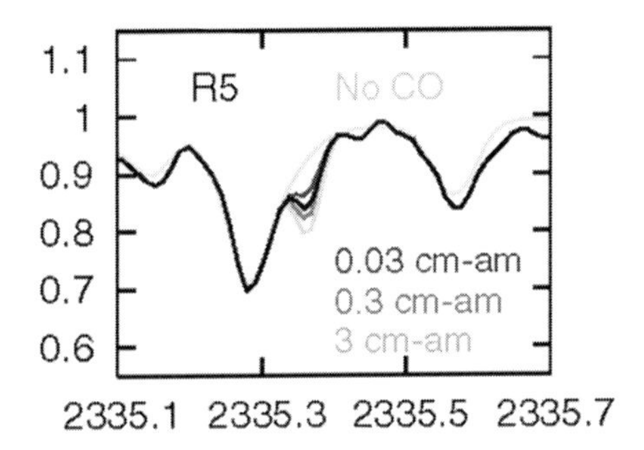


Figure 14.18 A part of the VLT/CRIRES spectrum of Triton (the brightest curve) that includes one of eight detected CO lines. The X-scale is in nanometers. The observed spectrum is compared to four synthetic spectra with CO abundances of 0, 0.03, 0.3, and 3 cm-atm and is between 0.03 and 0.3 in the figure. Using all eight CO lines, the CO abundance is 0.30 cm-atm with uncertainty of a factor of 3. From Lellouch et al. (2010).

methane surface density is $(1.9 \pm 0.7)\times10^{12}$ cm$^{-3}$. It may be compared with the mean density of the UVS ingress and egress, $4.7\times10^{11}$ cm$^{-3}$ (Herbert and Sandel 1991). Krasnopolsky and Cruikshank (1995) adopted high eddy diffusion in the troposphere, $K = 10^5$ cm$^2$ s$^{-1}$ below 8 km, suggested by Yelle et al. (1991), and got $[CH_4] = 3.1\times10^{11}$ cm$^{-3}$ near the surface, within uncertainties of the UVS data. The CRIRES data imply an increase in the methane surface density relative to the Voyager encounter by a factor of $\approx$5.

Eight CO lines were detected in the observed spectrum, and one of those is in Figure 14.18. The lines should be very narrow (because of the low pressure) and saturated. A measured CO column abundance is 0.30 cm-atm with uncertainty within a factor of 3.

Mixing ratios of the detected species are of great interest; however, there were no means to measure $N_2$. Mean atmospheric temperature retrieved from the $CH_4$ lines was $50^{+20}_{-15}$ K. Extrapolating the Voyager and stellar occultation data on the atmospheric pressure, the authors estimated it at $\approx$40 μbar in 2009. CO is uniformly mixed up to the ionospheric altitudes, and its mixing ratio is $\approx7\times10^{-4}$, while a $CH_4$ mixing ratio is $\approx2.6\times10^{-4}$ near the surface (slightly smaller values are in the paper). Actually the rather large $CH_4$ abundance favors its uniform mixing, say, in the lowest two scale heights; then the $CH_4$ mixing ratio is $2\times10^{-4}$. The CO and $CH_4$ mixing ratios are greater than those in the basic photochemical model (Table 14.6) for the conditions of the Voyager encounter by a factor of $\approx$2.

The pressure of 40 μbar corresponds to the $N_2$ ice temperature of 39 K. Saturated vapor of the pure $CH_4$ and CO would have mixing ratios of $10^{-4}$ and 0.15, respectively, using the data from Fray and Schmitt (2009). Raoult's law for ideal solutions gives partial vapor pressure as a product of saturated vapor pressure and mole fraction in the solution. The $CH_4$ and CO mole fractions in the $N_2$ ice on Triton are $10^{-3}$ and $5\times10^{-4}$ (Quirico et al. 1999; Table 14.2), and the expected $CH_4$ and CO mixing ratios are $10^{-7}$ and $7.5\times10^{-5}$, respectively, smaller than those measured by orders of magnitude.

Another limiting case is for fast evaporation, when slow diffusion in the ice may be neglected as well as condensation of gas onto the ice. Then a surface skin layer is enriched by

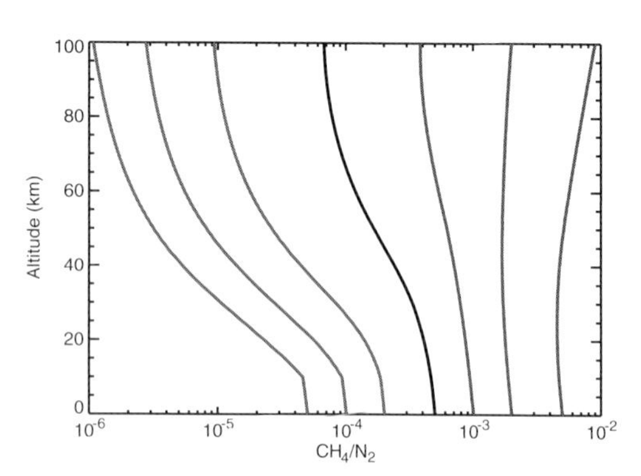


Figure 14.19 Calculated profiles of methane mixing ratio on Triton as functions of its ratio at $p = 14$ μbar ($h = 0$ km for the Voyager conditions). From Krasnopolsky (2012d).

less volatile species up to a level when mixing ratios in the released gas are equal to those in the bulk ice, that is, $10^{-3}$ for $CH_4$ and $5\times10^{-4}$ for CO. Actually these values are rather similar to those observed, taking into account their uncertainties. Seasonal variations with condensation and sublimation cycles may produce regions of pure ices as well as a case of so-called detailed balancing model (see Lellouch et al. 2010 and references therein) that provides intermediate solutions between the limiting cases considered above.

### *14.5.3 Methane Abundance and Triton's Photochemistry*

Photochemistry of Triton at the conditions of the Voyager encounter is very different from those of Titan and Pluto: molecular species are strongly depleted in the upper atmosphere where atomic species dominate, the ionosphere is dense, atomic nitrogen is more abundant than that on Titan and Pluto by orders of magnitude, production of HCN by the indirect photolysis of $N_2$ is much more effective as well. The latter is proportional to $[N_2]^2$ near a peak of the methane photolysis.

Abundance of methane is of crucial importance in these differences. Methane mixing ratios are $\approx10^{-4}$ in Triton's troposphere and $\approx10^{-2}$ in those of Titan and Pluto. However, the observed increase in the methane abundance since the Voyager flyby may generally stimulate transition from the Voyager-type photochemistry of Triton to those of Titan and Pluto.

Total loss of methane is mostly determined by the Lyman-alpha radiation and does not depend on methane abundance. Its effect is significant, if the abundance is low, and may be weak otherwise. Ionospheric processes add to the depletion of methane, while diffusive separation has the opposite effect and enhances methane in the thermosphere. Calculated profiles of methane mixing ratio for various abundances in the troposphere are shown in Figure 14.19. The atmosphere with a $CH_4$ fraction of $2\times10^{-4}$ still probably matches the basic features observed by Voyager and in the related studies.

# 15

# Pluto and Charon

## 15.1 Discovery and General Properties

U. Le Verrier analyzed perturbations in the orbit of Uranus and predicted existence of Neptune and its position on the sky. J. G. Galle searched for Neptune and discovered it in 1846 within 1 arcs off the prediction by Le Verrier. Later, P. Lowell calculated that the perturbations of Uranus by Neptune are insufficient to explain the Uranus orbit, and another planet should exist near the orbit of Neptune. That was Pluto discovered by C. Tombaugh in 1930. However, it became clear that Pluto is too small to fit Lowell's prediction. Therefore C. Tombaugh continued his search for planet X for 14 years, made some discoveries in this search, but could not find planet X.

Charon was discovered as a variable bulge of Pluto by J. W. Christy in 1978. Later it was observed as a separate object with a period of rotation of 6.387 days. The rotation plane was close to the Earth in 1985–1990, and that made it possible to observe Pluto–Charon mutual events (eclipses and transits). Those observations resulted in radii of both bodies with accuracies of $\approx$30 km for Pluto and $\approx$15 km for Charon. Even better radii of Pluto with uncertainties of $\approx$5 km were obtained by stellar occultations. However, the derived radii of Pluto referred to either its solid body or the top of the haze with a slant optical depth $\approx$1. The New Horizon radio occultations removed this ambiguity and gave Pluto's radius of 1190 km (Table 15.1).

Pluto's polar axis is close to its orbital plane; therefore seasonal changes on Pluto are very strong induced by the elliptic orbit and the low inclination, so that the subsolar latitude varies within $\pm 60^\circ$. For example, polar days/nights cover a quarter of Pluto's surface at solstices, being of 0.04 on the Earth.

A breakthrough in the studies of Pluto and Charon is related to the flyby of the New Horizons spacecraft at 12,500 km from Pluto's surface in July 2015. However, significant results were achieved in the ground-based studies and theory before this event. They include measurements of the radii, near-infrared spectroscopy of the ice composition, measurements of the ice temperature, stellar occultations with retrievals of atmospheric temperature and pressure, high-resolution spectroscopy of the atmospheric composition, and modeling of thermal balance and photochemistry.

Table 15.1 *Basic properties of Pluto and Charon*

| | |
|---|---|
| Perihelion (September 1989) | 29.66 AU |
| Aphelion (February 2114) | 49.30 AU |
| Orbital period | 248.0 years |
| Inclination | 17.2° |
| Obliquity | 122.5° |
| Radius | 1190 km |
| Mass | $1.30\times10^{25}$ g |
| Mean density | 1.86 g cm$^{-3}$ |
| Surface gravity | 62 cm s$^{-2}$ |
| Rotation period | 6.387 days |
| Charon rotation | synchronous |
| Charon orbit radius[a] | 19,571 km |
| Charon mass | $1.59\times10^{24}$ g |
| Charon radius | 606 km |
| Charon mean density | 1.71 g cm$^{-3}$ |
| New Horizons (July 14, 2015) | 32.9 AU, $F_{10.7}$ = 105 |

[a] Distance between the centers of Pluto and Charon.

## 15.2 Interior and Surface

### *15.2.1 Origin and Interior*

Along with Eris ($r$ = 1160 km, 96.3 AU), Pluto is the largest Kuiper belt or transneptunian object. Beside the big satellite Charon, it has a few small satellites. All of them are at the circular orbits in the equatorial plane of Pluto. Pluto's moons originated from a collision between Pluto and an other body of comparable size. Some of the released matter formed the moons.

Pluto's density of 1.86 g cm$^{-3}$ indicates that two-thirds of its mass is the solar-composition anhydrous rocks, and those rocks are three-fifths of the Charon mass (McKinnon et al. 2017). Radioactive heating could be sufficient for differentiation of Pluto's interior with a rocky core with radius of 850 km and a mantle of water ice above the core. Temperatures up to 1300 K are possible near the core center, and conditions near the core–mantle boundary may be favorable for a liquid ocean of $\approx$100 km deep.

### *15.2.2 Surface Composition and Temperature*

Near infrared spectroscopy is a standard tool to study composition of ices. Cruikshank et al. (1976) discovered methane ice bands, and later bands of $N_2$ and CO ices were detected as well (Figure 15.1). Spectral fits using Hapke's method (Hapke 1993) resulted in a proportion $N_2 : CH_4 : CO = 98 : 1.5 : 0.5\%$ (Owen et al. 1993). Traces of the ethane ice bands were found by Cruikshank et al. (1999) and in later studies (Figure 15.2). Spectra of Charon revealed $H_2O$ ice and ammonia hydrates (Figure 15.3), the latter with relative abundance of $\approx$3%. A review of the problem at the eve of the New Horizons flyby may be

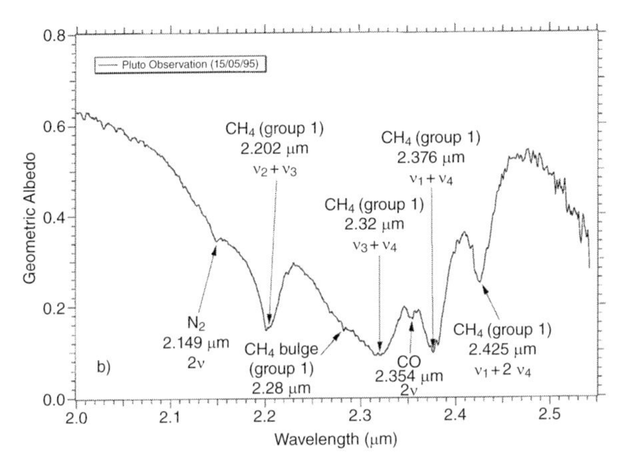


Figure 15.1 Spectrum of Pluto at 2.0–2.55 μm with identified bands of $CH_4$, $N_2$, and CO (Doute et al. 1999).

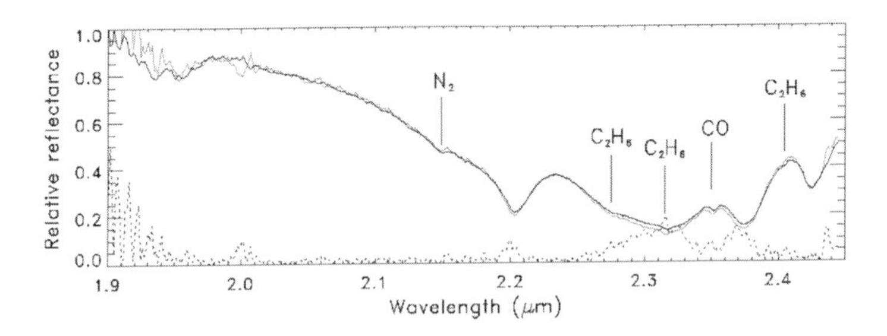


Figure 15.2 Two VLT spectra of Pluto and their difference (dotted line) indicating possible presence of ethane $C_2H_6$. From Merlin et al. (2010).

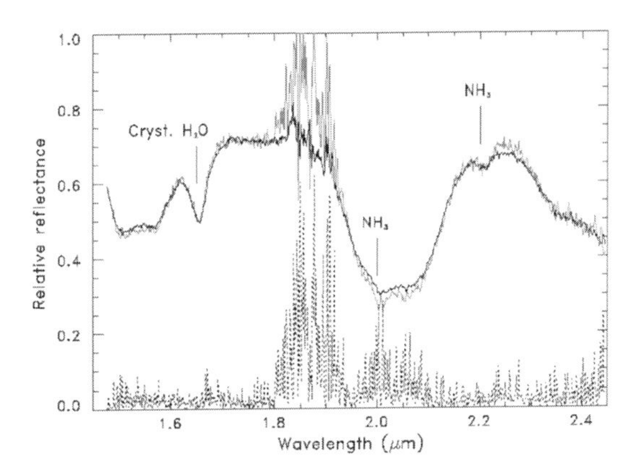


Figure 15.3 Two VLT spectra of Charon and their difference. The difference is large at 1.8–1.95 μm because of the telluric absorption. From Merlin et al. (2010).

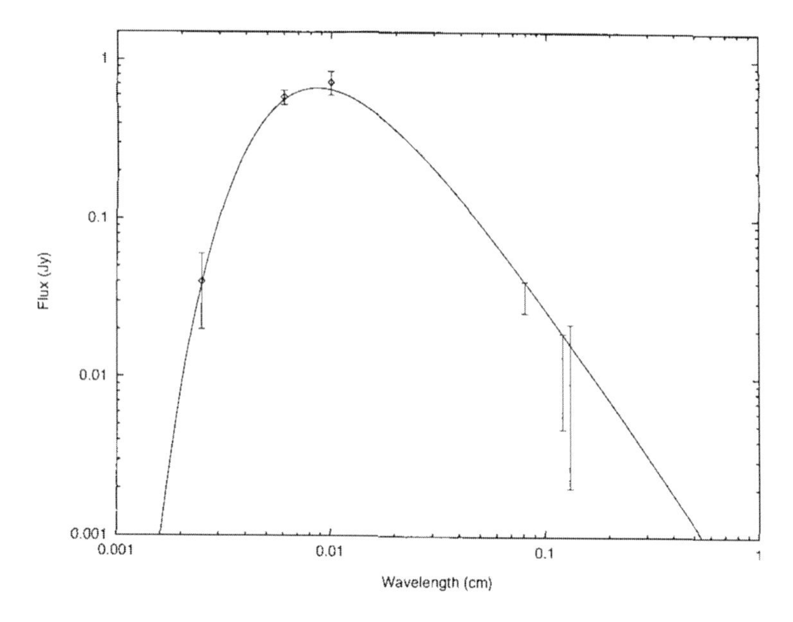


Figure 15.4 IRAS and ground-based observations of the Pluto–Charon system in the far infrared and millimeter range are fitted by a nonisothermal model of the system (Tryka et al. 1994).

found in Cruikshank et al. (2015). Tan and Kargel (2018) considered phase equilibria between the three components in the ice and the atmosphere on Pluto.

Nitrogen ice exists as the cubic phase $\alpha$ below 35.6 K and the hexagonal phase $\beta$ above this temperature. Therefore the $N_2$ ice band at 2.15 μm has a structure that varies with temperature at 30–50 K. Analysis of the observed band by Tryka et al. (1994) indicated the $N_2$ ice temperature of 40 $\pm$ 2 K on Pluto.

Measurements of the Pluto–Charon system were made at 25, 60, and 100 μm in 1983 using the Infrared Astronomical Satellite (IRAS). Ground-based observations in the millimeter range became available as well, and Tryka et al. (1994) created a nonisothermal model of the system to fit the data (Figure 15.4). The model involved the polar caps with $T = 40$ K and emissivity $\varepsilon = 0.6$, the equatorial belt with $\varepsilon = 0.95$, and Charon with $\varepsilon = 0.95$. Size of the polar caps and bolometric albedos of the equatorial belt and Charon were model parameters. The recommended model had the polar caps extending to 20° of latitude with albedos of 0.2 and 0.4 for the equatorial belt and Charon, respectively. Temperature is maximal at 51 K near local time of 14:00 at the equator.

Observations of the Pluto–Charon system using the Infrared Space Observatory (ISO) and Spitzer observatory (Figure 15.5) were analyzed by Lellouch et al. (2011b). The brightness temperature is steadily decreasing with wavelength from 53 K at 20 μm to 37 K at 150 μm. The anomaly of ISO at 150 μm did not repeat in the Spitzer data. Three types of surfaces on Pluto, those covered by $N_2$ ice, $CH_4$ ice, and a tholin/$H_2O$ mix, were considered with appropriate reflectivities and available spatial distributions. Thermal inertias of the $CH_4$ and tholin/$H_2O$ areas on Pluto and that of Charon were determined.

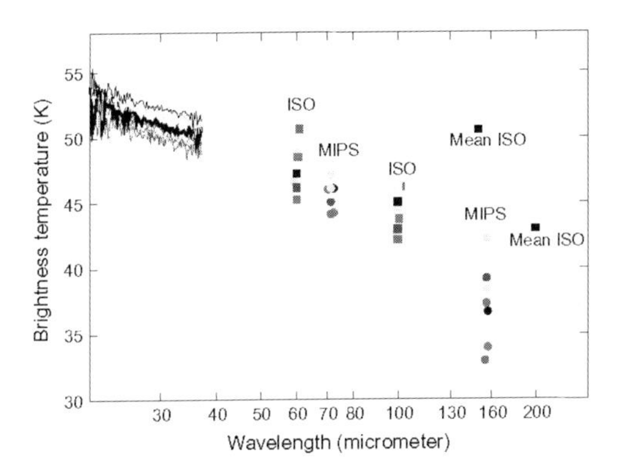


Figure 15.5 ISO and Spitzer (MIPS) and spectra at 20–37 μm) observations of the Pluto–Charon system. Different points at the same wavelength refer to different central longitudes of Pluto. The MIPS data near 150 μm do not confirm the mean ISO. From Lellouch et al. (2011b).

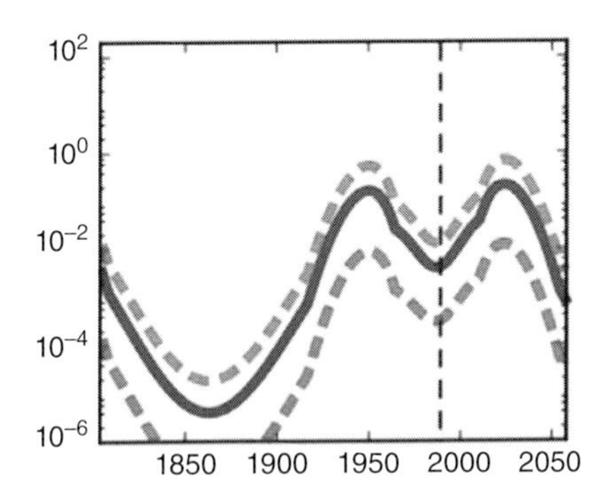


Figure 15.6 Calculated seasonal variations of the $N_2$ surface pressure (mbars). The $x$-axis is year, and the vertical line marks perihelion. This plot is for the polar caps extending to $\pm 45^\circ$, their emissivity of 0.6, and albedo of 0.1, 0.3, and 0.6 (three curves). From Stern et al. (2017a).

The derived values are much lower than those for compact ices, indicating high surface porosity. A mean surface temperature of Charon is $55.4 \pm 2.6$ K.

The New Horizons observations revealed rather different regions on the surface of Pluto. Most of $N_2$ and CO ices are found in Sputnik Planitia with a size of $\approx$1000 km near the anti-Charon point (longitude $\approx 180^\circ$E). This region does not have craters and therefore young ($<$10 Myr). Methane is abundant near 300°E. The observed mountains are of water ice. Some places are very dark, while a mean color is orange. Spatial distributions of the $N_2$, $CH_4$, CO, $H_2O$ ices and dark red material were studied by Schmitt et al. (2017) using the New Horizons NIR spectra. The reservoir of $N_2$ ice in Sputnik Planitia is estimated as a global-equivalent layer of 100 m thick (McKinnon et al. 2017). The Bond albedo is $0.72 \pm 0.07$ for Pluto and $0.25 \pm 0.03$ for Charon (Buratti et al. 2017) based on the LORRI camera at 0.35–0.85 μm.

Seasonal variations of the surface temperature and atmospheric pressure in the present and past epochs were modeled by Stern et al. (2017). Precession period of Pluto's obliquity is 3 Myr with the current subsolar latitude at perihelion near the equinox value of 17°N, being 75°N 0.9 Myr ago and 55°S 2.35 Myr ago. One of the calculated versions of the seasonal variations of atmospheric pressure in the current epoch is shown in Figure 15.6. These variations cover four orders of magnitude with a secondary minimum at perihelion. The pressure 0.9 Myr ago had a maximum of a few millibars at perihelion and a broad secondary maximum of ≈10 μbar near aphelion.

### *15.2.3 HST Spectroscopy in the UV Range*

The observed spectra (Figure 15.7) extend from 320 nm down to 200 nm, and the most interesting are those at 200–255 nm, because absorption features of some gases and ices are expected in this range. The spectrum from Krasnopolsky (2001) demonstrates a gradual decrease with wavelength that may be approximated by a power function $0.3\ (\lambda_0/\lambda)^{1.2}$ with $\lambda_0 = 200$ nm. This behavior is typical of nonabsorbing haze with particle radii comparable with the wavelength. The lack of the absorptions of $H_2O$, $CO_2$, $NH_3$, and $SO_2$ ices resulted in upper limits to their abundances that are less than 20%, 13%, 10%, and 4%, respectively. The lack of gaseous absorptions by di-, tri-, cyano-, and dicyano-acelylenes $C_4H_2$, $C_6H_2$, $HC_3N$, and $C_4N_2$ imposed upper limits that are smaller than calculated abundances of $C_4H_2 + C_6H_2$ in the photochemical model by Krasnopolsky and Cruikshank (1999) by a factor of 3 and agree with the calculated abundances of $HC_3N$ and $C_4N_2$.

The latest spectra by Stern et al. (2012) are more uncertain and indicate a weak absorption near 220 nm. Schindhelm et al. (2015) argued that the absorption is caused by and agrees with abundances of $C_4H_2$, $HC_3N$, and $C_2H_2$ from the model by Krasnopolsky and Cruikshank (1999).

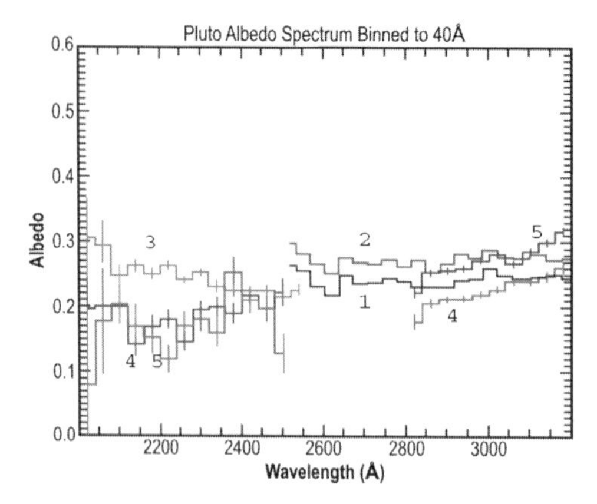


Figure 15.7 UV spectra of Pluto's geometric albedo observed using HST: (1, 2) Trafton and Stern (1996), longitudes 90° and 270°; (3) Krasnopolsky (2001); (4, 5) Stern et al. (2012), longitudes 95° and 273°. From Stern et al. (2012).

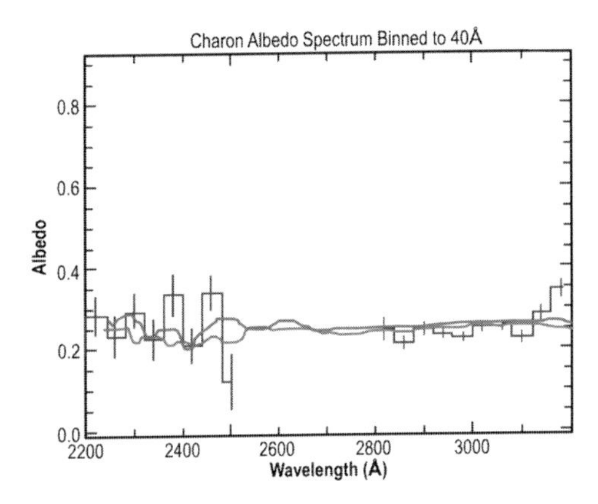


Figure 15.8 UV spectra of Charon's geometric albedo observed using HST. The continuous curves are from Krasnopolsky (2001) for longitudes of 90° and 270°, and the histogram is from Stern et al. (2012).

Spectra of Charon are shown in Figure 15.8, and no absorptions have been identified. The absorptions by $H_2O$ and ammonia ices begin below 220 nm and are not covered by the spectra.

## 15.3 Atmosphere before New Horizons and ALMA Observations

### *15.3.1 Stellar Occultations*

The atmosphere of Pluto was discovered in the stellar occultation that occurred on June 9, 1988. A previous occultation in 1985 was observed but not properly analyzed (see Sicardi et al. 2003). The occultation in 1988 was measured at a few sites including the Kuiper Airborne Observatory, where data of the highest quality were acquired and analyzed by Elliot and Young (1992). Composite analysis of the full set of the observational data was made by Millis et al. (1993).

The effect of the atmospheric refraction is dominant in the stellar occultation, and refraction is proportional to total gas abundance along the ray. Effective altitude is changing during the occultation, thus making possible to extract a profile of density in some altitude range. The technique of analysis and extraction was developed and discussed in detail (Figure 15.9) by Elliot and Young (1992). The measured densities can be converted into pressures and temperatures, if the mean molecular mass is known. Elliot and Young (1992) considered four versions of the atmospheric composition: $CH_4$, $N_2$, CO, and 0.5 Ar + 0.5 $CH_4$. The latter three versions have the same molecular mass $\mu = 28$ but slightly different refractivities. The retrieved temperature at a half-light radius $r_h$ = 1250 km (Figure 15.9) was equal to 60 K for $\mu = 16$ and 104 K for $\mu = 28$.

Yelle and Lunine (1989) calculated thermal balance of Pluto's $CH_4/N_2$ atmosphere at a level of ≈1 μbar for a various methane mixing ratio from 100% to 0.1%. The atmosphere is optically thin and heated mostly by the absorption of the solar light by the $CH_4$ band at 3.3 μm and cooled by emission of the $CH_4$ band at 7.6 μm. Collisional quenching of the $CH_4$ vibrational states is weaker by $N_2$ than by $CH_4$, and this is the only effect of the

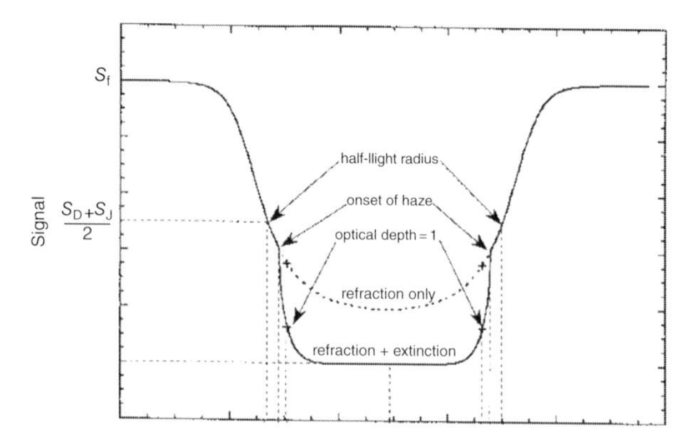


Figure 15.9 Pluto's stellar occultation curve observed using KAO in 1988 and its main points. The $x$-axis is time. From Elliot and Young (1992).

presence of $N_2$. Quenching parameters were poorly known at Pluto's temperatures, and the calculated atmospheric temperature at 1 μbar was stable at 106–100 K for $CH_4$ mixing ratios from 100% down to either 0.1% or 1% depending on the adopted quenching. Yelle and Lunine (1989) applied their results to preliminary data of the Pluto occultation and concluded that $N_2$ is the major species in the atmosphere of Pluto.

This conclusion was confirmed with better confidence by Elliot and Young (1992). The signal cutoff may be caused by either the solid surface or haze, and both possibilities were modeled by the authors. Evidently these data are more uncertain than the New Horizons (NH) observations. However, those data for the atmosphere near the half-light radius remain important to study the seasonal behavior of the atmosphere. Later five occultation events were observed and analyzed using the technique from Elliot and Young (1992), the mean half-light radius was ≈1275 km, and the retrieved pressures, temperatures, and temperature gradients are collected in Table 15.2.

Dias-Oliveira et al. (2015) observed occultations in 2012 and 2013, and their observing sites included VLT that acquired the data with very high signal-to-noise ratio. The occultation curves for the events in 2012 and 2013 were rather similar and fitted by a temperature profile (Figure 15.10) with four basic points: $T$ (1190 km) = 36 K, $T$ (1217 km) = 109.7 K, $T$ (1302 km) = 95.5 K, and $T$ (1392 km) = 80.6 K, being constant above. Sicardi et al. (2016) fitted the occultation on June 29, 2015, just a couple of weeks before the NH flyby, by this profile and gave pressures at 1275 km for the last three occultations (Table 15.2).

The measured seasonal variations of pressure at 1275 km are shown in Figure 15.11. They are compared with a value from the NH radio occultations (Hinson et al. 2017). The difference between the latest stellar occultations and the NH value looks reasonable, although exceeding the claimed uncertainties. The pressure approximately doubled for 27 years since the first occultation event to the NH flyby. This period is near one-tenth of Pluto's year with the first event close to the perihelion that was in September 1989. Young (2013) and Hansen et al. (2015) modeled seasonal variations of Pluto's atmosphere for various assumptions on the surface properties. Their studies were updated using the NH data.

Table 15.2 *Pressures, temperatures, and temperature gradients at* r = *1275 km from stellar occultations*

| Date | $p$ (μbar) | $T$ (K) | $dT/dr$ (K km$^{-1}$) | Reference |
|---|---|---|---|---|
| 1988-06-09 | 0.83 ± 0.11 | 103 ± 21 | −0.051 ± 0.070 | Elliot and Young (1992) |
| 2002-08-21 | 1.55 ± 0.2 | 97 ± 5 | – | Sicardy et al. (2003) |
| 2006-06-12 | 1.86 ± 0.10 | 104 ± 3 | −0.086 ± 0.033 | Young et al. (2008) |
| 2007-03-18 | 2.03 ± 0.20 | 98 ± 1 | −0.16 ± 0.01 | Person et al. (2008) |
| 2007-07-31 | 2.09 ± 0.09 | 103 ± 2 | −0.086 | Olkin et al. (2014) |
| 2009-04-21 | 2.59 ± 0.09 | – | – | Young (2013) |
| 2012-07-18[a] | 2.09 ± 0.02 | 100 | -0.17 | Dias-Oliveira et al. (2015), Sicardi et al. (2016) |
| 2013-05-04[a] | 2.27 ± 0.01 | 100 | −0.17 | |
| 2015-06-29[a] | 2.39 ± 0.03 | 100 | −0.17 | |
| *2015-07-14* | *1.95 ± 0.3* | *98 ± 5* | *−0.135* | *Hinson et al. (2017)* |

*Note.* The last line (in italic) is the NH radio occultations.
[a] These events were analyzed using the same temperature profile (Figure 15.10).

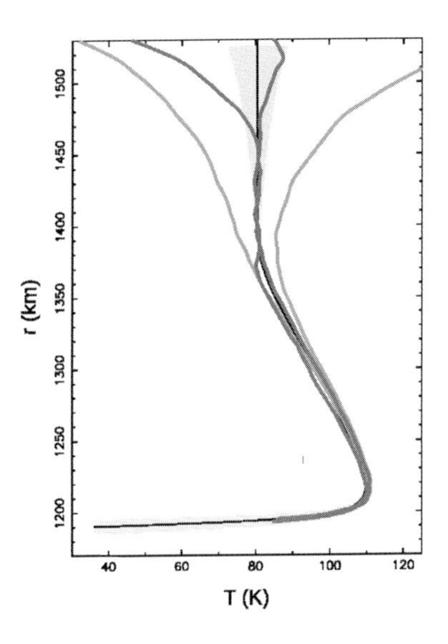


Figure 15.10 Temperature profiles from stellar occultation by Pluto observed using VLT on June 18, 2012: the mean profile with shadowed uncertainties, ingress and egress profiles, and two profiles with temperature at 1390 km changed by ±5 K. From Dias-Oliveira et al. (2015).

### *15.3.2 Ground-Based High-Resolution NIR Spectroscopy*

The detection of $CH_4$ and CO ices on Pluto was a clear indication of presence of these species in the atmosphere. Therefore attempts were made to search for these species and measure their abundances in the atmosphere. Young et al. (1997) detected gaseous

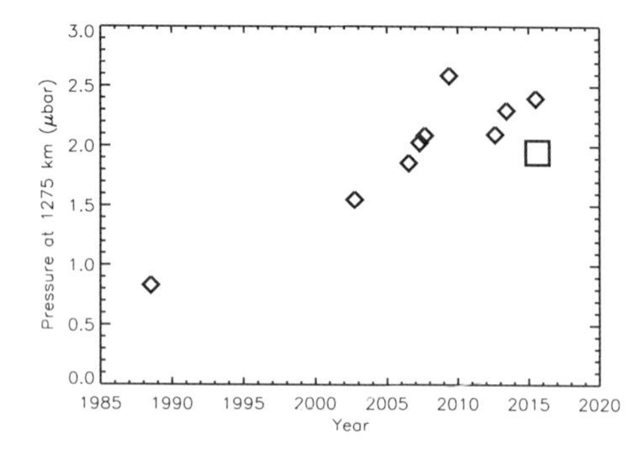


Figure 15.11 Variations of atmospheric pressure at $r$ = 1275 km ($h$ = 85 km) retrieved from stellar occultations and the NH radio occultation (square).

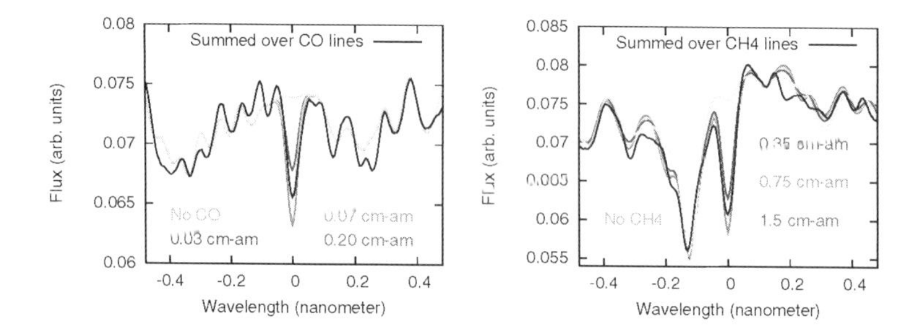


Figure 15.12 VLT/CRIRES observation of CO and $CH_4$ in Pluto's atmosphere at 2.35 μm. The observed lines are summed and fitted by synthetic spectra with various abundances of the species. The best fits are 0.07 cm-atm of CO and 0.75 cm-atm of $CH_4$. From Lellouch et al. (2011a).

methane using its lines at 1.67 μm observed by IRTF/CSHELL with a slit of 1.5 arcs and resolving power of $1.3\times10^4$. The retrieved abundance was $3^{+8}_{-2}\times10^{19}$ cm$^{-2}$. Therefore the abundance was uncertain within a factor of 3, and the uncertain total atmospheric pressure for either surface or tropopause at the bottom of the occultation curve added to uncertainties of the methane mixing ratio.

Later methane was observed at 1.65 μm by Lellouch et al. (2009) using VLT/CRIRES (Section 6.3) with a slit of 0.4 arcs and resolving power of $6\times10^4$. Then both $CH_4$ and CO were measured in Pluto's atmosphere with VLT/CRIRES at 2.35 μm (Lellouch et al. 2011a). Summing a few lines of each species and fitting the sums by synthetic spectra are shown in Figure 15.12. The retrieved abundances of $CH_4$ and CO were compared with a rather uncertain total gas abundance that was evaluated from 6.5 to 24 μbar. (The NH mean surface pressure of 11.5 μbar (Hinson et al. (2017) is just in the middle of this range.) The extracted column densities are greater for greater pressure, reducing the uncertainty that originates from the poorly known pressure. Mixing ratio of $CH_4$, $6^{+6}_{-3}\times10^{-3}$, and CO, $5^{+10}_{-2.5}\times10^{-4}$, are the final results.

### *15.3.3 Thermal Balance and Hydrodynamic Escape Modeling before the NH Flyby*

If effects of eddy diffusion and escape may be neglected in thermal balance, then the solution of the thermal balance equation is

$$\frac{a}{s+1}\left(T^{s+1} - T_0^{s+1}\right) = \int_{r_0}^{r} \frac{dr}{r^2} \int_{r}^{\infty} q(x)x^2 dx \tag{15.1}$$

(see Section 2.4.3). Here $aT^S$ is thermal conductivity, $T_0$ is the temperature at the lower boundary $r_0$, and $q(r)$ is the heating/cooling rate.

Strobel et al. (1996) upgraded the thermal balance calculations by Yelle and Lunine (1989) by detailed modeling with a non-LTE solar heating in the $CH_4$ bands at 2.3 and 3.3 μm, non-LTE radiative exchange and cooling in the $CH_4$ band at 7.6 μm, and LTE cooling by the CO rotational lines. Heating by the 2.3 μm band appeared stronger than that at 3.3 μm by a factor of 6, but referred to the deeper atmosphere near the surface. Strobel et al. (1996) considered a few versions of the atmospheric composition. A version with mixing ratios $f_{CH_4} = 3\times10^{-3}$ and $f_{CO} = 5\times10^{-4}$ was calculated as well. These mixing ratios correspond to the recent NH and ALMA data (see below). The calculated temperature for this model increased from 38 K adopted at the surface to 105 K at 60 km and then gradually decreased to 94 K at 200 km.

Zalucha et al. (2011) applied the model by Strobel et al. (1996) to calculate stellar occultation curves for conditions of the best observations and fitted the observed curves by changing the model parameters. Those parameters are the lower boundary conditions: surface radius, temperature, pressure, and $CH_4$ and CO mixing ratios. The most accurate fit to the surface radius gave 1180 ± 2 km in the occultation in 2006 (Table 15.2), and the surface radius was fixed at this value for all occultations. The results are weakly sensitive to surface temperature that was fixed at 37 K. The CO mixing ratio was fixed at $5\times10^{-4}$ as well, while surface pressure and methane mixing ratio were fitting parameters. The derived surface pressure was 9.2 μbar in 1988, and the mean of three occultations in 2002–2008 was 12.6 μbar. Therefore the increase in the surface pressure is smaller than that at 85 km (Figure 15.11). The retrieved $CH_4$ mixing ratios varied from $2.2\times10^{-3}$ to $9\times10^{-3}$ with a mean value of $5\times10^{-3}$, similar to that observed by Lellouch et al. (2011a). The retrieved temperature profiles showed an increase from 37 K at the surface to 110–125 K near 60 km, a decrease by ≈5 K with a shallow minimum near 250 km, and a weak growth to 113–128 K at 550 km.

Equation (15.1) diverges in the upper atmosphere because of the EUV and FUV heating. Therefore the approximation of a static atmosphere looked invalid for Pluto and this required hydrodynamic escape with adiabatic cooling of the expanding gas. Hydrodynamic escape (Section 2.5.3) from Pluto was under consideration for a few decades. Hunten and Watson (1982) applied the equations for hydrodynamic escape and established a EUV heating limit to this blow off. McNutt (1989) neglected two terms in the equations and found an analytic solution for the blow off, assuming that the EUV heating occurs in a narrow layer. Trafton et al. (1997) described a numerical solution for escape of a pure methane atmosphere and considered escape of $N_2$ dragged by the escaping methane. Krasnopolsky (1999) restored one of the neglected terms in McNutt's solution, included both the EUV heating by $N_2$ and the FUV heating by $CH_4$, and applied the method to a two-component ($N_2$ and $CH_4$) atmosphere. Strobel (2008) solved the equations numerically without the assumption of heat deposition in narrow layers. The assumption that thermal conductivity does not depend on density becomes questionable at extremely low densities, and this may restrict the applicability of the hydrodynamic escape equations. Erwin et al. (2013) created a hybrid fluid/kinetic model for Pluto's atmosphere that adopted the hydrodynamic approach below some level and direct Monte Carlo simulations based on the Boltzmann kinetic equation above that level. This level was at $r = 3000$ km for solar medium activity. Basic results of the models are in Table 15.3.

## 15.4 Atmosphere: New Horizons and ALMA Observations

### 15.4.1 NH Radio Occultations

Conditions and some results the NH radio occultations (Hinson et al. 2017) are in Table 15.4. There is a significant contrast between the entry and exit sites (Figure 15.13) that are the $N_2$ ice covered Sputnik Planitia and the sub-Charon area, which is free of $N_2$ ice. These data are

Table 15.3 *Basic results of modeling of hydrodynamic escape on Pluto*

| Model | $T_{max}$ (K) | $r_{ex}$ (km) | $T_{ex}$ (K) | $\Phi_{N2}$ ($10^{27}$ s$^{-1}$) | $\Phi_{CH_4}$ ($10^{27}$ s$^{-1}$) |
|---|---|---|---|---|---|
| Krasnopolsky (1999) | 114 | 3500 | 62 | 2.0 | 0.35 |
| Strobel (2008) | 117 | 4400 | 70 | 2.2 | – |
| Erwin et al. (2013) | 119 | 7700 | 79 | 2.6 | – |

*Note.* The chosen versions of the models refer to solar medium activity near Pluto's perihelion, $r_{ex}$ and $T_{ex}$ are the exobase radii and temperatures, and $\Phi$ is the escape rate.

Table 15.4 *Conditions and results of the NH radio occultations*

| Conditions | Sunset | Sunrise |
|---|---|---|
| Coordinates | 194°E 17°S | 16°E 15°N |
| Location | Sputnik Planitia | sub-Charon area |
| Pluto's radius, km | 1187.4 ± 3.6 | 1192.4 ± 3.6 |
| Surface pressure, μbar | 12.8 ± 0.7 | 10.2 ± 0.7 |
| $T$ near surface, K | 38.9 ± 2.1 | 51.6 ± 3.8 |
| $r_{max}$, km | 1215 | 1220 |
| $T_{max}$, K | 107 ± 6 | 106 ± 6 |
| $T$ at 100 km, K | 96.8 ± 6 | 96.3 ± 6 |

*Note.* From Hinson et al. (2017).

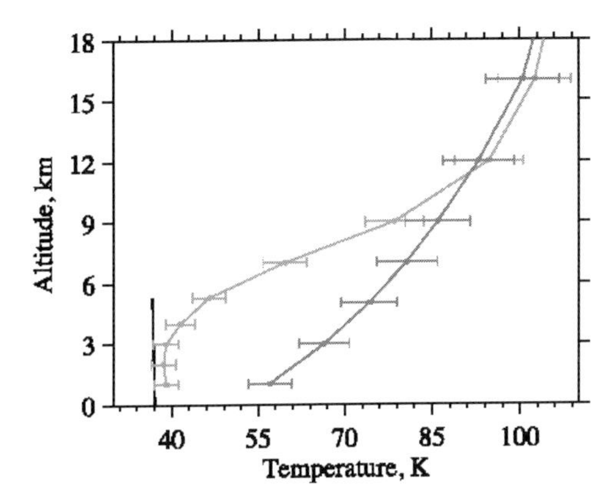


Figure 15.13 Temperature profiles in the lowest 18 km above Sputnik Planitia and the sub-Charon area observed by the NH radio occultations (Hinson et al. 2017). The vertical line is the $N_2$ condensation temperature.

helpful to study processes in the boundary layer on Pluto. The difference disappears above 12 km, and the temperature profiles become rather identical. The retrieved pressures and temperatures are tabulated by Hinson et al. (2017) up to 115 km. They are similar to the stellar occultation results at $r$ = 1200–1300 km (Sicardi et al. 2016) observed two weeks before the NH flyby.

### *15.4.2 NH UV Solar Occultations*

The most detailed data on the atmospheric composition were acquired using the NH solar UV occultations (Gladstone et al. 2016; Young et al. 2017). The UV spectrograph covered a spectral range of 52 to 187 nm with resolution of 0.35 nm. The Sun–Earth angle was small (16 arcmin) during the occultations, and the solar occultation sites were close to those of the radio occultations (Table 15.4). The solar diameter was ≈15 km and close to the pixel size; therefore it affected only slightly the spectral and spatial resolutions in the observation that took just ≈20 min.

The authors made significant efforts to get the most reliable analysis and interpretation of the observations, and their final results are shown in Figure 15.14. Absorption by $N_2$ dominates at 52–80 nm, $CH_4$ absorbs up to 140 nm, absorption by $C_2H_6$ extends to 145 nm, $C_2H_2$ has strong bands at 144, 148, and 152 nm, and $C_2H_4$ is measured by its absorption at 155–175 nm. Haze is evaluated by the absorption at 175–187 nm. A preferable vertical coordinate is the geopotential height

$$z = (r - r_S)\frac{r_S}{r},$$

while altitude is $h = r - r_S$; here $r_S$ = 1190 km is the surface radius. (The authors claim that logarithms of densities and pressure at hydrostatic equilibrium in an isothermal atmosphere

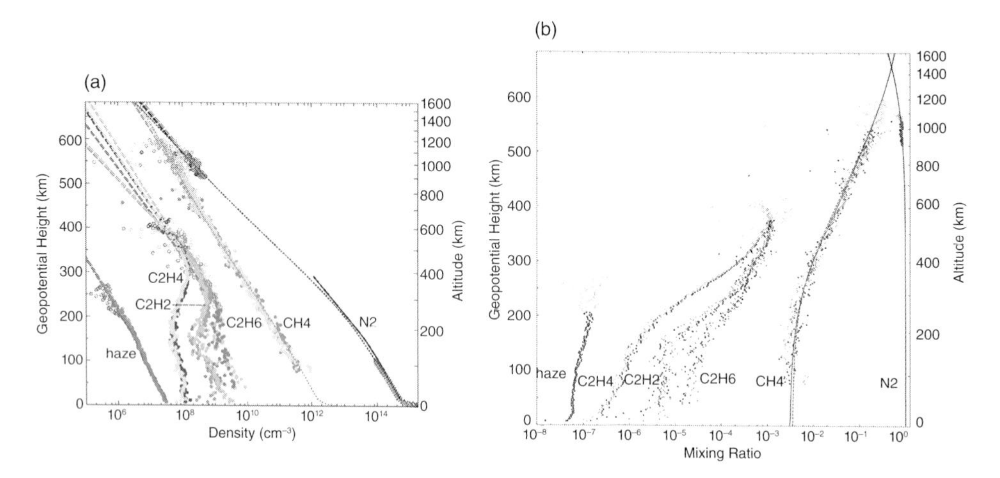


Figure 15.14 Mixing ratios and number densities of five species and haze retrieved from the NH solar UV occultations (Young et al. 2017). For convenience, number densities for haze are adopted as extinction coefficient times $10^{15}$ cm$^{-2}$. Two model profiles for $CH_4$ and one for $N_2$ are shown.

are straight lines relative to geopotential height. Actually this refers to the structure parameter $\lambda = \frac{\gamma Mm}{rkT}$ (2.38).)

Accurate retrievals of $N_2$ were possible in a narrow interval $h$ = 900 – 1100 km. The observed abundances of $CH_4$ extend from 100 to 1200 km, 50–450 km for $C_2H_6$, 0–600 km for $C_2H_2$ and $C_2H_4$, and 0–300 km for haze.

The measured $N_2$ density and temperature are $2\times10^8$ cm$^{-3}$ and 77 $\pm$ 16 K at 1000 km (Figure 15.15). They may be compared to $\approx 3\times10^9$ cm$^{-3}$ and $T$ = 97, 97, and 119 K at 1000 km in the models for slow hydrodynamic escape by Krasnopolsky (1999), Strobel (2008), and Erwin et al. (2013), respectively. Isothermal extension of the observed density and temperature at 1000 km confirms the existence of exobase with thermal escape and rules out hydrodynamic escape. The data are coupled with the radio occultations below 100 km and the stellar occultations below 200 km (Figure 15.10). The stellar occultation profile of temperature is isothermal at 80 K from 200 to 340 km. This temperature is close to that observed at 1000 km. However, the isothermal interpolation at 200–1000 km results in a density at 1000 km that significantly exceeds the observed value.

There is a variety of solutions that connect the density and temperature at 200 km with those at 1000 km, especially within the large uncertainty of temperature at 1000 km. A simple and reasonable approach is to extrapolate the temperature gradient observed by Dias-Oliveira et al. (2015; Figure 15.10) at 100–200 km up to a breakpoint with a constant temperature above the breakpoint. Then a region near the breakpoint may be smoothed (Figure 15.15).

The solid line in Figure 15.15 presents a nominal temperature profile that was fitted using a complicated multiparameter function. The dashed curve shows a profile that accounts for the observed $CH_4$ mixing ratio, diffusive separation, upward flow of methane

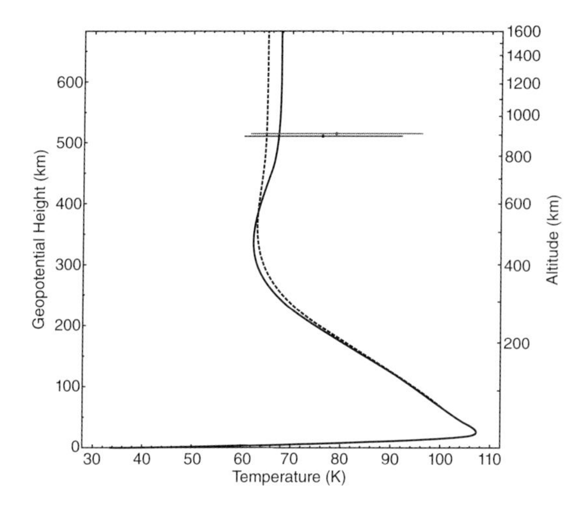


Figure 15.15 Temperatures near 1000 km derived from the $N_2$ densities at 900–1100 km measured by the solar UV occultations and two temperatures profiles that fit both radio and solar UV occultations (Young et al. 2017).

caused by its photolysis, charge exchange of $CH_4$ with $N_2^+$, and thermal escape. This solution gives a methane mixing ratio $3.5\times10^{-3}$ near the surface and constant eddy diffusion $K$ = 4000 cm$^2$ s$^{-1}$ with homopause at 12 km. The exobase is at 1700 km ($r_{ex} \approx$ 2900 km) with $T_{ex}$ = 65–68 K and thermal escape of (3–7) $\times$ 10$^{22}$ s$^{-1}$ for $N_2$ and (4–8) $\times$ 10$^{25}$ s$^{-1}$ for $CH_4$ that is smaller than the predictions for hydrodynamic escape by five and one orders of magnitude, respectively.

### *15.4.3 ALMA Observations*

The Atacama Large Millimeter/submillimeter Array (ALMA; Lellouch et al. 2017) is the largest interferometer that involves 66 dishes with diameters of 12 and 7 m. The observations of Pluto were conducted on June 12–13, 2015, just a month before the NH flyby. The detected CO and HCN rotational emission lines are shown in Figure 15.16. The CO line shape (Figure 15.17) was measured with resolving power of $2.5\times10^6$. Spatial resolution was 0.35 arcs, near one-half of the Pluto–Charon angular distance during the observations. Therefore Charon did not affect the observations of Pluto. Spectral fits to the observed CO line were made using the total pressure of 12 μbar and a CO mixing ratio and various temperature profiles as fitting parameters. Three fits are shown in Figure 15.17. Profile 2 is close to that from the NH radio and UV occultations (Figure 15.15), while the best fit (profile 3) indicates significant deviations from the NH radio occultation data. The NH profile is more uncertain at 200–600 km, and deviations at these altitudes are not ruled out. The recommended CO mixing ratio is 515 $\pm$ 40 ppm for the surface pressure of 12 μbar.

Fitting to the HCN line is sensitive to both temperature profile and altitude distribution of HCN. The best-fit profile of the HCN mixing ratio has a narrow peak of $\approx$0.03 ppm near

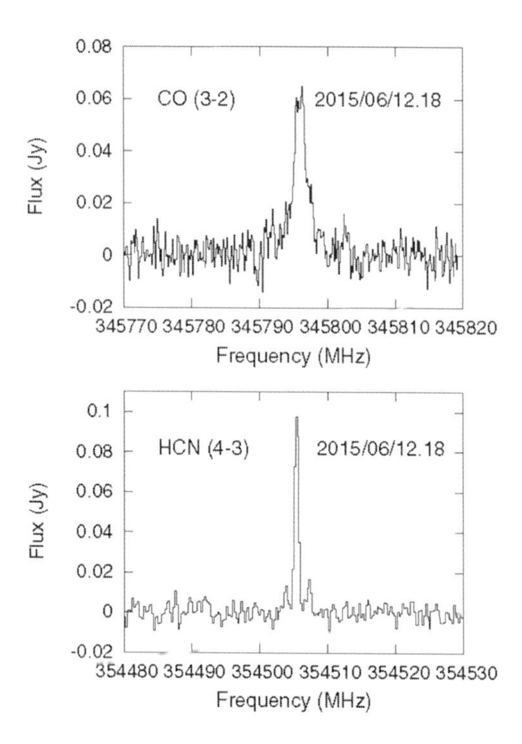


Figure 15.16 Lines of CO and HCN observed by Lellouch et al. (2017) using ALMA. Two weak lines near the main HCN line reflect its hyperfine structure.

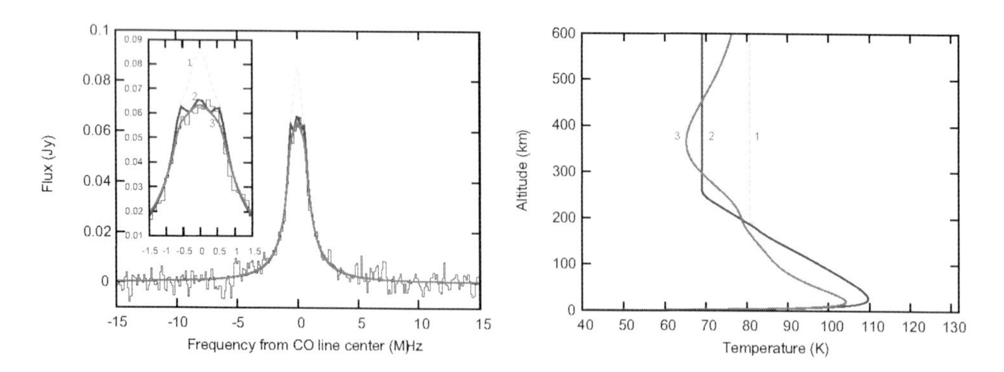


Figure 15.17 Fitting of the observed CO line shape by (1) temperature profile from Dias-Oliveira et al. (2015) and CO = 500 ppm, (2) the same but with temperature reduced to 69 K above 270 km, and (3) the best fit with CO = 507 ppm. From Lellouch et al. (2017).

40 km and a constant value of 15–40 ppm above 450 km. An alternative profile is for 40 ppm above 450 km and no HCN below 450 km. This means supersaturation by a few orders of magnitude above 450 km that may be possible if condensation nuclei are lacking. The HCN column abundance is $(1.6 \pm 0.4)\times10^{14}$ cm$^{-2}$.

Search for the $HC_3N$ and $HC^{15}N$ lines resulted in an upper limit of $2\times10^{13}$ cm$^{-2}$ to the $HC_3N$ column abundance and a lower limit $HC^{14}N/HC^{15}N > 125$ on Pluto. This ratio is equal to 60 on Titan. Mandt et al. (2017) explain the high lower limit by a more efficient aerosol trapping of $HC^{15}N$ relative to that of $HC^{14}N$.

### 15.4.4 Thermal Balance and the NH Data

This problem was studied by Strobel and Zhu (2017). They involved cooling of the atmosphere by rotational lines of HCN and $H_2O$ to explain the low temperatures of 65–70 K above 400 km. The model accounts for heating by absorption of the solar light by the $CH_4$ bands at 1.7, 2.3, 3.3, and 7.6 μm (Figure 15.18). The band at 7.6 μm absorbs thermal radiation in the atmosphere, and this is an additional source of heating below 20 km. The FUV heating by the methane absorption below 140 nm and the EUV heating by the $N_2$ absorption dominate above 400 km.

Processes of cooling are shown in the right panel of Figure 15.18. These are LTE cooling by rotational lines of $H_2O$, HCN, and CO, by the bands of $C_2H_2$ at 13.7 μm and $CH_4$ at 7.6 μm, and adiabatic cooling by a weak but nonnegligible thermal escape. Krasnopolsky (2018b) pointed out that the LTE conditions break for rotational lines near and above a level where $A \approx kn$; here $A$ is the transition probability, $k \approx 10^{-11}\, T^{1/2}$ cm$^3$ s$^{-1}$ is the collisional rate coefficient, and $n$ is the atmospheric number density. This means that the effect of cooling by the $H_2O$ and HCN rotational lines is significantly smaller above

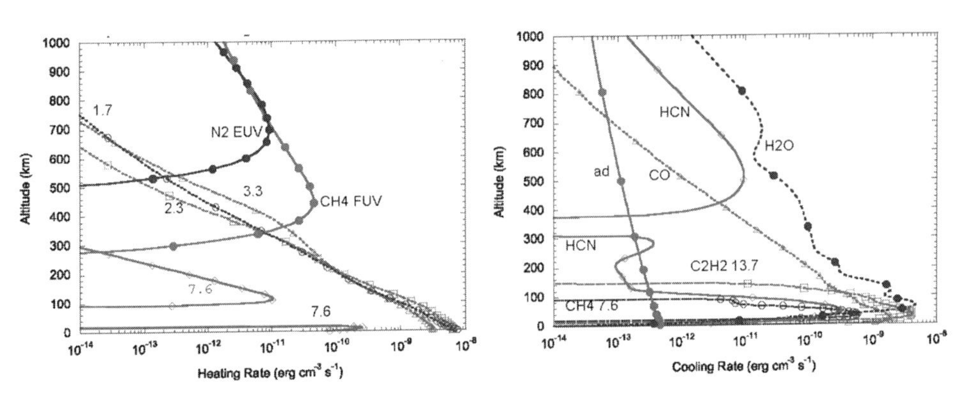


Figure 15.18 (a) Heating of Pluto's atmosphere by absorption of the solar light by the $CH_4$ bands at 1.7, 2.3, 3.3, and 7.6 μm and by the $CH_4$ FUV and $N_2$ EUV absorptions. (b) Cooling by the $H_2O$, HCN, and CO rotational lines, $C_2H_2$ band at 13.7 μm, $CH_4$ band at 7.6 μm, and adiabatic cooling. From Strobel and Zhu (2017).

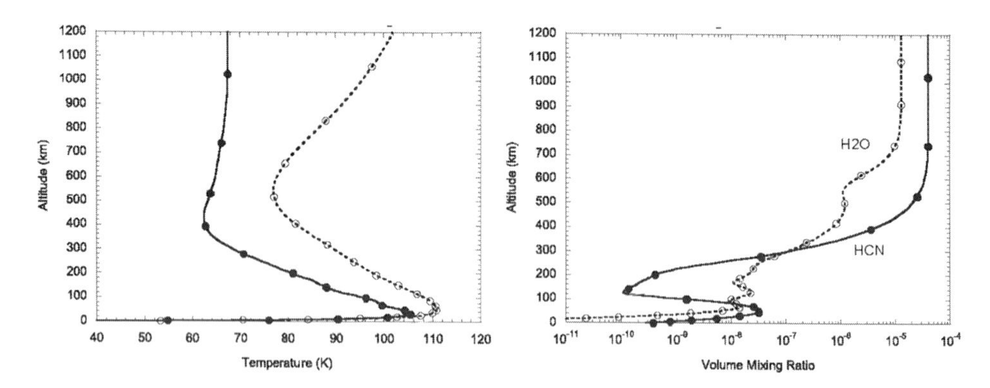


Figure 15.19 (a) Calculated temperature profiles with and without $H_2O$. (b) Adopted mixing ratios of HCN and $H_2O$. From Strobel and Zhu (2017).

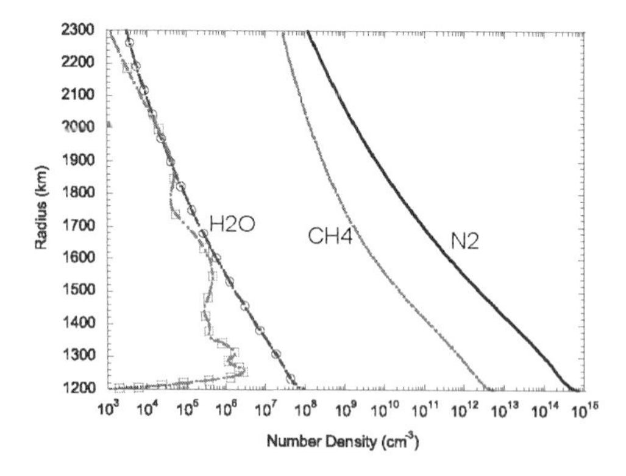


Figure 15.20 Profiles of number densities of $N_2$, $CH_4$, and $H_2O$ (calculated and best fit) in Strobel and Zhu (2017).

700 km (Figures 15.18 and 15,19) than that calculated by Strobel and Zhu (2017), and the observed cryosphere needs further study and explanation. That was made by Zhang et al. (2017; see Section 15.5.3).

The adopted HCN profile (Figure 15.19) is close to that retrieved from the ALMA observation (40 ppm above 450 km), and that of $H_2O$ (Figures 15.19 and 15.20) is compared with a profile, calculated for a production rate of $3.4\times10^{19}$ s$^{-1}$ (190 cm$^{-2}$ s$^{-1}$ or 90 g d$^{-1}$) released by ablation of the interplanetary dust near 500 km. The NH evaluation of the interplanetary dust flux is 200 kg d$^{-1}$ (Horanyi et al. 2016), very much greater than that of $H_2O$ in Strobel and Zhu (2017). The $H_2O$ profile (Figure 15.19) will be compared with that from a photochemical model by Wong et al. (2017) in Section 15.6.2.

## 15.5 Haze

### *15.5.1 NH Observations*

According to the solar UV occultations, the line-of-sight optical depth of the haze at 180 nm is $\tau_s \approx 2$ near the surface, and its scale height is $H \approx 70$ km. Then its vertical UV optical depth is

$$\tau_{UV} = \frac{\tau_s}{\sqrt{2\pi r/H}} = 0.2. \tag{15.2}$$

Pluto's haze was observed using the NH LORRI imager and MVIC camera at a few phase angles. Preliminary analysis of the haze observations was made by Gladstone et al. (2016). Brightness coefficient $\pi I/I_0$ at phase angles $\approx 167°$ was 0.2–0.3 in the red and 0.7–0.8 in the blue. (Here $I$ is the haze brightness, $I_0$ is the solar radiation, and phase angle is $\pi$ minus scattering angle.) The strong brightening to the blue indicates Rayleigh scattering by very small haze particles with radii $a \approx 0.01$ μm. However, the haze brightness is much smaller at moderate phase angle (0.02 in the red at 38°). Assuming a haze refractive index at $1.69 - 0.018i$ for tholins from Khare et al. (1984), the observed phase dependence results in the particle radius $a \approx 0.2$ μm based on the Mie formulas (Section 4.1). Therefore the haze particles are randomly shaped aggregates of small ($\approx$0.01 μm) spherical monomers, that is, fractals.

Brightness coefficient for optically thin conditions is

$$\frac{\pi I}{I_0} = P(\theta)\frac{\tau_s}{4}.$$

Here $P(\theta)$ is the scattering phase function that is equal to 5 in the red for the phase angle of 167° and $a = 0.2$ μm. The observed brightness coefficient was 0.2–0.3 in the red at 167°. Then the slant scattering optical depth is $\tau_s \approx 0.2$, and the vertical scattering optical depth in the red is $\tau_R \approx 0.02$. The Mie formulas give scattering efficiency of 2.7 for the observed haze particles with $a = 0.2$ μm, and their scattering cross section is $2.7\pi a^2 = 3.4\times10^{-9}$ cm$^2$. Then the particle column abundance is $6\times10^6$ cm$^{-2}$, and their mass is $1.3\times10^{-7}$ g cm$^{-2}$ for a particle density of 0.65 g cm$^{-3}$.

The observed haze extends in the UV and visible ranges up to 300 km (Figures 15.14a and 15.21), where its brightness decreases relative to that near the surface by factors of 150 and $10^4$, respectively. The steeper decrease in the visible indicates the particle size decreasing with altitude. The haze is structured with $\approx$20 layers, and the haze is denser in the northern hemisphere decreasing to equatorial and southern latitudes (Cheng et al. 2017). The particles may be approximated by spheres with radii of 0.5 μm near the surface and fractals near and above 45 km.

### *15.5.2 Models of the Haze*

Comparing the observed densities of the seven detected gases (five species in Figure 15.14a plus CO and HCN) with their saturated vapor densities (Table 13.8; see also Fray and Schmitt 2009) at Pluto's temperatures (Figure 15.15), one concludes that

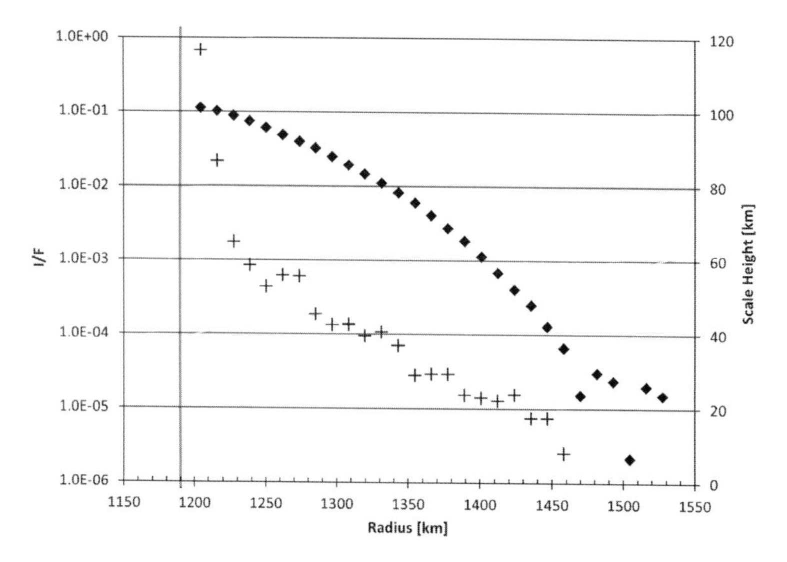


Figure 15.21 Azimuthally averaged brightness coefficient I/F and scale height of the haze observed at the limb using LORRI at phase angle of 166°. The bandpass is 350–850 nm, the effective mean wavelength is 608 nm. From Cheng et al. (2017).

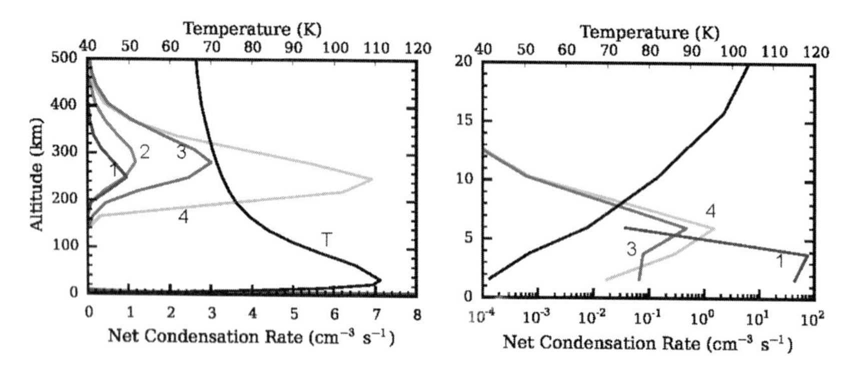


Figure 15.22 Condensation rates in Pluto's atmosphere from a photochemical model by Wong et al. (2017) adopted by Gao et al. (2017). Here 1, 2, 3, and 4 are $C_2H_6$, HCN, $C_2H_4$, and $C_2H_2$, respectively.

only HCN throughout the atmosphere and $C_2H_2$ at 300–500 km can condense on Pluto (Luspay-Kuti et al. 2017). All $C_2$ hydrocarbons can also condense in a thin layer near the surface with $T \approx 40$ K.

Gao et al. (2017) modeled the haze size distribution by downward transport and coagulation using the CARMA code to fit the haze extinction profile measured by the solar UV occultations. They applied condensation rates of four species (Figures 15.22 and 15.23) from a photochemical model by Wong et al. (2017) to calculate both spherical and purely fractal aggregate haze. The spherical haze requires production rates exceeding that in Figure 15.22 by a factor of 2–3, while the fractal aggregates with monomers of $\approx$0.01 μm agree with the

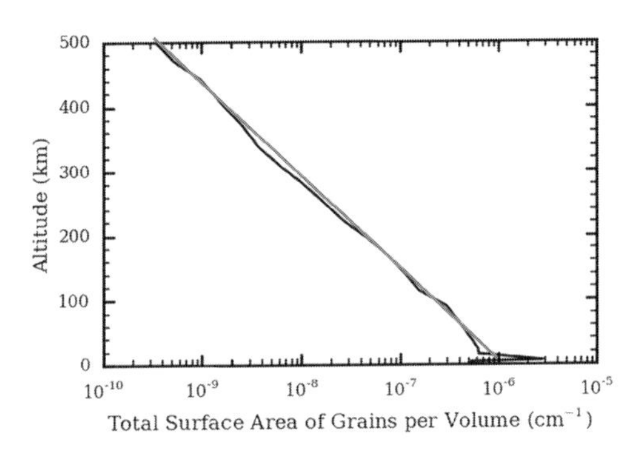


Figure 15.23 Volume surface area of the haze adopted by Wong et al. (2017) from the model by Gao et al. (2017). The straight line is the exponential fit to this function.

photochemistry. The fractal size distribution peaks at $a \approx 0.15$ μm near the surface and decreases to ≈0.02 μm at and above 200 km. To fit this distribution, a particle charge to radius ratio should be ≈30 $e^-$ $\mu m^{-1}$. The weighted-mean monomer radius increases in the model from 0.01 μm at 400 km to 0.0145 μm below 200 km.

However, only HCN and $C_2H_2$ can actually condense, and the condensation rate of $C_2H_2$ for the condensation efficiency $\alpha = 10^{-4}$ in the preferable model by Luspay-Kuti et al. (2017) is smaller than that in Figure 15.22 for $\alpha = 2 \times 10^{-5}$ by two orders of magnitude. Aerosol trapping, that is, irreversible sticking, is the only significant loss of the $C_2$ hydrocarbons below 500 km in Luspay-Kuti et al. (2017). It is equal to ≈1000 $cm^{-3}$ $s^{-1}$ for $C_2H_2$ at 200–300 km and exceeds the $C_2H_2$ condensation rate at 300–500 km by a factor of 3500 in their model and that in Figure 15.22 by two orders of magnitude. These data do not support the results of Wong et al. (2017) and Gao et al. (2017). Nevertheless, the exclusion of condensation of $C_2H_4$ and $C_2H_6$ from the model by Gao et al. (2017) would reduce the production of haze just by a factor of 1.5 (Figure 15.22) that looks within uncertainties of the model.

Bertrand and Forget (2017) included haze in their Pluto GCM. They modeled a global distribution of the solar and interplanetary Lyman-alpha radiation and photolysis of $CH_4$ by this radiation as the only source of haze in the model with a yield of 21 atomic units per photolysis event. They ran the model for a few assumptions on the particle radii and with/without South Pole $N_2$ condensation. Production of haze in 2015 was mostly in the sunlit northern hemisphere, and the haze is maximal at the North Pole without South Pole $N_2$ condensation. The haze latitudinal distribution is more homogeneous with a slight peak at the South Pole in the case of the South Pole $N_2$ condensation. The calculated mean haze mass is $2\text{–}4 \times 10^{-7}$ g $cm^{-2}$.

### *15.5.3 Haze and Thermal Balance of the Atmosphere*

Zhang et al. (2017) involved the haze to solve the problem of thermal balance of the atmosphere and explain the low temperature of ≈70 K in the upper atmosphere. They

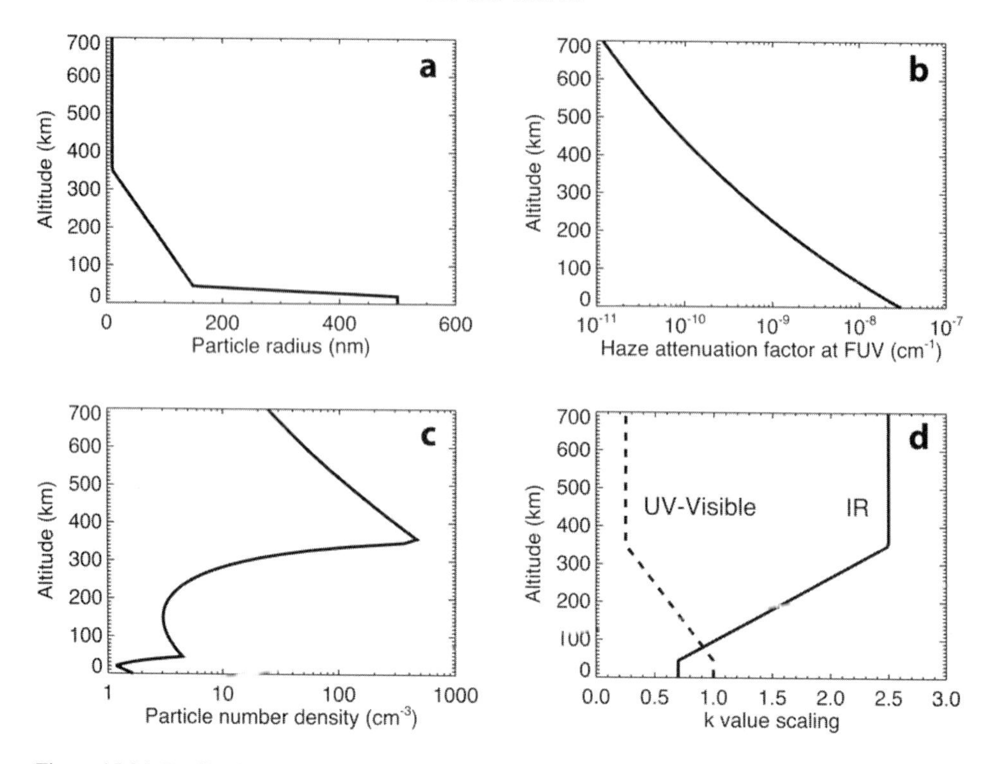


Figure 15.24 Profiles in (a) and (b) approximate the NH observations of the haze, (c) is deduced from (a) and (b), and (d) shows scaling factors to the tholin absorption coefficients from Khare et al. (1984, figure 246) to fit the observed temperature profile. From Zhang et al. (2017).

adopted a vertical profile of the haze particle radius (Figure 15.24a) and extrapolated the NH haze observations (Figure 15.14a) up to 700 km (Figure 15.24b). These profiles match fairly well the existing data. Then calculations using the Mie formulas result in the particle number density profile in Figure 15.24c.

Zhang et al. (2017) applied a few sources of the data on refractive index of tholins in their calculations. They found that heating of the atmosphere by the haze absorbing the sunlight mostly in the visible range is significantly larger below 700 km than heating by the gas absorption. The same refers to cooling by the blackbody radiation of the haze that is more effective than thermal emissions of the atmospheric gases. Therefore the temperature profile is mostly determined by the haze imaginary refractive indices at 0.3–1.0 μm and 10–300 μm that are responsible for heating and cooling, respectively. Using the refractive indices from Khare et al. (1984; Figure 13.2), Zhang et al. (2017) calculated scaling factors to the haze imaginary indices (Figure 15.24d) that are required to fit the observed temperature profile (Figure 15.25). The haze is significantly warmer than Pluto's surface, and thermal radiation of the haze could be detected in the middle infrared at 5–30 μm using, e.g., the James Webb Space Telescope.

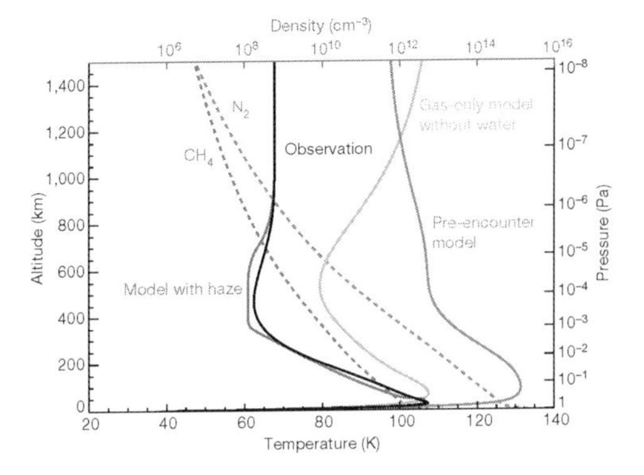


Figure 15.25 Observed density and temperature profiles and temperature profiles from three models (Zhang et al. 2017).

The required absorption coefficient of the haze in the visible range above 350 km is smaller than that in Khare et al. (1984) by a factor of 4 (Figure 15.24d), while that in the thermal infrared should be greater by a factor of 2.5. Both trends are not supported by the Cassini VIMS and CIRS observations on Titan, respectively (Figure 13.27). However, absorption coefficient in the very small haze particles above 350 km on Pluto may differ from that in Titan's haze.

### *15.5.4 Atmosphere of Charon*

The NH search for the atmosphere of Charon using solar UV occultations and airglow resulted in restrictive upper limits to abundances of 14 species (Stern et al. 2017a). The limit to $N_2$ is $3\times10^{15}$ cm$^{-2}$ and surface pressure of 4 picobar; the limit to $CH_4$ is $4\times10^{14}$ cm$^{-2}$ and 0.3 picobar. Hoey et al. (2017) calculated that Charon captures $2\times10^{24}$ s$^{-1}$ molecules from the escape flow from Pluto, and methane is 98% of this quantity. This capture supports an atmosphere with a total gas abundance of $1.5\times10^{13}$ cm$^{-2}$, that is, smaller than the upper limit to $CH_4$ by a factor of 27.

## 15.6 Photochemistry

### *15.6.1 Photochemical Models before the NH Flyby*

First models for Pluto were calculated by Summers et al. (1997). They adopted an isothermal atmosphere with $T$ = 104 K above ≈40 km and up to the exobase for all their models. That assumption was based on the stellar occultation by Elliot and Young (1992) and Triton's thermospheric profile. Their basic model was Triton's analog with mixing ratios $f_{CH4} = 4\times10^{-5}$ and $f_{CO} = 5\times10^{-4}$, that is, smaller than the current value for $CH_4$ by two orders of magnitude and equal to that for CO. The model results were presented in detail; it is not clear why atomic nitrogen is less abundant than that observed and modeled for Triton by three orders of magnitude and why, for example, the major ions are $C^+$ and $N_2^+$ instead of $C^+$ and $N^+$ (maybe this was a misprint). The authors also considered a few "high methane" models, and one of them with $f_{CH4} = 4\times10^{-3}$ and $f_{CO} = 5\times10^{-4}$ corresponds to the current values. Unfortunately, the only plot from that model is for the ionospheric composition with $C_2H_5^+$ as the most abundant ion. A model for the mean heliocentric distance of 39.4 AU was calculated as well.

Lara et al. (1997) made the first detailed photochemical model for Pluto's neutral atmosphere with a reasonable adopted methane abundance of 0.74% (Table 15.5), which is within a factor of 2 of the present value. They included a loss for condensation at a haze that is uniformly mixed in the atmosphere with a total visible optical depth $\tau = 0.07/\gamma$; here $\gamma$ is the sticking coefficient. (This is our interpretation of their condensation time.) The model involved 111 reactions of 29 neutral species. Some data of the model are in Table 15.5.

Krasnopolsky and Cruikshank (1999) developed a photochemical model of Pluto's atmosphere and ionosphere near perihelion. The model involved 191 reactions of 44 neutral and 23 ion species. Similar to that on Triton and Titan, ion reactions on Pluto significantly affect some neutral species. Thermal structure and gas flow in the model was adopted from the hydrodynamic escape calculations by Krasnopolsky (1999), and the flow velocity was included in the continuity equations for species. However, the flow velocity was much smaller than the diffusion velocity and weakly affected the solution.

Some results of the model for methane abundance of 0.9% are given in Table 15.5. The mixing ratios of $C_2H_2$ and $C_2H_4$ at the level $[N_2] = 10^{11}$ $cm^{-3}$ (500 km according to the NH solar UV occultations) are close to those observed. The smaller abundance of $C_2H_6$ is caused by a rate coefficient of the reaction

$$CH_3 + CH_3 + M \rightarrow C_2H_6 + M,$$

which was smaller than the present value by two orders of magnitude. Condensation at the haze particles was not considered in the model, and loss of all photochemical products (except $H_2$) was assumed at the cold surface with sticking coefficient $\gamma = 0.1$. Therefore abundances of $C_2H_2$ and $C_2H_4$ at the level $[N_2] = 2.5\times10^{13}$ $cm^{-3}$ (200 km according to the NH solar UV occultations) exceed those observed. The calculated dayside-mean peak electron density is 800 $cm^{-3}$, and $HCNH^+$ is the most abundant ion. (This agrees with an upper limit of 1000 $cm^{-3}$ to the peak electron density in the NH radio occultations near the terminator; Hinson et al. 2018.) Production and loss of the major species, their vertical profiles, column abundances, and precipitation and escape rates are discussed, shown, and tabulated in the paper.

### *15.6.2 Wong et al.'s Photochemical Model*

This is a model for Pluto's neutral atmosphere that is based on the NH findings (Wong et al. 2017). Luspay-Kuti et al. (2017) pointed out that this model adopts condensation of

Table 15.5 *Comparison of photochemical models with the NH solar UV occultations*

| Reference | $CH_4$ | $h_0$ (km) | $C_2H_2$ | $C_2H_4$ | $C_2H_6$ | HCN | $h_1$ (km) | $C_2H_2$ | $C_2H_4$ | $C_2H_6$ | HCN |
|---|---|---|---|---|---|---|---|---|---|---|---|
| NH Young et al. (2017) | 0.35% | 500 | 1000 | 1000 | 1000 | 40 | 200 | 15 | 2 | 50 | – |
| Wong et al. (2017) | 0.4% | 500 | 1150 | 650 | 650 | 40 | 200 | 5 | 0.2 | 175 | – |
| Krasnopolsky and Cruikshank (1999) | 0.9% | 520 | 800 | 500 | 25 | 350 | 130 | 240 | 40 | 23 | 50 |
| Lara et al. (1997) | 0.74% | 630 | 35 | 42 | 31 | 150 | 180 | 28 | 5 | 50 | 17 |

*Note.* $h_0$ and $h_1$ are the altitudes where $[N_2] = 10^{11}$ and $2.5 \times 10^{13}$ $cm^{-3}$, respectively; species mixing ratios are in ppm at $h_0$ and $h_1$. The HCN value of 40 ppm was observed by ALMA (Lellouch et al. 2017), and a corrected value is given for Wong et al. (2017) (see text).

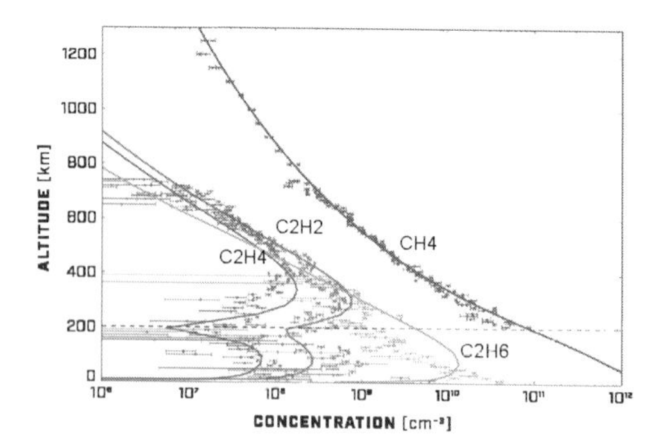


Figure 15.26 Number densities of $CH_4$, $C_2H_2$, $C_2H_4$, and $C_2H_6$ retrieved from the NH solar UV occultations (Gladstone et al. 2016) and calculated by Wong et al. (2017). (A black and white version of this figure will appear in some formats. For the color version, please refer to the plate section.)

$C_2H_4$ and $C_2H_6$ that contradicts the laboratory data, and this should significantly affect all conclusions of the model. Yet we describe it below as it is.

The model involves the NH temperature profile similar to that in Figure 15.15 and the measured surface pressure as the model initial data. The model has 40 levels up to 1300 km and includes 1600 reactions of 88 neutral species. The results are focused on the four hydrocarbons observed by the NH solar UV occultations, and their major production and loss processes are shown, while data on the other reactions are lacking in the paper.

Eddy diffusion is generally unknown on Pluto, and $K = 1000$ cm$^2$ s$^{-1}$ and $CH_4$ mixing ratio of 0.4% near the surface are the best combination that fits to the observed profile of methane (Figure 15.26). Photolysis of methane by the solar Lyman-alpha affects the calculated profile, and the solar Lyman-alpha is scaled by a factor of 1.43 to account for the local Lyman-alpha background.

The atmospheric temperature is $\approx$70 K at 200–400 km, and condensation at the haze particles is important at these altitudes. Densities of both haze and gases become very low above 400 km, while the atmosphere is warmer below 200 km; that is why condensation is the most effective in the altitude range of 200–400 km. The haze model by Gao et al. (2017) applies the hydrocarbon condensation rates (Figure 15.22) from Wong et al. (2017), while Wong et al. (2017) use an altitude profile of the haze volume surface area $A$ (Figure 15.23) from Gao et al. (2017). This profile from 0 to 500 km may be approximated by $A$ (cm$^{-1}$) = $10^{-6}\, e^{-h/H}$, where $H = 88$ km. Homogeneous nucleation is neglected, and excess of a condensable species over its saturation vapor is lost in collisions with the haze particles. This loss is proportional to the species sticking coefficient $\gamma$, which is a free parameter to fit the observed species abundances.

The column haze surface area in Figure 15.23 is $10^{-6}\, H = 8.8$, while the discussion in Section 15.5.1 gives $6\times10^6\times4\pi a^2 = 0.03$ for the scattering optical depth of 0.02 (0.013 in Gladstone et al. 2016) in the red. Evidently the volume surface area in Figure 15.23 refers to sums of the monomer surface areas. However, even the aggregate particles are much smaller than the mean free path in Pluto's atmosphere, and the monomer surface areas are inapplicable to the condensation process. Ratio of surface area of the haze particle to sum of the monomer areas may reach $\approx$0.2/0.01 $\approx$ 20, still much smaller than 8.8/0.03 $\approx$ 300. The remaining difference is unexplained (Krasnopolsky 2018b).

Radius of the equal volume sphere of the fractile particles on Titan is smaller than that of the sphere with equal projected area by a factor of 2.8 (Tomasko et al. 2008a). The latter is appropriate to the condensation process, and the sticking coefficients derived by Wong et al. (2017) should be scaled by at least a factor of $\approx$20/2.8 = 7. Furthermore, the scaled sticking coefficients look better, though sticking in condensation of hydrocarbons is generally unknown and was measured only for $H_2O$ and $HNO_3$ (Burkholder et al. 2015).

Acethylene $C_2H_2$ is formed by photolysis of $C_2H_4$ (two branches) and the reaction

$$C_2H_4 + CH \rightarrow C_2H_2 + CH_3.$$

The production maximum is near 400 km. The major loss is by condensation with $\gamma_{C2H2} = 3\times10^{-5}\times7 \approx 2\times10^{-4}$; other losses are reactions with CH, $^1CH_2$, and photolysis. The total loss peaks at $\approx$300 km.

Production of ethylene $C_2H_4$ is mostly by the reaction

$$CH_4 + CH \rightarrow C_2H_4 + H,$$

which peaks at 350 km. Its removal is by condensation, photolysis, and the reaction with CH.

Inversions in the observed number density profiles of $C_2H_2$ and $C_2H_4$ at 200 km (Figure 15.26) are explained by condensation at $\approx$70 K that begins at 200 km. The calculated inversion of $C_2H_2$ agrees with that observed. However, the saturated vapor of $C_2H_4$ is denser by orders of magnitude than that of $C_2H_2$ and cannot fit the observed inversion. Therefore Wong et al. (2017) assume that the saturated vapor pressure of ethylene is equal to that of acethylene and fits the observed densities of ethylene with $\gamma_{C2H4} = 10^{-4}\times7 = 7\times10^{-4}$ (Figure 15.26). This assumption is questionable and means that the model disagrees with the observed profile of ethylene.

Termolecular association of two $CH_3$ radicals is the only reaction that forms ethane. It is lost by condensation, reaction with CH, and photolysis. The best-fit sticking coefficient is $\gamma_{C2H6} = 3\times10^{-6}\times7 \approx 2\times10^{-5}$. The adopted condensation disagrees with the laboratory data.

A set of the HCN profiles for $\gamma_{HCN} = 10^{-5}$ to 1 results in variations of the HCN column abundance from $1.2\times10^{15}$ to $1.2\times10^{13}$ cm$^{-2}$, respectively. The authors fitted the calculated HCN profiles to a preliminary ALMA value of $5\times10^{13}$ cm$^{-2}$ with $\gamma_{HCN} = 10^{-2}$. The final HCN abundance in the ALMA observations is $1.6\times10^{14}$ cm$^{-2}$ (Section 15.4.3; Lellouch et al. 2017) and corresponds to $\gamma_{HCN} = 10^{-3}\times7 = 0.007$. This solution gives $f_{HCN}$ = 40 ppm at 500 km (Table 15.5), close to the preferable value in Lellouch et al. (2017).

Oxygen-bearing species (Figure 15.27) are calculated using the CO mixing ratio of $5\times10^{-4}$ at the surface and an exogenous $H_2O$ flux of $5\times10^5$ cm$^{-2}$ s$^{-1}$ at the upper boundary of 1300 km. This flux is equal to 1000 kg d$^{-1}$ of water, while the interplanetary

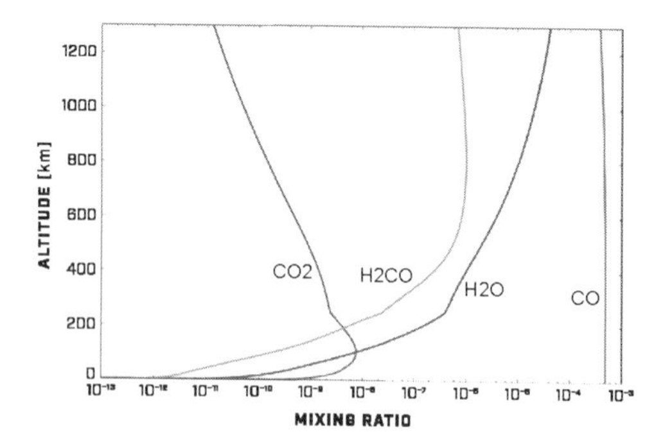


Figure 15.27 Oxygen-bearing species in the model by Wong et al. (2017).

dust delivery to Pluto is estimated at 200 kg $d^{-1}$ based on the NH and Pioneer 10/11 data (Horanyi et al. 2016). Poppe (2015) predicted the dust flux of $1.4\times10^{-17}$ g $cm^{-2}$ $s^{-1}$ for the NH flyby of Pluto. This flux is $4.7\times10^{5}$ $cm^{-2}$ $s^{-1}$ if it had completely comprised $H_2O$. On the other hand, Feuchgruber et al. (1997) evaluated influxes of $H_2O$ from the interplanetary dust to Saturn, Uranus, and Neptune at $\approx10^{6}$ $cm^{-2}$ $s^{-1}$, and the flux of $2\times10^{6}$ $cm^{-2}$ $s^{-1}$ at Titan's exobase is required to fit the observed $H_2O$ and other oxygen species (Krasnopolsky 2012d).

The calculated $H_2O$ abundances (Figure 15.27) are comparable to those required to explain the thermal structure of Pluto's atmosphere in the model by Strobel and Zhu (2017; Figures 15.19 and 15.20). However, Strobel and Zhu (2017) stated that the $H_2O$ influx is 190 $cm^{-2}$ $s^{-1}$ in their model. Condensation at the surface and the diffusion-limited thermal escape were the only loss for $H_2O$ in their model, and the neglect of chemical loss and condensation on the haze explains the difference by four orders of magnitude between the required fluxes of $H_2O$.

Wong et al. (2017) tabulated precipitation rates of 10 species in their model and compared them with those in M1 in Krasnopolsky and Cruikshank (1999). Currently the preferable model is M2, whose data are given in Table 15.6.

### *15.6.3 The Model by Luspay-Kuti et al.*

The model by Luspay-Kuti et al. (2017) indicates that $C_2H_4$ and $C_2H_6$ cannot condense in Pluto's atmosphere except a narrow layer near the surface with $T \approx 40$ K. To fit the observed NH profiles of the species, the authors propose irreversible sticking on the haze particles with sticking efficiency varying from zero near the surface to 0.019, 0.04, and 0.011 at 400 km for $C_2H_2$, $C_2H_4$, and $C_2H_6$, respectively. The authors understand that sticking efficiency should depend on temperature rather than altitude, but the latter is preferable. They succeed to fit the observed density profiles of the $C_2$ hydrocarbons. However, the calculated total loss by aerosol trapping is extremely high, $\approx3\times10^{11}$ $cm^{-2}$ $s^{-1}$ for $C_2H_2$ in their figure 9. Column production rates of the $C_2$ hydrocarbons are not given;

Table 15.6 *Precipitation rates (g $cm^{-2}$ $Byr^{-1}$) in the photochemical models*

| Model | $C_2H_2$ | $C_2H_4$ | $C_2H_6$ | $C_3H_4$ | HCN | $C_6H_6$ | $C_4H_2$ | $C_3H_6$ | $C_3H_3CN$ | $HC_3N$ |
|---|---|---|---|---|---|---|---|---|---|---|
| W2017 | 179 | 95 | 62 | 48 | 35 | 34 | 26 | 8 | 6 | 4 |
| KC1999 | 123 | 13 | 14 | 7 | 22 | – | 165 | – | – | 72 |

*Note.* W2017 is Wong et al. (2017), and KC1999 is Krasnopolsky and Cruikshank (1999).

they are typically $\approx 10^8$ cm$^{-2}$ s$^{-1}$ in the other models. The difference by three orders of magnitude between the column production and loss is striking, because usually the models tend to reduce this difference to be less than 1%. The extremely high aerosol production should be incompatible with the observed properties of the haze as well. Furthermore, the $C_2$ hydrocarbon densities at the surface are fitting parameters in the model, though the lower boundary conditions for photochemical products should reflect properties of the surface. Finally, the calculated peak total ion density of 30 cm$^{-3}$ is too low compared with the peak electron density of 2500 cm$^{-3}$ on the dayside and $\approx 1000$ cm$^{-3}$ near the terminator on Titan (Figure 13.45). Scaling the heliocentric distances of Titan and Pluto, the expected ratio of the peak electron densities is a factor of $\approx 3$.

Overall, the New Horizons flyby resulted in a breakthrough in the studies of Pluto and its atmosphere. The NH instruments and techniques for these studies were similar to those for the atmosphere of Triton in the Voyager 2 flyby: radio and solar UV occultations and imaging to observe haze. Both similarities and differences of the atmospheres of Pluto and Triton are exciting and stimulating further studies of these bodies.

# References


Adams, W.S., Dunham, T., 1932. Absorption bands in the infrared spectrum of Venus. *Publ. Astron. Soc. Pacific* 44, 243–247.

Agren, K., et al., 2009. On the ionospheric structure of Titan. *Planet. Space Sci.* 57, 1821–1827.

Ajello, J.M., et al., 2007. Titan airglow spectra from Cassini ultraviolet imaging spectrograph: EUV analysis. *Geophys. Res. Lett.* 34, L24204.

Ajello, J.M., et al., 2008. Titan airglow spectra from the Cassini ultraviolet imaging spectrograph: FUV disk analysis. *Geophys. Res. Lett.* 35, L06102.

Ajello, J.M., et al., 2012. Cassini UVIS observations of Titan nightglow spectra. *J. Geophys. Res.* 117, A12315.

Allen, D.A., Crawford, J.W., 1984. Cloud structure on the dark side of Venus. *Nature* 307, 222–224.

Allen, M., Lollar, B.S., Runnegar, B., Oehler, D.Z., Lyons, J.R., Manning, C.E., Summers, M.E., 2006. Is Mars alive? *EOS Tran. AGU* 87, 433–439.

Anderson, C.M., Samuelson, R.E., 2011. Titan's aerosol and stratospheric ice opacities between 18 and 500 lm: Vertical and spectral characteristics from Cassini CIRS. *Icarus* 212, 762–778.

Anderson, D.E., 1974. Mariner 6, 7, and 9 ultraviolet spectrometer experiment: analysis of hydrogen Lyman-alpha data. *J. Geophys. Res.* 79, 1513–1518.

Anderson, D.E., 1976. The Mariner 5 ultraviolet photometer experiment: Analysis of hydrogen Lyman-a data. *J. Geophys. Res.* 81, 1213–1216.

Anderson, D.E., Hord, C.W., 1971. Mariner 6 and 7 ultraviolet spectrometer experiment: Analysis of hydrogen Lyman alpha data. *J. Geophys. Res.* 76, 6666–6671.

Andreychikov, B.M., et al., 1987. X-ray radiometric analysis of the cloud aerosol of Venus by the Vega 1 and 2 probes. *Cosmic Res.* 25, 721–728.

Aoki, S., et al., 2015. Seasonal variations of the HDO/$H_2O$ ratio in the atmosphere of Mars at the middle of northern spring and beginning of northern summer. *Icarus* 260, 7–22.

Aoki, S., et al., 2017. Sensitive search of $CH_4$ on Mars by SOFIA/EXES. *EPSC Abstracts* 11, 460.

Aoki, S., et al., 2018. Mesospheric $CO_2$ ice clouds on Mars observed by Planetary Fourier Spectrometer onboard Mars Express. *Icarus* 302, 175–190.

Arney, G., Meadows, V., Crisp, D., Schmidt, S.J., Bailey, J., Robinson, T., 2014. Spatially resolved measurements of $H_2O$, HCl, CO, OCS, $SO_2$, cloud opacity, and acid concentration in the Venus near-infrared spectral windows. *J. Geophys. Res.* 119E, 1860–1891.

Atreya, S.K., 1986. *Atmospheres and Ionospheres of the Outer Planets and Their Satellites*. Springer, Berlin.

Bahou, M., Chung, C.Y., Lee, Y.P., Cheng, B.M., Yung, Y.L., Lee, L.C., 2001. Absorption cross sections of HCl and DCl at 135–232 nanometers: Implications for photodissociation on Venus. *Astrophys. J.* 559, L179–L182.

Bailey, J., Meadows, V.S., Chamberlain, S., Crisp, D., 2008. The temperature of the Venus mesosphere from $O_2(a^1\Delta_g)$ airglow observations. *Icarus* 197, 247–259.

Balsiger, H., et al., 2015. Detection of argon in the coma of comet 67P/Churyumov-Gerasimenko. *Sci. Adv.* 1, e1500377.

Banaszkiewicz, M., Lara, L.M., Rodrigo, R., Lopez-Moreno, J.J., Molina-Cuberos, G.J., 2000. A coupled model of Titan's atmosphere and ionosphere. *Icarus* 147, 386–404.

Barker, E.S., 1972. Detection of molecular oxyden in the Martian atmosphere. *Nature* 238, 447–448.

Barker, E.S., 1979. Detection of $SO_2$ in the UV spectrum of Venus. *Geophys. Res. Lett.* 6, 117–120.

Barstow, J.K., Tsang, C.C.C., Wilson, C.F., Irwin, P.G.J., Taylor, F.W., McGouldrick, K., Drossart, P., Piccioni, and Tellmann, G. 2012. Models of the global cloud structure on Venus derived from Venus Express observations. *Icarus* 217, 542–560.

Barth, C.A. and Hord, C.W., 1971. Mariner 6 and 7 ultraviolet spectrometer experiment: topography and polar cap. *Science* 173, 197–201.

Barth, C.A., Stewart, A.I., Hord, C.W., Lane, A.L., 1972. Mariner 9 ultraviolet spectrometer experiment: Mars airglow spectroscopy and variations in Lyman alpha. *Icarus* 17, 457–468.

Barth, C.A., et al., 1973. Mariner 9 ultraviolet spectrometer experiment: Seasonal variation of ozone on Mars. *Science* 179, 795–796.

Bell, J.M., et al., 2014. Developing a self-consistent descriptionof Titan's upper atmosphere without hydrodynamic escape. *J. Geophys. Res.* 119A, 4957–4972.

Belyaev, D., Montmessin, F., Bertaux, J.L., Mahieux, A., Fedorova, A., Korablev, O., Marcq, E., Yung, Y., Zhang, X., 2012. Vertical profiling of $SO_2$ and SO above Venus' clouds by SPICAV/SOIR solar occultations. *Icarus* 217, 740–751.

Belyaev, D.A., et al., 2017. Night side distribution of $SO_2$ content in Venus' upper atmosphere. *Icarus* 294, 58–71.

Benna, M., et al., 2015. First measurements of composition and dynamics of the Martian ionosphere by MAVEN's Neutral Gas and Ion Mass Spectrometer. *Geophys. Res. Lett.* 42, 8958–8965.

Bertaux, J.L., Blamont, J.E., Lepine, V.E., Kurt, V.G., Romanova, N.N., Smirnov, A.S., 1981. Venera 11 and Venera 12 observations of EUV emissions from the upper atmosphere of Venus. *Planet. Space Sci.* 29, 149–166.

Bertaux, J.-L., et al., 1982. Altitude profile of H in the atmosphere of Venus from Lyman alpha observations of Venera 11 and Venera 12 and origin of the hot exospheric component. *Icarus* 52, 221–244.

Bertaux, J.L., et al., 1996. VEGA 1 and VEGA 2 entry probes: An investigation of local UV absorption (220–400 nm) in the atmosphere of Venus ($SO_2$, aerosol, cloud structure). *J. Geophys. Res.* 101, 12709–12745.

Bertaux, J.L., et al., 2005a. Nightglow in the upper atmosphere of Mars and implications for atmospheric transport. *Science* 307, 566–569.

Bertaux, J.L., et al., 2005b. Discovery of an aurora on Mars. *Nature* 435, 790–794.

Bertaux, J.L., et al., 2006. SPICAM on Mars Express: Observing modes and overview of UV spectrometer data and scientific results. *J. Geophys. Res.* 111, E10S90.

Bertaux, J.L., Gondet, B., Lefèvre, F., Bibring, J.P., Montmessin, F., 2012. First detection of O2 1.27 μm nightglow emission at Mars with OMEGA/MEX and comparison with general circulation model predictions. *J. Geophys. Res.* 117, E00J04.

Bertaux, J.L., et al., 2014. Possible influence of Venus topography on the zonal wind and UV albedo at cloud top level. AGU Fall Meeting, abstract #P21B-3916.

Bertrand, T., Forget, F., 2017. 3D modeling of organic haze in Pluto's atmosphere. *Icarus* 287, 72–86.

Bezard, B., 2014. The methane mole fraction in Titan's stratosphere from DISR measurements during the Huygens probe's descent. *Icarus* 242, 64–73.

Bezard, B., de Bergh, C., Crisp, D., Maillard, J.P., 1990. The deep atmosphere of Venus revealed by high-resolution night-side spectra. *Nature* 345, 508–511.

Bezard, B., de Bergh, C., Fegley, B., Maillard, J.P., Crisp, D., Owen, T., Pollack, J.B., Grinspoon, D., 1993. The abundance of sulfur dioxide below the clouds of Venus. *Geophys. Res. Lett.* 20, 1587–1590.

Bézard, B., Fedorova, A., Bertaux, J.-L., Rodin, A., Korablev, O., 2011. The 1.10- and 1.18-μm nightside windows of Venus observed by SPICAV-IR aboard Venus Express, *Icarus*, 216, 173–183.

Bhattacharyya, D., Clarke, J.T., Bertaux, J.L., Chaufray, J.Y., Mayyasi, M., 2015. A strong seasonal dependence in the Martian hydrogen exosphere. *Geophys. Res. Lett.* 42, 8678–8685.

Biemann, K., Owen, T., Rushneck, D.R., et al., 1976. Search for organic and volatile inorganic components in two surface samples from the Chryse Planitia region of Mars. *J. Geophys. Res.* 82, 4641–4658.

Billebaud, F., Crovisier, J., Lellouch, E., Encrenaz, T., Maillard, J.P., 1991. High-resolution infrared spectrum of CO on Mars: Evidence for emission lines. *Planet. Space Sci.* 39, 213–218.

Billebaud, F., Maillard, J.P., Lellouch, E., Encrenaz, T., 1992. The spectrum of Mars in the (1–0) vibrational band of CO, *Astron. Astrophys.*, 261, 647–652.

Billebaud, F., et al., 2009. Observations of CO in the atmosphere of Mars with PFS onboard Mars Express. *Planet. Space Sci.* 57, 1446–1457.

Billmers, R.I., Smith, A.L., 1991. Ultraviolet–visible absorption spectra of equilibrium sulfur vapor: Molar absorptivity spectra of $S_3$ and $S_4$. *J. Phys. Chem.* 95, 4242–4245.

Bird, M.K., et al., 1997. Detection of Titan's ionosphere from Voyager 1 radio occultation observations. *Icarus* 130, 426–436.

Bird, M.K., et al., 2005. The vertical profile of winds on Titan. *Nature* 438, 800–802.

Bjoraker, G.L., Mumma, M.J., Larson, H.P., 1989. Isotopic abundance ratios for hydrogen and oxygen in the Martian atmosphere (abstract). *Bull. Am. Astron. Soc.* 21, 991.

Bjoraker, G.L., Larson, H.P., Mumma, M.J., Timmermann, R., Montani, J.L., 1992. Airborne observations of the gas composition of Venus above the cloud tops: Measurements of $H_2O$, HDO, HF and the D/H and $^{18}O/^{16}O$ isotope ratios. *Bull. Am. Astron. Soc.* 24, 995.

Bodmer, R., Bochsler, P., Geiss, J., von Steiger, R., Gloeckler, G., 1995. Solar wind helium isotopic composition from SWICS/ULYSSES. *Space Sci. Rev.* 72, 61–64.

Bogard, D.D., Clayton, R.N., Marti, K., Owen, T., Turner, G., 2001. Martian volatiles: Isotopic composition, origin, and evolution. *Climatol. Evol. Mars* 96, 425–458.

Borucki, W.J., Chameides, W.L., 1984. Lightning: Estimates of the rate of energy dissipation and nitrogen fixation. *Rev. Geophys. Space Phys.* 22, 363–372.

Borucki, W.J., Dyer, J.W., Phillips, J.R., Pham, P., 1991. Pioneer Venus orbiter search for Venusian lightning. *J. Geophys. Res.* 96, 11033–11043.

Borucki, W.J., McKay, C.P., Jebens, D., Lakkaraju, H.S., Vanajakshi, C.T., 1996. Spectral irradiance measurements of simulated lightning in planetary atmospheres. *Icarus* 123, 336–344.

Bougher, S.W., et al., 1990. The Venus nitric oxide night airglow: Model calculations based on the Venus thermospheric general circulation model. *J. Geophys. Res.* 95(A5), 6271–6284.

Bougher, S.W., Engel, S., Roble, R.G., Foster, B., 1999. Comparative terrestrial planet thermospheres. 2. Solar cycle variation of global structure and winds at equinox. *J. Geophys. Res.* 104, 16591–16611.

Bougher, S.W., Engel, S., Roble, R.G., Foster, B., 2000. Comparative terrestrial planet thermospheres. 3. Solar cycle variation of global structure and winds at solstices. *J. Geophys. Res.* 105, 17669–17692.

Bougher, S.W., Blelly, P.L., Combi, M., Fox, J.L., Mueller-Wodarg, I., Ridley, A., Roble, R.G., 2008. Neutral upper atmosphere and ionosphere modeling. *Space Sci. Rev.* 139, 107–141.

Bougher, S.W., McDunn, T.M., Zoldak, K.A., Forbes, J.M., 2009. Solar cycle variability of Mars dayside exospheric temperatures: Model evaluation of underlying thermal balance. *Geophys. Res. Lett.* 36, L05201.

Bougher, S.W., et al., 2015a. Mars Global Ionosphere-Thermosphere Model (MGITM): Solar cycle, seasonal, and diurnal variations of the Mars upper atmosphere. *J. Geophys. Res.* 120, 311–342.

Bougher, S.W., Brecht, A., Schulte, R., Fischer, J., Parkinson, C., Mahieux, A.,Wilquet, V., Vandaele, A.C., 2015b. Upper atmosphere temperature structure at the Venusian terminators: A comparison of SOIR and VTGCM results. *Planet. Space Sci.* 113–114, 337–347.

Bougher, S.W., et al., 2017. The structure and variability of Mars dayside thermosphere from MAVEN NGIMS and IUVS measurements: Seasonal and solar activity trends in scale heights and temperatures. *J. Geophys. Res.* 122, A1, 1296–1313.

Brace, L.H., et al., 1979. Empirical models of the electron temperature and density in the nightside Venus ionosphere. *Science* 205, 102–104.

Brecht, A.S., Bougher, S.W., 2012. Dayside thermal structure of Venus' upper atmosphere characterized by a global model. *J. Geophys. Res.* 117, E08002.

Brecht, A.S., Bougher, S.W., Gérard, J.-C., Parkinson, C., Rafkin, S., Foster, B., 2011. Understanding the variability of nightside temperatures, NO UV and $O_2$ IR nightglow emissions in the Venus upper atmosphere. *J. Geophys. Res.* 116, E08004.

Brinton, H.C., et al., 1980. Venus night-time hydrogen bulge. *Geophys. Res. Lett.* 7, 865–868.

Broadfoot, A.L., et al., 1989. Ultraviolet spectrometer observations of Neptune and Triton. *Science* 246, 1459–1466.

Buratti, B.J., et al., 2017. Global albedos of Pluto and Charon from LORRI New Horizons observations. *Icarus* 287, 207–217.

Burgdorf, M., Cruikshank, D.P., Dalle Ore, C.M., Sekiguchi, T., Nakamura, R., Orton, G., Quirico, E., Schmitt, B., 2010. A tentative identification of HCN ice on Triton. *Astrophys. J.* 718, L53–L57.

Burkholder, J.B., et al., 2015. Chemical kinetics and photochemical data for use in atmospheric studies. Evaluation Number 18. JPl Publication 15–10.

Butler, B.J., et al., 2001. Accurate and consistent microwave observations of Venus and their implications. *Icarus* 154, 226–238.

Capalbo, F.J., Benilan, Y., Yelle, R.V., Koskinen, T.T., 2015. Titan's upper atmosphere from Cassini/UVIS solar occultations. *Astrophys. J.* 814, 86–100.

Carleton, N.P., Traub, W.A., 1972. Detection of molecular oxygen on Mars. *Science* 177, 988–992.

Chaffin, M.S., et al., 2014. Unexpected variability of Martian hydrogen escape. *Geophys. Res. Lettt.* 41, 314–320.

Chamberlain, J.W., Hunten, D.M., 1987. *Theory of Planetary Atmospheres.* Academic Press, New York.

Chamberlain, S., Bailey, J., Crisp, D., Meadows, V., 2013. Ground-based near-infrared observations of water vapour in the Venus troposphere. *Icarus*, 222(1), 364–378.

Chase, M.W., Jr., 1998. NIST-JANAF thermodynamic tables, 4th ed. *J. Phys. Chem. Ref. Data. Monogr.* 9, 1–1951.

Chassefiere, E., 2009. Metastable methane clathrate particles as a source of methane to the Martian atmosphere. *Icarus* 204, 137–144.

Chassefiere, E., Bertaux, J.L., Kurt, V.G., Smirnov, A.S., 1986. Venus EUV measurements of helium at 58.4 nm from Venera 11 and Venera 12 and implications for the outgassing history. *Planet. Space Sci.* 34, 585–602.

Chaufray, J.Y., Bertaux, J.L., Leblanc, F., et al., 2008. Observation of the hydrogen corona with SPICAM on Mars Express. *Icarus* 195, 598–613.

Chaufray, J.Y., Leblanc, F., Quémerais, E., et al., 2009. Martian oxygen density at the exobase deduced from O I 130.4-nm observations by spectroscopy for the investigation of the characteristics of the atmosphere of Mars on Mars Express. *J. Geophys. Res.* 114, E02006.

Chaufray, J.Y., Bertaux, J.L., Leblanc, F., 2012a. First observation of the Venus UV dayglow at limb from SPICAV/VEX. *Geophys. Res. Lett.* 39, L20201.

Chaufray, J.-Y., et al., 2012b. Observation of the Venusian hydrogen corona with SPICAV on Venus Express. *Icarus* 217, 767–778.

Chaufray, J.Y., et al., 2015a. Variability of the hydrogen in the Martian upper atmosphere as similated by a 3D atmosphere-exosphere coupling. *Icarus* 245, 282–294.

Chaufray, J.Y., et al., 2015b. Observations of the nightside Venusian hydrogen corona with SPICAV/VEX. *Icarus* 262, 1–8.

Chen, Y., et al., 2015. Near infrared cavity ring-down spectroscopy for isotopic analyses of $CH_4$ on future Martian surface missions. *Planet. Space Sci.* 105, 117–122.

Cheng, A.F., et al., 2017. Haze in Pluto's atmosphere. *Icarus* 290, 112–133.

Christensen, P.R., 2003. Mars as seen from the 2001 Mars Odyssey Thermal Emission Imaging System experiment. *AGU Fall Meeting* P21A-02, abstract.

Christensen, P.R., et al., 1998. Results from the Mars Global Surveyor Thermal Emission Spectrometer. *Science* 279, 1692–1698.

Christian, H.J., et al., 2003. Global frequency and distribution of lightning as observed from space by the optical transient detector. *J. Geophys. Res.* 108, 4005.

Chung, C.Y., et al., 2001. Temperature dependence of absorption cross sections of $H_2O$, HDO, and $D_2O$ in the spectral region 140–193 nm. *Nucl. Instr. Meth. Phys. Res. A*, 467–468, 1572–1576.

Clancy, R.T., Lee, S.W., 1991. A new look at dust and clouds in the Mars atmosphere: Analysis of emissionphase function sequences from global Viking IRTM observations. *Icarus* 93,135–158.

Clancy R.T., Nair, H., 1996. Annual (aphelion-perihelion) cycles in the photochemical behavior of the global Mars atmosphère. *J. Geophys. Res.* 101, 12785–12790.

Clancy, R.T., Grossman, A.W., Wolf, M.J., James, P.B., Rudy, D.J., Billawala, Y.N., Sandor, B.J., Lee, S.W., Muhleman, D.O., 1996. Water vapor saturation at low altitudes around Mars aphelion: A key to Mars climate? *Icarus* 122, 36–62.

Clancy, R.T., Wolff, M.J., James, P.B., 1999. Minimal aerosol loading and global increases in atmospheric ozone during the 1996–1997 Martian northern spring season. *Icarus* 138, 49–63.

Clancy, R.T., Sandor, B.J., Moriarty-Schieven, G.H., 2004. A measurement of the 362 GHz absorption line of Mars atmospheric $H_2O_2$. *Icarus* 168, 116–121.

Clancy, R.T., Sandor, B.J., Moriarty-Schieven, G.H., 2012a. Thermal structure and CO distribution for the Venus mesosphere/lower thermosphere: 2001–2009 inferior conjunction sub-millimeter CO absorption line observations. *Icarus* 217, 779–793.

Clancy, R.T., et al., 2012b. Extensive MRO CRISM observations of 1.27 μm $O_2$ airglow in Mars polar night and their comparison to MRO MCS temperature profiles and LMD GCM simulations. *J. Geophys. Res.* 117, E00J10.

Clancy, R.T., et al., 2013. First detection of Mars atmospheric hydroxyl: CRISM Near-IR measurement versus LMD GCM simulation of OH Meinel band emission in the Mars polar winter atmosphere. *Icarus* 226, 272–281.

Clancy, R.T., et al., 2017. Vertical profiles of Mars 1.27 μm $O_2$ dayglow from MRO CRISM limb spectra: Seasonal/global behaviors, comparisons to LMDGCM simulations, and a global definition for Mars water vapor profiles. *Icarus* 293, 132–156.

Clarke, J.T., et al., 2014. A rapid decrease of the hydrogen corona of Mars. *Geophys. Res. Lett.* 41, 8013–8020.

Clarke, J.T., et al., 2017. Variability of D and H in the Martian upper atmosphere observed with the MAVEN IUVS echelle channel. *J. Geophys. Res.* 122(A2), 2536–2544.

Coates, A.J., et al., 2009. Heavy negative ions in Titan's ionosphere: Altitude and latitude dependence. *Planet. Space Sci.* 57, 1866–1871.

Combes, M, et al., 1988. The 2.5–12 micrometers spectrum of comet Halley from the IKS-VEGA experiment. *Icarus* 76, 404–436.

Connes, P., Connes, J., Benedict, W.S., Kaplan, L.D., 1967. Traces of HCl and HF in the atmosphere of Venus. *Astrophys. J.* 147, 1230–1237.

Connes, P., Connes, J., Kaplan, L.D., Benedict, W.S., 1968. Carbon monoxide in the Venus atmosphere. *Astrophys. J.* 152, 731–743.

Connes, J., Connes, P., Maillard, J.P., 1969. *Atlas des spectres dans le proche infrarouge de Venus, Mars, Jupiter et Saturn*. CNRS, Paris.

Connes, P., Noxon, J.F., Traub, W.A., Carleton, N.P., 1979. $O_2(^1\Delta)$ emission in the day and night airglow of Venus. *Astrophys. J.* 233, L29–L32.

Conrad, P.G., et al., 2016. In situ measurements of atmospheric krypton and xenon on Mars with Mars Science Laboratory. *Earth Planet. Sci. Lett.* 454, 1–9.

Conrath, B.J., et al., 1989. Infrared observations of the Neptunian system. *Science* 246, 1454–1459.

Cooray, V., 1997. Energy dissipation in lightning flashes. *J. Geophys. Res.* 102, 21401–21410.

Cordiner, M.A., et al., 2014. ALMA measurements of the HNC and $HC_3N$ distributions in Titan's atmosphere. *Astrophys. J. Lett.* 795, L30.

Cordiner, M.A., et al., 2015. Ethyl cyanide on Titan: Spectroscopic detection and mapping using ALMA. *Astrophys. J. Lett.* 800, L14.

Corlies, P., et al., 2017. Titan's topography and shape at the end of the Cassini mission. *Geophys. Res. Let.* 44, 11754–11761.

Cottini, V., Ignatiev, N.I., Piccioni, G., Drossart, P., Grassi, D., Markiewicz, W.J., 2012. Water vapor near the cloud tops of Venus from Venus Express/VIRTIS dayside data. *Icarus* 217, 561–569.

Cottini, V., Ignatiev, N.I., Piccioni, G., Drossart, P., 2015. Water vapor near Venus cloud tops from VIRTIS-H/Venus express observations 2006–2011. *Planet. Space Sci.* 113–114, 219–225.

Cotton, D.V., Bailey, J., Crisp, D., Meadows, V.S., 2012. The distribution of carbon monoxide in the lower atmosphere of Venus. *Icarus* 217, 570–584.

Courtin, R., Gautier, D., McKay, C.P., 2005. Titan's thermal emission spectrum: Reanalysis of the Voyager infrared measurements. *Icarus*, 114, 144–162.

Courtin, R., Sim, C.K., Kim, S.J., Gautier, D., 2012. The abundance of H2 in Titan's troposphere from the Cassini CIRS investigation. *Planet. Space Sci.* 69, 89–99.

Coustenis, A., Bezard, B., 1995. Titan's atmosphere from Voyager infrared observations. IV. Latitudinal variations of temperature and composition. *Icarus* 115, 126–140.

Coustenis, A., et al., 1991. Titan's atmosphere from Voyager infrared observations. III. Vertical distributions of hydrocarbons and nitrile near Titan's north pole. *Icarus* 89, 152–167.

Coustenis, A., et al., 1998. Evidence for water vapor in Titan's atmosphere from ISO/SWS data. *Astron. Astrophys.* 336, L85–L89.

Coustenis, A., et al., 2003. Titan's atmosphere from ISO mid-infrared spectroscopy. *Icarus* 161, 383–403.

Coustenis, A., et al., 2010. Titan trace gaseous composition from CIRS at the end of the Cassini-Huygens prime mission. *Icarus* 207, 461–476.

Coustenis, A., et al., 2013. Evolution of the stratospheric temperature and chemical composition over one Titanian year. *Astrophys. J.* 779, 177–186.

Coustenis, A., et al., 2018. Seasonal evolution of Titan's stratosphere near the poles. *Astrophys. J. Lett.* 854, L30.

Cox, C., et al., 2008. Distribution of the ultraviolet nitric oxide Martian night airglow: Observations from Mars Express and comparisons with a one-dimensional model. *J. Geophys. Res.* 113, E08012.

Cravens, T.E., et al., 2008. Energetic ion precipitation at Titan. *Geophys. Res. Lett.* 35, L03103.

Cravens, T.E., et al., 2009. Model-data comparison for Titan's nightside ionosphere. *Icarus* 199, 174–188.

Crisp, D., Meadows, V.S., Bezard, B., de Bergh, C., Maillard, J.P., Mills, F.P., 1996. Ground-based near-infrared observations of the Venus night side: 1.27-μm $O_2(a^1\Delta_g)$ airglow from the upper atmosphere. *J. Geophys. Res.* 101, 4577–4594.

Croteau, P., et al., 2011. Measurements of isotope effects in the photoionization of $N_2$ and implications for Titan's atmosphere. *Astrophys. J. Lett.* 728, L32.

Crovisier, J., Lellouch, E., de Bergh, C., Maillard, J.P., Lutz, B.L., Bezard, B., 2006. Carbon monoxide emissions at 4.7 lm from Venus' atmosphere. *Planet. Space Sci.* 54, 1398–1414.

Cruikshank, D.P. (ed.), 1995. *Neptune and Triton*. University of Arizona Press, Tucson.

Cruikshank, D.P., Apt, J., 1984. Methane on Triton: Physical state and distribution. *Icarus* 58, 306.

Cruikshank, D.P., Brown, R.H., Clark, R.N., 1984. Nitrogen on Triton. *Icarus* 58, 293.

Cruikshank, D.P., Pilcher, C.B., Morrison, D., 1976. Pluto: Evidence for methane ice. *Science* 194, 835–837.

Cruikshank, D.P., et al., 1993. Ices on the surface of Triton. *Science* 261, 742–745.

Cruikshank, D.P., et al., 2015. The surface composition of Pluto and Charon. *Icarus* 246, 82–92.

Cui, J., et al., 2009. Analysis of Titan's neutral upper atmosphere from Cassini Ion Neutral Mass Spectrometer measurements. *Icarus* 200, 581–615.

Cui, J., et al., 2012. The $CH_4$ structure in Titan's upper atmosphere revisited. *J. Geophys. Res.* 117, E11006.

Curry, S.M., et al., 2015. Comparative pick-up ion distribution at Mars and Venus: Consequences fro atmospheric deposition and escape. *Planet. Space Sci.* 115, 35–47.

Dalgarno, A., Babb, J.F., Sun, Y., 1992. Radiative association in planetary atmospheres. *Planet. Space Sci.* 40, 243–246.

Danielson, R.E., Moore, D.R., van der Hulst, H.C., 1969. The transfer of visible radiation through clouds. *J. Atmos. Phys.* 26, 1078–1083.

Davies, D.W., 1979. The vertical distribution of Mars water vapor. *J. Geophys. Res.* 84, 2875–2879.

De Bergh, C., Bezard, B., Owen, T., Crisp, D., Maillard, J.P., Lutz, B.L., 1991. Deuterium on Venus: Observations from Earth. *Science* 251, 547–549.

De Bergh, C., Bezard, B., Crisp, D., Maillard, J., Owen, T., Pollack, J., Grinspoon, D., 1995. Water in the deep atmosphere of Venus from high-resolution spectra of the night side. *Adv. Space Res.*, 15(4), 79–88.

De Bergh, C., et al., 2012. Application of a new set of methane line parameters to the modeling of Titan's spectrum in the 1.58 μm window. *Planet. Space Sci.* 61, 85–98.

Deighan, J., et al., 2015. MAVEN IUVS observations of the hot oxygen corona at Mars. *Geophys. Res. Lett.* 42, 9009–9014.

De Kok, R., et al., 2007. Oxygen compounds in Titan's stratosphere as observed by Cassini/CIRS. *Icarus* 186, 354–363.

De Kok, R., Irwin, P.G.J., Tsang, C.C., Piccioni, G., Drossart, P., 2011. Scattering particles in nightside limb observations of Venus' upper atmosphere by Venus Express VIRTIS. *Icarus* 211, 51–57.

De La Haye, V., et al., 2007. Cassini Ion and Neutral Mass Spectrometer data in Titan's upper atmosphere and exosphere: Observation of a suprathermal corona. *J. Geophys Res.* 112, A07309.

De La Haye, V., Waite, J., Jr., Cravens, T., Robertson, I., Lebonnois, S., 2008. Coupled ion and neutral rotating model of Titan's upper atmosphere. *Icarus* 197, 110–136.

Delitsky, M.L., Baines, K.H., 2015. Storms on Venus: Lightning-induced chemistry and predicted products. *Planet. Space Sci.* 113–114, 184–192.

DeMeo, F.E., et al., 2010. A search for ethane on Pluto and Triton. *Icarus* 208, 412–424.

Dennerl, K., et al., 2006. First observation of Mars with XMM-Newton. *Astron. Astrophys.* 451, 709–722.

Dias-Oliveira, A., Sicardy, B., Lellouch, E., et al., 2015. Pluto's atmosphere from stellar occultations in 2012 and 2013. *Astrophys. J.* 811, 53.

Dinelli, B.M., et al., 2013. An unidentified emission in Titan's upper atmosphere. *Geophys. Res. Lett.* 40, 1489–1493.

Dobrijevic, M., et al., 2016. 1D-coupled photochemical model of neutrals, cations and anions in the atmosphere of Titan. *Icarus* 268, 313–339.

Donahue, T.M., 1969. Deuterium in the upper atmospheres of Venus and Earth. *J. Geophys. Res.* 74, 1128–1137.

Donahue, T.M., Pollack, J.B., 1983. Origin and evolution of the atmosphere of Venus. In: Hunten, D.M., Colin, L., Donahue, T.M., Moroz, V.I. (Eds.), *Venus*, University of Arizona Press, Tucson, pp. 1003–1036.

Donahue, T.M., Hoffman, J.H., Hodges, R.R., Jr., 1981. Krypton and xenon in the atmosphere of Venus. *Geophys. Res. Lett.* 8, 513–516.

Donahue, T.M., Hoffman, J.H., Hodges, R.R., Watson, A.J., 1982. Venus was wet: A measurement of the ratio of D to H. *Science* 216, 630–633.

Donahue, T.M., Grinspoon, D.H., Hartle, R.E., Hodges, R.R., Jr., 1997. Ion/neutral escape of hydrogen and deuterium: Evolution of water. In: Bougher, S.W., Hunten, D.M., Phillips, R.J. (Eds.), *Venus II*, University of Arizona Press, Tucson, pp. 385–414.

Doose, L.R., et al., 2016. Vertical structure and optical properties of Titan's aerosols from radiance measurements made inside and outside the atmosphere. *Icarus* 270, 355–375.

Doute, S., et al., 1999. Evidence for methane segregation at the surface of Pluto. *Icarus* 142, 421–444.

Dreibus, G., Waenke, H., 1987. Volatiles on Earth and Mars – A comparison. *Icarus* 71, 225–240.

Durrance, S.T., 1981. The carbon monoxide fourth positive bands in the Venus dayglow. 1. Synthetic spectra. *J. Geophys. Res.* 86, 9115–9124.

Edberg, N.J.T., et al., 2013. Solar cycle modulation of Titan's ionosphere. *J. Geophys. Res.* 118A, 5255–5264.

Ekonomov, A.P., Golovin, Yu.M., Moroz, V.I., Moshkin, B.E., 1983a. Solar scattered radiation measurements by Venus probes. In: Hunten, D.M., Colin, L., Donahue, T.M., Moroz, V.I. (Eds.), *Venus*. University of Arizona Press, Tucson, pp. 632–649.

Ekonomov, A.P., Moshkin, B.E., Moroz, V.I., Golovin, Yu.M., Gnedykh, V.I., Grigoriev, A.V., 1983b. UV photometry at the Venera 13 and 14 landing probes. *Cosmic Res.* 21, 254–260.

Elliot, J.L., Young, L.A., 1992. Analysis of stellar occultation data for planetary atmospheres. I – Model fitting, with application to Pluto. *Astron. J.* 103, 991–1015.

Elliot, J.L., Veverka, J., Goguen, J., 1975. Lunar occultation of Saturn. I – the diameters of Tethus, Dione, Rhea, Titan, and Japetus. *Icarus* 26, 387–407.

Elliot, J.L., et al., 2000a. The thermal structure of Triton's middle atmosphere. *Icarus* 143, 425–428.

Elliot, J.L., et al., 2000b. The prediction and observation of the July 18, 1997 stellar occultation by Triton: More evidence for distortion and increasing pressure in Triton's atmosphere. *Icarus* 148, 347–369.

Elrod, M.K., et al., 2017. He bulge revealed: He and $CO_2$ diurnal and seasonal variations in the upper atmosphere of Mars as detected by MAVEN NGIMS. *J. Geophys. Res.* 122 (A2), 2564–2573.

Encrenaz, T., Lellouch, E., Rosenqvist, J., Drossart, P., Combes, M., Billebaud, F., de Pater, I., Gulkis, S. Maillard, J.P., Paubert, G., 1991. The atmospheric composition of Mars: ISM and ground-based observational data. *Ann. Geophys.* 9, 797– 803.

Encrenaz, Th., Lellouch, E., Cernicharo, J., Paubert, G., Gulkis, S., Spilker, T., 1995. The thermal profile and water abundance in the Venus mesosphere from H2O and HDO millimeter observations. *Icarus* 117, 162–172.

Encrenaz, T., et al., 2002. Astringent upper limit of the $H_2O_2$ abundance in the Martian atmosphere. *Astron. Astrophys.* 396, 1037–1044.

Encrenaz, T., Bézard, B. Greathouse, T.K., et al., 2004. Hydrogen peroxide on Mars: Evidence for spatial and seasonal variations. *Icarus* 170, 424–429.

Encrenaz, Th., et al., 2006. Seasonal variations of the Martian CO over Hellas as observed by OMEGA/Mars Express. *Astron. Astrophys.* 459, 265–270.

Encrenaz, T., et al., 2011. A stringent upper limit to $SO_2$ in the Martian atmosphere. *Astron. Astrophys.* 530, A37.

Encrenaz, T., et al., 2015a. Seasonal variations of hydrogen peroxide and water vapor on Mars: Further indications of heterogeneous chemistry. *Astron. Astrophys.* 578, A127.

Encrenaz, T., Moreno, R., Moullet, A., Lellouch, E., Fouchet, T., 2015b. Submillimeter mapping of mesospheric minor species on Venus with ALMA. *Planet. Space Sci.* 113–114, 275–291.

Encrenaz, T., et al., 2016. A map of D/H on Mars in the thermal infrared using EXES aboard SOFIA. *Astron. Astrophys.* 586, A62.

Engelke, C.W., Price, S.D., Kraemer, K.E., 2010. Spectral irradiance calibration in the infrared. XVII. Zero-magnitude broadband flux reference for visible-to-infrared photometry. *Astron. J.* 140, 1919–1928.

Erwin, J., Tucker, O.J., Johnson, R.E., 2013. Hybrid fluid/kinetic modeling of Pluto's escaping atmosphere. *Icarus* 226, 375–384.

Espenak, F., Mumma, M.J., Kostiuk, T., et al., 1991. Ground-based infrared measurements of the global distribution of ozone in the atmosphere of Mars. *Icarus* 92, 252–262.

Esposito, L.W., Winick, J.R., Stewart, A.I.F., 1979. Sulfur dioxide in the Venus atmosphere: Distribution and implications. *Geophys. Res. Lett.* 6, 601–604.

Esposito, L.W., Knollenberg, R.G., Marov, M.Y., Toon, O.B., Turco, R.P., 1983. The clouds and hazes of Venus. In: Hunten, D.M., et al. (Eds.), *Venus*. University of Arizona Press, Tucson, pp. 484–564.

Esposito, L.W., et al., 1988. Sulfur dioxide in the Venus cloud tops 1978–1986. *J. Geophys. Res.* 93, 5267–5276.

Esposito, L.W., et al., 1997. Chemistry of lower atmosphere and clouds. In: Bougher, S.W., Hunten, D.M., Phillips, R.J. (Eds.), *Venus II*. University of Arizona Press, Tucson, pp. 415–458.

Evans, K.F., 2007. SHDOMPPDA: A radiative transfer model for cloudy sky data assimilation. *J. Atmos. Sci.* 64, 3858–3868.

Farmer, C.B., Norton, R.H., 1989. Atlas of the infrared spectrum of the Sun and the Earth atmosphere from space. The Sun, vol. I. NASA Ref. Publication 1224.

Fast, K., Kostiuk, T., Espenak, F., et al., 2006. Ozone abundance on Mars from infrared heterodyne spectra. I. Acquisition, retrieval, and anticorrelation with water vapor. *Icarus* 181, 419–431.

Fast, K., Kostiuk, T., Lefèvre, F., et al., 2009. Comparison of HIPWAC and Mars Express SPICAM observations of ozone on Mars 2006–2008 and variation from 1993 IRHS observations. *Icarus* 203, 20–27.

Fedorov, A., et al., 2011. Measurements of the ion escape rates from Venus for solar minimum. *J. Geophys. Res.* 116, A07220.

Fedorova, A., Korablev, O., Perrier, S., Bertaux, J.L., Lefèvre, F., Rodin, A., 2006. Observation of $O_2$ 1.27 μm dayglow by SPICAM IR: Seasonal distribution for the first Martian year of Mars Express. *J. Geophys. Res.* 111, E09S07.

Fedorova, A., et al., 2008. HDO and $H_2O$ vertical distributions and isotopic ratio in the Venus mesosphere by Solar Occultation at Infrared spectrometer on board Venus Express. *J. Geophys. Res.* 113, E00B22.

Fedorova, A., et al., 2010. Viking observations of water vapor on Mars: Revision from up-to-date spectroscopy and atmospheric models. *Icarus* 208, 156–164.

Fedorova, A.A., et al., 2012. The $O_2$ nightglow in the Martian atmosphere by SPICAM onboard of Mars Express. *Icarus* 219, 596–608.

Fedorova, A.A, et al., 2014. Evidence for a bimodal size distribution for the suspended aerosol particles on Mars. *Icarus* 231, 239–260.

Fedorova, A., et al., 2016. Variations of water vapor and cloud top altitude in the Venus' mesosphere from SPICAV/VEX observations. *Icarus* 275, 143–162.

Fedorova, A.A., et al., 2018. Water vapor in the middle atmosphere of Mars during the 2007 global dust storm. *Icarus* 300, 440–457.

Fegley, B., Zolotov, M.Yu., Lodders, K., 1997. The oxidation state of the lower atmosphere and surface of Venus. *Icarus* 125, 416–439.

Feldman, P.D., Moos, H.W., Clarke, J.T., Lane, A.L., 1979. Identification of the UV nightglow from Venus. *Nature* 279, 221–223.

Feldman, P.D., Burgh, E.B., Durrance, S.T., Davidsen, A.F., 2000. Far-ultraviolet spectroscopy of Venus and Mars at 4 Å resolution with the Hopkins Ultraviolet Telescope on ASTRO-2. *Astrophys. J.* 538, 395–400.

Feldman, W.C., et al., 2004. Global distribution of near-surface hydrogen on Mars. *J. Geophys. Res.* 109, E09006.

Feofilov, A.G., Kutepov, A.A., Rezac, L., Smith, M.D., 2012. Extending MGS-TES temperature retrievals in the Martian atmosphere up to 90 km: Retrieval approach and results. *Icarus* 221, 949–959.

Feuchtgruber, H., Lellouch, E., de Graauw, T., Bezard, B., Encrenaz, T., Griffin, M., 1997. External supply of oxygen to the atmospheres of giant planets. *Nature* 389, 159–162.

Fink, U., Larson, H.P., Kuiper, G.P., Poppe, R.F., 1972. Water vapor in the atmosphere of Venus. *Icarus* 17, 617–631.

Flynn, G.J., 1996. The delivery of organic matter from asteroids and comets to the early surface of Mars. *Earth Moon Planets* 72, 469–474.

Folkner, W.M., et al., 2006. Winds on Titan from ground-based tracking of the Huygens probe. *J. Geophys. Res.* 111, E07S02.

Fonti, S., Marzo, G.A., 2010. Mapping the methane on Mars. *Astron. Astrophys.* 512, A51.

Forget, F., et al., 2009. Density and temperatures of the upper Martian atmosphere measured by stellar occultations with Mars Express SPICAM. *J. Geophys. Res.* 114, E01004,

Formisano, V., Atreya, S.K., Encrenaz, T., et al., 2004. Detection of methane in the atmosphere of Mars. *Science* 306, 1758–61.

Formisano, V., et al., 2005. The Planetary Fourier Spectrometer (PFS) onboard the European Mars Express mission. *Planet. Space Sci.* 53, 963–974.

Fox, J.L., 1982. Atomic carbon in the atmosphere of Venus. *J. Geophys. Res.* 87, 9211–9216.

Fox, J.L., 1993. The production and escape of nitrogen atoms on Mars, *J. Geophys. Res.* 98, 3297– 3310.

Fox, J.L., 2008. Morphology of the dayside ionosphere of Venus: Implication for ion outflows. *J. Geophys. Res.* 113, E1101

Fox, J.L., 2011. The post-terminator ionosphere of Venus. *Icarus* 216, 625–639.

Fox, J.L., 2012. The ionospheric source of the red and green lines of atomic oxygen in the Venus nightglow. *Icarus* 221, 787–799.

Fox, J.L., Black, J.H., 1989. Photodissociation of CO in the thermosphere of Venus. *Geophys. Res. Lett.* 16, 291–293.

Fox, J.L., Dalgarno, A., 1979a. Electron energy deposition in carbon dioxide. *Planet. Space Sci.* 27, 491–499.

Fox, J.L., Dalgarno, A., 1979b. Ionization, luminosity, and heating of the upperatmosphere of Mars. *J. Geophys. Res.* 84, 7315–7333.

Fox, J.L., Hac, A., 1997. Spectrum of hot O at the exobases of the terrestrial planets. *J. Geophys. Res.* 102, 24005–24011.

Fox, J.L., Hac, A., 2009. Photochemical escape of oxygen from Mars: A comparison of the exobase approximation to a Monte Carlo method. *Icarus* 204, 527–544.

Fox, J.L., Hac, A., 2010. Isotope fractionation in the photochemical escape of O from Mars. *Icarus* 208, 176–191.

Fox, J.L., Hac, A., 2018. Escape of O($^3P$), O($^1D$), and O($^1S$) from the Martian atmosphere. *Icarus* 300, 411–439.

Fox, J.L., Paxton, L.J., 2005. C and $C^+$ in the Venusian thermosphere/ionosphere. *J. Geophys. Res.* 110, A01311.

Fox, J.L., Sung, K.Y., 2001. Solar activity variations of the Venus thermosphere/ionosphere. *J. Geophys. Res.* 106, 21,305–21,335.

Fox, J.L., Victor, G.A., 1988. Electron energy deposition in $N_2$ gas. *Planet. Space Sci.* 36, 329–352.

Fox, J.L., Zhou, P., Bougher, S.W., 1996. The Martian thermosphere/ionosphere at high and low solar activities, *Adv. Space. Res.* 17(11), 203–218.

Fox, J.L., Benna, M., Mahaffy, P.R., Jakosky, B.M., 2015. Water and water ions in the Martian thermosphere/ionosphere. *Geophys. Res. Lett.* 42, 8977–8985.

Frandsen, B.N., Wennberg, P.O., Kjaergaard, H.G., 2016. Identification of OSSO as a near-UV absorber in the Venus atmosphere. *Geophys. Res. Lett.* 43, 11146–11155.

Franz, H.B., et al., 2017. Initial SAM calibration gas experiments on Mars: Quadrupole mass spectrometer results and implications. *Planet. Space Sci.* 138, 44–54.

Fray, N., Schmitt, B., 2009. Sublimation of ices of astrophysical interest: A bibliographic review. *Planet Space Sci.* 57, 2053–2080.

Fulchignoni, M., et al., 2005. In situ measurements of the physical characteristics of Titan's environment. *Nature* 438, 785–791.

Gacesa, M., Zhang, P., Kharchenko, V., 2012. Non-thermal escape of molecular hydrogen from Mars. *Geophys. Res. Lett.* 39, L10203.

Gagne, M.E., et al., 2012. Modeled $O_2$ nightglow distributions in the Venus atmosphere. *J. Geophys. Res.* 117, E12002.

Gagne, M.E., et al., 2013. New nitric oxide (NO) nightglow measurements with SPICAM/MEx as a tracer of Mars upper atmosphere circulation and comparison with LMD-MGCM model prediction: Evidence for asymmetric hemispheres. *J. Geophys. Res.* 118, 2172–2179.

Gao, P., et al., 2014. Bimodal distribution of sulfuric acid aerosol in the upper haze of Venus. *Icarus* 231, 83–98.

Gao, P., et al., 2017. Constraints on the microphysics of Pluto's photochemical haze from New Horizons observations. *Icarus* 287, 116–123.

García Muñoz, A., McConnell, J.C., McDade, I.C., Melo, S.M.L., 2005. Airglow on Mars: Some model expectations for the OH Meinel bands and the $O_2$ IR atmospheric band. *Icarus* 176, 75–95.

García Muñoz, A., Mills, F.P., Slanger, T.G., Piccioni, G., Drossart, P., 2009a. Visible and near-infrared nightglow of molecular oxygen in the atmosphere of Venus. *J. Geophys. Res.* 114, E12002.

García Muñoz, A., Mills, F.P., Piccioni, G., Drossart, P., 2009b. The near-infrared nitric oxide nightglow in the upper atmosphere of Venus. *Proc. Natl. Acad. Sci.* 106, 985–988.

Gelman, B.G., et al., 1979. Analysis of the chemical composition of the Venus atmosphere using the Venera 12 gas chromatograph. *Cosmic Res.* 17, 708–715.

Geminale, A., Formisano, V., Sindoni, G., 2011. Mapping methane in Martian atmosphere with PFS-MEX data. *Planet. Space Sci.* 59, 137–148.

Gerard, J.C., Cox, C., Saglam, A., Bertaux, J.L., Villard, E., Nehme, C., 2008. Limb observations of the ultraviolet nitric oxide nightglow with SPICAV on board Venus Express. *J. Geophys. Res.* 113, E00B03.

Gerard, J.C., et al., 2011. Measurements of the helium 584 Å airglow during the Cassini flyby of Venus. *Planet. Space Sci.* 59, 1524–1528.

Gerard, J.C., et al., 2011. EUV spectroscopy of Venus dayglow with UVIS on Cassini. *Icarus* 211, 70–80.

Gerard, J.C., Soret, L., Piccioni, G., Drossart, P., 2012. Spatial correlation of OH Meinel and $O_2$ infrared atmospheric nightglow emissions observed with VIRTIS-M on board Venus Express. *Icarus* 217, 813–817.

Giauque, W.F., Hornung, E.W., Kunzler, J.E., Rubin, T.R., 1960. The thermodynamic properties of aqueous sulfuric acid solutions and hydrates from 15 to 300 K. *J. Am. Chem. Soc.* 82, 62–67.

Gillett, F.C., 1975. Further observations of the 8–13 μm spectrum of Titan. *Astrophys. J.* 201, L41–L43.

Gilli, G., et al., 2015. Carbon monoxide and temperature in the upper atmosphere of Venus from VIRTIS/Venus Express non-LTE limb measurements. *Icarus* 248, 478–498.

Girazian, Z., et al., 2017. Ion densities in the nightside ionosphere of Mars: Effects of electron impact ionization. *Geophys. Res. Lett.* 44, 11,248–11,256.

Gladstone, G.R., et al., 2016. The atmosphere of Pluto as observed by New Horizons. *Science* 351(6279), aad8866.

Gnedykh, V.I., et al., 1987. Vertical structure of the cloud layer of Venus at the Vega 1 and 2 landing sites. *Cosmic Res.* 25, 707–712.

Gonzalez-Galindo, F., Forget, F., Lopez-Valverde, M.A., Angelats i Coll, M., Millour, E., 2009. A ground-to-exosphere Martian general circulation model: 1. Seasonal, diurnal, and solar cycle variation of thermospheric temperatures. *J. Geophys. Res.* 114, E04001.

Gonzalez-Galindo, F., et al., 2013. Three-dimensional Martian ionosphere model: 1. The photochemical ionosphere below 180 km. *J. Geophys. Res.* 118, 2105–2123.

Gordon, I.E., et al., 2017. The HITRAN 2016 molecular spectroscopic database. *J. Quant. Spec. Rad. Transfer* 203–69.

Gray, C.I., et al., 2014. The effects of solar flares, coronal mass ejections, and sloar wind streams on Venus' 5577 A oxygen green line. *Icarus* 233, 342–347.

Grebowsky, J.M., Strangeway, R.J., Hunten, D.M., 1997. Evidence for Venus lightning. In: Bougher, S.W., Hunten, D.M., Phillips, R.J. (Eds.), *Venus II*. University of Arizona Press, Tucson, pp. 125–157.

Greene, T.P., et al., 1993. CSHELL: A high spectral resolution echelle spectrograph for the IRTF. *Proc. SPIE* 1946, 313–323.

Griffith, C.A., Hall, J.L., Geballe, T.R., 2000. Detection of daily clouds on Titan. *Science* 290, 509–513.

Gringauz, K.I., Verigin, M.I., Breus, T.K., Gombosi, T., 1979. The interaction of electrons in the optical umbra of Venus with the planetary atmosphere: The origin of the nighttime ionosphere. *J. Geophys. Res.* 84, 2123–2127.

Grinspoon, D.H., Lewis, J.S., 1988. Cometary water on Venus: Implication of stochastic impacts. *Icarus* 74, 21–35.

Groeller, H., et al., 2010. Venus' atomic hot oxygen environment. *J. Geophys. Res.* 115, E12017.

Groeller, H., et al., 2015. Probing the Martian atmosphere with MAVEN/IUVS stellar occultations. *Geophys. Res. Lett.* 42, 9064–9070.

Gronoff, G., et al., 2008. Modeling the Venusian airglow. *Astron. Astrophys.* 482, 1015–1029.

Grundy, W.M., Young, L.A., Stansberry, J.A., Buie, M.W., Olkin, C.B., Young, E.F., 2010. Near-infrared spectral monitoring of Triton with IRTF/SpeX II: Spatial distribution and evolution of ices. *Icarus* 205, 594–604.

Gurrola, E.M. 1995. Interpretation of radar data from the Icy Galilean Satellites and Triton. PhD thesis, Stanford University.

Gurwell, M.A., 2004. Submillimeter observations of Titan: Global measures of stratospheric temperature, CO, HCN, $HC_3N$, and the isotopic ratios $^{12}C/^{13}C$ and $^{14}N/^{15}N$. *Astrophys. J. Lett.* 616, L7–L10.

Gurwell, M.A., Muhleman, D.O., Shah, K.P., Berge, G.L., Rudy, D.J., Grossman, A.W., 1995. Observations of the CO bulge on Venus and implications for mesospheric winds. *Icarus* 115, 141–158.

Gurwell, M.A., Melnick, G.J., Tolls, V., Bergin, E.A., Patten, B.M., 2007. SWAS observations of water vapor in the Venus mesosphere. *Icarus* 188, 288–304.

Guslyakova, S., et al., 2014. $O_2(a^1\Delta_g)$ dayglow limb observations on Mars by SPICAM IR on Mars-Express and connection to water vapor distribution. *Icarus* 239, 131–140.

Guslyakova, S., et al., 2016. Long-term nadir observations of the $O_2$ dayglow by SPICAM IR. *Planet. Space Sci.* 122, 1–12.

Gustaffson, T., Plummer, E.W., Eastman, D.E., Gudat, W., 1978. Partial photoionization cross sections of $CO_2$ between 20 and 40 eV studied with synchrotron radiation. *Phys. Rev.* A17, 175–180.

Gutcheck, R.A., Zipf, E.C., 1973. Excitation of the CO fourth positive system by the dissociative recombination of $CO_2^+$ ions. *J. Geophys. Res.* 78, 5429–5436.

Hanel, R., et al., 1972. Investigation of the Martian environment by infrared spectroscopy on Mariner 9. *Icarus* 17, 423–442.

Hansell, S.A., Wells, W.K., Hunten, D.M., 1995. Optical detection of lightning on Venus. *Icarus* 117, 345–351.

Hansen, C.J., Paige, D.A., 1992. A thermal model for the seasonal nitrogen cycle on Triton. *Icarus* 99, 273–288.

Hansen, C.J., Paige, D.A., Young, L.A., 2015. Pluto's climate modeled with new observational constraints. *Icarus* 246, 183–191.

Hansen, J.E., Hovenier, J.W., 1974. Interpretation of the polarization of Venus. *J. Atmos. Sci.* 31, 1137–1160.

Hansen, J.E, Travis, L.D., 1974. Light scattering in the planetary atmospheres. *Space Sci. Rev.* 16, 527–610.

Hanson, W.B., Santanini, S., and Zuccaro, D.R., 1977. The Martian ionosphere asobserved by Viking retarding potential analyzers, *J. Geophys. Res.*, 82, 4351–4363.

Hapke, B., 1993. *Theory of Reflectance and Emittance Spectroscopy*. Cambridge University Press, Cambridge, UK.

Hartle, R.E., Taylor, H.A., 1983. Identification of deuterium ions in the ionosphere of Venus. *Geophys. Res. Lett.* 10, 965–968.

Hartle, R.E., et al., 1996. Hydrogen and deuterium in the thermosphere of Venus: Solar cycle variations and escape. *J. Geophys. Res.* 101, 4525–4538.

Hartle, R.E., et al., 2006. Preliminary interpretation of Titan plasma interaction as observed by the Cassini Plasma Spectrometer: Comparisons with Voyager 1. *Geophys. Res. Lett.* 33, 8201, doi:10.10129/2005GL024817.

Hartogh, P., et al., 2010. Herschel/HIFI observations of HCl, $H_2O_2$, and $O_2$ in the Martian atmosphere: Initial results. *Astron. Astrophys.* 521, doi:10.1051/0004-6361/201015160.

Hase, F., Wallace, L., McLeod, S.D., Harrison, J.J., Bernath, P.F., 2010. The ACE-FTS atlas of the infrared solar spectrum. *J. Quant. Spectrosc. Radiat. Trans.* 111, 521–528.

Haus, R., Arnold, G., 2010. Radiative transfer in the atmosphere of Venus and application to surface emissivity retrieval from VIRTIS/VEX measurements. *Planet. Space Sci.* 58(12), 1578–1598.

Hebrard, E., Dobrijevic, M., Benilan, Y., Raulin, F., 2007. Photochemical kinetics uncertainties in modeling Titan's atmosphere: First consequences. *Planet. Space Sci.* 55, 1470–1489.

Hedin, A.E., Niemann, H.B., Kasprzak, W.T., Seiff, A., 1983. Global empirical model of the Venus thermosphere. *J. Geophys. Res.* 88, 73–83.

Herbert, F., Sandel, B.R., 1991. $CH_4$ and haze in Triton's lower atmosphere. *J. Geophys. Res.* 96, 19,241–19,252.

Hess, S.L., et al., 1980. The annual cycle of pressure on Mars measured by Viking Landers 1 and 2. *Geophys. Res. Lett.* 7, 197–200.

Hill, R.D., 1979. A survey of lightning energy estimates. *Rev. Geophys. Space Phys.* 17, 155–164.

Hillier, J., Veverka, J., Helfenstein, P., Lee, P., 1994. Photometric diversity of terrains on Triton. *Icarus* 109, 296–312.

Hinson, D.P., et al., 2017. Radio occultation measurements of Pluto's neutral atmosphere with New Horizons. *Icarus* 290, 96–111.

Hinson, D.P., et al., 2018. An upper limit on Pluto's ionosphere from radio occultation measurements with New Horizons. *Icarus* 307, 17–24.

Hodges, R.R., 1999. An exospheric perspective of isotopic fractionation of hydrogen on Venus. *J. Geophys. Res.* 104, 8463–8471.

Hodyss, R., et al., 2013. The solubility of $^{40}Ar$ and $^{84}Kr$ in liquid hydrocarbons: Implications for Titan's geological evolution. *Geophys. Res. Lett.* 40, 2935–2940.

Hoey, W.A., et al., 2017. Rarefied gas dynamic simulation of transfer and escape in the Pluto-Charon system. *Icarus* 287, 87–102.

Hoffman, J.H., Hodges, R.R., Jr., Donahue, T.M., McElroy, M.B., 1980. Composition of the Venus lower atmospherefrom the Pioneer Venus mass spectrometer, *J. Geophys. Res.* 85, 7882–7890.

Holler, B.J., Young, L.A., Grundy, W.M., Olkin, C.B., 2016. On the surface composition of Triton's southern latitudes. *Icarus* 267, 255–266.

Horanyi, M., Poppe, A., Sternovsky, Z., 2016. *Dust Ablation in Pluto's Atmosphere*. EGU General Assembly, Vienna, Austria.

Hord, C.W., et al., 1991. Galileo ultraviolet spectrometer experiment: Initial Venus and interplanetary cruise results. *Science* 253, 1548–1550.

Hörst, S.M., Vuitton, V., Yelle, R.V., 2008. The origin of oxygen species in Titan's atmosphere. *J. Geophys. Res.* 113, E10006.

Hubert, B., Gérard, J.-C., Gustin, J., Shematovich, V.I., Bisikalo, D.V., Stewart, A.I., Gladstone, G.R., 2010. UVIS observations of the FUV OI and CO 4P Venus dayglow during the Cassini flyby. *Icarus* 207, 549–557.

Hubert, B., et al., 2012. Cassini-UVIS observation of dayglow FUV emissions of carbon in the thermosphere of Venus. *Icarus* 220, 635–646.

Hueso, R., Peralta, J., Garate-Lopez, I., Bandos, T.V., Sanchez-Lavega, A., 2015. Six years of Venus winds at the upper cloud level from UV, visible and near infrared observations from VIRTIS on Venus Express. *Planet. Space Sci.* 113, 78–99.

Hunten, D.M., 1978. A Titan atmosphere with a surface temperature of 200 K. In: JPL (Ed.), *The Saturn System*, University of Arizona Press, Tucson, pp. 127–140.

Hunten, D.M., Watson, A.J., 1982. Stability of Pluto's atmosphere. *Icarus* 51, 665–667.

Ignatiev, N.I., Moroz, V.I., Moshkin, B.E., Ekonomov, A.P., Gnedykh, V.I., Grigoriev, A.V., Khatuntsev, I.V., 1997. Water vapour in the lower atmosphere of Venus: A new analysis of optical spectra measured by entry probes. *Planet. Space Sci.* 45, 427–438.

Ignatiev, N.I., Moroz, V.I., Zasova, L.V., Khatuntsev, I.V., 1999. Water vapor in the middle atmosphere of Venus: An improved treatment of the Venera 15 IR spectra. *Planet. Space Sci.* 47, 1061–1075.

Imanaka, H., Cruikshank, D.P., Khare, B.N., McKay, C.P., 2012. Optical constants of Titan tholins at mid-infrared wavelengths (2.5–25 lm) and the possible chemical nature of Titan's haze particles. *Icarus* 218, 247–261.

Iozenas, V.A., Krasnopolsky, V.A., 1970. Some ozonosphere characteristics from observational data of satellites. *Space Research* X, 215–222.

Irvine, W.M., 1968. Monochromatic phase curves and albedos for Venus. *J. Atmos. Sci.* 25, 610–616.

Irwin, P.G.J., de Kok, R., Negrao, A., Tsang, C.C.C., Wilson, C.F., Drossart, P., Piccioni, G., Grassi, D., Taylor, F.W., 2008. Spatial variability of carbon monoxide in Venus' mesosphere from Venus Express/visible and infrared thermal imaging spectrometer measurements. *J. Geophys. Res.* 113, E00B01.

Israel, G., et al., 2005. Complex organic matter in Titan's atmospheric aerosol from in situ pyrolysis and analysis. *Nature* 438, 796–799.

Istomin, V.G., Grechnev, K.V., Kochnev, V.A., 1983. Venera 13 and Venera 14: Mass spectrometry of the atmosphere. *Cosmic Res.* 21, 410–415.

Ivanov-Kholodny, G.S., et al., 1979. Daytime ionosphere of Venus as studied with Venera 9 and 10 dual-frequency radio occultation experiments. *Icarus* 39, 209–213.

Iwagami, N., et al., 2008. Hemispheric distributions of HCl above and below the Venus' clouds by ground-based 1.7 μm spectroscopy. *Planet. Space Sci.* 56, 1424–1434.

Jaffe, W., Caldwell, J., Owen, T., 1980. Radius and brightness temperature observations of Titan at centimeter wavelengths by the Very Large Array. *Astrophys. J.* 242, 806–811.

Jain, S.K., et al., 2015. The structure and variability of Mars upper atmosphere as seen in MAVEN/IUVS dayglow observations. *Geophys. Res. Lett.* 42, 9023–9030.

Jakosky, B.M., Farmer, C.B., 1982. The seasonal and global behavior of water vapor in Mars atmosphere: Complete global results of the Viking atmospheric water detector experiment. *J. Geophys. Res.* 87, 2999–3019.

Jakosky, B.M., Pepin, R.O., Johnson, R.E., Fox, J.L., 1994. Mars atmospheric loss and isotopic fractionation by solar-wind-induced sputtering and photochemical escape. *Icarus* 111, 271–288.

Jaquin, F., Gierasch, P., Kahn, R., 1986. The vertical structure of limb hazes in the Martian atmosphere. *Icarus* 68, 442–461.

Jenkins, J.M., et al., 2002. Microwave remote sensing of the temperature and distribution of sulfur compounds in the lower atmosphere of Venus. *Icarus* 158, 312–328.

Jennings, D.E., et al., 2008. Isotopic ratios in Titan's atmosphere from Cassini CIRS limb sounding: $HC_3N$ in the North. *Astrophys. J. Lett.* 681, L109.

Jensen, S.J.K., et al., 2014. A sink for methane on Mars? The answer is blowing in the wind. *Icarus* 236, 24–27.

Johnson, R.E., 2010. Thermally driven atmospheric escape. *Astrophys. J.* 716, 1573–1578.

Johnson, R.E., Tucker, O.J., Michael, M., Sittler, E.C., Smith, H.T., Young, D.T., Waite, J.H., 2009. Mass loss processes in Titan's atmosphere. In Brown, R.H., Lebreton, J.P., Waite, J.H. (Eds.), *Titan from Cassini–Huygens*, Springer, Dordrecht, pp. 373–392.

Jolly, A., et al., 2010. The $\nu_8$ bending mode of diacetylene: From laboratory spectroscopy to the detection of $^{13}C$ isotopologues in Titan's atmosphere. *Astrophys. J.* 714, 852–859.

Jolly, A., et al., 2015. Gas phase dicyanoacetylene ($C_4N_2$) on Titan: New experimental and theoretical spectroscopy results applied to Cassini CIRS data. *Icarus* 248, 340–346.

Kammer, J.A., Shemansky, D.E., Zhang, X., Yung, Y.L., 2013. Composition of Titan's upper atmosphere from Cassini UVIS EUV stellar occultations. *Planet. Space Sci.* 88, 86–92.

Kaplan, L.D., Munch, G., Spinrad, H., 1964. An analysis of the spectrum of Mars. *Astrophys. J.* 139, 237–242.

Kaplan, L.D., Connes, J., Connes, P., 1969. Carbon monoxide in the Martian atmosphere. *Astrophys. J.* 1457, L187– L192.

Kasting, J.F., Pollack, J.B., 1983. Loss of water from Venus. I. Hydrodynamic escape of hydrogen. *Icarus* 53, 479–508.

Käufl, H.U., et al., 2004. CRIRES: A high resolution infrared spectrograph for ESO's VLT. *SPIE* 5492, 1218.

Kawabata, K., et al., 1980. Cloud and haze properties from Pioneer Venus polarimetry. *J. Geophys. Res.* 85, 8129–8140.

Kaye, J.A., 1988. On the possible role of the reaction $O + HO_2 \rightarrow OH + O_2$ in OH airglow. *J. Geophys. Res.* 93, 285–288.

Keating, G.M., Nicholson, J.Y., Lake, L.R., 1980. Venus upper atmosphere structure. *J. Geophys. Res.* 85, 7941–7956.

Keating, G.M., et al., 1985. Models of Venus neutralupper atmosphere: Structure and composition. *Adv. Space Res.* 5(11), 117–172.

Kenner, R.D., Ogryzlo, E.A., Turley, S., 1979. On the excitation of the night airglow on Earth, Venus, and Mars. *J. Photochem.* 10, 199–203.

Kerzhanovich, V.V., Limaye, S.S., 1985. VIRA: Circulation of the atmosphere from the surface to 100 km. *Adv. Space Res.* 5(11), 59–84.

Khare, B.N., Sagan, C., 1973. Red clouds in reducing atmospheres. *Icarus* 20, 311.

Khare, B.N., et al., 1984. Optical constants of organic tholins produced in a simulated Titanian atmosphere: From X-rays to microwave frequencies. *Icarus* 60, 127–137.

Khatuntsev, I.V., Patsaeva, M.V., Titov, D.V., et al., 2013. Cloud level winds from the Venus Express monitoring camera imaging. *Icarus* 226, 140–158.

Khatuntsev, I.V., Patsaeva, M.V., Titov, D.V., Ignatiev, N.I., Turin, A.V., Fedorova, A.A., Markiewicz, W.J., 2017. Winds in the middle cloud deck from the near-IR imaging by the Venus Monitoring Camera onboard Venus Express. *J. Geophys. Res.* 122E, 2312–2327.

Kleinboehl, A., et al., 2015. No widespread dust in the middle atmosphere of Mars from Mars Climate Sounder observations. *Icarus* 261, 118–121.

Kliore, A.J., Luhmann, J.G., 1991. Solar cycle effects on the structure of the electron density profiles in the dayside ionosphere of Venus. *J. Geophys. Res.* 96, 21281–21289.

Kliore, A.J., Fjeldbo, G., Seidel, B.L., et al., 1973. S band radio occultation measurements of the atmosphere and topography of Mars with Mariner 9: Extended mission coverage of polar and intermediate latitudes, *J. Geophys. Res.* 78, 4331–4351.

Kliore, A.J., et al., 2008. First results from the Cassini radio occultations of the Titan ionosphere. *J. Geophys. Res.* 113, A09317.

Kliore, A.J., Nagy, A.F., Cravens, T.E., Richard, M.S., Rymer, A.M., 2011. Unusual electron density profiles observed by Cassini radio occultations in Titan's ionosphere: Effects of enhanced magnetospheric electron precipitation? *J. Geophys. Res.* 116, A11318.

Knollenberg, R.G., Hunten, D.M., 1980. Microphysics of the clouds of Venus: Results of the Pioneer Venus particle size spectrometer experiment. *J. Geophys. Res.* 85, 8039–8058.

Knudsen, W.C., Spenner, K., Michelson, P.F., Whitten, R.C., Miller, K.L., Novak, V., 1980. Suprathermal electron energy distribution within the dayside ionosphere of Venus. *J. Geophys. Res.* 85, 7754–7758.

Kolodner, M.A., Steffes, P.G., 1998. The microwave absorption and abundance of sulfuric acid vapor in the Venus atmosphere based on new laboratory measurements. *Icarus* 132, 151–169.

Kong, T.Y., McElroy, M.B., 1977a. Photochemistry of the Martian atmosphere. *Icarus* 32, 168–189.

Kong, T.Y., McElroy, M.B., 1977b. The global distribution of $O_3$ on Mars. *Planet. Space Sci.* 25, 839–857.

Korablev, O.I., Krasnopolsky, V.A., Rodin, A.V., Chassefiere, E., 1993. Vertical structure of Martian dust measured by solar infrared occultations from the Phobos spacecraft. *Icarus* 102, 76–87.

Korablev, O.I., et al., 2006. SPICAM IR acousto-optic spectrometer experiment on Mars Express. *J. Geophys. Res.* 111, E09S03.

Koskinen, T.T., Yelle, R.V., Snowden, D.S., Lavvas, P., Sandel, B.R., Capalbo, F.J., Benilan, Y., West, R.A., 2011. The mesosphere and thermosphere of Titan revealed by Cassini/UVIS stellar occultations. *Icarus* 216, 507–534.

Kostiuk, T., et al., 2001. Direct measurements of winds of Titan. *Geophys. Res. Lett.* 28, 2361–2364.

Krasnopolsky, V.A., 1966. Ultraviolet spectrum of the radiation reflected by the Earth's atmosphere and its use in determining the total abundance and vertical distribution of the atmospheric ozone. *Geomagn. Aeronomy* 6, 236–242.

Krasnopolsky, V.A., 1970. Nitric oxide at 110–220 km measured from the Cosmos 224 orbiter. *Geomagn. Aeronomy* 10, 660–663.

Krasnopolsky, V.A., 1974. Analysis of airglow observations at a planetary limb. *Geomagn. Aeronomy* 14, 567–571.

Krasnopolsky, V.A., 1975. On the structure of Mars' atmosphere at 120–220 km. *Icarus* 24, 28–32.

Krasnopolsky, V.A., 1979. Nightside ionosphere of Venus. *Planet. Space Sci.* 27, 1403–1408.

Krasnopolsky, V.A., 1980. Venera 9, 10: Spectroscopy of scattered radiation in the overcloud atmosphere. *Cosmic Res.* 18, 899–906.

Krasnopolsky, V.A., 1981. Spectroscopic evaluation of CO in the Martian upper atmosphere. *Cosmic Res.* 19, 902–906.

Krasnopolsky, V.A., 1982. Atomic carbon in the atmospheres of Mars and Venus. *Cosmic Res.* 20, 595–603.

Krasnopolsky, V.A., 1983a. Venus spectroscopy in the 3000–8000 Å region by Veneras 9 and 10. In: Hunten, D.M., Colin, L., Donahue, T.M., Moroz, V.I. (Eds.), *Venus*. University of Arizona Press, Tucson, pp. 459–483.

Krasnopolsky, V.A., 1983b. Lightning and nitric oxide on Venus. *Planet. Space Sci.* 31, 1363–1369.
Krasnopolsky, V.A., 1985. Total injection of water vapor into the Venus atmosphere. *Icarus* 62, 221–229.
Krasnopolsky, V.A., 1986. *Photochemistry of the Atmospheres of Mars and Venus*. Springer, Berlin.
Krasnopolsky, V.A., 1987. $S_3$ and $S_4$ absorption cross sections in the range of 340–600 nm and evaluation of the $S_3$ abundance in the lower atmosphere of Venus. *Adv. Space Res.* 7(12), 25–27.
Krasnopolsky, V.A., 1989. Vega mission results and chemical composition of Venusian clouds. *Icarus* 80, 202–210.
Krasnopolsky, V.A., 1993a. On the haze model for Triton. *J . Geophys. Res.* 98, 17123–17124.
Krasnopolsky, V.A., 1993b. Photochemistry of the Martian atmosphere (mean conditions). *Icarus* 101, 313–332.
Krasnopolsky, V.A., 1995. Uniqueness of a solution of a steady-state photochemical problem: Applications to Mars. *J. Geophys. Res.* 100, 3263–3276.
Krasnopolsky, V.A., 1997. Photochemical mapping of Mars. *J. Geophys. Res.* 102, 13313–13320.
Krasnopolsky, V.A., 1999. Hydrodynamic flow of $N_2$ from Pluto. *J. Geophys. Res.* 104, 5955–5962.
Krasnopolsky, V.A., 2000. On the deuterium abundance on Mars and some related problems. *Icarus* 148, 597–602.
Krasnopolsky, V.A., 2001. Middle ultraviolet spectroscopy of Pluto and Charon. *Icarus* 153, 277–284.
Krasnopolsky, V.A., 2002. Mars' upper atmosphere and ionosphere at low, medium, and high solar activities: implications for evolution of water. *J. Geophys. Res.* 107(E12), 5128.
Krasnopolsky, V.A., 2003. Spectroscopic mapping of Mars CO mixing ratio: Detection of north-south asymmetry. *J. Geophys. Res.* 108(E2), 5010, doi:10.1029/2002JE001926.
Krasnopolsky, V.A., 2005. A sensitive search for $SO_2$ in the Martian atmosphere: Implications for seepage and origin of methane. *Icarus* 178, 487–492.
Krasnopolsky, V.A., 2006a. A sensitive search for nitric oxide in the lower atmospheres of Venus and Mars: Detection on Venus and upper limit for Mars. *Icarus* 182, 80–91.
Krasnopolsky, V.A., 2006b. Photochemistry of the Martian atmosphere: Seasonal, latitudinal, and diurnal variations. *Icarus* 185, 153–170.
Krasnopolsky, V.A., 2006c. Some problems related to the origin of methane on Mars. *Icarus* 180, 359–367.
Krasnopolsky, V.A., 2007a. Long-term spectroscopic observations of Mars using IRTF/CSHELL: Mapping of $O_2$ dayglow, CO, and search for $CH_4$. *Icarus* 190, 93–102.
Krasnopolsky, V.A., 2007b. Chemical kinetic model for the lower atmosphere of Venus. *Icarus* 191, 25–37.
Krasnopolsky, V.A., 2008. High-resolution spectroscopy of Venus: Detection of OCS, upper limit to $H_2S$, and latitudinal variations of CO and HF in the upper cloud layer. *Icarus* 197, 377–385.
Krasnopolsky, V.A., 2009a. A photochemical model of Titan's atmosphere and ionosphere. *Icarus* 201, 226–256.
Krasnopolsky, V.A., 2009b. Seasonal variations of photochemical tracers at low and middle latitudes on Mars: Observations and models. *Icarus* 201, 564–569.

Krasnopolsky, V.A., 2010a. Venus night airglow: Ground-based detection of OH, observations of $O_2$ emissions, and photochemical model. *Icarus* 207, 17–27.

Krasnopolsky, V.A., 2010b. Solar activity variations of thermospheric temperature on Mars and a problem of CO in the lower atmosphere. *Icarus* 207, 638–647.

Krasnopolsky, V.A., 2010c. Spatially-resolved high-resolution spectroscopy of Venus. 1. Variations of $CO_2$, CO, HF, and HCl at the cloud tops. *Icarus* 208, 539–547.

Krasnopolsky, V.A., 2010d. Spatially-resolved high-resolution spectroscopy of Venus. 2. Variations of HDO, OCS, and $SO_2$ at the cloud tops. *Icarus* 209, 314–322.

Krasnopolsky, V.A., 2010e. The photochemical model of Titan's atmosphere and ionosphere: A version without hydrodynamic escape. *Planet. Space Sci.* 58, 1507–1515.

Krasnopolsky, V.A., 2011a. Excitation of the oxygen nightglow on the terrestrial planets. *Planet. Space Sci.* 59, 754–766.

Krasnopolsky, V.A., 2011b. A sensitive search for methane and ethane on Mars. *EPSC Abstracts* 6, 49.

Krasnopolsky, V.A., 2011c. Vertical profile of $H_2SO_4$ vapor at 70–110 km on Venus and some related problems. *Icarus* 215, 197–203.

Krasnopolsky, V.A., 2012a. Search for methane and upper limits to ethane and $SO_2$ on Mars. *Icarus* 217, 144–152.

Krasnopolsky, V.A., 2012b. A photochemical model for the Venus atmosphere at 47–112 km. *Icarus* 218, 230–246.

Krasnopolsky, V.A., 2012c. Observation of DCl and upper limit to $NH_3$ on Venus. *Icarus* 219, 244–249.

Krasnopolsky, V.A., 2012d. Titan's photochemical model: Further update, oxygen species, and comparison with Triton and Pluto. *Planet. Space Sci.* 73, 318–326.

Krasnopolsky, V.A., 2013a. Nighttime photochemical model and night airglow on Venus. *Planet. Space Sci.* 85, 78–88.

Krasnopolsky, V.A., 2013b. $S_3$ and $S_4$ abundances and improved chemical kinetic model for the lower atmosphere of Venus. *Icarus* 225, 570–580.

Krasnopolsky, V.A., 2013c. Night and day airglow of oxygen at 1.27 μm on Mars. Planet. *Space Sci.* 85, 243–249.

Krasnopolsky, V.A., 2014a. Observations of the CO dayglow at 4.7 μm on Mars: Variations of temperature and the CO mixing ratio at 50 km. *Icarus* 228, 189–196.

Krasnopolsky, V.A., 2014b. Chemical composition of Titan's atmosphere and ionosphere: Observations and the photochemical model. *Icarus* 236, 83–91.

Krasnopolsky, V.A., 2014c. Observations of CO dayglow at 4.7 μm, CO mixing ratios, and temperatures at 74 and 104–111 km on Venus. *Icarus* 237, 340–349.

Krasnopolsky, V.A., 2015a. CXO X-ray spectroscopy of comets and abundances of heavy ions in the solar wind. *Icarus* 247, 95–102.

Krasnopolsky, V.A., 2015b. Vertical profiles of $H_2O$, $H_2SO_4$, and sulfuric acid concentration at 45–75 km on Venus. *Icarus* 252, 327–333.

Krasnopolsky, V.A., 2015c. Variations of carbon monoxide in the Martian lower atmosphere. *Icarus* 253, 149–155.

Krasnopolsky, V.A., 2015d. Variations of the $HDO/H_2O$ ratio in the Martian atmosphere and loss of water from Mars. *Icarus* 257, 377–386.

Krasnopolsky, V.A., 2016a. Sulfur aerosol in the clouds of Venus. *Icarus* 274, 33–36.

Krasnopolsky, V.A., 2016b. Isotopic ratio of nitrogen on Titan: Photochemical interpretation. *Planet. Space Sci.* 134, 61–63.

Krasnopolsky, V.A., 2017a. On the iron chloride aerosol in the clouds of Venus. *Icarus* 286, 134–137.

Krasnopolsky, V.A., 2017b. Annual mean mixing ratios of $N_2$, Ar, $O_2$, and CO in the Martian atmosphere. *Planet. Space Sci.* 144, 71–73.
Krasnopolsky, V.A., 2018a. Disulfur dioxide and its near-UV absorption in the photochemical model of Venus atmosphere. *Icarus* 299, 294–299.
Krasnopolsky, V.A., 2018b. Some problems in interpretation of the New Horizons observations of Pluto's atmosphere. *Icarus* 301, 152–154.
Krasnopolsky, V.A., 2018c. On the carbon isotope ratio in Titan's atmosphere and interior. 42nd COSPAR Assembly, Pasadena, abstract id. B5.3-24-18.
Krasnopolsky, V.A., Forthcoming. Photochemistry of water in the Martian thermosphere and its effect on hydrogen escape. Icarus.
Krasnopolsky, V.A., Forthcoming. Venus nighttime photochemical model: Nightglow of $O_2$, NO, OH and abundances of $O_3$ and ClO. Icarus.
Krasnopolsky, V.A., Belyaev, D.A., 2017. Search for HBr and bromine photochemistry on Venus. *Icarus* 293, 111–118.
Krasnopolsky, V.A., Cruikshank, D.P., 1995. Photochemistry of Triton's atmosphere and ionosphere. *J. Geophys. Res.* 100, 21271–21286.
Krasnopolsky, V.A., Cruikshank, D.P., 1999. Photochemistry of Pluto's atmosphere and ionosphere near perihelion. *J. Geophys. Res.* 104, 21979–21996.
Krasnopolsky, V.A., Feldman, P.D., 2001. Detection of molecular hydrogen in the atmosphere of Mars. *Science* 294, 1914–1917.
Krasnopolsky, V.A., Feldman, P.D., 2002. Far ultraviolet spectrum of Mars. *Icarus* 160, 86–94.
Krasnopolsky, V.A., Gladstone, G.R., 1996. Helium on Mars: EUVE and Phobos data and implications for Mars' evolution. *J. Geophys. Res.* 101A, 15765–15772.
Krasnopolsky, V.A., Gladstone, G.R., 2005. Helium on Mars and Venus: EUVE observations and modeling. *Icarus* 176, 395–407.
Krasnopolsky, V.A., Krys'ko, A.A., 1976. On the night airglow of the Martian atmosphere. *Space Res.* 16, 1005–1008.
Krasnopolsky, V.A., Lefèvre, F., 2013. Chemistry of the atmospheres of Mars, Venus, and Titan. In: *Comparative Climatology of Terrestrial Planets*, Mackwell, S.J., et al. (Eds.), University of Arizona Press, Tucson, pp. 231–276.
Krasnopolsky, V.A., Mumma, M.J., 2001. Spectroscopy of comet Hyakutake at 80–700 Å: First detection of solar wind charge transfer emissions. *Astrophys. J.* 549, 629–634.
Krasnopolsky, V.A., Parshev, V.A., 1977. Altitude profile of water vapor on Mars. *Cosmic Res.* 15, 673–676.
Krasnopolsky, V.A., Parshev, V.A., 1979. Chemical composition of Venus' troposphere and cloud layer based on Venera 11, Venera 12, and Pioneer Venus measurements. *Cosmic Res.* 17, 630–637.
Krasnopolsky, V.A., Parshev, V.A., 1981. Chemical composition of the atmosphere of Venus. *Nature* 292, 610–613.
Krasnopolsky, V.A., Pollack, J.B., 1994. $H_2O$–$H_2SO_4$ system in Venus' clouds and OCS, CO, and $H_2SO_4$ profiles in Venus' troposphere. *Icarus* 109, 58–78.
Krasnopolsky, V.A., Kuznetsov, A.P., Lebedinsky, A.I., 1966. Ultraviolet spectrum of the Earth measured from Cosmos 65. *Geomagn. Aeronomy* 6, 145–148.
Krasnopolsky, V.A., Krysko, A.A., Rogachev, V.N., Parshev, V.A., 1976. Spectroscopy of the Venus night airglow from the Venera 9 and 10 orbiters. *Cosmic Res.* 14, 789–795
Krasnopolsky, V.A., Sandel, B.R., Herbert, F., 1992. Properties of haze in the atmosphere of Triton. *J. Geophys. Res.* 97, 11695–11700.
Krasnopolsky, V.A., Sandel, B.R., Herbert, F., Vervack, R.J., 1993. Temperature, $N_2$, and N density profiles of Triton's atmosphere: observations and model. *J. Geophys. Res.* 98, 3065–3078.

Krasnopolsky, V.A., Bjoraker, G.L., Mumma, M.J., Jennings, D.E., 1997. High-resolution spectroscopy of Mars at 3.7 and 8 μm: A sensitive search for $H_2O_2$, $H_2CO$, HCl, and $CH_4$, and detection of HDO. *J. Geophys. Res.* 102, 6525–6534.
Krasnopolsky, V.A., Mumma, M.J., Gladstone, G.R., 1998. Detection of atomic deuterium in the upper atmosphere of Mars. *Science* 280, 1576–1580.
Krasnopolsky, V.A., et al., 2002. X-ray emission from comet McNaught-Hartley (C/1999 T1). *Icarus* 160, 437–447.
Krasnopolsky, V.A., Maillard, J.P., Owen, T.C., 2004a. Detection of methane in the Martian atmosphere: Evidence for life. *Geophys. Res. Abstracts* 6, 06169.
Krasnopolsky, V.A., Maillard, J.P., Owen, T.C., 2004b. Detection of methane in the Martian atmosphere: Evidence for life? *Icarus* 172, 537–547.
Krasnopolsky, V.A., Maillard, J.P., Owen, T.C., Toth, R.A., Smith, M.D., 2007. Oxygen and carbon isotope ratios in the Martian atmosphere. *Icarus* 192, 396–403.
Krasnopolsky, V.A., Belyaev, D.A., Gordon, I.E., Li, G., Rothman, L.S., 2013. Observations of D/H ratios in $H_2O$, HCl, and HF on Venus and new DCl and DF line strengths. *Icarus* 224, 57–65.
Kuiper, G.P., 1944. Titan: A satellite with an atmosphere. *Astrophys. J.* 100, 378–383.
Kuiper, G.P., 1949. Survey of planetary atmospheres. In: *The Atmospheres of the Earth and Planets*, Kuiper, G.P. (Ed.), University of Chicago Press, Chicago, pp. 304–345.
Kumar, S., Broadfoot, A.L., 1975. He 584 Å airglow emission from Venus: Mariner 10 observations. *Geophys. Res. Lett.* 2, 357–360.
Kumar, S., Hunten, D.M., Broadfoot, A.L., 1978. Non-thermal hydrogen in the Venus exosphere: The ionospheric source and the hydrogen budget. *Planet. Space Sci.* 26, 1063–1075.
Kumar, S., Hunten, D.M., Pollack, J.B., 1983. Non-thermal escape of hydrogen and deuterium from Venus and implications for loss of water. *Icarus* 55, 369–375.
Kurokawa, H., et al., 2014. Evolution of water reservoirs on Mars: Constraints from hydrogen isotopes in Martian meteorites. *Earth Planet. Sci. Lett.* 394, 179–185.
Lacy, J.H., Richter, M.J., Greathouse, T.K., Jaffe, D.T., Zhu, Q., 2002. TEXES: A sensitive high-resolution grating spectrograph for the mid-infrared. *Pub. Astron. Soc. Pacific* 114, 153–168.
Lane, W.A., Opstbaum, R., 1983. High altitude Venus hazefrom Pioneer Venus limb scans. *Icarus* 54, 48–58.
Lara, L.M., Lellouch, E., Lopez-Moreno, J.J., Rodrigo, R., 1996. Vertical distributions of Titan's atmospheric neutral constituents. *J. Geophys. Res.* 101, 23262–23283.
Lara, L.M., Ip, W.-H., Rodrigo, R., 1997. Photochemical models of Pluto's atmosphere. *Icarus* 130, 16–35 .
Lavvas, P.P., Coustenis, A., Vardavas, I.M., 2008a. Coupling photochemistry with haze formation in Titan's atmosphere, Part I: Model description. *Planet. Space Sci.* 56, 27–66.
Lavvas, P.P., Coustenis, A., Vardavas, I.M., 2008b. Coupling photochemistry with haze formation in Titan's atmosphere, Part II: Results and validation with Cassini/Huygens data. *Planet. Space Sci.* 56, 67–99.
Lawrence, G.M., 1973. Production of $O(^1S)$ from photodissociation of $O_2$. *J. Geophys. Res.* 78, 8314–8318.
Lawrence, G.M., Barth, C.A., Argabright, V., 1977. Excitation of the Venus night airglow. *Science* 195, 573–574.
Leblanc, F., Johnson, R.E., 2002. Role of molecular species in pickup ion sputtering of the Martian atmosphere. *J. Geophys. Res.* 107, E25010.

Leblanc, F., et al., 2006. Martian dayglow as seen by the SPICAM UV spectrograph on Mars Express. *J. Geophys. Res.* 111, E05S11.

Leblanc, F., Chaufray, J.Y., Bertaux, J.L., 2007. On Martian nitrogen dayglow emission observed by SPICAM UV spectrograph/Mars Express. *Geophys. Res. Lett.* 34, L02206.

Lebonnois, S., 2005. Benzene and aerosol production in Titan and Jupiter's atmospheres: A sensitivity study. *Planet. Space Sci.* 53, 486–497.

Lebonnois, S., Quemerais, E., Montmessin, F., et al., 2006. Vertical distribution of ozone on Mars as measured by SPICAM/Mars Express using stellar occultations. *J. Geophys. Res.*, 111, E09S05.

Lebonnois, S., Burgalat, J., Rannou, P., Charnay, B., 2012. Titan global climate model: A new 3-dimensional version of the IPSL Titan GCM. *Icarus* 218, 707–722.

Lebonnois, S., Schubert, G., 2017. The deep atmosphere of Venus and the possible role of density-driven separation of $CO_2$ and $N_2$. *Nature Geoscience* 10, 473–477.

Lecuyer, C., et al., 2017. D/H fractionation during the sublimation of water ice. *Icarus* 285, 1–7.

Lee, L.C., Judge, D.L., 1972. Cross sections for the production of $CO_2^+(A^2\Pi_u, B^2\Sigma_u^+ \rightarrow X^2\Pi_g)$ fluorescence by vacuum ultraviolet radiation. *J. Chem. Phys.* 57, 4443–4447.

Lefèvre, F., Forget, F., 2009. Observed variations of methane on Mars unexplained by known atmospheric chemistry and physics. *Nature* 460, 720–723.

Lefèvre, F., et al., 2004. Three-dimensional modeling of ozone on Mars. *J. Geophys. Res* 109, E07004.

Lefèvre, F., Bertaux, J.L., Clancy, R.T., et al., 2008. Heterogeneous chemistry in the atmosphere of Mars. *Nature* 454, 971–975.

Lellouch, E., Paubert, G., Encrenaz, T., 1991. Mapping of CO millimeter-wave lines in Mars' atmosphere: The spatial variability of carbon monoxide on Mars. *Planet. Space Sci.* 39, 219–224.

Lellouch, E., Goldstein, J.J., Rosenqvist, J., Bougher, S.W., Paubert, G., 1994. Global circulation, thermal structure, and carbon monoxide distribution in Venus' mesosphere in 1991. *Icarus* 110, 315–339.

Lellouch, E., et al., 2009. Pluto's lower atmosphere structure and methane abundance from high-resolution spectroscopy and stellar occultations. *Astron. Astrophys.* 495, L17–L21.

Lellouch, E., de Bergh, C., Sicardy, B., Ferron, S., Kaufl, H.U., 2010. Detection of CO in Triton's atmosphere and the nature of surface-atmosphere interactions. *Astron. Astrophys.* 512, L8.

Lellouch, E., de Bergh, C., Sicardy, B., Kaufl, H.U., Smette, A., 2011a. High resolution spectroscopy of Pluto's atmosphere: Detection of the 2.3-lm CH4 bands and evidence for carbon monoxide. *Astron. Astrophys.* 530, L4.

Lellouch, E., Stansberry, J., Emery, J., Grundy, W., Cruikshank, D.P., 2011b. Thermal properties of Pluot's and Charon's surfaces from Spitzer observations. *Icarus* 214, 701–716.

Lellouch, E., et al., 2017. Detection of CO and HCN in Pluto's atmosphere with ALMA. *Icarus* 286, 289–307.

Leshin, L.A., 2000. Insights into Martian water reservoirs from analyses of Martian meteorite QUE94201. *Geophys. Res. Lett.* 27, 2017–2020.

Levine, J.S., Gregory, G.L., Harvey, G.A., Howell, W.E., Borucki, W.J., Orville, R.E., 1982. Production of nitric oxide by lightning on Venus. *Geophys. Res. Lett.* 9, 893–896.

Liang, M.C., et al., 2007. Source of nitrogen isotope anomaly in HCN in the atmosphere of Titan. *Astrophys. J. Lett.* 664, L115–L118.

Liaw, Y.P., Sisterson, D.L., Miller, N.L., 1990. Comparison of field, laboratory, and theoretical estimates of global nitrogen fixation by lightning. *J. Geophys. Res.* 95, 22489–22494.

Lillis, R., et al., 2017. Photochemical escape of oxygen from Mars: First results from MAVEN in situ data. *J. Geophys. Res.* 122(A3), 3815–3836.

Lindal, G.F., et al., 1983. The atmosphere of Titan: An analysis of the Voyager 1 radio-occultation measurements. *Icarus* 53, 348–363.

Liu, S.C., Donahue, T.M., 1976. The regulation of hydrogen and oxygen escape from Mars. *Icarus* 28, 231–246.

Loison, J.C., Dobrijevic, M., Hickson, K.M., Heays, A.N., 2017. The photochemical fractionation of oxygen isotopologues in Titan's atmosphere. *Icarus* 291, 17–30.

Lopez-Moreno, J.J., et al., 2008. Structure of Titan's low altitude ionized layer from the relaxation probe onboard Huygens. *Geophys. Res. Lett.* 35, L22104.

Lopez-Puertas, M., et al., 2013. Large abundances of polycyclic aromatic hydrocarbons in Titan's upper atmosphere. *Astrophys. J.* 770, 132–140.

Luginin, M., et al., 2016. Aerosol properties in the upper haze of Venus from SPICAV IR data. *Icarus* 277, 154–170

Luna, H., Michael, M., Shah, M.B., Johnson, R.E., Latimer, C.J., McConkey, J.W., 2003. Dissociation of $N_2$ in capture and ionization collisions with fast H+ and N+ ions and modeling of positive ions formation in the Titan atmosphere. *J. Geophys. Res.* 108 (E4), 5033. doi:10.1029/2002JE001950.

Luspay-Kuti, A., et al., 2017. Photochemistry on Pluto – I. Hydrocarbons and aerosols. *MNRAS* 472, 104–117.

Lutz, B.L., de Bergh, C., Owen, T., 1983. Titan: Discovery of carbon monoxide in its atmosphere. *Science* 220, 1374–1375.

Lyons, J.R., Yung, Y.L., Allen, M., 1992. Solar control of the upper atmosphere of Triton. *Science* 246, 1483–1485.

Lyons, J.R., Manning, C.E., Nimmo, F., 2005. Formation of methane on Mars by fluid-rock interaction in the crust. *Geophys. Res. Lett.* 32, L13201. doi:10.1029/2004GL022161.

Maguire, W.C., 1977. Martian isotopic ratios and upper limits for possible minor constituents as derived from Mariner 9 infrared spectrometer data. *Icarus* 32, 85–97.

Mahaffy, P.R., et al., 2013. Abundance and isotopic composition of gases in the Martian atmosphere from the Curiosity rover. *Science* 341, 263–266.

Mahaffy, P.R., et al., 2015. Structure and composition of the neutral upper atmosphere of Mars from the MAVEN NGIMS investigation. *Geophys. Res. Lett.* 42, 8951–8957.

Mahieux, A., et al., 2015a. Venus mesospheric sulfur dioxide measurement retrieved from SOIR on board Venus Express. *Planet. Space Sci.* 113–114, 193–204.

Mahieux, A., et al., 2015b. Hydrogen halides measurements in the Venus mesosphere retrieved from SOIR on board Venus Express. *Planet. Space Sci.* 113–114, 264–274.

Mahieux, A., et al., 2015c. Update of the Venus density and temperature profiles at high altitude measured by SOIR on board Venus Express. *Planet. Space Sci.* 113–114, 309–320.

Mahieux, A., et al., 2015d. Rotational temperatures of Venus upper atmosphere as measured by SOIR on board Venus Express. *Planet. Space Sci.* 113–114, 347–358.

Maiorov, B.S., et al., 2005. A new analysis of the spectra obtained by the Venera missions in the Venussian atmosphere. I. The analysis of the data received from the Venera 11 probe at altitudes below 37 km in the 0.44–0.66 lm wavelength range. *Solar Syst. Res.* 39, 267–282.

Majeed, T., McConnell, J.C., Strobel, D.F., Summers, M.E., 1990. The ionosphere of Triton. *Geophys. Res. Lett.* 17, 1721–1724.

Maltagliati, L., Montmessin, F., Korablev, O., Fedorova, A., Forget, F., Määttänen, A., Lefèvre, F., Bertaux, J.-L., 2013. Annual survey of water vapor vertical distribution and water-aerosol coupling in the Martian atmosphere observed by SPICAM/MEx solar occultations. *Icarus* 223, 942–962.

Maltagliati, L., et al., 2015. Titan's atmosphere as observed by VIMS/Cassini solar occultations: Gaseous components. *Icarus* 248, 1–24.

Manat, S.L., Lane, A., 1993. A compilation of the absorption cross sections of $SO_2$ from 106 to 403 nm. *J. Quant. Spectr. Rad. Transfer* 50, 267–276.

Mandt, K., et al., 2012a. The $^{12}C/^{13}C$ ratio on Titan from Cassini/INMS measurements and implications for the evolution of methane. *Astrophys. J.* 749, 160–174.

Mandt, K.E., et al., 2012b. Ion densities and composition of Titan's upper atmosphere derived from the Cassini Ion Neutral Mass Spectrometer: Analysis methods and comparison of measured ion densities to photochemical model simulations. *J. Geophys. Res.* 117, E10006.

Mandt, K.E., Mousis, O., Lunine, J., Gautier, D., 2014. Protosolar ammonia as the unique source of Titan's nitrogen. *Astrophys. J. Lett.* 788, L24.

Mandt, K., et al., 2017. Photochemistry on Pluto: Part II. HCN and nitrogen isotope fractionation. *MNRAS* 472, 118–128.

Marcq, E., Encrenaz, T., Bezard, B., Birlan, M., 2006. Remote sensing of Venus' lower atmosphere from ground-based spectroscopy: Latitudinal and vertical distribution of minor species. *Planet. Space Sci.* 54, 1360–1370.

Marcq, E., Bezard, B., Drossart, P., Piccioni, G., Reess, J.M., Henry, F., 2008. $S_3$ and $S_4$ absorption cross sections in the range of 340–600 nm and evaluation of the $S_3$ abundance in the lower atmosphere of Venus. *Geophys. Res.* 113, E00B07.

Marcq, E., Belyaev, D., Montmessin, F., Fedorova, A., Bertaux, J.L., Vandaele, A.C., Neefs, E., 2011. An investigation of the SO2 content of the Venusian mesosphere using SPICAV-UV in nadir mode. *Icarus* 211, 58–69.

Marcq, E., Bertaux, J.L., Montmessin, F., Belyaev, D., 2013. Variations of sulphur dioxide at the cloud top of Venus's dynamic atmosphere. *Nat. Geosci.* 6, 25–28.

Marcq, E., Lellouch, E., Encrenaz, Th., Widemann, Th., Birlan, M., Bertaux, J.L., 2015. Search for horizontal and vertical variations of CO in the day and night side lower mesosphere of Venus from CSHELL/IRTF 4.53 μm observations. *Planet. Space Sci.* 113–114, 256–263.

Marrero, T.R., Mason, E.A., 1972. Gaseous diffusion coefficients. *J. Phys. Chem. Ref. Data* 1, 3–118.

Marten, A., Hidayat, T., Biraud, Y., Moreno, R., 2002. New millimeter heterodyne observations of Titan: Vertical distribution of nitriles HCN, $HC_3N$, $CH_3CN$, and the isotopic ratio $^{15}N/^{14}N$ in its atmosphere. *Icarus* 158, 532–544.

Marty, B., et al., 2010. Nitrogen isotopes in the recent solar wind from the analysis of Genesis targets: Evidence for large scale isotope heterogeneityin the early Solar System. *Geochim. Cosmochim. Acta* 74, 340–355.

Masunaga, K., et al., 2013. Dependence of $O^+$ escape rate from the Venusian upper atmosphere on IMF directions. *Geophys. Res. Lett.* 40, 1682–1685.

McCleese, D.J., et al., 2010. Structure and dynamics of the Martian lower and middle atmosphere as observed by the Mars Climate Sounder: Seasonal variations in zonal mean temperature, dust, and water ice aerosols. *J. Geophys. Res.* 115, E12016.

McElroy, M.B., 1972. Mars: An evolving atmosphere. *Science* 175, 443–445.

McElroy, M.B., Donahue, T.M., 1972. Stability of the Martian atmosphere. *Science* 177, 986–988.

McElroy, M.B., Yung, Y.L., Nier, A.O., 1976. Isotopic composition of nitrogen: Implications for the past history of Mars' atmosphere. *Science* 194, 70–72.

McElroy, M.B., Prather, M.J., Rodriguez, J.M., 1982. Escape of hydrogen from Venus. *Science* 215, 1614–1615.

McEwan, M.J., Anicich, V.G., 2007. Titan's ion chemistry: A laboratory perspective. *Mass Spectrom. Rev.* 26, 281–319.

McGrath, M.A., et al., 1998. The ultraviolet albedo of Titan. *Icarus* 131, 382–392.

McKinnon, W.B., 2010. Radiogenic argon release from Titan: Sources, efficiency, and role of the ocean. AGU Fall Meeting, abstract P22A-01.

McKinnon, W.B., Zahnle, K.J., Ivanov, B.A., Melosh, H.J., 1997. Cratering on Venus: Models and observations. In: *Venus II*, Bougher, S.W., Hunten, D.M., Phillips, R.J. (Eds.), University of Arizona Press, Tucson, pp. 969–1014.

McKinnon, W.B., et al., 2017. Origin of the Pluto–Charon system: Constraints from the New Horizons flyby. *Icarus* 287, 2–11.

McLean, I.S., et al., 1998. The design and development of NIRSPEC: A near-infrared echelle spectrograph for the Keck II telescope. *SPIE* 3354, 566–578.

McNutt, R.L., 1989. Models of Pluto's upper atmosphere. *Geophys. Res. Lett.* 16, 1225–1228.

Mehr, F.J., Biondi, M.A., 1969. Electron temperature dependence of recombination of $O_2^+$ and $N_2^+$ ions with electrons. *Phys. Rev.* 181, 264–270.

Meier, R.R., 1991. Ultraviolet spectroscopy and remote sensing of the upper atmosphere. *Space Sci. Rev.* 58, 1–185.

Meier, R.R., Anderson, D.E., Stewart, A.I.F., 1983. Atomic oxygen emissions observed from Pioneer Venus. *Geophys. Res. Lett.* 10, 214–217.

Melosh, H.J., Vickery, A.M., 1989. Impact erosion of the primordial atmosphere of Mars. *Nature* 338, 487–489.

Merlin, F., et al., 2010. Chemical and physical properties of the variegated Pluto and Charon surfaces. *Icarus* 210, 930–940.

Merlivat, L., Nief, G., 1967. Fractionnement isotopique lors des changements d'e´tats solide-vapeur et liquide-vapeur de l'eau à des températures inférieures à 0°C, *Tellus* 19(1), 122– 127.

Migliorini, A., Grassi, D., Montabone, L., Lebonnois, S., Drossart, P., Piccioni, G., 2012. Investigation of air temperature on the nightside of Venus derived from VIRTIS-H on board Venus Express. *Icarus* 217, 640–647.

Migliorini, A., et al., 2013. The characteristics of the $O_2$ Herzberg II and Chamberlain bands observed with VIRTIS/Venus Express. *Icarus* 223, 609–614.

Miller, C.E., Yung, Y.L., 2000. Photo-induced isotope fractionation. *J. Geophys. Res.* D105, 29,039–29,051.

Millis, R.L., Wasserman, L.H., Franz, O.G., Nye, R.A., Elliot, J.L., Dunham, E.W., Bosh, A.S., Young, L.A., Slivan, S.M., Gilmore, A.C., 1993. Pluto's radius and atmosphere: Results from the entire 9 June 1988 occultation data set. *Icarus* 105, 282–297.

Mills, F.P., 1998. I. Observations and photochemical modeling of the Venus middle atmosphere. II. Thermal infrared spectroscopy of Europa and Callisto. PhD thesis, California Institute of Technology.

Mills, F.P., Allen, M., 2007. A review of selected issues concerning the chemistry in Venus' middle atmosphere. *Planet. Space Sci.* 55, 1729–1740.

Mischna, M.A., Allen, M., Richardson, M.I., et al., 2011. Atmospheric modeling of Mars methane surface release. *Planet. Space Sci.* 59, 227–237.

Mishchenko, M.I., 2000. Calculation of the amplitude matrix for a nonspherical particle in a fixed orientation. *Appl. Opt.* 39, 1026–1031.

Mishchenko, M.I., 2014. *Electromagnetic Scattering by Particles and Particle Groups: An Introduction*. Cambridge University Press, Cambridge, UK.

Mishra, A., et al., 2014. Revisited modeling of Titan's middle atmosphere electrical conductivity. *Icarus* 238, 230–234.

Mitrofanov, I.G., et al., 2007. Water ice permafrost on Mars: Layering structure and subsurface distribution according to HEND/Odyssey and MOLA/MGS data. *Geophys. Res. Lett.* 34, L18102.

Moinelo, A.C., et al., 2016. No statistical evidence of lightning in Venus night-side atmosphere from VIRTIS-Venus Express visible observations. *Icarus* 277, 395–400.

Molina-Cuberos, G.J., et al., 1999. Ionization by cosmic rays of the atmosphere of Titan. *Planet. Space Sci.* 47, 1347–1354.

Molina-Cuberos, G.J., et al., 2001. Ionospheric layer induced by meteoric ionization in Titan's atmosphere. *Planet. Space Sci.* 49, 143–153.

Molter, E.M., et al., 2016. ALMA observations of HCN and its isotopologues on Titan. *Astron. J.* 152, 42–49.

Montmessin, F., Fouchet, T., Forget, F., 2005. Modeling the annual cycle of HDO in the Martian atmosphere. *J. Geophys. Res.* 110, E03006.

Montmessin, F., et al., 2011. A layer of ozone detected in the nightside upper atmosphere of Venus. *Icarus* 216, 82–85.

Moreau, D., Esposito, L.W., Brasseur, G., 1991. The chemical composition of the dust-free Martian atmosphere: Preliminary results of a two-dimensional model. *J. Geophys. Res.* 96, 7933–7945.

Moreno, R., et al., 2012. The abundance, vertical distribution and origin of $H_2O$ in Titan's atmosphere: Hershel observations and photochemical modeling. *Icarus* 221, 753–767.

Moroz, V.I., 1964. New observations of Venus infrared spectrum (1.2–3.8 μm). *Astron. Zh.* 41, 711–715.

Moroz, V.I., Golovin, Yu.M., Moshkin, B.E., Ekonomov, A.P., 1981. Spectrophotometric experiment on the Venera 11 and 12 descent probes. 3. Results of the spectrophotometric measurements. *Cosmic Res.* 19, 599–612.

Moroz, V.I., et al., 1990. Water vapor and sulfur dioxide abundances at the Venus cloud tops from the Venera 15 infrared spectrometry data. *Adv. Space Res.* 10(5), 77–81.

Moudden, Y., 2007. Simulated seasonal variations of hydrogen peroxide in the atmosphere of Mars. *Planet. Space Sci.* 55, 2137–2143.

Moudden, Y., McConnell, J.C., 2007. Three-dimensional on-line modeling in a Mars general circulation model. *Icarus* 188, 18–34.

Mumma, M.J., Morgan, H.D., Mentall, J.E., 1975. Reduced absorption of the nonthermal CO ($A^1\Pi - X^1\Sigma$) fourth positive group by thermal CO and implications for the Mars upper atmosphere. *J. Geophys. Res.* 80, 168–172.

Mumma, M.J., Novak, R.E., DiSanti, M.A., Bonev, B.P., 2003. A sensitive search for methane on Mars. *Bull. Am. Astron. Soc.* 35, 937.

Mumma, M J., Villanueva, G.L., Novak, R.E., et al., 2009. Strong release of methane on Mars in northern summer 2003, *Science* 323, 1041–1045.

Munro, J.J., Harrison, S., Fujimoto, M.M., Tennyson, J., 2012. A dissociative electron attachment cross-section estimator. *J. Phys. Conf. Ser.* 388, 012013.

Murphy, D.M., Koop, T., 2005. Review of the vapour pressures of ice and supercooled water for atmospheric applications. *Q. J. R. Meteorol. Soc.* 131, 1539–1565.

Na, C.Y., Esposito, L.W., McClintock, W.E., Barth, C.A., 1994. Sulfur dioxide in the atmosphere of Venus: Modeling results. *Icarus* 112, 389–395.

Nagy, A.F., Cravens, T.E., 1988. Hot oxygen atoms in the upper atmospheres of Venus and Mars. *Geophys. Res. Lett.* 15, 433–435.

Nair, H., Allen, M., Anbar, A.D., Yung, Y.L., 1994. A photochemical model of the Martian atmosphere. *Icarus* 111, 124–150.

Nair, H., Summers, M.E., Miller, C.E., Yung, Y.L., 2005. Isotopic fractionation of methane in the Martian atmosphere. *Icarus* 175, 32–35.

Niemann, H.B., Kasprzak, W.T., Hedin, A.E., Hunten, D.M., Spencer, N.W., 1980. Mass spectrometric measurements of the neutral gas composition of the thermosphere and exosphere of Venus. *J. Geophys. Res.* 85, 7817–7827.

Niemann, H.B., et al., 2010. Composition of Titan's lower atmosphere and simple surface volatiles as measured by the Cassini-Huygens probe gas chromatograph mass spectrometer experiment. *J. Geophys. Res.* 115E, E12006.

Nier, A.O., McElroy, M.B., 1977. Composition and structure of Mars' upper atmosphère: Results from the neutral mass spectrometers on Viking 1 and 2. *J. Geophys. Res.* 82, 4341–4348.

Niles, P.B., Boynton, W.V., Hoffman, J.H., Ming, D.W., Hamara, D., 2010. Stable isotope measurements of Martian atmospheric $CO_2$ at the Phoenix landing site. *Science* 329, 1334–1337.

Nixon, C.A., et al., 2008a. The $^{12}C/^{13}C$ isotopic ratio in Titan hydrocarbons from Cassini/CIRS infrared spectra. *Icarus* 195, 778–791.

Nixon, C.A., et al., 2008b. Isotopic ratios in Titan's atmosphere from Cassini CIRS limb sounding: $CO_2$ at low and midlatitudes. *Astrophys. J. Lett.* 681, L109.

Nixon, C.A., et al., 2009. Infrared limb sounding of Titan with the Cassini composite infrared spectrometer: Effects of the mid-IR detector spatial responses. *Appl. Opt.* 48, 1912.

Nixon, C.A., et al., 2012. Isotopic ratios in Titan's methane: Measurements and modeling. *Astrophys. J.* 759, 159–174.

Nixon, C.A., et al., 2013. Detection of propylene in Titan's stratosphere. *Astrophys. J.* 776, L14.

Noland, M., et al., 1974. Six-color photometry of Iapetus, Titan, Rhea, Dione, and Tethys. *Icarus* 23, 334–354.

Norton, R.H., Beer, R., 1976. New apodizing functions for Fourier spectrometry. *J. Opt. Soc. Am.* 66, 259–264.

Noxon, J.F., Traub, W.A., Carleton, N.P., et al., 1976. Detection of $O_2$ dayglow emission from Mars and the Martian ozone abundance. *Astrophys. J.* 207, 1025–1030.

Ohtsuki, S., Iwagami, N., Sagawa, H., Ueno, M., Kasaba, Y., Imamura, T., Yanagisawa, K., Nishihara, E., 2008. Distribution of the Venus 1.27-μm $O_2$ airglow and rotational temperature. Planet. *Space Sci.* 56, 1391–1398.

Olkin, C.B., et al., 1997. The thermal structure of Triton's atmosphere: Results from the 1993 and 1995 occultations. *Icarus* 129, 178–201.

Olkin, C.B., et al., 2014. Pluto's atmospheric structure from the July 2007 stellar occultation. *Icarus* 239, 15–22.

Oschlisniok, J., et al., 2012. Microwave absorptivity by sulfuric acid in the Venus atmosphere: First results from the Venus Express radio science experiment VeRa. *Icarus* 221, 940–948.

Owen, T., 1964. A determination of the Martian $CO_2$ abundance. *Comm. Lunar Planet. Lab.* 2, 133.

Owen, T., Biemann, K., Rushneck, D.R., Biller, J.E., Homarth, D.W., Lafleur, A.L., 1977. The composition of the atmosphere at the surface of Mars. *J. Geophys. Res.* 82, 4635–4639.

Owen, T., Maillard, J.P., de Bergh, C., Lutz, B.L., 1988. Deuterium on Mars: The abundance of HDO and the value of D/H. *Science* 240, 1767–1771.

Owen, T.C., Roush, T.L., Cruikshank, D.P., et al., 1993. Surface ices and atmospheric composition of Pluto. *Science* 261, 745–748.

Oyama, V.I., et al., 1980. Pioneer Venus gas chromatography in the lower atmosphere of Venus. *J. Geophys. Res.* 85, 7891–7902.

Oze, C., Sharma, M., 2005. Have olivine, will gas: Serpentinization and the abiogenic production of methane on Mars. *Geophys. Res. Lett.* 32, L10203. doi:10.1029/2005GL022691.

Palmer, K.F., Williams, D., 1975. Optical constants of sulfuric acid: Applications to the cloud of Venus? *Appl. Opt.* 14, 208–219.

Pankine, A.A., et al., 2013. Retrievals of Martian atmospheric opacities from MGS TES nighttime data. *Icarus* 226, 708–722.

Parish, H.F., Schubert, G., Covey, C., Walterscheid, R.L., Grossman, A., Lebonnois, S., 2011. Decadal variations in a Venus general circulation model. *Icarus* 212, 42–65.

Parkinson, C.D., et al., 2015. Distribution of sulphuric acid aerosols in the clouds and upper haze of Venus using Venus Express VAST and VeRa temperature profiles. *Planet. Space Sci.* 113–114, 205–218.

Parkinson, T.D., Hunten, D.M. 1972. Spectroscopy and aeronomy of $O_2$ on Mars. *J. Atmos. Sci.* 29, 1380–1390.

Parkinson, W.H., Rufus, J., Yoshino, K., 2003. Absolute absorption cross section measurements of $CO_2$ in the wavelength region 163–200 nm and the temperature dependence. *Chem. Phys.* 290, 251–256.

Patsaeva, M.V., Khatuntsev, I.V., Patsaev, D.V., Titov, D.V., Ignatiev, N.I., Markiewicz, W.J., Rodin, A.V., 2015. The relationship between mesoscale circulation and cloud morphology at the upper cloud level of Venus. *Planet. Space Sci.* 113, 100–108.

Pätzold, M., Tellemann, S., Häusler, B., Bird, M.K., Tyler, G.L., Cristou, A.A., Withers, P., 2009. A sporadic layer in the Venus lower ionosphere of meteoric origin. *Geophys. Res. Lett.* 36, L05203.

Paxton, L.J., 1985. Pioneer Venus Orbiter Ultraviolet Spectrometer limb observations: Analysis and interpretation of the 166- and 156-nm data. *J. Geophys. Res.* 90, 5089–5096.

Paxton, L.J., Meier, R.R., 1986. Reanalysis of Pioneer orbiter ultraviolet spectrometer data OI 1304 intensities and atomic oxygen densities. *Geophys. Res. Lett.* 13, 229–232.

Paxton, L.J., Anderson, D.E., Stewart, A.I.F., 1985. The Pioneer Venus Orbiter ultraviolet spectrometer experiment: Analysis of hydrogen Lyman alpha data. *Adv. Space Res.* 5, 129–132.

Paxton, L.J., Anderson, D.E., Stewart, A.I.F., 1988. Pioneer Venus Orbiter ultraviolet spectrometer Lyman alpha data from near the subsolar region, 1988. *J. Geophys. Res.* 93, 1766–1772.

Penteado, P.F., et al., 2005. Measurements of $CH_3D$ and $CH_4$ in Titan from infrared spectroscopy. *Astrophys. J.* 629, L53–L56.

Penz, T., Lammer, H., Kulikov, Yu.N., Biernat, H.K., 2005. The influence of the solar particle and radiation environment on Titan's atmosphere evolution. *Adv. Space Res.* 36, 241–250.

Pereira, R.A., Baulch, D.L., Pilling, M.J., Robertson, S.H., Zeng, G., 1997. Temperature and pressure dependence of the multichannel rate coefficients for the CH3þOH system. *Journal of Physical Chemistry A* 101, 9681–9693.

Perrier, S., Bertaux, J.L., Lefèvre, F., et al., 2006. Global distribution of total ozone on Mars from SPICAM/MEX UV measurements, *J. Geophys. Res.* 111, E09S06.

Person, M.J., et al., 2008. Waves in Pluto's upper atmosphere. *Astron. J.* 136, 1510–1518.

Pertignani, A., van der Zande, W., Cosby, P.C., Hellberg, F., Thomas, R.D., Larsson, M., 2005. Vibrationally resolved rate coefficients and branching fractions in the dissociative recombination of $O_2^+$. *J. Chem. Phys.* 122, 014302.

Petrova, E.V., 2018. Glory of Venus and selection among the unknown UV absorber. *Icarus* 306, 163–170.

Petryanov, I.V., et al., 1981. Iron in the Venus clouds. *Dokl. Akad. Nauk SSSR* 260, 834–840.

Piccialli, A., Montmessin, F., Belyaev, D., Mahieux, A., Fedorova, A., Marcq, E., Bertaux, J.L., Vandaele, A.C., Korablev, O., 2015. Thermal structure of Venus upper aymosphere measured by stellar occultations with SPICAV/Venus Express. *Planet. Space Sci.* 113–114, 322–336.

Piccioni, G., et al., 2008. First detection of hydroxyl in the atmosphere of Venus. *Astron. Astrophys.* 483, L29–L33.

Piccioni, G., et al., 2009. Near-IR oxygen nightglow observed by VIRTIS in the Venus upper atmosphere. *J. Geophys. Res.* 114, E00B38.

Pieters, C.M., et al., 1986. The color of the surface of Venus. *Science* 234, 1379–1380.

Plaut, J.J., et al., 2007. Subsurface radar sounding of the south polar layered deposits of Mars. *Science* 316, 92–95.

Pollack, J.B., et al., 1977. Properties of aerosols in the Martian atmosphere, as inferred from Viking lander imaging data. *J. Geophys. Res.* 82, 4479–4496.

Pollack, J.B., et al., 1978. Properties of the clouds of Venus as inferred from airborne observations of its near infrared reflectivity spectrum. *Icarus* 34, 28–45.

Pollack, J.B., et al., 1979. Properties and effects of dust particles suspended in the Martian atmosphere. *J. Geophys. Res.* 84, 2929–2945.

Pollack, J.B., Schwartz, J.M., Rages, K., 1990. Scatterers in Triton's atmosphere: implications for the seasonal volatile cycle. *Science* 250, 440–443.

Pollack, J.B., et al., 1993. Near-infrared light from Venus' nightside: A spectroscopic analysis. *Icarus* 103, 1–42.

Poppe, A.R., 2015. Interplanetary dust influx to the Pluto–Charon system. *Icarus* 246, 352–359.

Porshnev, N.V., et al., 1987. Gas chromatographic analysis of products of thermal reactions of the cloud aerosol of Venus by the Vega 1 and 2 probes. *Cosmic Res.* 25, 715–720.

Price, C., Penner, J., Prather, M., 1997. NO*x* from lightning. 1. Global distribution based on lightning physics. *J. Geophys. Res.* 102, 5929–5941.

Prinn, R.G., 1971. Photochemistry of HCl and other minor constituents in the atmosphere of Venus. *J. Atmos. Sci.* 28, 1058–1068.

Quirico, E., Doute, S., Schmitt, B., de Bergh, C., Cruikshank, D.P., Owen, T.C., Geballe, T.R., Roush, T.L., 1999. Composition, physical state, and distribution of ices at the surface of Triton. *Icarus* 139, 159–178.

Ragent, B., Esposito, L.W., Tomasko, M.G., Marov, M.Ya., Shari, V.P., Lebedev, V.N., 1985. Particulate matter in the Venus atmosphere. *Adv. Space Res.* 5(11), 85–115.

Rages, K., Pollack, J.B., 1980. Titan aerosol: Optical properties and vertical distribution. *Icarus* 41, 119–130.
Rages, K., Pollack, J.B., 1983. Vertical distribution of scattering hazes in Titan's upper atmosphere. *Icarus* 55, 50–60.
Rages, K., Pollack, J.B., 1992. Voyager imaging of Triton's clouds and hazes. *Icarus* 99, 289–295.
Rannou, P., McKay, C.P., Botet, R., Cabane, M., 1999. Semi-empirical modelof absorption and scattering by isotropic fractal aggregates of spheres. *Planet. Space Sci.* 47, 385–396.
Rannou, P., Cours, T., Le Mouélic, S., Rodriguez, S., Sotin, C., Drossart, P., Brown, R., 2010. Titan haze distribution and optical properties retrieved from recent observations. *Icarus* 208, 850–867.
Rayner, J.T., Cushing, M.C., Vacca, W.D., 2009. The Infrared Telescope Facility (IRTF) spectral library: Cool stars. *Astrophys. J. Suppl. Series* 185, 289–432.
Rengel, M., et al., 2014. Hershel/PACS spectroscopy of trace gases of the stratosphere of Titan. *Astron. Astrophys.* 561, A4.
Richards, P.G., Fennelly, J.A., Torr, D.G., 1994. EUVAC: A solar EUV flux model for aeronomic calculations. *J. Geophys. Res.* 99, 8981–8992.
Robertson, I.P., et al., 2009. Structure of Titan's ionosphere: Model comparison with Cassini data. *Planet. Space Sci.* 57, 1834–1846.
Rodin, A., et al., 2014. High resolution heterodyne spectroscopy of the atmospheric methane NIR absorption. *Opt. Express* 22, 13825–13834.
Rothman, L.S., et al., 2013. The HITRAN 2012 molecular spectroscopic database. *J. Quant. Spec. Rad. Trans.* 130, 4–50.
Rousselot, P., et al., 2014. Toward a unique nitrogen isotopic ratio in cometary ices. *Astrophys. J. Lett.* 780, L17.
Royer, E.M., et al., 2016. Cassini UVIS observations of Titan ultraviolet airglow intensity dependence with solar zenith angle: Titan ultraviolet airglow variations. *Geophys. Res. Lett.* 44, 88–96.
Russell, C.T., Strangeway, R.J., Daniels, J.T.M., Zhang, T.L., Wei, H.Y., 2011. Venus lightning: Comparison with terrestrial lightning. *Planet. Space Sci.* 59, 965–973.
Rustad, D.S., Gregory, N.W., 1983. Vapor pressure of iron (III) chloride. *J. Chem. Eng. Data* 28, 151–155.
Safronov, V.S., 1969. Evolution of the Protoplanetary Cloud and Formation of the Earth and Planets [in Russian]. NASA TT-F-677, Nauka, Moscow.
Sagdeev, R.Z., et al., 1986. Overview of Vega Venus balloon in situ meteorological measurements. *Science* 231, 1411–1414.
Samuelson, R.E., Hanel, R.A., Kunde, V.G., Maguire, W.C., 1981. Mean molecular weight and hydrogen abundance of Titan's atmosphere. *Nature* 292, 688–693.
Sandel, B.R., et al., 2015. Altitude profiles of $O_2$ on Mars from SPICAM stellar occultations. *Icarus* 252, 154–160.
Sander, S.P., et al., 2011. Chemical Kinetics and Photochemical Data for Use in Atmospheric Studies. Evaluation 17, JPL Publication 10-6.
Sandor, B.J., Clancy, R.T., 2005. Water vapor variations in the Venus mesosphere from microwave spectra. *Icarus* 177, 129–143.
Sandor, B.J., Clancy, R.T., 2012. Observations of HCl altitude dependence and temporal variation in the 70–100 km mesosphere of Venus. *Icarus* 220, 618–626.
Sandor, B.J., Clancy, R.T., 2018. First measurements of ClO in the Venus atmosphere – Altitude dependence and temporal variation. *Icarus* 313, 15–24.
Sandor, B.J., Clancy, R.T., Moriarty-Schieven, G., Mills, F.P., 2010. Sulfur chemistry in the Venus mesosphere from $SO_2$ and SO microwave spectra. *Icarus* 208, 49–60.

Sandor, B.J., Clancy, R.T., Moriarty-Schieven, G., 2012. Upper limits for $H_2SO_4$ in the measosphere of Venus. *Icarus* 217, 839–844.

Sanko, N.F., 1980. Gaseous sulfur in the atmosphere of Venus. *Cosmic Res.* 18, 600–605.

Schindhelm, E., Stern, S.A., Gladstone, R., Zangari, A., 2015. Pluto and Charon's UV spectra from IUE to New Horizons. *Icarus* 246, 206–212.

Schinke, R., 1995. *Photodissociation Dynamics*. Cambridge University Press, Cambridge, UK.

Schmitt, B., et al., 2017. Physical state and distribution of materials at the surface of Pluto from New Horizons LEISA imaging spectrometer. *Icarus* 287, 229–260.

Schneider, N.M., et al., 2015. Discovery of diffuse aurora on Mars. *Science* 350, 0313.

Sebree, J.A., et al., 2016. $^{13}C$ and $^{15}N$ fractionation of $CH_4/N_2$ mixtures during photochemical aerosol formation: Relevance to Titan. *Icarus* 270, 421–428.

Seiersen, K, et al., 2003. Dissociative recombination of the cation and dication of $CO_2$. *Phys. Rev. A* 68(2), 022708.

Seiff, A., Schofield, J.T., Kliore, A., Taylor, F.W., Limaye, S.S., Revercomb, H.E., Sromovsky, L.A., Kerzhanovich, V.V., Moroz, V.I., Marov, M.Ya., 1985. Models of the structure of the atmosphere of Venus from the surface to 100 kilometers altitude. *Adv. Space Res.* 5(11), 3–58.

Serigano, J., et al., 2016. Isotopic ratios of carbon and oxygen in Titan's CO using ALMA. *Astrophys. J. Lett.* 821, L8.

Shaw, B.M., Lovell, R.J., 1969. Foreign-gas broadening of HF by CO2. *J. Opt. Soc, Am* 59, 1598–1601.

Shebanits, O., et al., 2013. Negative ion densities in the ionosphere of Titan: Cassini RPWS/LP results. *Planet. Space Sci.* 84, 153–162.

Shinnaka, Y., et al., 2014. $^{14}NH_2/^{15}NH_2$ ratio in comet C/2012 S1 (ISON) observed during its outburst in 2013 Novermber. *Astrophys. J. Lett.* 782, L16.

Sicardy, B., Widemann, T., Lellouch, E., et al., 2003. Large changes in Pluto's atmosphere as revealed by recent stellar occultations. *Nature* 424, 168–170.

Sicardy, B., Talbot, J., Meza, E., et al., 2016. Pluto's atmosphere from the 2015 June 29 ground-based stellar occultation at the time of the new horizons flyby. *Astrophys. J.* 819, L38 .

Sill, G.T., 1972. Sulfuric acid in the Venus clouds. *Comm. Lunar Planet. Lab.* 9, 191–198.

Sindoni, G., Formisano, V., Geminale, A., 2011. Observations of water vapour and carbon monoxide in the Martian atmosphere with the SWC of PFS/MEX. *Planet. Space Sci.* 59, 149–162.

Skrzypkowski, M.P., Gougousi, T., Johnsen, R., Golde, M.F., 1998. Measurement of the absolute yield of $CO(a^3\Pi)$ + O products in the dissociative recombination of $CO_2^+$ ions with electrons. *J. Chem. Phys.* 108, 8400–8407.

Slanger, T.G., Black,G., 1978. The $O_2(C^3\Delta_u \rightarrow a^1\Delta_g)$ bands in the nightglow spectrum of Venus. *Geophys. Res. Lett.* 5, 947–948.

Slanger, T.G., Huestis, D.L., Cosby, P.C., Chanover, N.J., Bida, T.A., 2006. The Venus nightglow ground-based observations and chemical mechanisms. *Icarus* 182, 1–9.

Smith, G.P., Robertson, R., 2008. Temperature dependence of oxygen atom recombination in nitrogen after ozone photolysis. *Chem. Phys. Lett.* 458, 6–10.

Smith, G.R., Strobel, D.F., Broadfoot, A.L., Sandel, B.R., Shemansky, D.F., Holberg, J.B., 1982. Titan's upper atmosphere: Composition and temperature from the EUV solar occultation results. *J. Geophys. Res.* 87, 1351–1359.

Smith, I.W.M., 1984. The role of electronically excited states in recombination reactions. *Int. J. Chem. Kinet.* 16, 423–443.

Smith, M.D., 2004. Interannual variability in TES atmospheric observations of Mars during 1999–2003. *Icarus* 167, 148–165.

Smith, M.D., 2009. THEMIS observations of Mars aerosol optical depth from 2002–2008. *Icarus* 202, 444–452.

Smith, M.D., Pearl, J.C., Conrath, B.J., Christensen, P.R., 2000. Mars Global Surveyor Thermal Emission Spectrometer (TES) observations of dust opacity during aerobraking and science phasing. *J. Geophys. Res.* 105, 9539–9552.

Smith, M.D., et al., 2006. One Martian year of atmospheric observations using MER Mini-TES. *J. Geophys. Res.* 111, E12S13.

Smith, M.D., et al., 2009. Compact Reconnaissance Imaging Spectrometer observations of water vapor and carbon monoxide. *J. Geophys. Res.* 114, E00D03.

Smith, M.D., et al., 2018. The climatology of carbon monoxide and water vapor on Mars as observed by CRISM and modeled by the GEM-Mars general circulation model. *Icarus* 301, 117–131.

Sonnabend, G., Sornig, M., Kroetz, P., Stupar, D., Schieder, R., 2008. Ultra high spectral resolution observations of planetary atmospheres using the Cologne tunable heterodyne infrared spectrometer. *J. Quant. Spec. Rad. Transfer* 109, 1016–1029.

Soret, L., Gerard, J.C., 2015. Is the $O_2(a^1\Delta_g)$ Venus nightglow emission controlled by solar activity? *Icarus* 262, 170–172.

Soret, L., et al., 2012a. Atomic oxygen on the Venus nightside: Global distribution deduced from airglow mapping. *Icarus* 217, 849–855.

Soret, L., Gerard, J.C., Piccioni, G., Drossart,P., 2012b. The OH Venus nightglow spectrum: Intensity and vibrational composition from VIRTIS Venus Express observations. *Planet. Space Sci.* 73, 387–396.

Soret, L., et al., 2016. SPICAM observations and modeling of Mars aurorae. *Icarus* 264, 398–406.

Spencer, J.R., Moore, J.M., 1992. The influence of thermal inertia on temperatures and frost stability on Triton. *Icarus* 99, 261–272.

Spenner, K., Knudsen, W.C., Lotze, W., 1996. Suprathermal electron fluxes in the Venus nightside ionosphere at moderate and high solar activity. *J. Geophys. Res.* 101, 4557–4563.

Spinrad, H., Münch, G., Kaplan, L.D., 1963. The detection of water vapor on Mars, *Astrophys. J.* 137, 1319–1321.

Sprague, A.L., et al., 2012. Interannual similarity and variation in seasonal circulation of Mars' atmospheric Ar as seen by the Gamma Ray Spectrometer on Mars Odyssey. *J. Geophys. Res.* 117, E04005.

Stenberg, G., et al., 2011. Observational evidence of alpha-particle capture at Mars. *Geophys. Res. Lett.* 38, L09101.

Stenberg, G., et al., 2015. Proton and alpha particle precipitation onto the upper atmosphere of Venus. *Planet. Space Sci.* 113–114, 369–377.

Stern, S.A., Cunningham, N.J., Hain, M.J., Spencer, J.R., Shinn, A., 2012. First ultraviolet reflectance spectra of Pluto and Charon by the Hubble Space Telescope Cosmic Origins Spectrograph: Absorption features and evidence for temporal change. *Astron. J.* 143(1), article 22.

Stern, S.A., et al., 2017a. Past epochs of significantly higher pressure atmospheres on Pluto. *Icarus* 287, 47–53.

Stern, S.A., et al., 2017b. New Horizons constraints on Charon's present day atmosphere. *Icarus* 287, 124–130.

Stevens, M.H., Strobel, D.F., Summers, M.E., 1992. On the thermal structure of Triton's thermosphere. *Geophys. Res. Lett.* 19, 669–672.

Stevens, M.H., et al., 2015. Molecular nitrogen and methane density retrievals from Cassini UVIS dayglow observations of Titan's upper atmosphere. *Icarus* 247, 301–312.

Stewart, A.I.F., 1972. Mariner 6 and 7 ultraviolet spectrometer experiment: Implications of $CO_2^+$, CO and O airglow. *J. Geophys. Res.* 77, 54–60.

Stewart, A.I.F., Barth, C.A., 1979. Ultraviolet night airglow of Venus. *Science* 205, 59–62.

Stewart, A.I.F., et al., 1980. Morphology of the Venus ultraviolet night airglow. *J. Geophys. Res.* 85(A13), 7861–7870.

Stewart, A.I.F., et al., 1992. Atomic oxygen in the Martian thermosphere. *J. Geophys. Res.* 97, 91–102.

Stiepen, A., et al., 2013. Venus nitric oxide nightglow mapping from SPICAV nadir observations. *Icarus* 226(1), 428–436.

Stiepen, A., et al., 2015. Ten years of Martian nitric oxide nightglow observations. *Geophys. Res. Lett.* 42, 720–725.

Stiepen, A., et al., 2017. Nitric oxide nightglow and Martian mesospheric circulation from MAVEN/IUVS observations and LMD-MGCM predictions. *J. Geophys. Res.* 122 (A5), 5782–5797.

Strobel, D.F., 2008. $N_2$ escape rates from Pluto's atmosphere. *Icarus* 193, 612–619.

Strobel, D.F., Summers, M.E., 1995. Triton's upper atmosphere and ionosphere. In: Cruikshank, D.P. (Ed.), *Neptune and Triton*, University of Arizona Press, Tucson, pp 1107 1150.

Strobel, D.F., Zhu, X., 2017. Comparative planetary nitrogen atmospheres: Density and thermal structures of Pluto and Triton. *Icarus* 291, 55–64.

Strobel, D.F., Cheng, A.F., Summers, M.E., Strickland, D.J., 1990. Magnetospheric interaction with Triton's ionosphere. *Geophys. Res. Lett.* 17, 1661–1664.

Strobel, D.F., Summers, M.E., Zhu, X., 1992. Titan's upper atmosphere: Structure and ultraviolet emissions. *Icarus* 100, 512–526.

Strobel, D.F., Zhu, X., Summers, M.E., Stevens, M.H., 1996. On the vertical structure of Pluto's atmosphere. *Icarus* 120, 266–289.

Summers, M.E., Strobel, D.F., 1991. Triton's atmosphere: A source of N and H for Neptune's magnetosphere. *Geophys. Res. Lett.* 18, 2309–2312.

Summers, M.E., Strobel, D.F., Gladstone, G.R., 1997. Chemical models of Pluto's atmosphere. In: Stern, S., Tholen, D. (Eds.), *Pluto and Charon*, University of Arizona Press, Tucson, pp. 391–434.

Sung, K., Varanasi, P., 2005. $CO_2$-broadened half-widths and $CO_2$-induced line shifts of $^{12}C^{16}O$ relevant to the atmospheric spectra of Venus and Mars. *J. Quant. Spec. Rad. Transfer* 91, 319–332.

Surkov, Yu.A., et al., 1983. Elemental composition of Venus' rocks. *Cosmic Res.* 21, 308–315.

Surkov, Yu.A., et al., 1987. Chemical composition of the cloud aerosol of Venus measured by the Vega 1 mass spectrometer. *Cosmic Res.* 25, 744–750.

Sylvestre, M., et al., 2018. Seasonal evolution of $C_2N_2$, $C_3H_4$, and $C_4H_2$ abundances in Titan's lower stratosphere. *Astron. Astrophys.* 609, A64.

Sze, N.D., McElroy, M.B., 1975. Some problems in Venus aeronomy. *Planet. Space Sci.* 23, 763–780.

Takacs, P.Z. et al., 1980. Mariner 10 observations of hydrogen Lyman alpha emission from the Venus exosphere: Evidence of complex structure. *Planet. Space Sci.* 28, 687–701.

Tan, S.P., Kargel, J.S., 2018. Solid-phase equilibria on Pluto's surface. *MNRAS* 474, 4254–4263.

Tanguy, L., Bézard, B., Marten, A., Gautier, D., Gérard, E., Paubert, G., Lecacheux, A., 1990. The stratospheric profile of HCN on Titan from millimeter observations. *Icarus* 85, 43–57.

Taylor, F.W., Crisp, D., Bezard, B., 1997. Near-infrared sounding of the lower atmosphere of Venus. In: Bougher, S.W., Hunten, D.M., Phillips, R.J. (Eds.), *Venus II*. University of Arizona Press, Tucson, pp. 325–352.

Taylor, H.A., Brinton, H.C., Bauer, S.J., Hartle, R.E., Cloutier, P.A., Daniell, R.E., 1980. Global observations of the composition and dynamics of the ionosphere of Venus: Implications for the solar wind interaction. *J. Geophys. Res.* 85, 7765–7777.

Teanby, N.A., et al., 2009. Titan's stratospheric C2N2, C3H4, and C4H2 abundances from Cassini/CIRS far-infrared spectra. *Icarus* 202, 620–631.

Teanby, N.A., et al., 2013. Constraints on Titan's middle atmosphere ammonia from Hershel/SPIRE submillimeter spectra. *Planet. Space Sci.* 75, 136–147.

Tellmann, S., Pätzold, M., Häusler, B., Bird, M.K., Tyler, G.L., 2009. Structure of the Venus neutral atmosphere as observed by the Radio Science experiment VeRa on Venus Express. *J. Geophys. Res.* 114, E00B36.

Teolis, B.D., et al., 2015. A revised sensitivity model for Cassini INMS: Results at Titan. *Space Sci. Rev.* 190, 47–84.

Theis, R.F., Brace, L.H., Mayr, H.G., 1980. Empirical models of the electron temperature and density in the Venus ionosphere. *J. Geophys. Res.* 85, 7787–7794.

Toigo, A.D., et al., 2013. High spatial and temporal resolution sampling of Martian gas abundances from CRISM spectra. *J. Geophys. Res.* 118E, 89–104.

Tokunaga, A.T., et al., 2008. Silicon immersion grating spectrograph design for the NASA Infrared Telescope Facility. *Proc. SPIE* 7014, 70146A.

Tomasko, M.G., et al., 2005. Rain, winds and haze during the Huygens probe's descent to Titan's surface. *Nature* 438, 765–778.

Tomasko, M.G., et al., 2008a. A model of Titan's aerosols based on measurements made inside the atmosphere. *Planet. Space Sci.* 56, 669–707.

Tomasko, M.G., et al., 2008b. Heat balance in Titan's atmosphere. *Planet. Space Sci.* 56, 648–659.

Toth, R.A., et al., 2008. Spectroscopic database of CO2 line parameters: 4300–7000 $cm^{-1}$. *J. Quan. Spec. Rad. Transfer* 109, 906–921.

Toublanc, D., Parisot, J.P., Brillet, J., Gautier, D., Raulin, F., McKay, C.P., 1995. Photochemical modeling of Titan's atmosphere. *Icarus* 113, 2–26.

Trafton, L.M., Stern, S.A., 1996. Rotationally resolved spectral studies of Pluto from 2500 to 4800 °A obtained with HST. *Astron. J.* 112, 1212–1224.

Trafton, L.M., Hunten, D.M., Zahnle, K.J., McNutt, R.L., Jr., 1997. Escape processes at Pluto and Charon. In: *Pluto and Charon*, Stern, S.A., Tholen, D.J. (Eds.), University of Arizona Press, Tucson, pp. 475–522 .

Trainer, M.G., Tolbert, M.A., McKay, C.P., et al., 2011. Limits on the trapping of atmospheric $CH_4$ in Martian polar ice analogs. *Icarus* 208, 192–197.

Traub, W.A., Carleton, N.P., Connes, P., et al., 1979. The latitude variation of $O_2$ dayglow and $O_3$ abundance on Mars. *Astrophys. J.* 229, 846–850.

Trauger, J.T., Lunine, J.I., 1983. Spectroscopy of molecular oxygen in the atmospheres of Venus and Mars. *Icarus* 55, 272–281.

Trokhimovsky, A., Fedorova, A., Korablev, O., Montmessin, F., Bertaux, J.L., Rodin, A., Smith, M.D., 2015. Mars' water vapor mapping by the SPICAM IR spectrometer: Five Martian years of observations. *Icarus* 251, 50–64.

Tryka, K.A., Brown, R.H., Anicich, V., Cruikshank, D.P., Owen, T.C., 1993. Spectroscopic determination of the phase composition and temperature of nitrogen ice on Triton. *Science* 261, 751–754.

Tryka, K.A., Brown, R.H., Cruikshank, D.P., et al., 1994. Temperature of nitrogen ice on Pluto and its implications for flux measurements. *Icarus* 112, 513–527.

Tsang, C.C.C., Wilson, C.F., Barstow, J.K., Irwin, P.G.J., Taylor, F.W., McGouldrick, K., Piccioni, G., Drossart, P., Svedhem, H., 2010. Correlations between cloud thickness and sub-cloud water abundance on Venus, *Geophys. Res. Lett.*, 37, L02202.

Tyler, G.L., et al., 1989. Voyager radio science observations of Neptune and Triton. *Science* 246, 1466–1473.

Upadhyay, H.O., Singh, R.P., Singh, R.N., 1994. Cosmic ray ionization of lower Venus atmosphere. *Earth Moon Planets* 65, 89–94.

Ustinov, E.A., 1977. Inverse problem of multiple scattering theory and interpretation of measurements of radiation scattered in the Venus cloud layer. *Cosmic Res.* 15, 768–775.

Vandaele, A.C., et al., 2008. Composition of the Venus mesosphere measured by solar occultation at infrared on board Venus Express. *J. Geophys. Res.* 113, E00B23.

Vandaele, A.C., Mahieux, A., Robert, S., Drummond, R., Wilquet, V., Bertaux, J.L., 2015. Carbon monoxide short term variability observed on Venus with SOIR/VEX. *Planet. Space Sci.* 113–114, 237–255.

Vandaele, A.C., et al., 2017. Sulfur dioxide in the Venus atmosphere. II. Spatial and temporal variability. *Icarus* 295, 1–15.

Van der Hulst, H.C., 1980. *Multiple Light Scattering: Tables, Formulas, and Applications.* Academic Press, New York.

Van der Hulst, H.C., 1981. *Light Scattering by Small Particles.* Dover, New York.

Vervack, R.J., Jr., 1997. Titan's upper atmospheric structure derived from Voyager ultraviolet spectrometer observations. PhD dissertation, University of Arizona.

Vervack, R.J., Sandel, B.R., Strobel, D.F., 2004. New perspectives on Titan's upper atmosphere from a reanalysis of the Voyager 1 UVS solar occultations. *Icarus* 170, 91–112.

Villanueva, G.L., et al., 2009. A sensitive search for deuterated water in comet 8P/Tuttle. *Astrophys. J. Lett.* 690, L5–L9.

Villanueva, G.I., et al., 2013. A sensitive search for organics ($CH_4$, $CH_3OH$, $H_2CO$, $C_2H_6$, $C_2H_2$, $C_2H_4$), hydroperoxyl ($HO_2$), nitrogen compounds ($N_2O$, $NH_3$, HCN) and chlorine species (HCl, $CH_3Cl$) on Mars using ground-based high-resolution infrared spectroscopy. *Icarus* 223, 11–27.

Villanueva, G.L., et al., 2015. Strong water isotopic anomalies in the Martian atmosphere: Probing current and ancient reservoirs. *Science* 348, 218–221.

Vinatier, S., Bezard, B., Nixon, C.A., 2007. The Titan $^{14}N/^{15}N$ and $^{12}C/^{13}C$ isotopic ratios in HCN from Cassini/CIRS. *Icarus* 191, 712–721.

Vinatier, S., et al., 2010. Analysis of Cassini/CIRS limb spectra of Titan acquired during the nominal mission. I. Hydrocarbons, nitriles and $CO_2$ vertical mixing ratio profiles. *Icarus* 205, 559–570.

Vinatier, S., et al., 2012. Optical constants of Titan's stratospheric aerosols in the 70–1500 $cm^{-1}$ spectral range constrained by Cassini/CIRS observations. *Icarus* 219, 5–12.

Vinatier, S., et al., 2015. Seasonal variations in Titan's middle atmosphere during the northern spring derived from Cassini/CIRS observations. *Icarus* 250, 95–115.

Vuitton, V., Yelle, R.V., McEwan, M.J., 2007. Ion chemistry and N-containing molecules in Titan's upper atmosphere. *Icarus* 191, 722–742.

Vuitton, V., Yelle, R.V., Cui, J., 2008. Formation and distribution of benzene on Titan. *J. Geophys. Res.* 113, E05007.

Vuitton, V., et al., 2009. Negative ion chemistry in Titan's upper atmosphere. *Planet. Space Sci.* 57, 1558–1572.

Von Zahn, U., Fricke, K.H., Hunten, D.M., Krankovsky, D., Mauersberger, K, Nier, A.O., 1980. The upper atmosphere of Venus during morning conditions. *J. Geophys. Res.* 85, 7829–7840.

von Zahn, U., Kumar, S., Niemann, H., Prinn, R., 1983. Composition of the Venus atmosphere. In: *Venus*, Hunten, D.M., Colin, L., Donahue, T.M., Moroz, V.I. (Eds.), University of Arizona Press, Tucson, pp. 299–430.

Waenke, H., Dreibus, G., 1988. Chemical composition and accretion history of terrestrial planets. *Philos. Trans. R. Soc. Lonon Ser. A* 325, 545–557.

Wahlund, J.E., et al., 2009. On the amount of heavy molecular ions in Titan's ionosphere. *Planet. Space Sci.*, 57, 1857–1865.

Watson, A.J., Donahue, T.M., Walker, J.C.G., 1981. The dynamics of a rapidly escaping atmosphere: Applications to the evolution of Earth and Venus. *Icarus* 48, 150–166.

Weaver, H.A., Feldman, P.D., Combi, M.R., Krasnopolsky, V.A., Lisse, C.M., Shemansky, D.E., 2002. A search for argon and O VI in three comets using the far Ultraviolet Spectroscopic Explorer. *Astrophys. J. Lett.* 576, L95–L98.

Webster, C.R., et al., 2013. Isotope ratios of H, C, and O in $CO_2$ and $H_2O$ of the Martian atmosphere. *Science* 341, 260–263.

Webster, C.R., et al., 2015. Mars methane detection and variability at Gale crater. *Science* 347, 415–417.

Webster, C.R., et al., 2018. Background levels of methane in Mars' atmosphere show strong seasonal variations. *Science* 360, 1093–1096.

Westlake, J.H., et al., 2012. Titan's ionospheric composition and structure: Photochemical modeling of Cassini INMS data. *J. Geophys. Res.* 117, E01003.

Westlake, J.H., et al., 2014a. The role of ion-molecule reactions in the growth of heavy ions in Titan's ionosphere. *J. Geophys. Res.* 119A, 5951–5963.

Westlake, J.H., et al., 2014b. Observed decline in Titan's thermospheric methane due to solar cycle drivers. *J. Geophys. Res.* 119A, 8586–8599.

Wiens, R.C., Bochshler, P., Burnett, D.S., Wimmer-Schweingruber, R.F., 2004. Solar and solar-wind isotopic compositions. *Earth Planet. Sci. Lett.* 222, 697–712.

Willacy, K., Allen, M., Yung, Y.L., 2016. A new astrobiological model of the atmosphere of Titan. *Astrophys. J.* 829, 79–90.

Wilquet, V., Fedorova, A., Montmessin, F., Drummond, R., Mahieux, A., Vandaele, A.C., Villard, E., Korablev, O., Bertaux, J.L., 2009. Preliminary characterization of the upper haze by SPICAV/SOIR solar occultation in UV to mid-IR onboard Venus Express. *J. Geophys. Res.* 114, E00B42.

Wilson, E.H., Atreya, S.K., 2004. Current state of modeling the photochemistry of Titan's mutually dependent atmosphere and ionosphere. *J. Geophys. Res.* 109, E06002.

Wilson, W.I., Klein, M.J., Kakar, R.K., Gulkis, S., Olsen, E.T., Ho, P.T.P., 1981. Venus. I. Carbon monoxide distribution and molecular line searches. *Icarus* 45, 624–637.

Wong, A.S., Morgan, C.G., Yung, Y.L., Owen, T.C., 2002. Evolution of CO on Titan. *Icarus* 155, 382–392.

Wong, M.H., et al., 2013. Isotopes of nitrogen on Mars: Atmospheric measurements by Curiosity's mass spectrometer. *Geophys. Res. Lett.* 40, 6033–6037.

Wong, M.L., et al., 2017. The photochemistry of Pluto's atmosphere as illuminated by New Horizons. *Icarus* 287, 110–115.

Woods, T.N., et al., 1996. Validation of the UARS solar ultraviolet irradiances: Comparison with the ATLAS 1 and 2 measurements. *J. Geophys. Res.* 101, 9541–9569.

Wraight, P.C., 1982. Association of atomic oxygen and airglow excitation mechanisms. *Planet. Space Sci.* 30, 251–259.

Wu, C.Y., Phillips, E., Lee, L.C., Judge, D.L., 1978. Atomic carbon emissions from photodissociation of $CO_2$. *J. Geophys. Res.* 83, 4869–4874.

Yelle, R.V., Lunine, J.I., 1989. Evidence for a molecule heavier than methane in the atmosphere of Pluto. *Nature* 339, 288–290 .

Yelle, R.V., Lunine, J.L., Hunten, D.M., 1991. Energy balance and plume dynamics in Triton's lower atmosphere. *Icarus* 89, 347–357.

Yelle, R.V., Lunine, J.L., Pollack, J.B., Brown, R.H., 1995. Lower atmospheric structure and surface-atmosphere interaction on Triton. In: *Neptune and Triton*, Cruikshank, D.P. (Ed.), University of Arizona Press, Tucson, pp. 1031–1106.

Yelle, R.V., Borggren, N., de la Haye, V., Kasprzak, W.T., Niemann, H.B., Mueller-Wodarg, I., Waite, J.H., Jr., 2006. The vertical structure of Titan's upper atmosphere from Cassini Ion Neutral Mass Spectrometer measurements. *Icarus* 182, 567–576.

Young, A.T., 1973. Are the clouds of Venus sulfuric acid? *Icarus* 18, 564–582.

Young, E.F., et al., 2008. Vertical structure in Pluto's atmosphere from the 2006 June 12 stellar occultation. *Astron. J.* 136, 1757–1769.

Young, L.A., 2013. Pluto's seasons: New predictions for New Horizons. *Astrophys. J. Lett.* 766, L22.

Young, L.A., Elliot, J.L., Tokunaga, A., de Bergh, C., Owen, T., 1997. Detection of gaseous methane on Pluto. *Icarus* 127, 258–262.

Young, L.A., et al., 2017. Structure and composition of Pluto's atmosphere from the New Horizons solar ultraviolet occultation. *Icarus* 300, 174–199.

Young, L.D.G., 1972. High-resolution spectra of Venus. *Icarus* 17, 632–658.

Yung, Y.L., 1987. An update of nitrile photochemistry on Titan. *Icarus* 72, 468–472.

Yung, Y.L., DeMore, W.B., 1982. Photochemistry of the stratosphere of Venus: Implications for atmospheric evolution. *Icarus* 51, 199–247.

Yung, Y.L., DeMore, W.B., 1999. *Photochemistry of Planetary Atmospheres*. Oxford University Press, Oxford, UK.

Yung, Y.L., Lyons, J.R., 1990. Triton: Topside ionosphere and nitrogen escape. *Geophys. Res. Lett.* 17, 1717–1720.

Yung, Y.L., Allen, M., Pinto, J.P., 1984. Photochemistry of the atmosphere of Titan: Comparison between model and observations. *Astrophys. J. Suppl.* 55, 465–506.

Yung, Y.L., et al., 2009. Evidence for carbonyl sulfide (OCS) conversion to CO in the lower atmosphere of Venus. *J. Geophys. Res.* 114, E00B34.

Zahnle, K., Kasting, J.F., Pollack, J.B., 1990. Mass fractionation of noble gases in diffusion-limited hydrodynamic hydrogen escape. *Icarus* 84, 502–527.

Zahnle, K., Haberle, R.M., Catling, D.C., Kasting, J.F., 2008. Photochemical instability of the ancient Martian atmosphere. *J. Geophys. Res.* 113, E11004.

Zalucha, X., Zhu, A.M., Gulbis, A .A .S., Strobel, D.F., Elliot, J.L., 2011. An analysis of Pluto occultation light curves using an atmospheric radiative-conductive model. *Icarus* 211, 804–818.

Zasova, L.V., Krasnopolsky, V.A., Moroz, V.I., 1981. Vertical distribution of $SO_2$ in the upper cloud layer of Venus and origin of UV absorption. *Adv. Space Res.* 1, 13–16.

Zasova, L.V., Moroz, V.I., Esposito, L.W., Na, C.Y., 1993. $SO_2$ in the middle atmosphere of Venus: IR measurements from Venera 15 and comparison to UV data. *Icarus* 105, 92–109.

Zasova, L.V., Moroz, V.I., Formisano, V., Ignatiev, N.I., Khatuntsev, I.V., 2004. Infrared spectrometry of Venus: IR Fourier spectrometer on Venera 15 as a precursor of PFS for Venus Express. *Adv. Space Res.* 34, 1655–1667.

Zasova, L.V., Moroz, V.I., Linkin, V.M., Khatountsev, I.A., Maiorov, B.S., 2006. Structure of the Venusian atmosphere from surface up to 100 km. *Cosmic Res.* 44, 364–383.

Zhang, M.H.G., Luhmann, J.G., Kliore, A.J., 1990. An observational study of the nightside ionospheres of Mars and Venus with radio occultation methods. *J. Geophys. Res.* 95, 17095–17102.

Zhang, X., Liang, M.C., Mills, F.P., Belyaev, D.A., Yung, Y.L., 2012. Sulfur chemistry in the middle atmosphere of Venus. *Icarus* 217, 714–739.

Zhang, X., Strobel, D.F., Imanaka, H., 2017. Haze heats Pluto's atmosphere yet explains its cold temperature. *Nature* 551, 352–355.

Zhu, X., Strobel, D.F., Erwin, J.T., 2014. The density and thermal structure of Pluto's atmosphere and associated escape processes and rates. *Icarus* 228, 301–314.

Zuber, M.T., et al., 1998. Observations of the north polar cap of Mars from the Mars orbiter laser altimeter. *Science* 282, 2053–2060.

# Index

$^{14}N/^{15}N$ in the Solar System, 421
absorption of the solar UV and EUV, 22
absorption, spontaneous and stimulated emissions, 112
acousto-optical tunable filter (AOTF), 78
activation energy, 127
adiabatic lapse rate, 13
aerosol on Mars
  Mars Express, 161
  MGS/TES observations, 160
  MRO/MCS, 163
  THEMIS, 161
  Viking data, 159
air and surface temperatures, 19
airmass factor, 58
allowed and forbidden transitions, 34
ALMA, 90
  Pluto, 481
  Titan, 379
ambipolar diffusion, 17
analysis of measured equivalent widths, 68
annual variations of pressure on Mars, 158
apodization, 83
Ar, variations on Mars, 179
argon and its isotopes on Venus, 249
asteroid size distribution, 7
asteroids, 6
astrobiology, 155
atomic carbon on Venus, 320
atomic deuterium on Venus, 318
atomic hydrogen
  from PV mass spectrometry, 315
  Venus, 313
atomic nitrogen, 40
atomic oxygen, 40
atomic oxygen on Venus, 318
atomic oxygen triplet at 130 nm
  on Mars, 208
attraction fields, 128
aurora on Mars, 219
Avogadro number, 11

barometric formula in exosphere, 25
basic $CO_2$–$H_2O$ chemistry, 197
basic ionospheric chemistry on Mars, 219
basic properties
  Mars, 156
  Pluto and Charon, 467
  Titan, 367
  Triton, 443
  Venus, 238
bimolecular, binary, two-body reactions, 122
blackbody radiation, 18
Bohr magneton, 34
boundary conditions, 142
boundary layer, Mars, 166
brightness coefficient, 58

Cassini–Huygens mission, 382
  Cassini instruments, 387
cavity ring-down spectrometer (CRDS), 84
centrifugal barrier model, 128
Chandra X-ray Observatory (CXO), 86
Chapman layer, 22, 219, 290
chemical composition of the Venus atmosphere below
    120 km, 279
chemical kinetic model for Venus below clouds
  model results, 345
  reactions and rate coefficients, 344
  sources of chemistry and boundary conditions, 343
chemical lifetime, 131
chlorine chemistry on Mars, upper limit, 196
chromosphere, 103
circular variable filters (CVF), 78
ClCO cycle, 350
cloud properties on Venus, 330
cloud variations on Venus, 332
CO ($A^1\Pi \rightarrow X^1\Sigma^+$) fourth positive system
  on Mars, 210
CO dayglow at 4.7 μm
  Mars, 212
  Venus, 322

CO detection and first observations, 173
CO in Venus upper atmosphere, 322
CO on Venus, 251
$CO_2$ and atmospheric pressure on Mars, 158
$CO_2$ and $N_2$, Venus, 242
$CO_2$ band at 15 μm, 87, 98
$CO_2$ band at 4.3 μm, 89
$CO_2$ emission at 15 μm, 21
$CO_2$ molecule, 51
$CO_2$ stability problem, 197
co-adding of atmospheric layers, 70
collisional or pressure (Lorentz) broadening, 65
column abundance, 12
column reaction rates, 148
comets, 8
complex numbers, 30
composition of Mars lower and middle atmosphere, 198
composition of Venus clouds based on the Vega probes, 335
condensation layer, 62
continuity equation, 141
cosines of incidence and viewing angles, 72
cosmic ray ionization
  Titan, 412
cosmic rays on Venus, 276
CRIRES, 81
cross-dispersed echelle spectrographs, 81
cryogenic echelle spectrograph CSHELL, 69
curves of growth, 67
cyclostrophic balance, 244

dayglow, 108
dayside ionosphere of Venus in the PV observations, 293
degrees of freedom for a gas molecule, 13
designation of term, 37
diffraction grating, 78
diffusion coefficient, 14
diffusion limit, 29, 147
discovery of Pluto and Charon, 467
dissociation energies of some molecules, 132
dissociation energy, 43
dissociative electron attachment, 116
dissociative recombination, 23, 130
dissociative recombination of $O_2^+$, $CO_2^+$, 118
double and triple collisions, 120

Earth's atmosphere, 20
echelle grating, 80
echelle spectrographs, 80
eddy diffusion, 16, 20, 29, 61–62, 64, 140–142, 144, 168, 200, 203, 205–206, 224, 248, 258, 279, 336, 338, 342–343, 345, 351, 353, 357–358, 406, 412, 427, 429, 431, 441, 450, 457, 465, 477
effective temperature of a planet, 19
effective temperatures of terrestrial planets, 106
Electron affinity and gas phase acidity of some molecules, 417
electron configuration, 36
electron energy loss function, 115
electronic states of $O_2$, 49
electronic, vibrational, and rotational energies, comparison, 43
elemental conservation, 18
emission rate factor, 113
enthalpies and entropies of formation, 122
equilibrium constant, 121
equivalent width, 67
escape and balance of helium on Venus, 329
escape of H
  Venus, 325
escape of H and $H_2$
  Mars, 224
escape of ions on Venus, 327
escape processes on Mars, 152
ethane on Mars, 194
E-type ionosphere, 23
evolution of nitrogen on Mars, 236
evolution of water on Venus, 327
excitation of oxygen nightglow on terrestrial planets, 301
exosphere, 25
exospheric temperature $T_\infty$
  Mars, 221
exospheric temperature versus solar activity, 24
extinction and scattering coefficients, 53
Extreme Ultraviolet Explorer (EUVE), 86

Far Ultraviolet Spectroscopic Explorer (FUSE), 86
Far UV (900–1200 Å) and EUV dayglow on Mars, 210
FeCl3 in Venus clouds, 334
Fermi resonance, 51
fine structure of hydrogen-like atoms, 32
finite difference analog of continuity equation, 141
flat field, 70
foreground spectrum, 70
formation of hydrogen bulge on Venus, 317
Fourier transform spectrometer (FTS), 82
fractionation factor, 153
fractionation factor for reservoir and buffer, 154
Franck–Condon principle, 48
F-type ionosphere, 23

gamma distribution, 55
gas chromatography, 102
Gaussian distribution, 55
general circulation model
  Titan, 394
Gibbs free energy, 121
global-mean model, 144
grating spectrographs, 79

H Lyman-alpha 122 nm
  on Mars, 205
$H_2O$ on Venus, 266
$H_2O$ vapor, 21

$H_2O$ vertical distribution on Mars, 168
$H_2O_2$ on Mars, 186
$H_2S$ on Venus, upper limit, 265
$H_2SO_4$ vapor on Venus, 265
haze
  photochemical production on Titan, 440
  Titan, 371
haze and production of $SO_X$ on Venus, 341
HBr and $Br_2$ on Venus, 274
HCl and DCl on Venus, 270
HDO and HDO/$H_2O$ ratio on Venus, 269
HDO/$H_2O$ ratio on Mars, 170
$He^+$ line at 30.4 nm, 36
heating efficiency, 22
heliosphere, heliopause, 104
helium atom, 38
helium bulge on Mars, 230
helium on Venus, 248
Henyey–Greenstein phase function, 58
heterogeneous reactions, 137
heterosphere, 16
HF and DF on Venus, 272
history of studies
  Mars, 155
  Venus, 238
HITRAN 2016 spectroscopic database, 71
H-like ions $C^{5+}$, $O^{7+}$, and $Ne^{9+}$, 36
homonuclear molecules, 46
homosphere, 16
Hubble Space Telescope (HST), 86
hydrodynamic escape, 27, 235, 477
hydrogen bulge on Venus, 315

immersion grating, 81
incondensable aerosol, 61
indirect photolysis of $N_2$, 458
infrared heterodyne spectrometers, 84
INMS Titan
  correction factors, 403
  escape of CH4 and H2, 406
  neutral upper atmosphere, 403
  nighttime ion mass spectra, 408
intensified CCD detector, 79
interval rule, 37
inversion of limb observations, 96
ion composition on Mars measured by MAVEN/NGIMS, 230
ionosphere, 22
  Mars, 219
  photochemical model for Titan, 438
  Pluto, 490, 496
  Titan, 375, 409
  Triton, 459
  Venus, 290
ionosphere of Venus
  radio occultations, 290
IRAM telescope, 90
iSHELL, 81
isotope fractionation by atmospheric escape, 152
isotope fractionation by condensation, 151
isotope fractionation in chemical reaction, 151
isotope ratios
  Titan, 420, 424
isotope ratios in molecular species, 96
Isotope ratios in the Venus and Earth atmospheres and in the solar wind, 251
isotopic ratios of C, N, O on Venus, 251

James Clerk Maxwell Telescope (JCMT), 90
Jupiter, 5

Kepler law, 3
krypton and xenon on Venus, 251
Kuiper belt, 7

Lamb shift, 36
lambda doubling, 47
Lambert reflection, 57
Langevin formula, 17, 130
latent heat of condensation, 13
Li 670.8 nm emission, 114
lightning on Venus, 276
local thermodynamic equilibrium, 21
log normal distribution, 56
loss of water from Mars, 234
low- and high-pressure limits of reactions, 132
lower boundary conditions, 143

magnetic dipole, 33
Mars Climate Database, 68, 74, 140, 171, 174, 182, 202–203, 221, 226
Mars photochemical GCMs, 203
Mars photochemistry, 197
  global-mean models, 200
  variations, 201
Mars Thermosphere General Circulation Model (MTGCM), 220
Martian polar caps, 157
mass of the asteroid belt, 7
mass of the Kuiper belt, 7
mass spectrometry, 101
  Mars, 178
mass, momentum, and energy conservation in hydrodynamic escape, 28
Maxwell–Boltzmann velocity distribution, 26
mean velocity of gas molecules, 26
mesopause, 21
mesosphere, 21
metastable states $O(^1D)$ and $N(^2D)$, 40
meteorite impact erosion on Mars, 232
meteorite ions on Venus, 292
methane, observations on Mars, 188
methane, origin on Mars, 193
methane, variability on Mars, 192
Mie formulas, 52
mixing ratios, 14

model for nighttime ozone on Venus, 364
model for OH band distribution on Venus, 365
model nightglow intensities of O2 1.27 μm, NO, OH on Venus, 364
models for sulfuric acid in Venus clouds, 338
mole fraction, 14
molecular diffusion, 14
molecular oxygen
  Mars, 177
  Venus, 256
monochromatic radiative equilibrium, 19
Morse function, 43
most abundant atmospheric gases, 5

Na 589.0/9.6 nm, 114
NASA Infrared Telescope Facility, 69
near UV absorber and iron chloride, 335
neon and its isotopes on Venus, 250
Neptune, 6
neutral composition and temperature on Mars from MAVEN/NGIMS, 228
$NH_3$ on Venus, upper limit, 274
nightside ionosphere
  Venus, 291
nightside ionosphere on Mars, 219
nighttime chemistry and nightglow, 362
nighttime ionosphere of Venus in PV observations and models, 294
NIRSPEC, 81
NIST Chemical Kinetics Database, 138, 147
nitrogen chemistry in the upper atmosphere, 222
NO nightglow
  Mars, 214
NO nightglow excitation, 119
NO on Mars, 196
NO on Venus, 275
nonspherical particles, 55
nonthermal escape, 27

O and C isotope ratios and evolution on Mars, 233
$O(^1S\text{-}^1D)$ line at 558 nm on Venus, 292
$O_2$ dayglow at 1.27 μm
  Mars, 183
OCS on Venus, 262
OH rovibrational bands, 119
operators of momentum and energy, 30
Öpik-Oort cloud, 8
optical constants of tholins, 368
orbitals, 46
ozone $O_3$, 90
  Mars, 180
  Venus, 256

parameters of targets for Solar System observations, 71
parity, 46
partial photochemical models, 140
particle size spectrometer at the PV large probe, 331
Pauli exclusion principle, 36, 47
periodic table of the chemical elements, 36
phosphorus in Venus clouds, 334
photochemical modeling
  Mars upper atmosphere, 223
photochemical modeling of Venus middle atmosphere, 350
photochemical models for Titan, 424
photodetachment, 419
photodissociation and photoionization of $CO_2$, 109
photodissociation of diatomic molecules
  $O_2$, $N_2$, CO, $H_2$, 132
photodissociation of polyatomic molecules, 135
photoelectron excitation rate, 116
photoelectron flux spectrum, 116
photoelectron production and loss, 115
photo-induced isotope fractionation, 151
photosphere, 103
Planck constant, 30
Planck law, 18
planetary atmospheres at the surface or 1 bar, 5
Pluto
  HST spectroscopy in UV range, 472
  origin and interior, 468
  surface composition and temperature, 468
Pluto New Horizons
  atmosphere of Charon, 489
  haze, 485
  haze and thermal balance of atmosphere, 487
  haze models, 485
  radio occultations, 478
  thermal balance, 483
  UV solar occultations, 479
Pluto's atmosphere
  high-resolution spectroscopy and search for CH4 and CO, 476
  pressures, temperatures, and temperature gradients at $r$ = 1275 km from stellar occultations, 475
  stellar occultations, 473
Pluto's atmosphere
  modeling of hydrodynamic escape, 477
  stellar occultations, 477
  thermal balance, 477
Pluto's photochemistry
  model by Luspay-Kuti et al. (2017), 494
  model by Wong et al. (2017), 490
  models before NH, 489
polar nightglow of $O_2$ at 1.27 μm on Mars, 216
polar nightglow of OH on Mars, 216
polarizability, 17
processing of observed spectra, 74
progressions, 47
properties of planets, 2
PV bus neutral mass spectrometer, 279
PV orbiter atmospheric drag, 279
PV orbiter neutral mass spectrometer, 282

quadrupole, 33
quantum defect, 38
quantum numbers, 31

R, P, Q-branches, 44
radiative recombination, 130
radiative transfer, direct problem, 60
radiative transfer, inverse problem, 59
radio occultations
  Titan, 371
radioactive decay of $^{238}U$, $^{235}U$, and $^{232}Th$, 149
radioactive decay of $^{40}K$, 151
radiogenic argon and helium on Mars, 237
Rayleigh distillation, 153
Rayleigh scattering, 54
reaction rate coefficient, 131
reaction rate coefficients, 147
requirements to boundary conditions, 143
Roche limit, 3
rotational instability, 9
rotational levels, 41
rotational temperature, 44, 96
RRKM theory, 128
Rydberg constant, 32

$S_3$ and $S_4$ on Venus, 263
$S_4$ cycle on Venus, 345
satellites with atmospheres, 4
saturated vapor pressures, 430
Saturn, 6
scale height, 11
scattering asymmetry factor, 58
Schroedinger wave equation, 30
selection rules, 34, 38, 44, 47
self-consistent photochemical models, 140
sequences, 48
single scattering approximation, 57
skin temperature, 19
$SO_2$ and SO, Venus, 257
$SO_2$ on Mars, upper limit, 195
solar activity, 24, 108, 135, 204–208, 210, 214–215, 220, 223–225, 229–230, 234–235, 279, 282, 289–290, 296, 301, 305, 326, 401–402, 407, 453
solar activity index, 24
solar and stellar occultations, 98
solar atmosphere, 103
solar constant, 105
solar corona, 103
solar infrared spectra, 71
solar spectrum, 105
solar wind, 103
some unsolved problems on Venus, 365
spatially resolved observations of CO on Mars, 174
spectra of isotopologues, 45
spherical albedo, 61
spin conservation, 122
standard isotope ratios, 150
statistical weight, 32
stellar occultations
  Mars upper atmosphere, 226
  $O_2$ in Mars upper atmosphere, 226
Stokes law, 61
stratosphere, 20
structure parameter, 25
submillimeter spectral range, 90
subsolar-to-antisolar circulation, 244
sulfur aerosol on Venus, 340
superfine structure, 36
superrotation, 244
  Titan, 393
  Venus, 244
symmetry of system, 41
synthetic spectra, 75

temperature gradient in troposphere, 12
temperature profile in thermosphere, 23
temperatures on Mars
  MGS/TES, 164
  MRO/MCS, 165
termolecular, three-body association, 136
TEXES, 81
thermal (Doppler) broadening, 65
thermal balance equation, 20
thermal balance of Pluto's $CH_4/N_2$ atmosphere, 473
thermal conductivity, 20
thermal effects of eddy diffusion, 24
thermal escape, 26–27, 152–153, 205, 325, 407, 425, 452, 457, 480–481, 483, 494
thermal wind, 244
thermochemical equilibrium, 121
thermosphere, 23
  Venus, 279
three-body association of oxygen, 119
time of mixing and diffusion, 17
time-dependent models, 148
Titan
  airglow, 401
  argon isotopes, 423
  carbon isotope ratio, 424
  chemical composition, 405
  dynamics, 393
  GCMS measurements, 383
  HSO observations, 382
  Huygens HASI, DWE, ACP measurements, 387
  Huygens/DISR observations, 384
  interior, 367
  ISO observations, 380
  microwave observations, 378
  NIR observations, 376
  nitrogen isotopes, 420, 422
  seasonal variations near poles, 394
  unsolved problems, 441
  UVIS occultations, 396
  UVS solar occultations, 372

VIMS observations, 388
Titan CIRS
haze properties, 392
nominal mission, 389
Titan stratosphere
Voyager/IRIS, 371
Titan, pre-Voyager studies, 368
Titan's atmosphere and ionosphere in the model by Dobrijevic et al. (2016), 425
Titan's ionosphere
daytime composition, 413
ionization of haze and embryos, 412
magnetospheric precipitation, 409
meteor ablation, 411
negative ions, 415
photodetachment of negative ions, 412
radio occultations, 409
RPWS observations, 409
Titan's photochemistry
hydrocarbons, Ar, and H2, 429
initial data, 427
ionosphere, 438
nitriles, 434
oxygen species, 437
production and loss rates, escape/precipitation flows, lifetimes, 431
production of haze, 440
Titius–Bode rule, 1
topography of Mars, 157
transition dipole moment, 33
transition probability, 33, 112
triple point of water, 158
Triton
composition of surface ice, 446
magnetospheric electrons, 453, 456
pre-Voyager studies, 443
seasonal variations, 448
spatial distribution of ices, 449
surface and interior, 445
surface temperature, 445
Triton's ionosphere
radio occultations, 449
Triton's atmosphere
atomic nitrogen, 451
CH4, 450
CH4 and CO abundances, 465
haze, 449
high-resolution spectroscopy, 464
N2, 452
stellar occultations, 460
thermal balance and temperature profile, 453
UVS solar occultations, 450
Voyager radio occultations, 449
Triton's photochemistry
CO, atomic species, ionosphere, 458
effects of methane abundance, 466
general features, 457
history of problem, 456
hydrocarbons, HCN, H2, 458
methane and eddy diffusion, 458
versions of the model, 459
Troe approximation, 137
tropopause, 20
troposphere of saturated vapor, 13
tunable laser spectrometer, 84
two-body association, 127
two-body problem, 1, 31

UMIST Database for Astrochemistry 2012, 139
uncertainty principle, 31
upper atmosphere, Venus Express, 283
upper boundary conditions, 143
uptake, or accommodation, or sticking coefficient, 138
Uranus, 6
UV dayglow
observations on Mars, 204
Triton, 455
UVS dayglow
Titan, 375

variability of long-living species on Mars, 174
variations of isotope ratios, 153
variations of the Martian exospheric temperature with solar activity, 221
Venus dayglow
180–300 nm (SPICAV), 311
80–180 nm (HUT, Cassini/UVIS), 310
EUVE, 310
Venus International Reference Atmosphere, 72
Venus model upper atmosphere, 284
Venus nightglow
CO UV, 306
in the visible range, 298
NO, 302
O2 at 1.27 μm, 299
OH, 305
Venus photochemical model
boundary conditions, 352
carbon monoxide, 353
chlorine and hydrogen chemistries, 359
input data, 350
odd nitrogen chemistry, 361
reactions, rate coefficients, column rates, 352
sulfur species, 357
water and oxygen allotropes, 359
Venus thermosphere general circulation model (VTGCM), 288
Venus upper atmosphere, variations with solar activity, 290
vertical and limb column abundances, 25
vertical flux, 14, 16
vertical step, 141
vibrational levels, 42
Viking retarding potential analyzers on Mars, 220

VIRA, 242
Voigt line shape, 66
volume emission rate, 108
VTGCM simulations of the $O_2$ 1.27 μm and NO UV nightglow, 307

water ice in Mars regolith, 172
water vapor column on Mars, 167
wave function, 30
weakly absorbing particles, 54
weighting function, 90, 93

X-ray fluorescent spectroscopy of cloud composition on Venus, 333
X-ray spectra of Mars and comets, 36

zero-point energy, 151
zonal and meridional wind on Venus, 246